DESIGN OF HIGHWAYS BRIDGES

DESIGN OF HIGHWAY BRIDGES

Based on AASHTO LRFD, Bridge Design Specifications

RICHARD M. BARKER
Professor of Civil Engineering
Virginia Tech
Blacksburg, Virginia

JAY A. PUCKETT
Professor of Civil Engineering
University of Wyoming
Laramie, Wyoming

A Wiley-Interscience Publication

JOHN WILEY & SONS, INC.

New York · Chichester · Weinheim · Brisbane · Singapore · Toronto

This text is printed on acid-free paper.

Copyright © 1997 by John Wiley & Sons, Inc.

All rights reserved. Published simultaneously in Canada.

Reproduction or translation of any part of this work beyond
that permitted by Section 107 or 108 of the 1975 United
States Copyright Act without the permission of the copyright
owner is unlawful. Requests for permission or further
information should be addressed to the Permissions Department,
John Wiley & Sons, Inc., 605 Third Avenue, New York, NY
10158-0012.

This publication is designed to provide accurate and
authoritative information in regard to the subject
matter covered. It is sold with the understanding that
the publisher is not engaged in rendering legal, accounting,
or other professional services. If legal advice or other
expert assistance is required, the services of a competent
professional person should be sought.

Library of Congress Cataloging in Publication Data:
Barker, R. M. (Richard M.)
    Design of highway bridges : based on AASHTO LRFD Bridge
Design Specifications / Richard M. Barker, Jay A. Puckett.
        p.   cm.
    Includes index.
    ISBN 0-471-30434-4 (cloth : alk. paper)
    1. Bridges—United States—Design and construction. I. Puckett,
J. A. (Jay Alan) II. Title.
TG300.B38   1997
624'.25—dc20   96-23741

Printed in the United States of America

10 9 8 7 6 5 4 3 2 1

This book is dedicated to our parents, our wives, and our children; who have shown us the way, been our constant support, and are our future.

CONTENTS

PREFACE

This book is written for senior level undergraduate or first year graduate students in civil engineering and for practicing civil engineers who have an interest in the design of highway bridges. The object of this book is to provide the student or practitioner a meaningful introduction to the design of medium- and short-span girder bridges. This objective is achieved by providing fundamental theory and behavior, background on the development of the specifications, procedures for design, and design examples.

This book is based on the American Association of State Highway and Transportation Officials (AASHTO) LRFD Bridge Design Specifications and System International (SI) units are used throughout. The general approach is to present theory and behavior upon which a provision of the specifications is based, followed by appropriate procedures, either presented explicitly or in examples. The examples focus on the procedures involved for a particular structural material and give reference to the appropriate article in the specifications. It is, therefore, essential that the reader have available a copy of the most recent edition of the AASHTO LRFD Bridge Design Specifications in SI units. (For those who have access to the World Wide Web, addendums to the specifications can be found at http://www2.epix.net/~modjeski.)

The scope of this book is limited to a thorough treatment of medium- and short-span girder bridges with a maximum span length of about 60 m. These bridge structures comprise approximately 80% of the U.S. bridge inventory and are the most common bridges designed by practitioners, illustrating the basic principles found in bridges of longer spans. Structure types included in this book are built of concrete, steel, and wood. Concrete cast-in-place slab, *T*-beam, and box-girder bridges and precast–prestressed systems are considered. Rolled steel beam and plate girder systems that are composite and noncomposite are included, as well as wood systems. This book concludes with a chapter on substructure design, which is a common component for all the bridge types.

Civil engineers are identified as primary users of this book because their formal education includes topics important to a highway bridge designer. These topics include studies in transportation systems, hydrodynamics of streams and channels, geotechnical engineering, construction management, environmental engineering, structural analysis and design, life cycle costing, material testing, quality control, professional and legal problems, and the people issues associated with public construction projects. This reference to civil engineers is not meant to exclude others from utilizing this book. However, the reader is expected to have one undergraduate course in structural design for each structural material considered. For example, if only the design of steel bridges is of interest, then the reader should have at least one course in structural analysis and one course in structural steel design.

Chapter 1 introduces the topic of bridge engineering with a brief history of bridge building and the development of bridge specifications in the United States. Chapter 2 emphasizes the need to consider aesthetics from the beginning of the design process and gives examples of successful bridge projects. Chapter 3 presents the basics on load and resistance factor design (LRFD) and indicates how these factors are chosen to obtain a desirable margin of safety.

Chapter 4 describes the nature, magnitude and placement of the various loads that act on a bridge structure. Chapter 5 presents influence function techniques for determining maximum and minimum force effects due to moving vehicle loads. Chapter 6 considers the entire bridge structure as a system and how it should be analyzed to obtain a realistic distribution of forces.

Chapters 7–9 are the design chapters for concrete, steel, and wood bridges. The organization of these three chapters is similar. A description of material properties is given first, followed by general design considerations. Then a discussion of the behavior and theory behind the member resistance expressions for the various limit states, and concluding with detailed design examples that illustrate the LRFD specification provisions.

Chapter 10 on substructure design completes the book. It includes general design considerations, an elastomeric bearing design example, and a stability analysis to check the geotechnical limit states for a typical abutment.

We suggest that a first course in bridges be based on Chapters 1–6, either Articles 7.1–7.6, 7.10.1, and 7.10.3 of Chapter 7 or Articles 8.1–8.4, 8.6–8.10, and 8.11.2, and conclude with Articles 10.1–10.3 of Chapter 10. It is assumed that some of this material will have been covered in prerequisite courses and can be referred to only as a reading assignment. How much of the material to present to a particular class is at the discretion of the professor, who is probably the best person to judge the background and maturity of the students. There is enough material in the book for more than one course in highway bridge design.

Practitioners who are entry level engineers will find the background material in Chapters 1–6 helpful in their new assignments and can use Chapters

7–10 for specific guidance on design of a particular bridge type. The same can be said for seasoned professionals, even though they would be familiar with the material in the loads chapter, they should find the other chapters of interest in providing background and design examples based on the AASHTO LRFD specifications.

ACKNOWLEDGMENTS

Acknowledgments to others who have contributed to the writing of this book is not an easy task because so many people have participated in the development of our engineering careers. To list them all is not possible, but we do recognize the contribution of our university professors at the University of Minnesota and Colorado State University; our engineering colleagues at Toltz, King, Duvall, Anderson & Associates, Moffatt & Nichol Engineers, and BridgeTech, Inc.; our faculty colleagues at Virginia Tech and the University of Wyoming; the government and industry sponsors of our research work; and the countless number of students who keep asking those interesting questions.

The contribution of John S. Kim, author of Chapter 10 on Substructure Design, is especially appreciated. We realize that many of the ideas and concepts presented in the book have come from reading the work of others. In each of the major design chapters, the influence of the following people is acknowledged: Concrete Bridges, Michael Collins, University of Toronto, Thomas T.C. Hsu, University of Houston, and Antoine Naaman, University of Michigan; Steel Bridges, Sam Easterling and Tom Murray, Virginia Tech, and Konrad Basler, Zurich, Switzerland; and Wood Bridges, Michael Ritter, USDA Forest Service.

We also wish to acknowledge those who have contributed directly to the production of the book. These include Elizabeth Barker who typed a majority of the manuscript, Jude Kostage who drafted most of the figures, and Brian Goodrich who made significant modifications for the conversion of many figures to SI units. Others who prepared figures, worked on example problems, handled correspondence, and checked page proofs were: Barbara Barker, Catherine Barker, Benita Calloway, Ann Crate, Scott Easter, Martin Kigudde, Amy Kohls, Kathryn Kontrim, Michelle Rambo-Roddenberry, and Cheryl Rottmann. Thanks also to the following state departments of transportation who supplied photographs of their bridges and offered encouragement: California, Minnesota, Pennsylvania, Tennessee, Washington, and West Virginia.

The patience and understanding that Charles Schmieg, Associate Editor, Minna Panfili, editorial program assistant, and Millie Torres, Associate Managing Editor at John Wiley & Sons, have shown us during the preparation and production of the manuscript are gratefully acknowledged. We also recognize the assistance provided by editors Dan Sayre and Robert Argentieri of John Wiley & Sons during the formative and final stages of this book.

Finally, on behalf of the bridge engineering community the authors wish to recognize John Kulicki of Modjeski & Masters and Dennis Mertz of the University of Delaware for their untiring leadership in the development of the LRFD Specification. The authors wish to thank these professionals for providing support and encouragement for the book and responding to many questions about the rationale and background of the specification. Others who contributed to the development of the LRFD Specification as members of the Code Coordinating Committee or as a Chair of a Task Group have also influenced the writing of this book. These include: John Ahlskog, Ralph Bishop, Ian Buckle, Robert Cassano, Paul Csagoly, J. Michael Duncan, Theodore Galambos, Andrzej Nowak, Charles Purkiss, Frank Sears, and James Withiam. A complete listing of the members of the task groups and the NCHRP panel that directed the project is given in Appendix D.

As with any new book, in spite of numerous proofreadings, errors do creep in and the authors would appreciate it if the reader would call them to their attention. You may write to us directly or, if you prefer, use our e-mail address: barker@vt.edu or puckett@uwyo.edu.

1

INTRODUCTION TO
BRIDGE ENGINEERING

Bridges are important to everyone. But they are not seen or understood in the same way by everyone, which is what makes their study so fascinating. A single bridge over a small river will be viewed differently by different people because the eyes each one sees it with are unique to that individual. Someone traveling over the bridge everyday while going to work may only realize a bridge is there because the roadway now has a railing on either side. Others may remember a time before the bridge was built and how far they had to travel to visit friends and to get the children to school. Civic leaders see the bridge as a link between neighborhoods and a way to provide fire and police protection and access to hospitals. In the business community, the bridge is seen as opening up new markets and expanding commerce. An artist will consider the bridge and its setting as a possible subject for a future painting. A theologian may see the bridge as symbolic of making a connection between God and human beings. While a boater on the river, looking up when passing underneath the bridge, will have a completely different perspective. Everyone is looking at the same bridge, but it produces different emotions and visual images in each one of them.

Bridges affect people. People use them, and engineers design them and later build and maintain them. Bridges do not just happen. They must be planned and engineered before they can be constructed. In this book, the emphasis will be on the engineering aspects of this process: selection of bridge type, analysis of load effects, resistance of cross sections, and conformance with bridge specifications. There is discussion of load factors, resistance and other factors of technical significance, but a good bridge engineer must never forget the *people* factor.

1.1 A BRIDGE IS THE KEY ELEMENT IN A TRANSPORTATION SYSTEM

A bridge is the key element in a transportation system for three reasons:

- It controls the capacity of the system.
- It is the highest cost per mile of the system.
- If the bridge fails, the system fails.

If the width of a bridge is insufficient to carry the number of lanes required to handle the traffic volume, the bridge will be a constriction to the flow of traffic. If the strength of a bridge is deficient and unable to carry heavy trucks, load limits will be posted and truck traffic will be rerouted. The bridge controls both the volume and weight of the traffic carried by the transportation system.

Bridges are expensive. The typical cost per mile of a bridge is many times that of the approach roads to the bridge. This investment must be taken seriously when the limited dollars available for a transportation system are distributed.

When a bridge is taken out of service, for whatever reason, the transportation system is restricted in its function. Traffic must be detoured over routes not designed to handle the increase in volume. Users of the system experience increased travel times and fuel expenses. Normalcy does not return until the bridge is repaired or replaced.

Because a bridge is the key element in a transportation system, balance must be achieved between handling future traffic volume and loads and the cost of a heavier and wider bridge structure. Strength must always be foremost, but so should measures to prevent deterioration. The designer of new bridges has control over these parameters and must make wise decisions so that capacity and cost are in balance, and safety is not compromised.

1.2 BRIDGE ENGINEERING IN THE UNITED STATES

Usually a discourse on the history of bridges begins with a log across a small stream or vines suspended above a deep chasm. This preamble is followed by the development of the stone arch by the Roman engineers of the second and first centuries BC and the building of beautiful bridges across Europe during the Renaissance period of the thirteenth, fourteenth, and fifteenth centuries. Next is the Industrial Revolution, which began in the last half of the eighteenth century and saw the emergence of cast iron, wrought iron, and finally steel for bridges. Such discourses are found in the books by Brown (1993), Gies (1963), Kirby et al. (1956), and are not repeated here. Instead a few of the bridges that are typical of those found in the United States are highlighted.

1.2.1 Stone Arch Bridges

When discussing stone arch bridges, the Roman bridge builders first come to mind. They utilized the semicircular arch and built elegant and handsome aqueducts and bridges, many of which are standing today. The oldest remaining Roman stone arch structure is from the seventh century BC and is a vaulted tunnel near the Tiber River. However, the oldest surviving stone arch bridge dates from the ninth century BC and is in Smyrna, Turkey, over the Meles River. In excavations of tombs and underground temples, archaeologists found arched vaults dating to the fourth millennium BC at Ur in one of the earliest Tigris–Euphrates civilizations (Gies, 1963). The stone arch has been around a long time and how its form was first discovered is unknown. But credit must be given to the Roman engineers, because they are the ones who saw the potential in the stone arch, developed construction techniques, built foundations in moving rivers, and left us a heritage of engineering works that we marvel at today.

Compared to these early beginnings, the stone arch bridges in the United States are relative newcomers. One of the earliest stone arch bridges is the Frankford Avenue Bridge over Pennypack Creek built in 1697 on the road between Philadelphia and New York. It is a three-span bridge, 23 m long, and is the oldest bridge in the United States that continues to serve as part of a highway system (Jackson, 1988).

Stone arch bridges were usually small scale and built by local masons. These bridges were never as popular in the United States as they were in Europe. Part of the reason for lack of popularity is that stone arch bridges are very labor intensive and expensive to build. However, with the development of the railroads in the mid-to-late nineteenth century, the stone arch bridge provided the necessary strength and stiffness for carrying heavy loads, and a number of impressive spans were built. One was the Starrucca Viaduct, Lanesboro, PA, which was completed in 1848 and another was the James J. Hill Stone Arch Bridge, Minneapolis, MN completed in 1883.

The Starrucca Viaduct is 317 m in overall length and is composed of 17 arches, each with a span of 15 m. The viaduct is located on what was known as the New York and Erie Railroad over Starrucca Creek near its junction with the Susquehanna River. Except for the interior spandrel walls being of brick masonry, the structure was of stone masonry quarried locally. The maximum height of the roadbed above the creek is 34 m (Edwards, 1959) and it still carries heavy railroad traffic.

The James J. Hill Stone Arch Bridge (Fig. 1.1) is 760 m long and incorporated 23 arches in its original design (later, two arches were replaced with steel trusses to provide navigational clearance). The structure carried Hill's

Fig. 1.1 James J. Hill Stone Arch Bridge, Minneapolis, MN [Hibbard Photo, Minnesota Historical Society, July 1905].

Great Northern Railroad (now the Burlington Northern) across the Mississippi River just below St. Anthony Falls. It played a key role in the development of the Northwest. The bridge was retired in 1982, just short of its 100th birthday, but it still stands today as a reminder of an era gone by and bridges that were built to last (Jackson, 1988).

1.2.2 Wooden Bridges

Early bridge builders in the United States (Timothy Palmer, Lewis Wernwag, Theodore Burr, and Ithiel Town) began their careers as millwrights or carpenter-mechanics. They had no clear conception of truss action and their bridges were highly indeterminate combinations of arches and trusses (Kirby and Laurson, 1932). They learned from building large mills how to increase clear spans by using the king-post system or trussed beam. They also appreciated the arch form and its ability to carry loads in compression to the abutments. This compressive action was important because wood joints can transfer compression more efficiently than tension.

   The long span wooden bridges built in the late eighteenth century and early nineteenth century incorporated both the truss and the arch. Palmer and Wernwag constructed trussed arch bridges in which arches were reinforced by trusses (Fig. 1.2). Palmer built a 74-m trussed arch bridge over the Piscataqua in New Hampshire in the 1790s. Wernwag built his "Colossus" in 1812 with a span of 104 m over the Schuylkill at Fairmount, PA.

In contrast to the trussed arch of Palmer and Wernwag, Burr utilized an arched truss in which a truss is reinforced by an arch (Fig. 1.3) and patented his design in 1817. An example of one that has survived until today is the Philippi Covered Bridge (Fig. 1.4) across the Tygant's Valley River, WV. It was completed in 1852 by Lemuel Chenoweth as a two-span Burr arched truss 176 m long. In later years, two reinforced concrete piers were added under each span to strengthen the bridge. As a result, it is able to carry traffic loads and is the nation's only covered bridge serving a federal highway.

One of the reasons many covered bridges have survived for well over a 100 years is that the wooden arches and trusses have been protected from the weather. (Another reason is that nobody has decided to set fire to them.) Palmer put a roof and siding on his "Permanent Bridge" (called permanent because it replaced a pontoon bridge) over the Schuylkill at Philadelphia in 1806, and the bridge lasted nearly 70 years before it was destroyed by fire in 1875.

Besides protecting the wood from alternating cycles of wet and dry that cause rot, other advantages of the covered bridge occurred. During winter blizzards, snow did not accumulate on the bridge. However, this presented another problem, bare wooden decks had to be paved with snow, because everybody used sleighs. Another advantage was that horses were not frightened by the prospect of crossing a rapidly moving stream over an open bridge because the covered bridge had a comforting barnlike appearance (so says the oral tradition). American folklore also says the covered bridges became favorite parking spots for couples in their rigs, out of sight except for the eyes of curious children who had climbed up and hid in the rafters (Gies, 1963). However, the primary purpose of covering the bridge was to prevent deterioration of the wood structure.

Another successful wooden bridge form first built in 1813 was the lattice truss, which Ithiel Town patented in 1820 (Edwards, 1959). This bridge consisted of strong top and bottom chords, sturdy end posts, and a web of lattice work (Fig. 1.5). This truss type was popular with builders because all of the web members were of the same length and could be prefabricated and sent

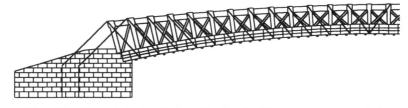

Fig. 1.2 Trussed Arch—designed by Lewis Wernwag, patented 1812.

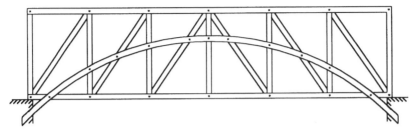

Fig. 1.3 Arched Truss—designed by Theodore Burr, patented 1817. From BRIDGES AND MEN by Joseph Gies. Copyright © 1963 by Joseph Gies. Used by permission of Doubleday, a division of Bantam Doubleday Dell Publishing Group, Inc.

to the job site for assembly. Another advantage is that it had sufficient stiffness by itself and did not require an arch to reduce deflections. This inherent stiffness meant that horizontal thrusts did not have to be resisted by abutments and a true truss, with only vertical reactions, had really arrived.

The next step toward simplicity in wooden bridge truss types in the United States is credited to an army engineer named Colonel Stephen H. Long who had been assigned by the War Department to the Baltimore and Ohio Railroad (Edwards, 1959). In 1829, Colonel Long built the first American highway–railroad grade separation project. The trusses in the superstructure had parallel chords that were subdivided into panels with counterbraced web members (Fig. 1.6). The counterbraces provided the necessary stiffness for the panels

Fig. 1.4 Philippi Covered Bridge [Photo by Larry Belcher, courtesy of West Virginia Department of Transportation].

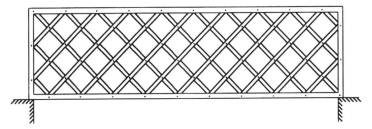

Fig. 1.5 Lattice Truss—designed by Ithiel Town, patented 1820. From BRIDGES AND MEN by Joseph Gies. Copyright © 1963 by Joseph Gies. Used by permission of Doubleday, a division of Bantam Doubleday Dell Publishing Group, Inc.

as the loading changed in the diagonal web members from tension to compression as the railroad cars moved across the bridge.

The development of the paneled bridge truss in wooden bridges enabled long span trusses to be built with other materials. In addition, the concept of web panels is important because it is the basis for determining the shear resistance of girder bridges. These concepts are called the modified compression field theory in Chapter 7 and tension field action in Chapter 8.

1.2.3 Metal Truss Bridges

Wooden bridges were serving the public well when the loads being carried were horse drawn wagons and carriages. Then along came the railroads with their heavy loads and the wooden bridges could not provide the necessary strength and stiffness for longer spans. As a result, wrought iron rods replaced wooden tension members and a hybrid truss composed of a combination of wood and metal members developed. As our understanding of which members were carrying tension and which were carrying compression increased, cast iron replaced wooden compression members, thus completing the transition to an all-metal truss form.

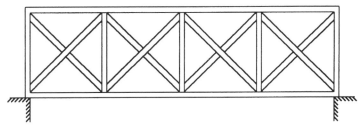

Fig. 1.6 Multiple King-Post Truss—designed by Colonel Stephen H. Long in 1829. From BRIDGES AND MEN by Joseph Gies. Copyright © 1963 by Joseph Gies. Used by permission of Doubleday, a division of Bantam Doubleday Dell Publishing Group, Inc.

In 1841, William Howe, uncle of Elias Howe, the inventor of the sewing machine, received a patent on a truss arrangement in which he took Long's panel system and replaced the wooden vertical members with wrought iron rods (Gies, 1963). The metal rods ran through the top and bottom chords and could be tightened by turnbuckles to hold the wooden diagonal web members in compression against cast iron angle blocks (Fig. 1.7). Occasionally, Howe truss bridges were built entirely of metal, but in general they were composed of both wood and metal components. These bridges have the advantages of the panel system as well as those offered by counterbracing.

A second variation on Long's panel system was patented by Thomas and Caleb Pratt (Caleb was the father of Thomas) in 1844 with wooden vertical members to resist compression and metal diagonal members, which resist only tension (Jackson, 1988). Most of the Pratt trusses built in the United States were entirely of metal and they became more commonly used than any other type. Simplicity, stiffness, constructability, and economy earned this recognition (Edwards, 1959). The distinctive feature of the Pratt truss (Fig. 1.8), and related designs, is that the main diagonal members are in tension.

In 1841, Squire Whipple patented a cast iron arch truss bridge (Fig. 1.9), which he used to span the Erie Canal at Utica, NY (*Note:* Whipple was not a country gentleman, his first name just happened to be Squire.) Whipple utilized wrought iron for the tension members and cast iron for the compression members. This bridge form became known as a bowstring arch truss, although some engineers considered the design to be more a tied arch than a truss (Jackson, 1988). The double-intersection Pratt truss of Figure 1.10, in which the diagonal tension members extended over two panels, was also credited to Whipple because he was the first to use the design when he built railroad bridges near Troy, NY.

To implement his designs, it is implied that Squire Whipple could analyze his trusses and knew the magnitudes of the tensile and compressive forces in the various members. He was a graduate of Union College, class of 1830, and in 1847 he published the first American treatise on determining the stresses produced by bridge loads and proportioning bridge members. It was

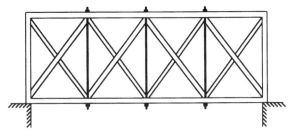

Fig. 1.7   Howe Truss, designed by William Howe, patented in 1841. From BRIDGES AND MEN by Joseph Gies. Copyright © 1963 by Joseph Gies. Used by permission of Doubleday, a division of Bantam Doubleday Dell Publishing Group, Inc.

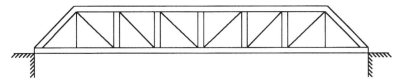

Fig. 1.8 Pratt Truss, designed by Thomas and Caleb Pratt, patented in 1844. From BRIDGES AND MEN by Joseph Gies. Copyright © 1963 by Joseph Gies. Used by permission of Doubleday, a division of Bantam Doubleday Dell Publishing Group, Inc.

titled *A Work on Bridge Building; consisting of two Essays, the one Elementary and General, the other giving Original Plans, and Practical Details for Iron and Wooden Bridges* (Edwards, 1959). In it he showed how one could compute the tensile or compressive stress in each member of a truss that was to carry a specific load (Kirby et al., 1956).

In 1851, Herman Haupt, a graduate of the United States Military Academy at West Point, class of 1835, authored a book titled *General Theory of Bridge Construction,* which was published by D. Appleton and Company, NY (Edwards, 1959). This book and the one by Squire Whipple were widely used by engineers and provided the theoretical basis for selecting cross sections to resist bridge dead loads and live loads.

One other development that was critical to the bridge design profession was the ability to verify the theoretical predictions with experimental testing. The tensile and compressive strengths of cast iron, wrought iron, and steel had to be determined and evaluated. Column load curves had to be developed by testing cross sections of various lengths. This experimental work requires large capacity testing machines.

The first testing machine to be made in America was built in 1832 to test a wrought iron plate for boilers by the Franklin Institute of Philadelphia (Edwards, 1959). Its capacity was about 90 kN, not enough to test bridge components. About 1862, William Sallers and Company of Philadelphia built a testing machine that had a rated capacity of 4500 kN and was specially designed for the testing of full-size columns.

Two testing machines were built by the Keystone Bridge Works, Pittsburgh, PA in 1869–1870 for the St. Louis Bridge Company to evaluate materials for

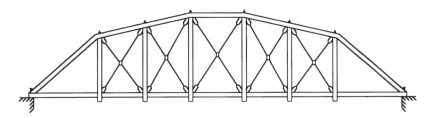

Fig. 1.9   Bowstring Arch—designed by Squire Whipple, patented in 1841.

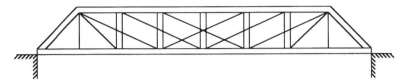

Fig. 1.10 Double-intersection Pratt—credited to Squire Whipple.

Eads' Bridge over the Mississippi River. One had a capacity of 900 kN while the other a capacity of 7200 kN. At the time it was built, the capacity of the larger testing machine was greater than any other in existence (Edwards, 1959).

During the last half of the nineteenth century, the capacity of the testing machines continued to increase until in 1904 the American Bridge Company built a machine having a tension capacity of 18 000 kN (Edwards, 1959) at its Ambridge, PA plant. These testing machines were engineering works in themselves, but they were essential to verify the strength of the materials and the resistance of components in bridges of ever increasing proportions.

1.2.4 Suspension Bridges

Suspension bridges capture the imagination of people everywhere. With their tall towers, slender cables, and tremendous spans, they appear as ethereal giants stretching out to join together opposite shores. Sometimes they are short and stocky, and seem to be guardians and protectors of their domain. Othertimes, they are so long and slender that they seem to be fragile and easily moved. Whatever their visual image, people react to them and remember how they felt when they first saw them.

Imagine the impression on a young child on a family outing in a state park and seeing the infamous "swinging bridge" across the raging torrent of a rock-strewn river (well, it seemed like a raging torrent) for the first time. And then to hear the jeers and challenge of the older children, daring you to cross as they moved side to side and purposely got the swinging bridge to swing. Well, it did not happen that first day, it felt more comfortable to stay with mother and the picnic lunch. But it did happen on the next visit, a year or two later. It was like a rite of passage. A child no longer, able to cross over the rock-strewn stream on the swinging bridge, not fighting it, but moving with it and feeling the exhilaration of being one with forces stronger than you were.

Suspension bridges also make strong impressions on adults and having an engineering education is not a prerequisite. People in the United States have enjoyed these structures on both coasts, where they cross bays and mouths of rivers; and the interior of the country, where they cross the great rivers, gorges, and straits. Most people understand that the cables are the tendons from which the bridge deck is hung, but they marvel at their strength and the ingenuity it took to get them in place. When people see photographs of workers on the towers of suspension bridges, they catch their breath, and then wonder at how small the workers are compared to the towers they have built. Suspension bridges bring out the emotions: wonder, awe, fear, pleasure; but mostly they are enjoyed for their beauty and grandeur.

In 1801, James Finley erected a suspension bridge with wrought iron chains of 21-m span over Jacob's Creek near Uniontown, PA. He is credited as the inventor of the modern suspension bridge with its stiff level floors and secured a patent in 1808 (Kirby and Laurson, 1932). In previous suspension bridges, the roadway was flexible and followed the curve of the ropes or chains. By stiffening the roadway and making it level, Finley developed a suspension bridge that was suitable, not only for footpaths and trails, but for roads with carriages and heavy wagons.

Most engineers are familiar with the suspension bridges of John A. Roebling: the Niagara River Bridge, completed in 1855 with a clear span of 250 m; the Cincinnati Suspension Bridge, completed in 1867 with a clear span of 322 m; and the Brooklyn Bridge, completed in 1883 with a clear span of 486 m. Of these three wire cable suspension bridges from the nineteenth century, the last two are still in service and are carrying highway traffic. However, there is one other long-span wire cable suspension bridge from this era that is noteworthy and still carrying traffic: the Wheeling Suspension Bridge, completed in 1849 with a clear span of 308 m (Fig. 1.11).

The Wheeling Suspension Bridge over the easterly channel of the Ohio River was designed and built by Charles Ellet who won a competition with John Roebling; that is, he was the low bidder. This result of a competition was also true of the Niagara River Bridge, except that Ellet walked away from it after the cables had been strung, saying that the $190,000 he bid was not enough to complete it. Roebling was then hired and he completed the project for about $400,000 (Gies, 1963).

The original Wheeling Suspension Bridge did not have the stiffening truss shown in Figure 1.11. This truss was added after a windstorm in 1854 caused the bridge to swing back and forth with increased momentum, the deck to twist and undulate in waves nearly as high as the towers, until it all came crashing down into the river (very similar to the Tacoma Narrows Bridge failure some 80 years later). The Wheeling Bridge had the strength to resist gravity loads, but it was aerodynamically unstable. Why this lesson was lost to the profession is unknown, but if it had received the attention it deserved, it would have saved a lot of trouble in the years ahead.

What happened to the Wheeling Suspension Bridge was not lost on John Roebling. He was in the midst of the Niagara River project when he heard of the failure, and immediately ordered more cable to be used as stays for the double-decked bridge. An early painting of the Niagara River Bridge shows the stays running from the bottom of the deck to the shore to provide added stability.

Fig. 1.11   Wheeling Suspension Bridge. [Photo by John Brunell, courtesy of West Virginia Department of Transportation].

In 1859 William McComas, a former associate of Charles Ellet, rebuilt the Wheeling Suspension Bridge. In 1872 Wilhelm Hildenbrand, an engineer with Roebling's company, modified the deck and added diagonal stay wires between the towers and the deck to increase the resistance to wind (Jackson, 1988) and to give the bridge the appearance it has today.

The completion of the Brooklyn Bridge in 1883 brought to maturity the building of suspension bridges and set the stage for the long-span suspension bridges of the twentieth century. Table 1.1 provides a summary of some of the notable long-span suspension bridges built in the United States and still standing.

Some comments are in order with regard to the suspension bridges in Table 1.1. The Williamsburg Bridge and the Brooklyn Bridge are of comparable span, but with noticeable differences. The Williamsburg Bridge has steel, rather than masonry towers. The deck truss is a 12.5-m deep lattice truss, compared to a 5.2-m deep stiffening truss of its predecessor. This truss gives the Williamsburg Bridge a bulky appearance, but it is very stable under traffic and wind loadings. Another big difference is that the wire in the steel cables of the Brooklyn Bridge was galvanized to protect it from corrosion in the briny atmosphere of the East River (Gies, 1963), while the wire in its successor was not. As a result, the cables of the Williamsburg Bridge have had

TABLE 1.1   Long-Span Suspension Bridges in the United States

Bridge	Site	Designer	Clear Span (m)	Date
Wheeling	West Virginia	Charles Ellet	308	1847
Cincinnati	Ohio	John Roebling	322	1867
Brooklyn	New York	John Roebling Washington Roebling	486	1883
Williamsburg	New York	Leffert Lefferts Buck	488	1903
Bear Mountain	Hudson Valley	C. Howard Baird	497	1924
Ben Franklin	Philadelphia	Ralph Modjeski Leon Moisseiff	533	1926
Ambassador	Detroit	Jonathon Jones	564	1929
George Washington	New York	Othmar Ammann	1067	1931
Golden Gate	San Francisco	Joseph Strauss Charles Ellis Leon Moisseiff	1280	1937
Verrazano-Narrows	New York	Ammann and Whitney	1298	1964

to be rehabilitated with a new protective system that cost $73 million (Bruschi and Koglin, 1996).

Another observation of Table 1.1 is the tremendous increase in clear span attained by the George Washington Bridge over the Hudson River in New York. It nearly doubled the clear span of the longest suspension bridge in existence at the time it was built, a truly remarkable accomplishment.

One designer, Leon Moisseiff, is associated with most of the suspension bridges in Table 1.1 that were built in the twentieth century. He was the designing engineer of the Manhattan and Ben Franklin bridges; participated in the design of the George Washington Bridge; and was a consulting engineer on the Ambassador, Golden Gate, and Oakland-Bay bridges (Gies, 1963). All of these bridges were triumphs and successes. He was a well-respected engineer who had pioneered the use of deflection theory, instead of the erroneous elastic theory, in the design of the Manhattan Bridge and those that followed. But Moisseiff will also be remembered as the designer of the Tacoma Narrows Bridge that self-destructed during a windstorm in 1940, not unlike that experienced by the Wheeling Suspension Bridge in 1854. The use of a plate girder to stiffen the deck undoubtedly contributed to providing a surface on which the wind could act, but the overall slenderness of the bridge gave it an undulating behavior under traffic even when the wind was not blowing. Comparing the ratio of depth of truss or girder to the span length for the Williamsburg, Golden Gate, and Tacoma Narrows bridges; we have 1:40, 1:164, and 1:350, respectively (Gies, 1963). The design had gone one step too far in making a lighter and more economical structure. The tragedy for bridge design professionals of the Tacoma Narrows failure was a tough lesson, but one that will not be forgotten.

1.2.5   Metal Arch Bridges

Arch bridges are aesthetically pleasing and can be economically competitive with other bridge types. Sometimes the arch can be above the deck, as in a tied-arch design, or as in the bowstring arch of Whipple (Fig. 1.9). Othertimes, when the foundation materials can resist the thrusts, the arch is below the deck. Restraint conditions at the supports of an arch can be fixed or hinged. And if a designer chooses, a third hinge can be placed at the crown to make the arch statically determinate or nonredundant.

The first iron arch bridge in the United States was built in 1839 across Dunlap's Creek at Brownsville in southwestern Pennsylvania on the National Road (Jackson, 1988). The arch consists of five tubular cast iron ribs that span 24 m between fixed supports. It was designed by Captain Richard Delafield and built by the U.S. Army Corps of Engineers (Edwards, 1959). It is still in service today.

The second cast iron arch bridge in this county was completed in 1860 across Rock Creek between Georgetown and Washington, DC. It was built by the Army Corps of Engineers under the direction of Captain Montgomery Meigs as part of a 24-km aqueduct, which brings water from above the Great Falls on the Potomac to Washington, DC. The two arch ribs of the bridge are 1.3-m diameter cast iron pipes that span 61 m with a rise of 6.1 m and carry water within its 38-mm thick walls. The arch supports a level roadway on open-spandrel posts that carried Washington's first horse-drawn street railway line (Edwards, 1959). The superstructure of the bridge was removed in 1916 and replaced by a concrete arch bridge. However, the pipe arches remain in place between the concrete arches and continue to carry water to the city today.

Two examples of steel deck arch bridges from the nineteenth century that still carry highway traffic are the Washington Bridge across the Harlem River in New York and the Panther Hollow Bridge in Schenely Park, Pittsburgh, PA (Jackson, 1988). The two-hinged arches of the Washington Bridge, completed in 1889, are riveted plate girders with a main span of 155 m. This bridge is the first American metal arch bridge in which the arch ribs are plate girders (Edwards, 1959). The three-hinged arch of the Panther Hollow Bridge, completed in 1896, has a span of 110 m.

One of the most significant bridges built in the United States is the steel deck arch bridge designed by James B. Eads across the Mississippi River at St. Louis. It took seven years to construct and was completed in 1874. The three-arch superstructure consisted of two 153-m side arches and one 159-m center arch that carried two decks of railroad and highway traffic (Fig. 1.12). The Eads Bridge is significant because of the very deep pneumatic caissons

Fig. 1.12 Eads Bridge, St. Louis, Missouri. [Photo courtesy of Kathryn Kontrim, 1996].

for the foundations, the early use of steel in the design, and the graceful beauty of its huge arches as they span across the wide river (Jackson, 1988).

Because of his previous experience as a salvage diver, Eads realized that the foundations of his bridge could not be placed on the shifting sands of the river bed but must be set on bedrock. The west abutment was built first with the aid of a cofferdam and founded on bedrock at a depth of 9 m. Site data indicated that bedrock sloped downward from west to east, with an unknown depth of over 30 m at the east abutment, presenting a real problem for cofferdams. While recuperating from an illness in France, Eads learned that European engineers had used compressed air to keep water out of closed caissons (Gies, 1963). He adapted the technique of using caissons, or wooden boxes, added a few innovations of his own, such as a sand pump, and completed the west and east piers in the river. The west pier is at a depth of 26 m and the east pier at a depth of 37 m.

However, the construction of these piers was not without cost. Twelve workman died in the east pier and one in the west pier from caisson's disease, or the bends. These deaths caused Eads and his physician, Dr. Jaminet, much anxiety because the east abutment had to go even deeper. Based on his own experience in going in and out of the caissons, Dr. Jaminet prescribed slow decompression and shorter working time as the depth increased. At a depth of 30 m, a day's labor consisted of two working periods of 45 min each, separated by a rest period. As a result of the strict rules, only one death occurred in the placement of the east abutment on bedrock at a depth of 42 m.

It is ironic that the lessons learned by Eads and Dr. Jaminet were not passed on to Washington Roebling and his physician, Dr. Andrew H. Smith, in the parallel construction of the Brooklyn Bridge. The speculation is that Eads and Roebling had a falling-out because of Ead's perception that Roebling had copied a number of caisson ideas from him. Had they remained on better terms, Roebling may not have been stricken by the bends, and partially paralyzed for life (Gies, 1963).

Another significant engineering achievement of the Eads Bridge was in the use of chrome steel in the tubular arches that had to meet, for that time, stringent material specifications. Eads insisted on an elastic limit of 345 MPa and an ultimate strength of 830 MPa for his steel at a time when the steel producers (one of which was Andrew Carnegie) questioned the importance of an elastic limit (Kirby et al., 1956). The testing machines mentioned in Section 1.2.3 had to be built, and it took some effort before steel could be produced that would pass the tests. The material specification of Eads was unprecedented in both its scale and quality of workmanship demanded, setting a benchmark for future standards (Brown, 1993).

The cantilever construction of the arches for the Eads Bridge was also a significant engineering milestone. Falsework in the river was not possible, so Eads built falsework on top of the piers and cantilevered the arches, segment by segment in a balanced manner, until the arches halves met at midspan (Kirby et al., 1956). On May 24, 1874, the highway deck was opened for pedestrians; on June 3 it was for vehicles; and on July 2 some 14 locomotives, 7 on each track, crossed side by side (Gies, 1963). The biggest bridge of any type ever built anywhere up to that time had been completed. (For a very complete description of the Eads Bridge and a narration of its building see Woodward, 1881).

Since the Eads Bridge, steel arch bridges longer than its 159 m center span have been constructed. These include the 298-m clear span Hell Gate Bridge over the East River in New York, completed in 1917; the 503 m clear span Bayonne Arch Bridge over the Kill van Kull between Staten Island and New Jersey, completed in 1931; and the world's longest 518 m clear span New River Gorge Bridge near Fayetteville, WV, completed in 1978 and designed by Michael Baker, Jr., Inc. (Fig. 1.13).

1.2.6   Reinforced Concrete Bridges

In contrast to wood and metal, reinforced concrete has a relatively short history. It was in 1824 that Joseph Aspdin of England was recognized for producing Portland cement by heating ground limestone and clay in a kiln. This cement was used to line tunnels under the Thames River because it was water resistent. In the United States, Portland cement was produced in Pennsylvania

Fig. 1.13   New River Gorge Bridge. [Photo by Terry Clark Photography, courtesy of West Virginia Department of Transportation].

in 1871 by D. O. Taylor and about the same time in Indiana by T. Millen of South Bend. It was not until the early 1880s that significant amounts were produced in the United States (McGregor, 1992).

In 1867, a French nursery gardener, Joseph Monier, received a patent for concrete tubs reinforced with iron. In the United States, Ernest Ransome of California was experimenting with reinforced concrete and in 1884 he received a patent for a twisted steel reinforcing bar. The first steel bar reinforced concrete bridge in the United States was built by Ransome in 1889: the Alvord Lake Bridge in Golden Gate Park, San Francisco, CA. This bridge has a modest span of 6 m, is 19.5 m wide, and is still in service (Jackson, 1988).

After the success of the Alvord Lake Bridge, reinforced concrete arch bridges were built in other parks because their classic stone arch appearance fit the surroundings. One of these that remains to this day is the 42-m span

Eden Park Bridge in Cincinnati, OH built by Fritz von Emperger in 1895. This bridge is not a typical reinforced concrete arch, but has a series of curved steel *I*-sections placed in the bottom of the arch and covered with concrete. This design was developed by Joseph Melan of Austria and, though it was used only for a few years, it played an important role in establishing the viability of reinforced concrete bridge construction (Jackson, 1988).

Begun in 1897, but not completed until 1907, was the high-level Taft Bridge carrying Connecticut Avenue over Rock Creek in Washington, DC. This bridge consists of five open-spandrel unreinforced concrete arches supporting a reinforced concrete deck. It was designed by George Morison and its construction was supervised by Edward Casey (Jackson, 1988). This bridge has recently been renovated and is prepared to give many more years of service.

Two reinforced concrete arch bridges in Washington, DC over the Potomac River are also significant. One is the Key Bridge (named after Francis Scott Key who lived near the Georgetown end of the bridge), completed in 1923, which connects Georgetown with Rosslyn, VA. It has seven open-spandrel three-ribbed arches designed by Nathan C. Wyeth and has recently been refurbished. The other is the Arlington Memorial Bridge, completed in 1932, which connects the Lincoln Memorial and Arlington National Cemetery. It has nine arches, eight are closed-spandrel reinforced concrete arches and the center arch, with a span of 66 m, is a double-leaf steel bascule bridge that has not been opened for several years. It was designed by the architectural firm of McKim, Mead, and White (Jackson, 1988).

Other notable reinforced concrete deck arch bridges still in service include: the nine-span, open-spandrel Colorado Street Bridge in Pasadena, CA, near the Rose Bowl, designed by Waddell and Harrington, and completed in 1913; the 30-m single-span, open-spandrel Shepperd's Dell Bridge across the Young Creek near Latourell, OR, designed by K. R. Billner and S. C. Lancaster, and completed in 1914; the 43-m single-span, closed-spandrel Canyon Padre Bridge on old Route 66 near Flagstaff, AZ, designed by Daniel Luten and completed in 1914; the ten-span, open-spandrel Tunkhannock Creek Viaduct near Nicholson, PA, designed by George Ray and completed in 1915 (considered to be volumetrically the largest structure of its type in the world); the thirteen-span, open-spandrel Mendota Bridge across the Minnesota River at Mendota, MN, designed by C. A. P. Turner and Walter Wheeler, and completed in 1926; the seven-span, open-spandrel Rouge River Bridge on the Oregon Coast Highway near Gold Beach, OR, designed by Conde B. McCullough and completed in 1931; the five-span, open-spandrel George Westinghouse Memorial Bridge across Turtle Creek at North Versailles, PA, designed by Vernon R. Covell and completed in 1931; and the 100-m single-span, open-spandrel Bixby Creek Bridge south of Carmel, CA, on State Route 1 amid the rugged terrain of the Big Sur (Fig. 1.14), designed by F. W. Panhorst and C. H. Purcell, and completed in 1933 (Jackson, 1988).

Reinforced concrete through-arch bridges were also constructed. James B. Marsh received a patent in 1912 for the Marsh rainbow arch bridge. This

Fig. 1.14 Bixby Creek Bridge, south of Carmel, California. [Roberts, 1990 (Used with permission of American Concrete Institute)].

bridge resembles a bowstring arch truss but uses reinforced concrete for its main members. Three examples of Marsh rainbow arch bridges still in service are the 27-m single-span Spring Street Bridge across Duncan Creek in Chippewa Falls, WI, completed in 1916; the eleven 27-m arch spans of the Fort Morgan Bridge across the South Platte River near Fort Morgan, CO, completed in 1923; and the 25-m single-span Cedar Creek Bridge near Elgin, KS, completed in 1927 (Jackson, 1988).

One interesting feature of the 1931 Rouge River Bridge, which is a precursor of things to come, is that the arches were built using the prestressing construction techniques first developed by the French engineer Eugene Freyssinet in the 1920s (Jackson, 1988). In the United States, the first prestressed concrete girder bridge was the Walnut Lane Bridge in Philadelphia, PA, which was completed in 1950. After the success of the Walnut Lane Bridge, prestressed concrete construction of highway bridges gained in popularity and is now used throughout the United States.

1.2.7 Girder Bridges

Girder bridges are the most numerous of all highway bridges in the United States. Their contribution to the transportation system often goes unrecognized, because the great suspension, steel arch, and concrete arch bridges are the ones people remember. The spans of girder bridges seldom exceed 150 m, with a majority of them less than 50 m, so they do not get as much attention as they perhaps should. Girder bridges are important structures because they are everywhere.

Girders are not as efficient as trusses in resisting loads over long spans. However, for short and medium spans the difference in material weight is small and girder bridges are competitive. In addition, the girder bridges have greater stiffness and are less subject to vibrations. This characteristic was important to the railroads and resulted in the early application of plate girders in their bridges.

A plate girder is an *I*-section assembled out of flange and web plates. The earliest ones were fabricated in England with rivets connecting double angles from the flanges to the web. In the United States, a locomotive builder, the Portland Company of Portland, ME, fabricated a number of railroad bridges around 1850 (Edwards, 1959). In early plate girders, the webs were often deeper than the maximum width of plate produced by rolling mills. As a result, the plate girders were assembled with the lengthwise dimension of the web plate in the transverse direction of the section from flange to flange. An example is a wrought iron plate girder span of 35 m built by the Elmira Bridge Company, Elmira, NY, in 1890 for the New York Central Railroad with a web depth of 2.7 m fabricated from plates 1.8 m wide (Edwards, 1959).

Steel plate girders eventually replaced wrought iron in the railroad bridge. An early example is the 457-m long Fort Sumner Railroad Bridge on concrete piers across the Pecos River, Fort Sumner, NM, completed in 1906 (Jackson, 1988). This bridge is still in service.

Other examples of steel plate girder bridges are the 2074-m long Knight's Key Bridge and the 1809-m long Pigeon Key Bridge, both part of the Seven Mile Bridge across the Gulf of Mexico from the mainland to Key West, FL (Jackson, 1988). Construction on these bridges began in 1908 and was completed in 1912. Originally they carried railroad traffic, but were converted to highway use in 1938.

Following the success of the Walnut Lane Bridge in Philadelphia, PA in 1950, prestressed concrete girders became popular as a bridge type for highway interchanges and grade separations. In building the interstate highway system, innumerable prestressed concrete girder bridges, some with single and multiple box sections, have been and continue to be built.

Some of the early girder bridges, with their multiple short spans and deep girders, were not very attractive. However, with the advent of prestressed concrete and the development of segmental construction, the spans of girder bridges have become longer and the girders more slender. The result is that the concrete girder bridge is not only functional but can also be designed to be aesthetically pleasing (Fig. 1.15).

1.2.8   Closing Remarks

Bridge engineering in the United States has come a long way since those early stone arch and wooden truss bridges. It is a rich heritage and much can

Fig. 1.15 Napa River Bridge. [Photo courtesy of California Department of Transportation].

be learned from the early builders in overcoming what appeared to be insurmountable difficulties. These builders had a vision of what needed to be done and, sometimes, by the sheer power of their will, completed projects that we view with awe today. The challenge for today's bridge engineer is to follow in the footsteps of these early designers and create and build bridges that other engineers will write about 100 and 200 years from now.

1.3 BRIDGE SPECIFICATIONS

For most bridge engineers, it seems that bridge specifications were always there. But that is not the case. The early bridges were built under a design-build type contract. A bridge company would agree, for some lump sum price, to construct a bridge connecting one location to another. There were no standard bridge specifications and the contract went to the low bidder. The bridge company basically wrote their own specifications when describing the bridge they were proposing to build. As a result, depending on the integrity, education, and experience of the builder, some very good bridges were constructed and at the same time some very poor bridges were built.

Of the highway and railroad bridges built in the 1870s, one out of every four failed, a rate of 40 bridges per year (Gies, 1963). The public was losing confidence and did not feel safe when traveling across any bridge. (The fear of crossing a bridge is a part of the gene pool that has been passed on to us today and it may have had its origin in the last half of the nineteenth century.) Something had to be done to improve the standards by which bridges were designed and built.

An event took place on the night of December 29, 1876, that attracted the attention of not only the public, but also the engineering profession. In a blinding snowstorm, an 11-car train with a doubleheader locomotive started across the Ashtabula Creek at Ashtabula, OH, on a 48-m iron bridge, when the first tender derailed, plowed up the ties, and caused the second locomotive to smash into the abutment (Gies, 1963). The coupling broke between the lead tender and the second locomotive, and the first locomotive and tender went racing across the bridge. The bridge collapsed behind them. The second locomotive, tender, and 11 cars plunged some 20 m into the creek. The wooden cars burst into flames when their pot-bellied stoves were upset, and a total of 80 passengers and crew died.

In the investigation that followed, a number of shortcomings in the way bridges were designed, approved, and built were apparent. The bridge was designed by an executive of the railroad who had limited bridge design experience. The acceptance of the bridge was by test loading with six locomotives, which only proved that the factor of safety was at least 1.0 for that particular loading. The bridge was a Howe truss with cast iron blocks for seating the diagonal compression members. These blocks were suspected of contributing to the failure. It is ironic that at a meeting of the American Society of Civil Engineers (ASCE), a statement was made that "the construction of the truss violated every canon of our standard practice" at a time when there were no standards of practice (Gies, 1963).

The American practice of using concentrated axle loads instead of uniformly distributed loads was introduced in 1862 by Charles Hilton of the New York Central Railroad (Edwards, 1959). It was not until 1894 that Theodore Cooper proposed his original concept of train loadings with concentrated axle loadings for the locomotives and tender followed by a uniformly distributed load representing the train. The Cooper series loading became the standard in 1903 when adopted by the American Railroad Engineering Association (AREA) and remain in use to the present day.

On December 12, 1914 the American Association of State Highway Officials (AASHO) was formed and in 1921 its Committee on Bridges and Allied Structures was organized. The charge to this committee was the development of standard specifications for the design, materials, and construction of highway bridges. During the period of development, mimeographed copies of the different sections were circulated to state agencies for their use. The first edition of the *Standard Specifications for Highway Bridges and Incidental Structures* was published in 1931 by AASHO.

The truck train load in the standard specifications is an adaptation of the Cooper loading concept applied to highway bridges (Edwards, 1959). The "H" series loading of AASHO was designed to adjust to different weights of trucks without changing the spacing between axles and wheels. These specifications have been reissued periodically to reflect the ongoing research and development in concrete, steel, and wood structures. They are now in their sixteenth edition, published in 1996 (AASHTO, 1996). In 1963, the AASHO became the American Association of State Highway and Transportation Officials (AASHTO). The insertion of the word "Transportation" was to recognize the officials responsibility for all modes of transportation (air, water, light rail, subways, tunnels, and highways).

In the beginning, the design philosophy utilized in the standard specification was working stress design (also known as allowable stress design). In the 1970s, variations in the uncertainties of loads were considered and load factor design was introduced as an alternative method. In 1986, the Subcommittee on Bridges and Structures initiated a study on incorporating the load and resistance factor design (LRFD) philosophy into the standard specification. This study recommended that LRFD be utilized in the design of highway bridges. The subcommittee authorized a comprehensive rewrite of the entire standard specification to accompany the conversion to LRFD. The result is the first edition of the AASHTO (1994) *LRFD Bridge Design Specifications* which is the document addressed in this book.

1.4   IMPLICATION OF BRIDGE FAILURES ON PRACTICE

On the positive side of the bridge failure at Ashtabula (OH) Creek in 1876 was the realization by the engineering profession that standards of practice for bridge design and construction had to be codified. Good intentions and a firm handshake were not sufficient to insure safety for the traveling public. Specifications, with legal ramifications if they were not followed, had to be developed and implemented. For railroad bridges, this task began in 1899 with the formation of the American Railway Engineering and Maintenance of Way Association and resulted in the adoption of Theodore Cooper's specification for loadings in 1903.

As automobile traffic expanded, highway bridges increased in number and size. Truck loadings were constantly increasing and legal limits had to be established. The original effort for defining loads, materials, and design procedures was made by the U.S. Department of Agriculture, Office of Public Roads in 1913 with the publication of its Circular No. 100, "Typical Specifications for the Fabrication and Erection of Steel Highway Bridges" (Edwards, 1959). In 1919, the Office of Public Roads became the Bureau of Public Roads (now the Federal Highway Administration) and a revised specification was prepared and issued.

The Committee on Bridges and Allied Structures of the AASHTO issued the first edition of *Standard Specifications for Highway Bridges* in 1931. It is interesting to note in the Preface of the sixteenth edition of this publication the listing of the years when the standard specifications were revised: 1935, 1941, 1944, 1949, 1953, 1957, 1961, 1965, 1969, 1973, 1977, 1983, 1989,

1992, and 1996. It is obvious that this document is constantly changing and adapting to new developments in the practice of bridge engineering.

In some cases, new information on the performance of bridges was generated by a bridge failure. A number of lessons have been learned from bridge failures that have resulted in revisions to the standard specifications. For example, changes were made to the seismic provisions after the 1971 San Fernando earthquake. Other bridge failure incidents that influence the practice of bridge engineering are given in the sections that follow.

1.4.1 Silver Bridge, Point Pleasant, WV, December 15, 1967

The collapse of the Silver Bridge over the Ohio River between Point Pleasant, WV and Kanauga, OH on December 15, 1967 resulted in 46 deaths and 9 injuries (NTSB, 1970).

Description. The Point Pleasant Bridge was a suspension bridge with a main span of 213 m and two equal side spans of 116 m. The original design was a parallel wire cable suspension bridge, but had provisions for a heat-treated steel eyebar suspension design that could be substituted if the bidders furnished stress sheets and specifications of the proposed materials. The eyebar suspension bridge design was accepted and built in 1927 and 1928.

Two other features of the design were also unique (Dicker, 1971): the eyebar cables were the top chord of the stiffening truss over a portion of all three spans and the base of each tower rested on rocker bearings. As a result, multiple load paths did not exist and the failure of a link in the eyebar cable would initiate rapid progressive failure of the entire bridge.

Cause of Collapse. The Safety Board found that the cause of the bridge collapse was a cleavage fracture in the eye of an eyebar of the north suspension chain in the Ohio side span (NTSB, 1970). The fracture was caused by development of a flaw due to stress corrosion and corrosion fatigue over the 40-year life of the bridge as the pin-connected joint adjusted its position with each passing vehicle.

Effect on Bridge Practice. The investigation following the collapse of the Silver Bridge disclosed the lack of regular inspections to determine the condition of existing bridges. Consequently, the National Bridge Inspection Standards were established under the 1968 Federal Aid Highway Act. This act requires that all bridges built with federal monies be inspected at regular intervals not to exceed two years. As a result, the state bridge agencies had to catalog all their bridges and the National Bridge Inventory was developed. There are over 577,000 bridges (100,000 are culverts) with spans greater than 6 m in the inventory.

It is ironic that even if the stricter inspection requirements had been in place, the collapse of the Silver Bridge probably could not have been prevented, because the flaw could not have been detected without disassembly of the eyebar joint. A visual inspection of the pin connections with binoculars from the bridge deck would not have been sufficient. The problem lies with using materials that are susceptible to stress corrosion and corrosion fatigue, and in designing structures without redundancy.

1.4.2 I-5 and I-210 Interchange, San Fernando, CA, February 9, 1971

At 6:00 a.m. (Pacific Standard Time), on February 9, 1971, an earthquake with a Richter magnitude of 6.6 occurred in the north San Fernando Valley area of Los Angeles, CA. The earthquake damaged approximately 60 bridges. Of this total, approximately 10% collapsed or were so badly damaged that they had to be removed and replaced (Lew et al., 1971). Four of the collapsed and badly damaged bridges were at the interchange of the Golden State Freeway (I-5) and Foothill Freeway (I-210). At this interchange, two men in a pickup truck lost their lives when the South Connector Overcrossing structure collapsed as they were passing underneath. These were the only fatalities associated with the collapse of bridges in the earthquake.

Description. Bridge types in this interchange included composite steel girders, precast prestressed *I*-beam girders, and prestressed and nonprestressed cast-in-place reinforced concrete box-girder bridges. The South Connector Overcrossing structure was a seven-span, curved, nonprestressed reinforced concrete box girder, carried on single column bents, with a maximum span of 39 m. The North Connector Overcrossing structure was a skewed four-span, curved, nonprestressed reinforced concrete box girder, carried on multiple column bents, with a maximum span of 55 m. A group of parallel composite steel girder bridges carried I-5 North and I-5 South over the Southern Pacific railroad tracks and San Fernando Road. Immediately to the east of this group, over the same tracks and road, was a two span cast-in-place prestressed concrete box girder, carried on a single bent, with a maximum span of 37 m.

When the earthquake struck, the South Connector collapsed on to the North Connector and I-5, killing the two men in the pickup truck. The North Connector superstructure held together, but the columns were bent double and burst their spiral reinforcement. One of the group of parallel bridges on I-5 was also struck by the falling South Connector and two others fell off their bearings. The bridge immediately to the east suffered major column damage and was removed.

Cause of Collapse. More than one cause contributed to the collapse of the bridges at the I-5 and I-210 Interchange. The bridges were designed for lateral seismic forces of about 4% of the dead load, which is equivalent to an acceleration of 0.04 g, and vertical seismic forces were not considered. From field measurements made during the earthquake, the estimated ground accelerations at the interchange were from 0.33 to 0.50 g laterally and from 0.17 to 0.25 g vertically. The seismic forces were larger than what the structures were designed for and placed an energy demand on the structures that could not be dissipated in the column-girder and column-footing connections. The connections failed, resulting in displacements that produced large secondary effects, which led to progressive collapse. Girders fell off their supports because the seat dimensions were smaller than the earthquake displacements. These displacement effects were amplified in the bridges that were curved or skewed and were greater in spread footings than in pile-supported foundations.

Effect on Bridge Practice. The collapse of bridges during the 1971 San Fernando earthquake pointed out the inadequacies of the lateral force and seismic design provisions of the specifications. Modifications were made and new articles were written to cover the observed deficiencies in design and construction procedures. The issues addressed in the revisions included the following: (1) seismic design forces include a factor that expresses the probability of occurrence of a high intensity earthquake for a particular geographic region, a factor that represents the soil conditions, a factor that reflects the importance of the structure, and a factor that considers the amount of ductility available in the design; (2) methods of analysis capable of representing horizontal curvature, skewness of span, variation of mass, and foundation conditions; (3) provision of alternative load paths through structural redundancy or seismic restrainers; (4) increased widths on abutment pads and hinge supports; and (5) dissipation of seismic energy by development of increased ductility through closely spaced hoops or spirals, increased anchorage and lap splice requirements, and restrictions on use of large diameter reinforcing bars. Research is continuing in all of these areas, and the specifications are constantly being revised as new information on seismic safety becomes available.

1.4.3   Sunshine Skyway, Tampa Bay, FL, May 9, 1980

The ramming of the Sunshine Skyway Bridge by the Liberian bulk carrier Summit Venture in Tampa Bay, FL on May 9, 1980 destroyed a support pier, and about 395 m of the superstructure fell into the bay. A Greyhound bus, a small pickup truck, and six automobiles fell 45 m into the bay. Thirty-five persons died and one was seriously injured (NTSB, 1981).

Description. The Sunshine Skyway was actually two parallel bridges across Lower Tampa Bay from Maximo Point on the south side of St. Petersburg to Manatee County slightly north of Palmetto, FL. The twin bridge structures are 6.82 km long and consist of post-tensioned concrete girder trestles, steel girder spans, steel deck trusses, and a steel cantilever through truss. The eastern structure was completed in 1954 and was one of the first bridges in the United States to use prestressed concrete. The western structure, which was struck by the bulk carrier, was completed in 1971. No requirements were made for structural pier protection.

The main shipping channel was spanned by the steel cantilever through truss with a center span of 263 m and two equal anchor spans of 110 m. The through truss was flanked on either end by two steel deck trusses with spans of 88 m. The bulk carrier rammed the second pier south of the main channel that supported the anchor span of the through truss and the first deck span. The collision demolished the reinforced concrete pier and brought down the anchor span and suspended span of the through truss, and one deck truss span.

Cause of Collapse. The Safety Board determined that the probable cause of the accident was the failure of the pilot of the Summit Venture to abort the passage under the bridge when the navigational references for the channel and bridge were lost in the heavy rain and high winds of an intense thunderstorm (NTSB, 1981). The lack of a structural pier protection system, which could have redirected the vessel and reduced the amount of damage, contributed to the loss of life. The collapse of the cantilever through truss and deck truss spans of the Sunshine Skyway Bridge was due to the loss of support of the pier rammed by the Summit Venture and the progressive instability and twisting failure that followed.

Effect on Bridge Practice. A result of the collapse of the Sunshine Skyway Bridge was the development of standards for the design, performance, and location of structural bridge pier protection systems. Provisions for determining vessel collision forces on piers and bridges are incorporated in the AASHTO (1994) LRFD Bridge Specifications.

1.4.4 Mianus River Bridge, Greenwich, CT, June 28, 1983

A 30-m suspended span of the eastbound traffic lanes of Interstate Route 95 over the Mianus River in Greenwich, CT collapsed and fell into the river on June 28, 1983. Two tractor-semitrailers and two automobiles drove off the edge of the bridge and fell 21 m into the river. Three persons died and three received serious injuries (NTSB, 1984).

Description. The Mianus River Bridge is a steel deck bridge of welded construction that has 24 spans, 19 of which are approach spans, and is 810 m long. The 5 spans over water have a symmetric arrangement about a 62.5-m main span, flanked by a 30-m suspended span and a 36.6-m anchor span on each side. The main span and the anchor span each cantilever 13.7 m beyond their piers to a pin and hanger assembly, which connects to the suspended span. The highway is six lanes wide across the bridge, but a lengthwise expansion joint on the centerline of the bridge separates the structure into two parallel bridges that act independent of each other. The bridge piers in the water are skewed 53.7° to conform with the channel of the Mianus River.

The deck structure over the river consists of two parallel haunched steel girders with floor beams that frame into the girders. The continuous five-span girder has four internal hinges at the connections to the suspended spans and is, therefore, statically determinate. The inclusion of hinges raises the question of redundancy and existence of alternative load paths. During the hearing after the collapse, some engineers argued that because there were two girders, if one pin and hanger assembly failed, the second assembly could provide an alternative load path.

The drainage system on the bridge had been altered by covering the curb drains with steel plates when the roadway was resurfaced in 1973 with bituminous concrete. With the curb drains sealed off, rainwater on the bridge ran down the bridge deck to the transverse expansion joints between the suspended span and the cantilever arm of each anchor span. During heavy rainfall, considerable water leaked through the expansion joint where the pin and hanger assemblies were located.

After the 1967 collapse of the Silver Bridge, the National Bridge Inspection Standards were established, which required regular inspections of bridges at intervals not exceeding two years. The Mianus River Bridge had been inspected by ConnDOTs Bridge Safety and Inspection Section 12 times since 1967 with the last inspection in 1982. The pin and hanger assemblies of the inside girders were observed from a catwalk between the separated roadways, but the pin and hanger assemblies connecting the outside girders were visually checked from the ground using binoculars. The inspectors noted there was heavy rust on the top pins from water leaking through the expansion joints.

Cause of Collapse. The eastbound suspended span that collapsed was attached to the cantilever arms of the anchor spans at each of its four corners. Pin and hanger assemblies were used to support the northeast (inside girder) and southeast (outside girder) corners of the eastern edge of the suspended span. The western edge was attached to the cantilever arms by a pin assembly without hangers. The pin and hanger assemblies consist of an upper pin in the cantilever arm and a lower pin in the suspended span connected by two hangers, one on either side of the web.

Sometime before the collapse of the suspended span, the inside hanger at the southeast corner came off the lower pin, which shifted all the weight on this corner to the outside hanger. With time, the outside hanger moved laterally outward on the upper pin. Eventually, a fatigue crack developed in the end of the upper pin, its shoulder fractured, the outside hanger slipped off, and the suspended span fell into the river.

The Safety Board concluded that the probable cause of the collapse of the Mianus River Bridge suspended span was the undetected lateral displacement of the hangers in the southeast corner suspension assembly by corrosion induced forces due to deficiencies in the State of Connecticut's bridge safety inspection and bridge maintenance program (NTSB, 1984).

Effect on Bridge Practice. A result of the collapse of the Mianus River Bridge was the development and enforcement of detailed and comprehensive bridge inspection procedures. The Mianus River Bridge was being inspected on a regular basis, but the inspectors had no specific directions as to what the critical elements were that could result in a catastrophic failure.

Another effect of this collapse was the flurry of activity in all the states to inspect all of their bridges with pin and hanger assemblies. In many cases, they found similar deterioration and were able to prevent accidents by repair or replacement of the assemblies. In designs of new bridges, pin and hanger assemblies have found disfavor and will probably not be used unless special provisions are made for inspectability and maintainability.

The investigation of the collapse also pointed out the importance of an adequate surface drainage system for the roadway on the bridge. Drains, scuppers, and downspouts must be designed to be self-cleaning and placed so that they discharge rainwater and melting snow with de-icing salts away from the bridge structure in a controlled manner.

1.4.5 Schoharie Creek Bridge, Amsterdam, NY, April 5, 1987

Three spans of the Schoharie Creek Bridge on I-90 near Amsterdam, NY fell 24 m into a rain-swollen creek on April 5, 1987 when two of its piers collapsed. Four automobiles and one tractor-semitrailer plunged into the creek. Ten persons died (NTSB, 1988).

Description. The Schoharie Creek Bridge consisted of five simply supported spans of lengths 30.5 m–33.5 m–36.6 m–33.5 m–30.5 m. The roadway width was 34.3 m and carried four lanes of highway traffic. The superstructure was composed of two main steel girders 3.66 m deep with transverse floor beams that spanned the 17.4 m between girders and cantilevered 7.8 m on either side. Stringers ran longitudinally between the floor beams and supported a noncomposite concrete deck. Members were connected with rivets.

The substructure consisted of four piers and two abutments. The reinforced concrete piers had two columns directly under the two girders and a tie beam near the top. Each pier was supported by a spread footing on dense glacial deposits. Piers 2 and 3 were located in the main channel of Schoharie Creek and were to be protected by riprap. Only the abutments were supported on piles. Unfortunately, in the early 1950s when this bridge was being designed, no satisfactory method was available to predict scour depth.

The bridge was opened to traffic on October 26, 1954 and on October 16, 1955, the Schoharie Creek experienced its flood of record (1900–1987) of 2170 m^3/s. The estimated discharge on April 5, 1987 when the bridge collapsed was 1840 m^3/s. The 1955 flood caused slight damage to the riprap, and in 1977 a consulting engineering firm recommended replacing missing riprap. This replacement was never done.

Records show that the Schoharie Creek Bridge had been inspected annually or biennially since the National Bridge Inspection Standards were required by the 1968 Federal Aid Highway Act. These inspections of the bridge were only of the above water elements and were usually conducted by maintenance personnel, not by engineers. At no time since its completion had the bridge received an underwater inspection of its foundation.

Cause of Collapse. The severe flooding of Schoharie Creek caused local scour to erode the soil beneath pier 3, which then dropped into the scour hole, and resulted in the collapse of spans 3 and 4. The bridge wreckage in the creek redirected the water flow so that the soil beneath pier 2 was eroded, and some 90 min later, it fell into the scour hole and caused the collapse of span 2. Without piles, the Schoharie Creek Bridge was completely dependent on riprap to protect its foundation against scour and it was not there.

The Safety Board determined that the probable cause of the collapse of the Schoharie Creek Bridge was the failure of the New York State Thruway Authority to maintain adequate riprap around the bridge piers, which led to the severe erosion of soil beneath the spread footings (NTSB, 1988). Contributing to the severity of the accident was the lack of structural redundancy in the bridge.

Effect on Bridge Practice. The collapse of the Schoharie Creek Bridge resulted in an increased research effort to develop methods for estimating depth of scour in a streambed around bridge piers and for estimating size of riprap to resist a given discharge rate or velocity. Methods for predicting depth of scour are now available.

An ongoing problem that needs to be corrected is the lack of qualified bridge inspection personnel. This problem is especially true for underwater inspections of bridge foundations, because there are approximately 300,000 bridges over water and 100,000 have unknown foundation conditions.

Once again the Safety Board recommends that bridge structures should be redundant and have alternative load paths. Engineers should finally be getting the message and realize that continuity is one key to a successful bridge project.

1.5 BRIDGE ENGINEER–PLANNER, ARCHITECT, DESIGNER, CONSTRUCTOR, AND FACILITY MANAGER

The bridge engineer is often involved with several or all aspects of bridge planning, design, and management. This situation is not typical in the building design profession where the architect usually heads a team of diverse design professionals consisting of architects, civil, structural, mechanical, and electrical engineers. In the bridge engineering profession, the bridge engineer works closely with other civil engineers who are in charge of the roadway design and alignment. After the alignment is determined, the engineer often controls the bridge type, aesthetics, and technical details. As part of the design process, the bridge engineer is often charged with reviewing shop drawing and other construction details.

Many aspects of the design affect the long-term performance of the system, which is of paramount concern to the bridge owner. The owner, who is often a department of transportation or other public agency, is charged with the management of the bridge, which includes routine inspections, rehabilitation and retrofits as necessary, and continual prediction of the life-cycle performance or deterioration modeling. Such bridge management systems (BMS) are beginning to play a large role in suggesting the allocation of resources to best maintain an inventory of bridges. A typical BMS is designed to predict the long-term costs associated with the deterioration of the inventory and recommend maintenance items to minimize total costs for a system of bridges. Because the bridge engineer is charged with maintaining the system of bridges, or inventory, his/her role differs significantly from the building engineer where the owner is often a real estate professional controlling only one, or a few, buildings, and then perhaps for a very short time.

In summary, the bridge engineer has significant control over the design, construction, and maintenance processes. With this control comes significant responsibility for public safety and resources. The decision the engineer makes in design will affect the long-term site aesthetics, serviceability, maintainability, and ability to retrofit for changing demands. In short, the engineer is (or interfaces closely with) the planner, architect, designer, constructor, and facility manager.

Many aspects of these functions are discussed in the following chapters where we illustrate both a broad-based approach to aid in understanding the global aspects of design, and also include many technical and detailed articles to facilitate the computational/validation of design. Often engineers become specialists in one or two of the areas mentioned in this discussion and interface with others who are expert in other areas. The entire field is so involved that near-complete understanding can only be gained after years of professional practice, and then, few individual engineers will have the opportunity for such diverse experiences.

REFERENCES

AASHTO (1994). *LRFD Bridge Design Specifications,* 1st ed., American Association of State Highway and Transportation Officials, Washington, DC.

AASHTO (1996). *Standard Specification for the Design of Highway Bridges,* 16th ed., American Association of State Highway and Transportation Officials, Washington, DC.

Brown, D. J. (1993). *Bridges,* Macmillan, New York.

Bruschi, M. G. and T. L. Koglin (1996). "Preserving Williamsburg's Cables," *Civil Engineering,* ASCE, Vol. 66, No. 3, March, pp. 36–39.

Dicker, D. (1971). "Point Pleasant Bridge Collapse Mechanism Analyzed," *Civil Engineering,* ASCE, Vol. 41, No. 7, July, pp. 61–66.

Edwards, L. N. (1959). *A Record of History and Evolution of Early American Bridges,* University Press, Orono, ME.

Gies, J. (1963). *Bridges and Men,* Doubleday, Garden City, NY.

Jackson, D. C. (1988). *Great American Bridges and Dams,* The Preservation Press, National Trust for Historic Preservation, Washington, DC.

Kirby, R. S. and P. G. Laurson (1932). *The Early Years of Modern Civil Engineering,* Yale University Press, New Haven, CT.

Kirby, R. S., S. Whithington, A. B. Darling, and F. G. Kilgour (1956). *Engineering in History,* McGraw-Hill, New York.

Lew, S. L., E. V. Leyendecker, and R. D. Dikkers (1971). "Engineering Aspects of the 1971 San Fernando Earthquake," *Building Science Series 40,* National Bureau of Standards, U.S. Department of Commerce, Washington, DC.

MacGregor, J. G. (1992). *Reinforced Concrete Mechanics and Design,* 2nd ed., Prentice Hall, Englewood Cliffs, NJ.

NTSB (1970). "Collapse of U.S. 35 Highway Bridge, Point Pleasant, West Virginia, December 15, 1967," *Highway Accident Report No. NTSB-HAR-71-1,* National Transportation Safety Board, Washington, DC.

NTSB (1981). "Ramming of the Sunshine Skyway Bridge by the Liberian Bulk Carrier SUMMIT VENTURE, Tampa Bay, Florida, May 9, 1980," *Marine Accident Report No. NTSB-MAR-81-3,* National Transportation Safety Board, Washington, DC.

NTSB (1984). "Collapse of a Suspended Span of Interstate Route 95 Highway Bridge Over the Mianus River, Greenwich, Connecticut, June 28, 1983," *Highway Accident Report No. NTSB-HAR-84/03,* National Transportation Safety Board, Washington, DC.

NTSB (1988). "Collapse of New York Thruway (I-90) Bridge Over the Schoharie Creek, Near Amsterdam, New York, April 5, 1987," *Highway Accident Report No. NTSB/HAR-88/02,* National Transportation Safety Board, Washington, DC.

Woodward, C. M. (1881). *A History of the St. Louis Bridge,* G. I. Jones and Company, St. Louis, MO.

2

AESTHETICS AND BRIDGE TYPES

2.1 INTRODUCTION

Oftentimes engineers deceive themselves into believing that if they have gathered enough information about a bridge site and the traffic loads, the selection of a bridge type for that situation will be automatic. Engineers seem to subscribe to the belief that once the function of a structure is properly defined, the correct form will follow. Furthermore, that form will be efficient and aesthetically pleasing. Perhaps we believe there is some great differential equation out there and if we could only describe the relationships and the gradients between the different parameters, apply the correct boundary conditions, and set the proper limits of integration, a solution of the equation will give us the best possible bridge configuration. Unfortunately, or perhaps fortunately, no such equation exists that will define the path we need to follow.

If we have no equation to follow, how is a conceptual design formulated? (In this context, the word design is meant in its earliest and broadest sense, it is the configuration one has before any calculations are made.) Without an equation and without calculations, how does a bridge get designed? An attempt is made in this chapter to answer this question by first examining the nature of the structural design process, then discussing aesthetics in bridge design, and finally, by presenting a description of different bridge types.

2.2 NATURE OF THE STRUCTURAL DESIGN PROCESS

The structural design process itself is probably different for every engineer because it is so dependent on personal experience. However, there are certain characteristics about the process that are common and can serve as a basis for discussion. For example, we know (1) that when a design is completed in our own mind, we must then be able to describe it to others; (2) that we all have different backgrounds and bring different knowledge into the design process; and (3) that the design is not completely open-ended, there are constraints that define an acceptable solution(s). These characteristics are part of the nature of structural design and influence how the process takes place.

A model of the design process incorporating these characteristics has been presented by Addis (1990) and includes the following components: output, input, regulation, and the design procedure. A schematic of this model is shown in Figure 2.1.

2.2.1   Description and Justification

The output component consists of description and justification. *Description* of the design will be drawings and specifications prepared by or under the direction of the engineer. Such drawings and specifications outline what is to be built and how it is to be constructed. *Justification* of the design requires the engineer to verify the structural integrity and stability of the proposed design.

In describing what is to be built, the engineer must communicate the geometry of the structure and the material from which it is made. At one time the engineer was not only the designer, but also the drafter and specification writer. We would sit at our desk, do our calculations, then turn around, maybe climb up on a stool, and transfer the results onto fine linen sheets with surfaces prepared to receive ink from our pens. It seemed to be a rite of passage that all young engineers put in their time on the "board." But then the labor was divided. Drafters and spec-writers became specialists, and the structural engineer began to lose the ability to communicate graphically and may have wrongly concluded that designing is mainly performing the calculations.

This trend toward separation of tasks has been somewhat reversed by the increased capabilities of personal computers. With computer-aided drafting

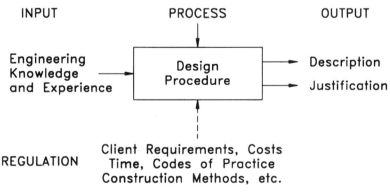

Fig. 2.1   Model of structural design process [Addis, 1990].

(CAD), structural analysis software, and word processing packages all on one system, the structural engineer is again becoming drafter, analyst, and spec-writer, that is, a more complete structural designer. In fact, it is becoming necessary for structural engineers to be CAD literate because the most successful structural analysis programs have CAD-driven preprocessors and postprocessors.

Justification of a proposed design is where most structural engineers excel. Given the configuration of a structure, its material properties, and the loads to which it is subjected; a structural engineer has the tools and responsibility to verify that a design satisfies all applicable codes and specifications. One note of caution: A structural engineer must not fall into the trap of believing that the verification process is infallible. To provide a framework for this discussion, we need to say a few words about deductive and inductive reasoning.

Deductive reasoning goes from broad general principles to specific cases. Once the general principles have been established, the engineer can follow a series of logical steps based on the rules of mathematics and applied physics and arrive at a unique answer that can be defended convincingly. An example would be the principle of virtual work, which can be used for a number of applications such as beam deformations and element stiffness matrices. Just follow the rules, put in the numbers, and the answer has to be correct. Wrong.

Inductive reasoning goes from specific cases to general principles. An example would be going from the experimental observation that doubling the load on the end of a wire doubled its elongation to the conclusion that there is a linear relationship between stress and strain. This conclusion may be true for some materials, and then only with restrictions, but it is not true for others. If experimental observations can be put into the form of an algebraic equation, this is often convenient. But it is also fallible.

It must be realized that deductive justification is based on quantities and concepts determined inductively. Consider, for example, a structural analysis and design program utilized to justify the adequacy of a reinforced concrete frame. Early on, screens will be displayed on the monitor asking the analyst to supply coordinates of joints, connectivity of the members, and boundary conditions. From this information, the computer generates a mathematical model of stick members that have no depth, joints that have no thickness, and supports modeled as rollers, hinges, or are completely restrained. Further, the mathematical model inductively assumes plane sections remain plane, distributed force values to be concentrated at nodes, and idealized boundary conditions at the supports. Next, the user is asked to supply constants or parameters describing material behavior, all of which have been determined inductively from experimental observations. Finally, the values of forces at the nodes determined by the equation solvers in the program must be interpreted as to their acceptance in the real world. This acceptance is based on inductively determined safety factors, load and resistance factors, or service-ability criteria. In short, what appears to be infallible deductive justification of a proposed design is, in fact, based on inductive concepts and is subject to possible error and, therefore, is fallible.

Oftentimes engineers select designs on the basis that they are easy to justify. If an engineer feels comfortable with the analysis of a particular bridge type, that bridge configuration will be used again and again. For example, statically determinate bridge structures of alternating cantilever spans and suspended spans were popular in the 1950s before the widespread use of computers because they were easy to analyze. The same could be said of the earlier railroad truss bridges whose analysis was made simple by graphical statics. One advantage of choosing designs that are easily justified is that those responsible for checking the design have no difficulty visualizing the flow of forces from one component to another. Now, with sophisticated computer software, an engineer must understand how forces are distributed throughout the members of more complex systems to obtain a completed design. The advantage of simple analysis of statically determinate structures is easily offset by their lack of redundancy or multiple load paths. Therefore, it is better to choose continuous beams with multiple redundancy even though the justification process requires more effort to ensure that it has been done properly.

Not only is there an interrelationship between deductive and inductive reasoning, there is also an interrelationship between description and justification. The configuration described for a bridge structure will determine its behavior. Triangles in trusses, continuous beams, arches, and suspension systems have distinctly different spacial characteristics, and therefore, behave differently. Description and justification are linked together, and it is important that a bridge engineer be proficient in both areas with an understanding of the interactions among them.

2.2.2   Public and Personal Knowledge

The input side of the design process shown in Figure 2.1 includes engineering knowledge and experience. An engineer brings both public and personal knowledge to the design process. Public knowledge is accumulated in books, databases, software, and so on, and can be passed on from generation to generation. Such knowledge bases include handbooks of material properties, descriptions of successful designs, standard specifications, theoretical mechanics, construction techniques, computer programs, cost data, and a multitude of other information too voluminous to describe here. Personal knowledge is what has been acquired by an individual through experience and is very difficult to pass on to someone else. People with experience seem to develop an intuitive understanding of structural action and behavior. They understand how forces are distributed and how elements can be placed to gather these forces together to carry them in a simple and efficient manner. And if you were to ask them how they do it, they may not be able to explain why they know that a particular configuration will work and another will not.

The linkage between judgment and experience has been explained this way: good judgment comes from experience and experience comes from bad judgment. Sometimes experience can be a tough teacher, but it is always increasing our knowledge base.

2.2.3 Regulation

Our bridge designs are not open-ended. There are may constraints that define the boundaries of an acceptable design. These constraints include: client's desires, architect's design, relevant codes, accepted practice, engineer's education, available materials, contractor's capabilities, economic factors, environmental concerns, legal factors, and last, but not least, political factors. For example, a bridge is to traverse coastal wetlands, the restrictions on how it can be built will often dictate the selection of the bridge type. If contractors in a particular region are not experienced in the construction method proposed by an engineer, then that may not be the proper design for that locality. Geometric constraints on alignment are quite different for a rural interstate overcrossing than for a densely populated urban interchange. Somehow a bridge designer must be able to satisfy all these restrictions and still have a bridge with pleasing appearance that remains personally and publically satisfying.

2.2.4 Design Process

The process of design is what occurs within the rectangular box of Figure 2.1. An engineer knows what the output has to be and what regulations govern the design, but because each person has accumulated different knowledge and experience, it is difficult to describe a procedure for design that will work in all cases. As Addis (1990) says, "Precisely how and why a structural engineer chooses or conceives a particular structure for a particular purpose is a process so nebulous and individual that I doubt if it is possible to study it at all."

It may not be possible to outline a procedure for the design process, but it is possible to identify its general stages. The first is the data gathering stage, followed by the conceptual, rhetorical, and schematic stages. In the data gathering stage, one amasses as much information as one can find about the bridge site, topography, functional requirements, soil conditions, availability of materials, hydrology, and temperature ranges. Above all, the designer must visit the bridge site, see the setting and its environment, and talk to people, many of whom have been thinking about the bridge project for a long time.

The conceptual or creative stage will vary from person to person because we all have different background, experience, and knowledge. But one thing is constant. It all begins with images in the mind. In the mind one can assimilate all of the information on the bridge site, and then mentally build the bridge, trying different forms, changing them, combining them, looking at them from different angles, driving over the bridge, walking under it, all in the mind's eye. Sometimes the configuration comes as a flash of inspiration, other times it develops slowly as a basic design is adjusted and modified in the mind of the designer.

Too often engineers associate solving problems with solving equations. So we are inclined to get out our calculation pad or get on the computer at our earliest convenience. That is not how the creative process works, in fact, putting ideas down on paper too early may restrict the process because the third spatial dimension and the feeling of spontaneity are lost.

Creative breakthroughs are not made by solving equations. Consider the words of Einstein in a letter to a friend,

> The words or the language, as they are written or spoken, do not seem to play any role in my mechanism of thought. The psychical entities which seem to serve as elements in thought are certain signs and more or less clear images which can be "voluntarily" reproduced and combined . . . this combinatory plan seems to be the essential feature in productive thought before there is any connection with logical construction in words or other kinds of signs which can be communicated to others. Albert Einstein in a letter to Jacques Hadamard (Friedhoff and Benzon, 1989)

So, if you thought the great physicist developed his theories using reams of paper and feverishly manipulating fourth-order tensors, that is wrong. Well, you may argue Einstein was a gifted person, very abstract, and what he did would not necessarily apply to ordinary people designing bridges.

Consider then the words of Leonhardt (1982) that follow the data gathering stage,

> The bridge must then take its initial shape in the imagination of the designer. . . . The designer should now find a quiet place and thoroughly think over the concept and concentrate on it with closed eyes. Has every requirement been met, will it be well-built, would not this or that be better looking . . . ?

These are words from a successful bridge designer, one of the family so to speak, that present what he has learned in more than 50 years of designing bridges. When in his distinguished career he realized this truth I do not know, but we should listen to him. The design of a bridge begins in the mind.

The rhetorical and schematic stages do not necessarily follow one another sequentially. They are simply stages that occur in the design process and may appear in any order and then reappear again. Once a design has been formulated in the mind, one may want to make some sketches to serve as a basis for discussion with one's colleagues. By talking about the design and in explaining it to others, the features of the design come into sharper focus. If there are any shortcomings, chances are they will be discovered and improved solutions will be suggested. In addition to being willing to talk about a design, we must also be willing to have it criticized.

In the design procedure outlined by Leonhardt (1982), he encourages a designer to seek criticism by posting sketches of the proposed design on the walls around the office so others can comment on them. It is surprising what additional pairs of trained eyes can see when they look at the sketches. Well, maybe it is no surprise, because behind every pair of eyes is a whole different set of experiences and knowledge, which brings to mind what de Miranda (1991) says about the three mentalities that must be brought to the design process,

> One should be creative and aesthetic, the second analytical, and the third technical and practical, able to give a realistic evaluation of the possibilities of the construction technique envisaged and the costs involved. If these three mentalities do not coexist in a single mind, they must always be present on terms of absolute equality in the group or team responsible for the design.

In short, make the sketches, talk about them, make revisions, let others critique them, defend the design, be willing to make adjustments, and keep interacting until the best possible design results. It can be a stimulating, challenging, and intellectually rewarding process.

The function of the design process is to produce a bridge configuration that can be justified and described to others. Now is the time to apply the equations for justification of the design and to prepare its description on plans and in specifications. Computers can help with the analysis and the drawings, but there are still plenty of tasks to keep engineers busy. The computer software packages will do thousands of calculations, but they must be checked. Computer-driven hardware can plot full-size plan sheets, but hundreds of details must be coordinated. Model specifications may be stored in a word processor file, but every project is different and has a unique description. A lot of labor follows the selection of the bridge configuration so it must be done right. As Leonhardt (1982) says, "The phase of conceptual and aesthetic design needs a comparatively small amount of time, but is decisive for the expressive quality of the work." In Section 2.3 we will look more closely at the aesthetic qualities of bridge design.

2.3 AESTHETICS IN BRIDGE DESIGN

If we recognize that the conceptual design of a bridge begins in the mind, we only need now to convince ourselves that the design we conceive in our mind is inherently beautiful. It is our nature to desire things that are lovely and appeal to our senses. We enjoy good music and soft lights. We furnish our homes with fine furniture, and select paintings and colors that please our eyes. We may say that we know nothing about aesthetics, yet our actions betray us. We do know what is tasteful, delights the eye, and is in harmony with its surroundings. It is just that we have not been willing to express it.

We need to realize that it is all right to have an opinion and put confidence in what has been placed within us. We simply need to carry over the love of beauty in our daily lives to our engineering projects.

When an engineer is comparing the merits of alternative designs, some factors are more equal than others. The conventional order of priorities in bridge design is safety, economy, serviceability, constructability, and so on. Somewhere down this list is aesthetics. There is no doubt in my mind that aesthetics needs a priority boost, and that it can be done without infringing upon the other factors.

Engineers have to get away from the belief that improved appearance increases the cost of bridges. Oftentimes the most aesthetically pleasing bridge is also the least expensive. Sometimes a modest increase in construction cost is required to improve the appearance of a bridge. Menn (1991) states that the additional cost is about 2% for short spans and only about 5% for long spans. This conclusion is seconded by Roberts (1992) in his paper on case histories of California bridges.

It is a mistake to believe that the public will not spend money on improved appearance. Given a choice, even with a modest increase in initial cost, the public will prefer the bridge that has the nicer appearance. Unfortunately, an engineer may realize this after it is too late. Gottemoeller (1991) tells of the dedication of a pedestrian bridge over a railroad track in the heart of a community in which speaker after speaker decried the ugliness of the bridge and how it had inflicted a scar on the city. They were not concerned about its function or its cost, only its appearance. Needless to say, they rejected a proposal for constructing a similar bridge nearby. It is unfortunate that an engineer has to build an ugly bridge that will remain long after its cost is forgotten to learn the lesson that the public is concerned about appearance.

It is not possible in this short chapter to completely discuss the topic of bridge aesthetics. Fortunately, there are good references dedicated to the subject, which summarize the thoughts and give examples of successful bridge designers throughout the world. Two of these topics are *Esthetics in Concrete Bridge Design,* edited by Watson and Hurd (1990); and *Bridge Aesthetics Around the World,* edited by Burke (1991). By drawing on the expertise in these references we will attempt to identify those qualities that most designers agree influence bridge aesthetics and to give practical guidelines for incorporating them into medium- and short-span bridges.

2.3.1   Definition of Aesthetics

The definition of the word aesthetics may vary according to the dictionary one uses. But usually it includes the words beauty, philosophy, and effect on the senses. A simple definition could be, *Aesthetics is the study of qualities of beauty of an object and of their perception through our senses.* Fernandez-Ordóñez (1991) has some wonderful quotations from the philosophers, such as

Love of beauty is the cause of everything good that exists on earth and in heaven.

Plato

and

Even if this particular aesthetic air be the last quality we seen in a bridge, its influence nonetheless exists and has an influence on our thoughts and actions.

Santayana

and

It is impossible to discover a rule that can be used to judge what is beautiful and what is not.

Hegel

The last quote from Hegel seems to contradict what we have set out to do in providing guidelines for aesthetically pleasing bridge designs. However, in another sense, it reinforces what was said earlier that we should not look for some magical equation that will tell us how to design a bridge. This lack of a magical equation should not discourage us from attempting to find basic principles for aesthetic design utilized by successful bridge designers.

From these philosophers we realize that it is difficult to argue against making something beautiful. It is not necessary that everyone agrees as to what makes a bridge beautiful, but it is important that designers are aware of the qualities of a bridge that influence the perception of beauty.

2.3.2   Qualities of Aesthetic Design

In reading the papers compiled by Watson and Hurd (1990) and Burke (1991), it quickly becomes apparent that there are number of qualities of successful aesthetic designs that are repeated. These are function, proportion, harmony, order, rhythm, contrast, texture, and use of light and shadow.

Some of these terms are familiar, others may not be, especially in their application to bridges. In an attempt to explain these terms, each one will be discussed along with illustrations of their application to bridges.

Function   For a bridge design to be successful, it must fulfill the purpose for which it is intended. Oftentimes the function of a bridge goes beyond the simple connection of points along a prescribed alignment with a given volume of traffic. For example, a bridge crossing a valley may have the function of safely connecting an isolated community with the schools and services of a larger community by avoiding a dangerous trip down and up steep and twisting roads. A bridge over a railroad track out on the prairie may have the function of eliminating a crossing at grade that claimed a number of lives. Sometimes a bridge has more than one function, such as the bridge across the Straits of Bosporus at Istanbul (Fig. 2.2). This bridge replaces a slow ferry boat trip, but it also serves the function of connecting two continents (Brown et al. 1976).

The function of a bridge must be defined and understood by the designer, client, and public. How that function is satisfied can take many forms, but it must always be kept in mind as the basis for all that follows. Implied with the successful completion of a bridge that fulfills its function, is the notion that it does it safely. If a bridge disappears in a flood, or other calamity, one does not take much comfort in the fact that it previously performed its function. A bridge must always safely perform its function.

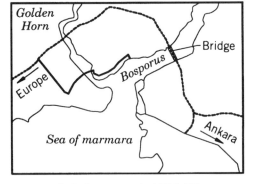

Fig. 2.2 Bosporus straits bridge at Istanbul (Brown et al. 1976) [Photo courtesy of Turkish Government Tourism Office, Washington, DC].

Proportion There was a time when proper proportions were thought to be represented by mathematics and that a rule existed that governed the correct proportions between the adjacent sides of a rectangle, that is the Golden Section with a 1:1.618 proportion. (It is amazing that they thought the eye could discern proportions to four significant figures.) There still are advocates (Lee, 1990) of geometric controls on bridge design and an illustration of the procedure is given in Figure 2.3. The proportioning of the Mancunian Way bridge cross section in Manchester, England, was carried out by making a layout of golden section rectangles in four columns and five rows. The three apexes of the triangles represent the eye-level position of drivers in the three lanes of traffic. The profile of the cross section was then determined by intersections of these triangles and the golden sections.

It may be that proportioning by golden sections is pleasing to the eye, but the usual procedure employed by successful designers has more freedom and arriving at a solution is often by trial and error. It is generally agreed that when a bridge is placed across a relatively shallow valley, as shown in Figure 2.4, the most pleasing appearance occurs when there are an odd number of spans with span lengths that decrease going up the side of the valley (Leonhardt, 1991).

When artists comment on the composition of a painting, they often talk about negative space. What they mean is the space-in-between, the empty spaces that contrast with and help define the occupied areas. Negative space highlights what is and what is not. In Figure 2.4, the piers and girders frame

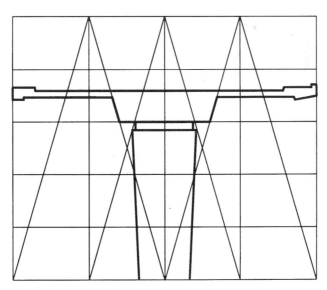

Fig. 2.3 Proportioning of Mancunian Way cross section (Lee, 1990). [Used with permission of American Concrete Institute].

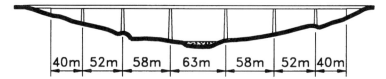

Fig. 2.4 Bridge in shallow valley: flat with varying spans; harmonious (Leonhardt, 1991). [From *Bridge Aesthetics Around the World,* copyright 1991 by the Transportation Research Board, National Research Council, Washington, DC. Reprinted with permission].

the negative space, and it is this space-in-between that must also have proportions that are pleasing to the eye.

The bridge over a deep valley in Figure 2.5 (Leonhardt, 1991) again has an odd number of spans, but they are of equal length. In this case, the negative spaces provide a transition of pleasing rectangular shapes from vertical to horizontal. Adding to the drama of the bridge is the slender continuous girder and the tall, tapered piers. An example of such a bridge is the Magnan Viaduct, near Nizza on the French Riviera, shown in Figure 2.6 (Muller, 1991).

Another consideration is the relative proportion between piers and girders. From a strength viewpoint, the piers can be relatively thin compared to the girders. However, when a bridge has a low profile, the visual impression can be improved by having strong piers supporting slender girders. This point is illustrated in Figure 2.7 (Leonhardt, 1991).

Slender girders can be achieved if the superstructure is made continuous. In fact, Wasserman (1991) says that superstructure continuity is the most important aesthetic consideration and illustrates this with the two contrasting photos in Figures 2.8 and 2.9. Most people would agree that the bridge in Figure 2.9 is awkward looking. It shows what can happen when least effort by a designer drives a project. It does not have to be that way. Consider this quotation from Gloyd (1990),

> "When push comes to shove, the future generation of viewers should have preference over the present generation of penny pinchers."

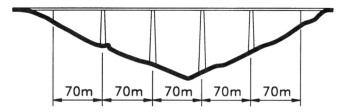

Fig. 2.5 Bridge in deep V-shaped valley: large spans and tapered piers (Leonhardt, 1991). [From *Bridge Aesthetics Around the World,* copyright 1991 by the Transportation Research Board, National Research Council, Washington, DC. Reprinted with permission].

Fig. 2.6 Magnan Viaduct near Nizza, France (Muller, 1991). [From *Bridge Aesthetics Around the World,* copyright 1991 by the Transportation Research Board, National Research Council, Washington, DC. Reprinted with permission].

A designer should also realize that the proportions of a bridge will change when viewed from an oblique angle as seen in Figure 2.10 (Menn, 1991). To keep the piers from appearing as a wall blocking the valley, Leonhardt (1991) recommends limiting the width of piers to about one-eighth of the span length (Fig. 2.11). He further recommends that if groups of columns are used as

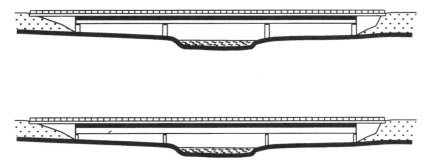

Fig. 2.7 Three-span beam: (*top*) pleasing appearance of slender beam on strong piers; (*bottom*) heavy appearance of deep beam on narrow piers (Leonhardt, 1991). [From *Bridge Aesthetics Around the World,* copyright 1991 by the Transportation Research Board, National Research Council, Washington, DC. Reprinted with permission].

Fig. 2.8 Example of Superstructure Continuity [Photo courtesy of Tennessee DOT—Geo. Hornal, photographer].

Fig. 2.9 Example of poor depth transitions and awkward configurations [Photo courtesy of Tennessee DOT].

Fig. 2.10   Single columns increase the transparency of tall bridge (Menn, 1991). [From *Bridge Aesthetics Around the World*, copyright 1991 by the Transportation Research Board, National Research Council, Washington, DC. Reprinted with permission].

piers, their total width should be limited to about one-third of the span length (Fig. 2.12).

Good proportions are fundamental to achieving an aesthetically pleasing bridge structure. Words can be used to describe what has been successful for some designers, but what works in one setting may not work in another. Rules and formulas will most likely fail. It finally gets down to the responsibility

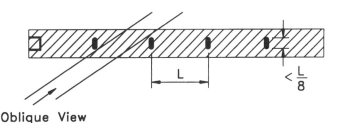

Oblique View

Fig. 2.11   Proportion for pier width not to exceed one-eighth of the span (Leonhardt, 1991). [From *Bridge Aesthetics Around the World*, copyright 1991 by the Transportation Research Board, National Research Council, Washington, DC. Reprinted with permission].

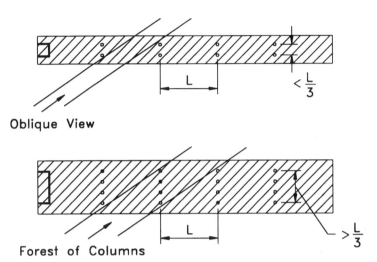

Fig. 2.12   Proportion for total width of groups of columns not to be larger than one-third of the span (Leonhardt, 1991). [From *Bridge Aesthetics Around the World,* copyright 1991 by the Transportation Research Board, National Research Council, Washington, DC. Reprinted with permission].

of each designer on each project to make the personal choices that will lead to a more beautiful structure.

Harmony   In this context, harmony means getting along well with others. The parts of the structure must be in agreement with each other and the whole structure must be in agreement with its surroundings.

Harmony between the elements of a bridge depends on the proportions between the span lengths and depth of girders, height and size of piers, and negative spaces and solid masses. The elements, spaces, and masses of the bridge in Figure 2.13 present a pleasing appearance because they are in harmony with one another. An example of lack of harmony between members and spaces is shown in Figure 2.14. This dissonance is caused by the placement of two dissimilar bridges adjacent to one another.

Harmony between the whole structure and its surroundings depends on the scale or size of the structure relative to its environment. A long bridge crossing a wide valley (Fig. 2.13) can be large because the landscape is large. But when a bridge is placed in an urban setting or used as an interstate overpass, the size must be reduced. Menn (1991) refers to this as integration of a bridge into its surroundings. Illustrations of bridges that are in harmony with their environment are the overpass in Figure 2.15 and the Linn Cove Viaduct of Figure 2.16. A bridge derives its size and scale from its surroundings.

Order and Rhythm   When discussing order and rhythm in bridge structures the same words and examples are often used to describe both. For example,

Fig. 2.13   A graceful long bridge over a wide valley, Napa River Bridge, California. [Permission granted by California DOT].

the bridge in Figure 2.17 illustrates both good order and rhythm. The eye probably first sees the repeating arches flowing across the valley with the regularity of a heartbeat. But also one perceives that all of the members are tied together in an orderly manner in an uninterrupted flow of beauty with a minimum change of lines and edges. If a girder was to replace one of the arch spans, the rhythm would be lost. Rhythm can bring about order, and good order can bring about a wholeness and unity of the structure.

The use of the same words to describe music and bridge aesthetics is not lost on the reader. Consider these comments by Grant (1990),

Fig. 2.14   Lack of harmony between adjacent bridges (Murray, 1991). [From *Bridge Aesthetics Around the World,* copyright 1991 by the Transportation Research Board, National Research Council, Washington, DC. Reprinted with permission].

Fig. 2.15 Well-proportioned concrete arch, West Lilac Road Overpass, I-15. [Permission granted by California DOT].

Fig. 2.16 Linn Cove Viaduct, North Carolina (Gottemoeller, 1991). [From *Bridge Aesthetics Around the World*, copyright 1991 by the Transportation Research Board, National Research Council, Washington, DC. Reprinted with permission].

Fig. 2.17 Tunkhannock Viaduct in Nicholson, Pennsylvania, designed by A. Burton Cohen. [Jet Lowe, HAER Collection, Historic American Engineering Record].

There is beauty and order in classical music—in the harmonies of different sounds, and in their disharmonies and rhythms. There is equal beauty in geometric and arithmetic relationships, similar or equal to those of the sounds.

There is a downside to this analogy with music when repetition and rhythm become excessive. Repeating similar spans too many times can become boring and monotonous, just as hearing the same music with a heavy beat that is repeated over and over again can be uncomfortable. It can also become quite aggravating to be driving down the interstate and seeing the same standard overcrossing mile after mile. The first one or two look just fine, but after a while one has to block out seeing the bridges to keep the mind from overloading. In these cases, something has to be done to break up the monotony.

Contrast and Texture   There is a place for contrast, as well as harmony, in bridge aesthetics. We see this in music and in paintings. There are bright sounds and bright colors, and then there are soft and subtle tones, all in the same composition. Incorporating these into the same composition of our bridges will keep them from becoming boring and monotonous.

All bridges do not have to blend in with their surroundings. Fernandez-Ordóñez (1991) quotes the following from Eduardo Torroja:

When a bridge is built in the middle of the country, it should blend in with the countryside, but very often, because of its proportions and dynamism, the bridge stands out and dominates the landscape.

This dominance seems to be especially true of cable-stayed and suspension bridges, such as seen in Figures 2.18 and 2.19. This dominance of the landscape does not subtract from their beauty.

There can also be contrast between the elements of a bridge to emphasize the slenderness of the girders and the strength of the piers and abutments. Texture can be used to soften the hard appearance of concrete and make certain elements less dominant. Large bridges seen from a distance must develop contrast through their form and mass, but bridges with smaller spans seen up close can effectively use texture. A good example of the use of texture is the I-82 Hinzerling Road Undercrossing near Prosser, Washington, shown in Figure 2.20. The textured surfaces on the solid concrete barrier and the abutments have visually reduced the mass of these elements and made the bridge appear to be more slender than it actually is.

Light and Shadow To use this quality effectively, the designer must be aware of how shadows occur on the structure throughout the day. If the bridge is running north and south the shadows will be quite different than if it is

Fig. 2.18 East Huntington Bridge in Huntington, WV [Photo by David Bowen, courtesy of West Virginia DOT].

Fig. 2.19   Brooklyn Bridge, New York City. [Jet Lowe, HAER Collection, Historic American Engineering Record].

Fig. 2.20   Texture reduces visual mass, I-82 Hinzerling Road Undercrossing, Prosser, WA [Photo courtesy of Washington State DOT].

running east and west. When sunlight is parallel to the face of a girder or wall, small imperfections in workmanship can cast deep shadows. Construction joints in concrete may appear to be discontinuous and hidden welded stiffeners may no longer be hidden.

One of the most effective ways to make a bridge girder appear slender is to put it partially or completely in shadow. Creating shadow becomes especially important with the use of solid concrete safety barriers that make the girders look deeper than they actually are (Fig. 2.21). Shadows can be accomplished by cantilevering the deck beyond the exterior girder as shown in Figure 2.22. The effect of shadow on a box girder is further improved by sloping the side of the girder inward.

Shadow and light have been used effectively in the bridges shown previously. The piers in the bridge of Figure 2.6 have ribs that cast shadows and make them look thinner. The deck overhangs of the bridges in Figures 2.8, 2.10, 2.15, and 2.20 cause changes in light and shadow that improve their appearance because the girders appear more slender and the harshness of a bright fascia is reduced.

2.3.3 Practical Guidelines for Medium- and Short-Span Bridges

The previous discussion on qualities of aesthetic design was meant to apply to all bridges in general. However, what works for a large bridge may not

Fig. 2.21 Concrete barrier wall and short-span overpass (Dorton, 1991). [From *Bridge Aesthetics Around the World,* copyright 1991 by the Transportation Research Board, National Research Council, Washington, DC. Reprinted with permission].

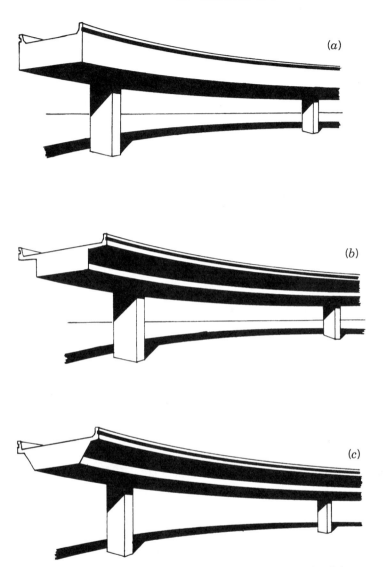

Fig. 2.22 (a) Vertical girder face without overhang presents a visual impact to the driver: structure looks deeper. (b) Increase in overhang creates more shade on face of girder, subduing the visual impact. (c) Sloping girders recede into shadow. Brightly lit face of barrier rail contrasts with shadow and stands out as a continuous, slender band of light, accentuating the flow of the structure. Structure appears subdued, inviting flow of traffic beneath. (Permission granted by the California DOT).

work for a small bridge. Medium- and short-span bridges have special problems. We will attempt to address those problems and offer practical solutions that have worked for other designers. Most of the illustrations used will be those of highway grade separations and crossings over modest waterways.

One word of caution is in order before presenting these guidelines and that is to reemphasize that rules and formulas will likely fail. Burke (1990) provides excerpts from the literature of bridge aesthetics warning against using them exclusively. However, for the inexperienced designer, or for one who does not feel particularly gifted artistically, the guidelines may be helpful. The following quotation taken from Burke (1990) is by Munro (1956) and presents a balanced approach:

> Although it is wise to report all past theories of aesthetics with some suspicion, it is equally wise to utilize them as suggestions.

Therefore, let us consider the guidelines that follow as simply recommendations or suggestions.

Resolution of Duality   Leonhardt (1991) makes the statement, "An odd number of spans is always better than an even number; this is an old and approved rule in architecture." He then goes on to illustrate the balance and harmony of odd numbered spans crossing a valley (Fig. 2.5) and a waterway (Fig. 2.7). So what is a designer to do with a grade separation over dual highways? If you are crossing two highways, the logical solution is to use a two-span layout. But this violates the principle of using odd numbered spans and causes a split composition effect (Dorton, 1991).

This problem is often called "unresolved duality" because the observer has difficulty in finding a central focal point when viewing two large voidal spaces. He suggests increasing the visual mass of the central pier to direct attention away from the large voidal spaces. This redirection has been done successfully in the design of the I-90 overpass near Olympia, WA, shown in Figure 2.23.

Another effective way to reduce the duality effect is to reduce the emphasis on the girder by increasing its slenderness relative to the central pier. This emphasis can be accomplished by increasing the spans and moving the abutments up the slope and has the added effect of opening up the traveled way and giving the feeling of free-flowing traffic. As shown in Figure 2.24, the use of sloping lines in the abutment face and pier top provide an additional feeling of openness. Proper proportions between the girder, pier, and abutment must exist as demonstrated in Figure 2.25. A fine example of applying these recommendations for resolving the duality effect is given by the Hinzerling Road bridge of Figure 2.20.

Generally speaking, the ideal bridge for a grade separation or highway interchange has long spans with the smallest possible girder depth and the smallest possible abutment size (Ritner, 1990). Continuity is the best way to

Fig. 2.23 Cedar Falls Road, Overpass, I-90, King County, WA [Photo courtesy of Washington State DOT].

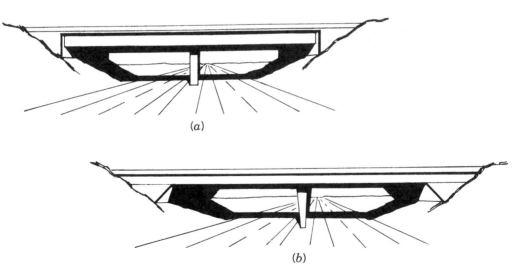

Fig. 2.24 (*a*) Vertical lines appear static. They provide interest and variety to the horizontal flow of the structure, but do not accentuate the flow. (*b*) Dynamic sloping lines provide interest and variety and accentuate flow. [Permission granted by the California DOT].

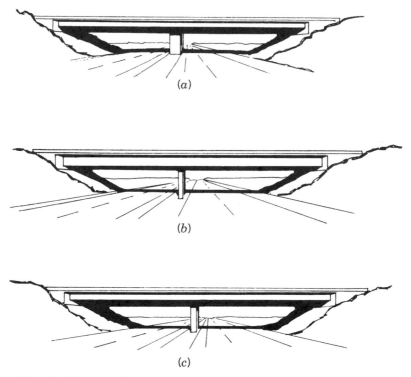

Fig. 2.25 (*a*) Massive columns overpower superstructure. (*b*) Massive superstructure overpowers spindly columns. (*c*) Substructure and superstructure are properly proportioned. [Permission granted by the California Department of Transportation].

minimize girder depth. In two-span applications, haunches can be used effectively, but as shown in Figure 2.26, proportions must be selected carefully. Leonhardt (1991) suggests that the haunch should follow a parabolic curve that blends in at midspan and is not deeper at the pier than twice the depth at midspan.

By utilizing these recommendations it is possible to overcome the duality effect and to design pleasing two-span and four-span highway bridges. Additional guidelines for the individual components of girders, overhangs, piers, and abutments, that will help the parts integrate into a unified, harmonious whole are given in the sections that follow.

Girder Span/Depth Ratio According to Leonhardt (1991), the most important criterion for the appearance of a bridge is the slenderness of the beam, defined by the L/d ratio, or span length/beam depth ratio. If the height of the opening is greater than the span, he suggests L/d can be as small as 10, while for long continuous spans L/d could be up to 45. The designer has a

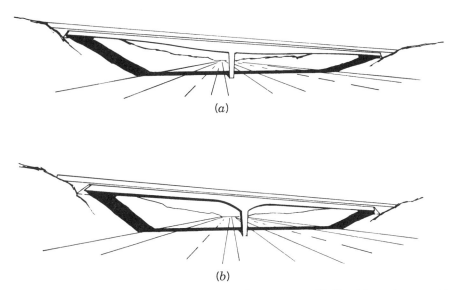

Fig. 2.26 (*a*) Long haunches give grace to the structure. (*b*) Short haunches appear awkward and abrupt, detracting from continuity of bridge. [Permission granted by the California Department of Transportation].

wide range of choices in finding the L/d ratio that best fits a particular setting. In light of the general objective of using a beam with the least possible depth, the L/d ratio selected should be on the high end of the range.

Because of structural limitations, the maximum L/d ratio will vary for different bridge types. Table 2.1 has been developed from recommendations given by ACI Committee 343 (1988), and those in Table 2.5.2.6.3-1 of the AASHTO Specifications (1994). The maximum values in Table 2.1 are traditional ratios given in previous editions of the AASHTO Specifications in an attempt to ensure that vibration and deflection would not be a problem. They should not be considered as absolute maximums, but only as guidelines. They compare well with L/d ratios that are desirable for a pleasing appearance.

Deck Overhangs It is not possible for many of the bridge types in Table 2.1 to have L/d ratios in excess of 30. However, it is possible to increase the apparent slenderness of the superstructure by placing part or all of the girder in shadow. Shadow can be created by cantilevering the deck slab beyond the exterior girder as shown in Figures 2.22 and 2.27.

When girders are spaced a distance S center to center in a multigirder bridge, a cantilevered length of the deck overhand w of about $0.4S$ helps balance the positive and negative moments in the deck slab. Another way to determine the cantilever length w is to proportion it relative to the depth of

TABLE 2.1 Typical and Maximum Span/Depth Ratios

Bridge Type	Typical	Maximum
Continuous Concrete Bridges	Committee 343	AASHTO
Nonprestressed slabs	20–24	
Nonprestressed girders		
T-beam	15±	15
Box girder	18±	18
Prestressed slabs		
Cast-in-place	24–40	37
Precast	25–33	
Prestressed girders		
Cast-in-place boxes	25–33	25
Precast I-beams	20–28	25
Continuous Steel Bridges	Caltrans	AASHTO
Composite I-beam		
Overall		31
I-beam portion		37
Composite welded girder	22	
Structural steel box	22	

the girder h. Leonhardt (1991) suggests a ratio of w/h of 2:1 for single-span, low-elevation bridges and 4:1 for long, continuous bridges high above the ground.

If the slope of the underside of the overhang is less than 1:4, that portion of the overhang will be in deep shadow (Murray, 1991). Both Leonhardt (1991) and Murray (1991) agree that the ratio of the depth of the fascia g to the depth of the girder h should be about 1:3 to give a pleasing appearance. By first selecting a cantilever length w, a designer can use these additional

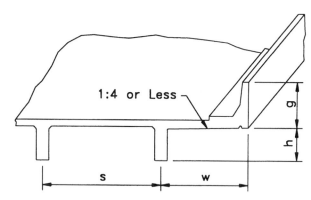

Fig. 2.27 Deck slab cantilevered over edge beam (after Leonhardt, 1991). [From *Bridge Aesthetics Around the World,* copyright 1991 by the Transportation Research Board, National Research Council, Washington, DC. Reprinted with permission].

proportions to obtain a visual effect of a more slender superstructure (Figs. 2.8 and 2.15).

When solid concrete barriers are used for safety rails, the fascia will appear to have greater depth. If a box girder with a sloping side is used, it is possible for the overhang to put the entire girder in shadow (Fig. 2.23) and improve the apparent slenderness. Also, it may be advantageous to change the texture (Fig. 2.20), or to introduce an additional shadow line that breaks up the flat surface at, say, the one-third point (Fig. 2.28).

Also shown in Figure 2.28 is a very important and practical detail—the drip groove. This drip groove somehow breaks the surface tension of rainwater striking the facia and prevents it from running in sheets and staining the side of the girder. It seems that architects have known about drip grooves forever, but in many cases engineers have been slow to catch on and we still see many discolored beams and girders. Perhaps all that is necessary is to point it out to them one time.

Piers In addition to having proper proportions between a pier and its superstructure (Fig. 2.25), a pier has features of its own that can improve the appearance of a bridge. As shown in Figure 2.29, piers come in many styles and shapes. The most successful ones are those that have some flare, taper, texture, or other feature that improves the visual experience of those who pass

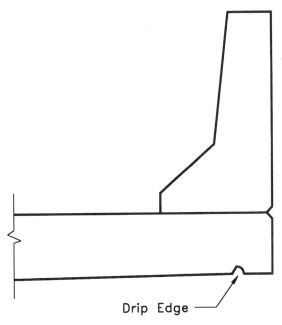

Fig. 2.28 Cantilevered overhang with drip groove (Mays, 1990) [Used with permission of the American Concrete Institute].

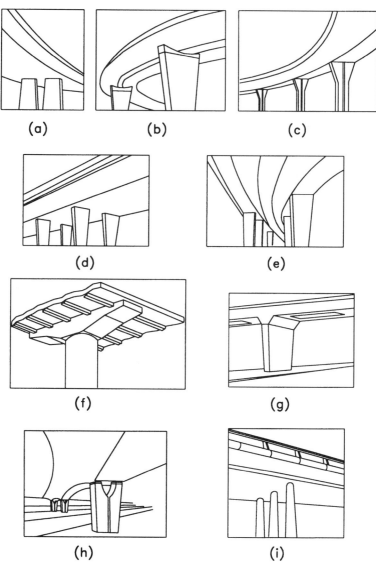

Fig. 2.29 Pier styles of contemporary bridges: wall type (a–e, g, h); T-type (f); and column type (i). (Glomb, 1991). [From *Bridge Aesthetics Around the World,* copyright 1991 by the Transportation Research Board, National Research Council, Washington, DC. Reprinted with permission].

by them. The key is that they fit in with the superstructure and its surroundings and that they express their structural process.

In general, tall piers should be tapered (Figs. 2.4, 2.5, and 2.30) to show their strength and stability in resisting lateral loads. Short piers can also be tapered (Fig,. 2.31) but in the opposite direction to show that less resistance

Fig. 2.30   Tall column with parabolic taper and raised edge (Menn, 1991). [From *Bridge Aesthetics Around the World,* copyright 1991 by the Transportation Research Board, National Research Council, Washington, DC. Reprinted with permission].

is desired at the bottom than at the top. And when the piers are of intermediate height (Fig. 2.32), they can taper both ways to follow a bending moment diagram that, in this case, has a point of inflection about one-third the height from the top.

An interesting example is given by Seim and Lin (1990) of a bridge pier that is tapered differently in two directions. The piers were designed for an elevated roadway at the San Francisco Airport (Fig. 2.33). They explain the design as follows:

At the San Francisco Airport, the Elevated Roadway is supported by piers with unusual tapers. Transversely and longitudinally, the tapers differ in order to express and to adjust to different structural requirements. In the longitudinal direction, continuity with the deck at the pier top requires moment resistance,

Fig. 2.31 Prestressed girders frame into the side of the supporting pier, eliminating from view the usual cap beam. [Permission granted by the California DOT].

while a hinge at the pier bottom relieved the stresses produced by the shrinkage and creep of the post-tensioned concrete deck. Hence, the pier widens at the top and narrows at the bottom. In the transverse direction, the pier behaves as a vertical cantilever, which requires a bottom wider than the top, and the two differing tapers are designed into the pier to present a natural solution to the structural requirements.

Note also in Figure 2.33 the narrowness of the piers relative to the width of the roadway they support.

There appears to be a preference among designers for piers that are integral with the superstructure, that is, they act together with the beams and girders to resist applied loads. Examples of integral piers are shown in Figures 2.31, 2.33, and 2.34. Where nonintegral substructures are used, Wasserman (1991) recommends hammerhead or *T*-piers, singly (Fig. 2.8) or joined (Fig. 2.35), over the more cluttered appearance of multiple-column bents (Fig. 2.36).

When interchanges are designed, multiple columns cannot be avoided, but they should be of similar form. In Figure 2.34, there are a variety of pier shapes and sizes, but they all belong to the same family. Contrast this example

Fig. 2.32   Piers with double taper (Seim and Lin, 1990) [Used with permission of the American Concrete Institute].

with the unfortunate mixture of supports in the bridge of Figure 2.37. Harmony between the elements of the bridge has been destroyed. The wall pier is too prominent because it has not been kept in the shade and its sloping front face adds to the confusion. All-in-all a good example of what not to do.

Abutments   Repeating what was said earlier, to obtain a pleasing appearance for a bridge, the girder should be as slender as possible. Large abutments may be needed to anchor a suspension bridge, but they are out of place for medium- and short-span bridges.

The preferred abutment is placed near the top of the bank, well out of the way of the traffic below (Fig. 2.24), which gives the bridge a feeling of openness and invites the flow of traffic. Some designers refer to this as a stub abutment, or if it is supported on columns or piling, a spill-through abutment, because the embankment material spills through the columns.

For a given length of an abutment, the flatter the slope of the embankment, the smaller the abutment appears, which can be seen in the comparisons of Figure 2.38. The preferred slope of the bank should be 1:2 or less.

Fig. 2.33   Elevated roadway at San Francisco airport (Seim and Lin, 1990) [Used with permission of the American Concrete Institute].

Another feature of the abutment that improves its appearance is to slope its face back into the bank from top to bottom. Elliot (1991) explains it this way:

> Sloping the face inward about 15 degrees, creates a magical illusion. Instead of seeming to suddenly stop against the vertical faces, the bridge now seems to flow smoothly into the supporting ground. This one feature will improve the appearance of a simple separation structure at virtually no increase in cost.

Examples of bridges with abutments illustrating this concept are shown in Figures 2.23, 2.24, and 2.26. The mass of the vertical-faced abutments of the bridges in Figures 2.35 and 2.36 could be reduced and the appearance improved if the faces were inclined inward.

The sloping ground from the abutment to the edge of the stream or roadway beneath the bridge is usually in the shade and vegetation does not easily grow on it (see Figs. 2.20 and 2.23). Whatever materials are placed on the slope to prevent erosion should relate to the bridge or the surrounding landscape. Concrete paving blocks or cast-in-place concrete relate to the abutment while rubble stone relates to the landscape. Proper selection of slope protection materials will give the bridge a neatly defined and finished appearance.

Fig. 2.34   Route 8/805 Interchange, San Diego, CA. [Permission granted by the California DOT].

2.3.4   Closing Remarks on Aesthetics

It is important for an engineer to realize that, whether intentional or not, a completed bridge becomes an aesthetic statement. Therefore, it is necessary to understand what qualities and features of a bridge tend to make that aesthetic statement a good one. This understanding will require training and time.

Suggestions have been made in this section on how to improve the appearance of medium- and short-span bridges. Some of these suggestions include numerical values for proportions and ratios, but most of them simply point out features that require a designer's attention. There are no equations, no computer programs or design specifications that can make our bridges beautiful. It is more an awareness of beauty on our part so that we can sense when we are in the presence of something good.

Aesthetics must be a part of the bridge design program from the beginning. It cannot be added on at the end to make the bridge look nice. At that time, it is too late. From the beginning, the engineer must consider aesthetics in the selection of spans, depths of girders, piers, abutments, and the relationship

Fig. 2.35 Hammerhead piers [Photo courtesy of Tennessee DOT—Geo. Hornal, photographer].

of one to another. It is a responsibility that cannot be delegated. We must demand it of ourselves, because the public demands it of us.

2.4 TYPES OF BRIDGES

There are any number of different ways in which to classify bridges. Bridges can be classified according to materials (concrete, steel, or wood), usage (pedestrian, highway, or railroad), span (short, medium, or long), or structural form (slab, girder, truss, arch, suspension, or cable-stayed). None of these classifications are mutually exclusive. They all seem to contain parts of one another within each other. For example, selection of a particular material does not limit the usage or dictate a particular structural form. On the other hand, there may be unique site characteristics that require a long-span bridge with high vertical clearance over navigable waters that will limit the choices of materials and structural form.

It would be a mistake to think that all an engineer has to do is to define the problem properly with regard to terrain, alignment and usage, and the selection of a particular bridge type will be determined. It will not happen. There are too many possibilities. The best that can be done is to describe the characteristics of the different bridge types, to realize that they overlap one

Fig. 2.36 Multiple-column bents [Photo courtesy of Tennessee DOT—Geo. Hornal, photographer].

another, and that no one bridge type has an exclusive advantage in a particular application.

Even as in the case of aesthetics, an engineer should not be looking for a magic formula or computer software package to select a bridge type. There is no substitute for experience. If a young engineer lacks experience in selecting a bridge type that is pleasing in appearance and efficient in structural form, there is nothing wrong in asking for advice. Much can be learned by interacting with those who have previously been through the process.

In this book, the design topics in concrete, steel, and wood highway bridges have been limited to medium- and short-spans, which obviously narrows the field of bridge types that are discussed in detail. However, to put this discussion in perspective, a brief overview of all bridge types commonly used will be presented.

The classification of bridge types in this presentation will be according to the location of the main structural elements relative to the surface on which the user travels, that is, whether the main structure is below, above, or coincides with the deck line.

2.4.1 Main Structure Below the Deck Line

Arched and truss-arched bridges are included in this classification. Examples are the masonry arch, the concrete arch (Fig. 2.17), the steel truss-arch, the

Fig. 2.37 Bridge with displeasing mixture of supports (Murray, 1991). [From *Bridge Aesthetics Around the World,* copyright 1991 by the Transportation Research Board, National Research Council, Washington, DC. Reprinted with permission].

steel deck truss, the rigid frame, and the inclined leg frame (Fig. 2.15) bridges. Striking illustrations of this bridge type are the New River Gorge Bridge (Fig. 2.39) in West Virginia and the Salginatobel Bridge (Fig. 2.40) in Switzerland.

With the main structure below the deck line in the shape of an arch, gravity loads are transmitted to the supports primarily by axial compressive forces. At the supports, both vertical and horizontal reactions must be resisted. The arch rib can be solid or it can be a truss of various forms. Xanthakos (1994) shows how the configuration of the elements affects the structural behavior of an arch bridge and gives methods for determining the force effects.

O'Connor (1971) summarizes the distinctive features of arch type bridges as

- The arch form is intended to reduce bending moments (and hence tensile stress) in the superstructure and should be economical in material compared with an equivalent straight, simply supported girder or truss.
- This efficiency is achieved by providing horizontal reactions to the arch rib. If these are external reactions, they can be supplied at reasonable cost only if the site is suitable. The most suitable site for this form of structure is a valley, with the arch foundations located on dry rock slopes.

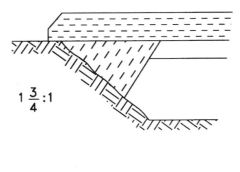

$1\frac{3}{4}:1$

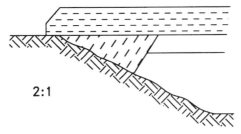

2:1

Fig. 2.38 Slope at abutment (Mays, 1990). [Used with permission of the American Concrete Institute].

- The conventional curved arch rib may have high fabrication and erection costs, although these may be controlled to some extent by the skill of the designer.
- The erection problem varies with the type of structure, being easiest for the cantilever arch and possibly most difficult for the tied arch. The difficulty with the latter arises from the fact that the horizontal reactions are not available until the deck is completed.
- The arch is predominantly a compression structure. For example, the open spandrel arch with the rib below the deck consists of deck, spandrel columns, and arch rib. The last two are compression members. The design must include accurate estimates of buckling behavior and should be detailed so as to avoid excessive reductions in allowable stress. The classic arch form tends to favor concrete as a construction material.
- The arch rib is usually shaped to take dead load without bending moments. This load is then called the form load. If the form load is large, the live load becomes essentially a small disturbance applied to a compressed member. Under these conditions, the normal first-order elastic theory is inadequate and errs on the unsafe side. Some form of deflection theory must be used for analysis. The effects of initial imperfections in the arch shape may become significant.

Fig. 2.39 New River Gorge Bridge [Photo by David Bowen, courtesy of West Virginia DOT].

- The conventional arch has two moment-resistant components—the deck and the arch rib. Undesirable and unanticipated distributions of moment may occur, particularly in regions where the spandrel columns are short, normally near the crown of the arch, which may be avoided by careful detailing; for example, by using pin-ended columns.
- The structure of most arches encroaches on the space bounded by the abutments and the deck. This encroachment may restrict clearance for passage beneath the arch and may involve the risk of collision with the arch rib.
- Aesthetically, the arch can be the most successful of all bridge types. It appears that through experience or familiarity, the average person regards the arch form as understandable and expressive. The curved shape is almost always pleasing. This aesthetic advantage is reduced for cases where the arch rises through the deck. However, even in these cases, the arch may be made particularly attractive.

2.4.2 Main Structure above the Deck Line

Suspension, cable-stayed, and through-truss bridges are included in this category. Both suspension and cable-stayed bridges are tension structures whose

Fig. 2.40 General view of Salginatobel bridge. [From M. S. Troitsky (1994). *Planning and Design of Bridges,* Copyright © 1994. Reprinted with permission of John Wiley & Sons, Inc.].

cables are supported by towers. Examples are the Brooklyn Bridge (Fig. 2.19) and the East Huntington Bridge (Fig. 2.18), respectively.

Suspension bridges (Fig. 2.41) are constructed with two main cables from which the deck, usually a stiffened truss is hung by secondary cables. Cable-stayed bridges (Fig. 2.42) have multiple cables that support the deck directly from the tower. Analysis of the cable forces in a suspension bridge must consider nonlinear geometry due to large deflections, while linear elastic analysis is usually sufficient for cable-stayed bridges.

O'Connor (1971) gives the following distinctive features for suspension bridges:

- The major element of the stiffened suspension bridge is a flexible cable, shaped and supported in such a way that it can transfer the major loads to the towers and anchorages by direct tension.
- This cable is commonly constructed from high strength wires, either spun in situ or formed from component, spirally formed wire ropes. In either case the allowable stresses are high, typically of the order of 600 MPa for parallel strands.
- The deck is hung from the cable by hangers constructed of high strength wire ropes in tension.

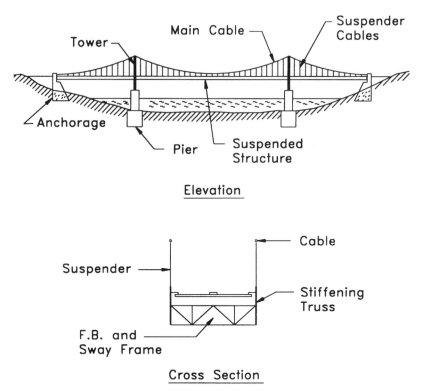

Fig. 2.41 Typical suspension bridge. [From M. S. Troitsky (1994). *Planning and Design of Bridges,* Copyright © 1994. Reprinted with permission of John Wiley & Sons, Inc.].

- This use of high strength steel in tension, primarily in the cables and secondarily in the hangers, leads to an economical structure, particularly if the self-weight becomes significant, as in the case of long spans.
- The economy of the main cable must be balanced against the cost of the associated anchorages and towers. The anchorage cost may be high in areas where the foundation material is poor.
- The main cable is stiffened either by a pair of stiffening trusses or by a system of girders at deck level.
- This stiffening system serves to (a) control aerodynamic movements and (b) limit local angle changes in the deck. It may be unnecessary in cases where the dead load is great.
- The complete structure can be erected without intermediate staging from the ground.
- The main structure is elegant and neatly expresses its function.
- The height of the main towers can be a disadvantage in some areas; for example, within the approach circuits for an airport.

(a)

(b)

(c)

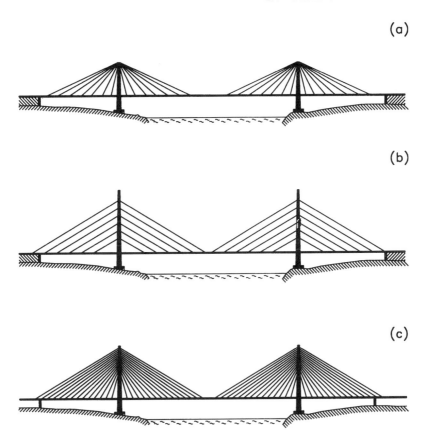

Fig. 2.42 Cable arrangements in cable-stayed bridges (Leonhardt, 1991). [From *Bridge Aesthetics Around the World,* copyright 1991 by the Transportation Research Board, National Research Council, Washington, DC. Reprinted with permission].

- It is the only alternative for spans over 600 m, and it is generally regarded as competitive for spans down to 300 m. However, even shorter spans have been built, including some very attractive pedestrian bridges.

and the following for cable-stayed bridges:

- The use of high strength cables in tension leads to economy in material, weight, and cost.
- As compared with the stiffened suspension bridge, the cables are straight rather than curved. As a result, the stiffness is greater. It will be recalled that the nonlinearity of the stiffened suspension bridge results from changes in the cable curvature and the corresponding change in bending moment taken by the dead-load cable tension. This phenomenon cannot occur in an arrangement with straight cables.

- The cables are anchored to the deck and cause compressive forces in the deck. For economical design, the deck system must participate in carrying these forces. In a concrete structure, this axial force compresses the deck.
- All individual cables are shorter than the full length of the superstructure. They are normally constructed of individual wire ropes, supplied complete with end fittings, prestretched and not spun. The cable erection problem differs greatly from that in the conventional suspension bridge.
- There is great freedom of choice in selecting the structural arrangement.
- Compared with the stiffened suspension bridge, the cable-braced girder bridge tends to be less efficient in supporting dead load, but more efficient under live load. As a result, it is not likely to be economical on the longest spans. It is commonly claimed to be economical over the range 100–350 m, but some designers would extend the upper bound as high as 800 m.
- The cables may be arranged in a single plane, at the longitudinal centerline of the deck. This arrangement capitalizes on the torsion capacity inherent in a tubular girder system, and halves the number of shafts in the towers. It also simplifies the appearance of the structure, and avoids cable intersections when the bridge is viewed obliquely.
- It is desirable to provide jacking details capable of modifying the cable forces. These can be arranged at the cable anchorages or at the tower tops. They are necessary to adjust for relaxation in the cables, errors in the cable lengths, or variations in their elastic modulus. They may also be used to modify the stress distribution due to dead load—for example, by prestressing the main span upwards.
- The presence of the cables facilitates the erection of a cable-stayed girder bridge. Temporary backstays of this type have been common in the cantilever erection of girder bridges. Adjustment of the cables provides an effective control during erection.
- Aerodynamic instability has not been found to be a problem in structures erected to date.
- The natural frequency of vibration differs from that of more conventional alternatives, such as the unbraced girder or the suspension bridge. In the case of the harp arrangement, the cables tends to balance a load on one side of the tower against a load on the other, causing a reduction in the dead-load moment in the deck and a possible reduction in the deck stiffness. However, the bridge may vibrate in a mode in which points at opposite ends of a cable have vertical movements in opposite senses. The contribution of the cables to the deck stiffness may be small, and this may lead to undesirable natural frequencies. The fan arrangements should be better in this regard.

A truss bridge (Fig. 2.43) consists of two main planar trusses tied together with cross girders and lateral bracing to form a three-dimensional truss that

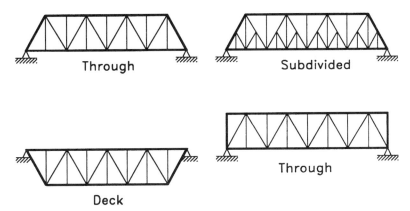

Fig. 2.43   Types of Bridge Trusses. [From M. S. Troitsky (1994). *Planning and Design of Bridges*, Copyright © 1994. Reprinted with permission of John Wiley & Sons, Inc.].

Fig. 2.44   Greater New Orleans Through-Truss Bridge. [Photo courtesy of Amy Kohls, 1996].

can resist a general system of loads. When the longitudinal stringers that support the deck slab are at the level of the bottom chord, this is a through-truss bridge as shown in Figure 2.44.

O'Connor (1971) gives the following distinctive features for truss bridges:

- A bridge truss has two major structural advantages: (1) the primary member forces are axial loads; (2) the open web system permits the use of a greater overall depth than for an equivalent solid web girder. Both these factors lead to economy in material and a reduced dead weight. The increased depth also leads to reduced deflections, that is, a more rigid structure.

- These advantages are achieved at the expense of increased fabrication and maintenance costs.

- The conventional truss bridge is most likely to be economical for medium spans. Traditionally, it has been used for spans intermediate between the plate girder and the stiffened suspension bridge. Modern construction techniques and materials have tended to increase the economical span of both steel and concrete girders. The cable-stayed girder bridge has become a competitor to the steel truss for the intermediate spans. These factors, all of which are related to the high fabrication cost of a truss, have tended to reduce the number of truss spans built in recent years. Nevertheless, economical solutions have been achieved for highway bridge spans in the range 150–500 m. The largest highway bridge truss span presently in service is the 480-m main span of the Greater New Orleans cantilever bridge. This value is exceeded by the 550-m span Quebec Bridge and the 520-m span Firth of Forth Bridge, both railway bridges. The economical threshold for the railway truss bridge may be as low as 75 m due to the loads that are significantly heavier than highway loads.

- The truss has become almost the standard stiffening structure for the conventional suspension bridge, largely because of its acceptable aerodynamic behavior.

- The relative light weight of a truss bridge is an erection advantage. It may be assembled member by member using lifting equipment of small capacity. Alternatively, the number of field connections may be reduced by fabricating and erecting the trusses bay by bay, rather than one member at a time.

- As in all bridge structures, it is important to achieve a compatible relationship between the deck and the main structure. This relationship is best achieved by causing the deck to act with the truss chords in taking axial loads. Alternatively, the deck may be isolated from the chords by a system of deck expansion joints.

- Compared with alternative solutions, the encroachment of a truss on the opening below is large if the deck is at the upper chord level, but is small if the traffic runs through the bridge, with the deck at the lower chord level. For railway overpasses carrying a railway above a road or another railway, the small construction depth of a through truss bridge is a major advantage. In some structures, it is desirable to combine both arrangements to provide a through truss over the main span with a small construction depth, and approaches with the deck at upper chord level.

- A truss bridge rarely looks aesthetically pleasing. This poor appearance is due partly to the complexity of the elevation, but also results from the awkward member intersections that appear in any oblique view. In a large-span bridge, these factors may become unimportant because of the visual impact of the large scale. In bridges of moderate span, it seems best to provide a simple and regular structure. For this reason, the Warren truss usually looks better than other forms.

2.4.3 Main Structure Coincides with the Deck Line

Girder bridges of all types are included in this category. Examples include slab (solid and voided), T-beam (cast-in-place), I-beam (precast or prestressed), wide-flange beam (composite and noncomposite), concrete box (cast-in-place and segmental, prestressed), steel box (orthotropic deck), and steel plate girder (straight and haunched) bridges.

Illustrations of concrete slab, T-beam, prestressed girder, and box-girder bridges are shown in Figure 2.45. A completed cast-in-place concrete slab bridge is shown in Figure 2.46. Numerous girder bridges are shown in the section on aesthetics. Among these are prestressed girders (Fig. 2.31), concrete box girders (Figs. 2.10, 2.23, and 2.34), and steel plate girders (Figs. 2.13, 2.35, and 2.36).

Girder type bridges carry loads primarily in shear and flexural bending. This action is relatively inefficient when compared to axial compression in arches and to tensile forces in suspension structures. A girder must develop both compressive and tensile forces within its own depth. These internal forces are separated by a lever arm sufficient to provide the internal resisting moment. Because the extreme fibers are the only portion of the cross section fully stressed, it is difficult to obtain an efficient distribution of material in a girder cross section. Additionally, stability concerns further limit the stresses and associated economy from a material utilization perspective. But from total economic perspective slab-girder bridges provide an economical and long-lasting solution for the vast majority of bridges. The U.S. construction industry is well tuned to provide this type of bridge. [As a result, girder bridges are typical for short-to-medium span lengths, say <50 m.]

In highway bridges, the deck and girders usually act together to resist the applied load. Typical bridge cross sections for various types of girders are

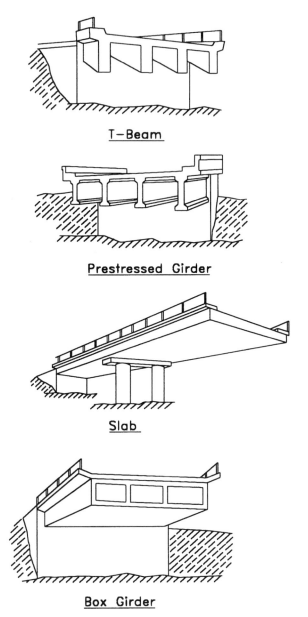

T–Beam

Prestressed Girder

Slab

Box Girder

Fig. 2.45   Types of concrete bridges. [Permission granted by the California DOT].

Fig. 2.46 Cast-in-place post-tensioned voided Slab Bridge (Dorton, 1991). [From *Bridge Aesthetics Around the World,* copyright 1991 by the Transportation Research Board, National Research Council, Washington, DC. Reprinted with permission].

shown in Table 2.2. They include steel, concrete, and wood bridge girders with either cast-in-place or integral concrete decks. These are not the only combinations of girders and decks, but represent those covered by the approximate methods of analysis in the AASHTO (1994) LRFD Specifications.

2.4.4 Closing Remarks on Bridge Types

For comparison purposes, typical ranges of span lengths for various bridge types are given in Table 2.3. In this book, the discussion will be limited to slab and girder bridges suitable for short-to-medium spans. For a general discussion in other bridge types, the reader is referred Xanthakos (1994).

2.5 SELECTION OF BRIDGE TYPE

One of the key submittals in the design process is the engineer's report to the bridge owner of the type, size, and location (TS & L) of the proposed bridge. The TS & L report includes a cost study and a set of preliminary bridge drawings. The design engineer has the main responsibility for the report, but

TABLE 2.2   Common Girder Bridge Cross Sections.[a]

SUPPORTING COMPONENTS	TYPE OF DECK	TYPICAL CROSS SECTION
Steel Beam	Cast-in-place concrete slab precast concrete slab, steel grid, glued/spiked panels, stressed wood	(a)
Closed Steel or Precast Concrete Boxes	Cast-in-place concrete slab	(b)
Open Steel or Precast Concrete Boxes	Cast-in-place concrete slab, precast concrete deck slab	(c)
Cast-in-Place Concrete Multi-cell Box	Monolithic concrete	(d)
Cast-in-Place Concrete Tee Beam	Monolithic concrete	(e)
Precast Solid, Voided or Cellular Concrete Boxes with Shear Keys	Cast-in-place concrete overlay	(f)

[a](AASHTO Table 4.6.2.2.1-1). [From *AASHTO LRFD Bridge Design Specifications,* Copyright © 1994 by the American Association of State Highway and Transportation Officials, Washington, DC. Used by permission].

opinions and advice will be sought from others within and without the design office. It is then submitted to all appropriate agencies, made available for public hearings, and must be approved before starting on the final design.

2.5.1   Factors To Be Considered

Selection of a bridge type involves consideration of a number of factors. In general, these factors are related to economy, safety, and aesthetics. It is

TABLE 2.2   (Continued)

SUPPORTING COMPONENTS	TYPE OF DECK	TYPICAL CROSS SECTION
Precast Solid, Voided or Cellular Concrete Box with Shear Keys and with or without Transverse Post-Tensioning	Integral concrete	(g)
Precast Concrete Channel Sections with Shesr Keys	Cast-in-place concrete overlay	(h)
Precast Concrete Double Tee Section with Shear Keys and with or without Transverse Post-Tensioning	Integral concrete	(i)
Precast Concrete Tee Section with Shear Keys and with or without Transverse Post-Tensioning	Integral concrete	(j)
Precast Concrete I or Bulb-Tee Sections	Cast-in-place concrete, prcast concrete	(k)
Wood Beams	Cast-in-place concrete or plank, glued/spiked panels or stressed wood	(l)

difficult to prepare a list of factors without implying an order of priority, but a list is necessary even if the priority changes from bridge to bridge. The discussion herein follows the outline presented by ACI-ASCE Committee 343 (1988) for concrete bridges, but the factors should be the same, regardless of the construction material.

Geometric Conditions of the Site   The type of bridge selected will often depend on the horizontal and vertical alignment of the highway route and on the clearances above and below the roadway. For example, if the roadway is on a curve, continuous box girders and slabs are a good choice because they have a pleasing appearance, can readily be built on a curve, and have a

TABLE 2.3   Span Lengths for Various Types of Superstructure

Structural Type	Material	Range of Spans (m)	Maximum Span in Service (m)
Slab	Concrete	0–12	
Girder	Concrete	12–250	240, Hamana-Ko Lane
	Steel	30–260	261, Sava I
Cable-stayed girder	Concrete	≤250	235, Maracaibo
	Steel	90–850	856, Normandy
Truss	Steel	90–550	550, Quebec (rail)
			480, Greater New Orleans, Nos. 1 and 2 (road)
Arch	Concrete	90–300	305, Gladesville
	Steel truss	240–500	510, New River Gorge
	Steel rib	120–360	365, Port Mann
Suspension	Steel	300–1400	1410, Humber

relatively high torsion resistance. Relatively high bridges with larger spans over navigable waterways will require a different bridge type than one with medium spans crossing a flood plain. The site geometry will also dictate how traffic can be handled during construction, which is an important safety issue and must be considered early in the planning stage.

Subsurface Conditions of the Site   The foundation soils at a site will determine whether abutments and piers can be founded on spread footings, driven piles, or drilled shafts. If the subsurface investigation indicates that creep settlement is going to be a problem, the bridge type selected must be one that can accommodate differential settlement over time. Drainage conditions on the surface and below ground must be understood because they influence the magnitude of earth pressures, movement of embankments, and stability of cuts or fills. All of these conditions influence the choice of substructure components which, in turn, influence the choice of superstructure. For example, an inclined leg rigid frame bridge requires strong foundation material that can resist both horizontal and vertical thrust. If it is not present, then another bridge type is more appropriate. The potential for seismic activity at a site should also be a part of the subsurface investigation. If seismicity is high, the substructure details will change, affecting the superstructure loads as well.

Functional Requirements   In addition to the geometric alignment that allows a bridge to connect two points on a highway route, the bridge must also function to carry present and future volumes of traffic. Decisions must be made on the number of lanes of traffic, inclusion of sidewalks and/or bike paths, whether width of the bridge deck should include medians, drainage of the surface waters, snow removal, and future wearing surface. In the case of stream and flood plain crossings, the bridge must continue to function during periods of high water and not impose a severe constriction or obstruction to the flow of water or debris. Satisfaction of these functional requirements will recommend some bridge types over others. For example, if future widening and replacement of bridge decks is a concern, multiple girder bridge types are preferred over concrete segmental box girders.

Aesthetics   The first part of this chapter emphasizes the importance of designing a bridge with a pleasing appearance. Whether intentional or not, every bridge makes an aesthetic statement. The fact that a highway bridge is usually out in the open, means it can be seen and reacted to by whoever passes by. Sometimes the reaction is neutral because the bridge did not stimulate the senses. But othertimes the senses respond and the viewer will be either delighted or offended. It should be the goal of every bridge designer to obtain a positive aesthetic response to the bridge type selected.

Economics and Ease of Maintenance   It is not possible to separate first cost and maintenance cost over the life of the bridge when comparing the economics of different bridge types. A general rule is that the bridge with the minimum number of spans, fewest deck joints, and widest spacing of girders will be the most economical. By reducing the number of spans in a bridge layout by one span, the construction cost of one pier is eliminated. Deck joints are a high maintenance cost item, so minimizing their number will reduce the life cycle cost of the bridge. When using the empirical design of bridge decks in the AASHTO (1994) LRFD Specifications, the same reinforcement is used for deck spans up to 4100 mm. Therefore, there is little cost increase in the deck for wider spacing of girders and fewer girders means less cost although at the "expense" of deeper sections.

Generally, concrete structures require less maintenance than steel structures. The cost and hazard of maintenance painting of steel structures should be considered in type selection studies (Caltrans, 1990).

One effective way to obtain the minimum construction cost is to prepare alternative designs and to allow contractors to propose an alternative design. The use of alternative designs permits the economics of the construction industry at the time of bidding to determine the most economical material and bridge type. By permitting the contractor to also submit an alternative design, the greatest advantage can be taken of new construction techniques to obtain less total project cost. The disadvantage of this approach is that a low initial cost may become the controlling criteria and life-cycle costs may not be considered.

Construction and Erection Considerations   The selection of the type of bridge to be built is often governed by construction and erection considerations. The length of time required to construct a bridge is important and will vary with bridge type. In general, the larger the prefabricated or precast members, the shorter the construction time. However, the larger the members, the more difficult they are to transport and lift into place.

Cast-in-place concrete bridges are generally economical for grade separations unless the falsework supporting the nonhardened concrete becomes a traffic problem. In that case, precast prestressed girders or welded steel plate girders would be a better choice.

The availability of skilled labor and specified materials will also influence the choice of a particular bridge type. For example, if there are no precast plants for prestressed girders within easy transport but there is a steel fabrication plant nearby that could make the steel structure more economical. However, other factors in the construction industry may be at work. The only way to determine which bridge type is more economical is to bid alternative designs.

Legal Considerations   In Figure 2.1, a model of the design process was presented. One of the components of the model was the constraint put on the design procedure by regulations. These regulations are usually beyond the control of the engineer, but they are real and must be considered.

Examples of regulations that will determine what bridge type can be built and where follow: Permits Over Navigable Waterways, National Environmental Policy Act, Department of Transportation Act, National Historic Preservation Act, Clean Air Act, Noise Control Act, Fish and Wildlife Coordination Act, The Endangered Species Act, Water Bank Act, Wild and Scenic Rivers Act, Prime and Unique Farmlands, and Executive Orders on Floodplain Management and Protection of Wetlands. Engineers who are not conscious of the effect the design of a bridge has on the environment will soon become conscious once they begin preparing the environmental documentation required by these acts.

In addition to the environmental laws and acts defining national policy, local and regional politics are also of concern. Commitments to officials or promises made to communities often must be honored and may preclude other nonpolitical issues. In general, engineers who tend to see things in black and white do not make good politicians, because politics is a gray area where solutions are usually derived by compromise. But even in this area good engineers are often successful and adjust to the political considerations.

2.5.2   Bridge Types Used for Different Span Lengths

Once a preliminary span length has been chosen, comparative studies are conducted to find the bridge type best suited to the site. For each group of bridge spans (small, medium, and large), experience has shown that certain bridge types are more appropriate than others. This experience is usually passed on from engineer to engineer and can be found in design aids prepared by state agencies and consulting firms. The comments that follow on common bridge types used for different span lengths are based on the experience of ACI-ASCE Committee 343 (1988), Caltrans (1990), and PennDOT (1993).

Small Span Bridges (up to 15 m)   The candidate structure types include: single or multicell culverts, slab bridges, *T*-beam bridges, wood beam bridges, precast concrete box beam bridges, precast concrete *I*-beam bridges, and composite rolled steel beam bridges.

Culvert Culverts are used as small span bridges to allow passage of small streams, livestock, vehicles, and pedestrians through highway embankments. These buried structures [A12.1]* are often the most economical solution for short spans. They are constructed of steel, aluminum, precast or cast-in-place reinforced concrete, and thermoplastics. Their structural form can be a pipe, pipe arch, plate arch, plate box, or rigid frame box. Either trench installations or embankment installations may be used. Minimum soil cover to avoid direct application of wheel loads is a function of the span length [Table A12.6.6.3-1] and is not less than 300 mm. It is often cited that there are 575,000 bridges over 6 m long in the National Bridge Inventory. What is seldom mentioned is that 100,000 of them are culverts.

Slab Slab bridges are the simplest and least expensive structure that can be built for small spans up to 12 m. These bridges can be built on ground supported falsework or constructed of precast elements. Construction details and formwork are the simplest of any bridge type. Their appearance is neat and simple, especially for low, short spans. Precast slab bridges constructed as simple spans require reinforcement in the topping slab to develop continuity over transverse joints at the piers, which is necessary to improve the riding quality of the deck and to avoid maintenance problems. Span lengths can be increased by use of prestressing. A design example of a simple span solid slab bridge is given in Chapter 7.

T-Beam *T*-Beam bridges (Table 2.2e) are generally economical for spans 10–20 m. These bridges usually are constructed on ground-supported falsework and require a good finish on all surfaces. Formwork may be complex, especially for skewed structures. Appearance of elevation is neat and simple, but not as desirable from below. Greatest use is for stream crossings, provided there is at least 2-m clearance above high water (floating debris may damage the girder stem). Usually, the *T*-beam superstructure is constructed in two

*References to AASHTO (1994) LRFD Specifications are enclosed in brackets and denoted by a letter A followed by the article number. A commentary is cited as the article number preceded by the letter C. Referenced figures and tables are enclosed in brackets to distinguish them from figures and tables in the text.

stages: first the stems and then the slabs. To minimize cracks at the tops of the stems, longitudinal reinforcement should be placed in the stem near the construction joint. To ease concrete placement and finishing, a longitudinal joint within the structure becomes necessary for bridges wider than about 20 m. A design example of a three-span continuous *T*-beam bridge is given in Chapter 7.

Wood Beam   Wood beam bridges (Table 2.2l) may be used for low speed, low truck volume roads, or in locations where a wood pile substructure can be constructed economically. Minimum width of roadway shall be 7.2-m curb to curb. The deck may be concrete, glued/spiked panels, or stressed wood. All wood used for permanent applications shall be impregnated with wood preservatives [A8.4.3.1]. The wood components not subject to direct pedestrian contact shall be treated with oil-borne preservatives [A8.4.3.2]. Main load carrying members shall be precut and drilled prior to pressure treatment. For a waterway crossing, abutments and piers shall be aligned with the stream and piers shall be avoided in the stream if debris may be a problem. A design example of a simple span glulam beam bridge is given in Chapter 9.

Precast Concrete Box Beam   Precast prestressed concrete box beam bridges can have spread boxes (Table 2.2b) or butted boxes (Table 2.2f and g) and can be used for spans from 10–50 m. These bridges are most suitable for locations where the use of falsework is impractical or too expensive. The construction time is usually shorter than that needed for cast-in-place *T*-beams. Precast box beams do not usually provide a comfortable ride because adjacent boxes will often have different camber and dead load deflections. Unreinforced grout keys often fail between adjacent units, allowing differential live load deflections to occur. A reinforced topping slab or transverse post-tensioning can alleviate this problem. Appearance of the spread box beam is similar to a *T*-beam while the butted box beam is similar to a cast-in-place box girder. For multiple spans, continuity should be developed for live load by casting concrete between the ends of the simple span boxes.

Precast Concrete I-Beam   Precast prestressed concrete *I*-beam bridges can be used for spans from 10–50 m and are competitive with steel girders. They have many of the same characteristics as precast concrete box beam bridges including the problems with different camber and ridability. The girders are designed to carry dead load and construction loads as simple span units. Live load and superimposed dead load design should use continuity and composite action with the cast-in-place deck slab. Appearance is like that of the *T*-beam: The elevation view is nice, but the underside looks cluttered. As in all concrete bridges, maintenance is low except at transverse deck joints.

Rolled Steel Beam   Rolled steel wide flange beam bridges are widely used because of their simple design and construction. These bridges are economical for spans up to 30 m when designing the deck as a composite and using cover plates in maximum moment regions. The use of composite beams is strongly recommended because they make a more efficient structure. Shear connectors, usually in the form of welded studs, must be designed to resist all forces tending to separate concrete and steel surfaces. The appearance of the multibeam bridge from underneath will be like that of the *T*-beam, but the elevation will be more slender (Table 2.1). The cost and environmental hazard of maintenance painting must be considered in any comparison with concrete bridges.

Medium Span Bridges (up to 50 m)   The candidate structure types include: precast concrete box beam bridges, precast concrete *I*-beam bridges, composite rolled steel beam bridges, composite steel plate girder bridges, cast-in-place concrete box girder bridges, and steel box-girder bridges.

Precast Concrete Box Beam and Precast Concrete I-Beam Characteristics of both of these precast prestressed concrete beams were discussed under small span bridges. As span lengths increase, transportation and handling may present a problem. Most state highway departments require a permit for any load over 24 m long and refuse permits for loads over 35 m long. Girders longer than 35 m may have to be brought to the site in segments and assembled there. The longer girders are heavy and firm ground is needed to store the girders and to provide support for the lifting cranes. The *I*-beams may be laterally unstable until incorporated into the structure and should be braced until the diaphragms are cast. A design example of a simple span precast pretensioned concrete *I*-beam is given in Chapter 7.

Composite Rolled Steel Beam Characteristics of composite rolled steel beams were discussed under small bridges. Composite construction can result in savings of 20–30% for spans over 15 m (Troitsky, 1994). Adding cover plates and providing continuity over several spans can increase their economic range to spans of 30 m. A design example of a simple span composite rolled steel beam bridge is given in Chapter 8.

Composite Steel Plate Girder Composite steel plate girders can be built to any desired size and consist of two flange plates welded to a web to form an asymmetrical *I*-section. These bridges are suitable for spans from 25 to 50 m and have been used for spans well over 100 m. Girders must be braced against each other to provide stability against overturning and flange buckling, to resist transverse forces, and to distribute concentrated vertical loads. Construction details and formwork are simple. Transportation of prefabricated girders over 35 m may be a problem. Composite steel plate girder bridges can be made to look attractive. Girders can be curved to follow alignment. This structure type has low dead load, which may be of value when foundation conditions are poor. A design example of a three-span continuous composite plate girder bridge is given in Chapter 8.

Cast-In-Place Reinforced Concrete Box Girder Nonprestressed reinforced concrete box-girder bridges (Table 2.2d) are adaptable for use in many locations. These bridges are used for spans of 15–35 m and are usually more economical than steel girders and precast concrete girders. Formwork is simpler than for a skewed *T*-beam, but it is still complicated. Appearance is good from all directions. Utilities, pipes, and conduits are concealed. High torsional resistance makes it desirable on curved alignment. They are an excellent choice in metropolitan areas.

Cast-In-Place Post-Tensioned Concrete Box Girder Prestressed concrete box-girder bridges afford many advantages in terms of safety, appearance, maintenance, and economy. These bridges have been used for spans up to 180 m. Because longer spans can be constructed economically, the number of piers can be reduced and shoulder obstacles eliminated for safer travel at overpasses. Appearance from all directions is neat and simple with greater slenderness than conventional reinforced concrete box girder bridges. High torsional resistance makes it desirable on curved alignment. Because of the prestress, the dead load deflections are minimized. Long-term shortening of the structure must be accommodated. Maintenance is very low, except that bearing and transverse deck joint details require attention. Addition of proper transverse and longitudinal post-tensioning greatly reduces cracking. Post-tensioned concrete box girders can be used in combination with conventional concrete box girders to maintain constant structure depth in long structures with varying span lengths. A design example of a three-span continuous cast-in-place post-tensioned concrete box-girder bridge is given in Chapter 7.

Composite Steel Box Girder Composite steel box-girder bridges (Table 2.2b and c) are used for spans of 20–150 m. These bridges are more economical in the upper range of spans and where depth may be limited. The boxes may be rectangular or trapezoidal and are effective in resisting torsion. They offer an attractive appearance and can be curved to follow alignment. Generally, multiple boxes would be used for spans up to 60 m and a single box for longer spans. Construction costs are often kept down by shop fabrication; therefore, designers should know the limitations placed by shipping clearances on the dimensions of the girders. Because of the many opportunities for welding and detail errors that can give rise to fatigue failures, the steel box should only be used in very special circumstances.

Large Span Bridges (50–150 m) The candidate structure types include: composite steel plate girder bridges, cast-in-place post-tensioned concrete box-girder bridges, post-tensioned concrete segmental bridges, concrete arch bridges, steel arch bridges, and steel truss bridges.

Composite Steel Plate Girder Characteristics of composite steel plate girder bridges are presented in medium span bridges. A design example of a medium span bridge is given in Chapter 8.

Cast-In-Place Post-Tensioned Concrete Box Girder Characteristics of cast-in-place post-tensioned concrete box-girder bridges are presented in medium span bridges. A design example of a medium span bridge is given in Chapter 7.

Post-Tensioned Concrete Segmental Construction (ACI-ASCE Committee 343, 1988) Various bridge types may be constructed in segments and post-tensioned to complete the final structure. The basic concept is to provide cost saving through standardization of details and multiple use of construction equipment. The segments may be cast-in-place or precast. If cast-in-place, it is common practice to use the balanced cantilever construction method with traveling forms. If precast, they may be erected by the balanced cantilever method, by progressive placement span by span, or by launching the spans from one end. Both the designer and the contractor have the opportunity to evaluate and choose the most cost efficient method. Table 2.4 from Troitsky (1994) indicates typical span length ranges for bridge types by conventional and segmented construction methods.

The analysis and design of prestressed concrete segmental bridges is beyond the scope of this book. The reader is referred to excellent reference books, such as Podolny and Muller (1982), on the design and construction of segmental bridges.

Concrete Arch and Steel Arch Characteristics of arch bridges are given in Section 2.4.1. Concrete arch bridges will usually be below the deck, but steel arch bridges can be both above and below the deck, sometimes in the same structure. Typical and maximum span lengths for concrete and steel arch bridges are given in Table 2.3. Arch bridges are pleasing in appearance and

TABLE 2.4 Range of Application of Bridge Type by Span Lengths Considering Segmental Construction[a]

Span (m)	Bridge Type
0–45	Precast pretensioned *I*-beam conventional
30–90	Cast-in-place post-tensioned box girder conventional
30–90	Precast-balanced cantilever segmental, constant depth
60–180	Precast-balanced cantilever segmental, variable depth
60–300	Cast-in-place cantilever segmental
240–450	Cable-stay with balanced cantilever segmental

[a]From M. S. Troitsky (1994). *Planning and Design of Bridges,* Copyright © 1994. Reprinted with permission of John Wiley & Sons, Inc.

are used largely for that reason even if a cost premium is involved. Arch bridges are not covered in this book, but they can be found in the books by Xanthakos (1994) and Troitsky (1994).

Steel Truss Characteristics of steel truss bridges are given in Section 2.4.2. Steel truss bridges can also be below the deck, and sometimes both above and below the deck in the same structure as seen in the cantilever truss bridge and its approach spans on the cover of the AASHTO (1994) LRFD Specifications. Truss bridges are not covered in this book, but they have a long history and numerous books, besides those already mentioned, can be found on their design and construction.

Extra Large (Long) Span Bridges (over 150 m)   An examination of Table 2.3 shows that all of the general bridge types, except slabs, have been built with span lengths greater than 150 m. These are the special bridges designed to meet special circumstances and are not covered in this book. Some of the bridge types in Table 2.2 were extended to their limit in attaining the long span lengths and may not have been the most economical choice.

Two of the bridge types, cable-stayed and suspension, are logical and efficient choices for long span bridges. Characteristics of cable-stayed bridges and suspension bridges are given in Section 2.4.2. These tension-type structures are graceful and slender in appearance and are well suited to long water crossings. Maintenance for both is above average because of the complexity of the hanger and suspension system. Construction is actually simpler than for the conventional bridge types for long spans, because falsework is usually not necessary. For additional information on the analysis, design, and construction of cable-stayed bridges the reader is referred to Podolny and Scalzi (1986) and Troitsky (1988), while O'Connor (1971) is a good reference for stiffened suspension bridges.

2.5.3   Closing Remarks on Selection of Bridge Type

In the selection of a bridge type, there is no unique answer. For each span length range there is more than one bridge type that will satisfy the design criteria. There are bound to be regional differences and preferences because of available materials, skilled workers, and knowledgeable contractors. For the same set of geometric and subsurface circumstances, the bridge type selected will probably be different in Pennsylvania than in California. And both would be the right answer for that place at that time.

Because of the difficulties in predicting the cost climate of the construction industry at the time of bidding, it is good policy to allow the contractor the option of proposing an alternative design. This design should be made whether or not the owner has required the designer to prepare alternative designs. This policy will improve the odds that the bridge type being built is the most economical.

In Section 2.2 on the design process, de Miranda (1991) was quoted as saying that for successful bridge design three mentalities must be present: (1) creative and aesthetic, (2) analytical, and (3) technical and practical. Oftentimes a designer will possess the first two mentalities and can select a bridge type that has a pleasing appearance and whose cross section has the theoretically least amount of material that the specifications will allow. But a designer may not be familiar with good, economical construction procedures and the third mentality is missing. By allowing the contractor to propose an alternative design, the third mentality is restored and the original design(s) are further validated or a better design may be proposed. Either way the design process is enhanced by incorporating the three mentalities.

REFERENCES

AASHTO (1994). *LRFD Bridge Design Specification,* 1st ed., American Association of State Highway and Transportation Officials, Washington, DC.

ACI-ASCE Committee 343 (1988). "Analysis and Design of Reinforced Concrete Bridge Structures," American Concrete Institute, Detroit, MI.

Addis, W. (1990). *Structural Engineering: The Nature of Theory and Design,* Ellis Horwood, London.

Brown, W. C., M. F. Parsons, and H. S. G. Knox, (1976). *Bosporus Bridge: Design and Construction,* The Institute of Civil Engineers, London.

Burke, M. P., Jr. (1990). "Bridge Aesthetics—Rules, Formulas, and Principles—the Negative View," *Esthetics in Concrete Bridge Design,* ACI MP1-7, American Concrete Institute, Detroit, MI, pp. 71–79.

Burke, M. P., Jr., Ed. (1991). *Bridge Aesthetics Around the World,* Transportation Research Board, National Research Council, Washington, DC.

Caltrans (1990). "Selection of Type," *Bridge Design Aids,* California Department of Transportation, Sacramento, CA, pp. 10–21 to 10–29.

de Miranda, F. (1991). "The Three Mentalities of Successful Bridge Design," *Bridge Aesthetics Around the World,* Transportation Research Board, National Research Council, Washington, DC, pp. 89–94.

Dorton, R. A. (1991). "Aesthetics Considerations for Bridge Overpass Design," *Bridge Aesthetics Around the World,* Transportation Research Board, National Research Council, Washington, DC, pp. 10–17.

Elliot, A. L. (1991). "Creating a Beautiful Bridge," *Bridge Aesthetics Around the World,* Transportation Research Board, National Research Council, Washington, DC, pp. 215–229.

Fernandez-Ordóñez, J. A. (1991). "Spanish Bridges: Aesthetics, History, and Nature," *Bridge Aesthetics Around the World,* Transportation Research Board, National Research Council, Washington, DC, pp. 205–214.

Friedhoff; R. M. and W. Benzon, (1989). *Visualization,* Harry N. Abrams, Inc., New York.

Glomb, J. (1991). "Aesthetic Aspects of Contemporary Bridge Design," *Bridge Aesthetics Around the World,* Transportation Research Board, National Research Council, Washington, DC, pp. 95–104.

Gloyd, C. S. (1990). "Some Thoughts on Bridge Esthetics," *Esthetics in Concrete Bridge Design,* ACI MP1-10, American Concrete Institute, Detroit, MI, pp. 109–117.

Gottemoeller, F. (1991). "Aesthetics and Engineers: Providing for Aesthetic Quality in Bridge Design," *Bridge Aesthetics Around the World,* Transportation Research Board, National Research Council, Washington, DC, pp. 80–88.

Grant, A. (1990). "Beauty and Bridges," *Esthetics in Concrete Bridge Design,* ACI MP1-6, American Concrete Institute, Detroit, MI, pp. 55–65.

Lee, David J. (1990). "Bridging the Artistic Gulf", *Esthetics in Concrete Bridge Design,* ACI MP1-9, American Concrete Institute, Detroit, MI, pp. 101–108.

Leonhardt, F. (1982). *Bridges: Aesthetics and Design,* The Architectural Press, London, and The MIT Press, Cambridge, MA.

Leonhardt, F. (1991). "Developing Guidelines for Aesthetic Design," *Bridge Aesthetics Around the World,* Transportation Research Board, National Research Council, Washington, DC, pp. 32–57.

Mays, R. R. (1990). "Aesthetic Rules Should Not Be Set in Concrete—A Bridge Architect's View on Bridge Design," *Esthetics in Concrete Bridge Design,* ACI MP1-16, American Concrete Institute, Detroit, MI, pp. 203–228.

Menn, C. (1991). "Aesthetics in Bridge Design," *Bridge Aesthetics Around the World,* Transportation Research Board, National Research Council, Washington, DC, pp. 177–188.

Muller, J. M. (1991). "Aesthetics and Concrete Segmental Bridges," *Bridge Aesthetics Around the World,* Transportation Research Board, National Research Council, Washington, DC, pp. 18–31.

Munro, T. (1956). *Towards Science in Aesthetics,* Liberal Arts Press, New York.

Murray, J. (1991). "Visual Aspects of Short- and Medium-Span Bridges," *Bridge Aesthetics Around the World,* Transportation Research Board, National Research Council, Washington, DC, pp. 155–166.

O'Connor, C. (1971). *Design of Bridge Superstructures,* Wiley-Interscience, New York.

PennDOT (1993). "Selection of Bridge Types," *Design Manual—Part 4,* Pennsylvania Department of Transportation, Vol. 1, Part A: Chapter 2, Harrisburg, PA.

Podolny, W., Jr. and J. M. Muller, (1982). *Construction and Design of Prestressed Concrete Segmental Bridges,* Wiley, New York.

Podolny, W., Jr. and J. B. Scalzi (1986). *Construction and Design of Cable-Stayed Bridges,* 2nd ed., Wiley, New York.

Ritner, J. C. (1990). "Creating a Beautiful Concrete Bridge," *Esthetics in Concrete Bridge Design,* ACI MP1-4, American Concrete Institute, Detroit, MI, pp. 33–45.

Roberts, J. E. (1992). "Aesthetic Design Philosophy Utilized for California State Bridges," *Journal of Urban Planning and Development,* ASCE, Vol. 118, No. 4, December, pp. 138–162.

Seim, C. and T. Y. Lin, (1990). "Aesthetics in Bridge Design: Accent on Piers," *Esthetics in Concrete Bridge Design,* ACI MP1-15, American Concrete Institute, Detroit, MI, pp. 189–201.

Troitsky, M. S. (1988). *Cable-Stayed Bridges, An Approach to Modern Bridge Design,* 2nd ed., Van Nostrand Reinhold, New York.

Troitsky, M. S. (1994). *Planning and Design of Bridges,* Wiley, New York.

Wasserman, E. P. (1991). "Aesthetics for Short- and Medium-Span Bridges," *Bridge Aesthetics Around the World,* Transportation Research Board, National Research Council, Washington, DC, pp. 58–66.

Watson, S. C. and M. K. Hurd, Eds. (1990). *Esthetics in Concrete Bridge Design,* ACI MP-1, American Concrete Institute, Detroit, MI.

Xanthakos, P. (1994). *Theory and Design of Bridges,* Wiley, New York.

3

GENERAL DESIGN
CONSIDERATIONS

3.1 INTRODUCTION

The justification stage of design can begin after the selection of possible alternative bridge types that satisfy the function and aesthetic requirements of the bridge location have been completed. As discussed in the opening pages of Chapter 2, justification requires that the engineer verify the structural safety and stability of the proposed design. Justification involves calculations to demonstrate to those who have a vested interest that all applicable specifications, design, and construction requirements are satisfied.

A general statement for assuring safety in engineering design is that the resistance of the materials and cross sections supplied exceed the demands put on them by applied loads, that is,

$$\text{Resistance} \geq \text{effect of the loads} \tag{3.1}$$

When applying this simple principle, it is essential that both sides of the inequality are evaluated for the same conditions. For example, if the effect of applied loads is to produce compressive stress on a soil, it is obvious that this should be compared to the bearing resistance of the soil, and not some other quantity. In other words, the evaluation of the inequality must be done for a specific loading condition that links together resistance and the effect of loads. This common link is provided by evaluating both sides at the same limit state.

When a particular loading condition reaches its limit, failure is the assumed result, that is, the loading condition becomes a failure mode. Such a condition is referred to as a limit state that can be defined as

A limit state is a condition beyond which a bridge system or bridge component ceases to fulfill the function for which it is designed.

95

Examples of limit states for girder type bridges include deflection, cracking, fatigue, flexure, shear, torsion, buckling, settlement, bearing, and sliding. Well-defined limit states are established so that a designer knows what is considered to be unacceptable.

An important goal of design is to prevent a limit state from being reached. However, it is not the only goal. Other goals that must be considered and balanced in the overall design are function, appearance, and economy. It is not economical to design a bridge so that none of its components could ever fail. Therefore, it becomes necessary to determine what is an acceptable level of risk or probability of failure. The determination of an acceptable margin of safety (how much greater the resistance should be compared to the effect of loads) is not based on the opinion of one individual but based on the collective experience and judgment of a qualified group of engineers and officials. In the highway bridge design community, AASHTO is such a group, since it relies on the experience of the state department of transportation engineers, research engineers, consultants, and engineers involved with design specifications outside the United States.

3.2   DEVELOPMENT OF DESIGN PROCEDURES

Over the years, design procedures have been developed by engineers to provide satisfactory margins of safety. These procedures were based on the engineer's confidence in the analysis of the load effects and the strength of the materials being provided. As analysis techniques improved and quality control on materials became better, the design procedures changed.

To understand where we are today, it is helpful to look at the design procedures of earlier AASHTO Specifications and how they have changed as technology changed.

3.2.1   Allowable Stress Design (ASD)

The earliest design procedures were developed with the primary focus on metallic structures kept in mind. Structural steels were observed to behave linearly up to a relatively well-defined yield point that was safely below the ultimate strength of the material. Safety in the design was obtained by specifying that the effect of the loads should produce stresses that were a fraction of the yield stress f_y: say one-half. This value would be equivalent to providing a safety factor F of 2, that is,

$$F = \frac{\text{resistance, } R}{\text{effect of loads, } Q} = \frac{f_y}{0.5f_y} = 2$$

Because the specifications set limits on the stresses, this became known as allowable stress design (ASD).

When ASD methods were first used, a majority of the bridges were open-web trusses or arches. By assuming pin-connected members and using statics, the analysis indicated members that were either in tension or compression. The required net area of a tension member under uniform stress was easily selected by dividing the tension force T by an allowable tensile stress f_t.

$$\text{required } A_{\text{net}} \geq \frac{\text{effect of load}}{\text{allowable stress}} = \frac{T}{f_t}$$

For compression members, the allowable stress f_c depended on whether the member was short (nonslender) or long (slender), but the rationale for determining the required area of the cross section remained the same; the required area was equal to the compressive force divided by an allowable stress value.

$$\text{required } A_{\text{gross}} \geq \frac{\text{effect of load}}{\text{allowable stress}} = \frac{C}{f_c}$$

These techniques were used as early as the 1860s to design many successful statically determinate long-span truss bridges. Similar bridges are built today, but they are no longer statically determinate because they are not pin-connected. As a result, the stresses in the members are no longer uniform.

The ASD method is also applied to beams in bending. By assuming plane sections remain plane, and linear stress–strain response, a required section modulus S can be determined by dividing the bending moment M by an allowable bending stress f_b.

$$\text{required } S \geq \frac{\text{effect of load}}{\text{allowable stress}} = \frac{M}{f_b}$$

Implied in the ASD method is the assumption that the stress in the member is zero before any loads are applied, that is, no residual stresses exist from forming the members. This assumption is seldom completely accurate, but is closer to being true for solid bars and rods than for thin open sections of typical rolled beams. The thin elements of rolled beams cool at different rates and residual stresses become locked into the cross section. Not only are these residual stresses highly nonuniform, they are also difficult to predict. Consequently, adjustments have to be made to the allowable bending stresses, especially in compression elements, to account for the effect of residual stresses.

Another difficulty in applying ASD to steel beams is that bending is usually accompanied by shear, and these two stresses interact. Consequently, it is not strictly correct to use tensile coupon tests (satisfactory for pin-connected trusses) to determine the yield strength f_y for beams in bending. Another definition of yield stress that incorporates the effects of shear stress would be more logical.

What is the point in discussing ASD methods applied to steel design in a book on bridge analysis and design? Simply this,

> ASD methods were developed for the design of statically determinate metallic structures. They do not necessarily apply in a straightforward and logical way to other materials and other levels of redundancy.

Designers of reinforced concrete structures have realized this for some time, and adopted strength design procedures many years ago. Wood designers are also moving toward strength design procedures. Both concrete and wood are nonlinear materials whose properties change with time and with changes in ambient conditions. In concrete, the initial stress state is unknown because it varies with placement method, curing method, temperature gradient, restraint to shrinkage, water content, and degree of consolidation. The only values that can be well defined are the strengths of concrete at its limit states. As described in Chapter 6, the ultimate strength is independent of prestrains and stresses associated with numerous manufacturing and construction processes, all of which are difficult to predict and are highly variable. In short, the ultimate strength is easier and more reliably predicted than behavior at lower load levels. That is the rationale for the adoption of the strength design procedures.

3.2.2 Variability of Loads

In regard to uncertainties in design, one other point concerning the ASD method needs to be emphasized. Allowable stress design does not recognize that different loads have different levels of uncertainty. Dead, live, and wind loads are all treated equally in ASD. The safety factor is applied to the resistance side of the design inequality, and the load side is not factored. In ASD, safety is determined by

$$\frac{\text{Resistance, } R}{\text{Safety Factor, } F} \geq \text{effect of loads, } Q \qquad (3.2)$$

For ASD, fixed values of design loads are selected, usually from a specification or design code. The varying degree of predictability of the different load types is not considered.

Finally, because the safety factor chosen is based on experience and judgment, quantitative measures of risk cannot be determined for ASD. Only the trend is known: If the safety factor is higher, the number of failures is lower. However, if the safety factor is increased by a certain amount, it is not known by how much this increases the probability of survival. Also, it is more meaningful to decision makers to say, "This bridge has a probability of 1 in 10,000 of failing in 75 years of service," than to say, "This bridge has a safety factor of 2.3."

3.2.3 Shortcomings of Allowable Stress Design

As just shown, ASD is not well suited for design of modern structures. Its major shortcomings can be summarized as follows:

1. The resistance concepts are based on elastic behavior of isotropic, homogeneous materials.
2. It does not embody a reasonable measure of strength, which is a more fundamental measure of resistance than is allowable stress.
3. The safety factor is applied only to resistance. Loads are considered to be deterministic (without variation).
4. Selection of a safety factor is subjective, and it does not provide a measure of reliability in terms of probability of failure.

What is needed to overcome these deficiencies is a method that is (a) based on the strength of material, (b) considers variability not only in resistance but also in the effect of loads, and (c) provides a measure of safety related to probability of failure. Such a method is incorporated in the AASHTO (1994) LRFD Bridge Specifications and is discussed in Section 3.2.4.

3.2.4 Load and Resistance Factor Design (LRFD)

To account for the variability on both sides of the inequality in Eq. 3.1, the resistance side is multiplied by a statistically based resistance factor ϕ, whose value is usually less than 1, and the load side is multiplied by a statistically based load factor γ, whose value is usually greater than 1. Because the load effect at a particular limit state involves a combination of different load types (Q_i) that have different degrees of predictability, the load effect side is represented by a summation of $\gamma_i Q_i$ values. If the nominal resistance is given by R_n, the safety criterion is

$$\phi R_n \geq \text{effect of } \Sigma \gamma_i Q_i \qquad (3.3)$$

Because Eq. 3.3 involves both load factors and resistance factors, the design method is called load and resistance factor design (LRFD). The resistance factor ϕ for a particular limit state must account for the uncertainties in

- Material properties.
- Equations that predict strength.
- Workmanship.
- Quality control.

- Consequence of a failure.

The load factor γ_i chosen for a particular load type must consider the uncertainties in

- Magnitudes of loads.
- Arrangement (positions) of loads.
- Possible combinations of loads.

In selecting resistance factors and load factors for bridges, probability theory has been applied to data on strength of materials, and statistics on weights of materials and vehicular loads.

Some of the pros and cons of the LRFD method can be summarized as follows:

Advantages of LRFD Method

1. Accounts for variability in both resistance and load.
2. Achieves fairly uniform levels of safety for different limit states and bridge types without involving complex probability or statistical analysis.
3. Provides a rational and consistent method of design.

Disadvantages of LRFD Method

1. Requires a change in design philosophy (from previous AASHTO methods).
2. Requires an understanding of the basic concepts of probability and statistics.
3. Requires availability of sufficient statistical data and probabilistic design algorithms to make adjustments in resistance factors to meet individual situations.

3.3   DESIGN LIMIT STATES

3.3.1   General

The basic design expression in the AASHTO (1994) LRFD Bridge Specifications that must be satisfied for all limit states, both global and local, is given as

$$\eta \sum \gamma_i Q_i \le \phi R_n \tag{3.4}$$

where Q_i is the force effect, R_n is the nominal resistance, γ_i is the statistically based load factor applied to the force effects, ϕ is the statistically based resistance factor applied to nominal resistance, and η is a load modification factor. For all nonstrength limit states, $\phi = 1.0$.

Equation 3.4 is Eq. 3.3 with the addition of the load modifier η. The load modifier is a factor that takes into account the ductility, redundancy, and operational importance of the bridge. It is given by the expression.

$$\eta = \eta_D \eta_R \eta_I \geq 0.95 \qquad (3.5)$$

where η_D is the ductility factor, η_R is the redundancy factor, and η_I is the operational importance factor. The first two factors refer to the strength of the bridge and the third refers to the consequence of a bridge being out of service. For all nonstrength limit states $\eta_D = \eta_R = 1.0$.

Ductility Factor η_D *[A1.3.3].** Ductility is important to the safety of a bridge. If ductility is present overloaded portions of the structure can redistribute the load to other portions that have reserve strength. This redistribution is dependent on the ability of the overloaded component and its connections to develop inelastic deformations without failure.

If a bridge component is designed so that inelastic deformations can occur, then there will be a warning that the component is overloaded. If it is reinforced concrete, cracking will increase and the component will show that it is in distress. If it is structural steel, flaking of mill scale will indicate yielding and deflections will increase. The effects of inelastic behavior are elaborated in Chapter 6.

Brittle behavior is to be avoided, because it implies a sudden loss of load carrying capacity when the elastic limit is exceeded. Components and connections in reinforced concrete can be made ductile by limiting the flexural reinforcement and by providing confinement with hoops or stirrups. Steel section can be proportioned to avoid buckling, which may permit inelastic behavior. Similar provisions are given in the specifications for other materials. In fact, if the provisions of the specifications are followed in design, experience has shown that the components will have adequate ductility [C1.3.3].

The values to be used for the strength limit state ductility factor are

$$\eta_D = 1.05 \text{ for nonductile components and connections}$$

$$\eta_D = 0.95 \text{ for ductile components and connections}$$

Redundancy Factor η_R *[A1.3.4].* Redundancy significantly affects the safety margin of a bridge structure. A statically indeterminate structure is redundant, that is, it has more restraints than are necessary to satisfy equilibrium. For

---

*The article numbers in the AASHTO (1994) LRFD Bridge Specifications are enclosed in brackets and preceded by the letter A if specifications and by the letter C if commentary.

example, a three-span continuous bridge girder in the old days would be classified as statically indeterminate to the second degree. Any combination of two supports, or two moments, or one support and one moment could be lost without immediate collapse, because the applied loads could find alternative paths to the ground. The concept of multiple load paths is the same as redundancy.

Single-load paths or nonredundant bridge systems are not encouraged. The Silver Bridge over the Ohio River between Pt. Pleasant, WV, and Gallipolis, OH, was a single-load path structure. It was constructed in 1920 as a suspension bridge with two main chains composed of eyebar links, much like large bicycle chains, strung between two towers. However, to make the structure easier to analyze, pin connections were made at the base of the towers. When one of the eyebar links failed in December 1967, there was no alternative load path, the towers were nonredundant, and the collapse was sudden and complete; 46 lives were lost.

In the 1950s a popular girder bridge system was the cantilever span, suspended span, cantilever span system. These structures were statically determinate and the critical detail was the linkage or hanger that supported the suspended span from the cantilevers. The linkage was a single-load path connection and if it failed, the suspended span would drop to the ground or water below. This failure is what happened to the bridge over the Mianus River in Connecticut, June 1983. Three lives were lost.

Redundancy in a bridge system will increase its margin of safety and this is reflected in the strength limit state redundancy factors given as

$$\eta_R = 1.05 \text{ for nonredundant members}$$

$$\eta_R = 0.95 \text{ for redundant members}$$

Operational Importance Factor η_I *[A1.3.5].* Bridges can be considered of operational importance if they are on the shortest path between residential areas and a hospital or school, or provide access for police, fire, and rescue vehicles to homes, businesses, and industrial plants. Bridges can also be considered essential if they prevent a long detour and save time and gasoline in getting to work and back home again. In fact, it is difficult to find a situation where a bridge would not be operationally important, because a bridge must be justified on some social or security requirement to have been built in the first place. One example of a nonimportant bridge could be on a secondary road leading to a remote recreation area that is not open year round. But then if you were a camper or backpacker and were injured or became ill, you'd probably consider any bridge between you and the more civilized world to be operationally important.

In the event of an earthquake, it is important that all lifelines, such as bridges, remain open. Therefore, the following requirements apply to the extreme event limit state as well as to the strength limit state:

$\eta_I \geq 1.05$ for a bridge of operational importance

$\eta_I \geq 0.95$ for a nonimportant bridge

For all other limit states

$$\eta_I = 1.0$$

Load Designation [A3.3.2]. Permanent and transient loads and forces that must be considered in a design are designated as follows:

Permanent Loads

DD	Downdrag
DC	Dead load of structural components and nonstructural attachments
DW	Dead load of wearing surfaces and utilities
EH	Horizontal earth pressure load
ES	Earth surcharge load
EV	Vertical pressure from dead load of earth fill

Transient Loads

BR	Vehicular braking force
CE	Vehicular centrifugal force
CR	Creep
CT	Vehicular collision force
CV	Vessel collision force
EQ	Earthquake
FR	Friction
IC	Ice load
IM	Vehicular dynamic load allowance
LL	Vehicular live load
LS	Live load surcharge
PL	Pedestrian live load
SE	Settlement
SH	Shrinkage
TG	Temperature gradient
TU	Uniform temperature
WA	Water load and stream pressure
WL	Wind on live load
WS	Wind load on structure

Load Combinations and Load Factors. The load factors for various load combinations and permanent loads are given in Tables 3.1 and 3.2, respectively. Explanations of the different limit states are given in the sections that follow.

TABLE 3.1   Load Combination and Load Factors[a]

Load Combination Limit State	DC DD DW EH EV ES	LL IM CE BR PL LS	WA	WS	WL	FR	TU CR SH	TG	SE	Use One of These at a Time			
										EQ	IC	CT	CV
STRENGTH-I	γ_p	1.75	1.00	-	-	1.00	0.50/1.20	γ_{TG}	γ_{SE}	-	-	-	-
STRENGTH-II	γ_p	1.35	1.00	-	-	1.00	0.50/1.20	γ_{TG}	γ_{SE}	-	-	-	-
STRENGTH-III	γ_p	-	1.00	1.40	-	1.00	0.50/1.20	γ_{TG}	γ_{SE}	-	-	-	-
STRENGTH-IV EH, EV, ES, DW DC ONLY	γ_p 1.5	-	1.00	-	-	1.00	0.50/1.20	- 	- 	-	-	-	-
STRENGTH-V	γ_p	1.35	1.00	0.40	0.40	1.00	0.50/1.20	γ_{TG}	γ_{SE}	-	-	-	-
EXTREME EVENT-I	γ_p	γ_{EQ}	1.00	-	-	1.00	-	-	-	1.00	-	-	-
EXTREME EVENT-II	γ_p	0.50	1.00	-	-	1.00	-	-	-	-	1.00	1.00	1.00
SERVICE-I	1.00	1.00	1.00	0.30	0.30	1.00	1.00/1.20	γ_{TG}	γ_{SE}	-	-	-	-
SERVICE-II	1.00	1.30	1.00	-	-	1.00	1.00/1.20	-	-	-	-	-	-
SERVICE-III	1.00	0.80	1.00	-	-	1.00	1.00/1.20	γ_{TG}	γ_{SE}	-	-	-	-
FATIGUE-LL, IM, AND CE ONLY	-	0.75	-	-	-	-	-	-	-	-	-	-	-

[a]AASHTO Table 3.4.1-1. [From *AASHTO LRFD Bridge Design Specifications,* Copyright © 1994 by the American Association of State Highway and Transportation Officials, Washington, DC. Used by Permission].

3.3.2   Service Limit State

The service limit state refers to restrictions on stresses, deflections, and crack widths of bridge components that occur under regular service conditions [A1.3.2.2]. For the service limit state, the resistance factors $\phi = 1.0$, and nearly all of the load factors γ_i are equal to 1.0. There are three different service limit state load combinations given in Table 3.1 to cover different design situations [A3.4.1].

Service I. This service limit state refers to the load combination relating to the normal operational use of the bridge with 90-km/h wind, and with all loads taken at their nominal values. It also relates to deflection control in buried structures and crack control in reinforced concrete structures.

Service II. This service limit state refers to the load combination relating only to steel structures and is intended to control yielding and slip of slip-critical connections due to vehicular live load. It corresponds to the overload provision for steel structures in past editions of the AASHTO Specifications.

Service III. This service limit state refers to the load combinations relating only to tension in prestressed concrete structures with the objective of crack control. The statistical significance of the 0.80 factor on live load is that the event is expected to occur about once a year for bridges with two traffic lanes,

TABLE 3.2 Load Factors for Permanent Loads, $\gamma_p{}^a$

Type of Load	Load Factor	
	Maximum	Minimum
DC: Component and Attachments	1.25	0.90
DD: Downdrag	1.80	0.45
DW: Wearing Surfaces and Utilities	1.50	0.65
EH: Horizontal Earth Pressure • Active • At-Rest	 1.50 1.35	 0.90 0.90
EV: Vertical Earth Pressure • Overall Stability • Retaining Structure • Rigid Buried Structure • Rigid Frames • Flexible Buried Structures other than Metal Box Culverts • Flexible Metal Box Culverts	 1.35 1.35 1.30 1.35 1.95 1.50	 N/A 1.00 0.90 0.90 0.90 0.90
ES: Earth Surcharge	1.50	0.75

^{a}AASHTO Table 3.4.1-2. [From *AASHTO LRFD Bridge Design Specifications,* Copyright © 1994 by the American Association of State Highway and Transportation Officials, Washington, DC. Used by Permission].

less often for bridges with more than two traffic lanes, and about once a day for the bridges with a single traffic lane. Service I is used to investigate compressive stresses in prestressed concrete components.

3.3.3 Fatigue and Fracture Limit State

The fatigue and fracture limit state refers to a set of restrictions on stress range caused by a design truck. The restrictions depend on the number of stress range excursions expected to occur during the design life of the bridge [A1.3.2.3]. They are intended to limit crack growth under repetitive loads and to prevent fracture due to cumulative stress effects in steel elements, components, and connections. For the fatigue and fracture limit state, $\phi = 1.0$.

Because the only load effect that causes a large number of repetitive cycles is the vehicular live load, it is the only load effect that has a nonzero load factor in the fatigue limit state (see Table 3.1). A load factor of 0.75 is applied to the vehicular live load, dynamic load allowance, and centrifugal force. Use of a load factor less than 1.0 is justified because statistics show that trucks at slightly lower weights cause more repetitive cycles of stress than those at the weight of the design truck [C3.4.1]. Incidentally, the fatigue design truck is different than the design truck used to evaluate other force effects. It is defined as a single truck with a fixed axle spacing [A3.6.1.4.1]. The truck load models are described in detail in Chapter 4.

Fracture due to fatigue occurs at stress levels below the strength measured in uniaxial tests. When passing trucks cause a number of relatively high stress excursions, cumulative damage will occur. When the accumulated damage is large enough, a crack in the material will start at a point of stress concentration. The crack will grow with repeated stress cycles, unless observed and arrested, until the member fractures. If fracture of a member results in collapse of a bridge, the member is called "fracture-critical." The eyebar chain in the Silver Bridge and the hanger link in the Mianus River Bridge were both fracture-critical members.

3.3.4 Strength Limit State

The strength limit state refers to providing sufficient strength or resistance to satisfy the inequality of Eq. 3.4 for the statistically significant load combinations that a bridge is expected to experience in its design life [A1.3.2.4]. Strength limit states include the evaluation of resistance to bending, shear, torsion, and axial load. The statistically determined resistance factors ϕ will usually be less than 1.0 and will have different values for different materials and strength limit states.

The statistically determined load factors γ_i are given in five separate load combinations in Tables 3.1 to address different design considerations. For force effects due to permanent loads, the load factors γ_p of Tables 3.2 shall be selected to give the most critical load combination for a particular strength limit state. Either the maximum or minimum value of γ_p may control the extreme effect and both must be investigated. Application of two different values for γ_p, could easily double the number of strength load combinations to be considered. Fortunately, not all of the strength limit states apply in every situation and some can be eliminated by inspection.

For all strength load combinations, a load factor of 0.50 is applied to *TU*, *CR*, and *SH* for nondisplacement force effects to represent the reduction in these force effects with time from the values predicted by an elastic analysis. In the calculation of displacements for these loads, a load factor of 1.2 is used to avoid undersized joints and bearings [C3.4.1].

Strength I. This strength limit state is the basic load combination relating to normal vehicular use of the bridge without wind [A3.4.1].

Strength II. This strength limit state is the load combination relating to the use of the bridge by permit vehicles without wind. If a permit vehicle is traveling unescorted, or if control is not provided by the escorts, the other lanes may be assumed to be occupied by the basic design vehicular live load.

For bridges longer than the permit vehicle, the presence of the design lane load, preceding and following the permit vehicle in its lane, should be considered [C3.4.1].

Strength III. This strength limit state is the load combination relating to the bridge exposed to wind velocity exceeding 90 km/h. The high winds prevent the presence of significant live load on the bridge [C3.4.1].

Strength IV. This strength limit state is the load combination relating to very high dead load/live load force effect ratios. The standard calibration process used to select load factors γ_i and resistance factors ϕ for the strength limit state was carried out for bridges with spans less than 60 m. For the primary components of large span bridges, the ratio of dead and live load force effects is rather high, and could result in a set of resistance factors different from those found acceptable for small and medium span bridges. To avoid using two sets of resistance factors with the load factors of the Strength *I* limit state, the Strength *IV* limit state load factors were developed for large span bridges [C3.4.1].

Strength V. This strength limit state is the load combination relating to normal vehicular use of the bridge with wind of 90-km/h velocity. The Strength *V* limit state differs from the Strength *III* limit state by the presence of live load on the bridge, wind on the live load, and reduced wind on the structure [A3.4.1].

3.3.5   Extreme Event Limit State

The extreme event limit state refers to the structural survival of a bridge during a major earthquake or flood, or when collided by a vessel, vehicle, or ice flow [A1.3.2.5]. The probability of these events occurring simultaneously is extremely low, therefore, they are specified to be applied separately. The recurrence interval of extreme events may be significantly greater than the design life of the bridge [C1.3.2.5]. Under these extreme conditions, the structure is expected to undergo considerable inelastic deformation by which locked-in force effects due to *TU, TG, CR, SH,* and *SE* are expected to be relieved [C3.4.1, see Chapter 6]. For the extreme event limit state, $\phi = 1.0$.

Extreme Event I. This extreme event limit state is the load combination relating to earthquake. This limit state also includes water load *WA* and friction *FR*. The probability of a major flood and an earthquake occurring at the same time is very small. Therefore, water loads and scour depths based on mean discharges may be warranted [C3.4.1].

Partial live load coincident with earthquake should be considered. The load factor for live load γ_{EQ} shall be determined on a project-specific basis [A3.4.1]. Suggested values for γ_{EQ} are 0.0, 0.5, and 1.0 [C3.4.1].

Extreme Event II. This extreme event limit state is the load combination relating to ice load, collision by vessels and vehicles, and to certain hydraulic events with reduced live load. The 0.50 live load factor signifies a low probability of the occurrence of the maximum vehicular live load, other than *CT*, and the extreme events [C3.4.1].

3.4   PRINCIPLES OF PROBABILISTIC DESIGN

To facilitate the use of statistics and probability, a brief primer on the basic concepts is given. This review provides the background for understanding how the LRFD code was developed. Probabilistic analyses are not necessary to apply the LRFD method in practice, except for rare situations that are not encompassed by the code.

There are several levels of probabilistic design. The fully probabilistic method (Level *III*) is the most complex, and requires a knowledge of the probability distributions of each random variable (resistance, load, etc.), and correlation between the variables. This information is seldom available, so it is rarely practical to implement the fully probabilistic method.

Level *II* probabilistic methods include the first-order second moment (FOSM) method, which uses simpler statistical characteristics of the load and resistance variables. Further, it is assumed that the load Q and resistance R are statistically independent.

The load and resistance factors employed in the AASHTO (1994) LRFD Bridge Specifications were determined by using Level *II* procedures, and other simpler methods for conditions where insufficient information was available to use the Level *II* methods. The following sections define and discuss the statistical and probabilistic terms that are involved in this Level *II* theory.

3.4.1   Frequency Distribution and Mean Value

Consider Figure 3.1, which is a histogram of the 28-day compressive strength distribution of 176 concrete cylinders, all intended to provide a design strength of 20.7 MPa. The ordinates represent the number of times a particular compressive strength (1.38 MPa intervals) was observed.

As is well known, the mean (average) value $\bar{x}$ of the N compressive strength values x_i is calculated by

Mean,

$$\bar{x} = \frac{\Sigma x_i}{N} \tag{3.6}$$

For the $N = 176$ tests, the mean value $\bar{x}$ is found to be 27.2 MPa.

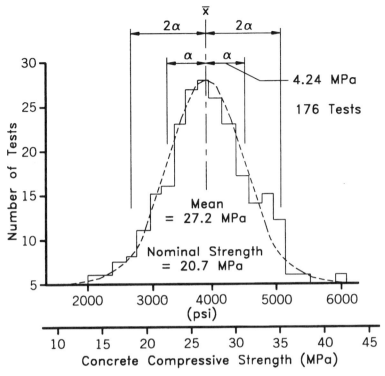

Fig. 3.1 Distribution of concrete strengths. [After J. G. MacGregor, *Reinforced Concrete: Mechanics and Design,* Copyright © 1992, p. 40. Reprinted by permission of Prentice Hall, Upper Saddle River, New Jersey].

Notice that the dashed smooth curve that approximates this histogram is the familiar bell-shaped distribution function that is typical of many natural phenomena.

3.4.2 Standard Deviation

The variance of the data from the mean is determined by summing up the square of the difference from the mean $\bar{x}$ (squared so that it is not sign dependent) and normalizing it with respect to the number of data points minus one.

$$\text{Variance} = \frac{\Sigma(x_i - \bar{x})^2}{(N - 1)} \qquad (3.7)$$

The standard deviation σ is a measure of the dispersion of the data in the same units as the data x_i. It is simply the square root of the variance.

Standard deviation,

$$\sigma = \sqrt{\frac{\Sigma(x_i - \bar{x})^2}{(N - 1)}} \qquad (3.8)$$

For the distribution of concrete compressive strength given in Figure 3.1, the standard deviation has been calculated as 4.24 MPa.

3.4.3 Probability Density Functions

The bell-shaped curve in Figure 3.1 can also represent the probability distribution of the data if the area under the curve is set to unity (probability = 1, includes all possible concrete strengths). To make the deviation $(x - \bar{x})$ for a particular point x nondimensional, it is divided by the standard deviation σ. The result is a probability density function, which shows the range of deviations and the frequency with which they occur. If the data are typical of those encountered in materials testing, the normal distribution curve of Figure 3.2a will often result. It is given by the function (Benjamin and Cornell, 1970)

$$f_x(x) = \frac{1}{\sigma\sqrt{2\pi}} \exp\left[-\frac{1}{2}\left\{\frac{x - m}{\sigma}\right\}^2 \right] \qquad -\infty \le x \le \infty \qquad (3.9)$$

where $f_x(x)$ gives the probable frequency of occurrence of the variable x as a function of the mean $m = \bar{x}$, and the standard deviation σ of the normal distribution. The frequency distributions need not be centered at the origin. The effect of changes in m and σ is shown in Figure 3.2.

The normal probability function has been studied for many years and its properties are well documented in statistics books. An important characteristic of the areas included between ordinates erected on each side of the center of the distribution curve is that they represent probabilities at a distance of one, two, and three standard deviations. These areas are 68.26, 95.44, and 99.73%, respectively.

EXAMPLE 3.1

Statistics indicate that the average height of the American male is 1.75 m (5 ft 9 in.) with a standard deviation of 0.076 m (3 in.). Table 3.3 shows the percentage of the male population in the United States in different height ranges. It is obvious that basketball players greater than 2.13 m (7 ft) tall are very rare individuals indeed.

There are other probability density functions besides the symmetric normal function shown in Figure 3.2. When the data distribution is unsymmetric, a

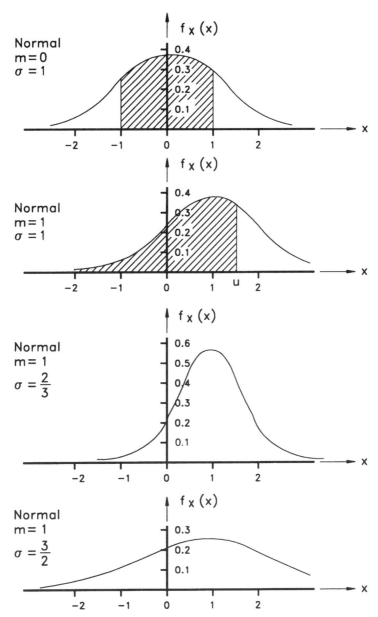

Fig. 3.2 Normal density functions. [From J. R. Benjamin and C. A. Cornell (1970). *Probability, Statistics, and Decisions for Civil Engineers.* Copyright © 1970. Reproduced with permission of the McGraw-Hill Companies].

TABLE 3.3 U.S. Males in Different Height Ranges[a]

Standard Deviation	Height Range	% of Male Population
1σ	1.67–1.83 m (5 ft 6 in.–6 ft 0 in.)	68.26
2σ	1.60–1.90 m (5 ft 3 in.–6 ft 3 in.)	95.44
3σ	1.52–1.98 m (5 ft 0 in.–6 ft 6 in.)	99.73
4σ	1.45–2.05 m (4 ft 9 in.–6 ft 9 in.)	99.997
5σ	1.37–2.13 m (4 ft 6 in.–7 ft 0 in.)	99.99997

[a]$\bar{x}$ = 1.75 m, σ = 0.076 m.

logarithmic normal (or simply lognormal) probability density function is often more suitable. Stated mathematically, if $Y = \ln(x)$ is normally distributed, then x is said to be lognormal. The lognormal function was used in calibrating the AASHTO (1994) LRFD Bridge Specification because it better represented the observed distribution of resistance data.

Lognormal probability density functions are shown in Figure 3.3 for different values of its standard deviation $\sigma = \zeta$. Notice that as the dispersion (the value of ζ) increases, the lack of symmetry becomes more pronounced.

Because $\ln(x)$ is a normal distribution, its mean λ_m and standard deviation ζ can be determined by a logarithmic transformation of the normal distribution function to give

Lognormal mean,

$$\lambda_m = \ln\left[\frac{\bar{x}}{\sqrt{(1 + V^2)}}\right] \tag{3.10}$$

and

Lognormal standard deviation,

$$\zeta = \sqrt{\ln(1 + V^2)} \tag{3.11}$$

where $V = \sigma/\bar{x}$ is the coefficient of variation and $\bar{x}$ and σ are defined by Eqs. 3.6 and 3.8, respectively. Thus, the mean and standard deviation of the lognormal function can be calculated from the statistics obtained from the standard normal function.

3.4.4 Bias Factor

In Figure 3.1, it was observed that the mean value $\bar{x}$ for the concrete compressive strength is 27.2 MPa. The design value, or nominal value x_n, of the concrete compressive strength for this population of concrete cylinders was specified as 20.7 MPa. There is a clear difference between what is specified and what is delivered. This difference is referred to as the "bias." It is common for the mean value to be larger than the nominal value because suppliers

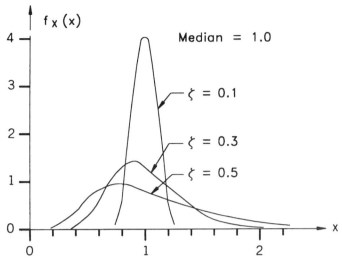

Fig. 3.3 Lognormal density functions. [From A. H-S. Ang and W. H. Tang (1975), *Probability Concepts in Engineering Planning and Design, Volume I—Basic Principles.* Copyright © 1975. Reprinted with permission of John Wiley & Sons, Inc.].

and manufacturers do not want their products rejected. Defining the bias factor λ as the ratio of the mean value $\bar{x}$ to the nominal value x_n, we have

Bias factor,

$$\lambda = \frac{\bar{x}}{x_n} \tag{3.12}$$

For the distribution of concrete compressive strength given in Figure 3.1, the bias factor is $27.2/20.7 = 1.31$.

3.4.5 Coefficient of Variation

To provide a measure of dispersion, it is convenient to define a value that is expressed as a fraction or percentage of the mean value. The most commonly used measure of dispersion is the coefficient of variation (V), which is the standard deviation (σ) divided by the mean value $(\bar{x})$

Coefficient of variation,

$$V = \frac{\sigma}{\bar{x}} \tag{3.13}$$

For the distribution of concrete compressive strength given in Figure 3.1, the coefficient of variation is $4.24/27.2 = 0.156$ or 15.6%.

Table 3.4 gives typical values of the bias factor and coefficient of variation for resistance of materials collected by Siu et al. (1975). Comparing the statistics in Table 3.4 for concrete in compression to those obtained from the tests reported in Figure 3.1, the bias factor is lower but the coefficient of variation is the same.

Table 3.5 gives the same statistical parameters for highway dead and live loads taken from Nowak (1993). The largest variation is the weight of the wearing surface placed on bridge decks. Also of interest, as indicated by the bias factor, is that the observed actual loads are greater than the specified nominal values.

3.4.6 Probability of Failure

In the context of reliability analysis, failure is defined as the realization of one of a number of predefined limit states. Load and resistance factors are selected to insure that each possible limit state is reached only with an acceptably small probability of failure. The probability of failure can be determined if the statistics (mean and standard deviation) of the resistance and load distribution functions are known.

To illustrate the procedure, first consider the probability density functions for the random variables of load Q and resistance R shown in Figure 3.4 for a hypothetical example limit state. As long as the resistance R is greater than the effects of the load Q, there is a margin of safety for the limit state under consideration. A quantitative measure of safety is the probability of survival given by

TABLE 3.4 Typical Statistics for Resistance of Materials[a]

Limit	Bias (λ_R)	COV (V_R)
Light-gage steel		
Tension and flexure	1.20	0.14
Hot-rolled steel		
Tension and flexure	1.10	0.13
Compression	1.20	0.15
Reinforced concrete		
Flexure	1.14	0.15
Compression	1.14	0.16
Shear	1.10	0.21
Wood		
Tension and flexure	1.31	0.16
Compression parallel to grain	1.36	0.18
Compression perpendicular to grain	1.71	0.28
Shear	1.26	0.14
Buckling	1.48	0.22

[a][Reproduced from W. W. C. Siu, S. R. Parimi, and N. C. Lind (1975). "Practical Approach to Code Calibration," *Journal of the Structural Division,* ASCE, 101(ST7), pp. 1469–1480. With permission].

TABLE 3.5 Statistics for Bridge Load Components[a]

Load Component	Bias (λ_Q)	COV (V_Q)
Dead load		
Factory-made	1.03	0.08
Cast-in-place	1.05	0.10
Asphaltic wearing surface	1.00	0.25
Live load (with dynamic load allowance)	1.10–1.20	0.18

[a]Nowak, 1993.

Probability of survival,

$$p_s = P(R > Q) \tag{3.14}$$

where the right-hand side represents the probability that R is greater than Q. Because the values of both R and Q vary, there is a small probability that the load effect Q may exceed the resistance R. This situation is represented by the shaded region in Figure 3.4. The complement of the probability of survival is the probability of failure, which can be expressed as

Probability of failure,

$$p_f = 1 - p_s = P(R < Q) \tag{3.15}$$

where the right-hand side represents the probability that $R < Q$.

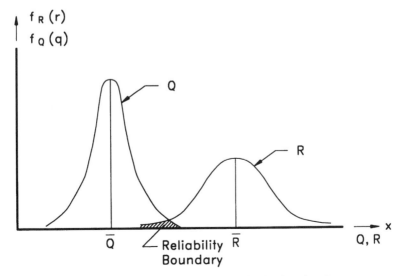

Fig. 3.4 Probability density functions for load and resistance.

The probability density functions for R and Q in Figure 3.4 have purposely been drawn to represent different coefficients of variation, V_R and V_Q, respectively. The areas under the two curves are both equal to unity, but the resistance R is shown with greater dispersion than Q. The shaded area indicates the region of failure, but the area is not equal to the probability of failure because it is a mixture of areas coming from distributions with different ratios of standard deviation to mean value. For quantitative evaluation of probability of failure p_f, it is convenient to use a single combined probability density function $g(R,Q)$ that represents the margin of safety. From this limit state function $g(R,Q)$, with its own unique statistics, the probability of failure and the safety index can be determined in a straightforward operation.

If R and Q are normally distributed, the limit state function $g(\ )$ can be expressed as

$$g(R,Q) = R - Q \tag{3.16}$$

For lognormally distributed R and Q, the limit state function $g(\ )$ can be written as

$$g(R,Q) = \ln(R) - \ln(Q) = \ln\left(\frac{R}{Q}\right) \tag{3.17}$$

In both cases, the limit state is reached when $R = Q$ and failure occurs when $g(R,Q) < 0$. From probability theory, when two normally distributed random variables are combined, then the resulting probability density function is also normal, that is, if R and Q are normally distributed, then the function $g(R,Q)$ is also normally distributed. Similarly, if R and Q are lognormal, then the function $g(R,Q)$ is lognormal. As a result, the statistics from the individual distributions can be used to calculate the statistics (mean and standard deviation) of the combined distribution.

Making use of these fundamental properties, the probability of failure for normally distributed R and Q can be obtained by

$$p_f = 1 - F_u\left(\frac{\overline{R} - \overline{Q}}{\sqrt{\sigma_R^2 + \sigma_Q^2}}\right) \tag{3.18}$$

and the probability of failure can be estimated for lognormally distributed R and Q by

$$p_f = 1 - F_u\left(\frac{\ln(\overline{R}/\overline{Q})}{\sqrt{V_R^2 + V_Q^2}}\right) \tag{3.19}$$

where $\overline{R}$ and $\overline{Q}$ are mean values, σ_R and σ_Q are standard deviations, V_R and V_Q are coefficients of variation of the resistance R and the load effect Q, and

$F_u(\ )$ is the standard normal cumulative distribution function. The cumulative distribution function $F_u(\ )$ is the integral of $f_x(x)$ between the limits $-\infty$ to u and will give the probability that x will be less than u. This integral is shown by the shaded area in Figure 3.2(b). (Note that u can be interpreted as the number of standard deviations by which x differs from the mean.) To determine the probability that a normal random variable lies in any interval, the difference between two values of $F_u(\ )$ will give this information. There is no simple expression for $F_u(\ )$, but it has been evaluated numerically and tabulated.

3.4.7   Safety Index, β

A simple alternative method for expressing the probability of failure is to use the safety index β. This procedure will be illustrated using the lognormal limit state function of Eq. 3.17. As noted previously, the lognormal distribution represents actual distributions of R and Q more accurately than the normal distribution. Also, numerical calculation of the statistics for the limit state function $g(\ )$ are more stable using the ratio R/Q than for using the difference R-Q because the difference R-Q is subject to loss of significant figures when R and Q are nearly equal.

If the function $g(R,Q)$ as defined by Eq. 3.17 has a lognormal distribution, its frequency distribution would have the shape of the curve shown in Figure 3.5. This curve is a single frequency distribution curve combining the uncertainties of both R and Q. The probability of attaining a limit state $(R < Q)$ is equal to the probability that $\ln(R/Q) < 0$ and is represented by the shaded area in Figure 3.5.

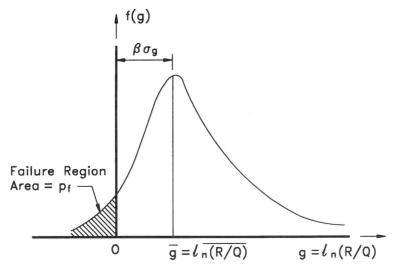

Fig. 3.5   Definition of safety index for lognormal R and Q.

The probability of failure can be reduced, and thus safety is increased by either having a tighter grouping of data about the mean $\bar{g}$ (less dispersion) or by moving the mean $\bar{g}$ to the right. These two approaches can be combined by defining the position of the mean from the origin in terms of the standard deviation σ_g of $g(R,Q)$. Thus, the distance $\beta\sigma_g$ from the origin to the mean in Figure 3.5 becomes a measure of safety and the number of standard deviations β in this measure is known as the "safety index."

> **Safety Index β** is defined as the number of standard deviations σ_g that the mean value $\bar{g}$ of the limit state function $g(\ )$ is greater than the value defining the failure condition $g(\ ) = 0$, that is, $\beta = \bar{g}/\sigma_g$.

Normal Distributions. If resistance R and load Q are both normally distributed random variables, and are statistically independent, the mean value $\bar{g}$ of $g(R,Q)$ given by Eq. 3.16 is

$$\bar{g} = \bar{R} - \bar{Q} \tag{3.20}$$

and its standard deviation is

$$\sigma_g = \sqrt{\sigma_R^2 + \sigma_Q^2} \tag{3.21}$$

where $\bar{R}$ and $\bar{Q}$ are mean values and σ_R and σ_Q are standard deviations of R and Q. If the horizontal axis in Figure 3.5 represented the limit state function $g(R,Q)$ and R-Q, equating the distances from the origin, $\beta\sigma_g = \bar{g}$, and substituting Eqs. 3.20 and 3.21, the relationship for the safety index β becomes

$$\beta = \frac{\bar{R} - \bar{Q}}{\sqrt{\sigma_R^2 + \sigma_Q^2}} \tag{3.22}$$

This closed-form equation is convenient because it does not depend on the distribution of the combined function $g(R,Q)$, but only on the statistics of R and Q individually.

If we compare Eqs. 3.18 and 3.22, the probability of failure p_f, written in terms of the safety index β, is

$$p_f = 1 - F_u(\beta) \tag{3.23}$$

By relating the safety index directly to the probability of failure, we have

$$\beta = F_u^{-1}(1 - p_f) \tag{3.24}$$

where F_u^{-1} is the inverse standard normal cumulative distribution function.

Values for the relationships in Eqs. 3.23 and 3.24 are given in Table 3.6 based on tabulated values for F_u and F_u^{-1} found in most statistics textbooks.

TABLE 3.6   Relationship between Probability of Failure and Safety Index for Normal Distributions

β	p_f	p_f	β
2.5	0.62×10^{-2}	1.0×10^{-2}	2.32
3.0	1.35×10^{-3}	1.0×10^{-3}	3.09
3.5	2.33×10^{-4}	1.0×10^{-4}	3.72
4.0	3.17×10^{-5}	1.0×10^{-5}	4.27
4.5	3.4×10^{-6}	1.0×10^{-6}	4.75
5.0	2.9×10^{-7}	1.0×10^{-7}	5.20
5.5	1.9×10^{-8}	1.0×10^{-8}	5.61

Roughly speaking, a change of 0.5 in β results in an order of magnitude change in p_f. As mentioned earlier, there is no comparable relationship between the safety factor used in ASD and the probability of failure, which is a major advantage of LRFD.

Lognormal Distributions. If the resistance R and load Q are both lognormally distributed random variables, and are statistically independent, it can be shown that the mean value of $g(R,Q)$ given by Eq. 3.17 is

$$\bar{g} = \ln \left(\frac{\bar{R}}{\bar{Q}} \right) \tag{3.25}$$

and its standard derivation is approximately

$$\sigma_g = \sqrt{V_R^2 + V_Q^2} \tag{3.26}$$

where $\bar{R}$ and $\bar{Q}$ are mean values and V_R and V_Q are coefficients of variation of R and Q, respectively. If we equate the distances from the origin in Figure 3.5, the relationship for the safety index β becomes

$$\beta = \frac{\ln(\bar{R}/\bar{Q})}{\sqrt{V_R^2 + V_Q^2}} \tag{3.27}$$

This closed-form equation is convenient because it does not depend on the distribution of the combined function $g(R,Q)$, but only on the statistics of R and Q individually.

The expression of σ_g in Eq. 3.26 is an approximation that is valid if the coefficients of variation, V_R and V_Q, are relatively small, less than about 0.20. This value can be seen by expressing the logarithmic function in Eq. 3.11 as an infinite series, that is,

$$\ln(1 + V^2) = V^2 - \tfrac{1}{2}(V^2)^2 + \tfrac{1}{3}(V^2)^3 - \tfrac{1}{4}(V^2)^4 + \cdots$$

For an infinite series with alternating signs, the error is no more than the first neglected term. Therefore, using only the first term a maximum relative error for $V = 0.20$ can be expressed as

$$\frac{\frac{1}{2}(V^2)^2}{V^2} = \frac{1}{2} V^2 = \frac{1}{2}(0.2)^2 = 0.02 \text{ or } 2\%$$

Therefore, $\ln(1 + V^2)$ can be replaced by V^2 without large error, and Eq. 3.11 then gives $\zeta \approx V$. The typical values given in Tables 3.4 and 3.5 for coefficient of variation of resistance of materials V_R and effect of loads V_Q are generally less than about 0.20; therefore, these variations can be used to represent the standard deviations of their respective lognormal distributions.

An approximate relationship between the safety index β and the probability of failure p_f has been developed by Rosenblueth and Esteva (1972) for lognormally distributed values of R and Q. It is given by the equation

$$p_f = 460 \exp(-4.3\beta) \qquad 2 < \beta < 6 \qquad (3.28)$$

The inverse function for this relationship is

$$\beta = \frac{\ln(460/p_f)}{4.3} \qquad 10^{-1} < p_f < 10^{-9} \qquad (3.29)$$

Values for both of these relationships are given in Table 3.7. By comparing Tables 3.6 and 3.7, the values for the normal and lognormal distributions are similar, but are not identical.

EXAMPLE 3.2

A prestressed concrete girder bridge with a simple span of 27 m and girder spacing of 2.4 m has the following bending moment statistics for a typical girder:

TABLE 3.7   Relationship between Probability of Failure and Safety Index for Lognormal Distributions

β	p_f	p_f	β
2.5	0.99×10^{-2}	1.0×10^{-2}	2.50
3.0	1.15×10^{-3}	1.0×10^{-3}	3.03
3.5	1.34×10^{-4}	1.0×10^{-4}	3.57
4.0	1.56×10^{-5}	1.0×10^{-5}	4.10
4.5	1.82×10^{-6}	1.0×10^{-6}	4.64
5.0	2.12×10^{-7}	1.0×10^{-7}	5.17
5.5	2.46×10^{-8}	1.0×10^{-8}	5.71

Effect of Loads (assumed normally distributed)

$$\overline{Q} = 4870 \text{ kN m} \qquad \sigma_Q = 415 \text{ kN m}$$

Resistance (assumed lognormally distributed)

$$R_n = 7040 \text{ kN m} \qquad \lambda_R = 1.05 \qquad V_R = 0.075$$

Determine the safety index for a typical girder using Eqs. 3.22 and 3.27. To use Eq. 3.22, the mean value and standard deviation of R must be calculated. From Eqs. 3.12 and 3.13.

$$\overline{R} = \lambda_R R_n = 1.05(7040) = 7390 \text{ kN m}$$

$$\sigma_R = V_R \overline{R} = 0.075(7390) = 554 \text{ kN m}$$

By substituting values into Eq. 3.22, we get

$$\beta = \frac{\overline{R} - \overline{Q}}{\sqrt{\sigma_R^2 + \sigma_Q^2}} = \frac{7390 - 4870}{\sqrt{554^2 + 415^2}} = 3.64$$

To use Eq. 3.27, the coefficient of variation of Q must be calculated from Eq. 3.13.

$$V_Q = \frac{\sigma_Q}{\overline{Q}} = \frac{415}{4870} = 0.085$$

If we substitute values into Eq. 3.27, we get

$$\beta = \frac{\ln(\overline{R}/\overline{Q})}{\sqrt{V_R^2 + V_Q^2}} = \frac{\ln(7390/4870)}{\sqrt{0.075^2 + 0.085^2}} = 3.68$$

The two results for β are nearly equal. The approximation used for σ_g in the development of Eq. 3.27 is reasonable. From either Table 3.6 or Table 3.7, the probability of failure of one of these girders is about 1:10,000.

3.5   CALIBRATION OF LRFD CODE

3.5.1   Overview of the Calibration Process

Several approaches can be used in calibrating a design code. Codes may be calibrated by use of judgment, fitting to other codes, use of reliability theory, or a combination of these approaches.

Calibration by judgment was the first approach used in arriving at code parameters. If the performance of a code was found to be satisfactory after many years, the parameter values were accepted as appropriate. Poor performance resulted in increasing safety margins. A fundamental disadvantage of this approach is that it results in nonuniform margins of safety, because excessively conservative code provisions will not result in problems and will therefore not be changed.

Calibration by fitting is usually done after there has been a fundamental change in either the design philosophy or the code format. In this type of calibration, the parameters of the new code are adjusted such that designs are obtained that are essentially the same as those achieved using the old code. The main objective of this type of calibration is to transfer experience from the old code to the new code.

Calibration by fitting is a valuable technique for ensuring that designs obtained with the new code do not deviate significantly from existing designs. It is also a relatively simple procedure since all that is involved is to match the code parameters from the old and new codes. The disadvantages of this type of calibration is that it does not necessarily result in more uniform safety margins or economy, because the new code essentially mimics the old code.

A code may also be calibrated by a more formal process using reliability theory. The formal process for estimating suitable values of load factors and resistance factors for use in bridge design, consists of the following steps (Barker et al., 1991):

Step 1. Compile the statistical database for load and resistance parameters.

Step 2. Estimate the level of reliability inherent in current design methods of predicting strengths of bridge structures.

Step 3. Observe the variation of the reliability levels with different span lengths, dead load to live load ratios, load combinations, types of bridges, and methods of calculating strengths.

Step 4. Select a target reliability index based on the margin of safety implied in current designs.

Step 5. Calculate load factors and resistance factors consistent with the selected target reliability index. It is also important to couple experience and judgment with the calibration results.

3.5.2 Calibration Using Reliability Theory

Calibration of the LRFD code for bridges using reliability theory followed the five steps outlined above. These steps are described in more detail in the following paragraphs.

Step 1

Compile a Database of Load and Resistance Statistics. Calibration using reliability theory requires that statistical data on load and resistance be available. The FOSM theories require the mean value and standard deviation to represent the probability density function. For a given nominal value, these two parameters are then used to calculate the companion nondimensional bias factor and coefficient of variation for the distribution.

The statistics for bridge load components given in Table 3.5 were compiled from available data and measurements of typical bridges. The live load statistics were obtained from surveys of truck traffic and weigh-in-motion data (Nowak, 1993).

Statistical data for resistance of materials given in Table 3.4 were obtained from material tests, component tests, and field measurements (Siu et al. 1975). Because these typical resistance statistics are 20 years old and were not developed specifically for bridges, data from current highway bridges were utilized in the LRFD calibration process.

The highway bridges selected for evaluation numbered about 200 and were from various geographic regions in the United States (California, Colorado, Illinois, Kentucky, Maryland, Michigan, Minnesota, New York, Oklahoma, Pennsylvania, Tennessee, and Texas). When selecting bridges representative of the nation's inventory, emphasis was placed on current and anticipated future trends in materials, bridge types, and spans. Steel bridges included composite and noncomposite rolled beams and plate girders, box girders, through trusses, deck trusses, pony trusses, arches, and a tied arch. Reinforced concrete bridges included slabs, T-beams, solid frame, and a box girder. Prestressed concrete bridges included a double tee, I-beams, and box girders. Wood bridges included sawn beam, glulam beam, a truss, and decks that were either nailed or prestressed transversely. Spans ranged from 9 m for a reinforced concrete slab bridge to 220 m for a steel arch bridge.

Resistance statistics were developed for a reduced set of the selected bridges, which included only the girder type structures. For each of the girder bridges, the load effects (moments, shears, tensions, and compressions) were calculated and compared to the resistance provided by the actual cross section. The statistical parameters of resistance for steel girders, reinforced concrete T-beams, and prestressed concrete girders are shown in Table 3.8 (Nowak, 1993). Comparing the resistance statistics of Tables 3.4 and 3.8, the bias factor and coefficient of variation for the girder bridges are slightly lower than those for the general population of structures.

Step 2

Estimate the Safety Index β in Current Design Methods. Risk levels implied in the existing specifications were determined by computing safety indexes for additional representative bridges. The additional bridges covered five span lengths from 9 to 60 m and five girder spacings from 1.2 to 3.6 m.

TABLE 3.8   Statistical Parameters of Resistance for Selected Bridges[a]

Type of Structure	Bias (λ_R)	COV (V_R)
Noncomposite steel girders		
Moment (compact)	1.12	0.10
Moment (noncompact)	1.12	0.10
Shear	1.14	0.105
Composite steel girders		
Moment	1.12	0.10
Shear	1.14	0.105
Reinforced concrete T-beams		
Moment	1.14	0.13
Shear w/steel	1.20	0.155
Shear w/o steel	1.40	0.17
Prestressed concrete girders		
Moment	1.05	0.075
Shear w/steel	1.15	0.14

[a]Nowak, 1993.

The reliability analysis was based on the FOSM methods, a normal distribution of the load, and a lognormal distribution of the resistance.

The mean value FOSM (MVFOSM) method used to derive Eqs. 3.22 and 3.27 is not the most accurate method that can be used to calculate values of the safety index β. While values of β determined from Eqs. 3.22 and 3.27 are sufficiently accurate to be useful for some purposes, it was considered worthwhile to use the more accurate advanced FOSM (AFOSM) method to derive values of β for the new AASHTO (1994) LRFD code.

An explicit expression for β cannot be written when the AFOSM method is used because the limit state function $g(\ )$ is linearized at a point on the failure surface, rather than at the mean values of the random variables. For the AFOSM method, an iterative procedure must be used in which an initial value of β is assumed and the process is repeated until the difference in calculated values of β on successive iterations is within a small tolerance. This iterative procedure is based on normal approximations to nonnormal distributions at the design point developed by Rackwitz and Fiessler (1978).

An estimate of the mean value of the lognormally distributed resistance R_n with bias factor λ_R and coefficient of variation V_R is given by Eq. 3.12 as

$$\overline{R} = \lambda_R R_n$$

and an assumed design point is

$$R^* = \overline{R}(1 - kV_R) = R_n\lambda_R(1 - kV_R) \tag{3.30}$$

where k is unknown. Because it is a modifier of the nominal resistance R_n, the term $\lambda_R(1 - kV_R)$ can be thought of as an estimate of a resistance factor ϕ^*, that is

$$\phi^* = \lambda_R(1 - kV_R) \tag{3.31}$$

and the parameter k is comparable to the number of standard deviations from the mean value. As an initial guess, k is often taken as 2.

An estimate of the standard deviation of R_n is obtained from Eq. 3.13 as

$$\sigma_R = V_R \overline{R}$$

and at an assumed design point becomes

$$\sigma'_R = R_n V_R \lambda_R(1 - kV_R) \tag{3.32}$$

For normally distributed R and Q, the safety index is given by Eq. 3.22 for the MVFOSM method. Substituting Eqs. 3.30 and 3.32 into Eq. 3.22 and transforming the lognormally distributed R_n into a normal distribution at the design point R^*, the safety index β can be expressed as (Nowak, 1993)

$$\beta = \frac{R_n \lambda_R(1 - kV_R)[1 - \ln(1 - kV_R)] - \overline{Q}}{\sqrt{[R_n V_R \lambda_R(1 - kV_R)]^2 + \sigma_Q^2}} \tag{3.33}$$

EXAMPLE 3.3

For the prestressed girder of Example 3.2, estimate the safety index at the design point R^* using Eq. 3.33 with $k = 2$. The statistics from Example 3.1 are

$$R_n = 7040 \text{ kN m} \qquad \lambda_R = 1.05 \qquad V_R = 0.075 \qquad \overline{Q} = 4870 \text{ kN m}$$

and

$$\sigma_Q = 415 \text{ kN m}$$

$$\phi^* = \lambda_R(1 - kV_R) = 1.05[1 - 2(0.075)] = 0.89$$

$$R^* = \phi^* R_n = 0.89(7040) = 6270 \text{ kN m}$$

$$\sigma'_R = V_R R^* = 0.075(6270) = 470 \text{ kN m}$$

Substitution into Eq. 3.33 gives

$$\beta = \frac{R^*[1 - \ln(1 - kV_R)] - \overline{Q}}{\sqrt{(\sigma'_R)^2 + \sigma_Q^2}} \tag{3.34}$$

$$\beta = \frac{6270[1 - \ln(1 - 2 \cdot 0.075)] - 4870}{\sqrt{470^2 + 415^2}} = \frac{7290 - 4870}{627}$$

$$\beta = 3.86$$

This estimate of the safety index β is slightly higher than the values calculated in Example 3.1. The value of β calculated at the design point on the failure surface by the iterative AFOSM method is considered to be more accurate than the value calculated by the MVFOSM method in Example 3.1.

Step 3

Observe the Variation of the Safety Indexes. Safety indexes were calculated by Nowak (1993) using the iterative AFOSM method for typical girder bridges. The study covered the full range of spans and girder spacings of simple span noncomposite steel, composite steel, reinforced concrete *T*-beam, and prestressed concrete *I*-beam bridges. For each of the bridge types five span lengths of 9, 18, 27, 36, and 60 m were chosen. For each span, five girder spacings of 1.2, 1.8, 2.4, 3.0, and 3.6 m were selected. For each case, cross sections were designed so that the actual resistance was equal to the required resistance of the existing code (AASHTO, 1989). In other words, the cross sections were neither overdesigned or underdesigned. It was not possible for one cross section to satisfy this criterion for both moment and shear, so separate designs were completed for both limit states.

Calculated safety indexes for prestressed concrete girders are shown in Figure 3.6 for simple span moment and in Figure 3.7 for shear. These results are typical of the other bridge types, that is, higher values of β for wider spacing of girders and lower values of β for shear than moment.

Observations of Figures 3.6 and 3.7 indicate for the moment a range of β from 2.0 to 4.5 with the lower value for small spans while for shear the range is 2.0–4.0 with the lower value for large spans. For these ranges of β, Tables 3.6 and 3.7 indicate that the probability of failure of designs according to AASHTO (1989) Standard Specifications varies from about 1:100 to 1:100,000. A uniform level of safety does not exist.

Step 4

Select a Target Safety Index β_T. A relatively large range of safety indexes were observed for moment and shear designs using the AASHTO (1989) Standard Specifications. These safety indexes varied mostly with span length and girder spacing and to a lesser extent with bridge type. What

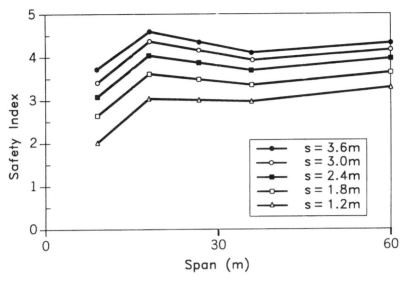

Fig. 3.6 Safety indexes for AASHTO (1989); simple span moment in prestressed concrete girders. [Nowak, 1993].

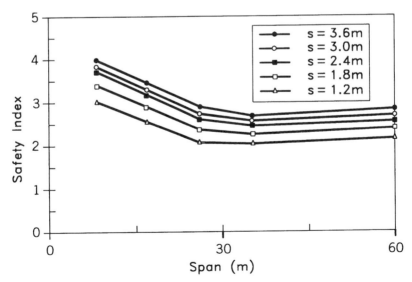

Fig. 3.7 Safety indexes for AASHTO (1989); simple span shear in prestressed concrete girders. [Nowak, 1993].

was desired in the calibration of the AASHTO (1994) LRFD code was a uniform safety index for all spans, spacings, and bridge types. To achieve this objective, a desired or target safety index is chosen and then load and resistance factors are calculated to give safety indexes as close to the target value as possible.

Based on the results of the parametric study by Nowak (1993), as well as calibrations of other specifications (OHBDC, 1992 and AISC, 1986), a target safety index $\beta_T = 3.5$ was selected. This value of β_T corresponds to the safety index calculated for moment in a simple span of 18 m with a girder spacing of 1.8 m using the AASHTO (1989) Standard Specifications. This calibration point can be seen in Figure 3.6 for prestressed concrete girders. Similar results were obtained for the other bridge types studied.

Step 5

Calculate Load and Resistance Factors. To achieve the desired or target safety index of $\beta_T = 3.5$, statistically based load and resistance factors must be calculated. Load factors must be common for all bridge types. The variation of β with span length is due to different ratios of dead load to live load. This effect can be minimized by proper selection of load factors for dead load and live load.

Resistance factors must account for the differences in reliability of the various limit states. For example, the safety indexes calculated for moment and shear shown in Figures 3.6 and 3.7 have different values and different trends.

It may not be possible to satisfy all of the conditions with $\beta_T = 3.5$. However, the objective of the calibration process is to select load and resistance factors that will generate safety indexes that are as close as possible to the target value. Acceptable sets of load factors and resistance factors occur when the calculated safety indexes cluster in a narrow band about the target value of $\beta_T = 3.5$.

To derive load factors γ and resistance factors ϕ from statistical considerations, assume that R and Q are lognormally distributed and that β is given by Eq. 3.22 so that

$$\overline{R} - \overline{Q} = \beta\sqrt{V_R^2 + V_Q^2} \tag{3.35}$$

It is desirable to separate the effects of R and Q, which can be done by using the approximation suggested by Lind (1971) for the value of the square-root term

$$\sqrt{\sigma_R^2 + \sigma_Q^2} \approx \alpha(\sigma_R + \sigma_Q) \tag{3.36}$$

where $\alpha = \sqrt{2}/2 = 0.707$ if $\sigma_R/\sigma_Q = 1.0$. Typical statistics for σ_R and σ_Q

indicate that the maximum range for σ_R/σ_Q is between $\frac{1}{3}$ and 3.0. Taking the extreme values for σ_R/σ_Q, then $\alpha = \sqrt{10}/4 = 0.79$. The maximum error in the approximation will only be 6% if $\alpha = 0.75$.

Substituting Eq. 3.36 into Eq. 3.35 yields

$$\overline{R} - \overline{Q} = \alpha\beta(\sigma_R + \sigma_Q)$$

which can be separated into

$$\overline{R} - \alpha\beta\sigma_R = \overline{Q} + \alpha\beta\sigma_Q \tag{3.37}$$

Recalling the definition of the bias factor (Eq. 3.12) and the coefficient of variation (Eq. 3.13), and setting $\beta = \beta_T$, we can write

$$R_n\lambda_R(1 - \alpha\beta_T V_R) = Q_n\lambda_Q(1 + \alpha\beta_T V_Q) \tag{3.38}$$

which can be written in the generic form of the basic design equation

$$\phi R_n = \gamma Q_n \tag{3.39}$$

where

$$\gamma = \lambda_Q(1 + \alpha\beta_T V_Q) \tag{3.40}$$

$$\phi = \lambda_R(1 - \alpha\beta_T V_R) \tag{3.41}$$

and the load and resistance factors are expressed only in terms of their own statistics and some fraction of the target safety index.

Establishing Load Factors. Trial values for the load factors γ_i can be obtained from Eq. 3.40 using the statistics from Table 3.5 for the different load components. By taking $\alpha = 0.75$ and $\beta_T = 3.5$, Eq. 3.40 becomes

$$\gamma_i = \lambda_{Q_i}(1 + 2.6V_{Q_i}) \tag{3.42}$$

and the trial load factors are

Factory-made	$\gamma_{DC1} = 1.03(1 + 2.6 \times 0.08) = 1.24$
Cast-in-place	$\gamma_{DC2} = 1.05(1 + 2.6 \times 0.10) = 1.32$
Asphalt overlay	$\gamma_{DW} = 1.00(1 + 2.6 \times 0.25) = 1.65$
Live load	$\gamma_{LL} = 1.10 \text{ to } 1.20 \ (1 + 2.6 \times .18) = 1.61 \text{ to } 1.76$

In the calibration conducted by Nowak (1993), the loads Q_i were considered to be normally distributed and the resistance R_n lognormally distributed. The expression used for trial load factors was

$$\gamma_i = \gamma_{Q_i}(1 + kV_{Q_i}) \tag{3.43}$$

where k was given values of 1.5, 2.0, and 2.5. The results for trial load factors were similar to those calculated using Eq. 3.42.

The final load factors selected for the Strength I Limit State (Table 3.1) were

$$\gamma_{DC1} = \gamma_{DC2} = 1.25 \qquad \gamma_{DW} = 1.50 \qquad \text{and } \gamma_{LL} = 1.75$$

Establishing Resistance Factors. Trial values for the resistance factors ϕ can be obtained from Eq. 3.41 using the statistics from Table 3.8 for the various bridge types and limit states. Because the chosen load factors represent values calculated from Eq. 3.43 with $k = 2.0$, the corresponding Eq. 3.41 becomes

$$\phi = \gamma_R(1 - 2.0V_R) \tag{3.44}$$

The calculated trial resistance factors and the final recommended values are given in Table 3.9. The recommended values were selected to give values of the safety index calculated by the iterative procedure that were close to β_T. Because of the uncertainties in calculating the resistance factors, they have been rounded off to the nearest 0.05.

Calibration Results. The test of the calibration procedure is whether or not the selected load and resistance factors develop safety indexes that are clustered around the target safety index and are uniform with span length and girder spacing. The safety indexes have been calculated and tabulated in Nowak (1993) for the representative bridges of span lengths from 9 to 60 m.

TABLE 3.9   Calculated Trial and Recommended Resistance Factors

Material	Limit State	Eq. 3.44	ϕ, Selected
Noncomposite steel	Moment	0.90	1.00
	Shear	0.90	1.00
Composite steel	Moment	0.90	1.00
	Shear	0.90	1.00
Reinforced concrete	Moment	0.85	0.90
	Shear	0.85	0.90
Prestressed concrete	Moment	0.90	1.00
	Shear	0.85	0.90

Typical calibration results are shown in Figures 3.8 and 3.9 for moment and shear in prestressed concrete girders. The two curves for γ of 1.60 and 1.70 in the figures show the effect of changes in the load factors for live load. Figures 3.8 and 3.9 both show uniform levels of safety over the range of span lengths, which is in contrast to the variations in safety indexes shown in Figures 3.6 and 3.7 before calibration.

The selection of the higher than calculated ϕ factors of Table 3.9 is justified because Figures 3.8 and 3.9 show that they result in reduced safety indexes that are closer to the target value of 3.5. Figure 3.9 indicates that for shear in prestressed concrete girders, a ϕ factor of 0.95 could be justified. However, it was decided to keep the same value of 0.90 from the previous code.

The selection of the live load factor of 1.75 was done after the calibration process was completed. With current highway truck traffic, this increased load factor provides a safety index greater than 3.5. This increase was undoubtedly done in anticipation of future trends in highway truck traffic.

3.5.3 Calibration of Fitting with ASD

The process of calibration with the existing ASD criteria avoids drastic deviations from existing designs. Calibration by fitting with ASD can also be used where there is insufficient statistical data to calculate ϕ from an expression like Eq. 3.44.

In the ASD format, nominal loads are related to nominal resistance by the safety factor F as stated previously in Eq. 3.2.

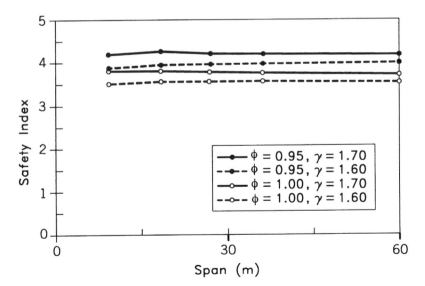

Fig. 3.8 Safety indexes for LRFD Code, simple span moments in prestressed concrete girders. [Nowak, 1993].

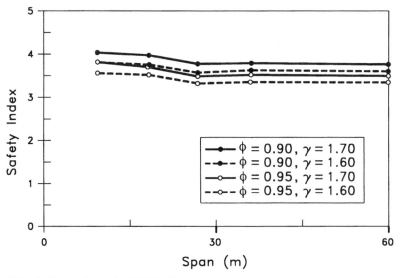

Fig. 3.9   Safety indexes for LRFD Code, simple span shears in prestressed concrete girders. [Nowak, 1993].

$$\frac{R_n}{F} \geq \Sigma Q_i \qquad (3.45)$$

Dividing Eq. 3.3 by Eq. 3.45 results in

$$\phi \geq \frac{\Sigma \gamma_i Q_i}{F \Sigma Q_i} \qquad (3.46)$$

If the loads consist of dead load Q_D and live load Q_L, then Eq. 3.46 becomes

$$\phi = \frac{\gamma_D Q_D + \gamma_L Q_L}{F(Q_D + Q_L)} \qquad (3.47)$$

Dividing both numerator and denominator by Q_L, Eq. 3.47 may be written as

$$\phi = \frac{\gamma_D \left(\dfrac{Q_D}{Q_L}\right) + \gamma_L}{F\left(\dfrac{Q_D}{Q_L} + 1\right)} \qquad (3.48)$$

EXAMPLE 3.4

Calculate the resistance factor ϕ for bending that is equivalent to an ASD safety factor $F = 1.6$ if the dead load factor γ_D is 1.25, the live load factor γ_L is 1.75, and the dead to live load ratio Q_D/Q_L is 3.0. Substituting values into Eq. 3.48

$$\phi = \frac{1.25(3.0) + 1.75}{1.6(3.0 + 1)} = 0.85$$

This value for ϕ is comparable to the values given in Table 3.9.

3.6 GEOMETRIC DESIGN CONSIDERATIONS

In water crossings or bridges over deep ravines or across wide valleys, the bridge engineer is usually not restricted by the geometric design of the highway. However, when two highways intersect at a grade separation or interchange, the geometric design of the intersection will often determine the span lengths and selection of bridge type. In this instance, collaboration between the highway engineer and the bridge engineer during the planning stage is essential.

The bridge engineer must be aware of the design elements that the highway engineer considers to be important. Both engineers are concerned about appearance, safety, cost, and site conditions. In addition, the highway engineer is concerned about the efficient movement of traffic between the roadways on different levels, which requires an understanding of the character and composition of traffic, design speed, and degree of access control so that sight distance, horizontal and vertical curves, superelevation, cross slopes, and roadway widths can be determined.

The document that gives the geometric standards is *A Policy on the Geometric Design of Highways and Streets,* AASHTO (1994a). The requirements in this publication are incorporated in the AASHTO (1994) LRFD Bridge Design Specification by reference [A2.3.2.2.3]. In the sections that follow, a few of the requirements that determine the roadway widths and clearances for bridges are given.

3.6.1 Roadway Widths

When traffic is crossing over a bridge there should not be a sense of restriction. To avoid a sense of restriction requires that the roadway width on the bridge be the same as that of the approaching highway. A typical overpass structure of a four-lane divided freeway crossing a secondary road is shown in Figure 3.10. The recommended minimum widths of shoulders and traffic lanes for the roadway on the bridge are given in Table 3.10.

Fig. 3.10   Typical overpass structure [Courtesy of Modjeski & Masters, Inc.].

For two-way elevated freeways in urban settings (Fig. 3.11), the traffic must be separated by a median barrier. The width of the barrier is 0.6 m. The minimum median width is obtained by adding the left shoulder widths in Table 3.10 to give 3.0 m for a four-lane and 6.6 m for six- and eight-lane roadways.

If a highway passes under a bridge, it is difficult not to notice the structure and to get a sense of restriction. As was discussed in the aesthetics section of Chapter 2, it is possible to increase the sense of openness by placing stub abutments on top of the slopes and providing an open span beyond the right shoulder. The geometric design requirements are stated in *A Policy on Geometric Design of Highways and Streets,* AASHTO (1994a) as follows:

Structures of this character should have liberal lateral clearances on the roadways at each level. All piers and abutment walls should be suitably offset from the

TABLE 3.10 Typical Roadway Widths for Freeway Overpasses[a]

Roadway	Width (m)
Lane width	3.6
Right shoulder width	
Four-lane	3.0
Six- and eight-lane	3.0
Left shoulder width	
Four-lane	1.2
Six- and eight-lane	3.0

[a](From *AASHTO A Policy on Geometric Design of Highways and Streets.* Copyright © 1994 by the American Association of State Highways and Transportation Officials, Washington, D.C. Used by Permission).

traveled way. The finished underpass roadway median and off-shoulder slopes should be rounded and there should be a transition to backslopes to redirect errant vehicles away from protected or unprotected structural elements.

In some areas it may be too costly to provide liberal lateral clearances and minimum dimensions are often used. The minimum lateral clearance from the edge of the traveled way to the face of the protective barrier should be the normal shoulder width given in Table 3.10. This clearance is illustrated

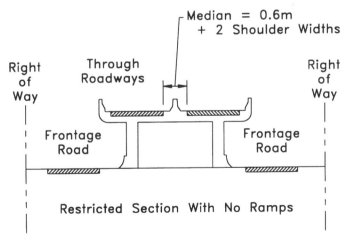

Fig. 3.11 Cross section for elevated freeways on structure with frontage roads (AASHTO Fig. VIII-9). [From *A Policy on Geometric Design of Highways and Streets,* Copyright © 1994 by the American Association of State Highway and Transportation Officials, Washington, DC. Used by Permission.]

in Figure 3.12 for a typical roadway underpass with a continuous wall or barrier. If the underpass has a center support, the same lateral clearance dimensions are applicable for a wall or pier on the left.

3.6.2 Vertical Clearances

For bridges over navigable waterways, the U.S. Coast Guard establishes the vertical clearance [A2.3.3.1]. For bridges over highways, the vertical clearances are given by *A Policy on Geometric Design of Highways and Streets,* AASHTO (1994a) [A2.3.3.2]. For freeways and arterial systems, the minimum vertical clearance is 4.9 m plus an allowance for several resurfacings of about 150 mm. For other routes, a lower vertical clearance is acceptable, but in no case should it be less than 0.5 m greater than the vehicle height allowed by state law. In general, a desired minimum vertical clearance of all structures above the traveled way and shoulders is 5.0 m.

3.6.3 Interchanges

The geometric design of the intersection of two highways depends on the expected volumes of through and turning traffic, the topography of the site, and the need to simplify signing and driver understanding to prevent wrong-way movements. There are a number of tested interchange designs and they vary in complexity from the simple two-level overpass with ramps shown in Figure 3.10 to the four-level directional interchange of Figure 3.13.

In comparing Figures 3.10 and 3.13, it is obvious that the bridge requirements for interchanges are dependent on the geometric design. In Figure 3.10, the bridges are simple overpasses with relatively linear ramps providing ac-

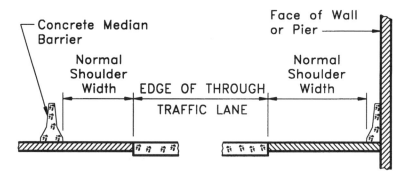

Fig. 3.12 Lateral clearances for major roadway underpasses (AASHTO Fig. X-5). [From *A Policy on Geometric Design of Highways and Streets,* Copyright © 1994 by the American Association of State Highway and Transportation Officials. Washington, DC. Used by Permission].

Fig. 3.13   Four-level directional interchange [Courtesy of Modjeski & Masters, Inc.].

cess between the two levels. In Figure 3.13, the through traffic is handled by an overpass at the lower two levels, but the turning movements are handled by sweeping curved elevated ramps at levels three and four. The geometric design of the highway engineer can strongly influence the structural design of the bridge engineer. These two engineers must work together during the planning phase and share one another's needs and desires for integrating the bridge structures into the overall mission of the highway system.

3.7   CLOSING REMARKS

In this chapter, we discussed general design considerations that range from the limit state philosophy of structural design to the calibration of the LRFD specification to the practical matter of horizontal and vertical clearances. All of these elements make up the design experience and must be understood by the successful bridge engineer. A bridge engineer is an analyst and must be able to justify a design by making calculations to show that the probable strength is greater than the probable effect of load by an acceptable safety margin. But a bridge engineer is more than an analyst and a number cruncher. A bridge engineer is also concerned about the appearance of the bridge and whether it can be safely traveled by an inexperienced driver who is using the bridge for the first time. As mentioned in Chapter 1, a bridge engineer is in a unique position of responsibility that affects not only the design project from its beginning to end, but also the safe operation of the structure throughout its life.

REFERENCES

AASHTO (1989). *Standard Specifications for Highway Bridges,* 14th ed., American Association of State Highway and Transportation Officials, Washington, DC.

AASHTO (1994). *LRFD Bridge Design Specifications,* 1st ed., American Association of State Highway and Transportation Officials, Washington, DC.

AASHTO (1994a). *A Policy on Geometric Design of Highways and Streets,* American Association of State Highway and Transportation Officials, Washington, DC.

AISC (1986). *Load and Resistance Factor Design,* Manual of Steel Construction, 1st ed., American Institute of Steel Construction, New York.

Ang, A. H-S. and W. H. Tang (1975). *Probability Concepts in Engineering Planning and Design, Volume I—Basic Principles,* Wiley, New York.

Barker, R. M., J. M. Duncan, K. B. Rojiani, P. S. K. Ooi, C. K. Tan, and S. G. Kim (1991). "Load Factor Design Criteria for Highway Structure Foundations," *Final Report, NCHRP Project 24-4,* Virginia Polytechnic Institute and State University, Blacksburg, VA, May.

Benjamin, J. R. and C. A. Cornell (1970). *Probability, Statistics, and Decisions for Civil Engineers,* McGraw-Hill, New York.

Lind, N. C. (1971). "Consistent Partial Safety Factors," *Journal of the Structural Division,* ASCE, Vol. 97, No. ST6, Proceeding Paper 8166, June, pp. 1651–1669.

MacGregor, J. G. (1992). *Reinforced Concrete Design,* 2nd ed, Prentice-Hall, Englewood Cliffs, NJ.

Nowak, A. S. (1993). "Calibration of LRFD Bridge Design Code," *NCHRP Project 12-33,* University of Michigan, Ann Arbor, MI.

OHBDC (1992). *Ontario Highway Bridge Design Code,* 3rd ed, Ministry of Transportation, Quality and Standards Division, Toronto, Ontario.

Rackwitz, R. and B. Fiessler (1978). "Structural Reliability under Combined Random Load Sequences," *Computers and Structures,* Vol. 9, No. 5, November, pp. 489–494, Pergamon Press, Great Britain.

Rosenblueth, E. and L. Esteva (1972). "Reliability Basis for Some Mexican Codes," *ACI Publication SP-31,* American Concrete Institute, Detroit, MI.

Siu, W. W. C., S. R. Parimi, and N. C. Lind (1975). "Practical Approach to Code Calibration," *Journal of the Structural Division,* ASCE, Vol. 101, No. ST7, Proceeding Paper 11404, July, pp. 1469–1480.

4

LOADS

4.1 INTRODUCTION

The engineer must consider all the loads that are expected to be applied to the bridge during its service life. Such loads may be divided into two broad categories: permanent loads and transient loads. The permanent loads remain on the bridge for an extended period, usually for the entire service life. Such loads include the self-weight of the girders and deck, wearing surface, curbs, parapets and railings, utilities, luminaries, and pressures from earth retainments. Transient loads typically include gravity loads due to vehicular, railway, and pedestrian traffic as well as lateral loads such as those due to water and wind, ice floes, ship collisions, and earthquakes. In addition, all bridges experience temperature fluctuations on a daily and seasonal basis and such effects must be considered. Depending on the structure type, other loads such as those from creep and shrinkage may be important, and finally, the superstructure supports may move, inducing forces in statically indeterminate bridges.

Transient loads, as the name implies, change with time and may be applied from several directions and/or locations. Typically, such loads are highly variable. It is the engineer's responsibility to anticipate which of these loads are appropriate for the bridge under consideration as well as the magnitude of the loads and how these loads are applied for the most critical load effect. Finally, it is reasonable to expect some loads to act in combination and such combinations must be considered for the appropriate limit state. A discussion of such considerations is presented in Chapter 3.

The loads appropriate for the design of short- and medium-span bridges are outlined in this chapter. The primary focus is on loads that are necessary for the superstructure design. Other loads are presented with only limited discussion. For example, ship impact is a very important and complex load

139

that must be considered for long-span structures over navigable waters. Similarly, seismic loads are of paramount importance in regions of high seismicity and must be considered for a bridge regardless of span length. Some bridges are an integral part of the life-line network that must remain intact after a seismic event. An understanding of such requirements requires prerequisite knowledge of structural dynamics combined with inelastic material response due to cyclic actions and is therefore discussed only briefly. This specialized topic is considered to be beyond the scope of this book.

Each type of load is presented individually with the appropriate reference to the AASHTO Specification including, where appropriate and important, a discussion regarding the development of the AASHTO provisions. The loads defined in this chapter are used in Chapter 5 to determine the load effects (shear and moment) for a girder line (single beam). In Chapter 6, the modeling of the three-dimensional system is discussed along with the reduction of the three-dimensional system to a girder-line. The primary purpose of this chapter is to define and explain the rationale of the AASHTO load requirements. Detailed examples using these loads are combined with structural analysis in the subsequent chapters.

4.2   GRAVITY LOADS

Gravity loads are those caused by the weight of an object on the bridge. Such loads are both permanent and transient and applied in a downward direction toward the center of the earth.

4.2.1   Permanent Loads

Permanent loads are those that remain on the bridge for an extended period of time, perhaps for the entire service life. Such loads include

- Dead load of structural components and nonstructural attachments (*DC*).
- Dead load of wearing surfaces and utilities (*DW*).
- Dead load of earth fill (*EV*).
- Earth pressure load (*EH*).
- Earth surcharge load (*ES*).
- Downdrag (*DD*).

The two letter abbreviations are those used by AASHTO and are also used in subsequent discussions and examples.

The dead load of the structural components and nonstructural attachments are definitely permanent loads and must be included in any analysis. Here components refer to those elements that are part of the load resistance system. Nonstructural attachments refer to such items as curbs, parapets, barrier rails,

TABLE 4.1   Unit Densities[a]

Material	Unit Weight (kg/m³)
Aluminum	2800
Bituminous wearing surfaces	2250
Cast iron	7200
Cinder filling	960
Compact sand, silt, or clay	1925
Concrete, lightweight (includes reinforcement)	1775
Concrete, sand-lightweight (includes reinforcement)	1925
Concrete, normal (includes reinforcement)	2400
Loose sand, silt, or gravel	1600
Soft clay	1600
Rolled gravel, macadam, or ballast	2450
Steel	7850
Stone masonry	2725
Hardwood	960
Softwood	800
Transit rails, ties and fastening per track	0.3[b]

[a]In AASHTO Table 3.5.1-1. [From *AASHTO LRFD Bridge Design Specifications.* Copyright © 1994 by the American Association of State Highway and Transportation Officials, Washington, DC. Used by Permission].
[b]In kilograms per cubic millimeter (kg/mm)

signs, illuminators, and guard rails. The weight of such items can be estimated by using the unit weight of the material combined with the geometry. For third-party attachments, for example, the guard rail, the manufacture's literature often contains weight information. In the absence of more precise information, the unit weights given in Table 4.1 may be used.

The dead load of the wearing surface (*DW*) is estimated by taking the unit weight times the thickness of the surface. This value is combined with the *DC* loads per Tables 3.1 and 3.2 (Tables 3.4.1-1 and 3.4.1-2).* Note that the load factors are different for the *DC* and *DW* loads. The maximum and minimum load factors for the *DC* loads are 1.25 and 0.90, respectively, and the maximum and minimum load factors for the *DW* loads are 1.50 and 0.65, respectively. The different factors are used because the *DW* loads have been determined to be more variable in load surveys than the *DC* loads. For example, Nowak (1993, 1995) noted the coefficients of variation (standard deviation per mean) for factory-made, cast-in-place (CIP), and asphalt surfaces are 0.08, 0.10, and 0.25, respectively. In short, it is difficult to estimate at the time of design how many courses and associated thicknesses of wearing surfaces may be applied by maintenance forces during the service life, but it is fairly easy to estimate the weight of other components.

---

*The article numbers in the AASHTO (1994) LRFD bridge specifications are enclosed in brackets and preceded by the letter A if specifications and by the letter C if commentary.

The dead load of earth fills (*EV*) must be considered for buried structures such as culverts. The *EV* load is determined by multiplying the unit weight times the depth of materials. Again the load factors per Table 3.1 and 3.2 (Tables A3.4.1-1 and A3.4.1-2) apply.

The earth surcharge load (*ES*) is calculated like the *EV* loads with the only difference being in the load factors. This difference is attributed to its variability. Note that part or all of this load could be removed at some time in the future, or perhaps the surcharge material (or loads) could be changed. Thus, the *ES* load has a maximum load factor of 1.50, which is higher than the typical *EV* factors that are about 1.35. Similarly, the minimum *ES* and *EV* factors are 0.75 and 0.90 (typical), respectively.

Soil retained by a structure such as a retaining wall, wing wall, or abutment creates a lateral pressure on the structure. The lateral pressure is a function of the geotechnical characteristics of the material, the system geometry, and the anticipated structural movements. Most engineers use models that yield a fluidlike pressure against the wall. Such a procedure is outlined in AASHTO Section 3.11 and is described in more detail in the AASHTO Section 10.

Downdrag is a force exerted on a pile or drilled shaft due to soil movement around the element. Such a force is permanent and typically increases with time. The details regarding the downdrag calculations are outlined in AASHTO Section 10, *Foundations*.

In summary, permanent loads must always be considered in the structural analysis. Some permanent loads are easily estimated, such as component self-weight, while others loads, such as lateral earth pressures, are more difficult due to the greater variabilities involved. Where variabilities are greater, higher load factors are used for maximum load effects and lower factors are used for minimum load effects.

4.2.2 Transient Loads

Although the automobile is the most common vehicular live load on most bridges, it is the truck that causes the critical load effects. In a sense, cars are "felt" very little by the bridge and come "free." More precisely, the load effects of the car traffic compared to the effect of truck traffic are negligible. Therefore, the AASHTO design loads attempt to model the truck traffic that is highly variable, dynamic, and may occur independent of, or in unison with, other truck loads. The principal load effect is the gravity load of the truck, but other effects are significant and must be considered. Such effects include impact (dynamic effects), braking forces, centrifugal forces, and the effects of other trucks simultaneously present. Furthermore, different design limit states may require slightly different truck load models. Each of these loads are described in more detail in the following sections. Much of the research involved with the development of the live load model and the specification calibration is presented in Nowak (1993, 1995). Readers interested in the details of this development are encouraged to obtain this reference for more background information.

Design Lanes. The number of lanes a bridge may accommodate must be established and is an important design criterion. Two such terms are used in the lane design of a bridge:

- Traffic lane.
- Design lane.

The traffic lane is the number of lanes of traffic that the traffic engineer plans to route across the bridge. A lane width is associated with a traffic lane and is typically 3600 mm. The design lane is the lane designation used by the bridge engineer for live-load placement. The design lane width and location may or may not be the same as the traffic lane. Here AASHTO uses a 3000-mm design lane and the vehicle is to be positioned within that lane for extreme effect.

The number of design lanes is defined by taking the integer part of the ratio of the clear roadway width divided by 3600 mm [A3.6.1.1.1]. The clear width is the distance between the curbs and/or barriers. In cases where the traffic lanes are less than 3600 mm wide, the number of design lanes shall be equal to the number of traffic lanes and the width of the design lane is taken as the width of traffic lanes. For roadway widths from 6000 to 7200 mm, two design lanes should be used and the design lane width should be one-half the roadway width.

The direction of traffic in the present and future design scenarios should be considered and the most critical cases should be used for design. Additionally, there may be construction and/or detour plans that cause traffic patterns to be significantly restricted or altered. Such situations may control some aspects of the design loading.

Vehicular Design Loads. A study by the Transportation Research Board (TRB) was used as the basis for the AASHTO loads (TRB, 1990). The TRB panel outlined many issues regarding the development (revision of) a national policy of truck weights. This document provides an excellent summary of history and policy alternatives and associated economic trade-offs. Loads that are above the legal weight and/or length limits but are regularly allowed to operate were cataloged. Although *all states in the Northeast allow such over-legal loads . . . ,* many others, from . . . *Florida to Alaska,* also routinely allow such loads. Typically, these loads are short-haul vehicles such as solid waste trucks and concrete mixers. Although above "legal" limits, these vehicles were allowed to operate routinely due to "grandfathering" provisions in state statutes. These vehicles are referred to as *exclusion vehicles.* It was felt by the engineers developing the load model that the exclusion trucks best represented the extremes involved in the present truck traffic (Kulicki, 1992).

Theoretically, one could use all the exclusion vehicles in each design and design for the extreme load effects (envelope of actions). As an analysis would be required for many vehicles, this is clearly a formidable task, even if automated. Hence, a simpler, more tractable model was developed. The objective of this model is to prescribe a set of loads such that the same extreme load effects of the model are approximately the same as the exclusion vehicles. This model consists of three distinctly different loads:

• Design truck.
• Design tandem.
• Design lane.

As illustrated in Figure 4.1(a), the design truck (the first of three separate live load configurations) is a model load that resembles the typical semitrailer truck [A3.6.1.2]. The front axle is 35 kN, located 4300 mm behind the drive axle is 145 kN, and the rear trailer axle is also 145 kN and is positioned at a variable distance ranging between 4300 and 9000 mm. The variable range means that the spacing used should cause critical load effect. The long spacing typically only controls where the front and rear portions of the truck may be positioned in adjacent structurally continuous spans such as for continuous

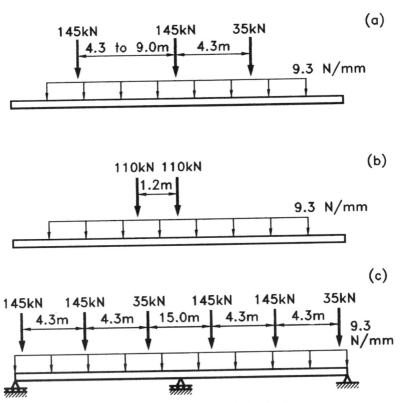

Fig. 4.1   The AASHTO design loads.

short-span bridges. The design truck is the same configuration that has been used by AASHTO (1996) Standard Specification since 1944 and is commonly referred to as HS20. The H denotes highway, the S denotes semitrailer, and the 20 is the weight of the tractor in tons (U. S. customary units). The new vehicle combinations as described in AASHTO (1994) LRFD Bridge Specifications are designated as HL-93 for highway loading accepted in 1993.

The second configuration is the design tandem and is illustrated in Figure 4.1(b). It consists of two axles weighing 110 kN each spaced at 1200 mm, which is similar to the tandem axle used in previous AASHTO Standard Specifications except the load is changed from 24 to 25 kips (110 kN).

The third load is the design lane load that consists of a uniformly distributed load of 9.3 N/mm and is assumed to occupy a region 3000 mm transversely. This load is the same as a uniform pressure of 64 lb/ft^2 (3.1 kPa) applied in a 10-ft (3000 mm) design lane. This load is similar to the lane load outlined in the AASHTO Standard Specifications for many years with the exception that the LRFD lane load does not require any concentrated loads.

The load effects of the design truck and the design tandem must each be *superimposed with* the load effects of the design lane load. This combination of lane and axle loads is a major deviation from the requirements of the earlier AASHTO Standard Specifications, where the loads were considered separately. It is important to understand that these loads are not designed to model any one vehicle or combination of vehicles, but rather the spectra of loads and their associated load effects.

Although the live load model was developed using the exclusion vehicles, it was also compared to other weigh-in-motion (WIM) studies. WIM studies gain truck weight data by using passive weighing techniques, so the operator is unaware that the truck is being monitored. Typically, bridges are instrumented to perform this task. Such studies include Goble (1991), Hwang et al. (1991), and Moses et al. (1985). These WIM studies were used as confirmation of the AASHTO live load by Kulicki (1992) and Nowak (1993).

Kulicki and Mertz (1991) compared the load effects (shear and moment) for one- and two-span continuous beams for the previous AASHTO loads and those presently prescribed. In their study, the HS20 truck and lane loads were compared to the maximum load effect of 22 trucks representative of today's traffic. The ratio of the maximum moments to the HS20 moment is illustrated in Figure 4.2. Similarly, the shear ratio is shown in Figure 4.3. Note that there is significant variation in the ratios and most ratios are greater than 1, indicating that the exclusion vehicle maximums are greater than the model load, a nonconservative situation. A perfect model would contain ordinates of unity for all span lengths. This model is practically not possible, but the combination of design truck with the design lane and the design tandem with the design lane gives improved results, which are illustrated in Figures 4.4 and 4.5. Note that the variation is much less as the ratios are more closely grouped over the span range, for both moment and shear, and for both simple and continuous spans. The implication is that the present model adequately rep-

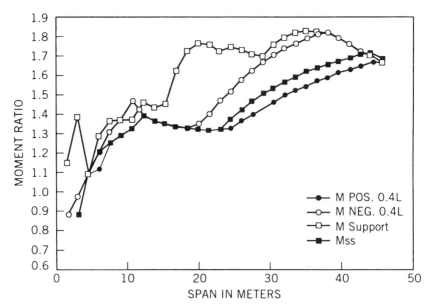

Fig. 4.2 Comparison of the exclusion vehicles to the traditional HS20 load effects—moment. (AASHTO Fig. C3.6.1.2.1-1). (From *AASHTO LRFD Bridge Design Specifications,* Copyright © 1994 by the American Association of State Highway and Transportation Officials, Washington, DC. Used by Permission).

resents today's traffic and a single-load factor may be used for all trucks. Note that in Figure 4.4, the negative moment of the model underestimates the effect of the exclusion vehicles. This underestimation occurs because the exclusion model includes only one vehicle on the bridge at a time, likely a nonconservative assumption for the negative moments and reactions at interior supports. As it is quite likely that an exclusion vehicle could be closely followed by another heavily load truck, it was felt that a third live-load combination was required to model this event. This live load combination is specified in AASHTO [A3.6.1.3.1]:

> . . . for negative moment over the interior supports (tension on top) 90 percent of the load effect of two design trucks spaced a minimum of 15 000 mm between the lead axle of one truck and the rear axle of the other truck, and the 4300 mm between the two 145 kN axles, combined with 90 percent of the effect of the design lane load.

Axles that do not contribute to the extreme force effect should be neglected (see Figure 4.1(c)). Nowak (1993) compared survey vehicles with others in the same lane to the AASHTO load model and the results are shown in Figures 4.6 and 4.7. The moments were chosen for illustration. The $M(75)$ moment represents the mean of the load effect due to the survey vehicles,

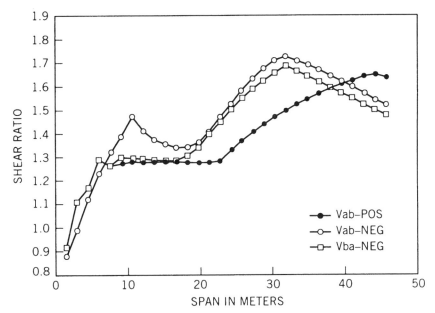

Fig. 4.3 Comparison of the exclusion vehicles to the traditional HS20 load effects—shear. (AASHTO Fig. C3.6.1.2.1-2). [From *AASHTO LRFD Bridge Design Specifications,* Copyright © 1994 by the American Association of State Highway and Transportation Officials, Washington, DC. Used by Permission].

HS20 is moment due to the traditional AASHTO (1996) Standard Specifications truck (same as the design truck), and LRFD is the moment due to the present AASHTO (1994) LRFD loads. Note that the present loads adequately represent the load survey with a bias of approximately 20%.

In summary, three design loads should be considered: the design truck, design tandem, and design lane. These loads are superimposed three ways to yield the live load effects, which are combined with the other load effects per Tables 3.1 and 3.2. These cases are illustrated in Table 4.2 where the number in the table indicates the appropriate *multiplier* to be used prior to superposition. Note that the term multiplier is used to avoid confusion with the load factors that are used to combine the various types of loads, for example, live and permanent loads in Tables 3.1 and 3.2.

Fatigue Loads. The strengths of various components of the bridge are sensitive to repeated stressing or fatigue. When the load is cyclic, the stress level that ultimately fractures the material can be significantly below the nominal yield strength. For example, depending on the details of the welds, steel could have a fatigue strength as low as 18 MPa [A6.5.3]. The fatigue strength is typically related to the range of live load stress and the number of stress cycles under service load conditions. As the majority of trucks do not exceed

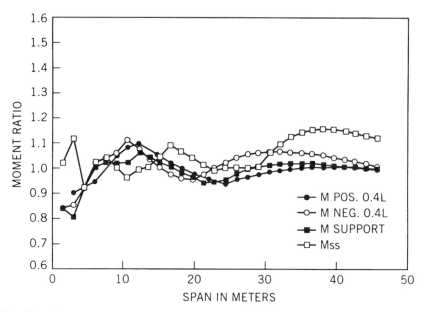

Fig. 4.4 Comparison of the design load effects with exclusion vehicle—moment. (AASHTO Fig. C3.6.1.2.1-3). [From *AASHTO LRFD Bridge Design Specifications,* Copyright © 1994 by the American Association of State Highway and Transportation Officials, Washington, DC. Used by Permission].

the legal weight limits, it would be unduly conservative to use the full live load model, which is based on exclusion vehicles to estimate this load effect. This means that a lesser load is used to estimate the live load stress range and is accommodated by using a single design truck with the variable axle spacing set at 9000 mm and a load factor of 0.75 as prescribed in Table 3.1 [Table A3.4.1-1]. The dynamic load allowance (*IM*) [A3.6.2] must be included and the bridge is assumed to be loaded in a single lane [A3.6.1.4.3b]. The average load effect due to the survey vehicles (used to calibrate the specification) was about 75% of the moment due to the design truck (Nowak, 1993), hence a load factor of 0.75 is used.

The number of stress-range cycles is based on traffic surveys. In lieu of survey data, guidelines are provided in AASHTO [A3.6.1.4.2]. The average daily truck traffic (ADTT) in a single lane may be estimated as

$$\text{ADTT}_{SL} = p(\text{ADTT})$$

where p is the fraction of traffic assumed to be in one lane as defined in Table 4.3.

Because the traffic patterns on the bridge are uncertain, the frequency of the fatigue load for a single lane is assumed to apply to all lanes.

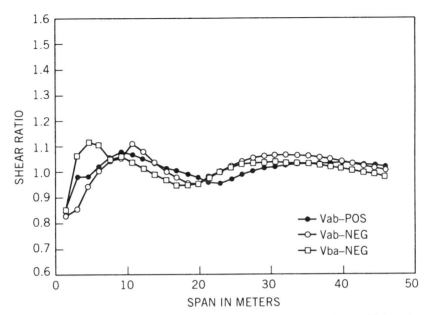

Fig. 4.5 Comparison of the design load effects with exclusion vehicle—shear. (AASHTO Fig. C3.6.1.2.1-4). [From *AASHTO LRFD Bridge Design Specifications,* Copyright © 1994 by the American Association of State Highway and Transportation Officials, Washington, DC. Used by Permission].

The ADTT is usually available from the bridge owner but in some cases only the average daily traffic (ADT) is available. In such cases, the percentage of trucks in the total traffic must be estimated. This percentage can vary widely with local conditions and the engineer should try to estimate this with a survey. For example, it is common for interstate roadways in the rural Western states to have the percentage of trucks exceed 45%. If survey data is not possible or practical, or if the fatigue limit state is not a controlling factor in the design, then AASHTO provides guidance. This guidance is illustrated in Table 4.4.

Note that the number of stress-range cycles is not used in the structural analysis directly. The fatigue truck is applied in the same manner as the other vehicles and the range of extreme stress (actions) are used. The number of stress-range cycles is used to establish the available resistance.

Pedestrian Loads. The AASHTO pedestrian load is 3.6×10^{-3} MPa, which is applied to sidewalks that are integral with a roadway bridge. If the load is applied to a bridge restricted to pedestrian and/or bicycle traffic, then a 4.1×10^{-3} MPa live load is used. These loads are comparable to the building corridor load of 4.8×10^{-3} MPa of the Uniform Building Code (UBC, 1994).

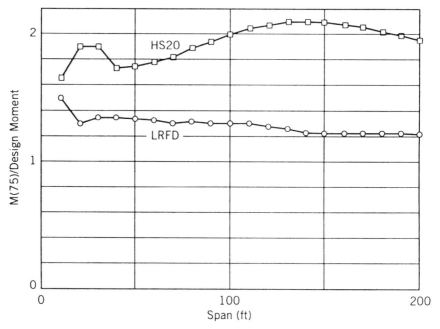

Fig. 4.6 Comparison of the design load effects with survey vehicles—moment. [Nowak, 1993].

The railing for pedestrian and/or bicycle must be designed for a load of 0.73 N/mm, both transversely and vertically on each longitudinal element in the railing system [A13.8 and A13.9]. In addition, as shown in Figure 4.8, railing must be designed to sustain a single concentrated load of 890 N applied to the top rail at any location and in any direction.

Deck and Railing Loads. The gravity loads for the design of the deck system are outlined in AASHTO [A3.6.1.3.3]. The deck must be designed for the load effect due to the design truck or the design tandem, whichever creates the most extreme effect. The two design vehicles should not be considered together in the same load case. For example, a design truck in a lane adjacent to a design tandem is not considered (consider all trucks of one kind). The design lane load is not considered in the design of the deck system, except in slab bridges where the load is carried principally in the longitudinal direction (see Chapter 3 on bridge types). Several methods are available for the analysis of decks subjected to these loads. A few of the more common methods are described in Chapter 6. The vehicular gravity loads for decks may be found in AASHTO [A3.6.1.3].

The deck overhang, located outside the facia girder and commonly referred to as the cantilever, is designed for the load effect of a uniform line load of 14.6 N/mm located 300 mm from the face of the curb or railing as shown

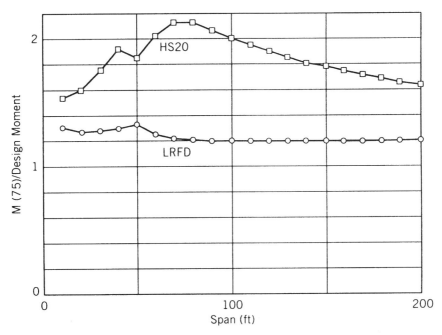

Fig. 4.7 Comparison of the Design Load Effects with Survey Vehicles—Shear. [Nowak, 1993].

in Figure 4.9. This load is derived by assuming that one-half of the 220-kN tandem (110 kN) is distributed over a length of 7600 mm. The rationale for this rather long length is that the barrier system is structurally continuous and periodically supported by cross beams or the cantilever slab that has been strengthened. In other words, the barrier behaves as another girder located on top of the deck and distributes the load over a longer length than if the barrier was not present. An illustration of a continuous barrier system is illustrated in Figure 4.10(a). The concrete curbs, parapets, barriers, and dividers should be made structurally continuous with the deck [A9.4.3]. The exception requires owner approval. If the barrier is not flexurally continuous, then the

TABLE 4.2 Load Multipliers for Live Loads

Live Load Combination	Design Truck	Design Tandem	Design Truck with 15 000-mm Headway[a]	Design Lane
1	1.0			1.0
2		1.0		1.0
3			0.9	0.9

[a]The two design truck and lane combination is for the moment at interior supports only.

TABLE 4.3   Fraction of Truck Traffic in a Single Lane, p^a

Number of Lanes Available to Trucks	p
1	1.00
2	0.85
3 or more	0.80

^{a}In AASHTO 3.6.1.4.2-1. [From *AASHTO LRFD Bridge Design Specifications,* Copyright © 1994 by the American Association of State Highway and Transportation Officials, Washington, DC. Used by Permission].

load should be distributed over a lesser length, increasing the cantilever moments. An example is illustrated in Figure 4.10(b). More details regarding deck design and analysis are presented in Chapter 6, 7, and 9.

The traffic barrier system and the deck overhang must sustain the infrequent event of a collision of a truck. The barrier is commonly referred to by many terms, such as parapet, railing, and barrier. The AASHTO uses the terms railing or railing system, and hereafter, this term is used in the same manner. The deck overhang and railing design is confirmed by crash testing as outlined in AASHTO [A13.7.2]. Here the rail/cantilever deck system is subjected to crash testing by literally moving vehicles of specified momentum (weight, velocity, and angle of attack) into the system. The momentum characteristics are specified as a function of performance levels that attempt to model various traffic conditions. The design loads crash worthiness are only used in the analysis and design of the deck and barrier systems. The design forces for the rail and deck design are illustrated in Table 4.5 for three performance levels (*PL*). The levels are described below [A13.7.2]:

TABLE 4.4   Fraction of Trucks in Traffic^a

Class of Highway	Fraction of Truck in Traffic
Rural interstate	0.20
Urban interstate	0.15
Other rural	0.15
Other urban	0.10

^{a}In AASHTO Table C3.6.1.4.2-1. [From *AASHTO LRFD Bridge Design Specifications,* Copyright © 1994 by the American Association of State Highway and Transportation Officials, Washington, DC. Used by Permission].

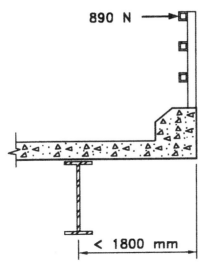

Fig. 4.8 Pedestrian rail loads.

PL-1 is used for short, low-level structures on rural highway systems, secondary expressways, and areas where a small number of heavy vehicles are expected and speeds are reduced.

PL-2 is used for high-speed mainline structures on freeways, expressways, highways, and area with a mixture of heavy vehicles and maximum tolerable speeds.

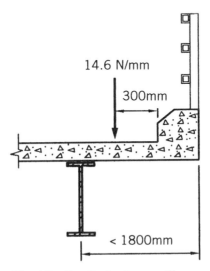

Fig. 4.9 Gravity load on cantilever.

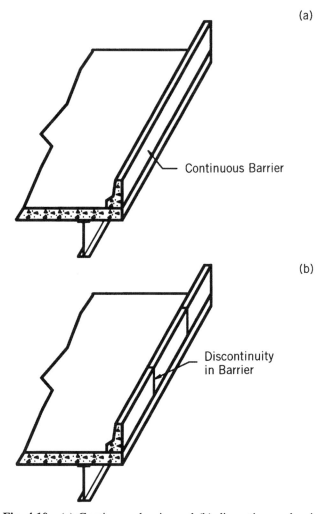

(a)

Continuous Barrier

(b)

Discontinuity
in Barrier

Fig. 4.10   (a) Continuous barrier and (b) discontinuous barrier.

PL-3 is used for freeways with variable cross slopes, reduced radius of
curvature, higher volume of mixed heavy vehicles and maximum tolerable speeds. Site specification justification shall be made for use of
this performance level.

Multiple Presence. Trucks will be present in adjacent lanes on roadways with
multiple design lanes but it is unlikely that three adjacent lanes will be loaded
simultaneously with the heavy loads. Therefore, some adjustments in the design loads are necessary. To account for this effect, AASHTO [A3.6.1.1.2]

TABLE 4.5 Design Forces for Traffic Railings[a]

Design Forces and Designations	Railing Performance Levels		
	PL-1	*PL*-2	*PL*-3
F_t Transverse (kN)	120	240	516
F_1 Longitudinal (kN)	40	80	173
F_v Vertical (kN)	20	80	222
L_t and L_1 (mm)	1220	1070	2 440
L_v (mm)	5550	5500	12 200
Minimum rail Height (mm)	510	810	1 020

[a]In AASHTO Table A13.2-1. [From *AASHTO LRFD Bridge Design Specifications,* Copyright © 1994 by the American Association of State Highway and Transportation Officials, Washington, DC. Used by Permission].

provides an adjustment factor for the multiple presence. Table 4.6, after AASHTO [Table A3.6.1.1.2-1], is provided.

Note that these factors should not be applied in situations where these factors have been implicitly included, such as in the load distribution factors outlined in AASHTO [A4.6.2]. If statical distribution factors are used or if the analysis is based on refined methods, then the multiple presence factors apply. The details of these analytical methods are described in Chapter 6. In addition, these factors apply in the design of bearings and abutments for the braking forces defined later. Lastly, the multiple presence factors should not be used in the case of the fatigue limit state.

Dynamic Effects. The roadway surface is not perfectly smooth, thus the vehicle suspension must react to roadway roughness by compression and extension of the suspension system. This oscillation creates axle forces that exceed the static weight during the time the acceleration is upward, and is less than

TABLE 4.6 Multiple Presence Factors[a]

Number of Design Lanes	Multiple Presence Factors "*m*"
1	1.20
2	1.00
3	0.85
More than 3	0.65

[a]In AASHTO Table 3.6.1.1.2-1. [From *AASHTO LRFD Bridge Design Specifications,* Copyright © 1994 by the American Association of State Highway and Transportation Officials, Washington, DC. Used by Permission].

the static weight when the acceleration is downward. Although commonly called impact, this phenomenon is more precisely referred to as dynamic loading.

There have been numerous experimental and analytical studies to determine the dynamic load effect. Paultre et al. (1992) provide an excellent review of analytical and experimental research regarding the effects of vehicle/bridge dynamics. In this paper, the writers outline the various factors used to increase the static load to account for dynamic effects. As illustrated in Figure 4.11, various bridge engineering design specifications from around the world use widely differing factors. The ordinate axis represents the load increase or dynamic load allowance (DLA) and the abscissa is the fundamental frequency of the structure. In cases where the specification value is a function of span length [e.g., AASHTO (1996)], the frequency is estimated using an empirically based formula. Note the wide variability for DLA. This variability indicates that the worldwide community has not reached a consensus about this issue.

One must carefully interpret and compare the results of such studies as the definitions of the dynamic effects are not consistent and is well portrayed by Bakht and Pinjarkar (1991) and Paultre et al. (1992). These writers describe the many definitions that have been used for dynamic load effects. Such

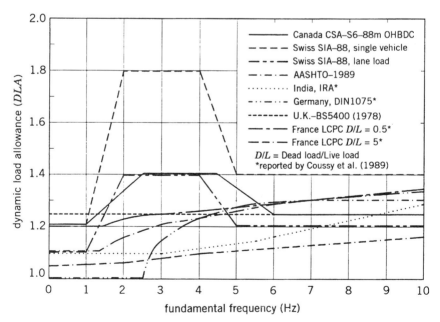

Fig. 4.11 International perspective of dynamic load allowance. [Paultre, 1992].

definitions have a significant effect on the magnitude of the DLA reported and consequently the profession's perception of dynamic effects. It is most common to compare the static and dynamic deflections as illustrated in Figure 4.12. A typical plot of a midspan deflection is shown as a function of vehicle position. The dynamic effect is defined herein as the amplification factor applied to the static response to achieve the dynamic load effect. This effect is called by many different terms: dynamic load factor, dynamic load allowance, and impact factor. Sometimes the factor includes the static load response (>1) and other times it includes only the dynamic response (<1). The term dynamic load allowance is used by AASHTO which is abbreviated *IM* (for impact). Although the terminology is inconsistent with the abbreviation, *IM* is traditionally used and some old habits will likely never die. When referring to Figure 4.12, the dynamic load allowance is

$$IM = \frac{D_{\text{dyn}}}{D_{\text{sta}}}$$

where D_{sta} is the maximum static deflection and D_{dyn} is the additional deflection due to the dynamic effects.

It is important to observe that this ratio varies significantly with different vehicle positions. Thus, it is quite possible to observe impact factors that greatly exceed those at the maximum deflections (and the AASHTO value).

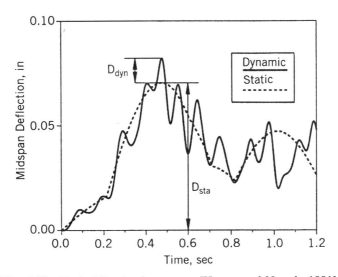

Fig. 4.12   Typical live load response. [Hwang and Nowak, 1991].

This characteristic is described by Bakht and Pinjarkar (1991), and Paultre et al. (1992). The DLA is of concern principally because it is used for the design and evaluation of bridges for the extreme load effects. Therefore, it is reasonable to define the DLA based on the extreme values.

The principal parameters that effect the impact factor are the dynamic characteristics of the truck, the dynamic characteristics of the bridge, and the roadway roughness. These characteristics are expected as all transient structural dynamic problems involve stiffness, mass, damping, and excitation. Hwang and Nowak (1991a, 1991b) present a comprehensive analytical study involving modeling a truck as rigid bodies interconnected with nonlinear suspension springs. The simply supported bridges were modeled using the standard equation for forced beam vibration, and the excitation was derived using actual roadway roughness data. Numerical integration was used to establish the response. The truck configurations were taken from weigh-in-motion studies of Moses and Ghosen (1985) and Ervin (1986). Simply supported steel and prestressed slab girder bridges were studied. The results offer insight into vehicle bridge dynamics. The dynamic and static components of midspan deflection for the steel girder bridges are illustrated in Figures 4.13 and 4.14. Note that the dynamic component remains almost unchanged with the truck weight while the static deflection increases linearly with weight, as expected. As the ratio of the two deflections is the DLA, it follows that the DLA decreases with truck weight, which is illustrated in Figure 4.15. The vertical dashed line was added by the authors to represent the weight of the 70 kip

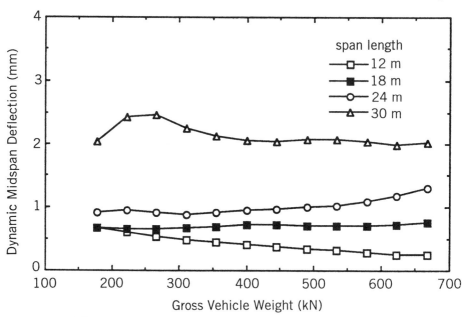

Fig. 4.13   Dynamic response. [Hwang and Nowak, 1991a].

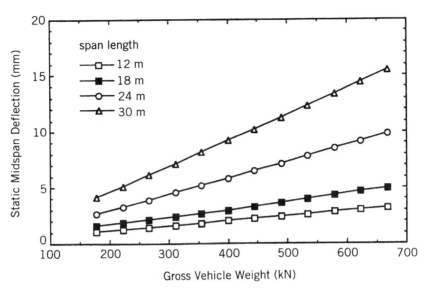

Fig. 4.14   Static response. [Hwang and Nowak, 1991a].

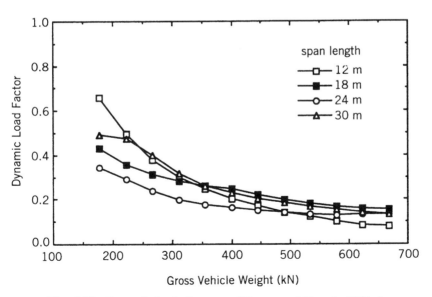

Fig. 4.15   Dynamic load allowance. [Hwang and Nowak, 1991a].

(312 kN) design truck combined with 44 ft (13.4 m) (typical truck length) of design lane load at 0.64 kip/ft (9.3 N/mm). Note that most of the data are below 0.3. Hwang and Nowak (1991a) also summarize their findings for various trucks, and roughness profiles, where four span lengths were considered. The average impact factors ranged from a low average of 0.09 (COV = 0.43) to a maximum of 0.21 (COV = 0.72), where COV is the coefficient of variation. These results indicate that the impact load effects are typically less than 30%, but with significant variation.

The global load dynamic effects are addressed in most studies regarding impact. Global means the load effect is due to the global system response such as the deflection, moment, or shear of a main girder. Local effects are the load effects that result from loads directly applied to (or in the local area of) the component being designed. These include decks and deck components. *In short, if a small variation in the live load placement causes a large change in load effect then the load effect should be considered local.*

The effects of impact on such components tend to be much greater than the effects on the system as a whole and are highly dependent on roadway roughness. First, this is because the load is directly applied to these elements, and second, their stiffness is much greater than that of the system as a whole. For many years the AASHTO used an impact formula that attempted to reflect this behavior by using the span length as a parameter. The shorter spans required increased impact to an upper limit of 0.3.

Other specifications, for example, the Ontario Highway Design Bridge Code (OHBDC) (1983), modeled this behavior as a function of the natural frequency of the system. This specification is illustrated in Figure 4.11. Although perhaps the most rational approach, it is problematic because the frequency must be calculated (or estimated) during the design process. Obtaining a *good* estimate of the natural frequency is difficult for an existing structure and certainly more difficult for a bridge being designed. This approach adds a level of complexity that is perhaps unwarranted. An empirical-based estimate can be obtained by a simple formula (Tilly, 1986).

The present AASHTO specification takes a very simplistic approach and defines the DLA as illustrated in Table 4.7 [A3.6.2].

TABLE 4.7[a]

Component	IM (%)
Deck joints—all limit states	75
All other components	
Fatigue and fracture limit states	15
All other limit states	33

[a]In AASHTO Table 3.6.2.1-1. [From *AASHTO LRFD Bridge Design Specifications,* Copyright © 1994 by the American Association of State Highway and Transportation Officials, Washington, DC. Used by Permission].

These factors are to be applied to the static load as

$$U_{L+I} = U_L(1 + IM) \tag{4.1}$$

where U_{L+I} is the live load effect plus allowance for dynamic loading, U_L is the live load effect of live load, and IM is the fraction given in Table 4.7.

The *all other components* include girders, beams, bearings (except elastomeric bearings), and columns. Clearly, the present specification does not attempt to model dynamic effects with great accuracy, but with sufficient accuracy and conservatism for design. Both experimental and analytical studies indicated that these values are reasonable estimates. Moreover, considering the variabilities involved, a flat percentage for dynamic load effect is practical, tractable for design, and reasonably based on research results. For the structural evaluation of existing bridges (rating), the engineer will likely use the criteria established in the AASHTO rating procedures that are a function of roadway roughness. At the time of design, the future roadway roughness and associated maintenance is difficult to estimate, thus more conservative values are appropriate.

Centrifugal Forces. Acceleration is the time derivative of the velocity vector and as such results from either a change of magnitude or direction of velocity. A truck can increase speed, decrease speed, and/or change directions as it moves along a curvilinear path. All of these effects require an acceleration of the vehicle that cause a force between the deck and the truck. Because its mass is large compared to the power available, a truck cannot increase its speed at a rate great enough to impose a significant force on the bridge. Conversely, a decrease in speed due to braking can create a significant acceleration (deceleration) that causes large forces on the bridge in the direction of the truck movement. The braking effect is described in the next section. Finally, as a truck moves along a curvilinear path, the change in direction of the velocity causes a centrifugal acceleration in the radial direction. This acceleration is

$$a_r = \frac{V^2}{r} \tag{4.2}$$

where V is the truck speed, and r is the radius of curvature of the truck movement.

The forces and accelerations involved are illustrated in Figure 4.16.

Newton's Second Law requires

$$F = ma \tag{4.3}$$

where m is the mass. Substitution of Eq. 4.2 into Eq. 4.3 yields

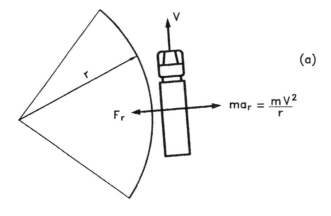

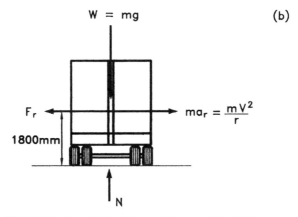

Fig. 4.16 Free body diagrams for centrifugal force.

$$F_r = \frac{mV^2}{r} \qquad (4.4)$$

where F_r is the force on the truck directed toward the center of the curve (outward on the bridge). The position of this force is at the center of mass, assumed to be at 1800 mm above the roadway surface [A3.6.3]. Note that the mass m is equal to

$$m = \frac{W}{g} \qquad (4.5)$$

where W is the weight of the vehicle, and g is the gravitational acceleration: 9.807 m/s^2.

Substitution of Eq. 4.5 into Eq. 4.4 yields

$$F_r = \left(\frac{V^2}{rg}\right) W \tag{4.6}$$

which is similar to the expression given in AASHTO [A3.6.3] where

$$F_r = CW \tag{4.7a}$$

where

$$C = \frac{4}{3}\left(\frac{v^2}{Rg}\right) \tag{4.7b}$$

v is the highway design speed (m/s), R is radius of curvature of traffic lane (m), and F_r is applied at the assumed center of mass at a distance 1800 mm above the deck surface.

Because the combination of the design truck with the design lane load gives a load approximately four-thirds of the effect of the design truck considered independently, a four-thirds factor is used to model the effect of a train of trucks. Equation 4.6 may be used with any system of consistent units. The multiple presence factors [A3.6.1.1.2] may be applied to this force as it is unlikely that all lanes will be fully loaded simultaneously.

Braking Forces. As described in the previous section, braking forces can be significant. Such forces are transmitted to the deck and must be taken into the substructure at the fixed bearings or supports. It is quite probable that all truck operators on a bridge will observe an event that causes the operators to apply the brakes. Thus, loading of multiple lanes should be considered in the design. Again, it is unlikely that all the trucks in all lanes will be at the maximum design level, therefore the multiple presence factors outlined previously may be applied [A3.6.1.1.2]. The forces involved are shown in Figure 4.17. The truck is initially at a velocity V and this velocity is reduced to zero over a distance s. The braking force and the associated acceleration are assumed to be constant. The change in kinetic energy associated with the truck is completely dissipated by the braking force. The kinetic energy is equated to the work performed by the braking force giving

$$\frac{1}{2}mV^2 = \int_0^s F_B\, ds = F_B\, s \tag{4.8}$$

where F_B is the braking force transmitted into the deck and m is the truck mass. Solving for F_B and substitution of the mass as defined in Eq. 4.5 yields

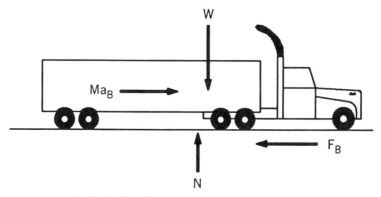

Fig. 4.17 Free body diagram for braking force.

$$F_B = \frac{1}{2}\left(\frac{W}{g}\right)\left(\frac{V^2}{s}\right) = \frac{1}{2}\left(\frac{V^2}{gs}\right) W = bW \qquad (4.9a)$$

where

$$b = \frac{1}{2}\left(\frac{V^2}{gs}\right) \qquad (4.9b)$$

b is the fraction of the weight that is applied to model the braking force. In the development of the AASHTO braking force fraction, it was assumed that the truck is moving at a velocity of 90 km/h = 25 m/s and a braking distance of 122 000 mm is required. Substitution of these values gives the braking force fraction

$$b = \frac{(25)^2}{2(9.807)(122)} = 0.26 \approx 25\%$$

The braking forces shall be taken as 25% of the axle weights of the design truck or the tandem truck placed in all lanes [A3.6.4]. The design lane is not included as it is assumed that the additional trucks "brake out of phase." Thus, the acceleration of the additional design lane trucks will pump the brakes and will not decelerate uniformly. This pumping action is assumed to occur at times different from when the design truck is at a maximum. Also implicit in the AASHTO value is that the coefficient of friction exceeds 0.25 for the tire-deck interface. The braking force is assumed to act horizontally at 1800 mm above the roadway surface in either longitudinal direction.

Permit Vehicles and Miscellaneous Considerations. Transportation agencies may include other vehicle loads to model load characteristic of their particular jurisdiction. For example, the Department of Transportation of Pennsylvania (PennDOT) uses series of trucks termed "umbrella loads." These loads represent vehicle loads that are actually used on Pennsylvania's highways (Koretzky et al. 1986). The moment is created on simple spans by the umbrella loads as illustrated in Figure 4.18. Presently, PennDOT uses these loads for design. Note that the largest load is 204 kips (907 kN) with a total length of 55 ft (16 800 mm).

Similarly, the Department of Transportation in California (Caltrans) uses a different load model for their structures (see Fig. 4.19). Caltrans' rationale is similar to PennDOTs. The load model should closely approximate the service conditions.

Other situations may dictate that a higher than average percentage of truck traffic is present that affects the fatigue limit state calculations. Circumstances that affect the flow of traffic, such as traffic signals, may effect the design loads. Lastly, a bridge location near an industrial site may cause the load characteristics to be significantly different than those prescribed by the specification. In all such cases, the characteristics of the truck loads should preferably be based upon survey data. If such data is not available or achievable, then professional judgment should be used.

4.3 LATERAL LOADS

4.3.1 Fluid Forces

The force on a structural component due to a fluid flow (water or air) around a component is established by Bernoulli's equation in combination with empirically established drag coefficients. Consider the object shown in an incompressible fluid in Figure 4.20. With the use of Bernoulli's equation, equating the upstream energy associated with the flow at point a with the energy associated with the stagnation point b where the velocity is zero yields

$$\tfrac{1}{2}\rho V_a^2 + p_a + \rho g h_a = \tfrac{1}{2}\rho V_b^2 + p_b + \rho g h_b \tag{4.10}$$

Assuming that points a and b are at the same elevation and that the reference upstream pressure at point a is zero, pressure at point b is

$$p_b = \tfrac{1}{2}\rho V_a^2 \tag{4.11}$$

The stagnation pressure is the maximum inward pressure possible as all the upstream kinetic energy is transferred to potential energy associated with the pressure. Because every point on the surface is not at stagnation, that is, some velocity exists and hence, the pressures at these points are less than the stagnation pressures. This effect is because the upstream energy is split between potential (pressure) and kinetic energies. The total pressure is integrated over

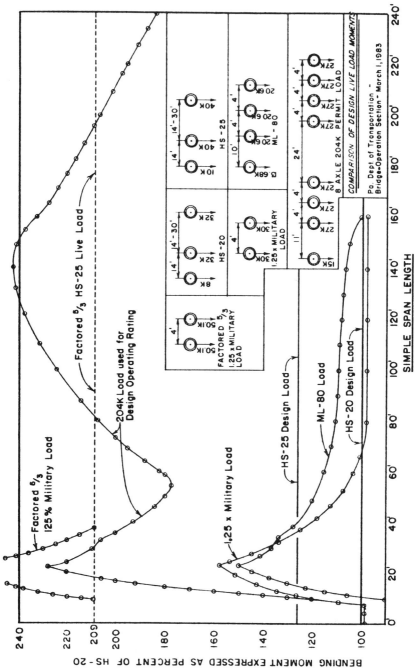

Fig. 4.18 PennDOT umbrella loads. [Koretzky et al., 1986].

166

Standard Permit Rating and Design Vehicles with Purple Loads

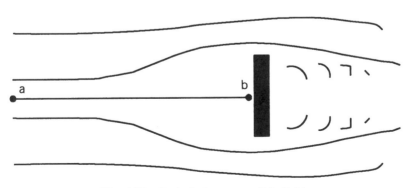

Load Rate	Axle Purple Loads (Unbonused)													
	#1	#2	#3	#4	#5	#6	#7	#8	#9	#10	#11	#12	#13
P5 [1]	26 k	24 k	24 k	24 k	24 k	—	—	—	—	—	—	—	—
P7	26 k	24 k	24 k	24 k	24 k	24 k	24 k	—	—	—	—	—	—
P9	26 k	24 k	24 k	24 k	24 k	24 k	24 k	24 k	24 k	—	—	—	—
P11	26 k	24 k	24 k	24 k	24 k	24 k	24 k	24 k	24 k	24 k	24 k	—	—
P13 [2]	26 k	24 k	24 k	24 k	24 k	24 k	24 k	24 k	24 k	24 k	24 k	24 k	24 k

[1] Minimum Vehicle
[2] Maximum Vehicle

P = Purple
G = Green = 0.867 × Purple
O = Orange = 0.667 × Purple

O = 13 Axle Orange Truck ◄───
G = 11 Axle Green Truck ◄───
G = 9 Axle Green Truck ◄───
P = 7 Axle Purple Truck ◄───
P = 5 Axle Purple Truck ◄───

Example Rating: P P G G O

10'-0" Clearance

Section A-A

2'-0"   6'-0"   2'-0"

Fig. 4.19   Caltrans umbrella loads. [Caltrans, 1995].

Fig. 4.20   Body in incompressible fluid.

the surface area and is used to obtain the fluid force. It is conventional to determine the integrated effect (or force) empirically and to divide the force by the projected area. This quotient establishes the average pressure on an object, which is a fraction of the stagnation pressure. The ratio of the average pressure to the stagnation pressure is commonly called the drag coefficient, C_d. The drag coefficient is a function of the object's shape and the characteristics of the fluid flow. With the use of a known drag coefficient, the average pressure on an object may be calculated as

$$p = C_d \tfrac{1}{2} \rho V^2 \tag{4.12}$$

It is important to note that the fluid pressures and associated forces are proportional to the velocity squared. For example, a 25% increase in the fluid velocity creates approximately a 50% increase in fluid pressure and associated force.

Wind Forces. The velocity of the wind varies with the elevation above the ground and the upstream terrain roughness, and therefore pressure on a structure is also a function of these parameters. Velocity increases with elevation, but at a decreasing rate. If the terrain is smooth then the velocity increases more rapidly with elevation. A typical velocity profile is illustrated in Figure 4.21, where several key parameters are shown. The parameter V_g is the geotropic velocity or the velocity independent of surface (boundary) effects, δ is the boundary layer thickness, usually defined as the height where the velocity of 99% of V_g, and V_{10} is the reference velocity at 10 m. Traditionally, this is the height at which wind velocity data is recorded. Since its introduction in 1916, the velocity profile has been modeled with a power function of the form

$$V_{DZ} = C V_{10} \left(\frac{Z}{10\ 000} \right)^\alpha$$

where C and α are empirically determined constants. This model is used in many building codes. Critics of the power law point out that its exponent is not a constant for a given upstream roughness but varies with height, that the standard heights used to establish the model were somewhat subjective, and lastly, that the model is purely a best-fit function and has no theoretical basis (Simiu, 1973, 1976). More recently, meteorologists and wind engineers are modeling the wind in the boundary layer with a logarithmic function. This function is founded on boundary layer flow theory and better fits experimental results. The general form of the logarithmic velocity profile is

$$V(Z) = \frac{1}{\kappa} V_0 \ln \left(\frac{Z}{Z_0} \right) \tag{4.13}$$

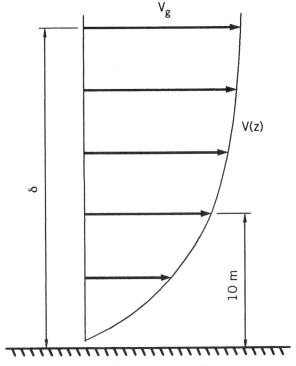

Fig. 4.21   Velocity profile.

where Z is the elevation above the ground, κ is von Karman's constant, ($\sim$0.4), Z_0 is the roughness of the ground upstream, and

$$V_0 = \sqrt{\frac{\tau_0}{\rho}} \tag{4.14}$$

where τ_0 is the shear stress at the ground surface and ρ is the density of air. The parameter V_0 is termed the shear friction velocity because it is related to the shear force (friction), and Z_0 is related to the height of the terrain roughness upstream. As expected, these parameters are difficult to mathematically characterize, so empirical values are used. Note that for a given upstream roughness, two empirical constants Z_0 and V_0 are required in Eq. 4.12. Therefore, two measurements of velocity at different heights can be used to establish these constants. This measurement has been done in the experiments of many investigators as reported by Simiu (1973, 1976) and Simiu and Scanlon (1986).

It is interesting that these constants are not independent and an expression can be formulated to relate them as shown below. The wind-generated shear stress at the surface of the ground is

$$\tau_0 = D_0 \rho V_{10}^2 \tag{4.15}$$

where D_0 is the surface drag coefficient and V_{10} is the wind speed at 10 000 mm above the low ground or water level, kilometers per hour (km/h).

With $Z = 10$ m, Eq. 4.13 is used to solve for V_0

$$V_0 = \kappa \frac{V_{10}}{\ln(10\ 000/Z_0)} \tag{4.16}$$

Equate the surface shear stress in Eq. 4.14 and Eq. 4.15 to obtain

$$D_0 = \left(\frac{V_0}{V_{10}}\right)^2 \tag{4.17a}$$

or

$$V_0 = \sqrt{D_0} V_{10} \tag{4.17b}$$

Substitution of Eq. 4.16 into Eq. 4.17a yields

$$D_0 = \left(\frac{\kappa}{\ln(10\ 000/Z_0)}\right)^2 \tag{4.18}$$

Finally, substitute the Eq. 4.18 into Eq. 4.17b to yield

$$V_0 = \left(\frac{\kappa}{\ln(10\ 000/Z_0)}\right) V_{10} \tag{4.19}$$

Equation 4.19 illustrates that for any reference velocity V_{10}, Z_0 and V_0 are related.

For unusual situations or for a more complete background, refer to Lui (1991). Lui outlines many issues in wind engineering in a format amenable to an engineer with a basic fluid mechanics background. Issues such as terrain roughness changes, local conditions, drag coefficients, and so on, are discussed in a manner relevant to the bridge/structural engineer.

The equation for velocity profile used by AASHTO [A3.8.1.1] is

$$V_{DZ} = 2.5 V_0 (V_{10}/V_B) \ln(Z/Z_0) \tag{4.20}$$

V_{DZ} is the design wind speed at design elevation Z (km/h) [same as $V(Z)$ in Eq. 4.13], V_B is the base wind velocity of 160 km/h yielding design pressures, V_0 is the "friction velocity," a meteorological wind characteristic taken as specified in Table 4.8 for upwind surface characteristics (km/h) and Z_0 is the "friction length" of the upstream fetch, a meteorological wind characteristic taken as specified in Table 4.8 (mm).

TABLE 4.8 Values of V_0 and Z_0 for Various Upstream Surface Conditions[a]

Condition	Open Country	Suburban	City
V_0 (km/h)	13.2	15.2	25.3
Z_0 (mm)	70	300	800

[a]In AASHTO Table 3.8.1.1-1. [From *AASHTO LRFD Bridge Design Specifications*, Copyright © 1994 by the American Association of State Highway and Transportation Officials, Washington, DC. Used by Permission].

The constant 2.5 is the inverse of the von Karman's constant 0.4. The ratio (V_{10}/V_B) is used to linearly proportion for a reference velocity other than 160 km/h.

Equation 4.19 may be used to illustrate the relationship between V_0 and Z_0. For example, use the city exposure

$$V_0 = \frac{0.4}{\ln\left(\dfrac{10\ 000}{800}\right)}\ (160\ \text{km/h}) = 25.3\ \text{km/h}$$

which agrees with Table 4.8.

The velocity at 10 m (V_{10}) may be established by fastest-mile-of-wind charts available in ASCE 7-88 for various recurrence intervals (ASCE, 1988), by site specific investigations, or in lieu of a better criterion, use 100 mph.

The wind pressure on the structure or component is established by scaling a basic wind pressure for $V_B = 160$ km/h. This procedure is

$$P_D = P_B \left(\frac{V_D}{V_B}\right)^2 = P_B \frac{V_D^2}{160^2} = P_B \frac{V_D^2}{25\ 600} \tag{4.21}$$

where the basic wind pressures are given in Table 4.9 [Table A3.8.1.2-1]. Table 4.9 includes the effect of gusts and the distribution of pressure on the surface (pressure coefficients). If we use Eq. 4.11 with a velocity of 160

TABLE 4.9 Base Pressures, P_B Corresponding to $V_B = 160$ km/h[a]

Structural Component	Windward Load (MPa)	Leeward Load (MPa)
Trusses, columns, and arches	0.0024	0.0012
Beams	0.0024	N/A
Large flat surfaces	0.0019	N/A

[a]In AASHTO Table 3.8.1.2-1. [From *AASHTO Bridge Design Specifications*, Copyright © 1994 by the American Association of State Highway and Transportation Officials, Washington, DC. Used by Permission].

km/h and a density of standard air (1000 kg/m³) we set a stagnation pressure of 1226 Pa = 0.00123 MPa. Therefore, Table 4.9 includes a large increase of about 100% for gusts. A discussion with the code writers established that the pressures used in the previous AASHTO specifications were reasonable, seldom controlled the design of short- or medium-span bridges, and conservative values were used. So, depending on the assumed gust response and pressure coefficients, the design wind speed is likely above 160 km/h.

Equation 4.21 uses the ratio of the design and base velocities squared because the pressure is proportional to the velocity squared. Additionally, the minimum wind loading shall not be less than 4.4 N/mm in the plane of the windward chord and 2.2 N/mm in the plane of the leeward chord on truss spans, and not less than 4.4 N/mm on girder spans [A3.8.1.2]. This wind load corresponds to the wind pressure on structures load combination (WS) as given in Table 4.9. This wind should be considered from all directions and the extreme values are used for design. Directional adjustments are outlined in AASHTO [A3.8.1.4], where the pressure is separated into parallel and perpendicular component pressures as a function of the attack angle. The details are not elaborated here.

The wind must also be considered on the vehicle. This load is 1.46 N/mm applied at 1800 mm above the roadway surface [A3.8.1.3].

For long-span structures, the possibility of aeroelastic instability exists. Here the wind causes a resonance situation with the structure, creating large deformations, actions, and possible failures. This phenomenon is best characterized by the famous Tacoma Narrows Bridge, which completely collapsed due to aeroelastic effects. This collapse brought attention to this important design consideration that is typically a concern in the analysis of long-span bridges. Due to the complexities involved, aeroelastic instability is considered beyond the scope of the AASHTO specification and this book.

Water Forces. Water flowing against and around the substructure creates a lateral force directly on the structure as well as debris that might accumulate under the bridge. Flood conditions are the most critical. As outlined above, the forces created are proportional to the square of velocity and to a drag coefficient. The use of Eq. 4.12 and the substitution of $\rho = 1000$ kg/m³ yields

$$p_b = \frac{1}{2}\rho C_d V_a^2 = 500 C_D V_a^2 \qquad (4.22)$$

where the AASHTO Equation [A3.7.3.1] is

$$p = 5.14 \times 10^{-4} C_D V^2 \qquad (4.23)$$

Here C_D is the drag coefficient given in Table 4.10, and V is the design velocity of the water for the design flood in strength and service limit states, and for the check flood in the extreme event limit state, m/s.

Note that C_D is the specific AASHTO value and C_d is a generic term.

TABLE 4.10 Drag Coefficient[a]

Type	C_D
Semicircular nosed pier	0.7
Square-ended pier	1.4
Drift lodged against pier	1.4
Wedged nosed pier with nose angle 90° or less	0.8

[a]In AASHTO Table 3.7.3.1-1. [From *AASHTO LRFD Bridge Design Specifications,* Copyright © 1994 by the American Association of State Highway and Transportation Officials, Washington, DC. Used by Permission].

If the substructure is oriented at an angle to the stream flow, then adjustments must be made. These adjustments are outlined in AASHTO [A3.7.3.2]. Where debris deposition is likely, the bridge area profile should be adjusted accordingly. Some guidance on this is given in AASHTO [A3.7.3.1 and its associated references].

Although not a force, the scour of the stream bed around the foundation can result in structural failure. The scour is the movement of the stream bed from around the foundation and this can significantly change the structural system, creating a situation that must be considered in the design. AASHTO [A2.6.4.4.1] outlines an extreme limit state for design. Because this issue is related to hydraulics, the substructure is not considered in detail here.

4.3.2 Seismic Loads

Depending on the location of the bridge site, the anticipated earthquake effects can be inconsequential or they can govern the design of the lateral load resistance system. The AASHTO Specifications have been developed to apply to all parts of the United States, so all bridges should be checked to determine if seismic loads are critical. In many cases the seismic loads are not critical and other lateral loads, such as wind, govern the design.

The provisions of the AASHTO Specifications are based on the following principles [C3.10.1]:

- Small-to-moderate earthquakes should be resisted within the elastic range of the structural components without significant damage.
- Realistic seismic ground motion intensities and forces are used in the design procedures.
- Exposure to shaking from large earthquakes should not cause collapse of all or part of the bridge. Where possible, damage should be readily detectable and accessible for inspection and repair.

The AASHTO provisions apply to bridges with conventional slab, girder, box girder, and truss superstructures whose spans do not exceed 150 m [A3.10.1]. Bridges with spans exceeding 150 m and other bridge types, such as suspension bridges, cable-stayed bridges, movable bridges and arches, are not applicable.

A discussion of the procedure used to determine when a bridge at a particular site requires a detailed seismic analysis is included in the next section. This section is followed by sections on minimum design forces and seismic load combinations.

Seismic Design Procedure. The six steps in the seismic design procedure are outlined in this section. The first step is to arrive at a preliminary design describing the type of bridge, the number of spans, the height of the piers, a typical roadway cross section, horizontal alignment, type of foundations, and subsurface conditions. The nature of the connections between the spans of the superstructure, between the superstructure and the substructure, and between the substructure and the foundation are also important. For example, if a bridge superstructure has no deck joints and is integral with the abutments, its response during a seismic event is quite different from one with multiple expansion joints. There are also innovative energy dissipating connections that can be placed below the superstructure at the abutments and pier caps to effectively isolate the superstructure from the effects of ground shaking. These devices can substantially reduce the magnitude of the inertial forces transmitted to a foundation component and can serve as a structural fuse that can be replaced or repaired if a larger earthquake occurs.

The second step is to determine the acceleration coefficient that is appropriate for the bridge site [A3.10.2]. Contours of horizontal acceleration in rock expressed as a percent of gravity are illustrated on the map of the continental United States shown in Figure 4.22.

At a given location, the acceleration coefficient from the map has a 90% probability of not being exceeded in 50 years. This value corresponds to a return period of about 475 years for the design earthquake. There is a 10% probability that an earthquake larger than the design earthquake implied by the acceleration coefficient from the map will occur. In some cases, such as for bridges on critical lifelines, a larger acceleration coefficient corresponding to the maximum probable earthquake, with a return period around 2500 years, must be used.

EXAMPLE 4.1

With the use of an enlarged version of the map of Figure 4.22 found at the end of Division I-A of the 16th edition of the AASHTO Specifications (1996), acceleration coefficients A can be determined for counties within each state. Some typical values for A are: Montgomery County, VA, $A = 0.075$; Albany County, WY, $A = 0.02$; and Imperial County, CA, $A > 0.80$.

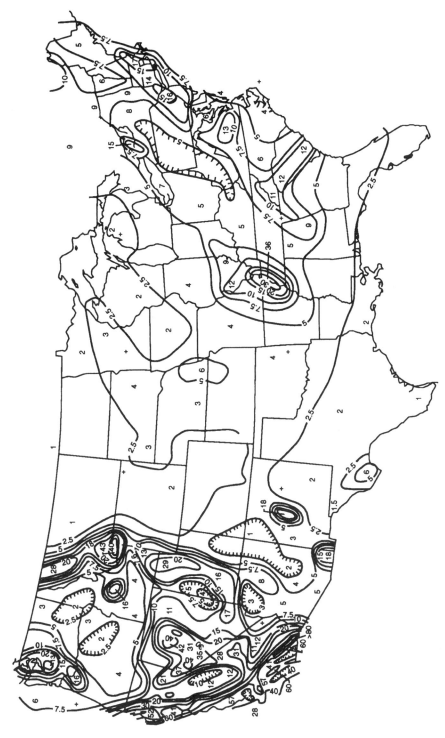

Fig. 4.22 Horizontal acceleration coefficient, 100 A, in rock for contiguous states. (AASHTO Figs. 3.10.2-1 and 3.10.2-2). [From *AASHTO LRFD Bridge Design Specifications*, Copyright © 1994 by the American Association of State Highway and Transportation Officials, Washington, DC. Used by Permission].

175

The third step is to determine the importance category of a bridge [A3.10.3]. Following a seismic event, transportation routes to hospitals, police and fire departments, communication centers, temporary shelters and aid stations, power installations, water treatment plants, military installations, major airports, defense industries, refineries, railroad and truck terminals must continue to function. Bridges on such routes should be classified as essential. In addition, a bridge that could collapse onto an essential route should also be classified as essential. Table 4.11 summarizes the characteristics of the three importance categories, one of which must be assigned to each bridge. Consideration should be given to possible future changes in the role of the bridge when assigning an importance category.

The fourth step is to determine the seismic performance zone for each bridge [A3.10.4]. These seismic zones group together regions of the United States that have similar seismic risk. The greater the acceleration coefficient, the greater the risk. The seismic zones are given in Table 4.12 and the higher the number the greater are the seismic performance requirements for the bridge in regard to the method of analysis, the length of bridge seats, and the strength of connections.

The fifth step is to determine a site coefficient S, which is dependent on the soil conditions at the bridge site [A3.10.5]. The acceleration coefficients given on the map of Figure 4.22 are in rock that may be at some depth below the surface where the bridge is located. Depending on the nature of the soil overlying the rock, the acceleration at the surface can be amplified to as much as double the acceleration in the rock. The four soil profiles given in Table 4.13 are used to select an approximate acceleration coefficient modifier from Table 4.14. In locations where the soil conditions are not known in sufficient detail or the soil profile does not fit any of the four types, the AASHTO Specifications [A3.10.5.1] state that a Type *II* Soil Profile shall be used. The use of this default site condition could be nonconservative and should not be used unless Type *III* and Type *IV* Soil Profiles have been ruled out by a geological or geotechnical engineer.

The sixth step is to determine the response modification factors (*R*-factors), which reduce the seismic force based on an elastic analysis of the bridge system [A3.10.7]. The force effects from an elastic analysis are to be divided

TABLE 4.11 Importance Categories

Importance Category	Description
Critical bridges	Must remain open to all traffic after the design earthquake (475-year return period) and open to emergency vehicles after a large earthquake (2500-year return period)
Essential bridges	Must be open to emergency vehicles after the design earthquake
Other bridges	May be closed for repair after a large earthquake

TABLE 4.12   Seismic Performance Zones[a]

Acceleration Coefficient	Seismic Zone
$A \leq 0.09$	1
$0.09 < A \leq 0.19$	2
$0.19 < A \leq 0.29$	3
$0.29 < A$	4

[a]In AASHTO Table 3.10.4-1. [From *AASHTO LRFD Bridge Design Specifications,* Copyright © 1994 by the American Association of State Highway and Transportation Officials, Washington, DC. Used by Permission].

by the response modification factors given in Table 4.15. The use of these R-factors, generally greater than 1, recognizes that when a design seismic event (475-year return period) occurs, energy is dissipated through inelastic deformation (hinging) in the substructure. This energy dissipation actually protects the structure from large shocks and allows it to be designed for reduced forces. In the event a large earthquake (2500-year return period) should occur, the hinging regions may have to be repaired, but if all of the components are properly tied together collapse does not occur. To insure that proper attention is given to the transfer of internal actions from one component to another, the R-factors for connections given in Table 4.16 do not reduce, and in some cases amplify, the force effects from an elastic analysis.

Based on the information obtained by completing the above steps, decisions can be made regarding the level of seismic analysis required, the design forces, and the design displacement requirements. For example, single-span bridges and bridges in Seismic Zone 1 do not have to be analyzed for seismic loads, while critical bridges in Seismic Zone 4 require a rigorous method of

TABLE 4.13   Soil Profiles

Type	Description
I	Rock of any description, either shale-like or crystalline in nature, or stiff soils where the soil depth is less than 60 000 mm, and the soil types overlying the rock are stable deposits of sands, gravels, or stiff clays
II	Stiff cohesive or deep cohesionless soils where the soil depth exceeds 60 000 mm and the soil types overlying the rock are stable deposits of sands, gravels, or stiff clays
III	Soft to medium-stiff clays and sands, characterized by 9000 mm or more of soft to medium-stiff clays with or without intervening layers of sand or other cohesionless soils
IV	Soft clays or silts greater than 12 000 mm in depth

TABLE 4.14 Site Coefficients[a]

	Soil Profile Type			
	I	*II*	*III*	*IV*
Site coefficient, *S*	1.0	1.2	1.5	2.0

[a]In AASHTO Table 3.10.5-1. [From *AASHTO LRFD Bridge Design Specifications,* Copyright © 1994 by the American Association of State Highway and Transportation Officials, Washington, DC. Used by Permission].

seismic analysis [A4.7.4]. Minimum analysis and displacement requirements for seismic effects are discussed in Chapter 6.

Minimum Seismic Design Connection Forces. When ground shaking due to an earthquake occurs and a bridge superstructure is set in motion, inertial forces equal to the mass times the acceleration are developed. These forces can be in any direction and must be restrained, or dissipated, at the connection between the superstructure and substructure. For a single-span bridge, the minimum design connection force in the restrained direction is to be taken as the product of the acceleration coefficient and the tributary dead load associated with that connection.

Bridges in Seismic Zone 1 do not require a seismic analysis and therefore nominal values are specified for the connection forces. To obtain the horizontal seismic forces in a restrained direction, the tributary dead load is multiplied by the value given in Table 4.17 [A3.10.9.2]. The tributary dead load to be used when calculating the longitudinal connection force at a fixed bearing of a continuous segment or simply supported span is the total dead load of the segment. If each bearing in a segment restrains translation in the trans-

TABLE 4.15 Response Modification Factors—Substructures[a]

	Importance Category		
Substructure	Other	Essential	Critical
Wall-type piers—larger dimension	2.0	1.5	1.5
Reinforced concrete pile bents			
a. Vertical piles only	3.0	2.0	1.5
b. One or more batter piles	2.0	1.5	1.5
Single Columns	3.0	2.0	1.5
Steel or composite steel and concrete pile bents			
a. Vertical piles only	5.0	3.5	1.5
b. One or more batter piles	3.0	2.0	1.5
Multiple column bents	5.0	3.5	1.5

[a]In AASHTO Table 3.10.7.1-1. [From *AASHTO LRFD Bridge Design Specifications,* Copyright © 1994 by the American Association of State Highway and Transportation Officials, Washington, DC. Used by Permission].

TABLE 4.16 Response Modification Factors—Connections[a]

Connection	All Importance Categories
Superstructure to abutment	0.8
Expansion joints within a span of the superstructure	0.8
Columns, piers, or pile bents to cap beam or superstructure	1.0
Columns or piers to foundations	1.0

[a]In AASHTO Table 3.10.7.1-2. [From *AASHTO LRFD Bridge Design Specifications,* Copyright © 1994 by the American Association of State Highway and Transportation Officials, Washington, DC. Used by Permission].

verse direction, the tributary dead load to be used in calculating the transverse connection force is the dead load reaction at the bearing. If each bearing supporting a segment is an elastomeric bearing, which offers little or no restraint, the connection is to be designed to resist the seismic shear forces transmitted through the bearing, but not less than the values represented by the multipliers in Table 4.17.

EXAMPLE 4.2

Determine the minimum longitudinal and transverse connection forces for a simply supported bridge span of 21 m in Montgomery County, VA (Zone 1), with a dead load of 115 kN/m. Assume that in the longitudinal direction one connection is free to move while the other is fixed, and that in the transverse direction both connections to the abutment are restrained.

SOLUTION

Acceleration coefficient $A = 0.075$
Total dead load $W_D = 115(21) = 2415$ kN

TABLE 4.17 Multiplier for Connection Force in Seismic Zone 1

Acceleration Coefficient	Soil Profile	Multiplier
$A \leq 0.025$	I or II	0.10
	III or IV	0.20
$0.025 < A \leq 0.09$	All	0.20

Connection force $F_C = ma = (W_D/g)(A \times g) = W_D A$
Longitudinal (min) $F_{CL} = (2415)(0.075) = 181$ kN   does not control
Table 4.17 $F_{CL} = 2415(0.20) = \underline{\underline{483 \text{ kN}}}$ at fixed end (controls)
Transverse $F_{CT} = \frac{1}{2}F_{CL} = \underline{\underline{241.5 \text{ kN}}}$ per abutment

Note that if the bridge was located in Albany County, WY, on good soil (Type *I* or Type *II*), the design connection forces would be cut in half.

Connections for bridges in Seismic Zone 2 are to be designed for the reaction forces determined by an elastic spectral analysis divided by the appropriate *R*-factor of Table 4.16. Connection forces for bridges in Seismic Zones 3 and 4 can be determined by either an elastic seismic analysis divided by *R*, or by an inelastic step-by-step time history analysis with $R = 1.0$ for all connections. The use of $R = 1.0$ assumes that the inelastic method properly models the material hysteretic properties and the accompanying energy dissipation. Further discussion on the seismic analysis requirements is given in Chapter 6.

Combination of Seismic Forces. Because of the directional uncertainty of earthquake motions, two load cases combining elastic member forces resulting from earthquakes in two perpendicular horizontal directions must be considered. The two perpendicular directions are usually the longitudinal and transverse axes of the bridge. For a curved bridge, the longitudinal axis is often taken as the line joining the two abutments. The two load cases are expressed as [A3.10.8]:

$$\text{Load Case 1} \quad 1.0 \, F_L + 0.3 \, F_T \tag{4.24a}$$

$$\text{Load Case 2} \quad 0.3 \, F_L + 1.0 \, F_T \tag{4.24b}$$

where

F_L = elastic member forces due to an earthquake in the direction of the longitudinal axis of the bridge.
F_T = elastic member forces due to an earthquake in the direction of the transverse axis of the bridge.

4.3.3 Ice Forces

Forces produced by ice must be considered when a structural component of a bridge, such as a pier or bent, is located in water and the climate is cold enough to cause the water to freeze. The usual sequence is that freeze-up occurs in late fall, the ice grows thicker in the winter and the ice breaks up in the spring. If the bridge is crossing a lake, reservoir, harbor, or other relatively quite body of water, the ice forces are generally static. These static forces can be horizontal when caused by thermal expansion and contraction or vertical if the body of water is subject to changes in water level. If the bridge is crossing a river with flowing water, the static forces exist throughout the winter months, but when the spring break-up occurs, larger dynamic forces are produced by floating sheets of ice impacting the bridge structure.

Effective Strength of Ice. Because the strength of ice is less than the strength of steel and concrete used in the construction of bridge piers, the static and dynamic ice forces on bridge piers are limited by the effective strength of the ice: the static thermal forces by the crushing strength and the dynamic forces by either the crushing strength or the flexural strength. The strength of the ice depends on the conditions that exist at the time it is formed, at the time it is growing in thickness, and at the time it begins to melt and break-up. If the ice is formed when the surface is agitated and freezes quickly, air will be entrapped within the structure of the ice and will give it a cloudy or milky appearance. This ice is not as strong as ice that is formed gradually and grows over a long period of time to be very solid and clear in appearance. The conditions during the winter months, when this ice is increasing in thickness, affects the strength of the ice. If snow cover is present and melts during a warming period and then freezes, weaker granular snow ice is formed. In fact, sections cut through ice sheets show varying layers of clear ice, cloudy ice, and snow ice. This ambiguity makes classification of the ice difficult. The conditions at the time of spring break-up also affect the strength. If the temperature throughout the thickness of the ice sheet is at the melting temperature when the ice breaks up, it will have less strength than when the average ice temperature is below the melting temperature.

An indication of the variation in crushing strength of ice at the time of break-up is given in AASHTO [A3.9.2.1] as shown in Table 4.18. These values are to be used in a semiempirical formula, discussed later, for determining dynamic ice forces on bridge piers.

Field Measurement of Ice Forces. Forces exerted by moving ice have been measured by Haynes et al. (1991) on a bridge pier in the St. Regis River in upstate New York. Other researchers who have measured ice forces in Canada, Alaska, and Vermont are listed in their report. The purpose of these studies

TABLE 4.18   Effective Ice Crushing Strength at Breakup

Average Ice Temperature	Condition of Ice	Effective Strength
At melting point	Substantially disintegrated	8.0 ksf (0.38 MPa)
	Somewhat disintegrated	16.0 ksf (0.77 MPa)
	Large pieces, internally sound	24.0 ksf (1.15 MPa)
Below melting point	Large pieces, internally sound	32.0 ksf (1.53 MPa)

is to provide data that can be used to calibrate design codes for changing local conditions.

In the Haynes study, a steel panel was instrumented and placed on the upstream nose of a pier (see Fig. 4.23). The panel pivots about its base and a load cell measures a reactive force when the panel is struck by moving ice. Whenever the signal from the load cell gets above a preset threshold level, the load cell force data along with the pressure transducer reading that determines the water depth are recorded. The ice force that produced the force in the load cell is then determined by balancing moments about the pin location.

In March 1990, a major ice run took place. Ice thickness was estimated to be about 6–8 in (152–203 mm) (nonuniform flow causes variations in ice cover thicknesses for most rivers). Plots of the ice force versus time for two of the largest ice force events during this run are shown in Figure 4.24. For the ice force record shown in Figure 4.24(a), the ice-structure interaction event lasted about 2.3 s and is believed to represent crushing failure of the ice because the force record has many oscillations without the force dropping to zero. The rapid increase and decrease of ice force shown in Figure 4.24(b) indicates an impact and possible rotation or splitting of the ice floe without much crushing. This impact event lasted only about 0.32 s and produced the maximum measured ice force of nearly 80,000 lb (356 kN). The largest ice force produced by the crushing failure of the ice was about 45,000 lb (200 kN).

One observation from these field measurements is the wide variation in ice forces against a pier produced in the same ice run by ice floes that were formed and broken up under similar conditions. Some of the ice sheets, prob-

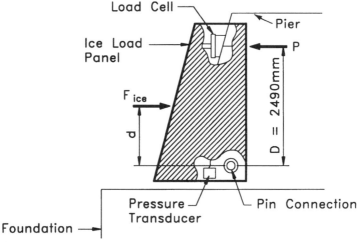

Fig. 4.23 Ice load panel on pier of the St. Regis River bridge. [From Haynes, *Cold Regions Research and Engineering Laboratory,* 1991].

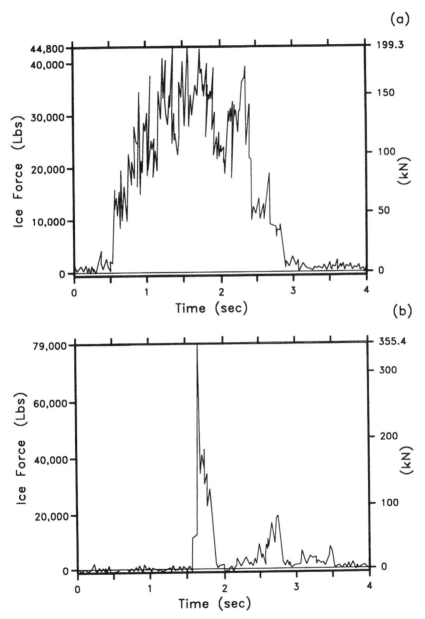

Fig. 4.24 Records of ice force versus time on 16–17 March 1990, (a) ice failure by crushing, (b) ice impact without much crushing. [From Haynes, *Cold Regions Research and Engineering Laboratory,* 1991].

ably the larger ones, were indented when they collided with the pier and failed by crushing. Other ice floes smaller in size and probably of solid competent ice banged into the pier with a larger force and then rotated and were washed past the pier. In light of this observation, it appears prudent to use only the last two categories for effective ice strength of 24 ksf (1.15 MPa) and 32 ksf (1.53 MPa) in Table 4.18, unless there is long experience with local conditions that indicate that ice forces are minimal.

Thickness of Ice. The formulas used to predict horizontal ice forces are directly proportional to the effective ice strength and to the ice thickness. The thicker the ice, the larger the ice force. Therefore, the thickness of ice selected by a designer is important and at the same time it is the parameter with the most uncertainty. It is usually thicker at the piers where cracking, flooding, freezing, and rafting (where one ice sheet gets under another) have occurred. It is usually thinner away from the pier where the water is flowing free. Ice not only grows down into the water, but also thickens on the top. Ice can thicken quite rapidly in cold weather, but can also be effectively insulated by a covering of snow. On some occasions the ice can melt out in midwinter and freeze-up has to begin again. And even if ice thickness has been measured over a number of years at the bridge site, this may not be the ice that strikes the bridge. It could come from as far away as 200 miles (320 km) upstream.

Probably the best way to determine ice thickness at a bridge site is to search the historical record for factual information on measured ice thickness and to talk to local people who have seen more than one spring break-up. These can be longtime residents, town or city officials, newspaper editors, state highway engineers, and representatives of government agencies. A visit to the bridge site is imperative because the locals can provide information on the thickness of ice and can also indicate what the elevation of the water level is at spring break-up.

If historical data on ice thickness is not available, a mathematical model based on how cold a region is can serve as a starting point for estimating thickness of ice. The following discussion is taken from Wortley (1984). The measure of "coldness" used is the freezing degree-day (FDD), which is defined as the departure of the daily mean temperature from the freezing temperature. For example, if the daily high was $-6.7°C$ and the low was $-12°C$, the daily average would be $-9.4°C$, which is $[0 - (-9.4)] = 9.4°C$ departure from the freezing temperature. The FDD would therefore be $9.4°C$. A running sum of FDDs (denoted by S_f) is a cumulative measure of winter's coldness. If this sum becomes negative due to warm weather, a new sum is started on the next freezing day.

An 80-year record of values of S_f at various sites around the Great Lakes, accumulated on a daily and weekly basis is given in Table 4.19. The daily basis is termed the mean S_f and weekly basis is termed the extreme S_f. The extreme sum is computed by accumulating the coldest weeks over the 80-year period.

TABLE 4.19   Eighty-year Mean and Extreme Freezing Degree Days (°F)[a]

Great Lake	Station	Mean	Extreme
Lake Superior	Thunder Bay, Ontario	2500	3300
	Houghton, Michigan	1650	2400
	Duluth, Minnesota	2250	3050
Lake Michigan	Escanaba, Michigan	1400	2400
	Green Bay, Wisconsin	1350	2300
	Chicago, Illinois	500	1400
Lake Huron	Parry Sound, Ontario	1500	2550
	Alpena, Michigan	1150	2000
	Port Huron, Michigan	600	1550
Lake Erie	Detroit, Michigan	500	1350
	Buffalo, New York	500	1200
	Erie, Pennsylvania	400	1100
	Cleveland, Ohio	300	1200
Lake Ontario	Kingston, Ontario	1150	2000
	Toronto, Ontario	600	1500
	Rochester, New York	600	1300

[a]After Assel, 1980.

Figure 4.25 is a map of the United States developed by Haugen (1993) from National Weather Service data covering the 30-year period from 1951–1980 giving contours of extreme freezing degree days in degrees Celsius (°C). For example, at Chicago, IL, the map contour gives 1260°F (700°C). This 30-year extreme is slightly less than the 80-year extreme value of 1400°F (777°C) given in Table 4.19.

Observations have shown that the growth of ice thickness is proportional to the square root of S_f. Neill (1981) suggests the following empirical equation for estimating ice thickness:

$$t = 0.083\alpha_t \sqrt{S_f(°F)} \quad \text{(in ft)} \qquad (4.25a)$$

$$t = 33.9\alpha_t \sqrt{S_f(°C)} \quad \text{(in mm)} \qquad (4.25b)$$

where α_t = coefficient for local conditions from Table 4.20 [C3.9.2.2] and S_f = sum of freezing degree days (°F or °C).

EXAMPLE 4.3

Use the map of Figure 4.25 and Eq. 4.25 to estimate the maximum thickness of ice on the St. Regis River, which flows into the St. Lawrence River in northern New York, assuming it is an average river with snow. The sum of

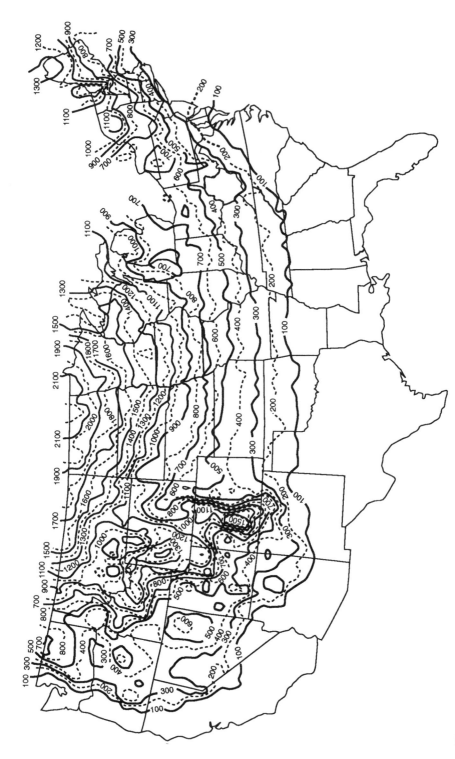

Fig. 4.25 Maximum sum of freezing degree days (FDD) in degrees celcius (°C). [From Haugen, *Cold Regions Research and Engineering Laboratory*, 1993].

TABLE 4.20   Locality Factors for Estimating Ice Thickness^a

Local Conditions	α_t
Windy lakes with no snow	0.8
Average lake with snow	0.5–0.7
Average river with snow	0.4–0.5
Sheltered small river with snow	0.2–0.4

^{a}From Neill, 1981.

freezing degree days is 1100°C, per Figure 4.25. Taking $\alpha_t = 0.5$, Eq. 4.25 yields

$$t = 33.9\alpha_t \sqrt{S_f} = 33.9 \times 0.5\sqrt{1100} = 560 \text{ mm}$$

In January 1990, the ice thickness was measured to be 14–26 (358–660 mm) inches near the bridge piers (Haynes et al., 1991). The calculated value compares favorably with the measured ice thickness. As a matter of interest, the designers of the bridge at this site selected an ice thickness of 36 in. (914 mm).

Dynamic Horizontal Ice Forces. When moving ice strikes a pier, the usual assumption is that the ice fails in crushing and the horizontal force on the pier is proportional to the width of the contact area, the ice thickness, and the effective compressive strength of the ice. During impact, the width of the contact area may increase from zero to the full width of the pier as the relative velocity of the ice floe with respect to the pier decreases. By equating the change in kinetic energy of a moving ice floe to the work done in crushing the ice, the critical velocity of the ice floe can be determined (Gershunov, 1986). The critical velocity is the velocity required to achieve full indentation of the structure into the ice. If the velocity of the ice floe is greater than the critical velocity, the ice floe continues to move and crush the ice on the full contact area.

The expressions for dynamic horizontal ice forces in AASHTO [A3.9.2.2] are independent of the velocity of the ice, which implies that the velocity of the approaching ice floe is assumed to be greater than the critical velocity. If $w/t > 6.0$, then the horizontal force F (N), due to moving ice is governed by crushing over the full width of the pier and is given by

$$F = F_c = C_a ptw \tag{4.26}$$

for which

$$C_a = (5t/w + 1)^{0.5} \tag{4.27}$$

where

p = effective ice crushing strength from Table 4.18 (MPa)
t = thickness of ice (mm)
w = pier width at level of ice action (mm)

When the pier nose is inclined at an angle greater than 15° from the vertical, an ice floe can ride up the inclined nose and fail in bending. If $w/t \leq 6$ the horizontal ice force F, in newtons, is taken as the lesser of the crushing force F_c from Eq. 4.26 or the bending failure force F_b given by

$$F = F_b = C_n p t^2 \tag{4.28}$$

for which

$$C_n = 0.5/\tan(\alpha - 15) \tag{4.29}$$

where

α = the inclination of the pier nose from the vertical, deg, but not less than 15°

EXAMPLE 4.4

Calculate the dynamic horizontal ice force predicted for the St. Regis River bridge pier at a water level where the pier width is 1220 mm. The pier nose is inclined only 5.7° from the vertical so the failure will be by crushing and Eq. 4.26 controls. Use an effective ice strength of 1150 kPa and the ice thickness of 8 in. (203 mm) observed on 16–17 March 1990.

$$C_a = (5t/w + 1)^{0.5} = \left(\frac{5 \times 203}{1220} + 1\right)^{0.5} = 1.35$$

$$F = F_c = C_a p t w = 1.35 \, (1150 \text{ kPa})(203 \text{ mm})(1220 \text{ mm}) = \underline{385 \text{ kN}}$$

The maximum ice force measured during the ice run of 16–17 March 1990 was 79.9 kips (355 kN) (Haynes et al., 1991), which is comparable to the predicted value.

The above ice forces are assumed to act parallel to the longitudinal axis of the pier. When an ice floe strikes the pier at an angle transverse forces are also developed. The magnitude of the transverse force F_t depends on the nose angle β of the pier and is given by [A3.9.2.4.1]

$$F_t = \frac{F}{2 \tan \left(\dfrac{\beta}{2} + \theta_f \right)} \tag{4.30}$$

where

F = horizontal ice force calculated by Eq. 4.26 or Eq. 4.28.

β = angle, deg, in a horizontal plane included between the sides of a pointed pier as shown in Figure 4.26. For a flat nose β is zero degrees. For a round nose β may be taken as 100°.

θ_f = friction angle between ice and pier nose.

EXAMPLE 4.5

Determine the transverse ice force corresponding to the dynamic horizontal ice force of Example 4.3 if the St. Regis River bridge pier has a pointed nose with an included angle of 90° and the friction angle is 10°.

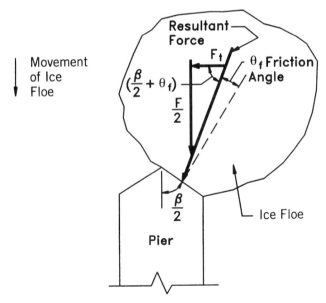

Fig. 4.26 Transverse ice force when a floe fails over a portion of a pier. (AASHTO Fig. C3.9.2.4.1-1). [From *AASHTO LRFD Bridge Design Specifications,* Copyright © 1994 by the American Association of State Highway and Transportation Officials, Washington, DC. Used by Permission].

$$F_t = \frac{F}{2\tan\left(\dfrac{\beta}{2} + \theta_f\right)} = \frac{385}{2\tan\left(\dfrac{90}{2} + 10\right)} = \underline{135\text{ kN}}$$

The longitudinal and transverse ice forces are assumed to act on the nose of the pier. When the ice movement is generally parallel to the longitudinal axis of the pier, two load combinations cases need to be investigated [A3.9.2.4]:

- A longitudinal force equal to F shall be combined with a transverse force of $0.15F$.
- A longitudinal force of $0.5F$ shall be combined with a transverse force of F_t.

If the longitudinal axis of the pier is skewed with respect to the flow, the total force on the pier is calculated on the basis of the projected pier width and resolved into components.

In regions where ice forces are significant, slender and flexible piers are not recommended. Ice-structure interaction can lead to amplification of the ice forces if the piers or pier components, including piles, are flexible.

Static Horizontal Ice Forces. When ice covers move slowly, the inertia can be neglected and the ice forces can be considered static. When ice is strained slowly it behaves in a ductile manner that tends to limit pressure. Additionally, ice creeps over time, which also decreases forces. The largest static ice forces are of thermal origin and occur when there is open water on one side of a structure and ice on the other.

Predictions of thermal ice pressures are difficult because they depend on the rate of change of temperature in the ice, the coefficient of thermal expansion $\sim0.000\,030/°F$ ($0.000\,054/°C$), the rheology of ice, the extent to which cracks have been filled with water, the thickness of the ice cover, and the degree of restrictions from the shores (Wortley, 1984). If thermal thrusts are calculated assuming the ice fails by crushing and using the strength values of Table 4.18, which neglect creep, the lateral loads determined will be too high.

Based on observations of ice in the Great Lakes and on stability calculations for rock-filled crib gravity structures, Wortley (1984) believes that reasonable ice thermal thrust values for this region are 5–10 kips/ft (73–146 kN/mm). If biaxial restraint conditions exist, such as in a harbor basin with a sheet piling bulkhead on most of its perimeter, the thermal thrusts can be doubled to 10–20 kips/ft (146–292 kN/mm).

Vertical Ice Forces. Changes in water level cause the ice sheet to move up and down and vertical loads result from the ice adhering to the structure. The vertical force on an embedded pile or pier is limited by the adhesive strength between the ice and the structure surface, by the shear strength of the ice, or by bending failure of the ice sheet some distance from the structure (Neill, 1981). Assuming there is no slippage at the ice-structure interface and no shear failure, a bending failure of the ice sheet will occur. If the pier is circular, this bending failure leaves a collar of ice firmly attached to the pier and a set of radial cracks in the floating ice sheet. When there is an abrupt water level fluctuation, the ice sheet will bend until the first circumferential crack occurs and a failure mechanism is formed. If the water level beneath ice sheet drops, the ice becomes a hanging dead weight (ice weighs 9.0 kN/m³). If the water level rises, a lifting force is transmitted to the pier or piling that could offset the dead load of a light structure.

The AASHTO Specifications give the following expressions for the maximum vertical force F_V on a bridge pier [A3.9.5]:

- For a circular pier, in newtons (N):

$$F_V = 0.3t^2 + 0.0169\ Rt^{1.25} \qquad (4.31)$$

- For a rectangular pier, in newtons per millimeter (N/mm) of pier perimeter

$$F_V = 2.3 \times 10^{-3}t^{1.25} \qquad (4.32)$$

where

t = ice thickness, in millimeters (mm)

R = radius of circular pier in millimeters (mm)

EXAMPLE 4.6

Calculate the vertical ice force on a 2000-mm diameter circular pier of a bridge crossing a reservoir that is subject to sudden changes in water level. Assume the ice is 600-mm thick.

$$F_V = 0.3t^2 + 0.0169\ Rt^{1.25} = 0.3(600)^2 + 0.0169(1000)(600)^{1.25}$$
$$= \underline{158\ 200\ N = 158.2\ kN}$$

Snow Loads on Superstructure. Generally snow loads are not considered on a bridge, except in areas of extremely heavy snowfall. In areas of significant snowfall, where snow removal is not possible, the accumulated snow loads may exceed the vehicle live loads. In some mountainous regions, snow loads up to 33.5 kPa may be encountered. In these areas, historical records and local experience should be used to determine the magnitude of the snow loads.

4.4 FORCES DUE TO DEFORMATIONS

4.4.1 Temperature

Two types of temperature changes must be included in the analysis of the superstructure (see [A3.12.2 and 3.12.3]). The first is a uniform temperature change where the entire superstructure changes temperature by a constant amount. This type of change lengthens or shortens the bridge, or if the supports are constrained it will induce reactions at the bearings and forces in the structure. This type of deformation is illustrated in Figure 4.27(a). The second type of temperature change is a gradient or nonuniform heating (or cooling) of the superstructure across its depth [see Fig. 4.27(b)]. Subjected to sunshine, the bridge deck heats more than the girders below. This nonuniform heating causes the temperature to increase more in the top portion of the system than in the bottom and the girder attempts to bow upward. If restrained by internal supports or by unintentional end restraints, compatibility actions are induced. If completely unrestrained, due to the piecewise linear nature of the imposed temperature distribution, internal stresses are introduced in the girder. In short, a statically determinate beam has internal stress due to the piecewise linear temperature gradient (even for a simply supported girder). This effect is discussed further in Chapter 6.

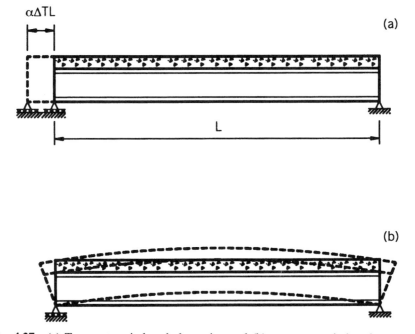

Fig. 4.27 (a) Temperature induced elongation and (b) temperature induced curvature.

As expected, the temperature range is considered a function of climate. Here AASHTO defines two climatic conditions: moderate and cold. A moderate climate is when the number of freezing days per year is less than 14. A freezing day is when the average temperature is less than 0°C. Table 4.21 gives the temperature ranges. The temperature *range* is used to establish the *change* in temperature used in analysis. For example, if the bridge is constructed at temperature of 20°C then the increase in a moderate climate for concrete is $\Delta T = 27 - 20 = 7°C$ and the decrease in temperature is $\Delta T = 27 - (-12) = 41°C$.

Theoretically, the range of climatic temperature is not a function of structure type, but the structure's temperature is a function of the climatic temperature record and specific heat of the material, mass, surface volume ratio, heat conductivity, wind conditions, shade, and so on. Because concrete bridges are more massive than steel and the specific heat of concrete is less than steel, an increase in climatic temperature causes a smaller temperature increase in the concrete structure than in the steel. Loosely stated, the concrete structure has more thermal inertia (systems with a large thermal inertia are resistant to changes in temperature) than its steel counterpart.

The temperature gradients are more sensitive to the bridge location than the uniform temperature ranges. The gradient temperature is a function of solar gain to the deck surface. In Western states, where solar radiation is greater, the temperature increases are also greater. The converse is true in the Eastern states. Therefore, the country is partitioned into the solar radiation zones shown in Figure 4.28 [A3.12.3]. The gradient temperatures outlined in Table 4.22 reference these radiation zones. The gradient temperature is considered in addition to the uniform temperature increase. Typically, these two effects are separated in the analysis and therefore are separated here. The AASHTO [A3.12.3] gradient temperatures are illustrated in Figure 4.29.

A temperature increase is considered positive in AASHTO. The temperature T_3 is zero unless determined from site specific study, but in no case is T_3 to exceed 3°C. In Figure 4.29, the dimension A is determined as follows:

TABLE 4.21 Temperature Ranges[a,b]

Climate	Steel or Aluminum	Concrete	Wood
Moderate	−18 to 50	−12 to 27	−12 to 24
Cold	−35 to 50	−18 to 27	−18 to 24

[a]In degrees celcius (°C).
[b]In AASHTO Table 3.12.2.1-1. [From *AASHTO LRFD Bridge Design Specifications,* Copyright © 1994 by the American Association of State Highway and Transportation Officials, Washington, DC. Used by Permission].

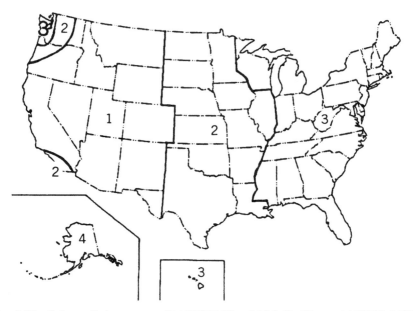

Fig. 4.28 Solar radiation zones. (AASHTO Fig. 3.12.3-1). [From *AASHTO LFRD Bridge Design Specifications*, Copyright © 1994 by the American Association of State Highway and Transportation Officials, Washington, DC. Used by Permission].

A = 300 mm for closed concrete structures that are 400 mm or more in depth. For shallower sections, A shall be 100 mm less than the actual depth.

$A = t - 100$ mm for steel superstructures,

where t = thickness of the concrete deck.

TABLE 4.22 Gradient Temperatures[a,b]

Zone	Concrete Surface		50-mm Asphalt		100-mm Asphalt	
	T_1	T_2	T_1	T_2	T_1	T_2
1	30	7.8	24	7.8	17	5
2	25	6.7	20	6.7	14	5.5
3	23	6	18	6	13	6
4	21	5	16	5	12	6

[a]In degrees celcius (°C).
[b]In AASHTO Table 3.12.3-1. [From *AASHTO LRFD Bridge Design Specifications*, Copyright © 1994 by the American Association of State Highway and Transportation Officials, Washington, DC. Used by Permission.]

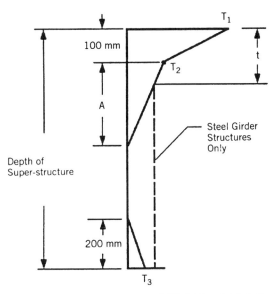

Fig. 4.29   Design temperature gradients. (AASHTO Fig. 3.12.3-2). [From *AASHTO LFRD Bridge Design Specifications,* Copyright © 1994 by the American Association of State Highway and Transportation Officials, Washington, DC. Used by Permission].

These temperature changes are used in the structural analyses described in Chapter 6.

4.4.2   Creep and Shrinkage

The effects of creep and shrinkage can have an effect on the structural strength, fatigue, and serviceability. Traditionally, creep is considered in concrete where its effect can lead to unanticipated serviceability problems that might subsequently lead to secondary strength problems. In addition, today, however, creep is also of concern in wooden structures. For example, in transversely prestressed wood, creep can lead to a loss in prestress and structural integrity leading to collapse. This creep effect is described in more detail in Chapter 9. Because creep and shrinkage are highly dependent on the material and the system involved, further elaboration is reserved for the chapters on design.

4.4.3   Settlement

Support movements may occur due to the elastic and inelastic deformation of the foundation. Elastic deformations include movements that affect the response of the bridge to other loads but do not lock-in permanent actions.

Such deformations may be modeled by approximating the stiffness of the support in the structural analysis model. This type of settlement is not a load but rather a support characteristic that should be included in the structural model. Inelastic deformations are movements that tend to be permanent and create locked-in permanent actions. Such movements may include the settlement due to consolidation, instabilities, or foundation failures. Some such movements are the result of loads applied to the bridge, and these load effects may be included in the modeling of the structural supports. Other movements are attributed to the behavior of the foundation independent of the loads applied to the bridge. These movements must be treated as a load and hereafter are called *imposed support deformations.*

The actions due to imposed support deformations in statically indeterminate structures are proportional to the stiffness. For example, for a given imposed deformation, a stiff structure develops larger actions than a flexible one. The statically determinate structures do not develop any internal actions due to settlement, which is one of the few inherent advantages of statically determinate systems. Imposed support deformations are estimated based on the geotechnical characteristics of the site and system involved. Detailed suggestions are given in AASHTO, Section 10.

4.5   COLLISION LOADS

4.5.1   Vessel Collision

On bridges over navigable waterways, the possibility of vessel collision with the pier must be considered. Typically, this is of concern for structures that are classified as long-span bridges, which are outside the scope of this book. Vessel collision loads are defined in AASHTO [A3.14].

4.5.2   Rail Collision

If a bridge is located near a railway, the possibility of a collision with the bridge as a result of a railway derailment exists. As this possibility is remote, the bridge must be designed for collision forces using the extreme limit state given in Tables 3.1 and 3.2. The abutments and piers within 9000 mm of the edge of the roadway, or within a distance of 15 000 mm of the centerline of the track must be designed for a 1800-kN force positioned at a distance of 1200 mm above the ground [A3.6.5.2].

4.5.3   Vehicle Collision

The collision force of a vehicle with the barrier rail and parapets is described previously in the section on Deck and Railing Loads, as well as in later chapters and will not be reiterated here.

4.6 SUMMARY

The various types of loads applicable to highway bridges are described with reference to the AASHTO specification. These loads are used in the subsequent chapters to determine the load effects and to explain the use of these effects in the proportioning of the structure. For loads that are particular to bridges, background is provided on the development and use of the loads, and in other cases, the AASHTO provisions are outlined with limited explanation leaving the detailed explanation to other references.

REFERENCES

AASHTO (1996). *Standard Specifications for Highway Bridges,* 16th ed., American Association of State Highway and Transportation Officials, Washington, DC.

AASHTO (1994). *LRFD Bridge Design Specification,* 1st ed., American Association of State Highway and Transportation Officials, Washington, DC.

ASCE (1988). *Minimum Design Loads for Buildings and Other Structures,* ASCE 7-88 American Society of Civil Engineers, New York.

Assel, R. A. (1980). Great Lakes Degree-Day and Temperature Summaries and Norms, 1897–1977, *NOAA Data Report ERL GLERL-15, January 1980,* Great Lakes Environmental Research Laboratory, Ann Arbor, MI.

Bakht, B. and S. G. Pinjarkar (1991). *Dynamic Testing of Highway Bridges—A Review,* Transportation Research Record 1223, National Research Council, Washington, DC.

Cassano, R. C. and R. J. Lebeau (1978). *Correlating Bridge Design Practice with Overload Permit Policy,* Transportation Research Record 664, National Research Council, Washington, DC.

Gershunov, E. M. (1986). "Collision of Large Floating Ice Feature With Massive Offshore Structure," *Journal of Waterway, Port, Coastal and Ocean Engineering,* ASCE, Vol. 112, No. 3, May, pp. 390–401.

Haugen, R. K. (1993). Meteorologist, U.S. Army Corps of Engineers, Cold Regions Research and Engineering Laboratory, Hanover, NH, personal communication.

Haynes, F. D., D. S. Sodhi, L. J. Zabilansky, and C. H Clark (1991). *Ice Force Measurements on a Bridge Pier in the St. Regis River, NY,* Special Report 91-14, U.S. Army Corps of Engineers, Cold Regions Research and Engineering Laboratory, Hanover, NH, October, 11*p.*

Hwang, E. S. and A. S. Nowak (1991a). "Simulation of Dynamic Load for Bridges," *Journal of Structural Engineering,* ASCE, Vol. 117, No. 5, May, pp. 1413–1434.

Hwang, E. S. and A. S. Nowak (1991b). *Dynamic Analysis of Girder Bridges,* Transportation Research Record 1223, National Research Council, Washington, DC.

Koretzky, H. P., K. R. Patel, R. M. McClure, D. A. VanHorn, (1986). *Umbrella Loads for Bridge Design,* Transportation Research Record 1072, National Research Council, Washington, DC.

Kulicki, J. M. (1992). personal communications.

Kulicki, J. M. and D. R. Mertz (1991). "A New Live Load Model for Bridge Design," *Proceedings of 8th Annual International Bridge Conference,* June. Pittsburgh, PA.

Liu, H. (1991). *Wind Engineering Handbook for Structural Engineers,* Prentice Hall, Englewood Cliffs, NJ.

Moses, F. and M. Ghosen (1985), *A Comprehensive Study of Bridge Loads and Reliability,* Final Report, FHWA/OH-85/005, Jan.

Neill, C. R., Ed. (1981). *Ice Effects on Bridges,* Roads and Transportation Association of Canada, Ottawa, Ontario, Canada.

Nowak, A. S. (1993). "Calibration of LRFD Bridge Design Code," *NCHRP Project,* 12–33, University of Michigan, Ann Arbor, MI.

Nowak, A. S. (1995). "Calibration of LRFD Bridge Code," *Journal of Structural Engineering,* ASCE, Vol. 121, No. 8, pp. 1245–1251.

Paultre, P., O. Challal, and J. Proulx (1992), "Bridge Dynamics and Dynamic Amplication Factors—A Review of Analytical and Experimental Findings," *Canadian Journal of Civil Engineering,* Vol. 19, pp. 260–278.

Simiu, E. (1973). "Logarithmic Profiles and Design Wind Speeds," *Journal of the Engineering Mechanics Division,* ASCE, Vol. 99, No. 10, October, pp. 1073.

Simiu (1976), "Equivalent Static Wind Loads for Tall Building Design," *Journal of the Structures Division,* ASCE, Vol. 102, No. 4, April.

Simiu, E. and R. H. Scanlan (1978). *Wind Effects on Structures: An Introduction to Wind Engineering,* Wiley-Interscience, New York, NY.

Tilly, G. P. (1986). "Dynamic Behavior of Concrete Structures," In *Developments in Civil Engineering,* Vol. 13 Report of the Rilem 65 MDB Committee, Elsevier, New York, NY.

TRB Committee for the Truck Weight Study (1990), *Truck Weights and Limits—Issues and Options,* Special Report 225, Transportation Research Board, National Research Council, Washington DC.

Uniform Building Code (1994). International Conference of Building Officials, Whitter, CA.

Wortley, C. A. (1984). Great Lakes Small-Craft Harbor and Structure Design for Ice Conditions: *An Engineering Manual, Sea Grant Advisory Report WIS-SG-84-426,* University of Wisconsin Sea Grant Institute, Madison, WI, 227p.

5

INFLUENCE FUNCTIONS AND GIRDER-LINE ANALYSIS

5.1 INTRODUCTION

As outlined in Chapter 4, bridges must carry many different types of loads, which may be present individually or in combination. The bridge engineer has the responsibility for analysis and design of the bridge subjected to these loads and for the placement of the loads in the most critical manner. For example, vehicular loads move, and hence the placement and analysis varies as the vehicle traverses the bridge. The engineer must determine the most critical load placement for all cross sections in the bridge. Frequently, this load placement is not obvious and the engineer must rely on systematic procedures to place the loads and to analyze the structure for this placement. Structural analysis using influence functions is the foundation of this procedure and is fundamental to the understanding of bridge analysis and design. The term influence function is used instead of the term influence line, because it is more general, that is, the function may be one-dimensional (1-D) (a line) or two-dimensional (2-D) (a surface).

The reader may have been exposed to influence functions, commonly called influence lines, in past coursework or professional practice. If so, then this chapter provides a review and perhaps a treatment unlike the previous exposure. To the novice, this chapter is intended to be comprehensive in both theory and application. The examples provide background and detailed analyses of all the structures used in the subsequent design chapters. Therefore, the reader should take careful note of the examples as they are referenced frequently throughout the remainder of this book.

Sign conventions are necessary to properly communicate the theory, procedures, and analytical results. Conventions are somewhat arbitrary and textbook writers use different conventions. Herein the following conventions are used for shear and moment diagrams:

- For a beam, moment causing compression on the top and tension on the bottom is considered positive as shown in Figure 5.1. Moment diagrams are plotted on the compression side of the element. For frames, the distinction of positive and negative is ambiguous.
- For a beam, positive shear is upward on the left face and downward on the right face as shown in Figure 5.1. The shear diagram is plotted so that the change in shear is in the direction of the applied load. Again, for frames, the distinction of positive and negative is ambiguous.
- Axial thrust is considered positive in tension. The side of the element on which to plot this function is arbitrary but must be consistent throughout the structure and labeled appropriately to avoid misinterpretation.

These conventions are summarized in Figure 5.1 and hereafter are referred to as the *designer sign convention* because they refer to those quantities (shears, moments, and axial load) that a designer uses to select and check member sizes. Additional sign conventions such as those used in analysis procedures are given as necessary.

5.2 DEFINITION

Influence function ≡ a function that represents the load effect (force or displacement) at a point in the structure as a unit action moves along a path or over a surface.

Consider the two-span beam illustrated in Figure 5.2(a). The unit action is a concentrated load that traverses the structures along the beam from left to right. The dimension x represents the location of the load. For the sake of this discussion, assume that an instrument that measures the flexural bending moment is located at point n and records this action as the unit load moves across the beam. The record of the moment as a function of load position is the influence function shown in Figure 5.2(b).

The load positioned in span AB causes a positive influence (positive moment) at n. Note that the maximum value occurs directly at point n. When the load is positioned in span BC, the influence of the load on the bending

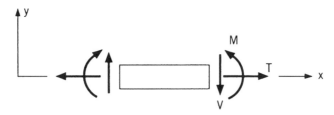

Fig. 5.1 Beam segment.

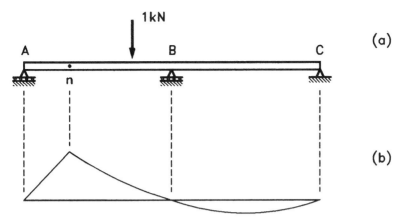

Fig. 5.2 Continuous beam.

moment at *n* is negative. A load in span *BC* causes the beam to deflect upward creating tension on top of the beam or a negative bending moment.

Consider the beam section shown in Figure 5.3(a) and the influence function for action *A* at point *m* shown in Figure 5.3(b). Assume that the influence function was created by a unit load applied downward in the same direction of the applied loads shown in Figure 5.3(a). Assuming that the structure

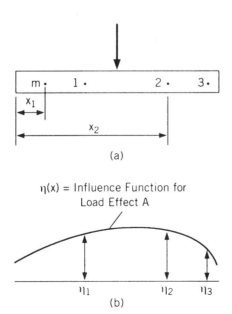

Fig. 5.3 (a) Concentrated loads on beam segment and (b) influence function for load effect *A*.

behaves linearly, the load P_1 applied at point 1 causes a load effect of P_1 times the function value $\eta(x_1) = \eta_1$. Similarly, the load P_2 applied at point 2 causes a load effect of P_2 times the function value $\eta(x_2) = \eta_2$, and so on. Superposition of all the load effects yields

$$\text{Action} \quad A = P_1\eta(x_1) + P_2\eta(x_2) + \cdots + P_n\eta(x_n) = \sum_{i=1}^{n} P_i\eta(x_i) = \sum_{i=1}^{n} P_i\eta_i$$

$$(5.1)$$

Linear behavior is a necessary condition for application of Eq. 5.1, that is, the influence coefficients must be based on a linear relationship between the applied unit action and the load effect. For example, if the unit load is applied at a specific point and then doubled, the resulting load effect will also double if the response is linear. For statically determinate structures, this relationship typically holds true except for cases of large deformation where consideration of deformed geometry must be considered in the equilibrium formulation. The unit action load effect relationship in statically indeterminate structures is a function of the relative stiffness of the elements. If stiffness changes are due to load application from either material nonlinearity and/or geometric nonlinearity (large deflections), then the application of the superposition implicit in Eq. 5.1 is negated. In such cases, the use of influence functions is not appropriate and the loads must be applied sequentially as expected in the real structure. Such an analysis is beyond the scope of this book and the reader is referred to books on structural and finite element analysis.

5.3   STATICALLY DETERMINATE BEAMS

The fundamentals of influence functions and their use are initially illustrated with statically determinate beams. Several examples are given.

5.3.1   Concentrated Loads

EXAMPLE 5.1

Use the beam shown in Figure 5.4(a) to determine the influence functions for the reaction at A, and the shear and moment at B. Point B is located at midspan.

Consider the unit load at position x on the beam AC of length ℓ. Because this system is statically determinate, the influence function may be based solely on static equilibrium. Use the free body diagram shown in Figure 5.4(b) to balance the moments about A and to determine the reaction R_C.

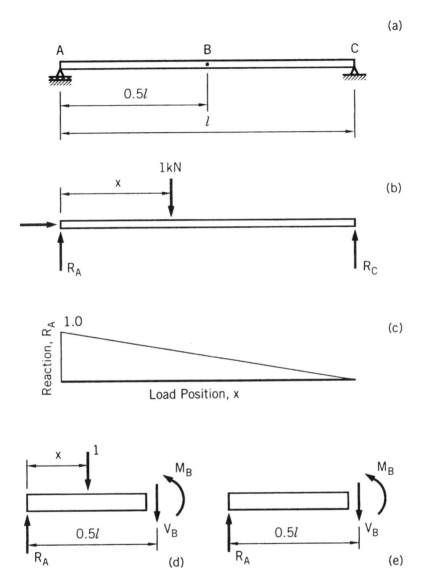

Fig. 5.4 (a) Simple beam, (b) moving unit load, (c) influence function for R_A, (d) Free body diagram AB with unit load at $x \leq 0.5\ell$, and (e) free body diagram AB with unit load at $x > 0.5\ell$.

(f)

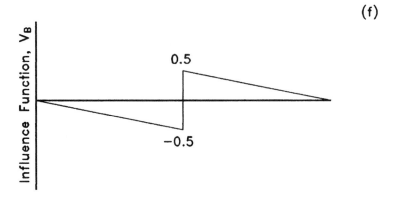

(g)

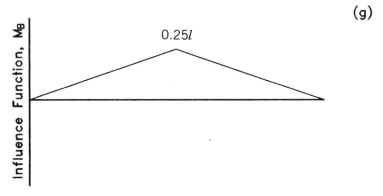

Fig. 5.4   (f) Influence diagram for V_B and (g) influence diagram for M_B.

$$\sum M_A = 0$$

$$(x) - R_C(\ell) = 0$$

$$R_C = \frac{x}{\ell}$$

Similarly, balancing moments about C yields the reaction at R_A.

$$\sum M_C = 1(\ell - x) - R_A(\ell) = 0$$

$$R_A = \frac{\ell - x}{\ell}$$

Summation of the vertical forces checks the previous moment summations.

If this check does not validate equilibrium, then there is an error in the calculation.

$$\Sigma F_V = 0$$

$$R_A + R_C - 1 = 0 \quad \text{OK}$$

The influence function for R_A is shown in Figure 5.4(c). Note the function is unity when the load position is directly over A and decreases linearly to zero when the load position is at C. Linearity is characteristic of influence functions (actions and reactions) for statically determinate structures. This point is elaborated on later. Next, the influence functions for the shear and moment at B are determined. Use the free body diagram shown in Figure 5.4(d) to sum the vertical forces yielding V_B.

$$\Sigma F_V = 0$$

$$R_A - 1 - V_B = 0$$

$$V_B = \frac{\ell - x}{\ell} - 1 = -\frac{x}{\ell}$$

Balancing moments about B gives the moment at B.

$$\Sigma M_B = 0$$

$$R_A(0.5\ell) - 1(0.5\ell - x) - M_B = 0$$

$$M_B = \frac{x}{2} \quad \text{when } 0 \le x \le 0.5\ell$$

Note that the functions for V_B and M_B are valid when $x \le \ell/2$. If $x > \ell/2$, then the unit load does not appear on the free body diagram. The revised diagram is shown in Figure 5.4(e). Again, by balancing forces and moments, the influence functions for V_B and M_B are established.

$$V_B = R_A = \frac{(\ell - x)}{\ell} \quad \text{when} \quad 0.5\ell \le x \le \ell$$

$$M_B = \frac{\ell - x}{2} \quad \text{when} \quad 0.5\ell \le x \le \ell$$

The influence functions for V_B and M_B are illustrated in Figures 5.4(f) and (g), respectively.

EXAMPLE 5.2

Use the influence functions developed in Example 5.1 to analyze the beam shown in Figure 5.5(a). Determine the reaction at A and the midspan shear and moment at B.

Use the influence function for R_A shown in Figure 5.4(c) to determine the influence ordinates at the load positions of P_1 and P_2 as shown in Figure 5.5(b). The equation developed in Example 5.1 may be used or the ordinates may be interpolated. As illustrated in Figure 5.5(b), the ordinate values are two-thirds and one-third for the positions of P_1 and P_2, respectively. Application of Eq. 5.1 yields

$$R_A = \sum_{i=1}^{2} P_i \eta_i = P_1(\tfrac{2}{3}) + P_2(\tfrac{1}{3})$$

The parameters V_B and M_B due to the applied loads may be determined in a similar manner. With the aid of Figures 5.5(c) and (d), application of Eq. 5.1 yields

$$V_B = P_1(-\tfrac{1}{3}) + P_2(\tfrac{1}{3})$$

and

$$M_B = P_1(\tfrac{\ell}{6}) + P_2(\tfrac{\ell}{6})$$

Comparison of these results with standard statics procedures is left to the reader.

5.3.2   Uniform Loads

Distributed loads are considered in a manner similar to concentrated loads. Consider the beam segment shown in Figure 5.6(a) that is loaded with a distributed load of varying magnitude. The influence function $\eta(x)$ for action A is illustrated in Figure 5.6(b). The load applied over the differential element Δx is $w(x)\Delta x$. This load is used in Eq. 5.1. In the limit as Δx goes to zero, the summation becomes the integration

$$\text{Action} \qquad A = \sum_{i=1}^{n} P_i \eta(x_i) = \sum_{i=1}^{n} w(x_i)\eta(x_i)\Delta x = \int_a^b w(x)\eta(x)dx \quad (5.2)$$

If the load is uniform then the load function $w(x) = w_0$ is a constant rather than a function of x and may be placed outside of the integral. Equation 5.2 becomes

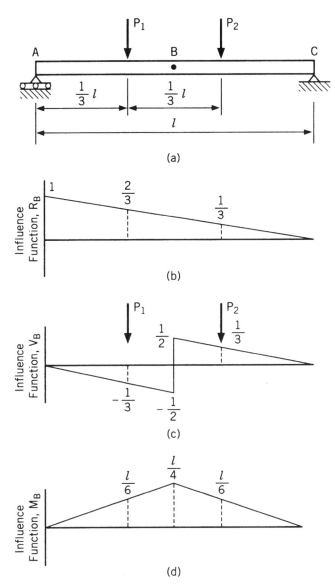

Fig. 5.5   (a) Simple beam, (b) influence function for V_B, (c) influence function for R_A, and (d) influence function for M_B.

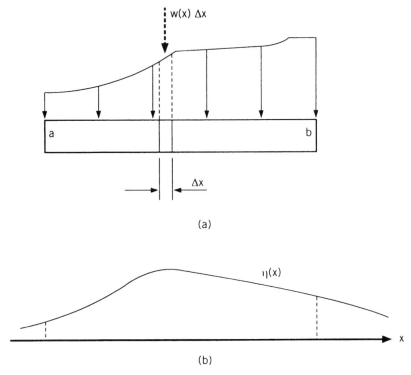

Fig. 5.6   (a) Beam segment with distributed load and (b) influence function.

$$\text{Action} \quad A = \int_a^b w(x)\eta(x)dx = w_0 \int_a^b \eta(x)dx \qquad (5.3)$$

Note that this integral is simply the area under the influence function over the range of load application.

EXAMPLE 5.3

Determine the reaction at A and the shear and moment at midspan for the beam shown in Example 5.1 [Figure 5.4(a)] subjected to a uniform load of w_0 over the entire span.

Application of Eq. 5.3 yields

$$R_A = \int_0^\ell w(x)\eta_{R_A}(x)dx = w_0 \int_0^\ell \eta_{R_A}(x)dx$$

$$R_A = \frac{w_0 \ell}{2}$$

$$V_B = \int_0^\ell w(x)\eta_{V_B}(x)dx = w_0 \int_0^\ell \eta_{V_B}(x)dx$$

$$V_B = 0$$

$$M_B = \int_0^\ell w(x)\eta_{M_B}(x)dx = w_0 \int_0^\ell \eta_{M_B}(x)dx$$

$$M_B = \frac{w_0 \ell^2}{8}$$

Again, comparison of these results with standard equilibrium analysis is left to the reader.

5.4 MULLER–BRESLAU PRINCIPLE

The analysis of a structure subjected to numerous load placements can be labor intensive and algebraically complex. The unit action must be considered at numerous locations requiring several analyses. The Muller–Breslau Principle allows the analyst to study one load case to generate the entire influence function. Because the function has the same characteristics whether generated by traversing a unit action or by the Muller–Breslau Principle, many of the complicating features are similar. The Muller–Breslau Principle has both advantages and disadvantages depending on the analytical objectives. These concerns are discussed in detail later.

The development of the Muller–Breslau Principle requires application of Betti's Theorem. This important energy theorem is prerequisite to the understanding of the Muller–Breslau Principle and is reviewed in Section 5.4.1.

5.4.1 Betti's Theorem

Consider two force systems P and Q associated with displacements p and q applied to a structure that behaves linear elastically. These forces and displacements are shown in Figures 5.7(a) and 5.7(b). Application of the $Q - q$ system to the structure and equating the work performed by gradually applied forces to the internal strain energy yields

$$\frac{1}{2}\sum_{i=1}^n Q_i q_i = U_{Qq} \tag{5.4}$$

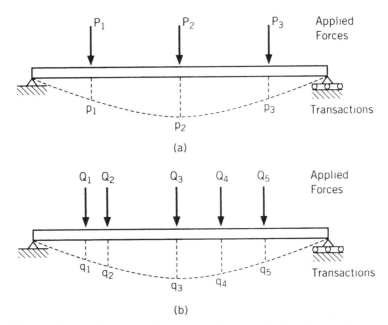

Fig. 5.7 (a) Displaced beam under system $P - p$ and (b) displaced beam under system $Q - q$.

where U_{Qq} is the strain energy stored in the beam when the loads Q are applied quasistatically* through displacement q.

Now apply the forces of the second system P with the Q forces remaining in place. Note that the forces Q are now at the full value and move through displacements p due to force P. The work performed by all the forces is

$$\frac{1}{2}\sum_{i=1}^{n} Q_i q_i + \sum_{i=1}^{n} Q_i p_i + \frac{1}{2}\sum_{i=1}^{n} P_i p_i = U_{\text{final}} \tag{5.5}$$

where U_{final} is the associated internal strain energy due to all forces applied in the order prescribed.

Use the same force systems to apply the forces in the reverse order, that is, P first and then Q. The work performed by all forces is

$$\frac{1}{2}\sum_{i=1}^{n} P_i p_i + \sum_{i=1}^{n} P_i q_i + \frac{1}{2}\sum_{i=1}^{n} Q_i q_i = U_{\text{final}} \tag{5.6}$$

*Because all loads are transient at some time, static load is technically a misnomer, but if the loads are applied slowly dynamic effects are small.

If the structure behaves linear elastically, then final displaced shape and internal strain energy are independent of the order of load application. Therefore, equating U_{final} in Eqs. 5.5 and 5.6 yields

$$\sum_{i=1}^{n} Q_i p_i = \sum_{i=1}^{n} P_i q_i \qquad (5.7)$$

In a narrative format, Eq. 5.7 states:

The product of the forces of the first system times the corresponding displacements due to the second force system is equal to the forces of the second system times the corresponding displacements of the first system.

Although derivation is performed with reference to a beam, the method is generally applicable to any linear elastic structural system.

5.4.2 Theory of Muller–Breslau Principle

Consider the beam shown in Figure 5.8(a), where the reaction R_A is of interest. Remove the support constraint and replace it with the reaction R_A as illustrated in Figure 5.8(b). Now replace the reaction R_A with a second force F and remove the applied forces P. Displace the released constraint a unit amount in the direction shown in Figure 5.8(c). Application of Betti's Theorem to the two systems (Eq. 5.7) yields

$$R_A(1) - P_1\delta_1 - P_2\delta_2 - \cdots - P_n\delta_n = F(0) \qquad (5.8)$$

which simplifies to

$$R_A = P_1\delta_1 + P_2\delta_2 + \cdots + P_n\delta_n = \sum_{i=1}^{n} P_i\delta_i \qquad (5.9)$$

A comparison of Eq. 5.1 to Eq. 5.9 reveals that application of Betti's Theorem yields the same result as direct application of superposition combined with the definition of an influence function. Hence, the ordinates δ in Eq. 5.9 must be the same as the ordinates η in Eq. 5.1. This observation is important because ordinates δ were generated by imposing a unit displacement at the released constraint associated with the action of interest. This constitutes the Muller–Breslau Principle, which is summarized below:

An influence function for an action may be established by removing the constraint associated with the action and imposing a unit displacement. The dis-

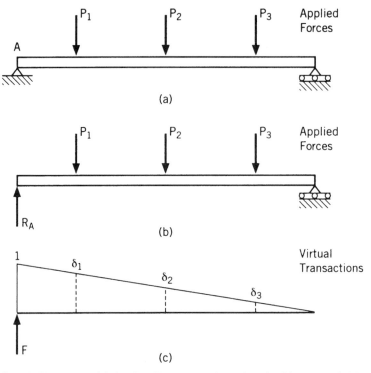

Fig. 5.8 (a) Structure with loads, (b) support A replaced with R_A, and (c) virtual displacement at A.

placement at every point in the structure is the influence function. In other words, the structure's displaced shape is the influence function.

The sense of the displacements that define the influence function must be considered. For concentrated or distributed forces, the translation colinear with the direction of action is used as the influence ordinate or function. If the applied action is a couple, then the rotation is the associated influence function. The latter can be established by application of Betti's Theorem in a similar manner.

EXAMPLE 5.4

Use the Muller–Breslau Principle to determine the influence function for the moment and shear at midspan of the beam shown in Example 5.1 (Fig. 5.4) and reillustrated for convenience in Figure 5.9(a).

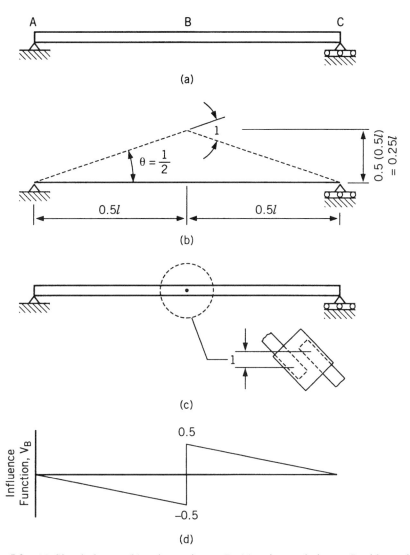

Fig. 5.9 (a) Simple beam, (b) unit rotation at B, (c) unit translation at B with mechanism for unit translation, and (d) influence function of V_B.

The moment is considered first. Release the moment at B by insertion of a hinge and apply a unit rotation at this hinge, that is, the relative angle between member AB and BC is one unit. Geometric and symmetry considerations require that the angle at the supports be one-half unit. Assuming small displacements (i.e., $\tan \theta = \sin \theta = \theta$), the maximum ordinate at B is determined by

$$\eta_{max} = \theta\left(\frac{\ell}{2}\right) = \frac{1}{2}\left(\frac{\ell}{2}\right) = \frac{\ell}{4} = 0.25\ell \qquad (5.10)$$

which is the same result as Example 5.1 [see Figs. 5.4(g) and 5.9(b)].

The shear influence function is determined in a similar manner. Release the translation continuity at B and maintain the rotational continuity. Such a release device is schematically illustrated in Figure 5.9(c). Apply a relative unit translation at B and maintain the slope continuity required on both sides of the release. This displacement gives the influence function shown in Figure 5.9(d), which is the same function given in Example 5.1 [Fig. 5.4(f)], as expected.

5.4.3 Qualitative Influence Functions

The displaced shape is not always as easily established as illustrated in Example 5.4. For example, the displaced shape of a statically indeterminate structure is more involved. Related procedures are described in Section 5.5. One of the most useful applications of the Muller–Breslau Principle is in the development of qualitative influence functions. Because most displaced shapes due to applied loads may be intuitively generated in an approximate manner, the influence functions may be determined in a similar fashion. Although exact ordinates and/or functions require more involved methods, a function can be estimated by simply releasing the appropriate restraint, inducing the unit displacement, and sketching the displaced shape. This technique is extremely useful in determining an approximate influence function that in turn aids the engineer in the placement of loads for the critical effect.

EXAMPLE 5.5

Use the qualitative method to establish the influence function for moment at point B for the beam shown in Figure 5.10(a).

Release the moment at B and apply a relative unit rotation. The resulting translation is illustrated in Figure 5.10(b). If a uniform live load is required, it is necessary to apply this load on spans AC and EF (location of positive influence) for the maximum positive moment at B, and on CE for the critical negative moment at B.

5.5 STATICALLY INDETERMINATE BEAMS

Primarily, two methods exist for the determination of influence functions:

- Traverse a unit action across the structure.

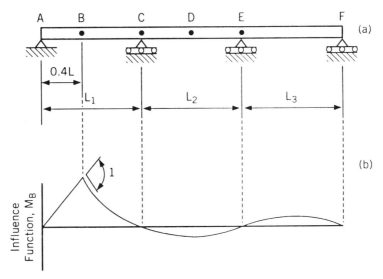

Fig. 5.10   (a) Continuous beam and (b) influence function M_B.

• Impose a unit translation at the released action of interest (Muller–Breslau).

Both of these methods are viable techniques for either hand or automated analysis. The principles involved with these methods have been presented in previous sections concerning statically determinate structures. Both methods are equally applicable to indeterminate structures, but are somewhat more involved. Both methods must employ either a flexibility approach such as consistent deformations, or stiffness techniques such as slope-deflection, moment distribution, and finite element analysis (matrix displacement analysis). Typically, stiffness methods are used in practice where slope-deflection and moment distribution are viable hand methods while the matrix approach is used in automated procedures. Both the unit load traverse and Muller–Breslau approaches are illustrated in the following sections using stiffness methods. Each method is addressed by first presenting the methodology and necessary tools reqired for the application, which is followed by examples. *These examples form the analytical basis for the design examples in the remainder of the book. Please pay careful attention to the examples.*

Throughout the remainder of this book a specialized notation is used to indicate a position on the structure. This notation, termed span point notation, is convenient for bridge engineering, and is illustrated by several examples in Table 5.1. Here the span point notation is described with the points that typically control the design of a continuous girder. For example, the shear is a maximum near the supports, the positive moment is a maximum in the span,

TABLE 5.1   Span Point Notation

Span Point Notation	Span	Percentage	Explanation	Critical Action (typical)
100	1	0	Left end of the first span	Shear
104	1	40	Forty percent of the way across the first span	Positive moment
110	1	100	Right end of the first span immediately left of the first interior support	Shear, negative moment
200	2	0	Left end of the second span immediately right of the first interior support	Shear, negative moment
205	2	50	Middle of the second span	Positive moment

and the negative moment is the largest, often called the *maximum negative moment,* at the supports. Mathematically, *maximum negative moment* is poor terminology but nevertheless it is conventional. Table 5.1 is provided for guidance in typical situations and for use in preliminary design. Final design calculations should be based on the envelope of all actions from all possible live load placements. Actions described with span point notation are in the designer's sign conventional outlined in Section 5.1 (see Fig. 5.1).

EXAMPLE 5.6

For the prismatic beam shown in Figure 5.11(a), determine the influence functions for the moments at the 104, 200, and 205 points and for shear at the 100, 104, 110, 200, and 205 points. Use span lengths of 100, 120, and 100 ft (30 480, 36 576, and 30 480 mm).

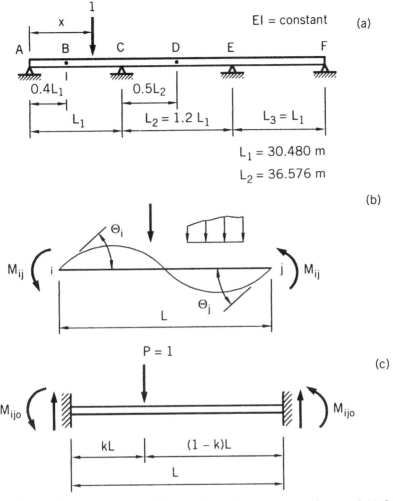

Fig. 5.11 (a) Continuous beam, (b) slope-deflection sign conventions, and (c) fixed-actions for concentrated loads.

A unit load is traversed across the beam and the slope-deflection method is used. This problem is repeated in Example 5.9 using the slope-deflection method combined with the Muller–Breslau Principle.

With the exception of the shear at the 104 and 205 points, these points were selected because the critical actions due to vehicular loads usually occur at or near these locations. These influence functions are subsequently used in Example 5.12 to determine the maximum load effect due to vehicular loads. The slope-deflection relationship between the end moments and rotations for a prismatic beam on nonsettling supports is given in Eq. 5.11. The subscripts reference the locations illustrated in Figure 5.11(b).

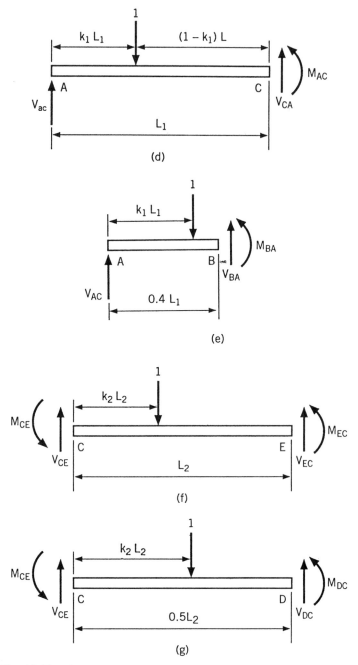

Fig. 5.11 (d) Free body diagram beam AC, (e) free body diagram beam segment AB, and (f) free body diagram beam CE, and (g) free body diagram beam segment CD.

$$M_{ij} = \frac{4EI}{L}\,\theta_i + \frac{2EI}{L}\,\theta_j + M_{ijo}$$

$$M_{ji} = \frac{2EI}{L}\,\theta_i + \frac{4EI}{L}\,\theta_j + M_{jio}$$

(5.11)

where

EI is the flexural rigidity.

L is the element length.

M_{ij} and M_{ji} are the moments at ends i and j, respectively.

M_{ijo} and M_{jio} are the fixed-end moments at ends i and j due the applied loads, respectively.

θ_i and θ_j are the rotations at end i and j, respectively.

Counterclockwise moments and rotations are considered positive in Eq. 5.11.

A fixed–fixed beam subjected to a concentrated load located at position kL is illustrated in Figure 5.11c. The end moments are

$$M_{ijo} = PL(k)(1 - k)^2$$ (5.12a)

$$M_{jio} = -PL(k^2)(1 - k)$$ (5.12b)

A single set of slope-deflection and equilibrium equations is desired for all locations of the unit load. Because the load must traverse all spans, the fixed-end moments must change from zero when the load is not on the span to moments based on Eq. 5.12 when the load is located on the span. To facilitate this discontinuity, a special form of MacCauley's notation (Pilkey and Pilkey, 1974) is used.

$\langle ij \rangle = \langle ji \rangle = 1$ if the unit load is located between i and j

$\langle ij \rangle = \langle ji \rangle = 0$ if the unit load is *not* located between i and j

(5.13)

Application of Eqs. 5.11, 5.12, and 5.13 to the continuous beam shown in Figure 5.11(a) gives

$$M_{AC} = \frac{4EI}{L_1}\,\theta_A + \frac{2EI}{L_1}\,\theta_C + k_1(1 - k_1)^2\,L_1\langle AC \rangle$$

$$M_{CA} = \frac{4EI}{L_1}\,\theta_C + \frac{2EI}{L_1}\,\theta_A + k_1^2(1 - k_1)L_1\langle AC \rangle$$

$$M_{CE} = \frac{4EI}{L_2}\,\theta_C + \frac{2EI}{L_2}\,\theta_E + k_2(1 - k_2)^2\,L_2\langle CE \rangle$$

$$M_{EC} = \frac{4EI}{L_2} \theta_E + \frac{2EI}{L_2} \theta_C + k_2^2(1 - k_2)L_2\langle CE \rangle$$

$$M_{EF} = \frac{4EI}{L_3} \theta_E + \frac{2EI}{L_3} \theta_F + k_3(1 - k_3)^2 L_3\langle EF \rangle$$

$$M_{FE} = \frac{4EI}{L_3} \theta_F + \frac{2EI}{L_3} \theta_E + k_3^2(1 - k_3) L_3\langle EF \rangle$$

where kL_i is the distance from the left end of the span i to the unit load.

The fixed-end moment terms with MacCauley's notation are zero except when the unit load is on the corresponding span. Equilibrium requires

$$M_{AC} = 0$$

$$M_{CA} + M_{CE} = 0$$

$$M_{EC} + M_{EF} = 0$$

$$M_{FE} = 0$$

The four rotations are determined by substitution of the slope-deflection equations into the four equilibrium equations, which can be solved for the four rotations. This yields a system of four linear algebraic equations. The resulting rotations are back-substituted into the slope-deflection equations to obtain the end moments. Conceptually, this process is straightforward, but as a practical matter, the solution process involves significant algebraic and numerical effort. A computer-based equation solver was employed where all the required equations were entered and unknowns were automatically determined and back-substituted to achieve the end moments.

The end shears and the internal shears and moments are determined from equilibrium considerations of each element. A free body diagram of span AC is illustrated in Figure 5.11(d). This diagram is valid if the unit load is on AC. For other cases, the diagram is valid without the unit load. Summation of moments about C [Fig. 5.11(d)] yields

$$V_{AC} = 1(1 - k_1)\langle AC \rangle + \frac{M_{CA}}{L_1}$$

Summation of moments about B [Fig. 5.11(e)] yields

$$M_{BA} = V_{AC}(0.4L_1) - (1)(0.4L_1 - k_1L_1)\langle AB \rangle$$

By using Figures 5.11(f) and (g), the end shears for span CE and the shear and moments at D are determined in a similar manner.

$$V_{CE} = (1)(1 - k_2)\langle CE \rangle + \frac{M_{EC} + M_{CE}}{L_2}$$

$$V_{EC} = 1\langle CE \rangle - V_{CE}$$

$$V_{DC} = 1\langle CD \rangle - V_{CE}$$

$$M_{DC} = V_{CE}(0.5L_2) - (1)(0.5L_2 - k_2L_2)\langle CD \rangle - M_{CE}$$

Conventional slope-deflection notation has been used (CCW positive). The results map to the span point notation described in Table 5.1 as $M_{CA} = M_{110} = M_{200}$, $M_{BA} = M_{104}$, $M_{DC} = M_{205}$, $V_{AC} = V_{100}$, $V_{BA} = -V_{104}$, $V_{CA} = -V_{110}$, $V_{CE} = V_{200}$, and $V_{DC} = -V_{205}$. Note the designer's sign convention defined in Section 5.1 is used with all actions described with span point notation, and the slope-deflection convention is used for the calculation of the end moments. The equations given are a function of load position, x. To generate the influence functions, a solution is necessary for each position considered. Typically, the load is positioned at the tenth points. This is done for the present system. The results are given in Table 5.2. Each column constitutes an influence function for the associated action. These functions are illustrated in Figure 5.12(a) (moments) and 5.12(b) (shears).

5.5.1 Integration of Influence Functions

As discussed previously, the integral or area under the influence function is useful for the analysis of uniformly distributed loads. As illustrated in Figures 5.12(a) and 5.12(b) these functions are discontinuous for either value or slope. These functions could be integrated in a closed-formed manner but this is extremely tedious. An alternate approach is to numerically integrate the influence functions. A piecewise straight linear approximation to influence function may be used and integration of this approximation results in the well-known trapezoidal rule. The integral approximation is

$$\text{Area} = b \sum_{i=1}^{n} \left(\frac{\eta_1}{2} + \eta_2 + \cdots + \eta_{n-1} + \frac{\eta_n}{2} \right) \tag{5.14}$$

where b is the regular distance between the available ordinates. A linear influence function is integrated exactly by Eq. 5.14. Generally, influence functions are nonlinear and a more accurate approach is desired. Simpson's Rule uses a piecewise parabolic approximation that typically approximates nonlinear functions more accurately than its linear counterpart, Eq. 5.14. Simpson's Rule requires that domains have uniformly spaced ordinates and an odd number of ordinates with an even number of spaces between them. Simpson's Rule is

TABLE 5.2 Influence Ordinates and Areas

Location	Position	M(104)	M(200)	M(205)	V(100)	V(104)	V(110)	V(200)	V(205)
100	0	0.00	0.00	0.00	1.00	0.00	0.00	0.00	0.00
101	3.048	1.53	-0.74	-0.27	0.88	-0.12	-0.12	0.03	0.03
102	6.096	3.08	-1.44	-0.52	0.75	-0.25	-0.25	0.05	0.05
103	9.144	4.67	-2.04	-0.74	0.63	-0.37	-0.37	0.07	0.07
104	12.192	6.31	-2.51	-0.91	0.52	0.52	-0.48	0.09	0.09
105	15.240	4.97	-2.81	-1.02	0.41	0.41	-0.59	0.10	0.10
106	18.288	3.73	-2.87	-1.05	0.31	0.31	-0.69	0.10	0.10
107	21.336	2.59	-2.67	-0.97	0.21	0.21	-0.79	0.09	0.09
108	24.384	1.58	-2.16	-0.78	0.13	0.13	-0.87	0.08	0.08
109	27.432	0.71	-1.28	-0.47	0.06	0.06	-0.94	0.04	0.04
110 or 200	30.480	0.00	0.00	0.00	0.00	0.00	-1.00	1.00	0.00
201	34.138	-0.62	-1.55	0.77	-0.05	-0.05	-0.05	0.93	-0.07
202	37.795	-1.02	-2.54	1.78	-0.08	-0.08	-0.08	0.84	-0.16
203	41.453	-1.22	-3.04	3.02	-0.10	-0.10	-0.10	0.73	-0.27
204	45.110	-1.26	-3.15	4.49	-0.10	-0.10	-0.10	0.62	-0.38
205	48.768	-1.18	-2.94	6.20	-0.10	-0.10	-0.10	0.50	0.50
206	52.426	-1.00	-2.49	4.49	-0.08	-0.08	-0.08	0.38	0.38
207	56.083	-0.76	-1.89	3.02	-0.06	-0.06	-0.06	0.27	0.27
208	59.741	-0.49	-1.22	1.78	-0.04	-0.04	-0.04	0.16	0.16
209	63.398	-0.23	-0.56	0.77	-0.02	-0.02	-0.02	0.07	0.07

210 or 300	67.056	0.00	0.00	0.00	1.00	0.00	0.00	0.00	0.00
301	70.104	0.14	0.35	-0.47	0.01	0.01	0.01	-0.04	-0.04
302	73.152	0.24	0.59	-0.78	0.02	0.02	0.02	-0.08	-0.08
303	76.200	0.29	0.73	-0.97	0.02	0.02	0.02	-0.09	-0.09
304	79.248	0.31	0.78	-1.05	0.03	0.03	0.03	-0.10	-0.10
305	82.296	0.31	0.77	-1.02	0.03	0.03	0.03	-0.10	-0.10
306	85.344	0.27	0.69	-0.91	0.02	0.02	0.02	-0.09	-0.09
307	88.392	0.22	0.56	-0.74	0.02	0.02	0.02	-0.07	-0.07
308	91.440	0.16	0.39	-0.52	0.01	0.01	0.01	-0.05	-0.05
309	94.488	0.08	0.20	-0.27	0.01	0.01	0.01	-0.03	-0.03
310	97.536	0.00	0.00	0.00	0.00	0.00	0.00	0.00	0.00
Pos. Area Span 1		27.03	0.00	0.00	13.37	4.18	0.00	1.98	1.98
Neg. Area Span 1		0.00	-57.03	-20.74	0.00	-2.99	-17.11	0.00	0.00
Pos. Area Span 2		0.00	0.00	95.56	0.00	0.00	0.00	18.29	4.16
Neg. Area Span 2		-28.67	-71.67	0.00	-2.35	-2.35	-2.35	0.00	-4.16
Pos. Area Span 3		1.90	4.74	0.00	0.51	0.51	0.51	0.00	0.00
Neg. Area Span 3		0.00	0.00	-20.74	0.00	0.00	0.00	-1.98	-1.98
Total Pos. Area		28.92	4.74	95.56	13.88	4.69	0.51	20.27	6.14
Total Neg. Area		-28.67	-128.70	-41.47	-2.35	-5.34	-19.46	-1.98	-6.14
Net Area		0.26	-123.96	54.08	11.53	-0.65	-18.95	18.29	0.00

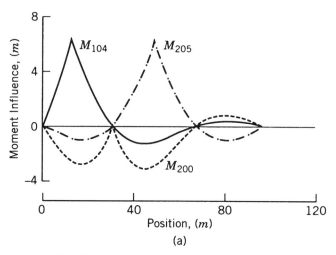

Fig. 5.12   (a) Moment influence functions.

$$\text{Area} = \frac{b}{3} \sum_{i=1}^{n} (\eta_1 + 4\eta_2 + 2\eta_3 + 4\eta_4 + \cdots + 2\eta_{n-2} + 4\eta_{n-1} + \eta_n) \quad (5.15)$$

Equation 5.15 was used to evaluate the positive, negative, and net areas for each function determined in Example 5.6. The results are given at the bottom of Table 5.2.

EXAMPLE 5.7

Use Simpson's rules to determine the positive, negative, and net areas of the influence function for M_{104} in Table 5.2.

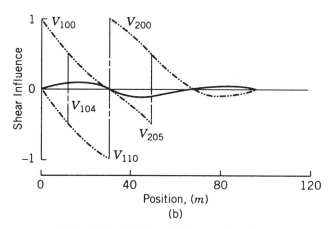

Fig. 5.12   (b) Shear influence functions.

$$A_{\text{Span1}} = [0 + 4(1.533) + 2(3.083) + 4(4.669) + 2(6.309) + 4(4.973)$$
$$+ 2(3.727) + 4(2.589) + 2(1.576) + 4(0.7074) + 0](3.048/3)$$
$$= 88.672$$
$$A_{\text{Span2}} = [0 + 4(-0.6209) + 2(-1.016) + 4(-1.218) + 2(-1.260)$$
$$+ 4(-1.1756) + 2(-0.9970) + 4(-0.7571) + 2(-0.4892)$$
$$+ 4(-0.2259) + 0](3.658/3) = -28.667$$
$$A_{\text{Span3}} = [0 + 4(0.1399) + 2(0.2350) + 4(0.2914) + 2(0.3136) + 4(0.3049)$$
$$+ 2(0.2743) + 4(0.2228) + 2(0.1567) + 4(0.0808) + 0](3.048/3)$$
$$= 6.221$$

These areas are added to give the positive, negative, and net areas.

$$A^+ = 88.672 + 6.221 = 94.893 \text{ m}^2 = 98.893 \times 10^6 \text{ mm}^2$$
$$A^- = -28.667 \text{ m}^2 = -28.667 \times 10^6 \text{ mm}^2$$
$$A^{\text{Net}} = 66.226 \text{ m}^2 = 66.226 \times 10^6 \text{ mm}^2$$

5.5.2   Relationship between Influence Functions

As illustrated in Example 5.6, the end actions, specifically end moments, are determined immediately after the displacements are established. This back-substitution process is characteristic of stiffness methods. The actions in the interior of the beam are based on static equilibrium considering the element loads and the end actions. Because the end actions and the associated influence functions are usually determined before other actions, it is convenient to establish relationships between the influence functions for the end actions and the functions for the actions in the interior portion of the span. The following discussion is rather detailed and requires techniques and associated notations that require careful study.

Consider the continuous beam shown in Figure 5.13(a) where the point of interest n is located at a distance βL from the end i. The actions at n may be determined by superposition of the actions at point n corresponding to a simple beam [Fig. 5.13(b)] and those corresponding to a simple beam with the end moment influence functions applied [Fig. 5.13(c)]. Note that the influence functions are actions that are applied on a free body diagram and treated in a manner similar to conventional actions. To illustrate, the influence functions are shown in Figure 5.13 (b-d) instead of their corresponding actions. With the use of superposition, the influence function for an action at n is determined by

$$\eta_n = \eta_s + \eta_e \qquad (5.16)$$

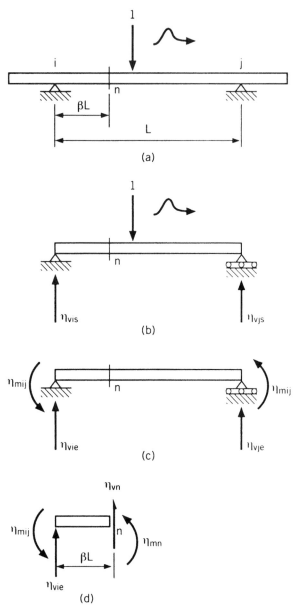

Fig. 5.13 (a) Continuous beam ij, (b) free body diagram simple beam ij with unit load, (c) free body diagram simple beam ij with end moments, and (d) free body diagram beam segment in βL.

where η_s is the influence function for the action at n for the unit action on the simple beam [Fig. 5.13(b)] and η_e is the influence function for the action at n due to the end actions on the simple beam [Fig. 13(c)].

By using the free body diagram shown in Figure 5.13(c), the shear influence function at i due to the end moments is determined from summation of moments about end j. The result is

$$\eta_{V_{ie}} = \frac{\eta_{M_{ij}} + \eta_{M_{ji}}}{L} \tag{5.17a}$$

where $\eta_{M_{ij}}$ and $\eta_{M_{ji}}$ are defined in Figure 5.13(c). Equation 5.16 is used to combine the shear influence function due to the unit action on the simple beam with the shear due the end moments. The result is

$$\eta_{V_n} = \eta_{V_s} + \frac{\eta_{M_{ij}} + \eta_{M_{ji}}}{L} \tag{5.17b}$$

By using Figure 5.13(d) and summing the moments about the point of interest n, the resulting influence function (end effects only) for moment at n is

$$\eta_{M_n} = \eta_{V_{ie}}\beta L - \eta_{M_{ij}} \tag{5.18a}$$

where βL is the distance from the left end of the span to the point of interest n.

Substitution of the left-end shear given in Eq. 5.17(a) into the moment expression given in Eq. 5.18(a) yields

$$\eta_{M_n} = (\beta - 1)\eta_{M_{ij}} + \beta\eta_{M_{ji}} \tag{5.18b}$$

Note the slope-deflection sign convention is used in the development of Eqs. 5.17 and 5.18. Any sign convention may be used as long as the actions are used consistently on the free body diagram, the sense of the action is correctly considered in the equilibrium equations, and results are properly interpreted with reference to the actions on the free body diagram.

EXAMPLE 5.8

Determine the influence ordinates for V_{104}, M_{104}, and M_{205} for the beam in Example 5.6 [Fig. 5.11(a)]. Use the influence functions for the end moments given in Table 5.2 and perform the calculations only for the ordinates at the 105 point.

Carefully note that the influence ordinates at 105 for actions at 104 and 205 are required. This means only *one ordinate* is established for the V_{104},

M_{104}, and M_{205} functions. The other ordinates may be determined in a similar manner.

The influence functions for the simple beam case are shown in Figure 5.14(b-c). The shear and moment ordinates at 105 are determined by linear interpolation. The sign convention used in Table 5.2 is the designer's sign convention and the slope-deflection sign convention is used in Eqs. 5.17 and 5.18. Therefore, the appropriate transformation must be performed. This calculation tends to be a bit confusing and requires careful study. To aid the reader, symbols have been added to reference the explanatory notes given below.

$$[\eta_{V_{104}}]_{@\,105} = \left[\eta_{V_{s_{104}}} + \frac{\eta_{M_{100}} + \eta_{M_{110}}}{L}\right]_{@\,105} = 0.6\left(\frac{5}{6}\right) + \left(\frac{0 + (-2.807)}{30.480}\right)$$

$$[\eta_{V_{104}}]_{@\,105} = 0.4079$$

$$[\eta_{M_{104}}]_{@\,105} = [\eta_{M_{s_{104}}} + (\beta - 1)\eta_{M_{100}} + \beta\eta_{M_{110}}]_{@\,105}$$

$$[\eta_{M_{104}}]_{@\,105} = \left[7.315\left(\frac{5}{6}\right) + (0.4^* - 1)(0) + 0.4(-2.807)\right]_{@\,105}$$

$$= 4.973 \text{ m} = 4973 \text{ mm}$$

$$[\eta_{M_{205}}]_{@\,105} = [\eta_{M_{s_{205}}} + (\beta - 1)\eta_{M_{200}} + \beta\eta_{M_{300}}]_{@\,105}$$

$$[\eta_{M_{205}}]_{@\,105} = 0\ddagger + (0.5\P - 1)(2.807) + (0.50)(0.7650\dagger)$$

$$[\eta_{M_{205}}]_{@\,105} = -1.020 \text{ m} = -1020 \text{ mm}$$

In summary, this method superimposes the effects of a unit load applied to the simple span with the effects of continuity (end moments). The unit load is applied only in the span containing the location of interest. Influence ordinates within this span are "affected" by the unit load and end effects. Function ordinates outside this span are affected only by the effects of continuity. Although a specific ordinate was used in this example, note that al-

---

*Here β is the fraction of the span length from the left end to the point of interest, that is, 104 is 40% from the left end. Do not confuse this with the location of where the ordinate is calculated, that is, 105.

†The sign on the M_{200} and M_{300} ordinates has been changed from the table to switch to the slope-deflection convention, for example, Table 5.2 gives −2.807 and +2.807 is used here.

‡The influence function is for the moment at 205 and the ordinate is being calculated for the ordinate of this function at 105. The simple beam function is superimposed only if the location where the ordinate calculation is being performed (105) is in the same span as the location of the point of interest (205). In this case, the two locations are in different spans.

¶Here β is the fraction of the span length from the left end to the point of interest. In this case, 205 is located at 50% of the second span.

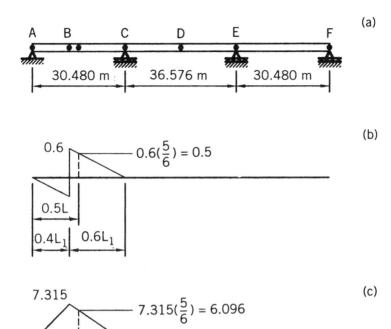

Fig. 5.14 (a) Continuous beam, (b) simple beam *AC* influence function for shear at 104, and (c) simple beam *AC* influence function for moment at 104.

gebraic functions may be used in a similar fashion, and perhaps what is more important, general algorithms may be developed using Eqs. 5.16, 5.17, and 5.18 and subsequently coded in computer programs. In addition, this calculation is quite amenable to spreadsheet calculation.

5.5.3 Muller–Breslau Principle for End Moments

The Muller–Breslau Principle may be conveniently used to establish the influence functions for the end moments. Subsequently, the end moments may be used with Eqs. 5.16 and 5.17 to establish all other influence functions.

The Muller–Breslau Principle requires that the displacements (in this case translation) be determined for the entire structure. The displacement of each element is solely a function of the end moments. The equation for the translation of a simple beam subjected to counterclockwise end moments M_{ij} and M_{ji} is

$$y = \frac{L^2}{6EI} [M_{ij}(2\varepsilon - 3\varepsilon^2 + \varepsilon^3) - M_{ji}(\varepsilon - \varepsilon^3)] \tag{5.19}$$

where $\varepsilon = x/L$, and y is the upward translation.

This equation can be derived many different ways, for example, direct integration of the governing equation. Verification is left to the reader. The Muller–Breslau procedure is described in Example 5.9.

EXAMPLE 5.9

Use the Muller–Breslau Principle to establish the influence function for the moment M_{104} for the beam of Example 5.6. Perform the calculations for the first span only.

The structure is reillustrated in Figure 5.15(a) for convenience. The influence function for the simple beam moment at the 104 point is illustrated in Figure 5.15(b). This function has been discussed previously and is not reiterated. Next determine the influence function for the end moments. Use the

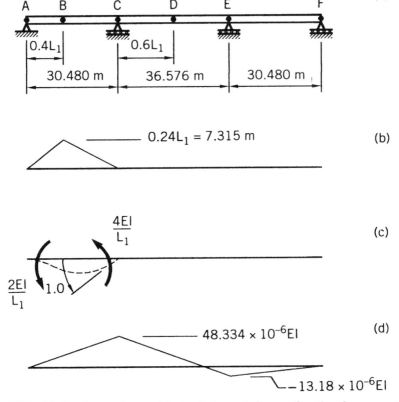

Fig. 5.15   (a) Continuous beam, (b) simple beam influence function for moment at 104, (c) unit rotation at C member CA, and (d) moment due to unit rotation at C.

Muller–Breslau Principle to release the moment at 110 and impose a unit displacement. The displaced shape is shown in Figure 5.15(c). The end moments for element AB are determined using the slope-deflection equations given in Eq. 5.11, Let $\theta_j = 1.0$, $\theta_i = 0$, and the fixed-end moments due to element loads are zero. The end moments are $2EI/L_1$ and $4EI/L_1$ for the left and right ends, respectively.

These moments are the fixed-end moments used in the slope-deflection equations, that is,

$$M_{AC} = \frac{4EI}{L_1}\theta_A + \frac{2EI}{L_1}\theta_C + \frac{2EI}{L_1}$$

$$M_{CA} = \frac{4EI}{L_1}\theta_C + \frac{2EI}{L_1}\theta_A + \frac{4EI}{L_1}$$

$$M_{CE} = \frac{4EI}{L_2}\theta_C + \frac{2EI}{L_2}\theta_E$$

$$M_{EC} = \frac{4EI}{L_2}\theta_E + \frac{2EI}{L_2}\theta_C$$

$$M_{EF} = \frac{4EI}{L_3}\theta_E + \frac{2EI}{L_3}\theta_F$$

$$M_{FE} = \frac{4EI}{L_3}\theta_F + \frac{2EI}{L_3}\theta_E$$

Equilibrium requires

$$M_{AC} = 0$$

$$M_{CA} + M_{CE} = 0$$

$$M_{EC} + M_{EF} = 0$$

$$M_{FE} = 0$$

The slope-deflection equations are substituted into the equilibrium equations and the four rotations are established. The resulting rotations are $\theta_A = -0.2455$, $\theta_C = -0.5089$, $\theta_E = 0.1339$, and $\theta_F = 0.0670$. These rotations are back-substituted into the slope-deflection equations to establish the end moments given below:

$$M_{AC} = 0 \text{ ft kips}$$

$$M_{CA} = 48.334 \times 10^{-6} \, EI$$

$$M_{CE} = -48.334 \times 10^{-6} \, EI$$

$$M_{EC} = -13.18 \times 10^{-6} \, EI$$

$$M_{EF} = 13.18 \times 10^{-6} \, EI$$

$$M_{FE} = 0$$

The moment diagram is shown in Figure 5.15(d).

Equation 5.19 is used to determine the translation due the end moments for the first span. This equation is the influence function $\eta_{M_{110}}$ and is given in Table 5.3. A sample calculation is given for the ordinate at location 103, $\varepsilon = x/L = 0.3$.

$$\eta_{M_{110}} = \frac{30\,480^2}{6EI} \{0.0EI[2(0.3) - 3(0.3^2) + 0.3^3]$$

$$- 48.334 \times 10^{-6} \, EI[0.3 - 0.3^3]\}$$

$$\eta_{M_{110}} = -2.042 \text{ m} = -2042 \text{ mm}$$

A comparison of the values in Tables 5.2 and 5.3 reveals that the influence

TABLE 5.3   Influence Ordinates for M_{104} for Span 1[a]

x (ft)	$\varepsilon = \dfrac{x}{L_1}$	η_{Ms104} (Unit Load on Simple Beam) (mm)	η_{M110} (Influence Function for M_{110}) (mm) (Eq. 5.19)	η_{M104} (Influence Function for Moment at 104) (mm)
0	0	0	0	0
10	0.1	1828	-740.7	1533
20	0.2	3658	-1436	3082
30	0.3	5486	-2042	4669
40	0.4	7315	-2515	6309
50	0.5	6096	-2807	4974
60	0.6	4877	-2874	3728
70	0.7	3658	-2678	2588
80	0.8	2438	-2155	1576
90	0.9	1219	-1280	787
100	1.0	0	0	0

[a]The parameter L_1 = 30,480 mm = 100 ft. The procedure is the same for the remaining span but the simple beam contribution is zero, which is left to the reader as an exercise.

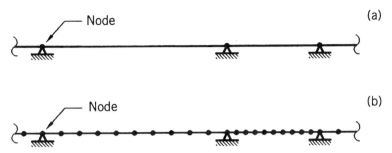

Fig. 5.16 (a) Discretized continuous beam—nodes at physical joints and (b) discretized continuous beam—nodes between physical joints.

ordinate $\eta_{M_{104}}$ is the same. Use Eq. 5.18 to combine the influence function $\eta_{M_{110}}$ and the simple beam function $\eta_{S_{104}}$. The result is shown in Table 3.3 and a sample calculation for the ordinate at the 103 point is given.

$$[\eta_{M_{104}}]_{@103} = 5486 + (0.4 - 1)(0) + (0.4)(-2042) = 4669 \text{ mm}$$

5.5.4 Automation by Matrix Structural Analysis

Traversing the unit action and the Muller–Breslau methods may be used in a stiffness-based matrix analysis. This unit action approach is conceptually straightforward and likely the easiest to implement in an existing stiffness-based code. Two approaches may be taken: (1) The structure is discretized with one element per span [Fig. 5.16(a)] or (2) The structure is discretized with several elements per span, often conveniently taken as 10 per span [Fig. 5.16(b)].

The use of one element per span requires special algorithms to

- Generate the equivalent joint loads for load placement at any position within an element.
- Determine the end actions after the displacements are known. This procedure involves adding the fixed-end actions to the actions from the analysis of the released structure.
- Calculate the actions and displacements at the required locations in the interior of the element.

With these tools available, one can use the standard matrix approach to place the unit actions at regular intervals along the load path and to calculate the actions at the required locations. This involves the solution of multiple load cases on a very small system of equations. The advantage is its computational efficiency. The disadvantages are its coding complexity and difficulties in including the nonprismatic affects that effect both stiffness and fixed-end action computations.

Alternatively, each span may be discretized into elements as illustrated in Figure 5.16(b). The node can be associated with the influence ordinates, eliminating the need for element load routines. The unit actions can be applied as joint loads and each load case generates one ordinate in the influence functions. The element end actions are available at regular locations. Although this is computationally more expensive than one element per span, it is simpler to code and the number of degrees of freedom required is relatively small by today's standards. Another advantage of this method is that the element cross-sectional properties may vary from element to element to account for the nonprismatic nature of the bridge. In addition, the displacements (e.g., influence function for translation at a point) are always available at every degree of freedom. The disadvantage is computational inefficiency, but this is minor.

The Muller–Breslau method may be used with either discretation scheme. The advantage of the Muller–Breslau approach is that only one load case is required for each influence function. Because the displacements establish the function, the back-substitution process is eliminated. This process is a minor computational advantage for standard beam and frame elements. The major disadvantage is that with an automated approach, one expects to develop complete action envelopes for locations along the beam for design, usually tenth points. Therefore, one load case is required for each action considered, likely one action for each degree of freedom, which is several times (beam = 2, frame = 3, etc.) the number of load cases for the unit action traverse. The computational saving is that the displacements are the influence ordinates and action recovery is not required. If one element per span is used, then algorithms must be developed to determine the influence ordinates (displacements) in the interior of the element (e.g., Eqs. 5.16–5.19). This increases code complexity, especially if nonprismatic beams are required.

In summary, designing an automated approach for the generation of influence function depends on the objectives and scope of the program. Typically, the combination of using multiple elements per span and applying a unit action as a joint load results in a code that is flexible, easy to maintain, and has the capability to generate influence functions for every end action in the system. Further, nonprismatic effects are naturally handled by changing the element properties. The computational efficacy for linear problems of this size becomes less important with ever increasing computational capability.

5.6   NORMALIZED INFLUENCE FUNCTIONS

Influence functions may be considered a type of structural property as they are independent of the load and dependent on the *relative* stiffness of each element. Consider the Muller–Breslau Principle, the displaced shape due to an imposed displacement is dependent on the *relative,* not *absolute,* values of stiffness. For a continuous prismatic beam, the cross section and material stiffness do not vary with location, therefore the influence functions are based on the only remaining parameter that affects stiffness, the span lengths. Note that in Figure 5.4, the influence functions for reaction and shear are independent of the span length, and the influence function for moment is proportional to span length. These relationships are similar for continuous beams, but here the shape is determined by the *relative* stiffness (in the case of a prismatic beam, the relative span lengths) and the ordinate values for moment are proportional to a characteristic span length.

For detailing and aesthetic reasons, bridges are often designed to be symmetrical about the center of the bridge, for example, the first and third spans of a three-span bridge are equal.

For economy and ease of detailing and construction, the engineer sets the span length to meet the geometric constraints and if possible, to have similar controlling actions in all spans. In a continuous structure, this is accommodated by making the outer spans shorter than the interior spans. Typical span ratios vary from 1.0 to 1.7. The spans for Example 5.6 are 30 480, 36 576, and 30 480 mm (100, 120, and 100 ft) that have a span ratio of 1.2. The shear influence functions from Example 5.6 can be used for a prismatic three-span continuous girder bridge with the same span ratio (i.e., 1.2). Similarly, the moment influence functions can be proportioned. For example, a bridge with spans of 10 668, 12 802, and 10 668 mm (35, 42, and 35 ft) has a span ratio of $42:35 = 1.2$. With the use of the first span as the characteristic span, the ordinates can be proportioned by $35:100 = 0.35$. For example, the smaller bridge has a maximum ordinate for the moment at 104 of

$$\eta_{M_{104}} = (0.35)6\,309 = 2208 \text{ mm}$$

The American Institute for Steel Construction (AISC) has published tables of normalized influence functions for various span configurations and span ratios (AISC, 1986). These were generated for a characteristic span length of 1.0. This format allows the engineer to use the tabulated values by multiplying by the actual characteristic span length. Table 5.2 (span ratio = 1.2) is normalized to a unit length for span one and the results are given in Table 5.4. This table is used in several examples that follow.

5.7 AASHTO VEHICLE LOADS

The AASHTO vehicle loads defined in Chapter 4 are used to determine the load effects for design. Because the vehicle loads are moving loads, load placement for maximum load effect may not be obvious. The influence function for a particular action is used in combination with the prescribed load to establish the load position for analysis. The engineer may place the load at

TABLE 5.4 Normalized Influence Functions (Three Span, Span Ratio = 1.2)[a]

Location	M(104)	M(200)	M(205)	V(100)	V(104)	V(110)	V(200)	V(205)
100	0.00000	0.00000	0.00000	1.00000	0.00000	0.00000	0.00000	0.00000
101	0.05028	-0.02431	-0.00884	0.87569	-0.12431	-0.12431	0.02578	0.02578
102	0.10114	-0.04714	-0.01714	0.75286	-0.24714	-0.24714	0.05000	0.05000
103	0.15319	-0.06703	-0.02437	0.63297	-0.36703	-0.36703	0.07109	0.07109
104	0.20700	-0.08250	-0.03000	0.51750	0.51750	-0.48250	0.08750	0.08750
105	0.16317	-0.09208	-0.03348	0.40792	0.40792	-0.59208	0.09766	0.09766
106	0.12229	-0.09429	-0.03429	0.30571	0.30571	-0.69429	0.10000	0.10000
107	0.08494	-0.08766	-0.03187	0.21234	0.21234	-0.78766	0.09297	0.09297
108	0.05171	-0.07071	-0.02571	0.12929	0.12929	-0.87071	0.07500	0.07500
109	0.02321	-0.04199	-0.01527	0.05801	0.05801	-0.94199	0.04453	0.04453
110 or 200	0.00000	0.00000	0.00000	1.00000	0.00000	-1.00000	-1.00000	0.00000
201	-0.02037	-0.05091	0.02529	-0.05091	-0.05091	-0.05091	0.92700	-0.07300
202	-0.03333	-0.08331	0.05829	-0.08331	-0.08331	-0.08331	0.83600	-0.16400
203	-0.03996	-0.09990	0.09900	-0.09990	-0.09990	-0.09990	0.73150	-0.26850
204	-0.04135	-0.10337	0.14743	-0.10337	-0.10337	-0.10337	0.61800	-0.38200
205	-0.03857	-0.09643	0.20357	-0.09643	-0.09643	-0.09643	0.50000	0.50000
206	-0.03271	-0.08177	0.14743	-0.08177	-0.08177	-0.08177	0.38200	0.38200
207	-0.02484	-0.06210	0.09900	-0.06210	-0.06210	-0.06210	0.26850	0.26850
208	-0.01605	-0.04011	0.05829	-0.04011	-0.04011	-0.04011	0.16400	0.16400
209	-0.00741	-0.01851	0.02529	-0.01851	-0.01851	-0.01851	0.07300	0.07300

210 or 300	0.00000	0.00000	0.00000	1.00000	0.00000	0.00000	0.00000	0.00000
301	0.00458	0.01145	-0.01527	0.01145	0.01145	0.01145	-0.04453	-0.04453
302	0.00771	0.01929	-0.02571	0.01929	0.01929	0.01929	-0.07500	-0.07500
303	0.00956	0.02391	-0.03188	0.02391	0.02391	0.02391	-0.09297	-0.09297
304	0.01029	0.02571	-0.03429	0.02571	0.02571	0.02571	-0.10000	-0.10000
305	0.01004	0.02511	-0.03348	0.02511	0.02511	0.02511	-0.09766	-0.09766
306	0.00900	0.02250	-0.03000	0.02250	0.02250	0.02250	-0.08750	-0.08750
307	0.00731	0.01828	-0.02437	0.01828	0.01828	0.01828	-0.07109	-0.07109
308	0.00514	0.01286	-0.01714	0.01286	0.01286	0.01286	-0.05000	-0.05000
309	0.00265	0.00663	-0.00884	0.00663	0.00663	0.00663	-0.02578	-0.02578
310	0.00000	0.00000	0.00000	1.00000	0.00000	0.00000	0.00000	0.00000
Pos. Area span 1	0.09545	0.00000	0.00000	0.43862	0.13720	0.0000	0.06510	0.06510
Neg. Area Span 1	0.00000	-0.06138	-0.02232	0.00000	-0.09797	-0.56138	0.00000	0.00000
Pos. Area Span 2	0.00000	0.00000	0.10286	0.00000	0.00000	0.00000	0.60000	0.13650
Neg. Area Span 2	-0.03086	-0.07714	0.00000	-0.07714	-0.07714	-0.07714	0.00000	-0.13650
Pos. Area Span 3	0.00670	0.01674	0.00000	0.01674	0.01674	0.01674	0.00000	0.00000
Neg. Area Span 3	0.00000	0.00000	-0.02232	0.00000	0.00000	0.00000	-0.06510	-0.06510
Total Pos. Area	0.10214	0.01674	0.10286	0.45536	0.15394	0.01674	0.66510	0.20160
Total Neg. Area	-0.03086	-0.13853	-0.04464	-0.07714	-0.17512	-0.63853	-0.06510	-0.20160
Net Area	0.07129	-0.12179	0.05821	0.37821	-0.02117	-0.62179	0.60000	0.00000

[a]Usage:

Multiply Influence Ordinates for Moment by Length of Span 1

Multiply Areas for Moment by Length of (Span 1)^2

Multiply Areas for Shear by Length of Span 1

Notes:

Area M(205)+ for Span 2 is 0.1036, 0.1052, and 0.1029 for trapezoidal, Simpson's and exact integration, respectively.

Areas V(205)+ and V(205)− for Span 2 were computed by Simpson's integration.

one or more positions by inspection and calculate the load effect for each load placement using Eq. 5.1. The maximum and minimum values are noted and used in subsequent design calculations.

Alternatively, the load is periodically positioned along the same path used to generate the influence function. For each placement, the load effect is calculated and compared to the previous one. The maximum and minimum load effects are recorded. This approach is most appropriate for automation and is the technique most often employed in computer programs that generate load effect envelopes.

The critical load placement is sometimes obvious when the influence function is available. As illustrated in Example 5.10, this is the case for the analysis of statically determinate beams.

Critical load placement on an influence function gives the maximum or minimum load effect for the particular action at the location associated with that function. Unfortunately, this location is likely not the location that gives the critical load effect in the span. For example, typical influence functions are generated at tenth points, but the critical location may be between the tenth point locations.

This critical location can be theoretically established for simple beams, and this formulation can be found in most elementary texts on structural analysis (e.g., McCormac and Elling, 1988; Laursen, 1988). We have chosen not to focus a great deal of attention on this aspect. From a practical perspective, the method only works for simple-span bridges. Automated approaches are written in a general way to accommodate both statically determinate and indeterminate systems with the same algorithms. Lastly, the absolute maximum or minimum load effect does not differ significantly from the tenth-point approximation. The two methods are compared in the following example.

EXAMPLE 5.10

Use the influence functions determined in Example 5.1 to calculate the maximum reaction R_{100}, shear V_{100}, and moment M_{105} for the AASHTO vehicle loads (AASHTO, 1994). Use a 10 668 mm (35 ft) span.

The influence lines for the actions required are shown in Figure 5.17(a–e). The critical actions for the design truck, design tandem, and the design lane loads are determined independently and are later superimposed as necessary. The design truck is used first, followed by the design tandem and finally the design lane load.

Design Truck Load. The critical load placement for R_{100} is shown in Figure 5.17(f). By using Eq. 5.1, this reaction is determined as

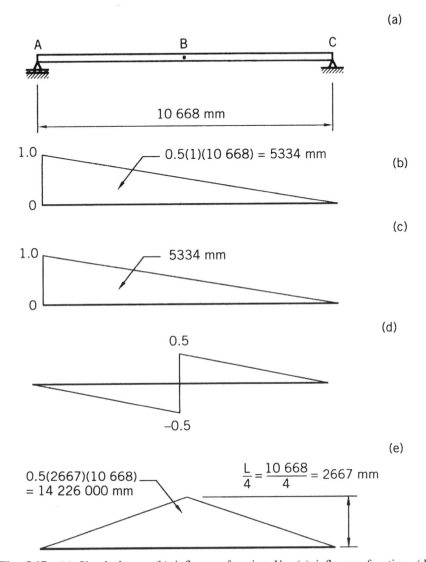

Fig. 5.17 (a) Simple beam, (b) influence function V_A, (c) influence function, (d) influence function V_A, and (e) Influence function M_A.

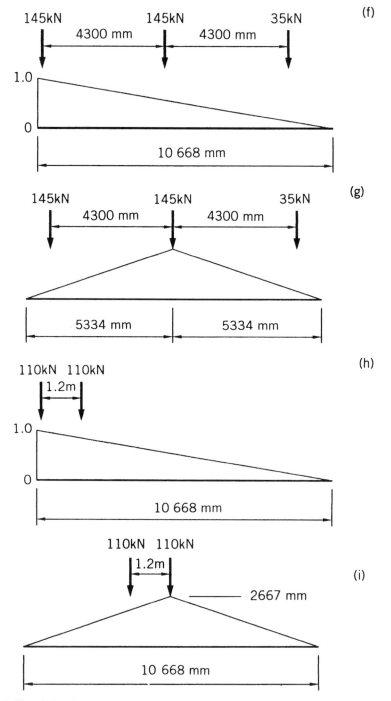

Fig. 5.17   (f) Design truck positioned for V_A, (g) design truck positioned for M_B, (h) tandem truck positioned for V_A, and (i) tandem truck positioned for M_B.

$$R_{100} = \sum_{i=1}^{3} P_i \eta_i = 145(1) + 145(6368/10\,668) + 35(2068/10\,668)$$

$$= 145.0 + 86.55 + 6.78 = 238.3 \text{ kN}$$

Note that $R_{100} = V_{100} = 238.3$ kN.

The critical load placement for M_{105} is illustrated in Figure 5.17(g). Multiplication of the loads times the ordinates gives

$$M_{105} = 2667[(145)(1) + 145(1034/5334) + 35(1034/5334)]$$

$$= 479.8 \times 10^3 \text{ mm kN} = 479.8 \text{ m kN}$$

Increasing the distance between the rear axles spreads the load and decreases the load effect. Thus, the 4300-mm (14 ft) variable axle spacing is critical and this will be the case for simple spans. The variable axle spacing can become critical for short multispan beams where the truck length is approximately the same as the span lengths.

Design Tandem Load. To determine R_{100}, the design tandem loads are placed as illustrated in Figure 5.17(h). The reaction is

$$R_{100} = 110(1) + 110(9468/10\,668) = 207.6 \text{ kN}$$

Again note, $V_{100} = R_{100} = 207.6$ kN.

The maximum moment at midspan is determined by placing the design tandem as shown in Figure 5.17(i). The result is

$$M_{105} = 2667[110(1) + 110(4134/5334)]$$

$$= 520.7 \times 10^3 \text{ mm kN} = 520.7 \text{ m kN}$$

Design Lane Load. Equation 5.3 is used to determine the shears and moments for the uniform lane load of 9.3 N/mm. This uniform load is multiplied by the appropriate area under the influence function. For example, the integral of the influence function for R_{100} is the area of a triangle or

$$\text{Area} = (1)(10\,668)/2 = 5334 \text{ mm}$$

Thus, the reaction R_{100} is calculated as

$$R_{100} = (9.3 \text{ N/mm})(5334 \text{ mm}) = 49.6 \text{ kN}$$

As before, $V_{100} = R_{100} = 49.6$ kN.

By using Figure 5.17(e), the moment at midspan is

$$M_{105} = (9.3\ \text{N/mm})(2667\ \text{mm})(10\ 668\ \text{mm})/2$$

$$= 132.3 \times 10^6\ \text{mm N} = 132.3\ \text{m kN}$$

The *absolute* maximum reaction and shear are as shown above, but the *absolute* maximum moments are slightly different. The actions for *simple beams* may be established with the following rules (AISC, 1986);

1. The maximum shear due to moving concentrated loads occurs at one support when one of the loads is at the support. With several moving loads, the location that will produce maximum shear must be determined by trial.
2. The maximum bending moment produced by moving concentrated loads occurs under one of the loads when that load is as far from one support as the center of gravity of all the moving loads on the beam is from the other support.

Position the design truck with the 145-kN kip wheel at 4259 mm. This position results in a maximum moment under the 145-kN wheel of 493.1×10^3 mm kN, which is slightly greater than the value at the 105 point (479.8×10^3 mm kN). Position the design tandem wheel at 4434 mm from the left, resulting in a maximum moment of 522.5×10^3 mm kN $= 522.5$ m kN under the wheel. The differences between these moments and the moments at the 105 point are approximately 2.7 and 0.4% for the design truck and tandem trucks, respectively. The absolute maximum moments are also given in Table 5.5. Note that the design lane load must be added to the design truck and to the design tandem loads [A3.6.1.3.1]. The maximums for these load cases occur at different locations, that is, the uniform lane load is at a maximum at midspan. This further complicates the analysis. A rigorous approach must determine the absolute maximum for the combined factored loads, which is only reasonable for simple spans. Adding the maximum design lane moments occurring at midspan to the absolute moment determined above gives

TABLE 5.5   Service Level Vehicle Design Loads[a,b,c]

Load	$R_{100} = V_{100}$ (kN)	M_{105} (m kN)	Absolute (M, m kN)
Design truck	238.3	479.8	493.1
Design tandem	207.6	520.7	522.5
Design lane	49.6	132.3	132.3
Truck + lane	**287.9**	612.1	625.4
Tandem + lane	257.2	**653.0**	**654.8**

[a] Simple span = 10 668 mm (35 ft).
[b] The critical values are in boldface.
[c] The impact factors are different for the truck and lane loads. These factors should be included for design.

a conservative estimate of the maximum. This calculation gives $493.1 + 132.3 = 625.4$ m kN for the design truck and lane loads, and $522.5 + 132.3 = 654.8$ m kN for the design tandem and lane loads. The differences between these moments and the moments at the 105 point are 2.1 and 0.3% for the truck and tandem plus lane loadings, respectively. Note that when the load effects are factored and added to the dead load effects, the percentage differences further decrease.

These calculations are summarized in Table 5.5.

EXAMPLE 5.11

Repeat Example 5.10 for a 30 480 mm (100 ft) span. The calculations are given below.

Design Truck Load.

$$R_{100} = 145(1) + 145(26\ 180/30\ 480) + 35(21\ 880/30\ 480) = 294.7\ \text{kN}$$

$$V_{100} = R_{100}$$

$$M_{105} = [30\ 480/4][145(1) + 145(10\ 940/15\ 240) + 35(10\ 940/15\ 240)]$$

$$= 2090 \times 10^3\ \text{mm kN} = 2090\ \text{m kN}$$

Design Tandem Load.

$$R_{100} = 110(1) + 110(29\ 280/30\ 480) = 215.7\ \text{kN}$$

$$V_{100} = R_{100}$$

$$M_{105} = [30\ 480/4][110(1) + 110(14\ 040/15\ 240)]$$

$$= 1610 \times 10^3\ \text{mm kN} = 1610\ \text{m kN}$$

Design Lane Load.

$$R_{100} = 9.3(\tfrac{1}{2})(1)(30\ 480) = 141.7 \times 10^3\ \text{N} = 141.7\ \text{kN}$$

$$V_{100} = R_{100}$$

$$M_{105} = 9.3(\tfrac{1}{2})(30\ 480/4)(30\ 480) = 1080 \times 10^6\ \text{mm N} = 1080\ \text{m kN}$$

The actions for the truck and tandem loads are combined with the lane load in Table 5.6.

The procedure for determining the actions in a continuous beam is similar to that illustrated for a simple beam. As illustrated previously, the influence diagrams are slightly more complicated as the functions are nonlinear with

TABLE 5.6   Service Level Design Loads[a]

Load	$R_{100} = V_{100}$ (kN)	M_{105} (m kN)
Design truck	294.7	2090
Design tandem	215.7	1610
Design lane	141.7	1080
Truck + lane	**436.4**	**3170**
Tandem + lane	357.4	2690

[a] Span = 30 480 mm (100 ft).

both positive and negative ordinates. To illustrate the calculation of the load effects for a continuous system, the three-span continuous beam of Example 5.6 is used. A few actions are used for illustration and the remaining actions required for design follow similar procedures.

EXAMPLE 5.12

Determine the shear V_{100}, the moment M_{104}, and the moment $M_{110} = M_{200}$ for the beam of Example 5.6 (Fig. 5.11). Use the normalized functions given in Table 5.4. The span lengths are 30 480, 36 576, and 30 480 mm (100, 120, and 100 ft). Use the AASHTO vehicle loads.

Design Lane Load.   Use the normalized areas at the bottom of Table 5.4 for the lane loads. Note that these areas require multiplication by the characteristic span length for shear and by the span length squared for moment. The positive and negative areas are used for the associated actions.

$$V_{100^-} = 9.3(-0.07714)(30\ 480) = -21\ 866\ N = -21.9\ kN$$

$$V_{100^+} = 9.3(0.455\ 36)(30\ 480) = 129\ kN$$

$$M_{104^+} = 9.3(0.102\ 14)(30\ 480^2) = 882.5\ m\ kN$$

$$M_{104^-} = 9.3(-0.030\ 86)(30\ 480^2) = -266.6\ m\ kN$$

$$M_{100^-} = 9.3(-0.138\ 53)(30\ 480^2) = -1197\ m\ kN$$

$$M_{110^+} = 9.3(0.016\ 74)(30\ 480^2) = 144.6\ m\ kN$$

Design Tandem Load.   The tandem axle is applied to the structure and the load effects are calculated with Eq. 5.1. The load placement is by inspection and noted below for each action.

For V_{100}, place the left axle at 100 and the second axle at 1200 mm (4 ft) from the left end. The influence ordinate associated with the second axle is determined by linear interpolation.

$$V_{100^+} = 110(1) + 110[1 - (1200/3048)(1 - 0.875\ 69)]$$

$$= 110 + 104.6 = 214.6\text{ kN}$$

For the most negative reaction at 100, position the right axle at 204.

$$V_{100^-} = 110(-0.103\ 37) + 110[-0.103\ 37 + (1200/3658)(0.103\ 37$$

$$- 0.099\ 90)]$$

$$V_{100^-} = -11.37 - 11.25 = -22.62\text{ kN}$$

For the positive moment at 104, position the left axle at 104. Again, determine the ordinate for the second axle by interpolation.

$$M_{104^+} = 110(0.207\ 00)(30\ 480) + 110[0.207\ 00 - (1200/3048)(0.207\ 00$$

$$- 0.163\ 17)](30\ 480)$$

$$M_{104^+} = 694.0 + 636.1 = 1330\text{ m kN}$$

Position the right axle at 204 for the most negative moment at 104. The result is

$$M_{104^-} = 110(-0.041\ 35)(30\ 480) + 110[-0.041\ 35$$

$$- (1200/3658)(-0.041\ 35 + 0.039\ 96)](30\ 480)$$

$$M_{104^-} = -138.6 - 140.2 = -278.8\text{ m kN}$$

Position the right axle at 204 for the most negative moment at 110.

$$M_{110^-} = 110(-0.103\ 37)(30\ 480) + 110[-0.103\ 37$$

$$- (1200/3658)(-0.103\ 37 + 0.099\ 90)](30\ 480)$$

$$M_{110^-} = -346.6 - 350.4 = -697.0\text{ m kN}$$

Design Truck Load. Position the rear axle at 100 for the maximum reaction:

$$R_{100^+} = 145(1) + 145(0.8266) + 35(0.6569) = 145.00 + 119.9 + 23.0$$

$$= 287.9\text{ kN}$$

Position the middle axle at 104 for the positive moment at the 104:

$M_{104^+} = 35[0.153\ 19 - (1252/3048)(0.153\ 19 - 0.101\ 44)](30\ 480)$

$\qquad + 145(0.207\ 00)(30\ 480) + 145[0.163\ 17 - (1252/3048)(0.163\ 17$

$\qquad - 0.122\ 29)](30\ 480)$

$\qquad\qquad M_{104^+} = 140.7 + 914.9 + 646.9 = 1702.5\ \text{m kN}$

Position the middle axle at 204 for the most negative moment at 104:

$M_{104^-} = 35[-0.038\ 57 - (642/3658)(-0.038\ 57 + 0.032\ 71)](30\ 480)$

$\qquad + 145(-0.041\ 35)(30\ 480) + 145[-0.039\ 96$

$\qquad - (642/3658)(-0.039\ 96 + 0.033\ 33)](30\ 480)$

$\qquad\qquad M_{104^-} = -40.0 - 182.8 - 171.5 = -394.3\ \text{m kN}$

Position the middle axle at 204 for the most negative moment at 110:

$M_{110^-} = 145[-0.099\ 9 - (642/3658)(-0.099\ 9 + 0.083\ 31)](30\ 480)$

$\qquad + 145(-0.103\ 37)(30\ 480) + 35[-0.096\ 43$

$\qquad - (642/3658)(-0.096\ 43 + 0.081\ 77)](30\ 480)$

$\qquad\qquad M_{110^-} = -428.6 - 456.9 - 100.1 = -985.6\ \text{m kN}$

Position the middle axle at 304 for the maximum positive moment at 110:

$M_{110^+} = 35[0.023\ 9 - (1252/3048)(0.023\ 9 - 0.019\ 29)](30\ 480)$

$\qquad + 145(0.025\ 71)(30\ 480) + 145[0.025\ 11 - (1252/3048)(0.025\ 11$

$\qquad - 0.022\ 50)](30\ 480)$

$\qquad\qquad M_{110^+} = 23.48 + 113.6 + 106.2 = 243.3\ \text{m kN}$

In the previous example, actions at selected points were determined. This procedure is generally permitted as long as the points and actions selected are representative of the extreme values (action envelope). These points are summarized in Table 5.1. Alternatively, all the extreme actions are determined at enough sections so that the envelope is represented. This is an extremely tedious process if performed by hand. Typically, all actions are determined at the tenth points. Therefore, this approach is most often automated.

A computer program called BT Beam—LRFD Analysis (BridgeTech, Inc. 1996) was used to develop the envelope of all actions at the tenth points. The automated procedure performs the calculations as presented in this chapter except that it uses a matrix formulation rather than a slope-deflection analysis. For a beam analysis, these analyses are identical. The results from this analysis are given in Table 5.7. A comparison of the values in this table with

TABLE 5.7 Action Envelope for Three-Span Continuous Beam 30 480, 36 576, and 30 480 mm (100, 120, and 100 ft)

	Positive Moment (m kN)							Negative Moment (m kN)						
Location	Truck	Tandem	Lane	Train	Truck Lane	Tandem + Lane	Critical M+	Truck	Tandem	Lane	Train	Truck + Lane	Tandem + Lane	Critical M−
100	0	0	0	N/A	0	0	0	0	0	0	N/A	0	0	0
101	755	570	351	N/A	1355	1109	1355	−99	−69	−66	N/A	−198	−158	−198
102	1274	977	615	N/A	2309	1915	2309	−198	−138	−132	N/A	−395	−315	−395
103	1572	1227	793	N/A	2884	2425	2884	−297	−207	−198	N/A	−593	−473	−593
104	1702	1330	884	N/A	3148	2653	3148	−396	−276	−264	N/A	−791	−631	−791
105	1669	1308	889	N/A	3109	2629	3109	−495	−345	−330	N/A	−989	−788	−989
106	1502	1179	808	N/A	2806	2376	2806	−594	−414	−396	N/A	−1186	−946	−1186
107	1184	951	640	N/A	2214	1905	2214	−693	−483	−462	N/A	−1384	−1104	−1384
108	752	649	386	N/A	1385	1249	1385	−792	−552	−528	N/A	−1582	−1262	−1582
109	257	303	184	N/A	526	587	587	−891	−621	−733	N/A	−1919	−1558	−1919
110	244	172	143	N/critical	467	371	467	−990	−689	−1185	−1857	−2502	−2102	−3290
200	243	172	143	N/critical	467	371	467	−990	−689	−1185	−1857	−2502	−2102	−3290
201	319	357	150	N/A	575	625	625	−779	−549	−632	N/A	−1668	−1363	−1668
202	875	741	347	N/A	1510	1332	1510	−666	−469	−393	N/A	−1278	−1017	−1278
203	1313	1047	647	N/A	2393	2040	2393	−552	−389	−382	N/A	−1116	−899	−1116
204	1586	1241	833	N/A	2943	2484	2943	−438	−309	−382	N/A	−965	−793	−965
205	1662	1303	896	N/A	3106	2629	3106	−325	−229	−382	N/A	−814	−686	−814
206	1586	1241	833	N/A	2943	2484	2943	−438	−309	−382	N/A	−965	−793	−965
207	1315	1047	647	N/A	2396	2039	2396	−552	−389	−382	N/A	−1116	−899	−1116
208	872	741	347	N/A	1506	1332	1506	−666	−469	−393	N/A	−1278	−1017	−1278
209	319	357	150	N/A	575	625	625	−779	−549	−632	N/A	−1668	−1362	−1668
210	244	172	143	N/critical	467	371	467	−991	−689	−1185	−1857	−2502	−2102	−3290
300	243	172	143	N/critical	467	371	467	−991	−689	−1185	−1857	−2502	−2102	−3290
301	230	303	184	N/A	491	587	587	−891	−620	−733	N/A	−1919	−1558	−1919
302	752	649	386	N/A	1385	1249	1385	−792	−552	−528	N/A	−1582	−1262	−1582
303	1184	951	640	N/A	2214	1904	2214	−693	−483	−462	N/A	−1384	−1104	−1384
304	1502	1179	808	N/A	2805	2375	2805	−594	−414	−396	N/A	−1186	−946	−1186
305	1669	1308	889	N/A	3108	2629	3108	−495	−345	−330	N/A	−989	−788	−989
306	1702	1330	884	N/A	3148	2653	3148	−396	−276	−264	N/A	−791	−631	−791
307	1572	1227	793	N/A	2884	2425	2884	−297	−207	−198	N/A	−593	−473	−593
308	1274	978	615	N/A	2309	1915	2309	−198	−138	−132	N/A	−395	−315	−395
309	755	571	351	N/A	1355	1110	1355	−99	−69	−66	N/A	−198	−158	−198
310	0	0	0	N/A	0	0	0	0	0	0	N/A	0	0	0

TABLE 5.7 Continued

	Positive Shear (kN)							Negative Shear (kN)						
Location	Truck	Tandem	Lane	Train	Truck + Lane	Tandem + Lane	Critical V+	Truck	Tandem	Lane	Train	Truck + Lane	Tandem + Lane	Critical V−
100	288	215	129	N/A	512	415	512	−33	−23	−22	N/A	−65	−52	−65
101	248	187	103	N/A	432	352	432	−33	−23	−23	N/A	−67	−54	−67
102	209	160	80	N/A	358	293	358	−46	−49	−29	N/A	−90	−94	−94
103	172	134	60	N/A	288	238	288	−83	−76	−37	N/A	−147	−138	−147
104	136	109	44	N/A	225	189	225	−121	−101	−49	N/A	−211	−184	−211
105	103	85	30	N/A	167	144	167	−158	−126	−65	N/A	−275	−232	−275
106	73	63	20	N/A	117	104	117	−194	−148	−83	N/A	−340	−280	−340
107	46	43	13	N/A	74	70	74	−226	−169	−104	N/A	−405	−329	−405
108	24	25	8	N/A	40	42	42	−256	−188	−127	N/A	−468	−377	−468
109	8	10	6	N/A	17	19	19	−283	−204	−153	N/A	−529	−425	−529
110	8	6	5	N/critical	15	12	15	−305	−217	−181	N/critical	−586	−469	−586
200	305	217	188	N/critical	594	477	594	−31	−22	−18	N/critical	−60	−47	−60
201	277	201	156	N/A	524	422	524	−31	−22	−20	N/A	−61	−49	−61
202	245	180	126	N/A	451	365	451	−33	−33	−24	N/A	−67	−67	−67
203	209	157	99	N/A	377	307	377	−62	−55	−31	N/A	−113	−104	−113
204	171	132	76	N/A	304	251	304	−96	−80	−42	N/A	−170	−148	−170
205	133	106	57	N/A	234	198	234	−133	−106	−57	N/A	−234	−198	−234
206	96	80	42	N/A	170	148	170	−171	−132	−76	N/A	−303	−251	−303
207	62	55	31	N/A	113	104	113	−209	−157	−99	N/A	−377	−308	−377
208	31	33	24	N/A	65	67	67	−245	−180	−126	N/A	−451	−365	−451
209	31	22	20	N/A	61	49	61	−277	−201	−156	N/A	−524	−422	−524
210	31	22	18	N/critical	60	47	60	−305	−217	−188	N/critical	−594	−477	−594
300	305	217	181	N/critical	587	470	587	−8	−6	−5	N/critical	−15	−12	−15
301	283	204	153	N/A	529	425	529	−8	−10	−6	N/A	−17	−19	−19
302	256	188	127	N/A	468	377	468	−24	−25	−8	N/A	−40	−42	−42
303	226	169	104	N/A	405	329	405	−46	−43	−13	N/A	−74	−70	−74
304	194	148	83	N/A	340	280	340	−73	−63	−20	N/A	−117	−104	−117
305	158	125	65	N/A	275	232	275	−103	−85	−30	N/A	−167	−144	−167
306	121	101	49	N/A	211	184	211	−136	−109	−44	N/A	−225	−189	−225
307	83	76	37	N/A	147	138	147	−172	−134	−60	N/A	−288	−238	−288
308	46	49	29	N/A	90	94	94	−209	−160	−80	N/A	−357	−293	−357
309	33	23	23	N/A	67	54	67	−248	−187	−103	N/A	−432	−352	−432
310	33	23	22	N/A	65	52	65	−288	−215	−129	N/A	−512	−415	−512

	Positive Reactions (kN)							Negative Reactions (kN)						
Location	Truck	Tandem	Lane	Train	Truck + Lane	Tandem + Lane	Critical R+	Truck	Tandem	Lane	Train	Truck + Lane	Tandem + Lane	Critical R−
100	288	215	129	N/A	512	415	512	−33	−23	−22	N/A	−65	−52	−65
110/200	320	219	369	516	795	661	797	−39	−28	−23	N/critical	−75	−60	−75
210/300	320	219	369	516	795	661	797	−39	−28	−23	N/critical	−75	−60	−75
310	288	215	129	N/A	512	415	512	−33	−23	−22	N/A	−65	−52	−65

[a]The truck, tandem, and train vehicle actions are multiplied by the dynamic load allowance of 1.33 prior to combining with the lane load.

those calculated previously shows minor differences. These differences are attributed to the load positioning procedures. The automated procedure moves the load along the influence diagram at relatively small intervals and the maxima/minima are stored. The hand calculations are based on a single load position estimating the maximum/minimum load effect.

The AASHTO vehicle loads are also applied to a three-span continuous beam with spans 10 668, 12 802, and 10 668 mm (35, 42, and 35 ft). The results are presented in Table 5.8. The values in Table 5.7 and 5.8 are used in the design examples presented in the remaining chapters.

5.8   INFLUENCE SURFACES

Influence functions (or surfaces) can represent the load effect as a unit action moves over a surface. The concepts are similar to those presented previously. A unit load is moved over a surface, an analysis is performed for each load placement, and the response of a specific action at a fixed location is used to create a function that is two dimensional, that is, $\eta(x,y)$ where x and y are the coordinates for the load position.

This function is generated by modeling the system with the finite element method (see Chapter 6). The function is used by employing superposition in a manner similar to Eq. 5.1. The analogous equation is

$$A = P_1\eta(x_1, y_1) + P_2\eta(x_2, y_2) + \cdots + P_n\eta(x_n, y_n) \qquad (5.20)$$
$$= \sum_{i=1}^{n} P_i\eta(x_i, y_i)$$

A distributed patch load is treated in a manner similar to distributive load in Eq. 5.3. The analogous equation is

$$A = \iint_{Area} w(x,y)\eta(x,y)dA \qquad (5.21)$$

where $w(x,y)$ is the distributive patch load and the integration is over the area where the load is applied. If the load is uniform, then $w(x,y)$ may be removed from the integration. For example, a uniform load such as the self-weight of a bridge deck is multiplied by the volume under the influence surface to determine the load effect. Numerical procedures similar to those described previously are used for the integration. Influence surfaces can be normalized and stored for analysis. Influence surfaces were used extensively in the development of the load distribution formulas contained in the AASHTO specification (Zokaie et al, 1991). This work is described in detail in Chapter 6.

TABLE 5.8 Action Envelope for Three-Span Continuous Beam 10 668, 12 802, and 10 668 mm (35, 42, and 35 ft)

Location	Positive Moment (kN m)							Negative Moment (kN m)						
	Truck	Tandem	Lane	Train	Truck + Lane	Tandem + Lane	Critical M+	Truck	Tandem	Lane	Train	Truck + Lane	Tandem + Lane	Critical M−
100	0	0	0	N/A	0	0	0	0	0	0	N/A	0	0	0
101	200	189	43	N/A	309	295	309	−28	−24	−8	N/A	−45	−40	−45
102	327	322	75	N/A	510	503	510	−56	−48	−16	N/A	−90	−80	−90
103	391	400	97	N/A	617	629	629	−84	−72	−24	N/A	−136	−119	−136
104	399	428	108	N/A	639	678	678	−112	−96	−32	N/A	−181	−159	−181
105	383	420	109	N/A	618	667	667	−140	−119	−40	N/A	−226	−199	−226
106	363	380	99	N/A	581	605	605	−167	−143	−48	N/A	−271	−239	−271
107	296	304	78	N/A	472	483	483	−195	−167	−57	N/A	−316	−279	−316
108	181	200	47	N/A	288	313	313	−223	−191	−65	N/A	−361	−319	−361
109	81	77	23	N/A	130	125	130	−251	−215	−90	N/A	−424	−376	−424
110	65	59	18	N/A	104	97	104	−327	−239	−145	N/A	−580	−463	−580
200	65	59	18	N/A	104	97	104	−327	−239	−145	N/A	−580	−463	−580
201	97	94	18	N/A	147	143	147	−208	−190	−77	N/A	−354	−330	−354
202	210	230	42	N/A	322	348	348	−178	−162	−48	N/A	−285	−264	−285
203	328	335	79	N/A	516	524	524	−147	−135	−47	N/A	−243	−226	−243
204	395	399	102	N/A	627	633	633	−117	−107	−47	N/A	−203	−189	−203
205	404	416	110	N/A	647	663	663	−87	−79	−47	N/A	−162	−152	−162
206	395	399	102	N/A	627	633	633	−117	−107	−47	N/A	−203	−189	−203
207	328	335	79	N/A	515	524	524	−147	−135	−47	N/A	−243	−226	−243
208	210	229	42	N/A	322	348	348	−178	−162	−48	N/A	−285	−264	−285
209	96	94	18	N/A	147	143	147	−208	−190	−77	N/A	−354	−330	−354
210	65	59	18	N/A	104	97	104	−327	−239	−145	N/A	−580	−463	−580
300	65	59	18	N/A	104	97	104	−327	−239	−145	N/A	−580	−463	−580
301	81	77	23	N/A	130	125	130	−261	−215	−90	N/A	−424	−376	−424
302	181	200	47	N/A	288	313	313	−223	−191	−65	N/A	−361	−319	−361
303	296	304	78	N/A	472	483	483	−195	−167	−57	N/A	−316	−279	−316
304	363	380	99	N/A	581	605	605	−167	−143	−48	N/A	−271	−239	−271
305	383	420	109	N/A	618	667	667	−139	−119	−40	N/A	−226	−199	−226
306	399	428	108	N/A	639	678	678	−112	−95	−32	N/A	−181	−159	−181
307	391	400	97	N/A	617	629	629	−84	−72	−24	N/A	−136	−119	−136
308	327	322	75	N/A	510	503	510	−56	−48	−16	N/A	−90	−80	−90
309	200	189	43	N/A	309	294	309	−28	−24	−8	N/A	−45	−40	−45
310	0	0	0	N/A	0	0	0	0	0	0	N/A	0	0	0

TABLE 5.8 Continued

Location	Positive Shear (kN)							Negative Shear (kN)						
	Truck	Tandem	Lane	Train	Truck + Lane	Tandem + Lane	Critical V+	Truck	Tandem	Lane	Train	Truck + Lane	Tandem + Lane	Critical V−
100	224	205	45	N/A	343	317	343	−26	−22	−8	N/A	−42	−37	−42
101	188	177	36	N/A	285	272	285	−26	−22	−8	N/A	−43	−38	−43
102	153	151	28	N/A	231	228	231	−42	−39	−10	N/A	−66	−62	−66
103	122	125	21	N/A	184	187	187	−65	−66	−13	N/A	−100	−101	−101
104	93	100	15	N/A	140	149	149	−85	−92	−17	N/A	−131	−139	−139
105	67	77	11	N/A	100	113	113	−104	−117	−23	N/A	−161	−178	−178
106	44	56	7	N/A	66	81	81	−136	−140	−29	N/A	−210	−215	−215
107	31	37	5	N/A	46	53	53	−168	−162	−36	N/A	−260	−251	−260
108	19	20	3	N/A	28	29	29	−198	−181	−45	N/A	−308	−286	−308
109	8	6	2	N/A	13	10	13	−226	−198	−54	N/A	−354	−317	−354
110	6	6	2	N/A	10	9	10	−254	−213	−63	N/A	−401	−347	−401
200	256	212	66	N/A	406	348	406	−24	−22	−6	N/A	−38	−35	−38
201	225	195	54	N/A	353	313	353	−24	−22	−7	N/A	−38	−36	−38
202	191	173	44	N/A	298	274	298	−27	−27	−8	N/A	−44	−44	−44
203	156	149	35	N/A	242	233	242	−43	−48	−11	N/A	−68	−75	−75
204	123	124	27	N/A	190	191	191	−66	−72	−15	N/A	−103	−111	−111
205	92	98	20	N/A	142	150	150	−92	−98	−20	N/A	−142	−150	−150
206	66	72	15	N/A	103	111	111	−123	−124	−27	N/A	−190	−191	−191
207	43	48	11	N/A	68	75	75	−156	−149	−35	N/A	−242	−233	−242
208	27	27	8	N/A	44	44	44	−191	−173	−44	N/A	−298	−274	−298
209	24	22	7	N/A	38	36	38	−225	−195	−54	N/A	−353	−313	−353
210	24	22	6	N/A	38	35	38	−256	−212	−66	N/A	−406	−348	−406
300	254	213	63	N/A	400	346	400	−6	−6	−2	N/A	−10	−9	−10
301	226	198	54	N/A	354	317	354	−8	−6	−2	N/A	−13	−10	−13
302	198	181	45	N/A	308	286	308	−19	−20	−3	N/A	−28	−29	−29
303	168	162	36	N/A	260	251	260	−31	−37	−5	N/A	−46	−53	−53
304	136	140	29	N/A	210	215	215	−44	−56	−7	N/A	−66	−81	−81
305	104	117	23	N/A	161	178	178	−67	−77	−11	N/A	−100	−113	−113
306	85	92	17	N/A	131	139	139	−93	−100	−15	N/A	−140	−149	−149
307	65	66	13	N/A	99	101	101	−122	−125	−21	N/A	−184	−187	−187
308	42	39	10	N/A	66	62	66	−153	−151	−28	N/A	−231	−228	−231
309	26	22	8	N/A	43	38	43	−188	−177	−36	N/A	−285	−272	−285
310	26	22	8	N/A	42	37	42	−224	−205	−45	N/A	−343	−318	−343

	Positive Reaction (kN)							Negative Reaction (kN)						
Location	Truck	Tandem	Lane	Train	Truck + Lane	Tandem + Lane	Critical R+	Truck	Tandem	Lane	Train	Truck + Lane	Tandem + Lane	Critical R−
100.0	224	205	45	N/A	343	317	343	−26	−22	−8	N/A	−42	−37	−42
110/200	295	218	129	N/A	521	419	521	−30	−27	−8	N/A	−48	−44	−48
210/300	295	218	129	N/A	521	419	521	−30	−27	−8	N/A	−48	−44	−48
310.0	224	205	45	N/A	343	318	343	−26	−22	−8	N/A	−42	−37	−42

[a]The truck, tandem, and train vehicle actions are multiplied by the dynamic load allowance of 1.33 prior to combining with the lane load.

5.9 SUMMARY

Influence functions are important for the structural analysis of bridges. They aid the engineer in the understanding, placement, and analysis of moving loads. Such loads are required to determine the design load effects. Several methods exist to generate influence functions. All methods have advantages and disadvantages for hand and automated methods. Several methods are illustrated in this chapter. Design trucks and lane loads have been used to generate the critical actions for four bridges. These envelopes are used in design examples presented in later chapters.

REFERENCES

AASHTO (1994). *LRFD Bridge Design Specification,* 1st ed., American Association for State Highway and Transportation Officials, Washington, D.C.

AISC (1986), *Moments, Shears, and Reactions for Continuous Bridges,* American Institute of Steel Construction, Chicago, IL.

BridgeTech, Inc. (1996), *LRFD Analysis Manual,* Laramie, WY, Version 1.4f.

Laursen, H.I. (1988), *Structural Analysis,* McGraw-Hill, 3rd ed., New York.

McCormac, J. and R.E. Elling (1988), *Structural Analysis—A Classical Approach,* 4th ed., Harper & Row, New York.

Pilkey, W.D. and O.H. Pilkey (1974), *Mechanics of Solids,* Quantum Publishers, New York.

Zokaie, T.L, T.A. Osterkamp, and R.A. Imbsen (1991), *Distribution of Wheel Loads on Highway Bridges, Final Report Project 12-26/1,* National Cooperative Highway Research Program, Transportation Research Board, National Research Council, Washington, DC.

6

SYSTEM ANALYSIS

6.1 INTRODUCTION

In order to design a complicated system such as a bridge, it is necessary to break the system into smaller, more manageable subsystems that are comprised of components. Subsystems include the superstructure, substructure, and foundation, while the components include beams, columns, deck slab, barrier system, cross frames, diaphragms, bearings, piers, footing, piles, and caps. The forces and deformations (load effects) within the components are necessary to determine the required size and material characteristics. It is traditional and implicit in the AASHTO Specification that design be performed on a component basis. Therefore the engineer requires procedures to determine the response of the structural system and ultimately its components.

In general, the distribution of the loads throughout the bridge requires equilibrium, compatibility, and that constitutive relationships (material properties) be maintained. These three requirements form the basis for all structural analysis, regardless of the level of complexity. Equilibrium requires that the applied forces, internal actions, and external reactions be statically in balance. Compatibility means that the deformations are internally consistent throughout the system (without gaps or discontinuities) and are consistent with the boundary conditions. Finally, the material properties, such as stiffness, must be properly characterized. Typically, the assumptions that are made regarding these three aspects of analysis determine the complexity and the applicability of the analysis model.

For example, consider the simply supported wide-flange beam subjected to uniform load shown in Figure 6.1. The beam is clearly a three-dimensional system because it has spatial dimension in all directions, but in mechanics of deformable bodies we learned that this system could be modeled by the familiar one-dimensional (1-D) equation

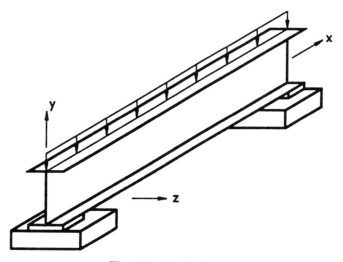

Fig. 6.1   Simple beam.

$$\frac{d^4y}{dx^4} = \frac{w(x)}{EI(x)} \qquad (6.1)$$

Several important assumptions were used in the development of Eq. 6.1. First, the material is assumed to behave linear elastically. Second, the strain (and stress) due to flexural bending is assumed to be linear. Third, the loads are concentrically applied such that the section does not torque, and finally, the beam is proportioned and laterally braced so that instability (buckling) does not occur. Although these assumptions are conventional and yield results comparable to laboratory results, often these conditions do not truly exist. First, for example, due to localized effects, some yielding most always occurs under reasonable service loads. Residual stresses from rolling result in some yielding at load levels below the predicted yield. These local effects do not significantly affect the global system response under service loads. Second, the bending stress profile is slightly nonlinear principally due to the load (stress) applied to the top of the beam and the reactions that create vertical normal stresses and strains that, due to Poisson's effect, also create additional horizontal stress. This effect is usually small. Third, concentric loading is difficult to achieve if the load is applied directly to the beam but if the beam is part of a slab system, then this assumption is perhaps more realistic. Finally, and importantly, Eq. 6.1 does not consider the local or global instability of the beam. It may be argued that other assumptions are also applicable but a discussion of these suffices for the purpose intended here.

The purpose for discussing such a seemingly simple system is to illustrate the importance of the modeling assumptions, and their relevance to the real system. It is the engineer's responsibility to understand the assumptions and their applicability to the system under study. When the assumptions do not adequately reflect the behavior of the real system, the engineer must be confident in the bounds of the error induced and the consequences of the error. Clearly, it is impossible to exactly predict the response of any structural system, but predictions can be of acceptable accuracy. The consequences of inaccuracies are a function of the mode of failure. This is elaborated in detail later.

The application of equilibrium, compatibility, and material response, in conjunction with the assumptions, constitutes the *mathematical model* for analysis. In the case of the simple beam, Eq. 6.1, with the appropriate boundary conditions, is the mathematical model. In other cases, the mathematical model might be a governing differential equation for a beam column or perhaps a thin plate, or it may be the integral form a differential equation expressed as an energy or variational principle. Whatever the mathematical model may be, the basis for the model and the behavior it describes must be understood.

In structural mechanics courses, there are numerous procedures presented to use the mathematical model represented by Eq. 6.1 to predict structural response. For example, direct integration, conjugate beam analogy, moment-area, slope-deflection, and moment distribution are all well-established methods. All of these methods either directly or indirectly involve the mathematical model represented by Eq. 6.1. The method used to solve the mathematical model is termed the *numerical model.* The selection of the numerical model depends on many factors including availability, ease of application, accuracy, computational efficiency, and the structural response required. In theory, numerical models based on the same mathematical model should yield the same response. In practice, this is generally true for simple mathematical models with one-dimensional elements such as beams, columns, and trusses. Where finite elements are used to model a continuum in two- or three-dimensions (2- or 3-D), features such as element types, mesh characteristics, and numerical integration, complicate the comparison of methods. This does not mean that several solutions exist to the same problem, but rather solutions should be comparable though not *exactly* identical, even though the mathematical models are the same.

Finally, the engineer should realize that even the simplest of systems are often mathematically intractable from a rigorous closed-form approach. It is rather easy to entirely formulate the mathematical model for a particular bridge, but the solution of the mathematical model is usually nontrivial and must be determined with approximate numerical models such as with the finite strip or finite element methods. It is important to realize that modeling approximations exist in *both* the mathematical and numerical models.

The modeling process is illustrated in Figure 6.2. At the top of the diagram is the real system as either conceptualized or as built. To formulate a mathematical model, the engineer must accept some simplifying assumptions that result in a governing equation(s) or formulation. Next, the engineer must

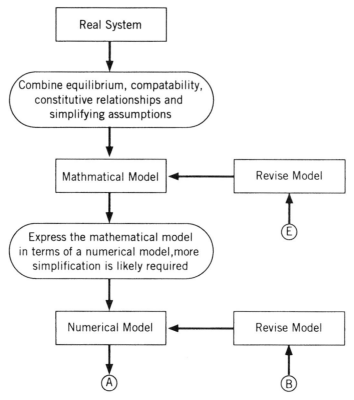

Fig. 6.2   Relationship of modeling to design.

translate the characteristics of the real system into the variables of the mathematical model. This includes definition of loads, material and cross-sectional properties, and boundary conditions. Likely, the engineer relies on more simplification here. The mathematical model is solved using a numerical model. Here some numerical approximation may be involved, or the model may solve the mathematical model "exactly." The results are then interpreted, checked, and used for component design. If the component properties vary significantly as a result of the design, then the numerical model should be altered and the revised results should be determined. Throughout the process, the engineer must be aware of the limitations and assumptions implicit in the analysis and should take precautions to insure that the assumptions are not violated, or that the consequences of the violations are acceptable.

Many model parameters are difficult to estimate and in such cases, the extreme conditions can be used to form an envelope of load effects to be used for design. For example, if a particular cross-sectional property is difficult to estimate due to complications such as composite action, concrete cracking, and creep effects, the engineer could model the section using the

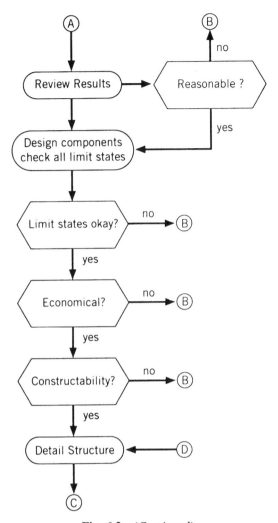

Fig. 6.2   (*Continued*).

upper and lower bounds and study the sensitivity of the procedure to the unknown parameters. This provides information about the importance of the uncertainty of parameters in the structural response.

6.2   SAFETY OF METHODS

As previously stated, it is important for the engineer to understand the limitations of the mathematical and numerical models, and the inaccuracies involved. As models are estimates of the actual behavior, it is important to

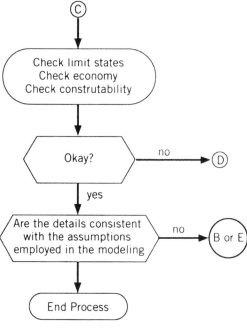

Fig. 6.2 (*Continued*)

clearly understand the design limit states and their relationship to the modes and consequences of failure. This finding is discussed in the sections that follow.

6.2.1   Equilibrium for Safe Design

An essential objective in any analysis is to establish a set of forces that satisfies equilibrium between internal actions and the applied loads at every point. The importance of equilibrium cannot be overstated and this is elaborated below.

Most of the analytical models described in this book are based on linear response, that is, the load effect is proportional to the load applied. Conversely, the resistance models used by AASHTO (and most other structural design specifications) implicitly assume nonlinear material response at the strength limit state. For example, the nominal flexural capacity of a braced compact steel section is $M_n = F_y Z$ and the flexural capacity of a reinforced concrete section involves the Whitney stress block where $f_c = 0.85f_c'$, and the steel stress is equal to the yield stress, and so on. Clearly, an inconsistency exists. The analysis is based on linear behavior and the resistance calculations are based on nonlinear behavior. The rationale for this is founded in the system behavior at and beyond yielding and is based in plasticity theory. The rationale is best explained by restating the *lower bound theorem* (Neal 1977, Horne, 1971):

> A load computed on the basis of an equilibrium moment distribution in which the moments are nowhere greater than M_p is less than or equal to the true plastic limit load.

Although stated in terms of bending moment, the theorem is valid for any type of action and/or stress. The essential requirements of the theorem are

- The calculated internal actions and applied forces should be in equilibrium everywhere.
- The materials and the section/member behavior must be ductile; that is, the material must be able to yield without fracture or instability (buckling).

In simpler terms, the theorem means that if a design is based on an analysis that is in equilibrium with the applied load and the structure behaves in a ductile manner, then the ultimate failure load will meet or exceed the design load. *This is one of the most important theorems in structural mechanics and is extremely relevant to design practice.*

This theorem offers wonderful assurance! How does it work? Consider the two beams shown in Figure 6.3(a) and 6.3(c). The beams are assumed to be designed such that any section can develop its full plastic moment capacity, which is $M_p = F_y Z$, and for the sake of simplicity it is assumed that the beam behaves elastic-plastic, that is, the moment that causes first yielding is the same as the plastic capacity (these two differ by ~ 10–15% for steel wide flange sections). The uniform load is applied monotonically to the simple beam of Figure 6.3(a) and the moment diagram is shown in Figure 6.3(b). When the moment at midspan reaches the capacity $M_p = w_u \ell^2 / 8$, a plastic hinge develops. This creates a loss in bending rigidity resulting in mechanism and collapse occurs. Now consider the beam shown in Figure 6.3(c) and its associated linear elastic moment diagram shown in Figure 6.3(d). The load is again applied monotonically up to the level where hinges form at the supports (negative moment). A loss in bending rigidity results and now the system becomes a simple beam with the plastic moment applied at the ends. Because the beam has not reached a mechanism, more load may be applied. A mechanism is finally reached when a hinge forms in the interior portion of the span (positive moment). This behavior illustrates redistribution of internal actions. Now assume the two beams have the same capacity M_p in positive and negative bending. For the simple beam, equate the capacities to the maximum moment at midspan yielding

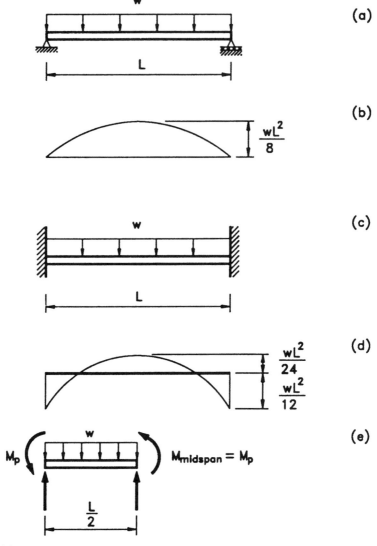

Fig. 6.3 (a) Uniformly loaded simple beam, (b) moment diagram, (c) uniformly loaded fixed–fixed beam, (d) moment diagram, and (e) free body diagram.

$$M_p = \frac{w_u \ell^2}{8}$$

$$w_u(\text{simple beam}) = \frac{8M_p}{\ell^2}$$

Use a similar procedure for the fixed–fixed beam to equate the capacity to the maximum elastic moment at the support yielding

$$M_p = \frac{w_u \ell^2}{12}$$

$$w_u(\text{fixed–fixed beam}) = \frac{12M_p}{\ell^2}$$

A free body diagram for the fixed–fixed beam is shown in Figure 6.3(e) for the state after the moment has reached the plastic moment capacity at the end. By equating the capacity M_p at midspan to the moment required by balancing the moment at midspan, one obtains

$$M_{\text{midspan}} = M_p + w_u\left(\frac{1}{2}\right)\left(\frac{1}{4}\right) = M_p$$

$$w_u = \frac{16M_p}{\ell^2}$$

Consider the relevance of this example to analysis and design. Suppose that the system is a fixed–fixed beam but the engineer designed for the simple beam moment. The design would have an additional capacity of $\frac{12}{8} = 1.5$ against initial yielding due to the neglected continuity moments and additional reserve against total collapse of $\frac{16}{8} = 2.0$ considering redistribution of internal actions. Even though the elastic moment diagram is used, the design is safe (but quite likely unnecessarily conservative). Now consider the more likely case where the engineer designs per the fixed–fixed elastic moment diagram shown in Figure 6.3(c). Here the additional reserve against collapse is $\frac{16}{12} = 1.33$. Note these factors should not be confused with the load factors of Tables 3.1 and 3.2 as these would be included in the elastic moment diagrams for the design. The AASHTO Specification allows the engineer to consider inelastic redistribution of internal actions in various articles. Because the amount of redistribution is related to the ductility of the component that is material and cross-section dependent, most of these provisions are outlined in the resistance articles of the specification.

The static or lower bound theorem implies that as long as equilibrium is maintained in the analytical procedures and adequate ductility is available, then the exact distribution of internal actions is not required. It is inevitable that the analysis and subsequent design overestimates the load effect in some locations while underestimates the effect in others. If the strength demands in the real structure are larger than the available resistance, yielding occurs and the actions redistribute to a location where the demands are less, and hence, more capacity exists to carry the redistributed actions. The requirement of ductility and equilibrium insures that redistribution occurs and that the system has the necessary capacity to carry the redistributed actions.

In summary, it is not required that the calculated system of forces be exact predictions of the forces that exist in the real structure. It is only necessary that the calculated system of forces satisfy equilibrium at every point. This requirement provides at least one load path. As illustrated for the fixed–fixed beam, redistribution of internal actions may also occur in statically redundant systems. As previously stated, in practice it is impossible to exactly predict the system of forces that exists in the real system, and therefore, the lower bound theorem provides a useful safety net for the strength limit state.

In the case where instability (buckling) may occur prior to reaching the plastic capacity, the static or lower bound theorem does *not* apply. If an instability occurs prior to complete redistribution, then equilibrium of the redistributed actions cannot be achieved and the structure may fail in an abrupt and dangerous manner.

6.2.2   Stress Reversal and Residual Stress

In Section 6.2.1, the ultimate strength behavior was introduced and it was assumed that the cross section achieved the full plastic moment capacity. There was no mention of how the section reaches this state nor what happens when the section is yielded and then unloaded. Both of these issues are important to understanding the behavior and design limit states for ductile materials.

Consider the cantilever beam shown in Figure 6.4(a), which has the cross section shown in Figure 6.4(b) with reference points of interest o, p, q, and r. Point o is located at the neutral axis, p is slightly above the o, r is located at the top, and q is midway between p and r. The section has residual stresses that are in self-equilibrium [see Fig. 6.4(c)]. In general, residual stresses come from the manufacturing process, construction process, temperature effects, intentional prestressing, creep, shrinkage, and so on. The beam is deflected at B and the load is measured. The product of the measured load and beam length is the moment at A, which is shown in the moment–curvature diagram illustrated in Figure 6.4(d). As the tip deflection increases, the moment increases with curvature and all points remain below the yield stress up to state a. Further increase causes initial yielding in the outer fibers and the yielding progresses toward the neutral axis until the section is in a fully plastic state. Figures 6.4(e–j) illustrate the elastic–plastic stress–strain curve for the material, the state of stress and strain of the cross-section points, and the section stress profiles. For example, at state b, points p, q, and r are all at the yield stress and the stress profile is uniform.

What happened to the residual stresses when progressing from point a to b? Because the section is initially stressed, the curvature at which yielding occurs and the rate at which the section reaches its full plastic capacity is affected, but the ultimate capacity is not. *This is an important aspect of struc-*

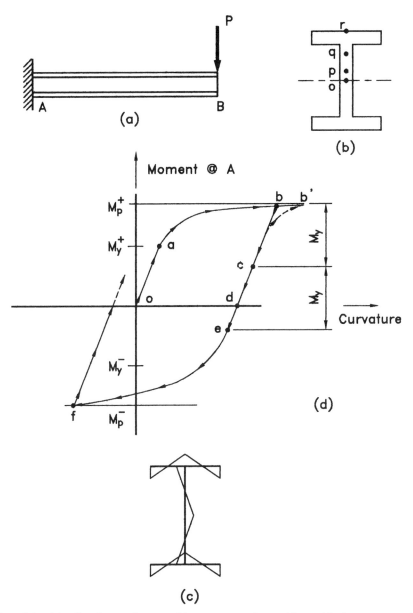

Fig. 6.4 (a) Cantilever beam, (b) cross section, (c) residual stresses, (d) moment–curvature diagram.

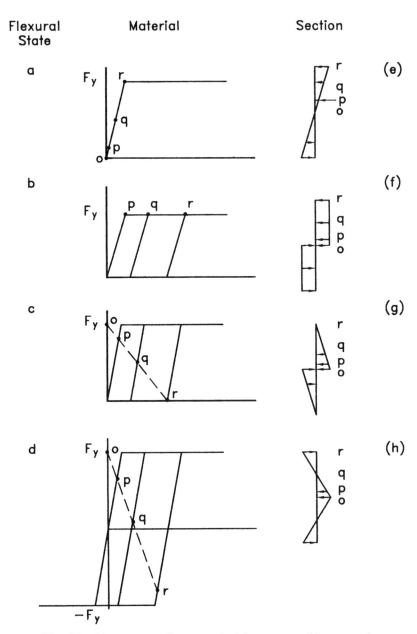

Fig. 6.4   (e) state at a, (f) state at b, (g) state at c, (h) state at d.

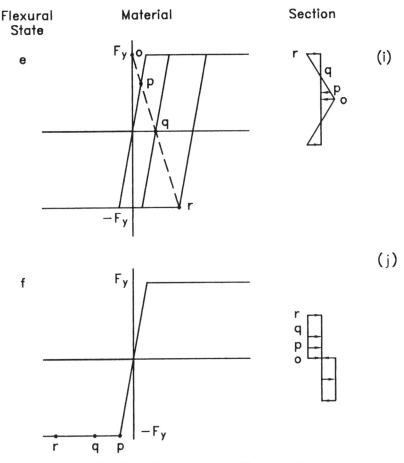

Fig. 6.4 (i) state at e, and (j) state at f.

tural design, as many residual stresses exist in a structure due to numerous reasons. Although such stresses may affect service level behavior and/or stability, they do not affect the capacity of ductile elements.

Upon load removal, the moment–curvature follows b-c-d in Figure 6.4(d). If reloaded in the initial direction, then the moment–curvature follows d-c-b'. The curved portion of the line is different from the o-a-b because the residual stresses have been removed. The curved portion must exist because the section yields incrementally with the outer fibers first and then progresses inward. Now start at state b and unload to c where the stress at r is zero. As illustrated in Figure 6.4(g), the change in stress is F_y, hence the change in moment required is M_y. Note the stress at o is still at yield because the point o is at the neutral axis and does not strain upon load removal. Points p and q have stresses that are proportionally less. Continue load removal until state d is achieved. Here all the load is removed and the structure is in self-equilibrium. The stresses are illustrated in Figure 6.4(h). Reverse the load by deflecting the beam upward at the tip until the yielding occurs at *r*. Because the stress at *r* was zero at state *c,* the change in moment required to produce a yield stress is again M_y [see Fig. 6.4(i)]. Further increases cause the section to reach its full plastic state at *f* shown in Figure 6.4(j). Note the nonlinear shape of the curve is again different because the initial stresses at state *e* are again different.

In summary, the initial or residual stresses do not affect the ultimate (ductile) capacity but they do affect the load-deflection characteristics in the post-yield region.

6.2.3   Repetitive Overloads

As the vehicular loading of a bridge is repetitive, the possibility of repeated loads that are above the service level are likely and their effects should be understood. In Section 6.2.2, the lower bound theorem is introduced for a single-load application that exceeds the yield strength of the beam at localized points. Consider the uniformly loaded prismatic beam shown in Figure 6.5(a) with the moment–curvature relationship shown in Figure 6.5(c). The uniform loading shown in Figure 6.5(a) is spatially invariant and the magnitude is cycled. Assume the load is applied slowly to a level where the moment at *A* and *C* exceeds the plastic moment, a hinge forms at *A,* and the section has yielded at *B* but just before the section is fully plastic at *B* [see Fig. 6.5(b)]. The resulting moment diagram is shown in Figure 6.5(d). Now the load is removed and the structure responds (unloads) elastically. The change in moment is the elastic moment diagram illustrated in Figure 6.5(e). The moment in the unloaded state is determined by superposing the inelastic (loading) and elastic (unloading) moment diagrams [Fig. 6.5(f)] and is termed the residual moment. The deflected shape of the beam is illustrated in Figure 6.5(g) where both the inelastic rotations at the beam ends and the elastic deflection due to the residual moments are shown.

To examine the effect of cyclic loads, consider the beam shown in Figure 6.6(a). The plastic collapse load for a single concentrated load is shown in Figure 6.6(b) and is used for reference. The loads W_1 and W_2 are applied independently. First, W_1 is increased to a level such that hinges form at *A* and *B* but not to a level such that the hinge forms at *D* [Figure 6.6(c)], which means that segment *CD* remains elastic and restrains collapse. Now remove load W_1 and the structure responses elastically, and residual moment and deflection remain. The residual deflection is illustrated in Figure 6.6(d). Next, apply W_2 to a level such that hinges form at *C* and *D* and segment *AC* remains elastic. The deflected shape is also illustrated in Figure 6.6(d). It is important to note that a complete mechanism has not formed but the deflections have increased. Now if the load cycle is repeated, the deflections may continue to increase and the resultant effect is a progressive buildup of permanent de-

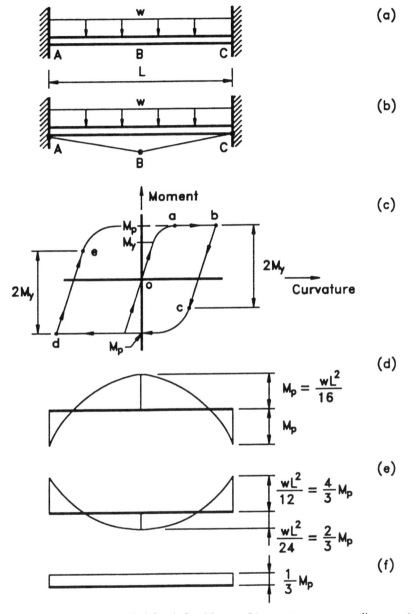

Fig. 6.5 (a) Uniformly loaded fixed–fixed beam, (b) moment-curvature diagram, (c) collapse mechanism, (d) moment diagram at collapse, (e) elastic (unloading) moment diagram, and (f) residual moment diagram.

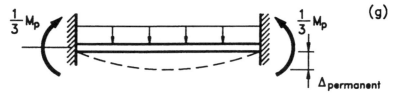

Fig. 6.5   (g) Residual displacement.

flections just as though the beam is deforming in the plastic collapse mechanism [Figure 6.6(e)]. This limit state is termed *incremental collapse.*

The residual moments and associated deflections can be determined by incrementally applying the loads as described and performing the analysis for each load step. Although a viable method, it tends to be tedious and is limited to simple structures and loads. A bridge system is much more complex than the beam previously described. The bridge must resist moving loads and the incremental approach must include the complexity of load position and movement, which greatly complicates the analysis. The important issue is that the repeated loads that cause incremental collapse are less than the static collapse loads. The load above which incremental collapse occurs is termed the *shakedown load.* If the load is below the shakedown load but above the load that causes inelastic action, then the structure experiences inelastic deformation in local areas but after a few load cycles, the structure behaves elastically under further loading. If the load exceeds the shakedown load but is less than the plastic collapse load, then incremental collapse occurs. Finally, if the load exceeds the plastic collapse load, the structure collapses. This discussion is summarized in Table 6.1.

Two theorems have been developed to determine the shakedown load: the lower and upper bound theorems. These theorems help to relate the elastic behavior to the inelastic behavior so complex systems can be analyzed without an incremental load analysis. The theorems are stated without proof. The interested reader is referred to an extensive reference by Horne (1971) and also Neal (1977).

The magnitude of variable repeated loading on a structure may be defined by a common load factor λ, where M_{max} and M_{min} represent the maximum and minimum elastic moments, and λM_{max} and λM_{min} represent the moments for load level λ, and $\lambda_s M_{max}$ and $\lambda_s M_{min}$ represent shakedown moments (Horne, 1971).

Lower Bound Theorem: The lower bound shakedown limit is given by a load factor λ for which residual moment m satisfies the inequalities:

$$m + \lambda M_{max} \leq M_p$$

$$m + \lambda M_{min} \geq -M_p \tag{6.2}$$

$$\lambda(M_{max} - M_{min}) \leq 2 M_y$$

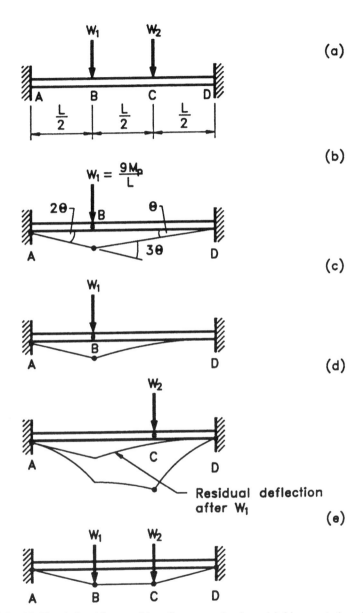

Fig. 6.6   (a) Fixed–fixed beam, (b) collapse mechanism, (c) hinges at A and B, (d) hinges at C and D, and (e) incremental collapse.

TABLE 6.1   Summary of Inelastic Behavior

Minimum Load	Maximum Load[a]	Result
0	Yield	Elastic behavior
Yield	Shakedown	Localized inelastic behavior initially Elastic behavior after shakedown
Shakedown	Plastic collapse	Incremental collapse
Plastic collapse	N/A	Collapse

[a]Not applicable = N/A.

The residual moment m does not necessarily have to be the exact residual moment field determined from incremental analysis but may be any self-equilibrating moment field. The third inequality is imposed to avoid an alternating plasticity failure where the material is yielded in tension and compression. Such a condition is unlikely in a bridge structure because the total change in moment at any point is far less than twice the yield moment.

Obviously, the residual moment m could be set to zero and the theorem simply implies that shakedown can be achieved if the moment is less than the plastic moment and the moment range is less than twice the yield moment. The former is similar to the ultimate strength limit state. The latter is seldom a problem with practical bridge structures (Horne, 1971).

Upper Bound Theorem: The upper bound shakedown load is determined by assuming an incremental collapse mechanism with hinges at locations j with rotations θ_j associated with the elastic moments M_{max} and M_{min}. The directions of the hinge rotations are consistent with the moments (Horne, 1971).

$$\lambda_{\text{shakedown}} \sum_j \{M_{j\text{max}} \text{ or } M_{j\text{min}}\}\theta_j = \sum_j M_{pj}|\theta_j| \qquad (6.3)$$

The elastic moment ($M_{j\text{max}}$ or $M_{j\text{min}}$) used is the one that causes curvature in the same sense as the hinge rotation.

EXAMPLE 6.1

Determine the yield, shakedown, and plastic collapse load for a moving concentrated load on the prismatic beam shown in Figure 6.7(a). Compare the shakedown load with the plastic collapse load and the initial yield load for the load at midspan. Assume that $M_p = M_y$, that is, neglect the spread of plasticity.

The elastic moment envelope is illustrated in Figure 6.7(b). The elastic envelopes may be established using any method appropriate for the solution of a fixed–fixed beam subjected to a concentrated load. The envelope values represent the elastic moments M_{max} and M_{min}. Use Eq. 6.3 and the mechanism shown in Figure 6.7(c) to obtain

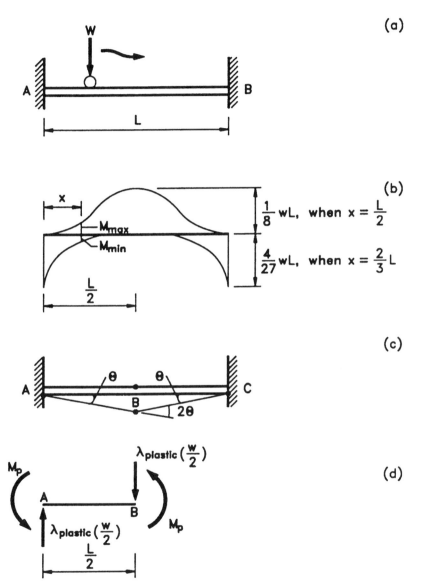

Fig. 6.7 (a) Fixed–fixed beam with moving load, (b) elastic moment envelope, (c) assumed incremental collapse mechanism, and (d) free body diagram.

$$\lambda_{\text{shakedown}} \left(\frac{4}{27} W\ell(\theta + \theta) + \frac{1}{8} W\ell(2\theta) \right) = M_p(4\theta)$$

$$\lambda_{\text{shakedown}} = \frac{432}{59} \frac{M_p}{W\ell} = 7.32 \frac{M_p}{W\ell}$$

To determine the plastic capacity, use the free body diagram shown in Figure 6.7(d) to balance the moment about B. The result is

$$\frac{\lambda_{\text{plastic}} W}{2} \left(\frac{\ell}{2} \right) = 2M_p$$

$$\lambda_{\text{plastic}} = \frac{8M_p}{W\ell}$$

Note that the shakedown load is approximately 92% of the plastic collapse load. The initial yield load is determined by equating the maximum elastic moment to $M_y = M_p$. The result is

$$\lambda_{\text{elastic}} = \frac{27}{4} \frac{M_p}{W\ell} = 6.75 \frac{M_p}{W\ell}$$

A comparison of the results is given in Table 6.2.

In summary, it is important to understand that plastic limit collapse may not be the most critical strength limit state, but rather incremental collapse should be considered. It has been demonstrated that plastic deformation occurs at load levels below the traditional plastic collapse for repeated loads. Some procedures outlined in the AASHTO Specification implicitly permit inelastic action and assume that shakedown occurs. Such procedures are discussed in later sections.

TABLE 6.2   Example 6.1 Summary

Load Level	Load Factor (λ)	$\dfrac{\lambda}{\lambda_{\text{elastic}}}$
Elastic	$6.75 \dfrac{M_p}{W\ell}$	1.00
Shakedown	$7.32 \dfrac{M_p}{W\ell}$	1.08
Plastic	$8.00 \dfrac{M_p}{W\ell}$	1.19

6.2.4 Fatigue and Serviceability

The static or lower bound theorem relates to the ultimate strength limit state. However, repetitive truck loads cause fatigue stresses that may lead to brittle fracture under service level loads. Because the loads creating this situation are at the service level and because the failure mode is brittle, there is little chance for load redistribution, hence the lower bound theorem does *not* apply. Thus, the only way to estimate the internal live load actions accurately and safely is to properly model the relative stiffness of all components and their connections. This aspect of the analysis and subsequent design is one characteristic that differentiates bridge engineers from their architectural counterparts. The building structural engineer is typically not concerned with a large number of repetitive loads at or near service load levels.

Service or working conditions can be the most difficult to model because bridges and the ground supporting them experience long-term deformation due to creep, shrinkage, settlement, and temperature change. The long-term material properties and deformations are difficult to estimate and can cause the calculated values to vary widely. It is best to try to bound the model parameters involved with an analysis and design for the envelope of extreme effects. Service limit states are important and should be carefully considered. A significant portion of a bridge manager's budget is spent for repair and retrofit operations. This effect is because the severe environment, including heavy loads, fluctuating temperature, and deicing chemicals cause serviceability problems that could ultimately develop into strength problems.

6.3 GRAVITY LOAD ANALYSIS

Most methods of analysis described in this chapter are based on three aspects of analysis: equilibrium, compatibility, and material properties, which are assumed to be linear elastic. The objective of these methods is to estimate the load effects based on the relative stiffness of the various components. The methods described vary from simplistic (beam-line) to rigorous (finite strip or finite element). Equilibrium is implicit in all methods and all methods attempt to achieve realistic estimates of the service level behavior. As the materials are assumed to behave linearly, these methods (as presented in this chapter) will not reflect the behavior after yielding occurs. As outlined earlier, the lower bound theorem prescribes that such analyses yield a conservative distribution of actions upon which to base strength design, and hopefully, a reasonable distribution of actions upon which to base service and fatigue limit states.

The discussion of specific analysis methods begins with the most common bridge types, the slab and slab-girder bridges. The discussion of these bridge types includes the most practical analytical procedures. As many books have been written on most of these procedures, for example, grillage, finite element, and finite strip methods, the scope must be restricted and is limited to address the basic features of each method and to address issues that are particularly relevant to the bridge engineer. Example problems are given to illustrate particular behavior or techniques. As in the previous chapter, the examples provide guidance for the analysis and design for the bridges presented in the resistance chapters. The reader is assumed to have the prerequisite knowledge of matrix structural analysis and/or the finite element method. If this is not the case, there are plenty of topics that are based on statics and/or the AASHTO Specification provision that should be read in detail. Other topics such as grillage, finite element, and finite strip analysis can be read with regard to observation of behavior rather than understanding the details of the analysis.

The discussion of slab and slab-girder bridges is followed by an abbreviated address of box systems. Many of the issues involved with the analysis of box systems are the same as slab-girder systems. Such issues are not reiterated and the discussion is focused on the behavioral aspects that are particular to box systems.

6.3.1   Slab-Girder Bridges

The slab and slab-girder bridges are the most common type of bridge in the United States. A few of these bridges are illustrated in Figures 2.8, 2.9, 2.26, 2.31, 2.35, and 2.36. These are made of several types and combinations of materials. Several examples are illustrated in Table 6.3.

A schematic illustration of a slab-girder bridge is shown in Figure 6.8(a). The principal function of the slab is to provide the roadway surface and to transmit the applied loads to the girders. This load path is illustrated in Figure 6.8(b). The load causes the slab-girder system to displace as shown in Figure 6.8(c). If linear behavior is assumed, the load to each girder is proportional to its displacement. As expected, the girder near the location of the load application carries more load than those away from the applied load. Compare the deflection of the girders in Figure 6.8(c). Equilibrium requires that the

TABLE 6.3   Examples of Slab-Girder Bridges[a]

Girder Material	Slab Material
Steel	CIP concrete
Steel	Precast concrete
Steel	Steel
Steel	Wood
CIP concrete	CIP concrete
Precast concrete	CIP concrete
Precast concrete	Precast concrete
Wood	Wood

[a]Cast in place = CIP.

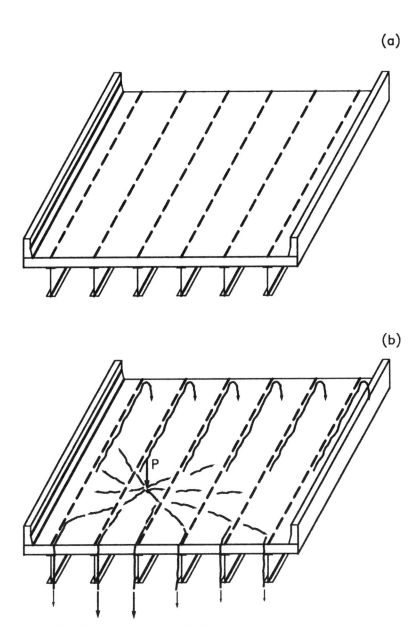

Fig. 6.8   (a) Slab-girder bridge, and (b) load transfer (boldface lines indicate larger actions).

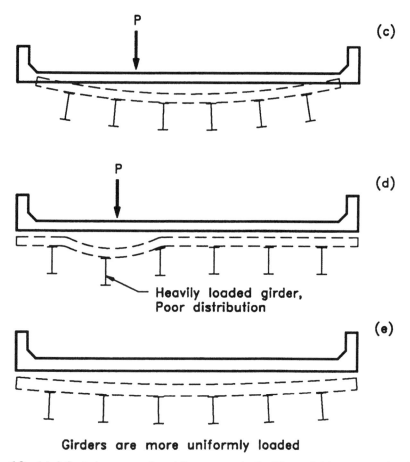

Fig. 6.8   (c) deflected cross section, (d) transversely flexible, and (e) transversely stiff.

summation of the load carried by all the girders equals the total applied load. The load carried by each girder is a function of the relative stiffness of the components that comprise the slab-girder system. The two principal components are the slab and the girders; other components include cross frames, diaphragms, and bearings. Only the slab and girder are considered here as the other components affect the behavior to a lesser extent.

The effect of relative stiffness is illustrated by considering the two slab-girder systems shown in Figures 6.8(d) and 6.8(e). The system shown in Figure 6.8(d) has a slab that is relatively flexible compared to the girder. Note the largest deflection is in the girder under the load and the other girder deflections are relatively small. Now consider the system shown in Figure 6.8(e) where the slab is stiffer than the previous case. Note the load (deflection) is distributed to the girders more evenly, therefore the load to each girder is less than shown in Figure 6.8(d).

The purpose of structural analysis is to determine the distribution of internal actions throughout the structure. Any method that is used should represent the relative stiffness of the slab and the girders. As outlined in the previous sections, the importance of accuracy of the analysis depends on the limit state considered and ductility available for the redistribution of actions after initial yielding. To illustrate, consider the simply supported slab-girder bridge shown in Figure 6.9(a). Assume the girders have adequate ductility for plastic analysis. Because of the simply supported configuration, this structure might traditionally be considered nonredundant, that is, one that does not have an alternative load path. Now assume the girder under the load yields and looses stiffness. Any additional load is then carried by the neighboring girders. If the load continues to increase, then the neighboring girders also yield, and additional load is carried by the nonyielded girders. If the slab has the capacity to transmit the additional load, then this process continues until all girders have reached their plastic capacities and a mechanism occurs in every girder. The ultimate load is obviously greater than the load that causes first yield. Note that this is predicted by the lower bound theorem described in Section 6.2.1. The shakedown theorems also apply.

So why should the engineer perform a complicated analysis to distribute the load to the girders? There are two principal reasons: (1) The failure mode may not be ductile, such as in a fatigue-related fracture or instability, and (2) The limit state under consideration may be related to serviceability and service level loads. Both reasons are important, and therefore, it is traditional to model the system as linear elastic to obtain reasonable distribution of internal actions for strength, service, and fatigue limit states. As the lower bound theorem may also apply, this approach is likely conservative and gives reasonable results for the strength limit states. In the case of the evaluation of an existing bridge where repair, retrofit, and/or posting is involved, it may be reasonable to use a linear elastic analysis for service load limit states and consider the nonlinear behavior for the strength limit states. Such a refinement could significantly influence the rehabilitation strategy or posting load.

Several methods for linear elastic analysis are described in the sections that follow. All are used in engineering practice and may be used for estimating the load effects for all limit states.

Behavior, Structural Idealization, and Modeling. Again consider the slab-girder system shown in Figure 6.9(a). The spatial dimensionality is a primary modeling assumption. The system may be modeled as a 1-, 1.5-, 2-, 2.5-, or 3-dimensional system. The 1-D system is shown in Figure 6.9(b). This system is a beam and may be modeled as such. Obviously, this is a simple model and is very attractive for design. The primary issue is how the load is distributed to the girder, which is traditionally done by using an empirically determined distribution factor to transform the 3-D system to a 1-D system. In short, the vehicle load (or load effect) from the beam analysis is multiplied by a factor that is a function of the relative stiffnesses of the slab-girder

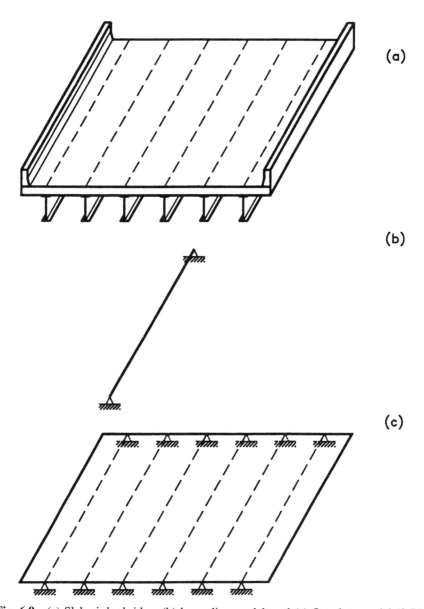

Fig. 6.9   (a) Slab-girder bridge, (b) beam-line model, and (c) flat plate model (2-D).

(d)

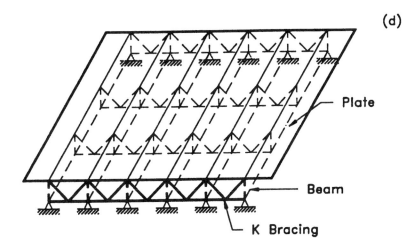

Plate

Beam

K Bracing

(e)

(f)

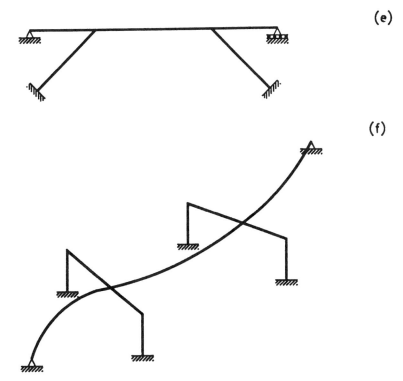

Fig. 6.9   (d) plane frame model (1.5-D), (e) 3-D model, and (f) space frame model (2.5-D).

system. This transforms the beam load effect to the estimated load effect in the system. Herein this procedure is called the beam-line method because only one girder is considered as opposed to modeling the entire bridge as a single beam.

A 2-D system is shown in Figure 6.9(c). This system eliminates the vertical dimension. What results is a system that is usually modeled with thin-plate theory for the deck combined with standard beam theory for the girders. The girder is brought into the plane of the deck (or plate) and supports are considered at the slab level. The eccentricity of both may be considered and included. The in-plane effects are usually neglected. Another type of 2-D system is the plane frame shown in Figure 6.9(e). Often the loads are distributed to the frame by distribution factors using the beam-line method. The analysis is performed on the plane frame.

In the 1.5-D system, the distribution factors are established by a 2-D system, but the girder actions are established using a 1-D system. This procedure is done because several computer programs exist for beam-line analysis and designs that are 1-D, but the designer wishes to use a refined procedure for the determination of the distribution factors, rather than that using the empirically based methods.

A 3-D system is shown in Figure 6.9(d). Here the full dimensionality is maintained. Minor components such as cross frames, diaphragms, and so on are often included. This model is the most refined and requires the most designer time and computer resources to perform.

The 2.5-D system typically uses a single-girder line in combination with other components and subsystems. Such a system is shown in Figure 6.9(f) where a curved box girder and its piers are modeled with space frame elements.

All of these methods are viable and have their place in engineering practice. It is not always appropriate, practical, or desirable to use the most refined method available. The complexity of the system, the load effects sought, the reason for the analysis, whether it be for design or evaluation, all are important considerations in the selection of the modeling procedures. The previous discussion is summarized in Table 6.4.

Beam-Line Method

Distribution Factor Method—Concepts. As previously described, the spatial dimensionality of the system can be reduced by using a distribution factor. This factor is established by analyzing the system with a refined method to establish the actions in the girders. For this discussion, bending moment is used for illustration but shear could also be used. The maximum moment at a critical location is determined with an analytical or numerical method and is denoted as M_{refined}. Next, the same load is applied to a single girder and a 1-D beam analysis is performed. The resulting maximum moment is denoted as M_{beam}. The distribution factor is defined as

TABLE 6.4 Spatial Modeling

Spatial Dimensionality	Mathematical Model	Numerical Model (Examples)	Figures
1	Beam theory	Stiffness (displacement) method Flexibility (force) method Consistent deformations Slope deflection Moment distribution	6.8(b)
2	Thin-plate theory Beam theory	Grillage Finite strip Finite element Harmonic analysis Classical plate solutions	6.8(c), 6.8(d)
3	Theory of elasticity Thin-plate theory Beam theory	Grillage Finite strip Finite element Classical solutions	6.8(e)
1.5	Thin-plate theory Beam theory	Grillage Finite strip Finite element Harmonic analysis Classical plate solutions	Not shown
2.5	Beam theory	Finite element	Fig. 6.9(f)

$$g = \frac{M_{\text{refined}}}{M_{\text{beam}}}$$

In the case of a 1.5-D analysis, this factor is used to convert the load effects established in the beam-line analysis to the estimated results of the entire system. For example, analyze the beam line for the live load and then multiply by the distribution factor g to obtain the estimated load effect in the system.

Alternatively, many analyses can be performed for numerous bridges and the effects of the relative stiffness of the various components, geometry effects, and load configuration may be studied. The results of these analyses are then used to establish empirically based formulas that contain the system parameters as variables. These formulas can then be used by designers to estimate the distribution factors *without* performing the refined analysis. Certainly, some compromise may be made in accuracy but this method generally gives good results. The AASHTO distribution factors are based on this concept and are presented in Table 6.5 where they are discussed in more detail.

Background. The AASHTO Specification has employed distribution factor methods for many years. In the most common case, the distribution factor was

TABLE 6.5 Vehicles per Girder for Concrete Deck on Steel or Concrete Beams; Concrete T-Beams; T- and Double T-Sections Transversely Post-Tensioned Together[a]

Action/Location	AASHTO Table	Distribution Factors (mg)[c]	Skew Correction Factor[b]	Range of Applicability
A. Moment interior girder	4.6.2.2b–1	One design lane loaded: $$mg^{SI}_{moment} = 0.06 + \left(\frac{S}{4300\ mm}\right)^{0.4} \left(\frac{S}{L}\right)^{0.3} \left(\frac{K_g}{Lt_s^3}\right)^{0.1}$$ Two or more (multiple) design lanes loaded: $$mg^{MI}_{moment} = 0.075 + \left(\frac{S}{2900\ mm}\right)^{0.6} \left(\frac{S}{L}\right)^{0.2} \left(\frac{K_g}{Lt_s^3}\right)^{0.1}$$	$1 - C_1(\tan\theta)^{1.5}$ $$C_1 = 0.25 \left(\frac{K_g}{Lt_s^3}\right)^{0.25} \left(\frac{S}{L}\right)^{0.5}$$ If $\theta < 30°$. then $C_1 = 0.0$ If $\theta > 60°$, then $\theta = 60°$	$1100 \leq S \leq 4900\ mm$ $110 \leq t_s \leq 300\ mm$ $6000 \leq L \leq 73\ 000\ mm$ No. of beams ≥ 4 $30° \leq \theta \leq 60°$
B. Moment exterior girder	4.6.2.2d–1	One design lane loaded: Use lever rule Two or more (multiple) design lanes loaded: $$mg^{ME}_{moment} = e*mg^{MI}_{moment}$$ $$e = 0.77 + \frac{d_e}{2800\ mm} \geq 1.0$$ d_e is positive if girder is inside of barrier, otherwise negative	N/A	$-300 \leq d_e \leq 1700\ mm$

284

C. Shear interior girder	4.6.2.2.3a–1	One design lane loaded $mg^{SI}_{\text{shear}} = 0.36 + \dfrac{S}{7600 \text{ mm}}$ Two or more (multiple) design lanes loaded $mg^{MI}_{\text{shear}} = 0.2 + \dfrac{S}{3600 \text{ mm}} - \left(\dfrac{S}{10\,700 \text{ mm}}\right)^2$	$1.0 + 0.20 \left(\dfrac{Lt^3_s}{K_g}\right)^{0.3} \tan\theta$	$1100 \le S \le 4900$ mm $110 \le t_s \le 300$ mm $6000 \le L \le 73\,000$ mm $4 \times 10^9 \le K_g \le 3 \times 10^{12}$ mm^4 No. of beams ≥ 4 $0° \le \theta \le 60°$
D. Shear exterior girder	4.6.2.2.3b–1	One design lane loaded Use lever rule Two or more (multiple) design lanes loaded $mg^{ME}_{\text{shear}} = e * mg^{MI}_{\text{shear}}$ $e = 0.6 + \dfrac{d_e}{3000 \text{ mm}}$ d_e is positive if girder is inside of barrier, otherwise negative	N/A	$-300 \le d_e \le 1700$ mm

[a]See Table 2.2 for applicable cross sections.
[b]Not applicable = N/A.
[c]Equations include multiple presence factor; for lever rule engineer must perform factoring by m.

$$g = \frac{S}{D}$$

where S is the girder spacing, (ft) and D is a constant depending on bridge type, the number of lanes loaded, and g may be thought of as the number of wheel lines carried per girder.

For example, for a concrete slab on a steel girder $D = 5.5$ was used for cases where two or more vehicles are present. Obviously, this is a very simplistic formula and easy to apply, but as expected, it does not always provide good estimates of the girder load in the full system. It has been shown by Zokaie et al. (1991) and Nowak (1993) that this formulation underestimates the load effects with close girder spacing and overestimates with wider spacing. To refine this approach, research was conducted to develop formulas that are based on more parameters and provide a better estimate of the true system response. This work was performed under NCHRP Project 12-26 (Zokaie et al., 1991) and provides the basis for the distribution factors presented in AASHTO [A4.6.2.2].*

AASHTO Specification—Distribution Factors. The distribution factors may be used for bridges with fairly regular geometry. As stated in AASHTO [A4.6.2.2], the method is limited to systems with

- A constant cross section.
- The number of beams is four or more.
- The beams are parallel and have approximately the same stiffness.
- The roadway part of the cantilever overhang does not exceed 3.0 ft (910 mm).
- The plan curvature is small [A4.6.1.2].
- The cross section is consistent with the sections shown in Table 2.2.

The provisions for load distribution factors are contained in several AASHTO articles and only a few are discussed here. These articles represent some of the most important provisions in Section 4 of the AASHTO Specification, and because of the many algebraically complex equations, these are not presented in the body of this discussion. For the sake of brevity, the most common bridge types, the slab and slab-girder bridge, are discussed here in detail. The analysis of other common types is discussed later. The distribution factors for slab-girder bridges are given in Table 6.5.

---

*The article number in AASHTO (1994) LRFD Bridge Specifications are enclosed in brackets and preceded by the letter A if specifications and by the letter C if commentary.

In Table 6.5

S is the girder spacing (mm).

L is the span length (mm).

t_s is the slab thickness (mm).

K_g is the longitudinal stiffness parameter (mm⁴).

$K_g = n(I_g + e_g^2 A)$, where

  n is the modular ratio, $(E_{\text{girder}}/E_{\text{deck}})$.

  I_g is the moment of inertia of the girder (mm⁴).

  e_g is the girder eccentricity which is the distance from the girder centroid to the middle centroid of the slab (mm).

  A is the girder area (mm²).

d_e is the distance from the center of the exterior beam and the inside edge of the curb or barrier (mm).

θ is the angle between the centerline of the support and a line normal to the roadway centerline.

The *lever rule* is a method of analysis. It involves a statical distribution of load based on the assumption that each deck panel is simply supported over the girder, except at the exterior girder that is continuous with the cantilever. Because the load distribution to any girder other than one directly next to the point of load application is neglected, the lever rule is a conservative method of analysis.

The equations in Table 6.5 were developed by Zokaie et al. (1991). Here investigators performed hundreds of analyses on bridges of different types, geometrics, and stiffnesses. Many of these structures were actual bridges that were taken from the inventories nationwide. Various computer programs were used for analysis and compared to experimental results. The programs that yielded the most accurate results were selected for further analysis in developing the AASHTO formulas. The database of actual bridges was used to determine "an average bridge" for each type. Within each type, the parametric studies were made to establish the distribution factor equations. Example results for the slab-girder bridge type are shown in Figure 6.10. Note that the most sensitive parameter for this type of bridge is the girder spacing. This observation is consistent with the traditional AASHTO distribution factor of $S/5.5$ ft. In fact, the division of the slope of this line, which is approximately 1.25, into the average girder spacing from the database, which is 7.5 ft yields $D = 6.0$, or approximately the value of $D = 5.5$ used by AASHTO for many years. It is important to note that the span length and girder stiffness affect the load distribution but to a lesser extent. This affect is reflected in the equations presented in Table 6.5. Unlike the previous AASHTO equations, the important parametric properties of the bridge were used to develop prediction models based on a power law. Each parameter was assumed to be independent of others in its affect on the distribution model. Although this is

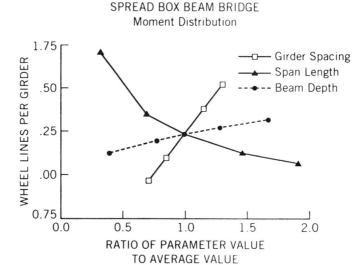

Fig. 6.10   Parametric studies. [After Zokaie et al., 1991].

probably not strictly true, the resulting equations seem to work well. The results of Table 6.5 are compared to finite element analysis (more rigorous and assumed to be more accurate) in Figure 6.11. In Figure 6.11(a), the rigorous analyses are compared to the old AASHTO procedures $[g = (S/D)]$, and in Figure 6.11(b), the rigorous analyses are compared to the equations of Table 6.5. Notice the great variability in the former and the decrease variability of the latter. Hence, the additional terms are necessary to better predict the system response. Traditionally, AASHTO has based analysis on the wheel line or half the axle weight. In the present specification, the analysis is assumed to be based on the entire vehicle weight. Thus, if one compares the distribution factors historically used by AASHTO to those presently used, then the traditional factors must be divided by 2, or the present factor must be multiplied by 2.

The single design lane formulas were developed with a single design truck and the multilane loaded formulas were developed with two *or* more trucks. Therefore, the most critical situation for two, three, or more vehicles was used in the development. The multiple presence factors given in Table 4.6 were included in the analytical results upon which the formulas are based. Thus, the *multiple presence factors are not to be used in conjunction with the factors given in Table 6.5, but rather the multiple presence is implicitly included in these factors.*

The development of the present AASHTO (1994) distribution factors was based on simply supported bridges. The investigators also studied systems to quantify the effect of continuity. Given the relative insensitivity of girder stiffness to the distribution factors (see Fig. 6.10), it is expected that continuity

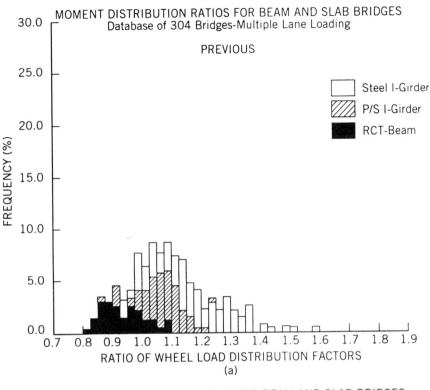

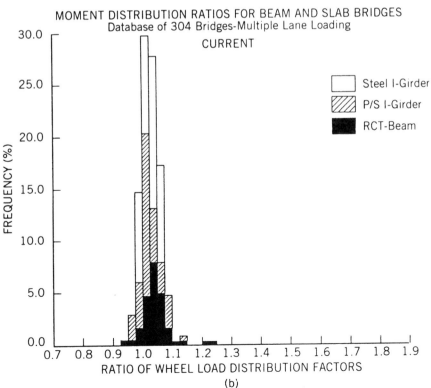

Fig. 6.11 Comparison of AASHTO distribution factor with rigorous analysis. [After Zokaie et al., 1991].

does not significantly affect the distribution factors. Zokaie determined that the effect of continuity was between 1.00 and 1.10 for most systems and suggested associated adjustments. The specification writers chose to eliminate this refinement because

- Correction factors dealing with 5% adjustments were thought to imply misleading levels of accuracy in an approximate method.
- Analysis carried out on a large number of continuous beam-slab-type bridges indicates that the distribution coefficients (factors) for negative moment exceed those obtained for positive moment by approximately 10%. On the other hand, it had been observed that stresses at or near internal bearings are reduced due to the fanning of the reaction force. This reduction is about the same magnitude as the increase in distribution factors, hence the two tend to cancel each other out.

READERS NOTE: Most of the examples developed in Chapter 6 were developed for a bridge with a span length of 35 ft (10 668 mm). During the writing of this book, the book was converted to SI units. The examples presented in this chapter involved numerous grillage, finite element, and finite strip analyses, which required many hours of development, sometimes required in order to report only a few numbers. Because the vehicle definitions are not precisely the same for the U.S. Customary and SI Specifications, the accurate SI conversion would require repeating all the analyses initially performed in U.S. Customary Units. This conversion was prohibitive and therefore the examples are described in SI and/or U.S. Customary Units with the analysis performing in this unit system. The intent of the problems is the distribution of load that is clearly independent of units.

EXAMPLE 6.1

The slab-girder bridge illustrated in Figure 6.12(a) with a simply-supported span of 35 ft (10 688 mm) is used in this example and several others that follow. Model the entire bridge as a single beam to determine the support reactions, shears, and bending moments for one- and two-lanes loaded using the AASHTO design truck.

A free body diagram is shown in Figure 6.12(b) with the design truck positioned near the critical location for flexural bending moment. Although this position does not yield the absolute maximum moment, which is 361.2 ft kps (498.7 kN m) (see Example 5.10), it is close to the critical location and this position facilitates analysis in later examples. The resulting moment diagram is shown in Figure 6.12(c). Note the maximum moment is 493.2 kN m (358.4 ft kips) for one-lane loaded, which is within 1% of the absolute maximum moment. This value is doubled for two trucks positioned on the

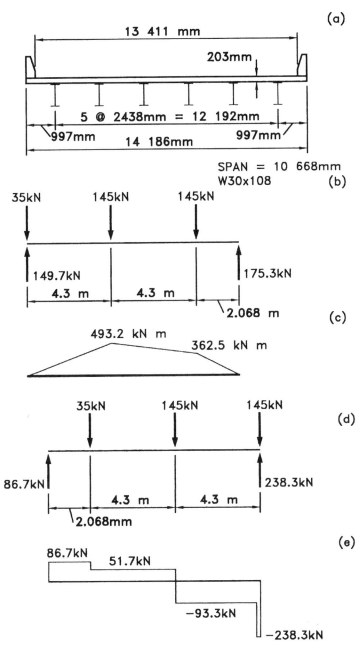

Fig. 6.12 (a) Cross section of a slab-girder bridge, (b) free body diagram—load for near critical flexural moment, (c) moment diagram, (d) free body diagram—load for near critical shear/reactions, and (e) shear diagram.

bridge giving a maximum of 986.4 kN m (716.8 ft kips). These values are used repeatedly throughout several examples that follow. The critical section for design is at the location of the maximum statical moment. This location is also used in several examples that follow.

A free body diagram is shown in Figure 6.12(d) with the design truck positioned for maximum shear/reaction force. The resulting maximum is 238.3 kN (52.8 kips) for one-lane loaded, and 476.6 kN (105.6 kips) for two-loaded lanes. See Figure 6.12(e). These values are also used in the examples that follow.

EXAMPLE 6.2

Determine the AASHTO distribution factors for bridge shown in Figure 6.12(a).

A girder section is illustrated in Figure 6.13 (see AISC 1994 for girder properties). The system dimensions and properties are as follows:

Girder spacing, $S = 8$ ft (2438 mm)
Span length, $L = 35$ ft (10 668 mm)
Deck thickness, $t_s = 8$ in. (203 mm)
Deck modulus of elasticity, $E_c = 3600$ ksi (24.82 GPa)
Girder modulus of elasticity, $E_s = 29\,000$ ksi (200.0 GPa)

Modular ratio, $n = \dfrac{E_s}{E_c} = \dfrac{29\,000}{3600} = 8.05$, use 8

Girder area, $A_g = 31.7$ in.^{2} (20 500 mm^2) Girder moment of inertia, $I_g = 4470$ in.^{4} (1860 $\times$ 10^6 mm^4)
Girder eccentricity, for example, $= t_s/2 + d/2 = 8/2 + 29.83/2 = 18.92$ in. (480 mm)

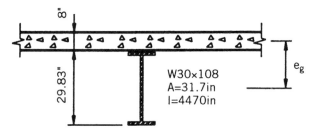

Fig. 6.13   Girder cross section.

Stiffness parameter, $K_g = n(I_g + e_g^2 A_g) = 8[4470 + (18.92^2)(31.7)] = $ 126 500 in.^{4} (52.6 $\times$ 10^9 mm^4)

$d_e = 3$ ft–$3\frac{1}{4}$ in. (cantilever) $- 1$ ft $3\frac{1}{4}$ in. (barrier) $= 2.0$ ft (610 mm)

The AASHTO distribution factors for moments are determined using rows A and B of Table 6.5.

The distribution factor for moment in the interior girder for one lane loaded is (Note the multiple presence factor m is included in the equations so this is denoted mg where m is included.)

$$mg^{SI}_{moment} = 0.06 + \left(\frac{S}{4300 \text{ mm}}\right)^{0.4} \left(\frac{S}{L}\right)^{0.3} \left(\frac{K_g}{Lt_s^3}\right)^{0.1}$$

$$mg^{SI}_{moment} = 0.06 + \left(\frac{2438}{4300}\right)^{0.4} \left(\frac{2438}{10\,668}\right)^{0.3} \left(\frac{52.6 \times 10^9}{(10\,668)(203^3)}\right)^{0.1} = 0.54$$

The distribution factor for moment in the interior girder for multiple lanes loaded is

$$mg^{MI}_{moment} = 0.075 + \left(\frac{S}{2900}\right)^{0.6} \left(\frac{S}{L}\right)^{0.2} \left(\frac{K_g}{Lt_s^3}\right)^{0.1}$$

$$mg^{MI}_{moment} = 0.075 + \left(\frac{2438}{2900 \text{ mm}}\right)^{0.6} \left(\frac{2438}{10\,668}\right)^{0.2} \left(\frac{52.6 \times 10^9}{(10\,668)(203^3)}\right)^{0.1} = 0.71$$

The distribution factor for moment in the exterior girder for multiple lanes loaded requires an adjustment factor

$$e = 0.77 + \frac{d_e}{2800 \text{ mm}} \geq 1.0$$

$$e = 0.77 + \frac{610}{2800} = 0.99 \quad \therefore \text{ use } e = 1.0$$

The adjustment factor for moment is multiplied by the factor for the interior girder and the result is

$$mg^{ME}_{moment} = e(mg^{MI}_{moment}) = 1.00(0.71) = 0.71$$

For the distribution factor for shear, rows C and D in Table 6.5 are used. The distribution factor for the interior girder with one lane loaded is

$$mg^{SI}_{shear} = 0.36 + \frac{S}{7600 \text{ mm}} = 0.36 + \frac{2438}{7600} = 0.68$$

Similarly, the factor for shear with multiple lanes loaded is

$$mg_{\text{shear}}^{\text{MI}} = 0.2 + \frac{S}{3600 \text{ mm}} - \left(\frac{S}{10\,700 \text{ mm}}\right)^2$$

$$= 0.2 + \frac{2438}{3600} - \left(\frac{2438}{10\,700}\right)^2 = 0.82$$

The adjustment for shear in the exterior girder is given in row D of Table 6.5. The calculation is shown below.

$$e = 0.6 + \frac{d_e}{3000 \text{ mm}} = 0.6 + \frac{610}{3000} = 0.80$$

The adjustment is multiplied by the interior distribution factor, the result is

$$mg_{\text{shear}}^{\text{ME}} = e(mg_{\text{shear}}^{\text{MI}}) = 0.80(0.82) = 0.66$$

The lever rule is used for the exterior girder loaded with one design truck. The details are addressed in the following example. The result is $mg_{\text{shear or moment}}^{\text{SI}} = 0.625$ times 1.2 (multiple presence factor) $= 0.75$ for both shear and moment. The AASHTO results are summarized in Table 6.6.

EXAMPLE 6.3

Use the lever method to determine the distribution factors for the bridge shown in Figure 6.12(a). Note U.S. customary units are used to establish the distribution factors because of clarity of presentation.

Exterior Girder. Consider Figure 6.14. The deck is assumed to be simply supported by each girder except over the exterior girder where the cantilever is continuous. If we consider truck No. 1, the reaction at A (exterior girder load) is established by balancing the moment about B,

TABLE 6.6   AASHTO Distribution Factor Method Results

Girder Location	Number of Lanes Loaded	Moment (kN m)	Moment Distribution Factor, (mg)	Girder Moment (kN m)	Simple Beam Reaction (kN)	Shear Distribution Factor (mg)	Girder Shear, (kN)
Exterior	1	498.7	$0.625 \times 1.2 = 0.75$	374.0	238.3	0.75	178.7
Exterior	2	498.7	0.71	354.1	238.3	0.66	157.3
Interior	1	498.7	0.54	269.3	238.3	0.68	162.0
Interior	2	498.7	0.71	354.1	238.3	0.82	195.4

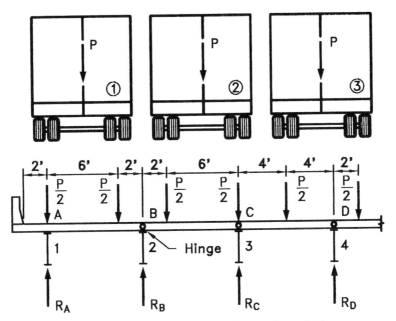

Fig. 6.14 Free body diagram—lever rule method.

$$R_A(8) = \left(\frac{P}{2}\right)(8) + \left(\frac{P}{2}\right)(2)$$

which reduces to

$$R_A = \left(\frac{P}{2}\right) + \left(\frac{P}{2}\right)\left(\frac{2}{8}\right) = 0.625P$$

The fraction of the truck weight P that is carried by the exterior girder is 0.625. The multiple presence factor of 1.2 (see Table 4.6) is applicable for the one-lane loaded case. Thus, the girder distribution factors are

$$mg_{\text{shear or moment}}^{\text{SE}} = (1.2)(0.625) = 0.75$$

and

$$mg_{\text{shear or moment}}^{\text{ME}} = (1.0)(0.625) = 0.625$$

This factor is "statically" the same for one- and two-lanes loaded because the wheel loads from the adjacent truck (No. 2) cannot be distributed to the exterior girder. This is because all the wheels lie inside the first interior girder and the effect of their load cannot be transmitted across the assumed hinge. As illustrated, the difference is due to the multiple presence factor.

Interior Girder. The distribution factor for the interior girder subjected to two or more loaded lanes is established by considering trucks numbered two and three, each of weight P, positioned with axles on deck panels BC and CD, as shown in Figure 6.14. Equilibrium requires that the reaction at C is

$$R_c = \left(\frac{2}{8}\right)\left(\frac{P}{2}\right) + \left(\frac{P}{2}\right) + \left(\frac{P}{2}\right)\left(\frac{4}{8}\right) + \left(\frac{P}{2}\right)(0) = 0.875P$$

and the distribution factor (multiple presence factor $= 1.0$) is

$$mg_{\text{shear or moment}}^{\text{MI}} = (1.0)(0.875) = 0.875$$

Only truck No. 2 is considered for the case of one loaded lane on an interior girder. This truck has one wheel line directly over girder No. 3 and one wheel line 6 ft from the girder. If we use statics, the girder reaction at C is

$$R_C = \left(\frac{P}{2}\right) + \left(\frac{2}{8}\right)\left(\frac{P}{2}\right) = 0.625P$$

and the distribution factor is

$$mg_{\text{shear or moment}}^{\text{SI}} = (1.2)(0.625) = 0.75$$

The distribution factors for shear and moment are the same under the pinned-panel assumption. The lever rule results are summarized in Tables 6.7. The format for these tables is consistently used in the remaining examples in this chapter, which permits the ready comparison of results from the various methods of analysis.

Grillage Method. Because the AASHTO and lever rule distribution factors are approximate, the engineer may wish to perform a more rigorous and accurate analysis. The advantages of more rigorous analysis include:

TABLE 6.7 Lever Rule Results

Girder Location	Number of Lanes Loaded	Moment (kN m)	Moment Distribution Factor (*mg*)	Girder Moment (kN m)	Simple Beam Reaction (kN)	Shear Distribution Factor (*mg*)	Girder Shear (kN)
Exterior	1	498.7	0.75	374.0	238.3	0.75	178.7
Exterior	2	498.7	0.625	311.7	238.3	0.625	148.9
Interior	1	498.7	0.75	374.0	238.3	0.75	178.7
Interior	2	498.7	0.875	436.4	238.3	0.875	208.5

- The simplifying factors/assumptions that are made in the development of distribution factors for beam-line methods may be obviated.
- The variability of uncertain structural parameters may be studied for their effect on the system response. For example, continuity, material properties, cracking, nonprismatic effects, and support movements may be of interest.
- More rigorous models are developed in the design process and can be used in the rating of permit (overweight) vehicles and determining a more accurate overload strength.

One of the best mathematical models for the deck is the thin plate that may be modeled with the biharmonic equation (Timoshenko and Woinowsky-Kreiger, 1959; Ugural, 1981)

$$\nabla^4 w = \frac{\partial^4 w}{\partial x^4} + 2\frac{\partial^4 w}{\partial x^2 \partial y^2} + \frac{\partial^4 w}{\partial y^4} = \frac{p(x)}{D} \tag{6.4}$$

where

w is the vertical translation
x is the transverse coordinate
y is the longitudinal coordinate
p is the vertical load,
D is the plate rigidity, equal to

$$D = \frac{Et^3}{12(1 - v^2)}$$

v is Poisson's ratio
t is the plate thickness
E is the modulus of elasticity

Equation 6.4 is for an isotropic (same properties in all directions) slab. Other forms are available for plates that exhibit significant orthotropy due to different reinforcement in the transverse and longitudinal directions (Timoshenko and Woinowsky-Kreiger, 1959; Ugural 1981). The development of Eq. 6.4 is based on several key assumptions: the material behaves linearly elastically, the strain profile is linear, the plate is isotropic, the vertical stresses due to the applied load are neglected, and the deformations are small relative to the dimensions of the plate.

Closed-form solutions to Eq. 6.4 are limited to a few cases that are based on simplified boundary conditions and loads. Even fewer solutions are available for girder-supported systems. Thus, approximate techniques or numerical models are used for the solution of Eq. 6.4; the most common methods include the grillage, finite element, and finite strip methods.

To gain a better understanding of the development and limitations of Eq. 6.4, the reader is referred to common references on the analysis of plates (Timoshenko and Woinowsky-Kreiger, 1959; Ugural, 1981). Due to the focus and scope of this work, it suffices here to take an abbreviated and very applied approach.

Consider the first term of Eq. 6.4 and neglect the transverse terms, then Eq. 6.4 becomes

$$\frac{\partial^4 w}{\partial x^4} = \frac{p(x)}{D} \tag{6.5}$$

which is the same as Eq. 6.1, the mathematical model for a beam. Now neglect only the middle term, the Eq. 6.4 becomes

$$\nabla^4 w = \frac{\partial^4 w}{\partial x^4} + \frac{\partial^4 w}{\partial y^4} = \frac{p(x)}{D} \tag{6.6}$$

which is the mathematical model for a plate system that has no torsional stiffness or associated torsional actions. In a practical sense such systems do not exist, and are merely mathematical models of a system where torsion exists but is neglected. This type of system would be similar to modeling a plate with a series of crossing beams where one element sits on top of the other as shown in Figures 6.15(a) and 6.15(b). Note that at the intersection of the beams the only interaction force between the element is a vertical force. This type of connection excessively simplifies the model of the deck, which is a continuum. In the continuum, a flexural rotation in one direction causes torsional rotation in an orthogonal direction. Consider the grillage joint shown in Figure 6.15(c). Here the joint is continuous for rotation in all directions, that is, the displacement of the joint is defined with the three displacements (degrees of freedom) shown in Figure 6.15(d), which includes vertical translation and two rotations. This type of joint, in combination with elements that have both flexural and torsional stiffness, is more like the continuum, and therefore, models it more accurately. This type of numerical model is called a grillage.

Grillage models became popular in the early 1960s with the advancement of digital computers. As the methodologies for the stiffness analysis (or displacement method) of frames were well known, researchers looked for convenient ways to model continua with frame elements. The grillage model is such a technique. Ideally the element stiffnesses in the grillage model would be such that when the continuum deck is subjected to a series of loads, the displacement of the continuum and the grillage are identical. In reality, the grillage can only approximate the behavior of the continuum described by

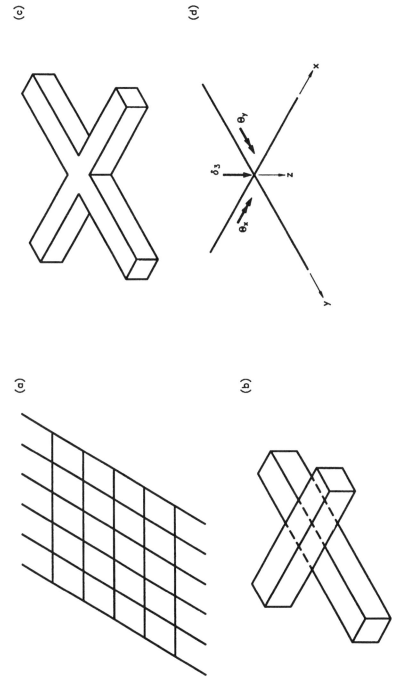

Fig. 6.15 (a) Grillage model, (b) crossing with translational continuity, (c) crossing with translational and rotational continuity, and (d) degrees of freedom in grillage (plane grid) modeling.

299

Eq. 6.4. The reason for this difference is twofold: (1) The displacement in the grillage tends to be more irregular (bumpy) than the continuum, and (2) The moment in the grillage is a function of the curvature along the beam. In the plate, the moment is a function of the curvatures in two orthogonal directions due to Poisson's effect. Fortunately, these effects are small and the grillage method has been shown to be a viable method of analysis.

Some advocates of the finite element and strip methods are quick to discount the grillage method because it is nonrigorous. But remember that such methods are used to obtain reasonable distribution of internal actions while accounting for equilibrium (recall the lower bound theorem discussed earlier). Both advocates and critics have valid points and a few of these are listed below:

- Grillages can be used with any program that has plane grid or space frame capabilities.
- The results are easily interpreted and equilibrium is easily checked by free body diagrams of the elements and system as a whole.
- Most all engineers are familiar with the analysis of frames.

The disadvantages are several:

- The method is nonrigorous and does not exactly converge to the exact solution of the mathematical model.
- Obtaining good solutions requires some experience with the grillage method. The mesh design and refinement can be somewhat of an art form.
- The assignment of the cross-sectional properties requires some discretion.

Hambly (1991) offers an excellent and comprehensive reference on modeling with grillages. The engineer interested in performing a grillage is encouraged to obtain this reference. Some of Hambly's suggestions regarding the design of meshes are paraphrased below:

- Consider how the designer wants the bridge to behave and place beam elements along lines of strength/stiffness, for example, parallel to girders, along edge beams and barriers, and along lines of prestress.
- The total number of elements can vary widely. It can be one element in the longitudinal direction if the bridge is narrow and behaves similarly to a beam, or can it be modeled with elements for the girders and other elements for the deck for wide decks where the system is dominated by the behavior of the deck. Elements need not be spaced closer than two to three times the slab thickness.
- The spacing of the transverse elements should be sufficiently small to distribute the effect of concentrated wheel loads and reactions. In the vicinity of such loads, the spacing can be decreased for improved results.

The element cross-sectional properties are usually based on the gross or un-cracked section and are calculated on a per unit length basis. These properties are multiplied by the center-to-center spacing of the elements to obtain the element properties, herein called the tributary length. Two properties are re-quired for the grillage model: flexural moment of inertia and the torsional constant. The moment of inertia is the familiar second moment of area, which is equal to

$$i_{deck} = \frac{bt^3}{12} \tag{6.7}$$

The torsional constant for grillage element is

$$j_{deck} = \frac{bt^3}{6} = 2i_{deck} \tag{6.8}$$

The moment of inertia I_{girder} for a beam element is determined in the usual way and its eccentricity e_g (for a composite beam) is accounted by

$$I = I_{girder} + e_g^2 A_{girder} \tag{6.9}$$

For noncomposite systems, e_g is zero and the beam is assumed to be at the middle surface of the deck and the beam's axial stiffness does not contribute.

For open sections that are comprised of thin rectangular shapes such as a wide flange or plate girder, the torsional constant is approximated by

$$J = \sum_{all\ rectangles} \frac{bt^3}{3} \tag{6.10}$$

where b is the long side and t is the narrower side ($b > 5t$). For open steel shapes, the torsional constant is usually small relative to the other parameters and has little affect on the response. For rectangular shapes that are not thin, the approximation is

$$J = \frac{3b^3t^3}{10(b^2 + t^2)} \tag{6.11}$$

The use of these properties is illustrated in the following example.

EXAMPLE 6.4

Use the grillage method to determine the end shear (reactions) and maximum bending moments in the girders in Figure 6.12(a), which is illustrated in

Example 6.1. In addition, determine the distribution factors for moment and shear for girders for one- and two-lanes loaded.

The slab-girder bridge is discretized by a grillage model with the two meshes shown in Figures 6.16(a) and 6.16(b). The section properties are calculated below.

Girder Properties

$$E_s = 29\,000 \text{ ksi (200.0 GPa)}$$

$$A_g = 31.7 \text{ in.}^2 \text{ (20 453 mm}^2)$$

$$d = 29.83 \text{ in. (4536 mm)}$$

$$e_g = \left(\frac{t_s}{2}\right) + \left(\frac{d}{2}\right) = \left(\frac{8}{2}\right) + \left(\frac{29.83}{2}\right) = 18.92 \text{ in. (481 mm)}$$

$$I_g = 4470 \text{ in.}^4 \text{ (noncomposite girder)} = 1.860 \times 10^9 \text{ mm}^4$$

$$J_g = 4.99 \text{ in.}^4 \text{ (noncomposite girder)} = 2.077 \times 10^6 \text{ mm}^4$$

$$I_g \text{ (composite girder)} = I_g + e_g^2 A_g = 4470 + 18.92^2 \text{ (31.7)}$$

$$= 15\,810 \text{ in.}^4 \text{ (steel)} = 6.58 \times 10^9 \text{ mm}^4$$

Deck Properties

$$E_c = 3600 \text{ ksi (24.82 GPa)}$$

$$t_s = 8 \text{ in. (203 mm)}$$

$$v = 0.15$$

$$i_s = \frac{1}{12}(12)(8^3) = 512 \text{ in.}^4 \text{ (per ft)} = 700\,000 \text{ mm}^4 \text{ (per mm)}$$

$$j_s = \frac{1}{6}(12)(8^3) = 1024 \text{ in.}^4 \text{ (per ft)} = 1\,400\,000 \text{ mm}^4 \text{ (per mm)}$$

Element Properties. The elements that model the girders have the same properties as indicated above. Note that only the moment of inertia and the torsional constant are required in the grillage. The element properties for the deck are a function of the mesh size. For the coarse mesh in Figure 6.16(a), the elements oriented in the transverse (*x*-direction) are positioned at 7 ft (2134 mm) center to center. Therefore, the properties assigned to these elements are

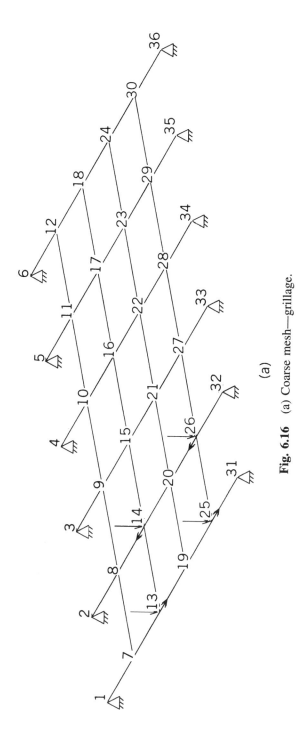

Fig. 6.16 (a) Coarse mesh—grillage.

(a)

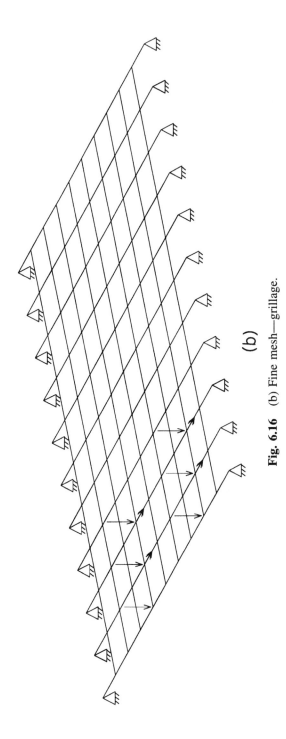

Fig. 6.16 (b) Fine mesh—grillage.

(b)

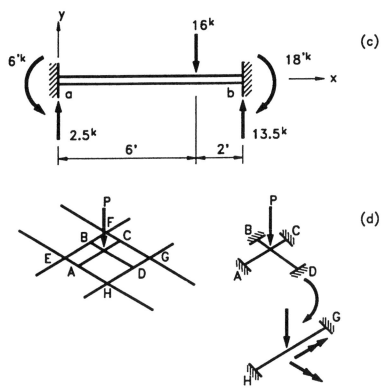

Fig. 6.16 (c) Fixed–fixed beam with wheel load (equivalent joint loads) and (d) load positioned between elements.

$$I_s = i_s(\text{tributary length}) = 512 \frac{\text{in.}^4}{\text{ft}} (7 \text{ ft})$$

$$= 3584 \text{ in.}^4 \text{ (transverse)} = 1.49 \times 10^9 \text{ mm}^4$$

$$J_s = j_s(\text{tributary length}) = 1024 \frac{\text{in.}^4}{\text{ft}} (7 \text{ ft})$$

$$= 7168 \text{ in.}^4 \text{ (transverse)} = 2.98 \times 10^9 \text{ mm}^4$$

The properties for the portion of the deck above the girders (at 8-ft centers) are

$$I_s = i_s(\text{tributary length}) = 512 \frac{\text{in.}^4}{\text{ft}} (8 \text{ ft})$$

$$= 4096 \text{ in.}^4 \text{ (longitudinal)} = 1.70 \times 10^9 \text{ mm}^4$$

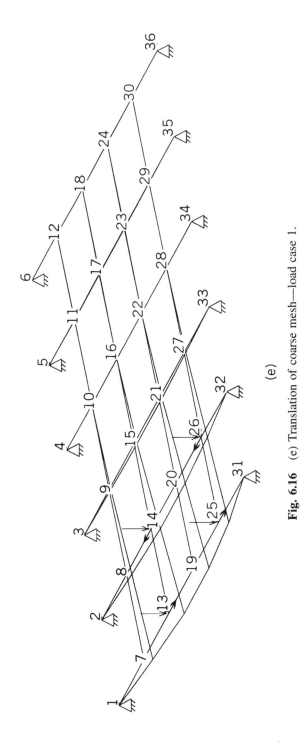

Fig. 6.16 (e) Translation of coarse mesh—load case 1.

(e)

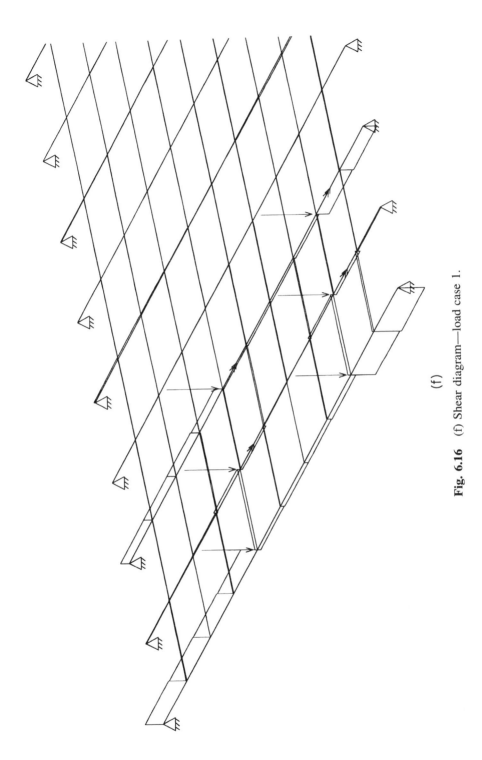

Fig. 6.16 (f) Shear diagram—load case 1.

(f)

307

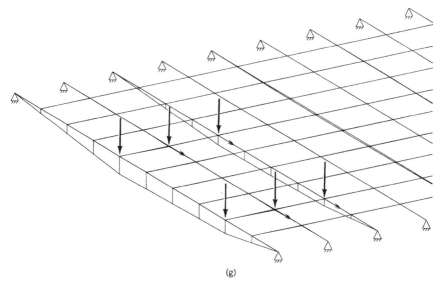

(g)

Fig. 6.16 (g) Moment diagram—load case 1.

$$J_s = j_s(\text{tributary length}) = 1024 \, \frac{\text{in.}^4}{\text{ft}} \, (8 \text{ ft})$$
$$= 8192 \text{ in.}^4 \text{ (longitudinal)} = 3.41 \times 10^9 \text{ mm}^4$$

For the fine mesh, the tributary width of the deck elements oriented in the transverse and longitudinal directions are 3.5 ft (1067 mm) and 4.0 ft (1219 mm), respectively. The associated element properties are

$$I_s = i_s(\text{tributary length}) = 512(3.5) = 1792 \text{ in.}^4 \text{ (transverse)}$$
$$= 746 \times 10^6 \text{ mm}^4$$
$$J_s = j_s(\text{tributary length}) = 1024(3.5) = 3584 \text{ in.}^4 \text{ (transverse)}$$
$$= 1.49 \times 10^9 \text{ mm}^4$$

and

$$I_s = i_s(\text{tributary length}) = 512(4) = 2048 \text{ in.}^4 \text{ (longitudinal)}$$
$$= 852 \times 10^6 \text{ mm}^4$$
$$J_s = j_s(\text{tributary length}) = 1024(4) = 4096 \text{ in.}^4 \text{ (longitudinal)}$$
$$= 1.70 \times 10^9 \text{ mm}^4$$

For the girder element properties, the associated properties of the beam and the slab contributions are added. The steel girder is transformed to concrete using the modular ratio of $n = 8$. The result for the fine mesh is

$$I_g = I_g(\text{composite beam})n + I_s$$

$$= 15\,410(8) + 2048 = 125\,300 \text{ in.}^4 = 52.1 \times 10^9 \text{ mm}^4$$

$$J_g = J_g(\text{composite beam})n + J_s$$

$$= 4.99(8) + 4096 = 4136 \text{ in.}^4 = 1.72 \times 10^9 \text{ mm}^4$$

The support boundary conditions are assumed to be restrained against translation in all directions at the girder ends. Although some torsional restraint may be present, it is difficult to estimate. By comparing the analysis of the system with both ends torsionally restrained and without, this effect was observed to be small and the torsionally unrestrained case is reported.

Eight load cases were used and are described below.

1. The design truck is positioned for near critical maximum midspan moment and end shear in the exterior girder (No. 1) for one-lane loaded [see Figs. 6.16(a) and 6.16(b)].
2. Number 1 is repeated for two-lanes loaded [see Fig. 6.16(b)].
3. The design truck is positioned for near critical maximum midspan moment in the interior girder (No. 3) for one-lane loaded [see Fig. 6.16(b)].
4. Number 3 is repeated for two-lanes loaded [see Fig. 6.16(b)].
5–8. Numbers 1–4 are repeated with the design vehicles moved so that the rear 32 kip (145 kN) axle is near the support to create critical shears and reactions.

Because some of the concentrated wheel loads lie between nodes, their statical equivalence must be determined. For example, the load that lies between nodes 13 and 14 in the coarse mesh is illustrated in Figure 6.16(a). The statical equivalent actions are determined from the end actions associated with this load applied on a fixed-ended beam as shown in Figure 6.16(c). The negative, or opposite, actions are applied to the grillage. The applied joint loads for the coarse mesh are illustrated in Table 6.8 for load case No. 1. Again, U.S. units are presented for clarity.

The nodal loads for the other load cases and for the fine mesh are established in a similar manner. It is common to neglect the joint load moments and assign the loads based on a simple beam distribution, hence the moments are not included. Although all the loads have been assigned to a node, the distribution of the load is not correct and may lead to errors. The effect of the applied moments decreases with finer meshing. Thus, the finer mesh not only reduces the errors in the stiffness model but also reduces the unnecessary errors due to modeling the load. If the load is applied directly to elements as member loads, then the algorithm inherent in the software should correctly

TABLE 6.8 Nodal Loads—Coarse Mesh

Load Case 1:	Exterior Girder–One Lane Loaded	
Node	Load, P_y, (kips)	Moment, M_z (ft kips)
13	−18.5	−6
14	−13.5	18
25	−18.5	−6
26	−13.5	18
Sum	−64	

determine the joint load forces and moments. The software should correctly superimpose the fixed end actions with the actions from the analysis of the released (joint-loaded) system to yield the correct final action accounting for the effect of the load applied directly to the member. If the load is applied within a grillage panel, then the statical equivalence becomes more difficult, as loads must be assigned to all of these nodes (this was conveniently and purposefully avoided in this example). The easiest approach in this case is to add another grillage line under the load. If this is not viable, then the load may be assigned by using the subgrillage *A-B-C-D* shown in Figure 6.16(d). Next assign subgrillage end actions to the main grillage element *HG*, and proceed as previously illustrated. The main difference is that the torque must also be considered. An alternative to this tedious approach is to refine the mesh to a point where the simple beam nodal load assignments are viable because the fixed-end torsion and bending moment are relatively small.

Analysis Results. The translations for load case No. 1 for the coarse mesh is shown in Figure 6.16(e). Note that the translations are greater near the point of load application and the supports are restraining the translations as expected. The shear and moment diagrams for load case No. 2 are shown in Figures 6.16(f) and 6.16(g). Tables 6.9 and 6.10 summarize the maximum midspan moments and end reactions (maximum shears) for the four load cases. The simple beam actions are given for this position (see Example 6.1) and are illustrated in Figures 6.12(c) and 6.12(e). The associated actions are illustrated in Figures 6.16(f) and 6.16(g). The distribution factors are also given in the tables. The critical distribution factors are highlighted. These distribution factors are compared with the AASHTO factors in addition to those derived from the finite element and finite strip methods in later examples.

The critical values for flexural moment (using the fine mesh) are highlighted in Table 6.10. The critical moment for the exterior girder with one-lane loaded is 1.2 (multiple presence) × 221.2 ft kips = 265.4 ft kips with a distribution factor of $mg^{SE}_{moment} = 1.2 \times 0.62 = 0.74$ and the exterior girder moment for two-lanes loaded is 1.0 × 232.8 = 232.8 ft kips with a distribution factor of $mg^{ME}_{moment} = 0.65$. The maximum interior girder moments are 1.2 × 157.4 = 188.8 ft kips ($mg^{SI}_{moment} = 0.53$) and 1.0 × 258.8 = 258.8 ft

TABLE 6.9 Summary of Moments—Grillage Analysis

Load Case[a]	Girder	Beam Analysis Moment, (ft kips)	Max. Moment, (ft kips) (Coarse Mesh)	Distribution Factor (mg)	Max. Moment, (ft kips) (Fine Mesh)	Distribution Factor (mg)
1	1	358.4	202.5	0.57	**221.2**	**0.62**
1	2	358.4	123.2	0.34	116.7	0.33
1	3	358.4	37.4	0.10	18.5	0.05
1	4	358.4	0.0	0.00	−4.5	−0.01
1	5	358.4	−3.8	−0.01	−3.0	−0.01
1	6	358.4	−0.6	0.00	0.0	0.00
	Sum	Total Moment = 358.4	358.7	1.00	348.9	0.98
2	1	358.4	236.2	0.66	**232.8**	**0.65**
2	2	358.4	257.5	0.72	240.6	0.67
2	3	358.4	182.3	0.51	183.2	0.51
2	4	358.4	49.2	0.14	46.4	0.13
2	5	358.4	−2.3	−0.01	−1.6	0.00
2	6	358.4	−6.3	−0.02	−5.4	−0.02
	Sum	Total Moment = 2(358.4) = 716.8	716.7	2.00	695.0	1.94

311

TABLE 6.9 (*Continued*)

Load Case[a]	Girder	Beam Analysis Moment, (ft kips)	Max. Moment, (ft kips) (Coarse Mesh)	Distribution Factor, (mg)	Max. Moment, (ft kips) (Fine Mesh)	Distribution Factor (mg)
3	1	358.4	36.7	0.10	19.7	0.06
3	2	358.4	148.1	0.41	123.9	0.35
3	3	358.4	132.3	0.37	**157.4**	**0.44**
3	4	358.4	45.5	0.13	48.8	0.14
3	5	358.4	1.2	0.00	1.4	0.00
3	6	358.4	−5.3	−0.01	−5.2	−0.01
	Sum	Total Moment = 358.4	358.5	1.00	346.0	0.98
4	1	358.4	26.4	0.07	11.8	0.03
4	2	358.4	167.6	0.47	141.6	0.40
4	3	358.4	255.1	0.72	**258.8**	**0.72**
4	4	358.4	203.7	0.57	204.3	0.57
4	5	358.4	69.0	0.19	75.4	0.21
4	6	358.4	4.9	0.01	−6.6	0.02
	Sum	Total Moment = 2(358.4) = 716.8	716.9	2.03	685.2	1.95

[a]Load Cases:
1. One-lane loaded for the maximum exterior girder actions (Girder No. 1).
2. Two-lanes loaded for the maximum exterior girder actions (Girder No. 1).
3. One-lane loaded for the maximum interior girder actions (Girder No. 3).
4. Two-lanes loaded for the maximum interior girder actions (Girder No. 3).

TABLE 6.10 Summary of Reactions—Grillage Analysis

Load Case[a]	Girder	Beam Analysis Reaction (kips)	Max. Reactions, kips (Coarse Mesh)	Distribution Factor (*mg*)	Max. Reactions, (Fine Mesh) (kips)	Distribution Factor (*mg*)
5	1	52.8	30.3	0.57	**34.4**	**0.65**
5	2	52.8	19.8	0.38	19.3	0.37
5	3	52.8	3.2	0.06	−0.4	−0.01
5	4	52.8	−0.1	−0.00	−0.4	−0.01
5	5	52.8	−0.3	−0.01	−0.2	−0.00
5	6	52.8	−0.0	−0.00	0.1	0.00
	Sum	52.8[b]	52.9	1.00	52.8	1.00
6	1	52.8	32.8	0.62	**33.4**	**0.63**
6	2	52.8	41.2	0.78	39.0	0.74
6	3	52.8	29.1	0.55	31.7	0.60
6	4	52.8	3.2	0.06	2.3	0.04
6	5	52.8	−0.3	−0.01	−0.4	−0.01
6	6	52.8	−0.4	−0.01	−0.4	−0.01
	Sum	105.6[b]	105.6	1.99	105.6	1.99
7	1	52.8	2.3	0.04	0.5	0.01
7	2	52.8	26.3	0.50	19.9	0.38
7	3	52.8	21.0	0.40	**30.4**	**0.58**
7	4	52.8	3.7	0.07	2.6	0.05
7	5	52.8	−0.1	−0.00	−0.2	−0.00
7	6	52.8	−0.4	−0.01	−0.4	−0.01
	Sum	52.8[b]	52.8	1.00	52.8	1.01
8	1	52.8	1.9	0.04	0.1	0.00
8	2	52.8	24.2	0.46	19.5	0.37
8	3	52.8	40.7	0.77	**46.0**	**0.87**
8	4	52.8	32.5	0.62	33.3	0.63
8	5	52.8	6.6	0.13	8.2	0.16
8	6	52.8	−0.3	−0.01	−1.3	−0.02
	Sum	105.6[b]	105.6	2.01	105.6	2.01

[a]Load Cases:
  5. One-lane loaded for the maximum exterior girder actions (Girder No. 1).
  6. Two-lanes loaded for the maximum exterior girder actions (Girder No. 1).
  7. One-lane loaded for the maximum interior girder actions (Girder No. 3).
  8. Two-lanes loaded for the maximum interior girder actions (Girder No. 3).
[b]Beam reaction for entire bridge.

kips (mg^{ME}_{moment} = 0.72) for one- and two-lanes loaded, respectively. Note the coarse mesh yields approximately the same results as the fine mesh, hence convergence is deemed acceptable. The total moment at the critical section is 358.4 ft kips for one-lane loaded and 716.8 ft kips for two-lanes loaded. Note the summation of moments at the bottom of each load case. The differences are due to the presence of the nominal deck elements located between the girders. These elements are not shown in the table. Because of their low stiffness, they attract a small amount of load that causes the slight difference between the sum of the girder moments and the statical moment. Inclusion of these elements in the summation eliminates this discrepancy. The distribution factors do not sum to 1.0 (one-lane loaded) or 2.0 (two-lanes loaded) for the same reason. Small differences between the reported values and these values are due to rounding.

The critical reaction/shears are highlighted. The multiple presence factors (Table 4.6) are used to adjust the actions from analysis. The maximum reaction for the exterior girder with one lane loaded is 1.2 × 34.4 = 41.4 kips (mg^{SE}_{shear} = 0.78) and 1.0 × 33.4 = 33.4 kips (mg^{ME}_{shear} = 0.63) with two-lanes loaded. For the interior girder the reactions are 1.2 × 30.4 = 36.5 kips (mg^{SI}_{shear} = 0.69) and 1.0 × 46.0 kips (mg^{ME}_{shear} = 0.87) for one- and two-lanes loaded, respectively. The summation of the end reactions is equal (within rounding) to maximum system reaction of 52.8 (one lane) and 105.6 (two lanes). The nominal longitudinal deck elements in the fine mesh were not supported at the end, hence the total load must be distributed to the girders at the ends and the reactions check as expected.

The result of these analyzes are compared to those from other methods in a later example. The results presented in Tables 6.9 and 6.10 are summarized in Table 6.11. This tabular format is consistent with that used previously and permits ready comparison of the results from the various methods.

Finite Element Method. The finite element method is one of the most general and powerful contemporary numerical methods. It has the capability to model many different mathematical models and to combine these models as necessary. For example, finite element procedures are available to model Eq. 6.1 for the girders and Eq. 6.4 for the deck, and combine the two models into

TABLE 6.11   Grillage Method Summary—Fine Mesh

Girder Location	Number of Lanes Loaded	Moment (ft kips)	Distribution Factor (*mg*)	Reactions (kips)	Distribution Factor (*mg*)
Exterior	1	265.2	0.74	41.2	0.78
Exterior	2	232.8	0.65	33.4	0.63
Interior	1	190.0	0.53	37.0	0.69
Interior	2	258.8	0.72	46.0	0.87

one that simultaneously satisfies both equations and the associated boundary conditions. Like the grillage method, the most common finite element models are based on a stiffness (or displacement approach), that is, a system of equilibrium equations is established and solved for the displacements at the degrees of freedom. The scope of this method seems unending with many texts and reference books, research papers, and computer programs to address and use it. Here, only the surface is scratched and the reader is strongly encouraged to gain more information by formal and/or self-study. The method is easily used and abused. With modern software it is very easy to generate thousands of equations and still have an inappropriate model. The discussion herein is a brief overview of the finite element method as related to the engineering of slab-girder bridges and it is assumed that the reader has had a course and/or experience with the method.

The finite element formulation is commonly used in two ways: 2-D and 3-D models.

The 2-D model is the simplest and involves fewer degrees of freedom. Here plate elements that usually contain three degrees of freedom per node are used to model the deck on the basis of the mathematical model described by Eq. 6.4. The girders are modeled with grillage or plane grid elements with three degrees of freedom per node. Examples of these elements are illustrated in Figure 6.17(a) and 6.17(b). The girder properties may be based on Eqs. 6.9–6.11. The deck properties typically include the flexural rigidities in orthogonal directions or the deck thickness and material properties upon which the rigidities can be based. The nodal loads and/or element loads are determined in the usual manner.

Because there are many different elements with differing number of degrees of freedom and response characteristics, it is difficult to provide general guidance mesh characteristics, other than those usually addressed in standard references. It is important to suggest that at least two meshes be studied to obtain some knowledge of the convergence characteristics. If the response changes significantly with refinement, a third (or fourth) mesh should likely be studied.

Because of the importance of maintaining equilibrium, the analytical results should be checked for global equilibrium. It is very easy to mistakenly apply the loads in the wrong direction or in the wrong location. It is strongly suggested that global equilibrium be checked by hand. We have caught numerous errors in input files and in computer code by this simple check. If the program being used does not have a way to obtain reactions, then perhaps the stiff boundary spring elements can be used at the supports and the element forces are the reactions. If the program does not produce reactions, or they cannot be deduced from the element forces, then the use of another program that does is recommended. In short, no matter how complex the model: **Always check statics.**

This simple check ensures that ductile elements designed on the basis of the analysis provides at least one viable load path and likely an opportunity

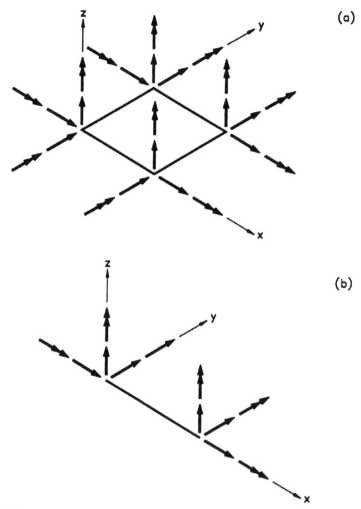

Fig. 6.17 (a) Example of shell element and (b) example of space frame element.

for redistribution should yielding occur. A statics check is necessary for any method of analysis.

As an alternative to the 2-D model, the bridge may be modeled as a 3-D system. Here Eq. 6.4 is used to mathematically model the out-of-plane behavior of the deck and the in-plane effects are modeled using a similar fourth order partial differential equation (Timoshenko and Goodier, 1970). In-plane effects arise from the bending of the system, which produces compression in the deck and tension in the girder under the influence of positive bending moments. The in- and out-of-plane effects are combined into one element, commonly called a shell element. A typical shell element is shown in Figure

6.17(a) where in- and out-of-plane degrees of freedom are illustrated. Typically, the in- and out-of-plane effects are considered uncoupled, which results in a linear formulation. The girders are usually modeled with space frame elements that have six degrees of freedom per node, the same as the shell element. The girder eccentricity (composite girder) is modeled by placing the elements at the centroidal axis of the girder, which creates many additional degrees of freedom. To avoid additional computational effort, the degrees of freedom at the girders may be related to the degrees of freedom of the plate by assuming that a rigid linkage exists between these two points. This linkage can be easily accommodated in the element formulation for the space frame element. This capability is typically included in commercial software and is denoted by several terms: rigid links, element offset, or element eccentricities. An alternative approach is to use the additional degrees of freedom at the girder level but to declare these nodes to be slaves to the deck nodes directly above. A last alternative is to be lazy in the refinement of the model and just include the girder nodes, which produces a much larger model, but of course one can complain (boast) about how large the model is and how long it takes to execute! (This action is not recommended.)

EXAMPLE 6.5

Use the finite element method to determine the end shear (reactions) and midspan flexural bending moments in the girders in Figure 6.12(a) as illustrated in Example 6.1. In addition, determine the distribution factors for moment and shear in girders for one- and two-lanes loaded.

The system is discretized with the 2-D meshes shown in Figures 6.18(a) 6.18(b). The girder properties are the same as in Example 6.4 with the exception that the deck properties are not added as before because the deck is modeled with the shell element as shown in Figure 6.17(a). Here the in-plane effects are neglected and the plate bending portion is retained. The deck rigidities are calculated internal to the finite element program on the basis of $t = 203$ mm $= 8$ in. and $E = 24\,800$ MPa $= 3600$ ksi, and $v = 0.15$. The girder properties and nodal loads are calculated as in the previous example.

The maximum moments and reactions are summarized in Table 6.12. A table similar to Tables 6.9 and 6.10 could be developed and the results would be quite similar. For the sake of brevity, such tables are not shown and only the maximum actions are reported. The multiplication indicates the application of the multiple presence factors.

Finite Strip Method. The finite strip method is a derivative of the finite element method. The mathematical models described previously are the usual basis for analysis so that converged finite element and strip models should yield the same "exact" solutions. The finite strip method employs strips to

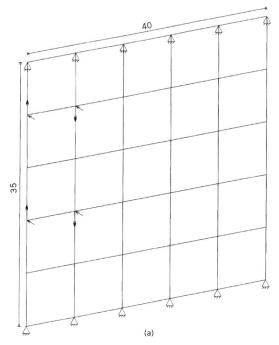

Fig. 6.18   (a) Finite element coarse mesh.

discretize the continuum as shown in Figure 6.19(a). A strip is an element that runs the entire length of the deck. With the typical polynomial shape function used in the finite element method, this type of mesh would be unacceptable. However, the finite strip method uses a special shape function that considers the boundary conditions at the ends to be simply supported. This condition permits the use of a Fourier sine series for the displacement in the longitudinal direction while a third order polynomial is used in the transverse direction. A typical lower order shape function is

$$w(x,y) = \sum_{m=1}^{r} f_m(x)Y_m = \sum_{m=1}^{r} (A_m + B_m x + C_m x^2 + D_m x^3)\sin\left(\frac{m\pi y}{\ell}\right) \quad (6.12)$$

where

$f_m(x)$ is a third order polynomial with coefficients A_m, B_m, C_m, and D_m.
Y_m is the sine function.
ℓ is the span length.
y is the longitudinal coordinate.
m is the series index that has a maximum value of r.

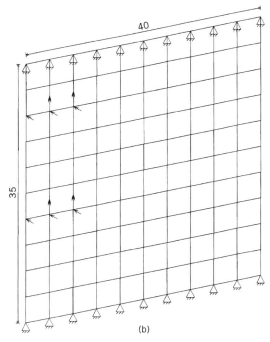

Fig. 6.18 (b) Finite element fine mesh.

TABLE 6.12 Finite Element Results, Critical Actions[a]

Girder Location	Number of Lanes Loaded	Moment (ft kips)	Distribution Factor (*mg*)	Reactions (kips)	Distribution Factor (*mg*)
Exterior	1	(1.2)(206.0) = 247.2	0.68	(1.2)(31.4) = 37.7	0.71
		(1.2)(196.9) = 236.3	0.66	(1.2)(29.8) = 35.8	0.68
Exterior	2	(1.0)(220.8) = 220.8	0.62	(1.0)(30.6) = 30.6	0.58
		(1.0)(219.4) = 219.4	0.61	(1.0)(30.4) = 30.4	0.58
Interior	1	(1.2)(154.9) = 186.9	0.52	(1.2)(30.2) = 36.2	0.69
		(1.2)(154.8) = 185.8	0.52	(1.2)(30.2) = 36.2	0.69
Interior	2	(1.0)(258.8) = 258.8	0.72	(1.0)(44.2) = 44.2	0.84
		(1.0)(249.0) = 249.0	0.69	(1.0)(44.9) = 44.9	0.85

[a]Coarse mesh on first line, fine mesh on the second line.

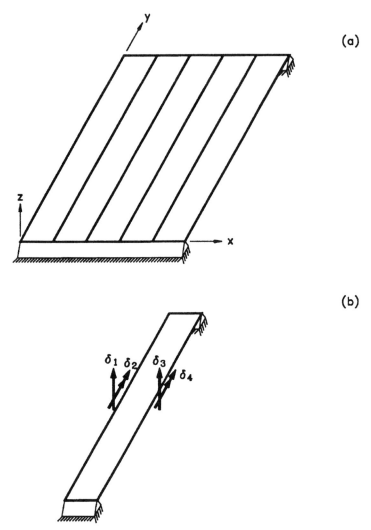

Fig. 6.19   (a) Example of a finite strip model and (b) finite strip element.

It is important to note that the polynomial function is the same one typically used in standard beam elements and may be rewritten in terms of the four degrees of freedom at the strip edges [see Figure 6.19(b)]. The degrees of freedom include two translations and two rotations per harmonic considered (value of m). The total number of degrees of freedom is the number of nodal lines times two, for example, if 50 strips are used with 50 terms, the total number of unknowns is 51(nodal lines) $\times$ 2(unknowns per nodal line) $\times$ 50(terms) = 5100(unknowns). The mathematics of the element formulation involves a procedure very similar to the finite element method. For example, the element stiffness matrix involves

$$[S] = \int_{vol} [B]^T [D][B]\{\delta\}dV \qquad (6.13)$$

where B contains the curvatures or generalized strain, D contains the plate rigidities, and δ contains the four degrees of freedom. Equation 6.13 is presented to remind the reader that differentiation and integration are involved with the element formulation.

An important feature of the finite strip method is its efficiency. When the shape function in Eq. 6.12 is twice differentiated to obtain curvatures, the polynomial function may change but the sine function remains a sine function. Upon substitution into the strain matrix B in Eq. 6.13, the summations remain. A term-by-term expansion of the series in combination with the necessary matrix multiplication yields terms with the following integrals:

$$I = \int_0^{\ell} \sin\left(\frac{m\pi y}{\ell}\right) \sin\left(\frac{n\pi y}{\ell}\right) dy$$

$$I = \frac{1}{2} \quad \text{when } m = n \qquad (6.14)$$

$$I = 0 \quad \text{when } m \neq n$$

This integration is zero when the terms in the series are not the same (termed orthogonality). This important feature causes all terms where n is not equal to m to be zero, which permits the programmer to consider each term separately, and completely uncouples the equations to be solved. For example, if 50 strips are used with 50 terms, then the total number of degrees of freedom is 5100, as before, but this size system is never assembled or solved. Instead, the system is solved for one term at a time or 51(nodal lines) × 2 degrees of freedom per nodal line, which results in 102 degrees of freedom per mode. Thus, this system is solved repetitively for the 50 modes and the results are appropriately superimposed. Hence, a very small problem (the same as a continuous beam with 51 nodes) is solved numerous times. This approach is vastly more efficient than considering the full 5100 degrees of freedom in one solution. A typical finite strip model runs in about 10% of the time as a finite element model with a similar number of degrees of freedom using solvers that account for the small bandwidth and symmetry of the stiffness matrix.

A brief treatise of the finite strip method is provided in this section and the main objective was to introduce the reader to the rationale for it use.

Complete details can be found in books by Cheung (1976) and Loo and Cusens (1978).

EXAMPLE 6.6

Use the finite strip method to determine the end shear (reactions) and midspan flexural bending moments in the girders in Figure 6.14(a) as illustrated in Example 6.1. In addition, determine the distribution factors for moment and shear in girders for one- and two-lanes loaded.

The system is modeled with 20 uniform strips and 100 terms. Studies showed that this discretization is adequate for slab-girder systems (Finch and Puckett, 1992). A large number of terms is required to accurately determine the shear forces near the concentrated forces and girder ends. If only flexural effects are required near midspan, then only about 10 terms are required. The girder, deck properties, and load positioning are the same as in the previous example. The results are summarized in Table 6.13.

EXAMPLE 6.7

Compare the results from the AASHTO, lever, grillage, finite element, and finite strip methods.

Tables 6.6, 6.7, and 6.11–6.13 have been combined for comparison of the methods and the results are given in Tables 6.14 and 6.15 for moment and shear, respectively.

Recall that the basis for the AASHTO multilanes loaded formulas includes the possibility of three or more lanes being loaded and creating a situation more critical than the two-lane case. Therefore, the AASHTO values are influenced by this and are generally, but not always, slightly higher than the two-lane numerical results. Most values compare within 10% except the lever

TABLE 6.13 Finite Strip Results

Girder Location	Number of Lanes Loaded	Moment (ft kips)	Distribution Factor (*mg*)	Reaction (kips)	Distribution Factor (*mg*)
Exterior	1	(1.2)(204.3) = 245.2	0.68	(1.2)(29.6) = 35.5	0.67
Exterior	2	(1.0)(218.6) = 218.6	0.61	(1.0)(30.6) = 30.6	0.58
Interior	1	(1.2)(154.1) = 184.9	0.52	(1.2)(26.4) = 31.7	0.60
Interior	2	(1.0)(250.8) = 250.8	0.70	(1.0)(41.7) = 41.7	0.79

TABLE 6.14   Summary of Analysis Methods—Moment (ft kips)

Girder	Number of Lanes Loaded	AASHTO	Lever	Grillage	Finite Element	Finite Strip
Exterior	1	268.8	268.8	265.2	236.3	245.2
		0.75	0.75	0.74	0.66	0.68
Exterior	2	254.5	224.0	232.3	219.4	218.6
		0.71	0.625	0.65	0.61	0.61
Interior	1	193.5	268.8	190.0	185.8	184.9
		0.54	0.75	0.53	0.52	0.52
Interior	2	254.5	313.6	258.8	249.0	250.8
		0.71	0.875	0.72	0.69	0.70

method, which tends to be conservative because load sharing is limited to neighboring girders.

6.3.2   Slab Bridges

The slab bridge is another common bridge type frequently used for short spans, usually less than 15 240 mm (50 ft). The slab bridge does not have any girders, and therefore, the load must be carried principally by flexure in the longitudinal direction. A very simplistic approach (perhaps valid for the ultimate strength limit states) is to divide the total statical moment by the bridge width to achieve a moment per unit width for design. This type of analysis is valid by the lower bound theorem for consideration of the strength limit state assuming adequate transverse strength and ductility is available. The results of this procedure are most certainly underestimates of the localized moments near the application of the load under linear elastic conditions, that is, service and fatigue limits states. Hence, it is necessary to determine the moments under service conditions. The moments are determined by es-

TABLE 6.15   Summary of Analysis Methods—Reactions (kips)

Girder	Number of Lanes Loaded	AASHTO	Lever	Grillage	Finite Element	Finite Strip
Exterior	1	39.6	39.6	41.2	35.8	35.5
		0.75	0.75	0.78	0.68	0.67
Exterior	2	34.8	33.0	33.4	30.4	30.6
		0.66	0.625	0.63	0.58	0.58
Interior	1	35.9	39.6	37.0	36.2	31.7
		0.68	0.75	0.70	0.69	0.60
Interior	2	43.3	46.2	46.0	44.9	41.7
		0.82	0.875	0.87	0.85	0.79

tablishing the width of the bridge that is assigned to carry one vehicle, or in other words, the structural width per design lane. The width for one-lane loaded is [A4.6.2.3]

$$E^S = 10.00 + 5.0\sqrt{L_1 W_1} \qquad \text{(6.15a-US)}$$

$$E^S = 250 + 0.42\sqrt{L_1 W_1} \qquad \text{(6.15a-SI)}$$

and the width for multilanes loaded is

$$E^M = 84.00 + 1.44\sqrt{L_1 W_1} \leq \frac{W}{N_L} \qquad \text{(6.15b-US)}$$

$$E^M = 2100 + 0.12\sqrt{L_1 W_1} \leq \frac{W}{N_L} \qquad \text{(6.15b-SI)}$$

where

$E^{S \text{ or } M}$ is the structural width per design lane [in. (mm)], for single and multiple lanes loaded,

L_1 is the modified span length taken equal to the lesser of the actual span or 60.0 ft (18 000 mm),

W_1 is the modified edge-to-edge width of bridge taken equal to the lesser of the actual width or 60.0 ft (18 000 mm) for multilane loading, or 30 ft (9000 mm) for single-lane loading,

W is the physical edge-to-edge width of the bridge [ft (mm)] and

N_L is the number of design lanes [A3.6.1.1.1].

The adjustment for skew is

$$r = 1.05 - 0.25\tan \theta \leq 1.00 \qquad \text{(6.15c)}$$

where θ is the skew angle define previously in Table 6.5. Note that skew reduces the longitudinal load effects.

EXAMPLE 6.8

Determine the slab width that is assigned to a vehicle (design lane) for the bridge described in Example 6.1 [see Fig. 6.12(a)] *without* the girders. Use a 20 in. (508 mm) deck thickness. Assume three design lanes are possible.

By using Eq. 6.15(a) for one-lane loaded, the width is

$$E^S = 10.00 + 5.0\sqrt{L_1 W_1} = 10.00 + 5.0\sqrt{(35)(44)}$$

$$= 206 \text{ in./lane} = 17.2 \text{ ft/lane}$$

and by using Eq. 6.15(b) for multiple lanes loaded, the width is

$$E^M = 84.00 + 1.44\sqrt{L_1 W_1} = 84.00 + 1.44\sqrt{(35)(44)} = 140.5 \text{ in./lane}$$

$$= 11.7 \text{ ft/lane} \le \frac{W}{N_L} = \frac{44}{3} = 14.7 \text{ ft/lane} \therefore E^M = 11.7 \text{ ft}$$

The bending moment is determined for a design lane that is divided by the width E to determine the moment per unit length for design.

From the simple-beam analysis given in Example 6.1, the maximum bending moment for one lane is 358.4 ft kips. Using this moment, the moments per ft are

$$M_{LL}^S = \frac{M_{\text{beam}}}{E^S} = \frac{358.4 \text{ ft kips/lane}}{17.6 \text{ ft/lane}} = 20.8 \text{ ft kips/ft}$$

and

$$M_{LL}^M = \frac{M_{\text{beam}}}{E^M} = \frac{358.4 \text{ ft kips/lane}}{11.7 \text{ ft/lane}} = 30.6 \text{ ft kips/ft}$$

Because the slab bridge may be properly modeled by Eq. 6.4, all the methods described earlier may be used. To illustrate, brief examples of the grillage and finite element methods are given below. Most of the modeling details remain the same as previously presented. The girders are obviously omitted and the loading is the same as the previous examples. Shear is not a problem with the slab bridge and this limit state need not be considered [A4.6.2.3]. Zokaie (1991) reexamined this long-time AASHTO provision and confirmed the validity of this approach. Only flexural bending moment is presented.

EXAMPLE 6.9

Use the grillage method to model the slab bridge described in Example 6.8. Use the fine mesh used in Example 6.4 and consider two-lanes loaded for bending moment. The deck may be modeled as an isotropic plate.

All the deck section properties are proportional to the thickness cubed. Hence, for the 20-in (508 mm) slab, the properties determined in Example 6.4 are multiplied by $(20/8)^3 = 2.5^3 = 15.625$. The distribution of internal actions is not a function of the actual thickness, but rather the relative rigidities in the transverse and longitudinal directions. Because isotropy is assumed in this example, any uniform thickness may be used for determining the actions. The displacements are proportional to the actual stiffness (thickness cubed as noted above).

The loads are positioned as shown in Figure 6.20(a). The moments in the grillage elements are divided by the tributary width associated with each longitudinal element, 4.0 ft. The moments M_y (beam-like) are illustrated in Figure 6.20(b) and are summarized in Table 6.16.

The critical values (highlighted) are 20.56 ft kips/ft with an associated width of 358.4/20.56 = 17.4 ft. The total moment across the critical section is the summation of the grillage moments. Equilibrium dictates that this moment be 2 lanes × 358.4 ft kips = 716.8 ft kips, which is the summation of the moments in the grillage elements at the critical section, validating equilibrium. The AASHTO value is 30.6 ft kips/ft for two-lanes loaded, which is approximately 50% greater than the maximum grillage value of 20.56 ft kip/ft. This difference is discussed in more detail later.

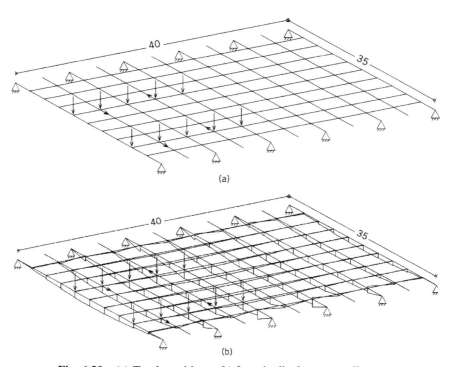

Fig. 6.20   (a) Truck positions, (b) Longitudinal moment diagram.

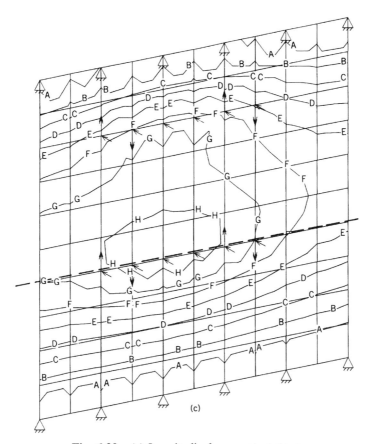

Fig. 6.20   (c) Longitudinal moment contour.

EXAMPLE 6.10

Use the finite element method to model the slab bridge described in Example 6.8. Use the fine mesh used in Example 6.5 and consider two- and three-lanes loaded for maximum bending moment. The deck may be modeled as an isotropic plate.

The deck thickness is increased to 20 in. (508 mm) and the girders are removed from the model presented in Example 6.5. The nodal loads are the same as in the previous example.

The moments that cause flexural stress in the longitudinal direction are illustrated in Table 6.17 and Figure 6.20(b). Contour plots of the flexural moments are illustrated in Fig. 6.20(c). Note the values in Table 6.18 are the contour values for the bridge at the dashed line. As expected, the longitudinal moments are significantly greater than the transverse moments. These figures

TABLE 6.16 Analysis Results for the Grillage Model at the Critical Section (Two-Lanes Loaded)

Element	1	2	3	4	5	6	7	8	9	10	11	Total[a]
Moment (ft kips)	58.65	62.46	69.57	75.05	82.26	81.46	72.30	65.04	55.65	48.91	45.46	716.81
Tributary Length (ft)	4.0	4.0	4.0	4.0	4.0	4.0	4.0	4.0	4.0	4.0	4.0	44.0
Moment (ft kips/ft)	14.66	15.62	17.39	18.76	**20.56**	20.37	18.08	16.26	13.91	12.23	11.37	N/A
Statical Moment (ft kips)	358.4	358.4	358.4	358.4	358.4	358.4	358.4	358.4	358.4	358.4	358.4	N/A
Width per Lane (ft)	24.4	22.9	20.6	19.1	**17.4**	17.6	19.8	22.0	25.8	29.3	31.5	N/A

[a]Not applicable = N/A.

TABLE 6.17 Analysis Results for the Finite Element Model Near the Critical Section (Two-Lanes Loaded)

Tenth Point Across Section	1 Edge	2	3	4	5	6	7	8	9	10	11 Edge	Total[a]
Moment (ft kips/ft) (fine mesh)	14.43	14.73	18.05	17.89	**20.54**	20.24	18.29	15.46	12.78	11.37	11.06	N/A
Element width (ft)	2.2	4.4	4.4	4.4	4.4	4.4	4.4	4.4	4.4	4.4	2.2	44.0
Moment (ft kips per element)	31.7	64.8	79.4	78.7	90.4	89.1	80.5	68.0	56.2	50.0	24.3	713.1
Statical Moment (ft kips)	358.4	358.4	358.4	358.4	358.4	358.4	358.4	358.4	358.4	358.4	358.4	358.4
Width per Lane (ft)	24.8	24.3	19.9	20.0	**17.4**	17.7	19.6	23.2	28.0	31.5	32.4	N/A

[a]Not applicable = N/A.

are provided to give the reader a sense of the distribution of internal actions in a two-way system that is traditionally modeled as a one-way system, that is, as a beam.

The maximum moment in the finite element method is 20.54 ft kip/ft and is associated with a width of 358.4/20.54 = 17.4 ft. These values compare very well with the grillage moment of 20.56 ft kip/ft and the associate width of 17.4 ft. The finite element moments are reported at the nodes along the critical moment section for the entire system. Therefore, the total moment at the section reported is 713.1 ft kips and is slightly less than the total statical moment of 2 × 358.4 = 716.8 ft kips for two-loaded lanes. Note that the width per lane is used only if a beam-line analysis is required, that is, a 1.5-D analysis where the load distribution is developed by a numerical model and the design is based on analysis of a beam. Alternatively, the entire design could be based on the mathematical/numerical model. Note: The other load cases are also required for the design.

Both the grillage and finite element methods do not compare well with the AASHTO value of 30.6 ft kips/ft. Recall the AASHTO multilane formulas implicitly include two-, three-, or more lanes loaded. Because this bridge has a curb-to-curb width of 44 ft, likely three 10-ft design lanes should be considered for design and is considered below.

Why is it important to initially present the two-lane loaded case rather than the three-lane case for the refined methods? There are three reasons: (1) to highlight the assumptions included in the AASHTO distribution formulas, (2) to illustrate that the two-lane load does not always give most critical results, and (3), if the results from an analytical approach differ significantly (more than 15%) from the AASHTO value, then the differences should be understood and justified. Zokaie et al. (1991) presented numerous histograms similar to Figure 6.11 where the results of the simplified AASHTO formulas are within a 15% of results based on more rigorous methods. This result suggests that significant deviation should be carefully investigated.

The two-lane loaded finite element model is modified to include an additional vehicle placed adjacent to the others and located near the edge of the deck [see Fig. 6.20(a)]. The results are given in Table 6.18 and are plotted in Figure 6.20(b).

Note that the moment of 34.08 ft kips and the associated distribution width of 10.5 ft are critical. Now, AASHTO [A3.6.1.1.2] provides a multiple presence factor of 0.85 for bridges with three design lanes (see Table 4.6). Hence, the moment of 34.08 ft kips may be multiplied by 0.85 yielding a critical value of (0.85)(34.08) = 29.0 ft kips and the associated distribution width is 12.4 ft. These values compare reasonably well with the AASHTO values of 30.6 ft kips and 11.7 ft. On the basis of the preceding analysis, it is likely that the AASHTO equation for distribution width is governed by three-lanes loaded for this bridge width.

TABLE 6.18 Analysis Results for the Finite Element Model Near the Critical Section (Three-Lanes Loaded)

Tenth Point Across Section	1 Edge	2	3	4	5	6	7	8	9	10	11 Edge	Total[a]
Moment (ft kips/ft) (fine mesh)	16.45	16.82	20.64	21.16	24.80	25.87	25.72	25.30	27.75	30.59	**34.08**	N/A
Tributary width (ft)	2.2	4.4	4.4	4.4	4.4	4.4	4.4	4.4	4.4	4.4	2.2	44.0
Tributary Moment (ft kips)	36.2	74.0	90.8	93.1	109.1	113.8	113.1	111.3	122.1	134.6	75.0	1073.1
Statical Moment (ft kips)	358.4	358.4	358.4	358.4	358.4	358.4	358.4	358.4	358.4	358.4	358.4	358.4
Width per Lane (ft)	21.8	21.3	17.4	16.9	14.5	13.9	13.9	14.2	12.9	11.7	**10.5**	N/A

[a]Not applicable = N/A.

6.3.3 Slabs in Slab-Girder Bridges

The slab design may be accomplished by three methods: (1) the analytical strip method approach, (2) the empirical approach, and (3) the yield-line method. The analytical method requires a linear elastic analysis upon which to proportion the slab to satisfy the strength and service limit states. The empirical approach requires that the designer satisfy a few simple rules regarding the deck thickness and reinforcement details, and limit states are assumed to be automatically satisfied without further design validations. The empirical approach is elaborated in more detail in the following chapters on design. The third method is the yield-line method and is based on inelastic yielding of the deck, and therefore, is appropriate for the strength and extreme-event limit states. All three methods may be used to proportion the slab. All three methods yield different designs that are generally viable and reasonable. In this section, the strip method is first outlined with a discussion of the AASHTO provisions and an illustrative example. A brief discussion of the yield-line method follows, also reinforced with an example. The empirical approach is outlined in Chapter 7 on concrete design.

Linear Elastic Methods. A deck slab may be considered as a one-way slab system because its aspect ratio (panel length divided by the panel width) is large. For example, a typical panel width (girder spacing) is 2400–3600 mm (8–11 ft) and a typical girder length from 9100 to 61 000 mm (30–200 ft). The associated aspect ratios vary from 3.75 to 10. Deck panels with an aspect ratio of 1.5 or larger may be considered one-way systems [A4.6.2.1.4]. Such systems are assumed to carry the load effects in the short-panel direction, that is, in a beamlike manner. Assuming the load is carried to the girder by one-way action, then the primary issue is the width of strip (slab width) used in the analysis and subsequent design. Guidance is provided in AASHTO [A4.6.2], *Approximate Methods.*

The strip width SW [in. (mm)] for a CIP section is

$$M^+: SW^+ = 26.0 + 6.6S \qquad \text{(6.16a-US)}$$

$$M^+: SW^+ = 660 + 0.55S \qquad \text{(6.16a-SI)}$$

$$M^-: SW^+ = 48.0 + 3.0S \qquad \text{(6.16b-US)}$$

$$M^-: SW^- = 1220 + 0.25S \qquad \text{(6.16b-SI)}$$

$$\text{Overhang} \quad SW^{\text{Overhang}} = 45 + 10.0X \qquad \text{(6.16c-US)}$$

$$\text{Overhang} \quad SW^{\text{Overhang}} = 1140 + 0.833X \qquad \text{(6.16c-SI)}$$

where S is the girder spacing [ft (mm)], and X is the distance from the load point to the support [ft (mm)].

Strip widths for other deck systems are given in AASHTO [Table A4.6.2.1.3-1]. A model of the strip on top of the supporting girders is shown in Figure 6.21(a). A design truck is shown positioned for near critical positive moment. The slab-girder system displaces as shown in Figure 6.21(b). This displacement may be considered as the superposition of the displacements associated with the local load effects [Fig. 6.21(c)] and the global load effects [Fig. 6.21(d)]. The global effects consist of bending of the strip due to the displacement of the girders. Here a small change in load position does not significantly affect these displacements, hence this is a global effect. The local effect is principally attributed to the bending of the strip due to the application of the wheel loads on this strip. A small movement, for example, one foot transversely, significantly affects the local response. For decks, usually the local effect is significantly greater than the global effect. The global effects

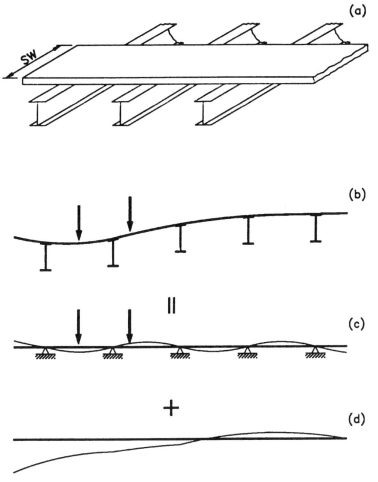

Fig. 6.21 (a) Idealized design strip, (b) transverse section under load, (c) rigid girder model, and (d) displacement due to girder translation.

may be neglected and the strip may be analyzed with classical beam theory assuming that the girders provide rigid support [A4.6.2.1.5]. Because the lower bound theorem is applicable and because this distribution of internal actions accounts for equilibrium, the strip method yields adequate strength and should, in general, yield a reasonable distribution of reinforcement. To account for the stiffening effect of the support (girder) width, the design shears and moments may be taken as critical at the face of the support for monolithic construction and at one-quarter flange width for steel girders [A4.6.2.1.5].

> Sign convention for slabs: A positive slab moment creates compression on the top and a negative moment creates compression on the bottom. Where plotted, the moment is plotted on the compression face.

EXAMPLE 6.11

Determine the shear and moments required in the transverse direction for the slab shown in Figure 6.22(a).
The strip widths are [A4.6.2.1.3]

$$SW^+ = 26.0 + 6.6S = 26.0 + 6.6(8) = 78.8 \text{ in.} = 6.6 \text{ ft}$$

and

$$SW^- = 48.0 + 3.0S = 48.0 + 3.0(8) = 72.0 \text{ in.} = 6.0 \text{ ft}$$

The strip model of the slab consists of the continuous beam shown in Figure 6.22(b). Here influence functions may be used to position the design truck transverse for the most critical actions. Because this approach was taken earlier, an alternative approach, moment distribution is used here but any beam analysis method may be employed (based on Eq. 6.1). The near-critical truck position for moment in span BC is shown in Figure 6.22(a). Although the beam has eight spans including the cantilevers, it may be simplified by terminating the system at joint E with a fixed support and neglecting the cantilever because it is not loaded and contributes no rotational stiffness. This simplification has little affect on the response. The analysis results are shown in Table 6.19.

The most negative moment is approximately -21.9 ft kips (nearest 0.1 ft kip). This end moment is used in the free body diagram shown in Figure 6.22(c) to determine the end shears for element BC. The end-panel moment diagram is shown in Figure 6.22(d). The critical moments are divided by the strip width to obtain the moments per foot. The results are

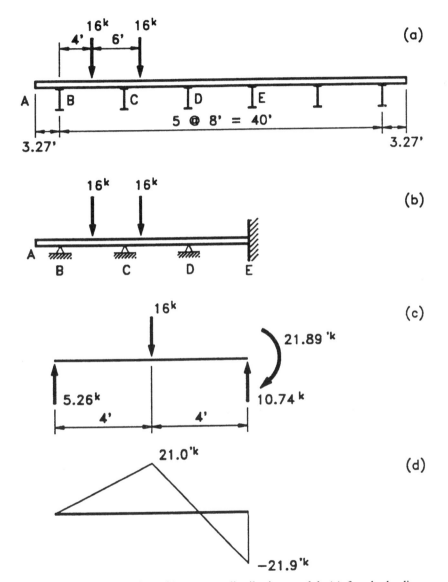

Fig. 6.22 (a) Cross section, (b) moment distribution model, (c) free body diagram for *BC*, (d) moment diagram for *BC*

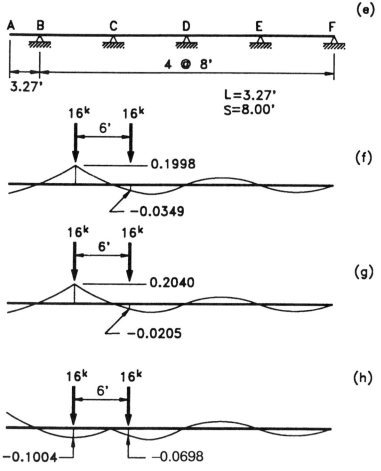

Fig. 6.22 (e) transverse beam, (f) position for moment 205, (g) position for moment 204, and (h) position for moment 300.

$$m^+ = \frac{21.0 \text{ ft kips}}{6.6 \text{ ft}} = 3.18 \text{ ft kips/ft}$$

and

$$m^- = \frac{-21.9 \text{ ft kips}}{6 \text{ ft}} = -3.65 \text{ ft kips/ft}$$

The m^+ and m^- indicate moments at the middle of the panel and over the

TABLE 6.19   Moment Distribution Analysis

	BC^a	CB	CD	DC	DE	ED
Stiffness	0	0.75	1.00	1.00	1.00	Fixed
Distribution Factor	N/A	0.429	0.571	0.5	0.5	0.0
Fixed-end Moment	16	-16	18	-6	0	0
Adjustment	-16					
Carry-over		-8				
Fixed-end Moment		-24	18	-6	0	0
Distribution		2.574	3.426	3.000	3.000	0.000
Carry-over			1.500	1.713		1.500
Distribution		-0.643	-0.857	-0.857	-0.857	
Carry-over			-0.429	-0.429		-0.429
Distribution		0.184	0.245	0.215	0.215	
TOTAL		-21.9	21.9	-2.36	2.36	1.07

^{a}Not applicable = N/A.

girder, respectively. These moments may be considered representative and used for the other panels as well.

As an alternative to the moment distribution, the beam model for transverse moments may be modeled with influence function developed especially for slab analysis. Such functions are described in Appendix A where influence functions for four equal interior spans with length S and cantilever span with length L. This configuration is illustrated in Figure 6.22(e). The influence function for moment at 205, 204, and 300 are shown in Figures 6.22(f)–(h). The near critical load positions are also illustrated in these figures. The calculation of the beam moments are based on Eq. 5.1 and are given below. The influence ordinates are from Table A1 in Appendix A.

$$M_{205} = 16(0.1998)(8) + 16 \left(\frac{-0.0317 - 0.0381}{2} \right) (8)$$

$$M_{205} = 16(0.1998)(8) + 16(-0.0349)(8) = 21.1 \text{ ft kips}$$

$$m_{205}^+ = \frac{21.1 \text{ ft kips}}{6.6 \text{ ft}} = 3.20 \text{ ft kips/ft}$$

$$M_{204} = 16(0.2040)(8) + 16\left(\frac{-0.0155 - 0.0254}{2}\right)(8)$$

$$M_{204} = 16(0.2040)(8) + 16(-0.0205)(8) = 23.5 \text{ ft kips}$$

$$m_{204}^+ = \frac{23.5 \text{ ft kips}}{6 \text{ ft}} = 3.92 \text{ ft kips/ft}$$

$$M_{300} = 16(-0.1004)(8) + 16\left(\frac{-0.0634 - 0.0761}{2}\right)(8)$$

$$M_{300} = 16(-0.1004)(8) + 16(-0.0698)(8) = -21.8 \text{ ft kips}$$

$$m_{300}^- = \frac{-21.8 \text{ ft kips}}{6 \text{ ft}} = -3.63 \text{ ft kips/ft}$$

Note that the moment at 205 is essentially the same as the moment distribution results. Repositioning the load slightly to the left, at the 204, the panel moment is increased to 23.5 ft kips (3.92 ft kips/ft). The negative moment remains essentially the same. The critical panel moments are $m^+ = 3.92$ ft kips/ft and $m^- = -3.65$ ft kips/ft. These moments are compared with the rigorous methods in the example that follows.

The grillage, finite element, and finite strip methods may be used to model the deck actions. The procedures outlined earlier in this chapter are generally applicable. The joint loads must be positioned transversely in the most critical position. The longitudinal positioning affects the response of the system, and to illustrate, two positions are used in the following example. The first is near the support and the second is at midspan. The results from each are compared with the AASHTO strip method.

EXAMPLE 6.12

Use the grillage, finite element, and finite strip methods to determine the moments in the first interior panel of the system shown in Figure 6.12(a). Position the design truck axle at 3.5 ft from the support and at midspan with the wheel positioned transversely as shown in Figure 6.23(a).

The fine meshes for the grillage, finite element, and finite strip methods are used as in Examples 6.4–6.6. The equivalent joint loads are determined for the truck position described and the resulting moments are given in Table 6.20. To obtain the moment from the grillage model the element moment must be divided by the associated tributary length. The resulting flexural bending moment diagrams for the load positioned at midspan and near the support are shown in Figures 6.23(b) and 6.23(c), respectively. The transverse moment per unit length (lower case) at midpanel is

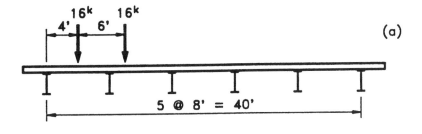

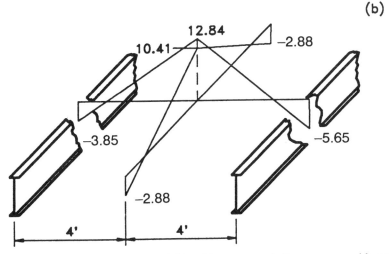

Fig. 6.23 (a) Cross section, and (b) grillage moment diagram near midspan.

$$m^+_{transverse} = \frac{12.84 \text{ ft kips}}{3.5 \text{ ft}} = 3.67 \text{ ft kips/ft}$$

$$m^+_{longitudinal} = \frac{10.41 \text{ ft kips}}{4 \text{ ft}} = 2.60 \text{ ft kips/ft}$$

and transverse moment over the girder is

$$m^-_{transverse} = \frac{-5.65 \text{ ft kips}}{3.5 \text{ ft}} = -1.61 \text{ ft kips/ft}$$

The slab moments near the girder support are

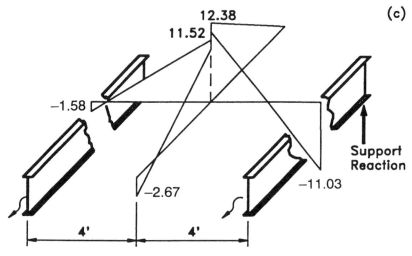

Fig. 6.23 (c) Grillage moment diagram near support.

$$m^+_{\text{transverse}} = \frac{11.52 \text{ ft kips}}{3.5 \text{ ft}} = 3.29 \text{ ft kips/ft}$$

$$m^+_{\text{longitudinal}} = \frac{12.38 \text{ ft kips}}{4 \text{ ft}} = 3.10 \text{ ft kips/ft}$$

and

$$m^-_{\text{transverse}} = \frac{-11.03 \text{ ft kips}}{3.5 \text{ ft}} = -3.15 \text{ ft kips/ft}$$

The finite element analysis gives results directly in terms of moment per foot and therefore no intermediate calculations are required and the results are presented in Table 6.20. The moments from the AASHTO strip method are also given for comparison.

The grillage, finite element, and finite strip methods include both the local and the global load effects. Note that there is significant difference between these results that has not been exhibited in the previous examples. Also note that the grillage method gives results that are in better agreement with the AASHTO moments. It is also interesting to compare the moments in the transverse and longitudinal directions. For example, at midspan of the grillage, the ratio of the positive transverse to longitudinal moment is 3.67/2.60 = 1.4 and near the support the ratio is 3.29/3.10 = 1.06. The same ratios for the finite strip method are 4.38/5.78 = 0.76 and 4.84/5.21 = 0.93 at the

TABLE 6.20   Finite Element and AASHTO Moments (ft kips/ft) (transverse moments only)

	AASHTO Strip Method Transverse Only	Grillage Transverse Longitudinal	Finite Strip Transverse Longitudinal	Finite Element Transverse Longitudinal
m^+ @ Support	3.92	3.29	4.84	4.45
		3.10	5.21	4.05
m^- @ Support	−3.65	−3.15	−3.28	−1.26
m^+ @ Midspan	3.92	3.67	4.38	4.26
		2.60	5.78	3.63
m^- @ Midspan	−3.65	−1.61	−2.18	−0.31

midspan and near-support values, respectively. This result indicates that the rigorous analysis gives significant longitudinal moments, that is, moments that are considered small in the AASHTO strip method. Near the support, the behavior is affected by the boundary conditions, that is, the support assumptions at the end of the bridge. In this example, the deck is assumed to be supported across the full-width (deck and girders). This boundary condition creates significant longitudinal stiffness that attracts the longitudinal moment, which is equal to or exceeds the transverse moment. If only the girders are supported, the transverse moments increase by about 50% and significantly exceed the AASHTO values. The longitudinal moments also decrease significantly. The finite strip moments are higher than all other moments because the positive moments were taken directly under the concentrated load, in an area where the curvature is increased locally due to the presence of the load (frequently called dishing). This effect decreases rapidly away from the load to values that are similar to the grillage and AASHTO values. The finite element values are similar to the finite strip values when the values directly under the load are considered. The moments were significantly lower (~one-half) at the element centroids located approximately 2.5 ft (760 mm) away from the load position. Finally, it should be noted that the AASHTO strip method significantly overestimates the negative moment over the girders in the middle of the longitudinal span because of the assumption that the girders do not translate. The results are better near the support where the girder translation is small.

It is relevant to note the differences and difficulties that arise in modeling the deck with the various methods. What if only one method was used with one mesh? How would the engineer know if the answers are correct? Further, is the maximum moment directly under the load the proper moment for design or should the actions be spread over a larger area? What effect does modeling the wheel load as a patch rather than a concentrated load have? How is the patch load properly modeled in the grillage, finite strip, and finite element models? Is a flat plate (or shell) element appropriate to represent the ultimate limit state where significant arching action has been observed in experimental research? The answers to these questions are best established by studying the system under consideration. These questions and many others are beyond the scope of this chapter, but it suffices to note that there are many important issues that must be addressed to properly model the localized effects in structures. The best way to answer the many questions that arise in modeling is to modify the model and to observe the changes. Modeling local effects takes judgment, skill, and usually significant time. It is more difficult than modeling the global response, for example, determining distribution factors.

Again place analysis in perspective. The lower bound theorem requires that the one-load path be established for safe design, which makes the AASHTO method viable because the load is distributed transversely and nominal "distribution" steel is used in the longitudinal direction (see Chapter 7). The remaining limit states are associated with service loads such as cracking and fatigue and if these can be assured by means other than rigorous analysis, then so much the better. More information on this topic is given in Chapter 7.

In summary, the distribution of internal actions in a bridge deck is complex and not easily modeled. The AASHTO method seems to give reasonable results for this example and as shown in the subsequent design of this deck, the AASHTO moments result in a reasonable distribution of reinforcement.

The ultimate limit state can be modeled with the yield-line method. This method can be used to gain additional insight into the behavior of deck systems under ultimate loading conditions.

Yield-Line Analysis. The yield-line method is a procedure where the slab is assumed to behave inelastically and exhibits adequate ductility to sustain the applied load until the slab reaches a plastic collapse mechanism. Because the reinforcement proportioning required by AASHTO gives underreinforced or ductile systems, this assumption is realistic. The slab is assumed to collapse at a certain ultimate load through a system of plastic hinges called yield lines. The yield lines form a pattern in the slab creating the mechanism. Two methods are available for determining the ultimate load by the yield-line method: The equilibrium approach and the energy approach. The energy approach is described here because it is perhaps the simplest to implement. The energy approach is an upper-bound approach, which means that the ultimate load established with the method is either equal to or greater than the actual (i.e., nonconservative). If the exact mechanism or yield-line pattern is used in the energy approach, then the solution is theoretically exact. Practically, the yield pattern can be reasonably estimated and the solution is also reasonable for design. Patterns may be selected by trial or a systematic approach may be used. Frequently, the yield-line pattern can be determined in terms of a few (sometimes one) characteristic dimensions. These dimensions may be used in a general manner to establish the ultimate load, and then the load is minimized with respect to the characteristic dimensions to obtaining the lowest value. Simple differentiation is usually required.

The fundamentals and the primary assumptions of the yield-line theory are as follows (Ghali and Neville, 1989):

- In the mechanism, the bending moment per unit length along all yield lines is constant and equal to the moment capacity of the section.
- The slab parts (area between yield lines) rotate as rigid bodies along the supported edges.
- The elastic deformations are considered small relative to the deformation occurring in the yield lines.
- The yield lines on the sides of two adjacent slab parts pass through the point of intersection of their axes of rotation.

Consider the reinforcement layout shown in Figure 6.24(a) and the free body diagram shown in Figure 6.24(b). The positive flexural capacities in the two directions are m_t and m_ℓ. Here the axis labels t and ℓ are introduced for the transverse and longitudinal directions, usually associated with a bridge. In general, the orthogonal directions align with the reinforcement. Assume a yield line crosses the slab at an angle α relative to the direction of longitudinal reinforcement as shown in Figure 6.24(a). Equilibrium requires that

$$m_a = m_\ell \cos^2 \alpha + m_t \sin^2 \alpha$$

$$m_{\text{twist}} = (m_\ell - m_t) \sin \alpha \cos \alpha$$

(6.17)

If the slab is isotopically reinforced, then $m_t = m_\ell = m$ and Eq. 6.17 simplifies to

$$m_a = m(\cos^2 \alpha + \sin^2 \alpha) = m$$

$$m_{\text{twist}} = 0$$

(6.18)

Therefore, for isotropic reinforcement, the flexural capacity is independent of the angle of the yield line and may be uniformly assigned the value of the capacity in the direction associated with the reinforcement.

Virtual work may be used to equate the energy associated with the internal yielding along the yield lines and the external work of the applied loads. Consider the slab segment shown in Figure 6.25 where a yield line is positioned at an angle to the axis of rotation of the slab segment. By definition of work, the internal energy for yield line i is the dot product of the yield-line moment and the rotation, or mathematically stated

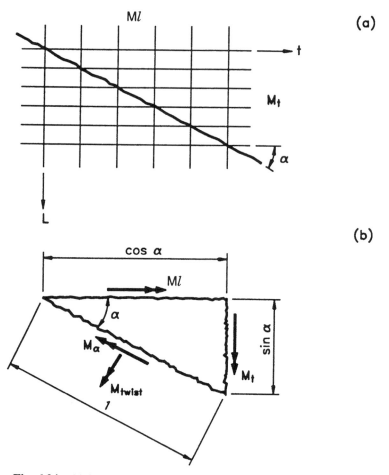

Fig. 6.24   (a) Deck reinforcement layout and (b) free body diagram.

$$U_i = \int m_i \cdot \theta_i \, d\ell = m_i \ell_i (\cos\alpha_i)\theta_i \tag{6.19a}$$

and, the total system energy is

$$U = \sum_{i=1}^{NL} m_i \ell_i (\cos\alpha_i)\theta_i \tag{6.19b}$$

where NL is the number of yield lines in the system and θ_i is the associated rotation.

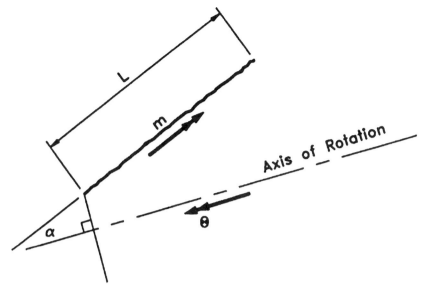

Fig. 6.25 Slab part.

The total system energy is the summation of the contributions from all the slab segments. It is perhaps simpler to think of the dot product as the projection of moment on the axes of rotation times the virtual rotation. To facilitate this, both moment and rotation may be broken into orthogonal components (usually associated with the system geometry) and the dot product becomes

$$U_{\text{int}} = \sum_{i=1}^{NL} (m_i \ell_i) \cdot \theta_i = \sum_{i=1}^{NL} M_{ti}\theta_{ti} + \sum_{i=1}^{NL} M_{\ell i}\theta_{\ell i} \tag{6.20}$$

where M_{ti} and $M_{\ell i}$ are the components of $m_i \ell_i$ and θ_{ti} and $\theta_{\ell i}$ are the components of θ.

The virtual external work for uniform and concentrated loads is

$$W_{\text{ext}} = \int pw dA + \sum P_i w_i \tag{6.21}$$

where p is the distributed load, w is the virtual translation, and w_i is the translation at concentrated load P_i.

In the examples that follow, a typical slab on a slab-girder bridge is studied with the yield-line method. This method is illustrated for a concentrated load applied in the middle portion of the bridge, near the end of the bridge, and on the cantilever. Several of the important features of the method are illustrated in this analysis.

EXAMPLE 6.13

Determine the required moments due to the concentrated loads positioned as shown in Figure 6.26(a) in combination with uniformly distributed loads.

The assumed yield line patterns are also illustrated in Figure 6.26(a). The girder spacing is S, the cantilever overhang is H, and G is the wheel spacing (gage), usually 6 ft (1800 mm), or the spacing between the wheels of adjacent trucks, usually 4 ft (1200 mm). First consider the load positioned in the center of a panel near midspan as illustrated by point A in Figure 6.26(a).

Because the system is axisymmetric, the load is distributed evenly to all sectors ($d\alpha$). The analysis may be performed on the sector as shown in Figure 6.26(b) and the total energy is determined by integration around the circular path. By using Eq. 6.20, the internal work associated with yield-line rotation is

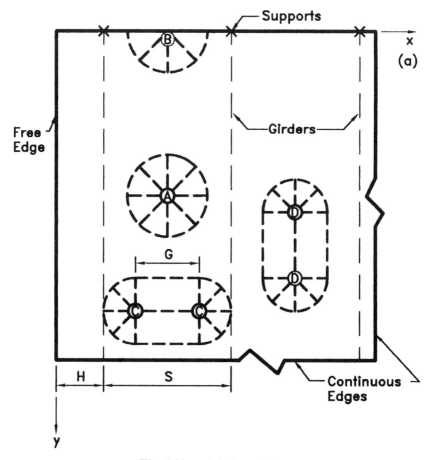

Fig. 6.26 (a) Axle positions.

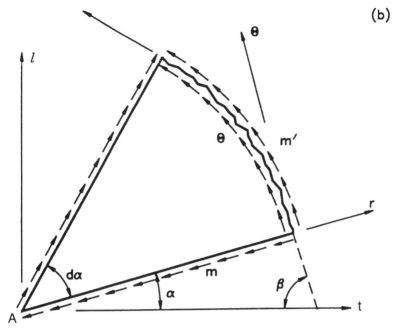

Fig. 6.26   (b) Sector (part).

$$U_{int} = U_{perimeter} + U_{radial\ fans} = \int_0^{2\pi} m'\theta r d\alpha + \int_0^{2\pi} m\theta r d\alpha$$

where α is the orientation of the radial yield line, θ is the virtual rotation at the ring of the yield line pattern, m is the positive moment capacity for the orientation α, and m' is the negative moment capacity for the orientation α. The moment capacity for a general orientation α is given in Eq. 6.17. By using Eq. 6.18, the internal strain energy becomes

$$U_{int} = U_{perimeter} + U_{radial\ fans} = \int_0^{2\pi} (m'_\ell \cos^2 \alpha + m'_t \sin^2 \alpha)\theta r d\alpha$$

$$+ \int_0^{2\pi} (m_\ell \cos^2 \beta + m_t \sin^2 \beta)\theta r d\beta$$

where β is the compliment of α. Note that

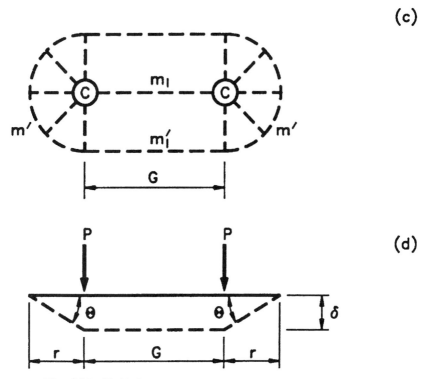

(c)

(d)

Fig. 6.26   Yield lines for axles. (c) Plan view; (d) elevation view.

$$\int_0^{2\pi} \sin^2 \alpha \; d\alpha = \int_0^{2\pi} \cos^2 \alpha \; d\alpha = \pi$$

U_{int} simplifies to

$$U_{\text{int}} = U_{\text{perimeter}} + U_{\text{radial fans}} = (m'_\ell + m'_t)\pi r\theta + (m_\ell + m_t)\pi r\theta$$

The energies for the perimeter and the fans are kept separate to facilitate further manipulation. The virtual rotation and translation δ at the load are related by

$$\delta = \theta r$$

The external virtual work due to the concentrated load P, in combination with a uniform load q, is established by using Eq. 6.21, resulting in

$$W_{ext} = P\delta + q \left(\frac{\pi r^2}{3}\right) \delta = Pr\theta + q \left(\frac{\pi r^2}{3}\right) r\theta$$

By equating the external work and internal energy, one obtains

$$\pi(m'_\ell + m'_t) + \pi(m_\ell + m_t) = P + \frac{q\pi r^2}{3} \tag{6.22}$$

The moment summations may be thought of as double the average moment capacity, or

$$m' = \tfrac{1}{2}(m'_\ell + m'_t)$$

and

$$m = \tfrac{1}{2}(m_\ell + m_t)$$

Substitution of the average moments into Eq. 6.22 results in

$$(m' + m) = \frac{P}{2\pi} + \frac{qr^2}{6} \tag{6.23}$$

The average capacities m and m' are used for convenience in subsequent calculations. For a comparison with the elastic analysis to be presented later, neglect the uniform load and assume that the positive and negative capacities are equal, the required moment capacity is

$$m = m' = \frac{P}{4\pi} \tag{6.24}$$

Now consider the load positioned near the edge of the slab at point B or at point E as shown in Figure 6.26(a), where the yield-line pattern is also shown. Note that due to symmetry, the length of the yield lines in this system constitute one-half the length of the previous system (point A). Thus, the internal energy is one-half of that given for the yield-line pattern for point A (Eq. 6.22), the associated uniform load is also one-half of the previous value but the concentrated load is full value, and the required moment is doubled (Eq. 6.23) giving

$$(m' + m) = \frac{P}{\pi} + \frac{qr^2}{6} \tag{6.25}$$

Next consider the two loads positioned at point C as illustrated in Figures 6.26(a) and 6.26(c). The internal energy is

$$U_{int} = U_{radial\ fan} + U_{straight} = 2\pi r\theta(m' + m) + \frac{2G\delta}{r}(m'_\ell + m_\ell)$$

The work due to the two concentrated loads plus the uniform load is

$$W_{ext} = 2P\delta + qr\delta\left(\frac{\pi r}{3} + G\right)$$

Again, equating the internal and external energies, one obtains

$$P + \frac{qr}{2}\left(\frac{\pi r}{3} + G\right) = \pi(m' + m) + \frac{G}{r}(m'_\ell + m_\ell) \tag{6.26}$$

Now consider the load at point D in Figure 6.26(a). The only difference between the analysis of this position and that of point C is that the moment capacity associated with the straight lines is now the capacity of the transverse reinforcement. Equation 6.26 becomes

$$P + \frac{qr}{2}\left(\frac{\pi r}{3} + G\right) = \pi(m' + m) + \frac{G}{r}(m'_t + m_t) \tag{6.27}$$

Note that as G goes to zero, Eqs. 6.26 and 6.27 reduce to Eq. 6.25, as expected. These equations are used to estimate the ultimate strength of slabs designed with the AASHTO procedures in the design chapters that follow.

It is interesting to compare the moments based on the yield-line analysis to those based on the elastic methods.

EXAMPLE 6.14

Determine the moments required for a wheel load of AASHTO design truck [P = 16 kips (72.5 kN)] position in the interior panel. Compare these moments to those obtained from the AASHTO strip and finite element methods. By neglecting the uniform load, Eq. 6.23 can be used to determine the required moment

$$(m' + m) = \frac{P}{2\pi} + \frac{(q = 0)r^2}{6} = \frac{16}{2\pi} = 2.55 \text{ ft kips/ft}$$

If we assume that the positive and negative moment capacities are the same, the required capacity is

$$m = m' = 1.27 \text{ ft kips/ft}$$

From Table 6.20 the AASHTO strip method moments are $m^+ = 3.50$ and m^- $= -3.65$ ft kips/ft, and the finite element transverse moments are $m^+ = 4.26$ and $m^- = -0.31$ ft kips/ft near midspan and $m^+ = 4.45$ and $m^- = -1.26$ ft kips/ft near the support. Thus the elastic distribution is quite different from the inelastic, which is consistent with test results where slabs typically test at a minimum eight times the service-level loads (16 kips).

In summary, several methods have been described for proportioning the moment and steel in a bridge deck. The AASHTO strip method is permitted by the specification and offers a simple method for all limit states. In light of the lower bound theorem, this is a conservative method. The yield-line method uses inelastic analysis techniques and is pertinent only to the strength and extreme limit states. Other methods are required in conjunction with this to insure serviceability. Finally, the empirical design method is not an analytical approach, but rather a set of rules upon which to proportion the deck. The discussion of this method is presented in following chapters.

6.3.4 Box-Girder Bridges

Behavior, Structural Idealization, and Modeling. The box-girder bridge is a common structural form in both steel and concrete. The multicell box girder may be thought of as slab-girder bridge with a bottom slab that encloses the section [see Fig. 6.27(a)]. This closure creates a "closed section" that is torsionally much stiffer than its open counterpart. This characteristic makes the box-girder ideal for bridges that have significant torsion induced by horizontal curvature resulting from roadway alignments. The characteristic is illustrated in Figures 2.29, 2.33, and 2.34. For example, the box-girder bridge is often used for tightly spaced interchanges that require curved alignments, because of its torsional resistance and fine aesthetic qualities.

Box systems are built in a wide variety of configurations, most are illustrated in Table 2.2. Examples include: Closed steel or precast boxes with a cast-in-place (CIP) deck (b), open steel or precast boxes with CIP deck (c), CIP multicell box (d), precast boxes with shear keys (f), and transversely posttensioned precast boxes (g). These systems can be separated into three primary categories of box systems: single- and double-cell [Fig. 6.27(a)], multicell (three or more cells, Fig. 6.27(b), and spread box systems [the boxes noncontiguous, Fig. 6.27(c)]. As expected, the behavior of these systems is distinctly different within each category and the design concern varies widely depending on construction methods. Due to the large number of systems and construction methods, selected systems are described with limited detail.

The single- and two-cell box systems are usually narrow compared to the span and behave similar to a beam and are often modeled with space frame

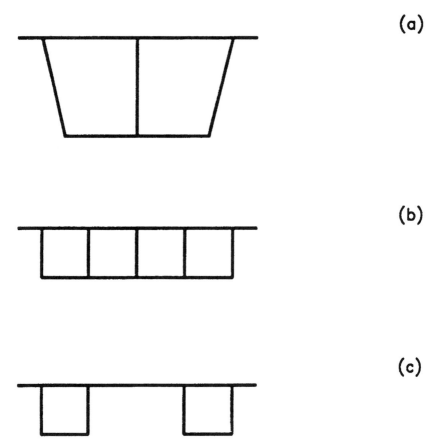

(a)

(b)

(c)

Fig. 6.27   (a) Two-cell box section, (b) multicell box section, and (c) spead-box section.

elements. Such systems are designed for the critical combinations of bending moment, shear, and torsion created due to global effects and the local effect of the vehicle applied directly to the deck. As stated in Chapter 4, global means the load effect is due to the global system response such as the deflection, moment, or shear of a main girder. Local effects are the actions and displacements that result from loads directly applied to (or in the local area of) the component being designed. Recall that if a small spatial variation in the live load placement causes a large change in load effect then the load effect is considered local.

The various displacement modes for a two-cell box-girder cross section are illustrated in Figure 6.28(a–e). Here the total displacement is decomposed into four components: global bending, global torsion, local flexure, and local distortion due to global displacements. The global bending is due to the girder behaving as a single beam, that is, the strain profile is assumed to be linear

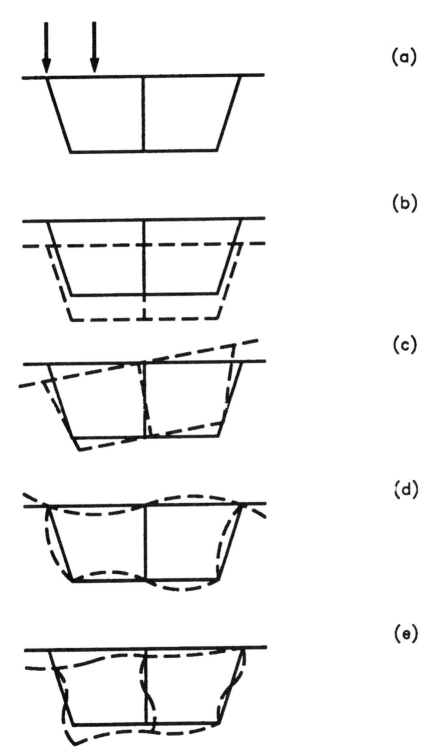

Fig. 6.28 (a) Eccentric loading, (b) global flexural deformation, (c) global torsional deformation, (d) local transverse bending deformation, and (e) distortional transverse deformation due to global displacements.

TABLE 6.21a Distribution Factors for Multicell Box Beams, and Box Sections Transversely Post-Tensioned Together—U.S. Customary Units[a]

Action/Location	AASHTO Table	Distribution Factors $(mg)^c$	Skew Correction Factor[b]	Range of Applicability
A. Moment interior girder	4.6.2.2b-1	Single design lane loaded $$mg_{\text{moment}}^{SI} = \left(1.75 + \frac{S}{3.6}\right)\left(\frac{1}{L}\right)^{0.35}\left(\frac{1}{N_c}\right)^{0.45}$$ Two or more (multiple) design lanes loaded $$mg_{\text{moment}}^{MI} = \left(\frac{13}{N_C}\right)^{0.3}\left(\frac{S}{18}\right)\left(\frac{90}{L}\right)^{0.25}$$	N/A	$7.0 \le S \le 13.0$ ft $60 \le L \le 240$ ft $N_C \ge 3$ If $N_C > 8$ Use $N_C = 8$
B. Moment exterior girder	4.6.2.2d-1	$$mg_{\text{moment}}^{ME} = \frac{W_e}{14}$$	N/A	$W_e \le S$
C. Shear interior girder	4.6.2.2.3a-1	Single design lane loaded $$mg_{\text{shear}}^{SI} = 0.5\left[\left(\frac{S}{3\text{ ft}}\right)^{0.6}\left(\frac{d}{12\,L}\right)^{0.1}\right]$$ Two or more (multiple) design lanes loaded $$mg_{\text{shear}}^{MI} = 0.5\left[\left(\frac{S}{3.4\text{ ft}}\right)^{0.9}\left(\frac{d}{12\,L}\right)^{0.1}\right]$$	N/A	$6.0 \le S \le 13.0$ ft $20 \le L \le 240$ ft $3 \le d \le 9$ ft $N_C \ge 3$

D. Shear exterior girder	4.6.2.2.3b-1	One design lane loaded Use lever rule		$-2.0 \le d_e \le 5.0$ ft $0 \le \theta \le 60°$
		Two or more (multiple) design lanes loaded $mg^{ME}_{shear} = e * mg^{MI}_{shear}$ $e = \dfrac{8\ \text{ft} + d_e}{12.5\ \text{ft}}$	$1.0 + \left[0.25 + \dfrac{12\,L}{70d} \right] \tan\theta$ for shear in the obtuse corner	$6.0 \le S \le 13.0$ ft $20 \le L \le 240$ ft $3 \le d \le 9$ ft
		d_e is positive if girder web is inside of barrier, otherwise negative		$N_c \ge 3$

[a] S is the girder spacing [ft (mm)].
L is the span length [ft(mm)].
N_c is the number of cells.
N_b is the number of beams.
W_e is half the web spacing, plus the total overhang spacing [ft(mm)].
d is the overall depth of a girder [ft(mm)].
d_e is the distance from the center of the exterior beam to the inside edge of the curb or barrier [ft (mm)].
θ is the angle between the centerline of the support and a line normal to the roadway centerline.
The *lever rule* is in Example 6.3.
[b] Not applicable = N/A.
[c] Equations include multiple presence factor; for lever rule the engineer must perform factoring by m.

356

TABLE 6.21b Distribution Factors for Multicell Box Beams, and Box Sections Transversely Post-Tensioned Togther—SI Units

Action/Location	AASHTO Table	Distribution Factors (mg)	Skew Correction[a] Factor	Range of Applicability
A. Moment interior girder	4.6.2.2b-1	Single design lane loaded $mg^{SI}_{moment} = \left(1.75 + \dfrac{S}{1100}\right)\left(\dfrac{300}{L}\right)^{0.35}\left(\dfrac{1}{N_c}\right)^{0.45}$ Two or more (multiple) design lanes loaded $mg^{MI}_{moment} = \left(\dfrac{13}{N_c}\right)^{0.3}\left(\dfrac{S}{430}\right)\left(\dfrac{1}{L}\right)^{0.25}$	N/A	$2100 \leq S \leq 4000$ mm $18\,000 \leq L \leq 73\,000$ mm $N_c \geq 3$ If $N_c > 8$ Use $N_c = 8$
B. Moment exterior girder	4.6.2.2d-1	$mg^{ME}_{moment} = \dfrac{W_e}{4300 \text{ mm}}$	N/A	$W_e \leq S$
C. Shear interior girder	4.6.2.3a-1	Single design lane loaded $mg^{SI}_{shear} = \left(\dfrac{S}{2900 \text{ mm}}\right)^{0.6}\left(\dfrac{d}{L}\right)^{0.1}$ Two or more (multiple) design lanes loaded $mg^{MI}_{shear} = \left(\dfrac{S}{2200 \text{ mm}}\right)^{0.9}\left(\dfrac{d}{L}\right)^{0.1}$	N/A	$1800 \leq S \leq 4000$ mm $6000 \leq L \leq 73\,000$ mm $890 \leq d \leq 2800$ mm $N_c \geq 3$
D. Shear exterior girder	4.6.2.3b-1	One design lane loaded Use lever rule Two or more (multiple) design lanes loaded $mg^{ME}_{shear} = e*mg^{MI}_{shear}$ $e = 0.64 + \dfrac{d_e}{3800 \text{ mm}}$ d_e is positive if girder web is inside of barrier, otherwise negative	$1.0 + \left[0.25 + \dfrac{L}{70d}\right]\tan\theta$ for shear in the obtuse corner	$-600 \leq d_e \leq 1500$ mm $0 \leq \theta \leq 60°$ $1800 \leq S \leq 4000$ mm $6000 \leq L \leq 73\,000$ mm $900 \leq d \leq 2700$ mm $N_c \geq 3$

and there is no twisting or distortion, that is, the section shape remains unaltered [Fig. 6.28(b)]. The global torsional rotation is illustrated in Figure 6.28(c). As in the bending mode, the section shape remains unaltered by the load and the section is twisted due to eccentrically applied loads. The local bending mode is shown in Figure 6.28(d). Here the loads create out-of-plane bending in the deck. Because the girder webs are continuous with the slab, the webs and the bottom slab also bend. The intersection of elements (physical joints) rotate, but do not translation, in this mode. Finally, the distortion mode is illustrated in Figure 6.28(e). The slab and webs flex due to the translation and rotation of the physical joints, that is, the displacements shown in Figure 6.28(b) plus those shown in Figure 6.28(c). Superpose all of these modes to establish the system response.

There are numerous analytical methods available for the analysis of box-girder systems, ranging from the rigorous and complex to the simplistic and direct. The selection of the method depends on the response sought and its use. The box system may be modeled with finite elements, finite strips, and beams. All approaches are viable and the one selected depends on several factors including: the number of cells, the geometry (width/length, skew, diaphragms, cross bracing), construction method, type of box system (single, multi, or spread boxes), and, of course, the reason for and application of the results.

In general, the one- and two-cell box systems have spans that are much greater than their section widths and can be modeled as beams, usually with space frame elements. The beam is modeled so that the global flexural and torsional response are considered. These actions are then used with the resistance provisions in the usual manner. For the case of steel boxes, the web thicknesses tend to be thin and local stiffeners are required. The local bending effects are modeled by considering the box as a frame in the transverse direction, and obtaining reasonable distribution of shear and bending moments (due to the out-of-plane deformations). This model can be based on the distribution width outlined in, for example, Eq. 6.16.

The distortional deformation is modeled by imposing the resulting beam displacements at the joints of the transverse frame (plane frame), which creates bending of the deck and web elements. The results of the local bending and distortional deformation are superposed to establish the local out-of-plane actions.

As with the single- and two-cell system, multicell (three or more) box systems can be accurately modeled with the finite element and finite strip methods. Both formulations can simultaneously model the in- and out-of-plane load effects associated with global and local behavior. These methods are certainly the most common rigorous methods used in engineering practice. The principle difference between the slab-girder and the multicell box bridge is that the box section has significantly more torsional stiffness, which enables better load distribution.

Beam-Line Methods. The AASHTO Specification has equations for distribution factors for multicell box beams. These are applied in a fashion similar to the slab-girder systems and are summarized in Table 6.21.

Other box systems, such as spread box beams and shear-keyed systems have similar formulas but due to space limitations are not presented. These formulas are given in AASHTO [A4.6.2.2]. An example is given to illustrate the use of the AASHTO formulas for a multicell system.

EXAMPLE 6.15

Determine the distribution factors for one- and multilane loaded for the concrete cast-in-place box system shown in Figure 6.29. The bridge has no skew.

The AASHTO distribution factors are used as in the previous example with slab-girder bridges. The factors for one-, two-, and three-lane loadings are established for the interior and exterior girder moments and shears. Each case is considered separately. Table 6.21 is used exclusively for all calculations. Note that the multiple presence factor is included.

Interior girder moment for one-lane loaded:

$$mg_{moment}^{SI} = \left(1.75 + \frac{S}{3.6}\right)\left(\frac{1}{L}\right)^{0.35}\left(\frac{1}{N_c}\right)^{0.45}$$

$$mg_{moment}^{SI} = \left(1.75 + \frac{13}{3.6}\right)\left(\frac{1}{100}\right)^{0.35}\left(\frac{1}{3}\right)^{0.45} = 0.65$$

Interior girder moment for multiple-lanes loaded.

$$mg_{moment}^{MI} = \left(\frac{13}{N_C}\right)^{0.3}\left(\frac{S}{18}\right)\left(\frac{90}{L}\right)^{0.25}$$

$$mg_{moment}^{MI} = \left(\frac{13}{3}\right)^{0.3}\left(\frac{13}{18}\right)\left(\frac{90}{100}\right)^{0.25} = 1.09$$

Exterior girder moment for one-lane and multiple-lane loaded.

$$W_e = \tfrac{1}{2}S + \text{overhang} = \tfrac{1}{2}(13) + 3.75 = 10.25 \text{ ft}$$

$$mg_{moment}^{(S \text{ or } M)E} = \frac{W_e}{14} = \frac{10.25}{14} = 0.73$$

Interior girder shear for one-lane loaded.

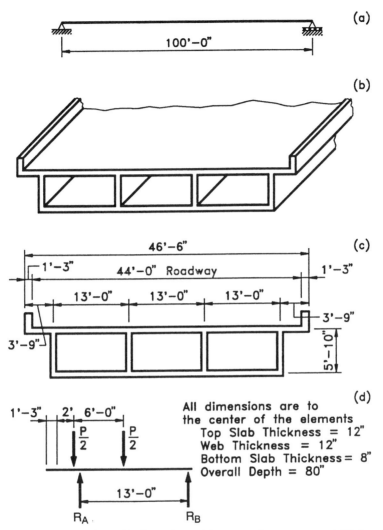

Fig. 6.29 (a) Box section, (b) dimensions, (c) span and supports, and (d) free body diagram.

$$mg^{SI}_{\text{shear}} = 0.5\left[\left(\frac{S}{3\text{ ft}}\right)^{0.6}\left(\frac{d}{12L}\right)^{0.1}\right]$$

$$mg^{SI}_{\text{shear}} = 0.5\left[\left(\frac{13}{3\text{ ft}}\right)^{0.6}\left(\frac{(6.66)(12)}{(12)(100)}\right)^{0.1}\right] = 0.92$$

Interior girder shear for multiple-lanes loaded.

$$mg_{\text{shear}}^{MI} = 0.5\left[\left(\frac{S}{3.4\text{ ft}}\right)^{0.9}\left(\frac{d}{12L}\right)^{0.1}\right]$$

$$mg_{\text{shear}}^{MI} = 0.5\left[\left(\frac{13}{3.4\text{ ft}}\right)^{0.9}\left(\frac{(12)(6.66)}{(12)(100)}\right)^{0.1}\right] = 1.28$$

Exterior girder shear for one-lane loaded: The lever rule is used for this calculation. Refer to the free body diagram illustrated in Figure 6.29(d). Balance the moment about B to determine the reaction R_A

$$\sum M_B = 0$$

$$\frac{P}{2}(7.5) + \frac{P}{2}(13.5) - R_A(13) = 0$$

$$R_A = 0.81P$$

$$g_{\text{shear}}^{SE} = 0.81$$

$$mg_{\text{shear}}^{SE} = 1.2(0.81) = 0.97$$

Exterior girder shear for two-lanes loaded: The interior distribution factor is used with an adjustment factor based on the overhang length.

$$mg_{\text{shear}}^{ME} = e * mg_{\text{shear}}^{MI} = (0.84)(1.28) = 1.08$$

where

$$e = \frac{8\text{ ft} + d_e}{12.5\text{ ft}} = \frac{8\text{ ft} + 2.5}{12.5\text{ ft}} = 0.84$$

The AASHTO distribution factors are compared with those based on a finite element analysis in the next section. The application of these factors is similar to that of the slab-girder bridge. Details regarding the application are presented in Chapter 7.

Finite Element Method. The box-girder bridge may be modeled with the finite element method by using shell elements. These elements must have the capability to properly model the in-plane and out-of-plane effects. One of the principal characteristics of the box girder is shear lag. This phenomenon is the decrease in stress (flange force) with increased distance away from the web. The mesh must be sufficiently fine to model this effect. A mesh that is too coarse tends to spread the flange force over a larger length, therefore decreasing the peak forces. Another important characteristic is the proper modeling of the diaphragms and support conditions. The diaphragms are transverse walls periodically located within the span and at regions of concentrated load such as supports. The diaphragms tend to stiffen the section torsionally and reduce the distortional deformation, which produces a stiffer structure with improved load distribution characteristics. Because the supports are typically located at a significant distance below the neutral axis, the bearing stiffness is important in the modeling. For example, the bottom flange force can change significantly if the support conditions are changed from pin–roller to pin–pin. Finally, because the box section has significant torsional stiffness, the effects of skew can also be significant and must be carefully considered. For example, it is possible for an eccentrically loaded box-girder web to lift completely off its bearing seat. This effect can greatly increase the reactions and associated shear forces in the area of the support. Such forces can create cracks and damage bearings.

A simply supported three-cell box girder is modeled in the example below.

EXAMPLE 6.16

For the bridge illustrated in Figure 6.29, use the finite element method to determine the distribution factors for the bending moment at midspan.

The system was modeled with the two meshes shown in Figures 6.30(a) and 6.30(b). Both meshes produced essentially the same results, hence convergence was achieved for the parameters under study. The load effect of each web was based on the longitudinal force per unit length, f, at the bottom. This quantity is readily available from the analysis. The sum of the forces for all the webs is divided by the number of lanes loaded. This ratio is then divided into the force in each girder. The distribution factor for girder i may be mathematically expressed as

$$g = \frac{f_i}{\left(\sum_{\text{No. Girders}} f_i / \text{No. Lanes Loaded} \right)} \tag{6.28}$$

where f_i is the load effect in the girder web i.

The flexural bending moment for the entire web could be used as in the slab-girder bridge, but this quantity is not readily available. To determine the moment, the force per unit length must be numerically integrated over the web depth, that is, $M = \int f y \, dy$. If the end supports are not restrained, and therefore, do not induce a net axial force in the section, then the force per length is nominally proportional to the moment. The loads are positioned for the maximum flexural effect in the exterior and interior webs for one-, two- and three-lanes loaded. For shear/reactions, the loads were positioned for the maximum reaction and the distribution factors were determined by Eq. 6.28.

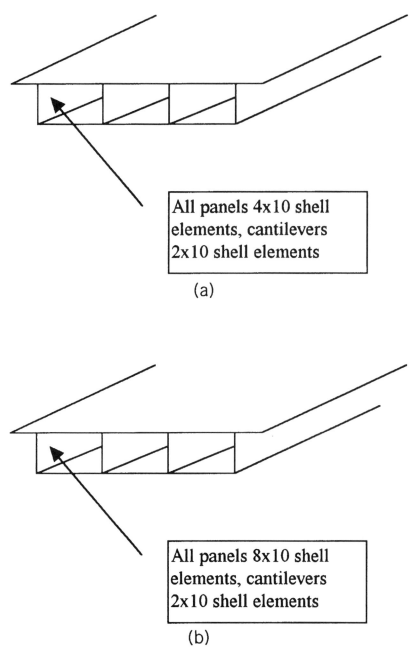

All panels 4x10 shell
elements, cantilevers
2x10 shell elements

(a)

All panels 8x10 shell
elements, cantilevers
2x10 shell elements

(b)

Fig. 6.30   (a) Coarse mesh and (b) fine mesh.

The distribution factors based on the finite element methods are shown in Table 6.22. The results from the AASHTO distribution formulas are also shown for comparison.

Note that there are significant differences between the AASHTO formulas and finite element results. In general, the finite element model exhibits better load distribution than the AASHTO values. Recall that the AASHTO values are empirically determined based solely on statistical observation. Therefore, it is difficult to explain the differences for this particular structure. It is interesting to note that unlike the slab-girder and slab bridges, the analytical results can vary significantly from the AASHTO method and perhaps this highlights the need for rigorous analysis of more complex systems. Again, note that the comparison of the distribution factors does not suggest that the only reason to perform a rigorous analysis is to establish the distribution factors. Because the finite element-based actions are available directly, the designer may wish to proportion the bridge based on these actions. Other load cases must be included and the envelope of combined actions is used for design.

6.4 EFFECTS OF TEMPERATURE, SHRINKAGE, AND PRESTRESS

6.4.1 General

The effects of temperature, shrinkage, and prestress are treated in a similar manner. All of these create a state where the structure is prestrained prior to the application of gravity and/or lateral loads. It is unlikely that the effects of these loads exceed any strength limit state, but these loads can certainly be of concern regarding serviceability. As discussed earlier in this chapter, for ductile systems, prestrains and prestress are eliminated when the ultimate limit states are reached (see Section 6.2.2). Therefore, the combination of these is typically not of concern for the ultimate limit state. They may be of concern regarding the determination of the load that creates first yield, deformation of the structure for the design of bearings and joints, and other service-level phenomena.

Because the stiffness-based methods are frequently used in the analysis of bridge systems, the discussion of prestraining is limited to these methods. The finite element analysis of the system subjected to prestrain is similar to the stiffness method, and most finite element textbooks address this issue. The force or flexibility method is a viable technique, but is seldom used in contemporary computer codes and, therefore, is not addressed.

The general procedure for the analysis of frame elements subjected to prestraining effect is illustrated in Figure 6.31. The structure subjected to the effects of temperature and prestressing forces is shown in Figure 6.31(a). Each element in the system may be separated from the structure as illustrated in

TABLE 6.22 Distribution Factors Based on the Finite Element Method

Girder Location	Lanes Loaded	Finite Element Moment Dist. Factor (mg)	AASHTO Moment Distribution Factor (mg)	Finite Element Shear or Reaction Distribution Factor (mg)	AASHTO Shear or Reaction Distribution Factor (mg)
Exterior	1	$0.53(1.2) = 0.64$	0.73	$0.72(1.2) = 0.86$	0.81
Exterior	2	$0.85(1.0) = 0.85$	0.73	$0.84(1.0) = 0.84$	1.08
Exterior	3	$1.00(0.85) = 0.85$	0.73	$1.34(0.85) = 1.14$	1.08
Interior	1	$0.38(1.2) = 0.46$	0.65	$0.71(1.2) = 0.85$	0.92
Interior	2	$0.62(1.0) = 0.62$	1.09	$1.21(1.0) = 1.21$	1.28
Interior	3	$0.80(0.85) = 0.68$	1.09	$1.22(0.85) = 1.04$	1.28

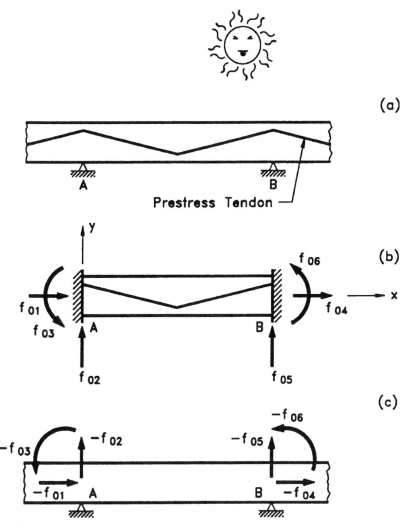

Fig. 6.31   (a) Girder subjected to prestrain, (b) restrained system subjected to prestrain, and (c) equivalent joint loads.

Figure 6.31(b). Here the joints are locked against rotation and translation with restraining actions located at the end of the element. The opposite (negative) of the restraining actions are applied on the joint-loaded structure and the analysis proceeds in the same manner as with any other joint-load effect. The displacements from the joint-loaded system [Fig. 6.31(c)] are the displacements in the entire prestrained system. However, the actions from the joint-loaded system must be superposed with the actions from the restrained system to achieve the actions in the entire system. Again, the main difference between the analysis for effects of prestrain/prestress and the analysis for directly applied load is the analysis of the restrained system. In the sections that follow, the effects of prestressing forces and temperature effects are discussed. The effects of shrinkage may be determined similar to temperature effects and is not explicitly included.

6.4.2 Prestressing

The element loads for various tendon paths are tabularized and available to aid in determining equivalent element loads that are subsequently used in establishing the restraining actions. The equivalent element loads for several commonly used tendon configurations are illustrated and are found in textbooks on advanced analysis, for example, see Ghali et al. (1989). These loads may be applied just as one would apply any other loads.

6.4.3 Temperature Effects

Most bridges experience daily and seasonal temperature variations causing material to shorten with decreased temperatures and lengthen with increased temperatures. It has been observed that these temperature fluctuations can be separated into two components: a uniform change and a gradient. The uniform change is the effect due to the entire bridge changing temperature by the same amount. The temperature gradient is created when the top portion of the bridge gains more heat due to direct radiation than the bottom. Because the strains are proportional to the temperature change, a nonuniform temperature strain is introduced. In this section, the axial strain and curvature formulas are presented for the effect of temperature change. These formulas are given in discrete form. The formulas may be implemented in stiffness and flexibility formulations. An example is presented to illustrate the usage of the formulas.

AASHTO Temperature Specifications. Uniform temperature increases and temperature gradients are outlined in AASHTO [A3.12.1 and A3.12.2]. The uniform change is prescribed is Table 4.21 and the temperature gradient is defined in Table 4.22 and Figures 4.22, and 4.23. The temperature change creates a strain of

$$\varepsilon = \alpha \, (\Delta T)$$

where α is the coefficient of thermal expansion and ΔT is the temperature change.

This strain may be used to determine a change in length by the familiar equation

$$\Delta \ell = \varepsilon \ell = \alpha (\Delta T) \ell$$

where ℓ is the length of the component, and the stress in a constrained system of

$$\sigma = \alpha(\Delta T)E$$

The response of a structure to the AASHTO multilinear temperature gradient is more complex than its uniform counterpart and can be divided into two effects: (1) gradient induced axial strain, and (2) gradient induced curvature. The axial strain is described first.

Temperature Gradient Induced Axial Strain. The axial strain ε due to the temperature gradient is (Ghali et al., 1989)

$$\varepsilon = \frac{\alpha}{A} \int_A T(y)dA \qquad (6.29)$$

where

α is the coefficient of thermal expansion.
$T(y)$ is the gradient temperature as shown in Figure 4.29.
y is the distance from the neutral axis.
dA is the differential cross-sectional area.
The integration is over the entire cross section.

If the coefficient of thermal expansion is the same for all cross-section materials, standard transformed section analysis may be used to establish the cross-sectional properties. For practical purposes, steel and concrete may be assumed to have the same expansion coefficients.

By discretization of the cross section into elements, Eq. 6.29 simplifies this to a discrete summation. Consider the single element shown in Figure 6.32. Although the element shown is rectangular, the element shape is arbitrary. The element's elastic centroidal axis is located at a distance $\bar{y}_i$, and y is the location of the differential area element dA. The area of the element and second moment of area are denoted by A_i and I_i, respectively. Note that $y_i = y - \bar{y}_i$. The temperature at location y is

$$T(y) = T_{ai} + \frac{\Delta T_i}{d_i} y_i = T_{ai} + \frac{\Delta T_i}{d_i} (y - \bar{y}_i) \qquad (6.30)$$

where T_{ai} is the temperature at the element centroid, ΔT_i is the temperature difference from the bottom of the element to the top, and d_i is the depth of the element.

Substitution of Eq. 6.30 into Eq. 6.29 yields

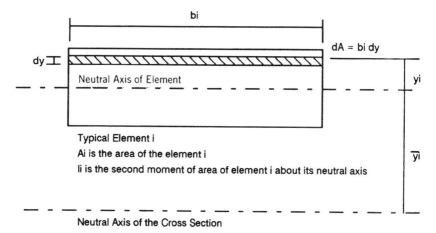

Fig. 6.32 Cross section with discrete element. Example cross section.

$$\varepsilon = \frac{\alpha}{A}\sum\int \left[T_{ai} + \frac{\Delta T_i}{d_i}(y - \bar{y}_i) \right] dA_i \qquad (6.31)$$

where the summation is over all elements in the cross section and the integration is over the domain of the discrete element. Integration of each term in Eq. 6.31 yields

$$\varepsilon = \frac{\alpha}{A}\sum \left[T_{ai}\int dA_i + \frac{\Delta T_i}{d_i}\int y\,dA_i - \frac{\Delta T_i \bar{y}_i}{d_i}\int dA_i \right] \qquad (6.32)$$

Substitution of $A_i = \int dA_i$ and $\bar{y}_i A_i = \int y\,dA_i$ in Eq. 6.32 yields

$$\varepsilon = \frac{\alpha}{A}\sum \left[T_{ai}A_i + \frac{\Delta T_i}{d_i}\bar{y}_i A_i - \frac{\Delta T_i}{d_i}\bar{y}_i A_i \right] \qquad (6.33)$$

Note the second and third term sum to zero, and Eq. 6.33 simplifies to

$$\varepsilon = \frac{\alpha}{A}\sum T_{ai}A_i \qquad (6.34)$$

Equation 6.34 is the discrete form of Eq. 6.29, which is given in AASHTO [A4.6.6]. Note that only areas of the cross section with gradient temperature contribute to the summation.

Temperature Gradient Induced Curvature. Temperature induced curvature is the second deformation that must be considered. The curvature, ψ, due to the gradient temperature is (Ghali et al., 1989)

$$\psi = \frac{\alpha}{I} \int T(y)y\,dA \tag{6.35}$$

where I is the second moment of area of the entire cross section about the elastic centroidal axis.

Substitution of Eq. 6.30 into Eq. 6.35, and expansion yields

$$\psi = \frac{\alpha}{I} \sum \left[T_{ai} \int y_i\,dA_i + \frac{\Delta T_i}{d_i} \int (y^2 - \bar{y}_i y)\,dA_i \right] \tag{6.36}$$

where the summation is over all elements in the cross section and the integration is over the domain of the discrete element. Performing the required integration in Eq. 6.36 yields

$$\psi = \frac{\alpha}{I} \sum \left[T_{ai}\bar{y}_i A_i + \frac{\Delta T_i}{d_i} I_i - \frac{\Delta T_i}{d_i} \bar{y}_i^2 A_i \right] \tag{6.37}$$

The parallel axis theorem is used to relate the cross-section properties in Eq. 6.37, that is,

$$I_i = \bar{I}_i + \bar{y}_i^2 A_i \tag{6.38}$$

which can be rearranged as

$$\bar{I}_i = I_i - \bar{y}_i^2 A_i \tag{6.39}$$

The combination of Eq. 6.39 with Eq. 6.37 yields

$$\psi = \frac{\alpha}{I} \sum \left[T_{ai}\bar{y}_i A_i + \frac{\Delta T_i}{d_i} \bar{I}_i \right] \tag{6.40}$$

which is the discrete form of the integral equation given in Eq. 6.35 and in AASHTO [A4.6.6].

Using Strain and Curvature Formulas. The axial strain and curvature may be used in both flexibility and stiffness formulations for frame elements. In the former, ε may be used in place of P/AE, and ψ may be used in place of M/EI in traditional displacement calculations. The flexibility method requires the analysis of the released statically determine and stable system. The analysis of the released system is conceptually straightforward, but this is not the case for the multilinear temperature distribution. Although the distribution does not create external reactions, it does create internal self-equilibriating stresses. These stresses must be superimposed with actions created from the redundants. The complete details are presented elsewhere (Ghali et al., 1989).

In the stiffness method, the fixed-end actions for a prismatic frame element may be calculated as

$$N = E A \, \varepsilon \tag{6.41}$$

$$M = E I \, \psi \tag{6.42}$$

where N is the axial force (constant with respect to length), M is the flexural bending moment (again constant), and E is Young's modulus.

These actions may be used to determine the equivalent joint loads in the usual manner and the resulting displacements may be used to recover the actions in the joint-loaded systems. These actions must be superimposed with the actions (usually stresses in this case) in the restrained (fixed) system. The temperature-related stresses in the restrained system, *not the fixed-end actions* in Eq. 6.41 and 6.42, must be used because the temperature gradient is not constant across the section. This complication forces algorithm modifications in existing stiffness codes because the fixed-end actions do not superimpose directly with the actions of the joint-loaded structure. The following example provides guidance on this issue.

EXAMPLE 6.17

The transformed composite cross section shown in Figure 6.33(a) is subjected to the temperature gradients associated with Zone 1 (Table 4.22) for a plain concrete surface. This temperature variation is also shown in Figure 6.33(a). The dimensions of the cross section were selected for ease of computation and illustration rather than based on typical proportions. A modular ratio of 8 is used. The cross section is used in the simple beam shown in Figure 6.33(b). The beam subjected to the temperature change has been divided into two sections labeled 1 and 2 in Figure 6.33(a). The cross sections and material properties are listed in Table 6.23 with reference to the labeled sections. All other areas of the cross section do not have a temperature gradient and therefore are not included in the summations. These section and material properties are used to calculate the axial strain, curvature, fixed-end axial force, and flexural bending moment.

By using Eqs. 6.34 and 6.41, the gradient induced axial strain and fixed-end axial force are

$$\varepsilon = (11.7 \times 10^{-6}/96\,780)\,[19(25\,800) + 4(25\,800)] = 71.74 \times 10^{-6}$$
$$N = 200\,000(96\,780)(71.74 \times 10^{-6}) = 1390 \text{ kN}$$

By using Eqs. 6.40 and 6.42, the curvature and fixed-end flexural moments are calculated below.

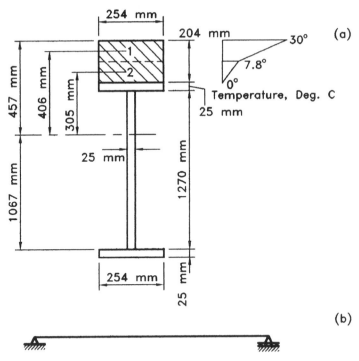

Fig. 6.33 (a) Temperature stress distribution and (b) simple span girder.

$$\psi = (11.7 \times 10^{-6}/23.9 \times 10^{9}) \, [19(406)(25\,800) + (22/102)(22.2 \times 10^{6})$$
$$+ \, 4(304)(25\,800) + (8/102)(22.2 \times 10^{6})]$$

$$\psi = 115 \times 10^{-9}$$

$$M = 200\,000(23.9 \times 10^{9})(115 \times 10^{-9}) = 550 \times 10^{6} \text{ N mm} = 550 \text{ kN m}$$

TABLE 6.23 Cross-Section Properties

Properties	Reference Section 1	Reference Section 2	Total Section[a]
A_i (mm)^2	25 800	25 800	96 780
$\bar{y}_i$ (mm)	406	304	1067
I_i (mm^4)	22.2×10^6	22.2×10^6	23.9×10^9
α	11.7×10^{-6}	11.7×10^{-6}	11.7×10^{-6}
E(MPa)	200 000 ($n = 8$)	200 000 ($n = 8$)	200 000
T_{ai} (°C)	19	4	N/A
ΔT_i (°C)	22	8	N/A

[a]Not applicable = N/A.

The restrained temperature stresses are determined by $\sigma = \alpha \Delta TE$. The restrained stress in the top of the section is

$$\sigma = 11.7 \times 10^{-6}\,(30)(200\,000) = 70.2\ \text{MPa}$$

and the stress in the bottom is zero because ΔT is zero as shown in Figure 6.34(b). The moment is constant in the joint-loaded system and the associated flexural stress distribution is shown in Figure 6.34(c). The axial force in the unrestrained (joint-loaded) system is also constant and the associated axial stress is shown in Figure 6.34(d). Superimposing the stresses in the restrained and joint-loaded systems gives the stress distribution shown in Figure 6.34(e). Note, the external support reactions are zero, but the internal stresses are not. The net internal force in each section is zero, that is, there is no axial force nor flexural moment. The equivalent joint loads shown at the top of Figure 6.34(c) are used to calculate the displacements {0.00175 rad, −0.00175 rad, 2.2 mm} referenced in Figure 6.34(a).

Summary of Temperature Effects. The AASHTO (1994) LRFD Bridge Design Specification requires that a prescribed temperature gradient be used to model temperature effects in girders. The prescribed multilinear gradients are used to develop a method involving discrete summations that are used to determine the axial strain and curvature. These formulas may be implemented into stiffness and flexibility programs. An example is used to illustrate the usage of the formulas presented.

6.4.4   Shrinkage and Creep

Concrete creep and shrinkage are difficult to separate as these two effects occur simultaneously in the structural system. The load effect may be estimated by an analysis similar to the procedure previously described for temperature effects. The temperature strain $\alpha(\Delta T)$ may be replaced with a shrinkage or creep strain. The strain profile is obviously different than that produced by a temperature gradient, but the appropriate strain profile may be used in a similar manner.

6.5   LATERAL LOAD ANALYSIS

As with the gravity loads, the lateral loads must also be transmitted to the ground, that is, a load path must be provided. Lateral loads may be imposed from wind, water, ice floes, and seismic events. The load due to ice floes and water is principally a concern of the substructure designer and discussion of the system for resisting these loads is presented in Chapter 10. The system analysis for wind loads is discussed in Section 6.5.1 and the analysis for seismic loads is briefly introduced in Section 6.5.2.

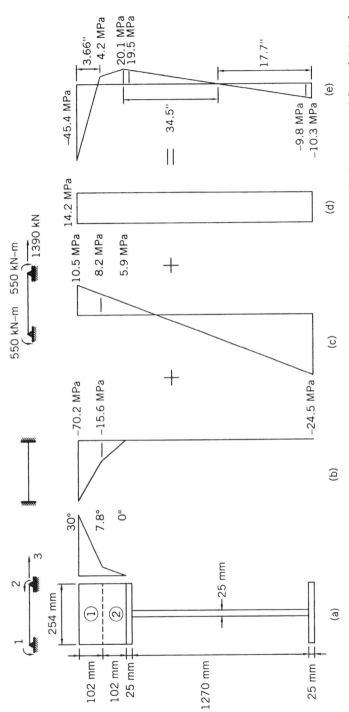

Fig. 6.34 (a) Cross section, (b) restrained system, (c) unrestrained system (bending), (d) unrestrained system (axial), and (e) total stress.

373

6.5.1 Wind Loads

The wind pressure is determined by the provisions in AASHTO [A3.8], which are described in Chapter 4 (Wind Forces). This uniform pressure is applied to the superstructure as shown in Figure 6.35(a). The load is split between the upper and lower wind-resisting systems. If the deck and girders are composite or are adequately joined to support the wind forces, then the upper system is considered to be a diaphragm where the deck behaves as a very stiff beam being bent about the y–y axis as shown in Figure 6.35(a). Note that this is a common and reasonable assumption given that the moment of inertia of the deck about the y–y axis is quite large. Wind on the upper system can be considered transmitted to the bearings at the piers and the abutments via the diaphragm acting as a deep beam. It is traditional to distribute the wind load to the supporting elements on a tributary area basis [see Fig.

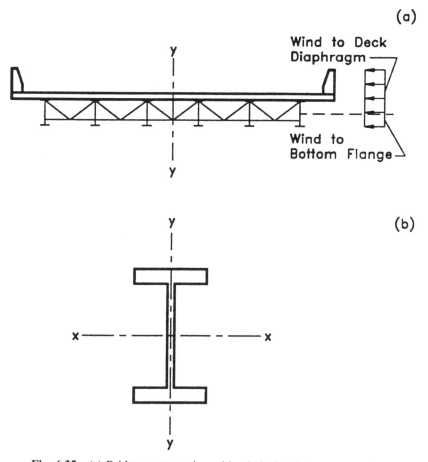

Fig. 6.35 (a) Bridge cross section with wind, (b) girder cross section.

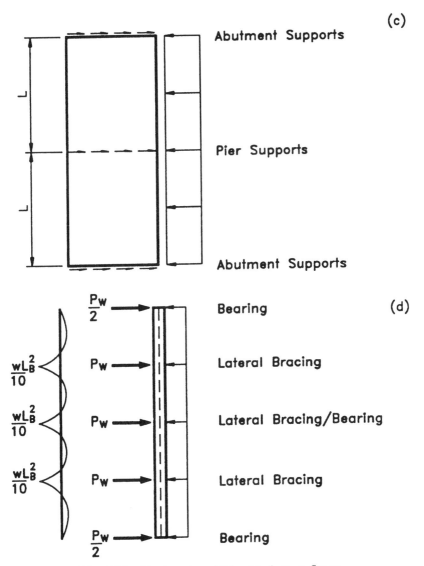

Fig. 6.35   (c) plan view, (d) load to bottom flange.

6.35(c)]. If there are no piers, or if the bearing supports at the piers do not offer lateral restraint, then all the diaphragm loads must be transmitted to abutment bearings, one-half to each. If the bearings restrain lateral movement and the pier support is flexible (e.g., particularly tall), then a refined model of the system might be warranted to properly account for the relative stiffness of the piers, the bearings, and the abutments. The in-plane deformation of the deck may usually be neglected.

(e)

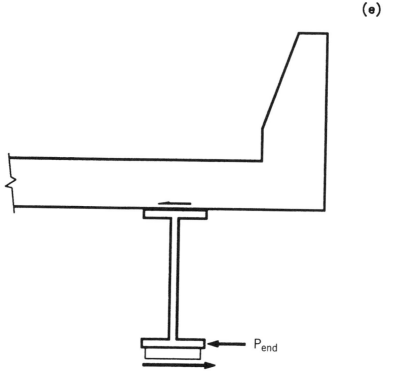

Fig. 6.35   (e) load to end external bearing.

The wind load to the lower system is carried by the girder in weak-axis bending [y–y in Fig. 6.35(b)]. Most of the girder's strength and stiffness in this direction are associated with the flanges. Typically, the bottom flange is assumed to carry the lower system load as shown in Figure 6.35(d). The bottom flange is usually supported by intermediate bracing provided by a cross frame [see Fig. 6.35(a)], steel diaphragm element (transverse beams), or in the case of a concrete beam a concrete diaphragm (transverse). These elements provide: the compression-flange bracing required for lateral torsional buckling while the concrete is being placed; the transverse elements also aid in the gravity load distribution, to a minor extent; and finally, the bracing periodically supports the bottom flange which decreases the effective span length for the wind loading. The bracing forces are illustrated in Figure 6.35(d) where the free body diagram is shown with the associated approximate moment diagram. In AASHTO [A4.6.2.7] an approximate analysis is permitted, where the bracing receives load on the basis of its tributary length. The moment may also be approximated with $WL_B^2/10$ [C4.6.2.7.1]. In place

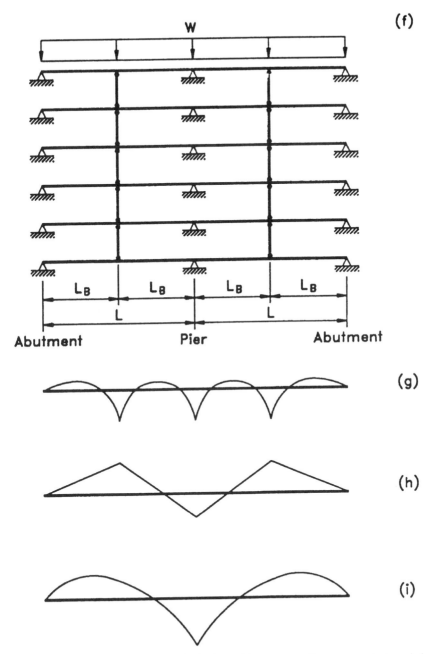

Fig. 6.35 (f) girder flanges—load sharing, (g) moment diagram—exterior girder flange, (h) moment diagram—interior girder flange, and (i) moment diagram—interior girder flange, uniform load.

of the approximate analysis, a more exact beam analysis may be performed, but this refinement is seldom warranted. Once the load is distributed into the bracing elements, it is transmitted into the deck diaphragm by the cross frames, or by frame action in the case of diaphragm bracing. Once the load is distributed into the diaphragm, it combines with the upper system loads that are transmitted to the supports. If the deck is noncomposite or the deck-girder connection is not strong enough to transmit the load, then the path is different. This case is described later.

At the supports, the load is path designed so that the load can be transmitted from the deck level into the bearings. The cross-framing system must be designed to support these loads. The deck diaphragm loads may be uniformly distributed to the top of each girder. Note that the end supports receive the additional load from the bottom flange for the tributary length between the first interior bracing and the support. This load is shown as P_{end} in Figure 6.35(e).

In the case of insufficient diaphragm action (or connectors to the diaphragm), the upper system load must be transmitted into the girder in weak-axis bending. The load distribution mechanism is shown in Figure 6.35(f). The girders translate together because they are coupled by the transverse elements. The cross frames (or diaphragms) are very stiff axially and may be considered rigid. This system may be modeled as a plane frame, or more simply, the load may be equally shared between all girders and the load effect of the wind directly applied to the exterior girder may be superimposed. The local and global effects are shown in Figures 6.35(g–h) and 6.35(i), respectively. For longer spans, the bracing is periodically spaced and the global response is more like a beam subjected to a uniform load. Approximating the distribution of the load as uniform, permits the global analysis to be based on the analysis of a beam supported by bearing supports. The load is then split equally between all of the girders. Mathematically stated,

$$M_{Total} = M_{Local} + \frac{M_{Global}}{\text{No. of girders}}$$

This analysis is indeed approximate, but adequate ductility should be available and the lower bound theorem applies. The procedures described account for all the load and, therefore, equilibrium can be maintained. The procedures also parallel those outlined in the commentary in AASHTO [C4.6.2.7.1]. Similar procedures may be used for box systems.

The procedures outlined herein do not, nor do those of AASHTO, address long-span systems where the aeroelastic wind effects are expected.

6.5.2 Seismic Load Analysis

The load path developed to resist lateral loads due to wind is the same load path followed by the seismic loads. The nature of the applied load is also similar. The wind loads acting on the superstructure are uniformly distributed along the length of the bridge and the seismic loads are proportional to the distributed mass of the superstructure along its length. What is different is the magnitude of the lateral loads and the dependence of the seismic loads on the period of vibration of various modes excited during an earthquake and the degree of inelastic deformation which tends to limit the seismic forces.

Because of the need to resist lateral wind loads in all bridge systems, designers have already provided the components required to resist the seismic loads. In a typical superstructure cross section, the bridge deck and longitudinal girders are tied together with struts and bracing to form an integral unit acting as a large horizontal diaphragm. The horizontal diaphragm action distributes the lateral loads to the restrained bearings in each of the segments of the superstructure. A segment may be a simply supported span or a portion of a multispan bridge that is continuous between deck joints.

During an earthquake, a segment is assumed to maintain its integrity, that is, the deck and girders move together as a unit. In a bridge with multiple segments, they often get out-of-phase with each other and may pound against one another if gaps in the joints are not large enough. In general, the deck and girders of the superstructure within a segment are not damaged during a seismic event, unless they are pulled off their supports at an abutment or internal hinge. Thus, an analysis for seismic loads must provide not only the connection force at the restrained supports, but also an estimate of the displacements at unrestrained supports. Procedures for determining these seismic forces and displacements are discussed in the sections that follow.

Minimum Analysis Requirements. The analysis should be more rigorous for bridges with higher seismic risk and greater importance. Also, more rigorous seismic analysis is required if the geometry of a bridge is irregular. A regular bridge does not have changes in stiffness or mass that exceed 25% from one segment to another along its length. A horizontally curved bridge may be considered regular if the subtended angle at the center of curvature, from one abutment to another, is less than 60° and does not have an abrupt change in stiffness or mass.

Minimum analysis requirements based on seismic zone, geometry, and importance are given in Table 6.24.

Single-span bridges do not require a seismic analysis regardless of seismic zone. The minimum design connection force for a single-span bridge is the product of the acceleration coefficient and the tributary area as discussed in Chapter 4 (Minimum Seismic Design Connection Forces).

Bridges in Seismic Zone 1 do not require a seismic analysis. The minimum design connection force for these bridges is the tributary dead load multiplied by the coefficient given in Table 4.17.

Either a single-mode or a multi-mode spectral analysis is required for bridges in the other seismic zones depending on their geometry and importance classification. The single-mode method is based on the first or fundamental mode of vibration and is applied to both the longitudinal and

TABLE 6.24 Minimum Analysis Requirements for Seismic Effects[a]

Seismic Zone	Single-Span Bridges	Multispan Bridges					
		Other Bridges		Essential Bridges		Critical Bridges	
		Regular	Irregular	Regular	Irregular	Regular	Irregular
1	None[b]	None	None	None	None	None	None
2	None	SM[c]	SM	SM	MM	MM	MM
3	None	SM	MM[d]	MM	MM	MM	TH
4	None	SM	MM	MM	MM	TH[e]	TH

[a]In AASHTO Table 4.7.4.3.1-1. [From *AASHTO LRFD Design Bridge Specification*. Copyright © 1994 by the American Association of State Highway and Transportation Officials, Washington, DC. Used by permission.]
[b]None = no seismic analysis is required.
[c]SM = single mode elastic method.
[d]MM = multimode elastic method.
[e]TH = time history method.

transverse directions of the bridge. The multimode method includes the effects of all modes equal in number to at least three times the number of spans in the model [A4.7.4.3.3].

A time–history analysis is required for critical bridges in Seismic Zones 3 and 4. This analysis involves a step-by-step integration of the equations of motion for a bridge structure when it is subjected to ground accelerations that vary with time. Historical records of the variation in acceleration, velocity, and displacement due to ground shaking are referred to as time histories, hence the name for the analysis method. Careful attention must be paid to the modeling of the structure and the selection of the time step used in the analysis. If elastic material properties are used, the R factors of Tables 4.15 and 4.16 apply to the substructures and connection forces, respectively. If inelastic material properties are used, all of the R factors are 1.0 because the inelastic analysis accounts for the energy dissipation and redistribution of seismic forces and no further modification is needed. Oftentimes, when selecting a time history for a specific bridge site, a historical record may not be available for the soil profile that is present. In this case, artificial time histories are generated that include the magnitude, frequency content, and duration of the ground shaking anticipated at the bridge site. Obviously, a time–history analysis requires considerable skill and judgment, and an analyst experienced in inelastic, dynamic, numerical analysis should be consulted.

Elastic Seismic Response Spectrum. Both of the spectral methods of analysis require that a seismic response spectrum be given for the bridge site. The purpose of a response spectrum analysis is to change a problem in dynamics into an equivalent problem in statics. The key to this method of analysis is to construct an appropriate response spectrum for a particular soil profile. A response spectrum can be defined as a graphical representation of the maximum response of single-degree-of-freedom elastic systems to earthquake ground motions versus the periods or frequencies of the system (Imbsen, 1981). The response spectrum is actually a summary of a whole series of time–history analysis.

A response spectrum is generated by completing the steps illustrated in Figure 6.36. The single-degree-of-freedom (SDOF) system is shown as an inverted pendulum oscillator that could represent the lumped mass of a bridge superstructure supported on columns or piers. The undamped natural period of vibration (s) of the SDOF system is given by

$$T = 2\pi \sqrt{\frac{m}{k}} \tag{6.43}$$

where

m = mass of the system = W/g
W = contributing dead load for the structure (kips)
g = acceleration of gravity = 386.4 in./s^2
k = stiffness of the structure supporting the mass (kips/in.)

When damping is introduced into the system, it is usually expressed as a ratio of critical damping given by

$$\xi = \frac{c}{c_c} \tag{6.44}$$

where

c = coefficient of viscous damping
c_c = critical damping coefficient, minimum amount of damping required to prevent a structure from oscillating = $4\pi m/T$

It can be shown that the period of vibration of a damped structure T_D is related to the undamped period T by

$$T_D = \frac{T}{\sqrt{1 - \xi^2}} \tag{6.45}$$

where ξ = ratio of actual damping to critical damping, Eq. 6.44.

In an actual structure, damping due to internal friction of the material and relative moment at connections seldom exceeds 20% of critical damping.

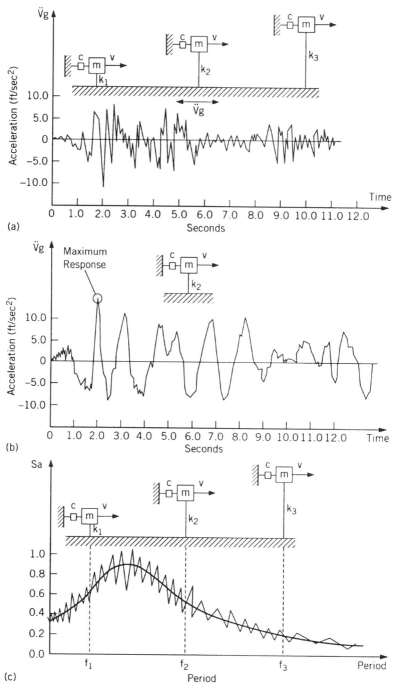

Fig. 6.36   (a) Time history record of ground acceleration applied to a damped single-degree-of-freedom system, (b) time history record of structure response, and (c) maximum responses of single-degree-of-freedom systems. [From Imbsen, 1981.]

Substituting this value into Eq. 6.45 increases the undamped period of vibration by only 2%. In practice, this difference is neglected and the damped period of vibration is assumed to be equal to the undamped period.

In Figure 6.36(a), a time–history record of ground acceleration is shown below three SDOF systems with identical mass and damping, but with different structural stiffness to give three different periods of vibration. The acceleration response from an elastic step-by-step time–history analysis of the second SDOF system is shown in Figure 6.36(b), and the maximum response is indicated. This procedure is repeated for a large number of SDOF systems with different stiffnesses until the maximum response is determined for a whole spectrum of periods of vibration. A response spectrum for acceleration is plotted in Figure 6.36(c) and gives a graphical representation of the variation of maximum response with period of vibration. For design purposes, a curve is drawn through the average of the maximum response to give the elastic response spectrum shown by the smooth line in Figure 6.36(c). This response spectrum was developed for a single earthquake under one set of soil conditions. When a response spectrum is used for design purposes, it is usually based on more than one earthquake and includes the effects of different soil conditions.

Seismic Design Response Spectra. It is well known that local geologic and soil conditions influence the intensity of ground shaking and the potential for damage during an earthquake. The 1985 earthquake in Mexico is an example of how the soils overlying a rock formation can modify the rock motions dramatically. The epicenter of the earthquake was near the west coast of Mexico, not too far from Acapulco. However, most of the damage was done some distance away in Mexico City. The difference in the ground shaking at the two sites was directly attributable to their soil profile. Acapulco is quite rocky with thin overlying soil, while Mexico City is sitting on an old lake bed overlain with deep alluvial deposits. When the earthquake struck, Acapulco took a few hard shots of short duration that caused only moderate damage. In contrast, the alluvial deposits under Mexico City shook like a bowl of gelatin for some time and extensive damage occurred. The response spectra characterizing these two sites are obviously different and these differences due to soil conditions must be recognized in the response spectra developed for the seismic analysis of bridges.

In order to describe the characteristics of response spectra for different soil profiles statistical studies of a number of accelerometer records have been conducted. These studies defined soil profiles similar to those in Table 4.13 as a reasonable way to differentiate the characteristics of surface response. For each of the accelerometer records within a particular soil profile, an elastic response spectrum was developed as previously illustrated in Figure 6.36. An average response spectrum was obtained from the individual response system at different sites with the same soil profile but subjected to different earthquakes. This procedure was repeated for the four soil profiles that had been

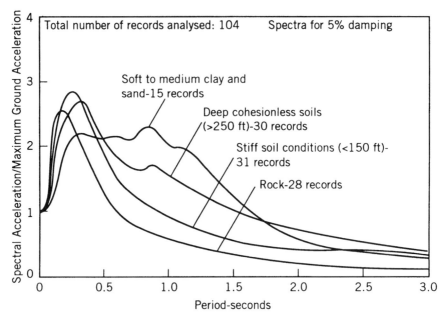

Fig. 6.37 Average acceleration spectra for different site conditions. Normalized with respect to maximum ground acceleration. [From Seed et al., 1976.]

defined. The results of the study by Seed et al. (1976), which included the analysis of over 100 accelerometer records, are given in Figure 6.37. The elastic response spectra in this figure were developed for 5% of critical damping and the accelerations have been normalized with respect to the maximum ground acceleration.

The shape of the average spectra in Figure 6.37 first ascends, levels off, and then decays as the period of vibration increases. As the soil profile becomes more flexible, the period at which decay begins is delayed, so that at larger periods the softer soils have larger accelerations than the stiffer soils. These variations in acceleration with period and soil type are expressed in AASHTO [A3.10.6] by an elastic seismic response coefficient C_{sm} defined as

$$C_{sm} = \frac{1.2AS}{T_m^{2/3}} \leq 2.5A \qquad (6.46)$$

where

    T_m = period of vibration of the mth mode (s)
    A = acceleration coefficient from Figure 4.22
    S = site coefficient from Table 4.14

The seismic response coefficient is a modified acceleration coefficient that is multiplied times the effective weight of the structure to obtain an equivalent lateral force to be applied to the structure. Because C_{sm} is based on an elastic response, the member forces resisting the equivalent lateral force used in design are divided by the appropriate R factors given in Tables 4.15 and 4.16.

The shape of the seismic response spectra defined by Eq. 6.46 does not have an ascending branch, but simply levels off at 2.5A. This characteristic can be seen in Figure 6.38 in the plot of C_{sm}, normalized with respect to the acceleration coefficient A, for different soil profiles versus the period of vibration. For Soil Profile Types *III* and *IV*, the maximum value of 2.5A is overly conservative, as can be seen in Figure 6.37, so that in areas where $A \geq 0.10$, C_{sm} need not exceed 2.0A (Fig. 6.38). Also, for Soil Profiles III and IV, for short periods an ascending branch is defined as (for modes other than the first mode):

$$C_{sm} = A(0.8 + 4.0T_m) \leq 2.0A \tag{6.47}$$

For intermediate periods, $0.3 \text{ s} \leq T_m \leq 4.0 \text{ s}$, a characteristic of earthquake response spectra is that the average velocity spectrum for larger earthquakes of magnitudes 6.5 or greater is approximately horizontal. This characteristic implies that C_{sm} should decrease as $1/T_m$. However, because of the concern for high ductility requirements in bridges with longer periods, it was decided to reduce C_{sm} at a slower rate of $1/T_m^{2/3}$. For bridges with long periods ($T_m > 4.0 \text{ s}$), the average displacement spectrum of large earthquakes becomes horizontal. This implies that C_{sm} should decay as $1/T_m^2$ and Eq. 6.46 becomes

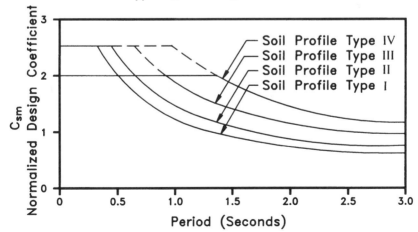

Fig. 6.38 Design response spectra for various soil profiles. Normalized with respect to acceleration coefficient A. (AASHTO Fig. C3.10.6.1-1). [From *AASHTO LRFD Bridge Design Specifications,* Copyright © 1994 by the American Association of State Highway and Transportation Officials, Washington, DC. Used by Permission.]

$$C_{sm} = \frac{3AS}{T_m^{4/3}} \qquad T_m \geq 4.0 \text{ s} \qquad (6.48)$$

6.6 SUMMARY

This chapter includes numerous topics related to the structural analysis of bridge systems. It is the intended to provide a broad-based perspective for analysis. It is important to understand the concepts involved with the plastic and shakedown limits and how they relate to the AASHTO design specification which is primarily based on elastic analysis. It is also important to understand the development and usage of the AASHTO distribution factors, the meaning of one- and multiple-lanes loaded and comparisons and modeling techniques associated with advanced methods such as grillage, finite element, and finite strip methods. Several examples were presented to address the elastic and inelastic analysis of deck systems. Here it is important to note the variation in results of these methods that are sensitive to local effects. Finally, several miscellaneous topics were presented including prestress and temperature effects, and an introduction to seismic analysis. Seismic analysis and design is beyond the scope of this book and the reader is referred to other references. It is unlikely that this chapter is comprehensive on any one topic but will provide a useful introduction and the background necessary for better analysis of bridges.

REFERENCES

AASHTO (1994). *LRFD Bridge Design Specification* 1st ed., American Association of State Highway and Transportation Officials, Washington, DC.

AISC (1994). *Manual of Steel Construction—LFRD,* 1st ed., American Institute for Steel Construction, Chicago, IL.

Cheung, Y. K. (1976). *Finite Strip Method in Structural Analysis,* 1st ed., Pergamon Press, New York.

Finch, T. R. and J. A. Puckett (1992). *Transverse Load Distribution of Slab-Girder Bridges,* Report 92-9, Vol. No. 5, Mountain Plains Consortium, Fargo, ND.

Ghali, A. and A. M. Neville (1989). *Structural Analysis—A Unified Classical and Matrix Approach,* 3rd ed., Chapman & Hall, New York.

Hambly, E. C. (1991). *Bridge Deck Behavior,* 2nd ed., E & FN SPON, An imprint of Chapman & Hall, London, UK.

Horne, M. R. (1971). *Plastic Theory of Structures,* Willan Clowes & Sons, London, UK.

Imbsen, R. A. (1981). *Seismic Design of Highway Bridges,* Report No. FHWA-IP-81-2, Federal Highway Administration, Washington, DC.

Loo, Y. C. and A. R. Cusens (1978). *The Finite Strip Method in Bridge Engineering,* View Point Publications, Great Britain, UK.

Neal, N. G. (1977). *The Plastic Methods of Structural Analysis,* 3rd ed., Chapman & Hall, London, UK.

Nowak, A. S. (1993). *Calibration of LRFD Bridge Design Code—NCHRP Project 12-33,* University of Michigan, Ann Arbor, MI.

Seed, H. B., C. Ugas, and J. Lysmer (1976). *Site Dependent Spectra for Earthquake Resistant Design,* Bulletin of the Seismological Society of America, Vol. 66, No. 1, February, pp. 221–244.

Timoshenko, S. and S. Woinowsky-Kreiger (1959). *Theory of Plates and Shells,* 2nd ed., McGraw-Hill, New York.

Timoshenko, S. P. and J. N. Goodier (1970). *Theory of Elasticity,* 3rd ed., McGraw-Hill, New York.

Ugural, A. C. (1981). *Stresses in Plates and Shells,* McGraw-Hill, New York.

Zokaie, T., T. A. Osterkamp, and R. A. Imbsen (1991). *Distribution of Wheel Loads on Highway Bridges,* Final Report Project 12-26/1, National Cooperative Highway Research Program, Transportation Research Board, National Research Council, Washington, DC.

7

CONCRETE BRIDGES

7.1 INTRODUCTION

Concrete is a versatile building material. It can be shaped to conform to almost any alignment and profile. Bridge superstructures built of reinforced and prestressed concrete can be unique one-of-a-kind structures formed and constructed at the job site or they can be look-alike precast girders and box beams manufactured in a nearby plant. The raw materials of concrete— cement, fine aggregate, coarse aggregate, and water—are found in most areas of the world. In many countries without a well-developed steel industry, reinforced concrete is naturally the preferred building material. However, even in North America with its highly developed steel industry, bridges built of concrete continue to be competitive.

Concrete bridges can be designed to satisfy almost any geometric alignment from straight to curved to doubly curved as long as the clear spans are not too large. Cast-in-place (CIP) concrete box girders are especially suited to curved alignment because of their superior torsional resistance and the ability to keep the cross section constant as it follows the curves. With the use of post-tensioning, clear spans of 45 m are common. When the alignment is relatively straight, precast prestressed girders can be utilized for multispan bridges, especially if continuity is developed for live load. For relatively short spans, say less than 12 m, flat slab bridges are often economical. Cast-in-place (CIP) girders monolithic with the deck slab (*T*-beams) can be used for clear spans up to about 20 m, longer if continuity exists. Some designers do not like the underside appearance of the multiple ribs, but if the bridge is over a small waterway rather than a traveled roadway, there should be no objection.

Cast-in-place concrete bridges may not be the first choice if speed of construction is of primary importance. Also, if formwork cannot be suitably supported, such as in a congested urban setting where traffic must be maintained, the design of special falsework to provide a construction platform may be a disadvantage.

Longer span concrete bridges have been built using match-cast and cable-supported segmental construction. These structural systems require analysis and construction techniques that are relatively sophisticated and are beyond the scope of this book. In this chapter, short-to-medium-span bridges constructed of reinforced and prestressed concrete are discussed.

After a review of the behavior of the materials in concrete bridges, the resistance of cross sections to bending and shear is presented. A relatively detailed discussion of these two topics is given because of the introduction in the AASHTO (1994) LRFD Bridge Specification of a unified flexural theory for reinforced and prestressed concrete beams and the modified compression field theory for shear resistance. In the development of the behavior models, the sign convention adopted for strains and stresses is that tensile values are positive and compressive values are negative. This results in stress–strain curves for concrete that are drawn primarily in the third quadrant instead of the familiar first quadrant.

It is not necessary to go through each detailed step of the material response discussion. The information is given so that a reader can trace the development of the provisions in the specification. At the end of this chapter, a number of example problems are given to illustrate the application of the resistance equations that are derived. A concrete bridge deck with a barrier wall is designed followed by design examples of a solid slab, a T-beam, a prestressed beam, and a box-girder bridge.

7.2 REINFORCED AND PRESTRESSED CONCRETE MATERIAL RESPONSE

In order to predict the response of a structural element subjected to applied forces, three basic relationships must be established: (1) equilibrium of forces, (2) compatibility of strains, and (3) constitutive laws representing the stress–strain behavior of the materials in the element. For a two-dimensional (2-D) element without torsion that is subjected to bending by transversely applied forces, there are three equilibrium equations between the applied external forces and the three-internal resisting forces: moment, shear, and axial load. When the external forces are applied, the cross section deforms and internal longitudinal, transverse, and shear strains are developed. These internal strains must be compatible. Longitudinal strains throughout the depth of a section are related to one another through the familiar assumption of plane sections before bending remain plane sections after bending. The longitudinal strains are related to the transverse, shear, and principal strains through the relationships described in Mohr's circle of strain. The stress–strain relationships provide the key link between the internal forces (which are integrations over an area of the stresses) and the deformations of the cross section. These interrelationships are shown schematically in Figure 7.1, and will be described in more specific terms in the sections that follow.

On the left of Figure 7.1 is a simple model used in psychology to illustrate that the manner in which individuals or groups respond to certain stimuli depends on their psychological makeup. In individuals, we often speak of one's constitution; in groups, the response depends on the constituents; in concrete, it depends on constitutive laws. The analogy to concrete elements may be imperfect, but the point is that knowledge of the behavior of the material is essential to predicting the concrete response of the element to external loads.

Another point in regard to the relationships in Figure 7.1 is that they involve both deductive and inductive reasoning. The equilibrium and compatibility equations are deductive in that they are based on general principles of physics and mechanics that are applied to specific cases. If the equations are properly written, they will lead to a set of unique correct answers. On the other hand, the constitutive equations are inductive as they are based on specific observations from which expressions are written to represent general behavior. If the trends exhibited by the data are not correctly intepreted or an

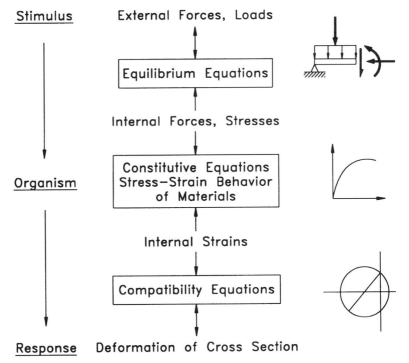

Fig. 7.1   Interrelationship between equilibrium, material behavior, and compatibility.

important parameter is overlooked, the predicted response will not be verified by experimental tests. As more experimental data become available the constitutive equations change and the predicted response improves. The AASHTO (1994) Bridge Specification incorporates the current state-of-practice regarding material response; however, one should expect that changes will occur in the constitutive equations in the future as additional test data and/or new materials become available.

7.3 CONSTITUENTS OF FRESH CONCRETE

Concrete is a conglomerate artificial stone. It is a mixture of large and small particles held together by a cement paste that hardens and will take the shape of the formwork in which it is placed. The proportions of the coarse and fine aggregate, Portland cement, and water in the mixture influence the properties of the hardened concrete. The design of concrete mixes to meet specific requirements can be found in concrete materials textbooks (Troxell et al., 1968). In most cases a bridge engineer will select a particular class of concrete from a series of predesigned mixes, usually on the basis of the desired 28-day compressive strength, f_c'. A typical specification for different classes of concrete is shown in Table 7.1.

- Class A concrete is generally used for all elements of structures and specifically for concrete exposed to saltwater.
- Class B concrete is used in footings, pedestals, massive pier shafts, and gravity walls.
- Class C concrete is used in sections under 100 mm in thickness such as reinforced railings and for filler in steel grid floors.
- Class P concrete is used when strengths in excess of 28 MPa are required. For prestressed concrete, consideration should be given to limiting the nominal aggregate size to 20 mm.

A few brief comments on the parameters in Table 7.1 and their influence on the quality of concrete selected are in order. Air entrained (*AE*) concrete improves durability when subjected to freeze–thaw cycles and exposure to deicing salts. This improvement is accomplished by adding a detergent or vinsol resin to the mixture that produces a very even distribution of finely divided air bubbles. This even distribution of pores in the concrete prevents large air voids from forming and breaks down the capillary pathways from the surface to the reinforcement.

The water–cement ratio (W/C) by weight is the single most important strength parameter in concrete. The lower the W/C ratio, the greater the strength of the mixture. Obviously, increasing the cement content will increase the strength for a given amount of water in the mixture. For each class of concrete, a minimum amount of cement in kilograms per cubic meter (kg/

TABLE 7.1 Concrete Mix Characteristics by Class[a]

Class of Concrete	Minimum Cement Content (kg/m³)	Maximum Water/Cement Ratio (kg/kg)	Air Content Range (%)	Coarse Aggregate Per AASHTO M43 Square Size of Openings	28-Day Compressive Strength (MPa)
A	362	0.49		25–4.75	28
A(AE)	362	0.45	6.0 ± 1.5	25–4.75	28
B	307	0.58		50–4.75	17
B(AE)	307	0.55	5.0 ± 1.5	50–4.75	17
C	390	0.49		12.5–4.75	28
C(AE)	390	0.45	7.0 ± 1.5	12.5–4.75	28
P	334	0.49	As specified elsewhere	25–4.75 or 19–4.75	As specified elsewhere
Lightweight	334		As specified in the contract documents		

[a]AASHTO Table C5.4.2.1-1. [From *AASHTO LRFD Bridge Design Specifications*, Copyright © 1994 by the American Association of State Highway and Transportation Officials, Washington, DC. Used by permission.]

m³) is specified. By increasing the cement above these minimums, it is possible to increase the water content and still obtain the same W/C ratio. This increase of water content may not be desirable because excess water, which is not needed for the chemical reaction with the cement and for wetting the surface of the aggregate, will eventually evaporate and cause excessive shrinkage and less durable concrete. As a result, AASHTO [A5.4.2.1]* places an upper limit of 475 kg/m³ on the denominator of the W/C ratio to limit the water content of the mixture.

To obtain quality concrete that is durable and strong, it is necessary to limit the water content, which may produce problems in workability and placement of the mixture in the forms. To increase workability of the concrete mix without increasing the water content, chemical additives have been developed. These admixtures are called high-range water reducers (superplasticizers) and are very effective in improving both wet and hardened concrete properties. They must be used with care, and the manufacturer's directions must be followed, because they can have unwanted side effects such as accelerated setting times. Laboratory testing should be performed to establish both the wet and hardened concrete properties using aggregates representative of the construction mix.

In recent years, very high strength concretes with compressive strengths approaching 200 MPa have been developed in laboratory samples. The key to obtaining these very high strengths is the same as for obtaining very durable concrete and that is having an optimum graded mixture so that all of the gaps between particles are filled with finer and finer material until in the limit no voids exist. In the past, attention has been given to providing a well-graded mixture of coarse and fine aggregate so that the spaces between the maximum aggregate size would be filled with smaller particles of gravel or crushed stone, which in turn would have their spaces filled with fine aggregate or sand. Filling the spaces between the fine aggregate would be the powderlike Portland cement particles which, when it reacted with water, bonded the whole conglomerate together. In very high strength concretes this has been taken one step further where a finer cementitious material is introduced to fill the gaps between the Portland cement particles. These finely divided mineral particles are typically pozzolans, fly ash, or silica fume. They can replace some of the Portland cement in satisfying the minimum cement content and must be added to the weight of the Portland cement in the denominator of the W/C ratio.

7.4 PROPERTIES OF HARDENED CONCRETE

The 28-day compressive strength f_c' is the primary parameter, which affects

---

*The article number in the AASHTO (1994) LRFD Bridge Specifications are enclosed in brackets and preceded by the letter A if specifications and by the letter C if commentary.

a number of the properties of hardened concrete such as tensile strength, shear strength, and modulus of elasticity. A standard 150-mm diameter × 300-mm high cylinder is placed in a testing machine and loaded to a compressive failure to determine the value of f_c'. It should be noted that this is an unconfined compression test. When concrete is placed in a column or beam with lateral or transverse reinforcement, the concrete is in a state of triaxial or confined stress. The confined concrete stress state increases the peak compressive stress and the maximum strain over that of the unconfined concrete. It is necessary to include this increase in energy absorption or toughness when examining the resistance of reinforced concrete cross sections.

7.4.1 Short-Term Properties of Concrete

Concrete properties determined from a testing program represent short-term response to loads because these tests are usually completed in a matter of minutes, in contrast to a time period of months or even years over which load is applied to concrete when it is placed in a structure. These short-term properties are useful in assessing the quality of concrete and the response to short-term loads such as vehicle live loads. However, these properties must be modified when they are used to predict the response due to sustained dead loads such as self-weight of girders, deck slabs, and barrier rails.

Concrete Compressive Strength and Behavior. In AASHTO [A5.4.2.1] a minimum 28-day compressive strength of 16 MPa for all structural applications is recommended and a maximum compressive strength of 70 MPa unless additional laboratory testing is conducted. Bridge decks should have a minimum compressive strength of 28 MPa to provide adequate durability.

When describing the behavior of concrete in compression, a distinction has to be made between three possible stress states: uniaxial, biaxial, and triaxial. Illustrations of these three stress states are given in Figure 7.2. The uniaxial stress state of Figure 7.2(a) is typical of the unconfined standard cylinder test used to determine the 28-day compressive strength of concrete. The biaxial stress state of Figure 7.2(b) will occur in the reinforced webs of beams subjected to shear, bending, and axial load. The triaxial state of stress of Figure 7.2(c) will occur within the core of an axially load column that is confined by lateral ties or spirals.

The behavior of concrete in uniaxial compression [Fig. 7.2(a)] can be described by defining a relationship between normal stress and strain. A simple relationship for concrete strengths less than 40 MPa is given by a parabola as

$$f_c = f_c' \left[2\left(\frac{\varepsilon_c}{\varepsilon_c'}\right) - \left(\frac{\varepsilon_c}{\varepsilon_c'}\right)^2 \right] \tag{7.1}$$

where f_c is the compressive stress corresponding to the compressive strain ε_c,

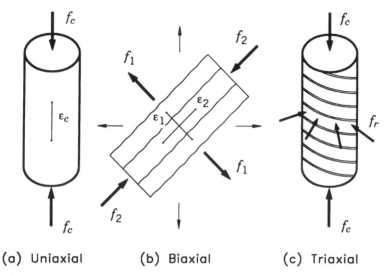

(a) Uniaxial   (b) Biaxial   (c) Triaxial

Fig. 7.2 Compressive stress states for concrete. (a) Uniaxial, (b) biaxial, and (c) triaxial.

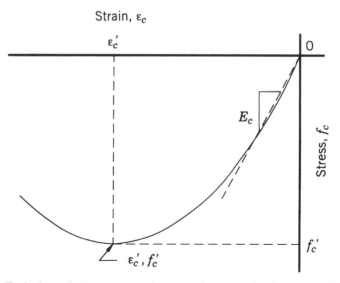

Fig. 7.3 Typical parabolic stress–strain curve for unconfined concrete in uniaxial compression.

and f'_c is the peak stress from a cylinder test, and ε'_c is the strain corresponding to f'_c. This relationship is shown graphically in Figure 7.3. The sign convention adopted is that compressive stresses and compressive strains are negative values.

The modulus of elasticity given for concrete in AASHTO [A5.4.2.4] is an estimate of the slope of a line from the origin drawn through a point on the stress–strain curve at $0.4 f'_c$. This secant modulus E_c in megapascals (MPa) is shown in Figure 7.3 and is given by the expression

$$E_c = 0.043\gamma_c^{1.5}\sqrt{f'_c} \tag{7.2}$$

where γ_c is the mass density of concrete in kilograms per cubic meter (kg/m^3) and f'_c is the absolute value of the specified compressive strength of concrete in megapascals. For $\gamma_c = 2300$ kg/m^3 and $f'_c = 28$ MPa,

$$E_c = 0.043(2300)^{1.5}\sqrt{f'_c} = 4800\sqrt{f'_c} = 4800\sqrt{28} = 25 \text{ GPa}$$

When the concrete is in a state of biaxial stress, the strains in one direction affect the behavior in the other. For example, the ordinates of the stress strain curve for principal compression f_2 in the web of a reinforced concrete beam [Fig. 7.2(b)] are reduced when the perpendicular principal stress f_1 is in tension. This phenomenon was quantified by Vecchio and Collins (1986) and results in a modification of Eq. 7.1 as follows:

$$f_2 = f_{2 \text{ max}}\left[2\left(\frac{\varepsilon_2}{\varepsilon'_c}\right) - \left(\frac{\varepsilon_2}{\varepsilon'_c}\right)^2\right] \tag{7.3}$$

where f_2 is the principal compressive stress corresponding to ε_2 and $f_{2 \text{ max}}$ is a reduced peak stress given by

$$f_{2 \text{ max}} = \frac{f'_c}{0.8 + 170\varepsilon_1} \leq f'_c \tag{7.4}$$

where ε_1 is the average principal tensile strain of the cracked concrete. These relationships are illustrated in Figure 7.4. Hsu (1993) refers to this phenomenon as compression softening and presents mathematical expressions that are slightly different than Eqs. 7.3 and 7.4 because he includes both stress and strain softening (or reduction in peak values). When cracking becomes severe, the average strain ε_1 across the cracks can become quite large and, in the limit, causes the principal compressive stress f_2 to go to zero. A value of $\varepsilon_1 = 0.004$ results in a one-third reduction in f_2.

When the concrete within a beam or column is confined in a triaxial state of stress by lateral ties or spirals [Fig. 7.2(c)], the out-of-plane restraint provided by the reinforcement increases the peak stress and peak strain above the unconfined values. For confined concrete in compression, the limiting

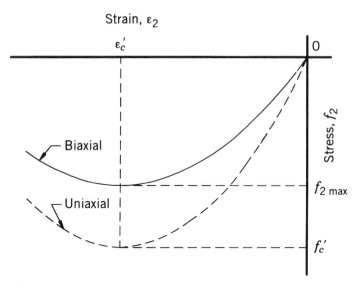

Fig. 7.4   Comparison of uniaxial and biaxial stress–strain curves for unconfined concrete in compression.

ultimate strain is dramatically increased beyond the 0.003 value often used for unconfined concrete. This increased strain on the descending branch of the stress–strain curve adds ductility and toughness to the element and provides a mechanism for dissipating energy without failure. As a result the confinement of concrete within closely spaced lateral ties or spirals is essential for elements located in seismic regions in order to absorb energy and allow the deformation necessary to reduce the earthquake loads.

Confined Concrete Compressive Strength and Behavior. Figure 7.5 shows a comparison of typical stress–strain curves for confined and unconfined concrete in compression for an axially loaded column. The unconfined concrete is representative of the concrete in the shell outside of the lateral reinforcement, which is lost due to spalling at relatively low compressive strains. The confined concrete exhibits a higher peak stress f'_{cc} and a larger corresponding strain ε_{cc} than the unconfined concrete strength f'_{co} and its corresponding strain ε_{co}.

One of the earliest studies to quantify the effect of lateral confinement was by Richart et al. (1928) in which hydrostatic fluid pressure was used to simulate the lateral confining pressure f_r. The model used to represent the strength of confined concrete was similar to the Coulomb shear failure criterion used for rock (and other geomaterials).

$$f'_{cc} = f'_{co} + k_1 f_r \tag{7.5}$$

where f'_{cc} is the peak confined concrete stress, f'_{co} is the unconfined concrete

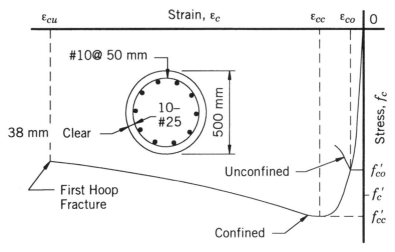

Fig. 7.5 Comparison of unconfined and confined concrete stress–strain curves in compression.

strength, f_r is the lateral confining pressure, and k_1 is a coefficient that depends on the concrete mix and the lateral pressure. From these tests, Richart et al. (1928) determined that the average value of $k_1 = 4$. Setunge et al. (1993) propose that a simple lower bound value of $k_1 = 3$ be used for confined concrete of any strength below 120 MPa.

Richart et al. (1928) also suggested a simple relationship for the strain ε_{cc} corresponding to f'_{cc} as

$$\varepsilon_{cc} = \varepsilon_{co}\left(1 + k_2\frac{f_r}{f'_{co}}\right) \tag{7.6}$$

where ε_{co} is the strain corresponding to f'_{cc} and $k_2 = 5k_1$. Also, f'_{cc} is commonly taken equal to $0.85 f'_c$ to account for the lower strength of the concrete placed in a column compared to that in the control cylinder.

The lateral confining pressure f_r in Eqs. 7.5 and 7.6, produced indirectly by lateral reinforcement, needs to be determined. Mander, et al. (1988), following an approach similar to Sheikh and Uzumeri (1980), derive expressions for the effective lateral confining pressure f'_r for both circular hoop and rectangular hoop reinforcement. Variables considered are the spacing, area, and yield stress of the hoops; the dimensions of the confined concrete core, and the distribution of the longitudinal reinforcement around the core perimeter. It is convenient to use f'_{cc} on the area of the concrete core A_{cc} enclosed within centerlines of the perimeter hoops. However, not all of this area is effectively confined concrete, and f_r must be adjusted by a confinement effectiveness coefficient k_e to give an effective lateral confining pressure of

$$f'_r = k_e f_r \tag{7.7}$$

in which

$$k_e = A_e/A_{cc} \tag{7.8}$$

where A_e is the area of effectively confined concrete,

$$A_{cc} = A_c - A_{st} = A_c(1 - \rho_{cc}) \tag{7.9}$$

where A_c is the area of the core enclosed by the centerlines of the perimeter hoops or ties, A_{st} is the total area of the longitudinal reinforcement, and

$$\rho_{cc} = A_{st}/A_c \tag{7.10}$$

EXAMPLE 7.1

Determine the confinement effectiveness coefficient k_e for a circular column with spiral reinforcement of diameter d_s between bar centers if the arch action between spirals with a clear vertical spacing of s' has an amplitude of $s'/4$ (see Fig. 7.6). Midway between the spirals, A_e will be the smallest with a diameter of $d_s - s'/4$, that is

$$A_e = \frac{\pi}{4}\left(d_s^2 - \frac{s'}{4}\right)^2 = \frac{\pi}{4}d_s^2\left(1 - \frac{s'}{4d_s}\right)^2 = \frac{\pi}{4}d_s^2\left[1 - \frac{s'}{2d_s} + \left(\frac{s'}{4d_s}\right)^2\right]$$

Neglecting the higher order term, which is much less than one, yields

$$A_e \approx \frac{\pi}{4}d_s^2\left(1 - \frac{s'}{2d_s}\right)$$

and with $A_c = \pi d_s^2/4$, Eqs. 7.8 and 7.9 yield

$$k_e = \frac{1 - (s'/2d_s)}{1 - \rho_{cc}} \le 1.0 \tag{7.11}$$

(Note that the definition of d_s for this model is different than the outside diameter of the core d_c often used in selecting spiral reinforcement.)

If we consider the half-section of depth s in Figure 7.7, which is confined by a spiral with hoop tension at yield exerting a uniform lateral pressure f_r (tension shown is positive) on the concrete core, the equilibrium of forces requires

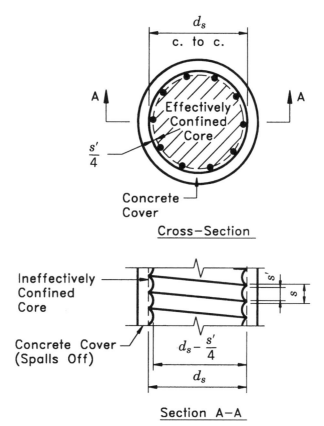

Fig. 7.6 Effectively confined core for circular spirals. [Reproduced from J. B. Mander, M. J. N. Priestley, and R. Park (1988). Theoretical Stress–Strain Model for Confined Concrete *Journal of Structural Engineering,* ASCE, 14(8), pp. 1804–1826. With permission.]

$$2A_{sp}f_{yh} + f_r s d_s = 0 \qquad (7.12)$$

where A_{sp} is the area of the spiral, f_{yh} is the yield strength of the spiral, and s is the center to center spacing of the spiral. Solve Eq. 7.12 for the lateral confining pressure

$$f_r = \frac{-2A_{sp}f_{yh}}{sd_s} = -\frac{1}{2}\rho_s f_{yh} \qquad (7.13)$$

where ρ_s is the ratio of the volume of transverse confining steel to the volume of confined concrete core, that is,

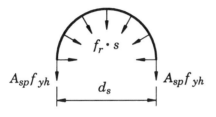

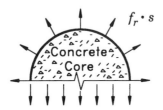

Fig. 7.7 Half-body diagrams at interface between spiral and concrete core.

$$\rho_s = \frac{A_{sp}\pi d_s}{\frac{\pi}{4}sd_s^2} = \frac{4A_{sp}}{sd_s} \tag{7.14}$$

Mander et al. (1988) give expressions similar to Eqs. 7.11, 7.13, and 7.14 for circular hoops and rectangular ties.

EXAMPLE 7.2

Determine the peak confined concrete stress f'_{cc} and corresponding strain ε_{cc} for a 500-mm diameter column with 10 No. 25 longitudinal bars and No. 10 round spirals at 50-mm pitch. The material strengths are $f'_c = -28$ MPa and $f_{yh} = 400$ MPa. Assume that $\varepsilon_{co} = -0.002$ and that the concrete cover is 38 mm. Use the lower bound value of $k_1 = 3$ and the corresponding value of $k_2 = 15$.

$$s = 50 \text{ mm}, \ s' = 50 - 11 = 39 \text{ mm}$$

$$d_s = 500 - 2(38) - 2(\tfrac{1}{2})(11) = 413 \text{ mm}$$

$$A_c = \frac{\pi}{4}(d_s)^2 = \frac{\pi}{4}(413)^2 = 134\,000 \text{ mm}^2$$

$$\rho_{cc} = \frac{A_{st}}{A_c} = \frac{10(500)}{134\,000} = 0.0373$$

$$\rho_s = \frac{4A_{sp}}{d_s s} = \frac{4(100)}{50(413)} = 0.0194$$

$$k_e = \frac{1 - \frac{s'}{2d_s}}{1 - \rho_{cc}} = \frac{1 - \frac{39}{2(413)}}{1 - 0.0373} = 0.990$$

$$f'_r = -\tfrac{1}{2}k_e\rho_s f_{yn} = -\tfrac{1}{2}(0.990)(0.0194)(400)$$

$$= -3.8 \text{ MPa} = \underline{3.8\text{-MPa compression}}$$

$$f'_{cc} = f'_{co} + k_1 f'_r = 0.85(-28) + 3(-3.8)$$

$$= -35.2 \text{ MPa} = \underline{35.2\text{-MPa compression}}$$

$$\varepsilon_{cc} = \varepsilon_{co}\left(1 + k_2\frac{f'_r}{f'_{co}}\right) = -0.002\left[1 + 15\left(\frac{-3.8}{-23.8}\right)\right]$$

$$= -0.0068 = \underline{0.0068 \text{ shortening}}$$

Again, note the negative signs indicate compression.

Over the years, researchers developed stress–strain relationships for the response of confined concrete in compression that best fits their experimental data. Sheikh (1982) presented a comparison of seven models used by investigators in different research laboratories. All but one of these models use different equations for the ascending and descending branches of the stress–strain curve. The model considered to best fit the experimental data is one he and a colleague developed earlier (Sheikh and Uzumeri, 1980).

The stress–strain model proposed by Mander et al. (1988) for monotonic compression loading up to first hoop fracture is a single equation relating the longitudinal compressive stress f_c as a function of the corresponding longitudinal compressive strain ε_c

$$f_c(x) = \frac{f'_{cc}rx}{r - 1 + x^r} \tag{7.15}$$

where

$$x = \frac{\varepsilon_c}{\varepsilon_{cc}} \tag{7.16}$$

$$r = \frac{E_c}{E_c - E_{sec}} \tag{7.17}$$

and the secant modules of confined concrete at peak stress is

$$E_{\text{sec}} = \frac{f'_{cc}}{\varepsilon_{cc}} \tag{7.18}$$

This curve continues until the confined concrete strain reaches an ε_{cu} value large enough to cause the first hoop or spiral to fracture. Based on an energy balance approach and test results, Mander et al. (1988) present an integral equation that can be solved numerically for ε_{cu}.

EXAMPLE 7.3

Determine the parameters and plot the stress–strain curves for the confined and unconfined concrete of the column section in Example 7.2. Assume concrete strain at first hoop fracture $\varepsilon_{cu} = 8\varepsilon_{cc} = 8(-0.0068) = -0.0544$.

$$E_c = 0.043\gamma_c^{1.5}\sqrt{f'_c} = 0.043(2300)^{1.5}\sqrt{28} = 25 \text{ GPa}$$

$$E_{\text{sec}} = \frac{f'_{cc}}{\varepsilon_{cc}} = \frac{-35.2}{-0.0068} = 5.2 \text{ GPa}$$

$$r = \frac{E_c}{E_c - E_{\text{sec}}} = \frac{25}{25 - 5.2} = 1.26$$

$$f_c(\varepsilon_c) = 1.26f'_{cc}\frac{(\varepsilon_c/\varepsilon_{cc})}{0.26 + (\varepsilon_c/\varepsilon_{cc})^{1.26}} \qquad 0 \le \varepsilon_c \le 8\varepsilon_{cc}$$

This last expression for f_c is Eq. 7.15 and has been used to plot the curve shown in Figure 7.5.

From the above discussion, it is apparent that the behavior of concrete in compression is different when it has reinforcement within and around the concrete than when it is unreinforced. A corollary to this concrete behavior is that the response in tension of reinforcement embedded in concrete is different than the response of bare steel alone. The behavior of the tension reinforcement is discussed later after a brief discussion about the tensile behavior of concrete.

Concrete Tensile Strength and Behavior. Concrete tensile strength can be measured either directly or indirectly. A direct tensile test [Fig. 7.8(a)] is preferred for determining the cracking strength of concrete but requires special equipment. Consequently, indirect tests, such as the modulus of rupture test and the split cylinder test, are often used. These tests are illustrated in Figure 7.8.

The modulus of rupture test [Fig. 7.8(b)] measures the tensile strength of concrete in flexure with a plain concrete beam loaded as shown. The tensile

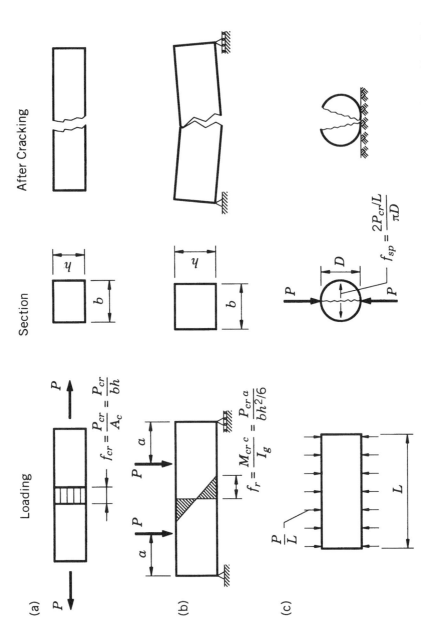

Fig. 7.8 Direct and indirect concrete tensile tests. (a) Direct tension test, (b) modulus of rupture test, and (c) split cylinder test.

stress through the depth of the section is nonuniform and is maximum at the bottom fibers. A flexural tensile stress is calculated from elementary beam theory for the load that cracks (and fails) the beam. This flexural tensile stress is called the modulus of rupture f_r. For normal weight concrete, AASHTO [A5.4.2.6] gives the following expression for f_r (MPa).

$$f_r = 0.63\sqrt{f'_c} \tag{7.19}$$

where f'_c is the absolute value of the cylinder compressive strength of concrete (MPa).

In the split cylinder test [Fig. 7.8(c)], a standard cylinder is laid on its side and loaded with a uniformly distributed line load. Nearly uniform tensile stresses are developed perpendicular to the compressive stresses produced by opposing line loads. When the tensile stresses reach their maximum strength, the cylinder splits in two along the loaded diameter. A theory of elasticity solution (Timoshenko and Goodier, 1951) gives the splitting tensile stress f_{sp} as

$$f_{sp} = \frac{2P_{cr}/L}{\pi D} \tag{7.20}$$

where P_{cr} is the total load that splits the cylinder, L is the length of the cylinder, and D is the diameter of the cylinder.

Both the modulus of rupture (f_r) and the splitting stress (f_{sp}) overestimate the tensile cracking stress (f_{cr}) determined by a direct tension test [Fig. 7.8(a)]. If they are used, nonconservative evaluations of resistance to restrained shrinkage and splitting in anchorage zones can result. In these and other cases of direct tension, a more representative value must be used. For normal weight concrete, Collins and Mitchell (1991) and Hsu (1993) estimate the direct cracking strength of concrete, f_{cr}, as

$$f_{cr} = 0.33\sqrt{f'_c} \tag{7.21}$$

where f'_c is the cylinder compressive strength (MPa).

The direct tension stress–strain curve (Fig. 7.9) is assumed to be linear up to the cracking stress f_{cr} at the same slope given by E_c in Eq. 7.2. After cracking and if reinforcement is present, the tensile stress decreases but does not go to zero. Aggregate interlock still exists and is able to transfer tension across the crack. The direct tension experiments by Gapalaratnam and Shah (1985), using a stiff testing machine, demonstrate this behavior. This response is important when predicting the tensile stress in longitudinal reinforcement and the shear resistance of reinforced concrete beams. Collins and Mitchell (1991) give the following expressions for the direct tension stress–strain curve shown in Figure 7.9:

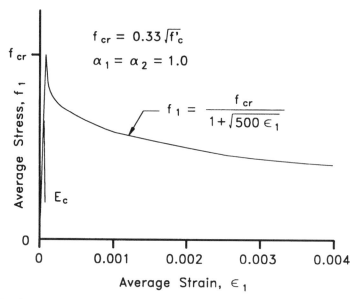

Fig. 7.9   Average stress versus average strain for concrete in tension. (Collins and Mitchell, 1991). [Reprinted by permission of Prentice Hall, Upper Saddle River, NJ.]

Ascending Branch $(\varepsilon_1 \leq \varepsilon_{cr} = f_{cr}/E_c)$

$$f_1 = E_c \varepsilon_1 \tag{7.22}$$

where ε_1 is the average principle tensile strain in the concrete and f_1 is the average principle tensile stress.

Descending Branch $(\varepsilon_1 > \varepsilon_{cr})$

$$f_c = \frac{\alpha_1 \alpha_2 f_{cr}}{1 + \sqrt{500 \varepsilon_1}} \tag{7.23}$$

where α_1 is a factor accounting for bond characteristics of reinforcement:

$\alpha_1 = 1.0$ for deformed reinforcing bars

$\alpha_1 = 0.7$ for plain bars, wires, or bonded strands

$\alpha_1 = 0$ for unbonded reinforcement and α_2 is a

factor accounting for sustained or repeated loading

$\alpha_2 = 1.0$ for short-term monotonic loading

$\alpha_2 = 0.7$ for sustained and/or repeated loads

If no reinforcement is present, there is no descending branch and the concrete tensile stress after cracking is zero. However, if the concrete is bonded to reinforcement, concrete tensile stresses do exist. Once again it is apparent that the behavior of concrete with reinforcement is different than that of plain concrete.

7.4.2 Long-Term Properties of Concrete

At times it appears that concrete is more alive than it is dead. If compressive loads are applied to concrete for a long period of time, it wants to get away from them. Concrete generally gains strength with age unless a deterioration mechanism, such as that caused by the intrusion of the chloride ion, saps its strength. And if concrete does not do anything else, it will shrink and crack. But even this behavior can be reversed by immersing the concrete in water and refilling the voids and closing the cracks. It appears that concrete never completely dries and there is always some gelatinous material that has not hardened and provides resiliency between the particles. These time-dependent properties of concrete are influenced by the conditions at time of placement and the environment that surrounds it throughout its service life. It is difficult to predict the exact effect of all of the conditions, but estimates can be made of the trends and changes in behavior.

Compressive Strength of Aged Concrete. If a concrete bridge has been in service for a number of years and a strength evaluation is required, the compressive strength of a core sample is a good indication of the quality and durability of the concrete in the bridge. There are also nondestructive methods for determining the compressive strength, often in an indirect way by first determining the modulus of elasticity and then back-calculating to find the compressive strength. Another device measures the rebound of a steel ball that has been calibrated against the rebound on concrete of known compressive strength.

In general, the trend is that the compressive strength of concrete increases with age. However, to determine the magnitude of the increase, there is no substitute for conducting a field investigation.

Shrinkage of Concrete. Shrinkage of concrete is a decrease in volume under constant temperature due to loss of moisture after concrete has hardened. This time-dependent volumetric change depends on the water content of the fresh concrete, the type of cement and aggregate used, the ambient conditions (temperature, humidity, and wind velocity) at the time of placement, the curing procedure, the amount of reinforcement, and the volume/surface area ratio. In AASHTO [A5.4.2.3.3], an empirical equation from Collins and Mitchell (1991) is presented that is used to evaluate the shrinkage strain ε_{sh} based on the drying time, the relative humidity, and the volume/surface area ratio.

$$\varepsilon_{sh} = -k_s k_h \left(\frac{t}{35 + t} \right) 0.51 \times 10^{-3} \qquad (7.24)$$

where t is the drying time in days, k_s is a size factor taken from Figure 7.10, and k_h is a humidity factor given in Table 7.2. If humidity at the site is not known, an annual average value depending on the geographic location can be taken from Figure 7.11. Equation 7.24 assumes that the concrete is moist cured and that low-absorption aggregates are used. In spite of these limitations, Eq. 7.24 does indicate the trend and relative magnitude of the shrinkage strains, which are illustrated in Example 7.4.

EXAMPLE 7.4

Estimate the shrinkage strain in a 200-mm thick concrete bridge deck whose top and bottom surfaces are exposed to drying conditions in an atmosphere with 70% relative humidity. The volume/surface area ratio for 1 mm² of deck area is

$$\frac{\text{volume}}{\text{surface area}} = \frac{200(1)(1)}{2(1)(1)} = 100 \text{ mm}$$

From Figure 7.10 for $t = 5$ years (≈ 2000 days), $k_s = 0.73$, and from Table 7.2 for $H = 70\%$, $k_h = 1.0$. Thus Eq. 7.24 gives

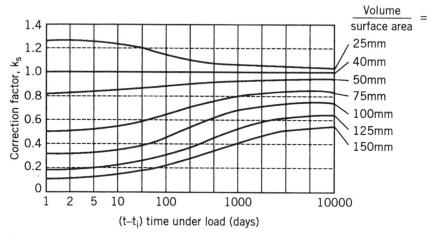

Fig. 7.10 Factor k_s for volume/surface area ratio (AASHTO Fig. 5.4.2.3.3-2). [From *AASHTO LRFD Bridge Design Specifications*. Copyright © 1994 by the American Association of State Highway and Transportation Officials, Washington, DC. Used by permission.]

TABLE 7.2 Factor k_h for Relative Humidity H^a

Average Ambient Relative Humidity H (%)	k_h
40	1.43
50	1.29
60	1.14
70	1.00
80	0.86
90	0.43
100	0.00

[a]In AASHTO Table 5.4.2.3.3-1. [From AASHTO LRFD Bridge Design Specifications, Copyright © (1994) by the American Association of State Highway and Transportation Officials, Washington DC. Used by permission.]

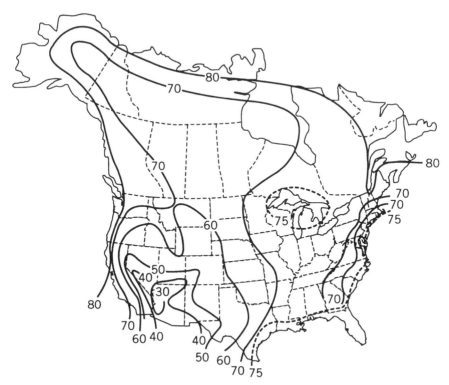

Fig. 7.11 Annual average ambient relative humidity in percent (AASHTO Fig. 5.4.2.3.3-1). [From *AASHTO LRFD Bridge Design Specifications.* Copyright © 1994 by the American Association of State Highway and Transportation Officials, Washington, DC. Used by permission.]

$$\varepsilon_{sh} = -(0.73)(1.0)\left(\frac{2000}{35 + 2000}\right)0.51 \times 10^{-3} = -0.000\,37$$

where the negative sign indicates shortening.

The variation of shrinkage strain with drying time for these conditions is shown in Figure 7.12. Because this empirical equation does not include all of the variables affecting shrinkage, the commentary in AASHTO [C5.4.2.3.1] indicates that the results may be off by $\pm 50\%$ and the actual shrinkage strains could be larger than -0.0008 [C5.4.2.3.3]. Even if the values are not exact, the trend shown in Figure 7.12 of increasing shrinkage strain at a diminishing rate as drying time increases is correct. When specific information is not available on the concrete and the conditions under which it is placed, AASHTO [A5.4.2.3.1] recommends values of shrinkage strain to be taken as -0.0002 after 28 days and -0.0005 after 1 year of drying.

There are a number of measures that can be taken to control the amount of shrinkage in concrete structures. One of the most effective is to reduce the water content in the concrete mixture, because it is the evaporation of the excess water that causes the shrinkage. A designer can control the water content by specifying both a maximum water/cement ratio and a maximum cement content. Use of hard, dense aggregates with low absorption results in less shrinkage because they require less moisture in the concrete mixture to wet their surfaces. Another effective method is to control the temperature in the concrete before it hardens so that the starting volume for the beginning of shrinkage has not been enlarged by elevated temperatures. This temperature control can be done by using a low heat of hydration cement and by cooling

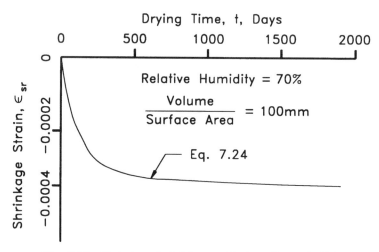

Fig. 7.12   Variation of shrinkage with time. Example 7.4.

the materials in the concrete mixture. High outdoor temperatures during the summer months need to be offset by shading the aggregate stockpiles from the sun and by cooling the mixing water with crushed ice. It has often been said by those in the northern climates that the best concrete (fewest shrinkage cracks) is placed during the winter months if kept sufficiently warm during cure.

Creep of Concrete. Creep of concrete is an increase in deformation with time when subjected to a constant load. In a reinforced concrete beam, the deflection continues to increase due to sustained loads. In reinforced concrete beam-columns, axial shortening and curvature increase under the action of constant dead loads. Prestressed concrete beams lose some of their precompression force because the concrete shortens and decreases the strand force and associated prestress. The creep phenomenon in concrete influences the selection and interaction of concrete elements and an understanding of its behavior is important.

Creep in concrete is associated with the change of strain over time in the regions of beams and columns subjected to sustained compressive stresses. This time-dependent change in strains relies on the same factors that affect shrinkage strains plus the magnitude and duration of the compressive stresses, the compressive strength of the concrete, and the age of the concrete when the sustained load is applied. Creep strain ε_{CR} is determined by multiplying the instantaneous elastic compressive strain due to the sustained loads ε_{ci} by a creep coefficient ψ, that is,

$$\varepsilon_{CR}(t,t_i) = \psi(t,t_i)\varepsilon_{ci} \tag{7.25}$$

where t is the age of the concrete in days from the time it is first placed in a form, and t_i is the age of the concrete in days when the permanent load is applied. In AASHTO [A5.4.2.3.2], an empirical equation taken from Collins and Mitchell (1991) is given for the creep coefficient. It is expressed as

$$\psi(t,t_i) = 3.5k_ck_f\left(1.58 - \frac{H}{120}\right)t_i^{-0.118}\left[\frac{(t - t_i)^{0.6}}{10 + (t - t_i)^{0.6}}\right] \tag{7.26}$$

where H is the relative humidity (%), k_c is the creep factor for the effect of the volume/surface area ratio taken from Figure 7.13, and

$$k_f = \frac{62}{42 + f'_c} \tag{7.27}$$

where f'_c is the absolute value of the 28-day compressive strength of the concrete (MPa). If H is not known for the site, a value can be taken from Figure 7.11.

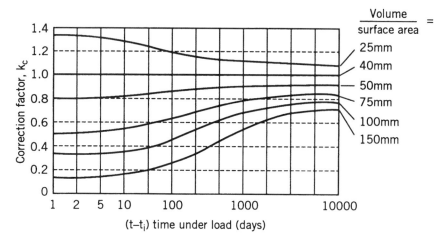

Fig. 7.13   Factor k_c for volume/surface area ratio (AASHTO Fig. 5.4.2.3.2-1). [From *AASHTO LRFD Bridge Design Specifications.* Copyright © 1994 by the American Association of State Highway and Transportation Officials, Washington, DC. Used by permission.]

The *H* factor may be higher than ambient for a water crossing due to evaporation in the vicinity of the bridge.

EXAMPLE 7.5

Estimate the creep strain in the bridge deck of Example 7.4 after 1 year if the compressive stress due to sustained loads is 10 MPa, the 28-day compressive strength is 31 MPa, and $t_i = 15$ days. The modulus of elasticity from Eq. 7.2 is

$$E_c = 0.043(2300)^{1.5}\sqrt{31} = 26.4 \text{ GPa}$$

and the initial compressive strain becomes

$$\varepsilon_{ci} = \frac{f_{cu}}{E_c} = \frac{-10}{26\,400} = -0.000\,38$$

For a *volume/surface area ratio* = *100 mm* and $(t - t_i) = (365 - 15) = 350$ *days*, Figure 7.13 gives a correction factor $k_c = 0.68$. The concrete strength factor k_f from Eq. 7.27 is

$$k_f = \frac{62}{42 + 31} = 0.85$$

The creep coefficient in an environment where $H = 70\%$ is given by Eq. 7.26 as

$$\psi(365,15) = 3.5(0.68)(0.85)\left(1.58 - \frac{70}{120}\right)15^{-0.118}\frac{350^{0.6}}{10 + 350^{0.6}} = 1.13$$

Thus, the estimated creep strain after 1 year is (Eq. 7.25)

$$\varepsilon_{CR}(365,15) = 1.13(-0.000\,38) = -0.000\,43$$

which is of the same order of magnitude as the shrinkage strain. Again, this estimate could be in error by $\pm 50\%$. For the same conditions as this example, the variation of total compressive strain with time after application of the sustained load is shown in Figure 7.14. The total compressive strain $\varepsilon_c(t,t_i)$ is the sum of the initial elastic strain plus the creep strain and the rate of increase diminishes with time. This total strain can be expressed as

$$\varepsilon_c(t,t_i) = \varepsilon_{ci} + \varepsilon_{CR}(t,t_i) = [1 + \psi(t,t_i)]\varepsilon_{ci} \tag{7.28}$$

For this example, the total compressive strain after 1 year is

$$\varepsilon_c(365,15) = (1 + 1.13)(-0.000\,38) = -0.000\,81$$

over two times the elastic value.

Creep strains can be reduced by the same measures taken to reduce shrinkage strains, that is, by using a low water content in the concrete mixture and

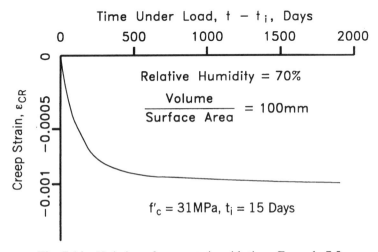

Fig. 7.14   Variation of creep strain with time. Example 7.5.

keeping the temperature relatively low. Creep strain can also be reduced by using steel reinforcement in the compression zone because the portion of the compressive force it carries is not subject to creep. By delaying the time at which permanent loads are applied, creep strains will be reduced because the more mature concrete will be drier and less resilient. This trend is reflected in Eq. 7.26, where larger values of t_i for a given age of concrete t result in a reduction of the creep coefficient $\psi(t,t_i)$.

Finally, not all effects of creep deformation are harmful. When differential settlements occur in a reinforced concrete bridge, the creep property of concrete actually decreases the stresses in the elements from those that would be predicted by an elastic analysis.

Modulus of Elasticity for Permanent Loads. To account for the increase in strain due to creep under permanent loads, a reduced long-term modulus of elasticity $E_{c,LT}$ can be defined as

$$E_{c,LT} = \frac{f_{ci}}{[1 + \psi(t,t_i)]\varepsilon_{ci}} = \frac{E_{ci}}{1 + \psi(t,t_i)}$$

where E_{ci} is the modulus of elasticity at time t_i. Assuming that E_{ci} can be represented by the modulus of elasticity E_c from Eq. 7.2, then

$$E_{c,LT} = \frac{E_c}{1 + \psi(t,t_i)} \tag{7.29}$$

When transforming section properties of steel to equivalent properties of concrete for service limit states, the modular ratio n is used. It is defined as

$$n = \frac{E_s}{E_c} \tag{7.30}$$

A long-term modular ratio n_{LT} for use with permanent loads can be similarly defined, assuming that steel does not creep,

$$n_{LT} = \frac{E_s}{E_{c,LT}} = n[1 + \psi(t,t_i)] \tag{7.31}$$

EXAMPLE 7.6

For the conditions of Example 7.5, estimate the long-term modular ratio using $t = 5$ years.

From Figure 7.13 for $(t - t_i) = 5(365) - 15 = 1810$ days, $k_c = 0.75$. Thus,

$$\psi(1825,15) = 3.50(0.75)(0.85)\left(1.58 - \frac{70}{120}\right)15^{-0.118}\frac{1810^{0.6}}{10 + 1810^{0.6}} = 1.45$$

and

$$n_{LT} = 2.45n$$

In evaluating designs based on service and fatigue limit states, an effective modular ratio of $2n$ for permanent loads is assumed [A5.7.1]. In AAAHTO [A5.7.3.6.2], which is applicable to the calculation of deflection and camber, the long-time deflection is estimated as the instantaneous deflection multiplied by the factor:

$$3.0 - 1.2\frac{A'_s}{A_s} \geq 1.6 \qquad (7.32)$$

where A'_s is the area of the compression reinforcement and A_s is the area of nonprestressed tension reinforcement. This factor is essentially $\psi(t,t_i)$ and if $A'_s = 0$, Eq. 7.31 gives a value of $n_{LT} = 4n$. For calculating stresses due to permanent loads on a long-term composite section, AASHTO [A6.10.5.1.1b] specifies that the concrete slab shall be transformed by using $n_{LT} = 3n$. Based on these recommendations and the calculations made for the creep coefficient, it is reasonable to use the following simple expression for the modulus of elasticity for permanent loads.

$$E_{c,LT} = \frac{E_c}{3} \qquad (7.33)$$

7.5 PROPERTIES OF STEEL REINFORCEMENT

Reinforced concrete is simply concrete with embedded reinforcement, usually steel bars or tendons. Reinforcement is placed in structural members at locations where it will be of the most benefit. It is usually thought of as resisting tension, but it is also used to resist compression. If shear in a beam is the limit state that is being resisted, longitudinal and transverse reinforcement are placed to resist diagonal tension forces.

The behavior of nonprestressed reinforcement is usually characterized by the stress–strain curve for bare steel bars. The behavior of prestressed steel tendons is known to be different for bonded and unbonded tendons, which suggests that we should reconsider the behavior of nonprestressed reinforcement embedded in concrete.

7.5.1 Nonprestressed Steel Reinforcement

Typical stress–strain curves for bare steel reinforcement are shown in Figure 7.15 for steel grades 280, 420, and 520. The response of the bare steel can be broken into three parts, elastic, plastic, and strain hardening. The elastic portion *AB* of the curves respond in a similar straight-line manner with a constant modulus of elasticity E_S = 200 000 MPa up to a yield strain of ε_y = f_y/E_S. The plastic portion *BC* is represented by a yield plateau at constant stress f_y until the onset of strain hardening. The length of the yield plateau is a measure of ductility and it varies with the grade of steel. The strain hardening portion *CDE* begins at a strain of ε_h and reaches maximum stress f_u at a strain of ε_u before dropping off slightly at a breaking strain of ε_b. The three portions of the stress–strain curves for bare steel reinforcement can be characterized symbolically as

Elastic Portion *AB*

$$f_s = \varepsilon_s E_s \qquad 0 \le \varepsilon_s \le \varepsilon_y \tag{7.34}$$

Plastic Portion *BC*

$$f_s = f_y \qquad \varepsilon_y \le \varepsilon_s \le \varepsilon_h \tag{7.35}$$

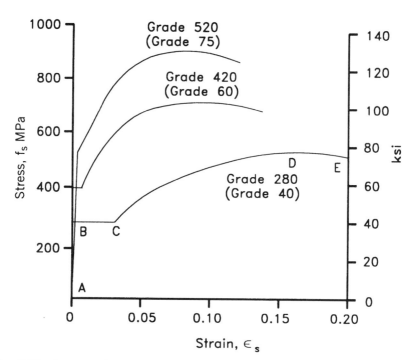

Fig. 7.15 Stress–strain curves for bare steel reinforcement. (From Holzer et al., 1975.)

Strain-Hardening Portion *CDE*

$$f_s = f_y \left[1 + \frac{\varepsilon_s - \varepsilon_h}{\varepsilon_u - \varepsilon_h} \left(\frac{f_u}{f_y} - 1 \right) \exp\left(1 - \frac{\varepsilon_s - \varepsilon_h}{\varepsilon_u - \varepsilon_h} \right) \right] \qquad \varepsilon_h \le \varepsilon_s \le \varepsilon_b \quad (7.36)$$

Equation 7.36 and the nominal limiting values for stress and strain in Table 7.3 are taken from Holzer et al. (1975). The curves shown in Figure 7.15 are calibrated to pass through the nominal yield stress values of the different steel grades. The actual values for the yield stress from tensile tests average about 15% higher. The same relationship is assumed to be valid for both tension and compression. When steel bars are embedded in concrete, the behavior is different than for the bare steel bars. The difference is due to the fact that concrete has a finite, though small, tensile strength, which was realized early in the development of the mechanics of reinforced concrete as described by Collins and Mitchell (1991) in this quote from Morsch (1908):

> Because of friction against the reinforcement, and of the tensile strength which still exists in the pieces lying between the cracks, even cracked concrete decreases to some extent the stretch of the reinforcement.

Concrete that adheres to the reinforcement and is uncracked reduces the tensile strain in the reinforcement. This phenomenon is called "tension stiffening."

An experimental investigation by Scott and Gill (1987) confirmed the decrease in tensile strain in the reinforcement between cracks in the concrete. To measure the strains in the reinforcement without disturbing the bond characteristics on the surface of the bars, they placed the strain gages inside the bars. This internal placement of strain gages was done by splitting a bar in half, machining out a channel, placing strain gages and their lead wires, and then gluing the halves back together. The instrumented bar was then encased in concrete, except for a length at either end that could be gripped in the jaws of a testing machine. Tensile loads were then applied and the strains along the bar at 12.5-mm increments were recorded. Figure 7.16 presents the strains

TABLE 7.3   Nominal Limiting Values for Bare Steel Stress–Strain Curves[a]

f_y, (MPa) (ksi)	f_u, (MPa) (ksi)	ε_y	ε_h	ε_u	ε_b
280 (40)	550 (80)	0.001 38	0.0230	0.140	0.200
420 (60)	730 (106)	0.002 07	0.0060	0.087	0.136
520 (75)	900 (130)	0.002 59	0.0027	0.073	0.115

[a]Holzer et al., 1975.

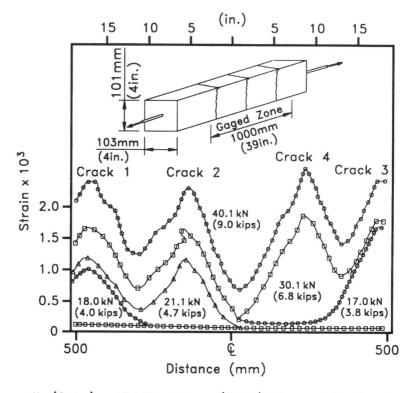

f'$_c$ (Cube) = 45 MPa, 12mm (0.5in) Dia. reinforcing bar

Fig. 7.16 Variation of steel strain along the length of a tension specimen tested by Scott and Gill (1987). (From Collins and Mitchell, 1991). [Reprinted by permission of Prentice Hall, Upper Saddle River, NJ.]

in one of their bars over the 1000-mm long section at increasing levels of tensile load.

To represent the behavior of reinforcement embedded in concrete as shown in Figure 7.16, it is convenient to define an average stress and average strain over a length long enough to include at least one crack. The average stress–strain behavior for concrete stiffened mild steel reinforcement is shown in Figure 7.17 and compared to the response of a bare bar (which is indicative of the bar response at a crack where the concrete contribution is lost). The tension stiffening effect of the concrete is greatest, as would be expected, at low strains and tends to round off the sharp knee of the elastic-perfectly plastic behavior. This tension stiffening effect results in the average steel stress showing a reduced value of apparent yield stress f_y^* and its accompanying apparent yield strain ε_y^*. At higher strains, the concrete contribution diminishes and the embedded bar response follows the strain-hardening portion of the bare steel curve.

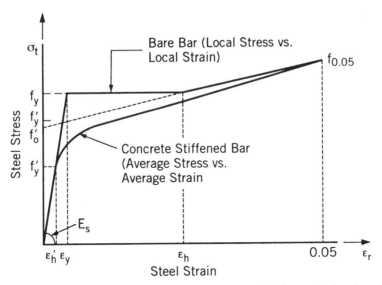

Fig. 7.17 Stress–strain curve for mild steel. [Reprinted with permission from T. T. C. Hsu (1993). *Unified Theory of Reinforced Concrete,* CRC Press, Boca Raton, FL. Copyright CRC Press, Boca Raton, FL © 1993.]

Also shown by the dashed line in Figure 7.17 is a linear approximation to the average stress–strain response of a mild steel bar embedded in concrete. A derivation of this approximation and a comparison with experimental data are given in Hsu (1993). The equation for these two straight lines are given by

Elastic Portion

$$f_s = E_s \varepsilon_s \qquad \text{when} \qquad f_s \leq f_y' \qquad (7.37)$$

Postyield Portion

$$f_s = (0.91 - 2B)f_y + (0.02 + 0.25B)E_s \varepsilon_s \qquad \text{when} \qquad f_s > f_y' \quad (7.38)$$

where

$$f_y' = \text{intersection stress level} = (0.93 - 2B)\, f_y \qquad (7.39)$$

$$B = \frac{1}{\rho}\left(\frac{f_{cr}}{f_y}\right)^{1.5} \qquad (7.40)$$

ρ = steel reinforcement ratio based on the net concrete section
  = $A_s/(A_g - A_s)$

f_{cr} = tensile cracking strength of concrete, taken as $0.33\sqrt{f'_c}$(MPa)

f_y = steel yield stress of bare bars (MPa)

Figure 7.18 compares the bilinear approximation of an average stress–strain curve ($\rho = 0.01, f_{cr} = 2$ MPa, $f_y = 400$ MPa) with a bare bar and test results by Tamai et al. (1987). This figure illustrates what was stated earlier and shows how the response in tension of reinforcement embedded in concrete is different than the response of bare steel alone.

7.5.2 Prestressing Steel

The most common prestressing steel is seven-wire strand, which is available in stress–relieved strand and low-relaxation strand. During manufacture of the strands, high carbon steel rod is drawn through successively smaller diameter dies, which tends to align the molecules in one direction and increases the strength of the wire to over 1700 MPa. Six wires are then wrapped around one central wire in a helical manner to form a strand. The cold drawing and twisting of the wires creates locked-in or residual stresses in the strands. These residual stresses cause the stress–strain response to be more rounded and to exhibit an apparently lower yield stress. The apparent yield stress can

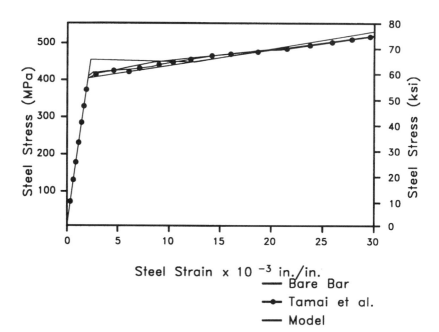

Fig. 7.18 Average Stress–Strain Curves of Mild Steel: Theories and Tests. [Reprinted with permission from T. T. C. Hsu (1993). *Unified Theory of Reinforced Concrete,* CRC Press, Boca Raton, FL. Copyright CRC Press, Boca Raton, FL © 1993.]

be raised by heating the strands to about 350°C and allowing them to cool slowly. This process is called stress relieving. Further improvement in behavior by reducing the relaxation of the strands is achieved by putting the strands into tension during the heating and cooling process. This process is called strain tempering and produces the low-relaxation strands. Figure 7.19 compares the stress–strain response of seven-wire strand manufactured by the different processes.

High strength deformed bars are also used for prestressing steel. The deformations are often like raised screw threads so that devices for post-tensioning and anchoring bars can be attached to their ends. The ultimate tensile strength of the bars is about 1000 MPa.

A typical specification for the properties of the prestressing strand and bar is given in Table 7.4. Recommended values for modulus of elasticity, E_p, for prestressing steels are 197 000 MPa for strands and 207 000 MPa for bars.

The stress–strain curves for the bare prestressing strand shown in Figure 7.19 have been determined by a Ramberg–Osgood function to give a smooth transition between two straight lines representing elastic and plastic behavior. Constants are chosen so that the curves pass through a strain of 0.01 when the yield strengths of Table 7.4 are reached. Collins and Mitchell (1991) give the following expression for low-relaxation strands with

$$f_{pu} = 1860 \text{ MPa}$$

$$f_{ps} = E_p\varepsilon_{ps}\left\{0.025 + \frac{0.975}{[1 + (118\varepsilon_{ps})^{10}]^{0.10}}\right\} \leq f_{pu} \qquad (7.41)$$

while for stress-relieved strands with $f_{pu} = 1860$ MPa

$$f_{ps} = E_p\varepsilon_{ps}\left\{0.03 + \frac{0.97}{[1 + (121\varepsilon_{ps})^{6}]^{0.167}}\right\} \leq f_{pu} \qquad (7.42)$$

and for untreated strands with $f_{pu} = 1655$ MPa

$$f_{ps} = E_p\varepsilon_{ps}\left\{0.03 + \frac{1}{[1 + (106\varepsilon_{ps})^{2}]^{0.5}}\right\} \leq f_{pu} \qquad (7.43)$$

These curves are based on the minimum specified strengths. The actual stress–strain curves of typical strands will probably have higher yield strengths and be above those shown in Figure 7.19.

A tendon can be either a single strand or bar, or it can be a group of strands or bars. When the tendons are bonded to the concrete the *change* in strain of the prestressing steel is equal to the change in strain of the concrete. This condition exists in pretensioned beams where the concrete is cast around the tendons and in post-tensioned beams where the tendons are pressure grouted after they are prestressed. At the time the concrete or grout is placed,

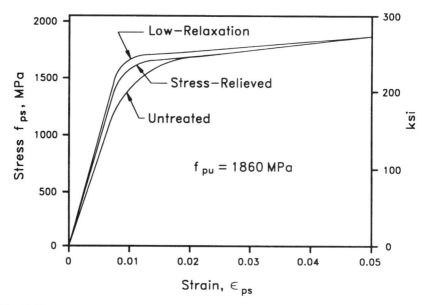

Fig. 7.19 Stress–strain response of seven-wire strand manufactured by different processes. (After Collins and Mitchell, 1991). [Reprinted by permission of Prentice Hall, Upper Saddle River, NJ.]

the prestressing tendon will have been stretched and will have a difference in strain of $\Delta\varepsilon_{pe}$ when the two materials are bonded together. The strain in the prestressing tendon ε_{ps} can be determined at any stage of loading from the strain in the surrounding concrete ε_{cp} as

$$\varepsilon_{ps} = \varepsilon_{cp} + \Delta\varepsilon_{pe} \qquad (7.44)$$

where ε_{cp} is the concrete strain at the same location as the prestressing tendon, and $\Delta\varepsilon_{pe}$ is

TABLE 7.4 Properties of Prestressing Strand and Bar[a]

Material	Grade or Type	Diameter (mm)	Tensile Strength, f_{pu} (MPa)	Yield Strength, f_{py} (MPa)
Strand	1725 MPa (Grade 250)	6.35–15.24	1725	85% of f_{pu} except
	1860 MPa (Grade 270)	9.53–15.24	1860	90% of f_{pu} for low-relaxation strand
Bar	Type 1, Plain	19–35	1035	85% of f_{pu}
	Type 2, Deformed	15–36	1035	80% of f_{pu}

[a]In AASHTO Table 5.4.4.1-1. [From *AASHTO LRFD Bridge Design Specifications,* Copyright © 1994 by the American Association of State Highway and Transportation Officials, Washington, DC. Used by permission.]

$$\Delta\varepsilon_{pe} = \varepsilon_{pe} - \varepsilon_{ce} \tag{7.45}$$

where ε_{pe} is the strain corresponding to the effective stress in the prestressing steel after losses f_{pe} expressed as

$$\varepsilon_{pe} = \frac{f_{pe}}{E_p} \tag{7.46}$$

and ε_{ce} is the strain in the concrete at the location of the prestressing tendon resulting from the effective prestress. If the tendon is located along the centroidal axis, then

$$\varepsilon_{ce} = \frac{A_{ps}f_{pe}}{E_c A_c} \tag{7.47}$$

where A_{ps} is the prestressing steel area and A_c is the concrete area. This strain is always small and is usually ignored (Loov, 1988) so that $\Delta\varepsilon_{pe}$ is approximately equal to

$$\Delta\varepsilon_{pe} \approx f_{pe}/E_p.$$

In the case of an unbonded tendon, slip results between the tendon and the surrounding concrete and the strain in the tendon becomes uniform over the distance between anchorage points. The total change in length of the tendon must now equal the total change in length of the concrete over this distance, that is,

$$\varepsilon_{ps} = \bar{\varepsilon}_{cp} + \Delta\varepsilon_{pe} \tag{7.48}$$

where $\bar{\varepsilon}_{cp}$ is the average strain of the concrete at the location of the prestressing tendon, averaged over the distance between anchorages of the unbonded tendon.

7.6 LIMIT STATES

Reinforced concrete bridges must be designed so that their performance under load does not go beyond the limit states prescribed by AASHTO. These limit states are applicable at all stages in the life of a bridge and include service, fatigue, strength, and extreme event limit states. The condition that must be met for each of these limit states is that the factored resistance is greater than the effect of the factored load combinations, or simply, supply must exceed demand. The general inequality that must be satisfied for each limit state can be expressed as

$$\phi R_n \geq \eta \sum \gamma_i Q_i \tag{7.49}$$

where ϕ is a statistically based resistance factor for the limit state being examined; R_n is the nominal resistance; η is a load multiplier relating to ductility, redundancy, and operational importance; γ_i is a statistically based load factor applied to the force effects as defined for each limit state in Table 3.1; and Q_i is a force effect. The various factors in Eq. 7.49 are discussed more fully in Chapter 3, and are repeated here only for convenience.

7.6.1 Service Limit States

Service limit states related to how a bridge performs when subjected to the forces it sees when it is put into service. Actions to be considered are cracking, deformations, and stresses for concrete and prestressing tendons under regular service conditions. Because the provisions for service limit states are not derived statistically, but rather are based on experience and engineering judgment, the resistance factors ϕ and load factors γ_i are usually taken as unity. There are some exceptions for vehicle live loads and wind loads as shown in Table 3.1.

Control of Flexural Cracking in Beams [A5.7.4.3]. The width of flexural cracks in reinforced concrete beams is controlled by distributing the reinforcement over the region of maximum concrete tension. The width of a crack is influenced by the tensile stress and the detailing of the reinforcement. At the Service *I* limit state, the tensile stress in the nonprestressed reinforcement f_s based on a cracked section analysis shall not exceed f_{sa} given by

$$f_s \leq f_{sa} = \frac{Z}{(d_c A)^{1/3}} \leq 0.6 f_y \tag{7.50}$$

where Z is a crack width parameter, d_c is the concrete cover measured from the extreme tension fiber to the center of the closest bar, but not to be taken greater than 50 mm; and A is the area of concrete having the same centroid as the principal tensile reinforcement, divided by the number of bars. The quantity Z is selected from Table 7.5 for different exposure conditions and indirectly provides a limit on crack width.

Once Z has been selected, the most effective way to increase f_{sa} is to use more bars. Thus, Eq. 7.50 encourages using several smaller bars at moderate spacing rather than a few larger bars of equivalent area. This procedure will distribute the reinforcement over the region of maximum concrete tension and provide good crack control. New provisions are expected for crack control in box culverts in the 1996 AASHTO Interim Specifications.

To guard against excessive spacing of bars when flanges of *T*-beams and box girders are in tension, the flexural tension reinforcement is to be distrib-

TABLE 7.5 Crack Width Parameter Z^a

Exposure Condition	Z (N/mm)	Crack Width (mm)
Moderate	30 000	0.41
Severe	23 000	0.30
Buried structures	17 000	0.23

^{a}In AASHTO [A5.7.3.4]. [From *AASHTO LRFD Bridge Design Specifications,* Copyright © 1994 by the American Association of State Highway and Transportation Officials, Washington, DC. Used by permission.]

uted over the lesser of the effective flange width or a width equal to one-tenth of the span. If the effective flange width exceeds one-tenth the span, additional longitudinal reinforcement, with area not less than 0.4% of the excess slab area, is to be provided in the outer portions of the flange.

For relatively deep flexural members, reinforcement should also be distributed in the vertical faces in the tension region to control cracking in the web. If the web depth exceeds 900 mm, longitudinal skin reinforcement is to be uniformly distributed over a height of $d/2$ nearest the tensile reinforcement. The area of skin reinforcement A_{sk} in square millimeters per millimeter (mm²/mm) of height required on each side face is

$$A_{sk} \geq 0.001(d_e - 760) \leq \frac{A_s + A_{ps}}{1200} \qquad (7.51)$$

where d_e is the effective depth from the extreme compression fiber to the centroid of the tensile reinforcement, A_s is the area of the nonprestressed steel, and A_{ps} is the area of the prestressing tendons. The maximum spacing of the skin reinforcement is not to exceed either $d/6$ or 300 mm.

Deformations. Service load deformations may cause deterioration of wearing surfaces and local cracking in concrete slabs. Vertical deflections and vibrations due to moving vehicle loads can cause motorists concern. To limit these effects, optional deflection criteria are suggested [A2.5.2.6.2] as

- Vehicular load, general . . . SPAN/800.
- Vehicular load on cantilever arms . . . SPAN/300.

where the vehicle load includes the impact factor *IM* and the multiple presence factor *m*.

When calculating the vehicular deflection, it should be taken as the larger of that resulting from the design truck alone or that resulting from 25% of the design truck taken together with the design lane load. All of the design lanes should be loaded and all of the girders are assumed to share equally in supporting the load. This statement is equivalent to a deflection distribution factor g equal to the number of lanes divided by the number of girders.

Calculated deflections of bridges have been difficult to verify in the field because of additional stiffness provided by railings, sidewalks, and median barriers not usually accounted for in the calculations. Therefore, it seems reasonable to estimate the instantaneous deflection using the elastic modulus for concrete E_c from Eq. 7.2 and the gross moment of inertia I_g [A5.7.3.6.2]. This estimate is much simpler, and probably just as reliable, as using the effective moment of inertia I_e based on a value between I_g and the cracked moment of inertia I_{cr}. It also makes the calculation of the long-term deflection more tractable because it can be taken as simply 4.0 times the instantaneous deflection [A5.7.3.6.2].

Stress Limitations for Concrete. Service limit states still apply in the design of reinforced concrete members that have prestressing tendons, which precompress the section so that concrete stresses f_c can be determined from elastic uncracked section properties and the familiar equation

$$ f_c = -\frac{P}{A_g} \pm \frac{Pey}{I_g} \mp \frac{My}{I_g} \tag{7.52} $$

where P is the prestressing force, A_g is the cross-sectional area, e is the eccentricity of the prestressing force, M is the moment due to applied loads, y is the distance from the centroid of the section to the fiber, and I_g is the moment of inertia of the section. If the member is a composite construction, it is necessary to separate the moment M into the moment due to loads on the girder M_g and the moment due to loads on the composite section M_c, because the y and I values are different, that is,

$$ f_c = -\frac{P}{A_g} \pm \frac{Pey}{I_g} \mp \frac{M_g y}{I_g} \mp \frac{M_c y_c}{I_c} \tag{7.53} $$

where the plus and minus signs for the stresses at the top and bottom fibers must be consistent with the chosen sign convention where tension is positive. These linear elastic concrete stress distributions are shown in Figure 7.20.

Limits on the concrete stresses are given in Tables 7.6 and 7.7 for two load stages: (1) prestress transfer stage—immediately after transfer of the prestressing tendon tensile force to the concrete but prior to the time-dependent losses due to creep and shrinkage, and (2) service load stage—after allowance for all prestress losses. The concrete compressive strength at time of initial loading f'_{ci}, the 28-day concrete compressive strength f'_c, and the resulting stress limits are all in megapascals. A precompressed tensile zone is a region that was compressed by the prestressing tendons but has gone into tension

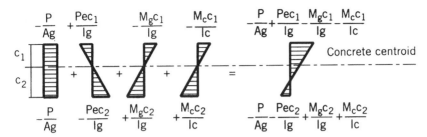

Fig. 7.20   Linear-elastic concrete stress distributions in composite prestressed beams.

when subjected to dead load and live load moments. The stress limits in the tables are for members with prestressed reinforcement only and do not include those for segmentally constructed bridges.

For the components that include both prestressed and nonprestressed reinforcement (often called partially prestressed because only a part of the reinforcement is prestressed), the compressive stress limits are those given in Tables 7.6 and 7.7, but because cracking is permitted, the tensile stress limit is given by Eq. 7.50, where f_{sa} is to be interpreted as the change in stress after decompression.

Stress Limitations for Prestressing Tendons.   The tendon stress, due to prestress operations or at service limit states, shall not exceed the values as specified by AASHTO in Table 7.8 or as recommended by the manufacturer of the tendons and anchorages. The tensile strength f_{pu} and yield strength f_{py} for prestressing stand and bar can be taken from Table 7.4.

TABLE 7.6   Stress Limits for Concrete at Prestress Transfer Stage[a]

Compressive stresses	
Pretensioned components	$0.60 f'_{ci}$
Post-tensioned components	$0.55 f'_{ci}$
Tensile stresses	
Precompressed tensile zone without bonded reinforcement	N/A[b]
Other tensile zones without bonded reinforcement	$0.25 \sqrt{f'_{ci}} \leq 1.38$
Tensile zones with bonded reinforcement sufficient to resist 120% of the tension force in the cracked concrete computed on the basis of an uncracked section	$0.58 \sqrt{f'_{ci}}$
Handling stresses in prestressed piles	$0.415 \sqrt{f'_{ci}}$

[a]In AASHTO [A5.9.4.1]. [From *AASHTO LRFD Bridge Design Specifications,* Copyright © 1994 by the American Association of Highway and Transportation Officials, Washington, DC. Used by permission.]
[b]Not applicable = N/A

TABLE 7.7 Stress Limits for Concrete at Service Load Stage[a]

Compressive stress—load combination Service I	
Due to permanent loads	$0.45 f'_c$
Due to permanent and transient loads and during shippng and handling	$0.60 f'_c$
Tensile Stresses—load combination Service III for components with bonded prestressing tendons.	
Precompressed tensile zone assuming uncracked sections.	
Components with bonded prestressing tendons other than piles	$0.50 \sqrt{f'_c}$
Components subjected to severe corrosive conditions	$0.25 \sqrt{f'_c}$
Components with unbonded prestressing tendons	No tension
Other tensile zone stresses are limited by those given for the prestress transfer stage in Table 7.6.	

[a]In AASHTO [A5.9.4.2]. [From *AASHTO LRFD Bridge Design Specifications,* Copyright © 1994 by the American Association of State Highway and Transportation Officials, Washington, DC. Used by permission.]

TABLE 7.8 Stress Limits for Prestressing Tendons[a]

	Tendon Type		
	Stress Relieved Strand and Plain High-Strength Bars	Low Relaxation Strand	Deformed High-Strength Bars
At jacking: (f_{pi})			
Pretensioning	$0.72 f_{pu}$	$0.78 f_{pu}$	
Post-tensioning	$0.76 f_{pu}$	$0.80 f_{pu}$	$0.75 f_{pu}$
After transfer (f_{pt})			
Pretensioning	$0.70 f_{pu}$	$0.74 f_{pu}$	
Post-tensioning			
At anchorages and			
couplers immediately	$0.70 f_{pu}$	$0.70 f_{pu}$	$0.66 f_{pu}$
after anchor set	$0.70 f_{pu}$	$0.74 f_{pu}$	$0.66 f_{pu}$
Post-tensioning—General			
At service limit state: (f_{pe})			
After losses	$0.80 f_{py}$	$0.80 f_{py}$	$0.80 f_{py}$

[a]In AASHTO Table 5.9.3-1. [From *AASHTO LRFD Bridge Design Specification,* Copyright © 1994 by the American Association of State Highway and Transportation Officials, Washington, DC. Used by permission.]

7.6.2 Fatigue Limit State

Fatigue is a characteristic of a material in which damage accumulates under repeated loadings so that failure occurs at a stress level less than the static strength. In the case of highway bridges, the repeated loadings that cause fatigue are the trucks that pass over them. An indicator of the fatigue damage potential is the stress range f_f of the fluctuating stresses produced by the moving trucks. A second indicator is the number of times the stress range is repeated during the expected life of the bridge. In general, the higher the ratio of the stress range to the static strength, the fewer the number of loading cycles required to cause fatigue failure.

In calculating the fatigue stress range f_f, the fatigue loading described in Chapter 4 is used. This loading consists of a special fatigue truck with constant axle spacing of 9000 mm to the rear axle of the trailer, applied to one lane of traffic without multiple presence, and with an impact factor IM of 15%. The Fatigue Load Combination of Table 3.1 has only a load factor of 0.75 applied to the fatigue truck, all the other load factors are zero. Elastic-cracked section properties are used to calculate f_f, except gross section properties can be used for members with prestress where the sum of the stresses due to unfactored permanent loads and prestress plus 1.5 times the unfactored fatigue load does not exceed a tensile stress of $0.095\sqrt{f_c'}$.

For reinforced concrete components, AASHTO [A5.5.3] does not include the number of cycles of repeated loading as a parameter in determining the fatigue strength. What is implied is that the values given for the limits on the stress range are low enough so that they can be considered as representative of infinite fatigue life. Background on the development of fatigue stress limits for concrete, reinforcing bars, and prestressing strands can be found in the report by ACI Committee 215 (1992), which summarizes over 100 references on analytical, experimental, and statistical studies of fatigue in reinforced concrete. In their report, which serves as the basis for the discussion that follows, the fatigue stress limits appear to have been developed for 2–10 million cycles.

Fatigue of Plain Concrete. When plain concrete beams are subjected to repetitive stresses that are less than the static strength, accumulated damage due to progressive internal microcracking eventually results in a fatigue failure. If the repetitive stress level is decreased, the number of cycles to failure N increases. This effect is shown by the S–N curves in Figure 7.21, where the ordinate is the ratio of the maximum stress S_{max} to the static strength and the abscissa is the number of cycles to failure N, plotted on a logarithmic scale. For the case of plain concrete beams, S_{max} is the tensile stress calculated at the extreme fiber assuming an uncracked section and the static strength is the rupture modulus stress f_r.

The curves a and c in Figure 7.21 were obtained from tests in which the stress range between a maximum stress and a minimum stress equal to 75

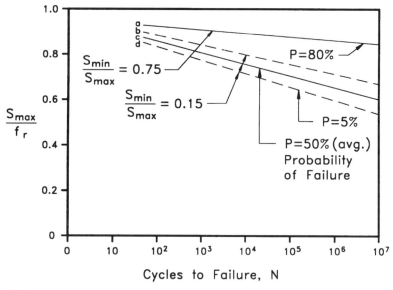

Fig. 7.21 Fatigue strength of plain concrete beams. [ACI Committee 215, 1992.] (Used with permission of American Concrete Institute).

and 15% of the maximum stress, respectively. It can be observed that an increase of the stress range results in a decreased fatigue strength for a given number of cycles. Curves *b* and *d* indicate the amount of scatter in the test data. Curve *b* corresponds to an 80% chance of failure while curve *d* represents a 5% chance of failure. Curves *a* and *c* are averages representing 50% probability of failure.

The *S–N* curves for concrete in Figure 7.21 are nearly linear from 100 cycles to 10 million cycles and have not flattened out at the higher number of cycles to failure. It appears that concrete does not exhibit a limiting value of stress below which the fatigue life is infinite. Thus, any statement on the fatigue strength of concrete must be given with reference to the number of cycles to failure. ACI Committee 215 (1992) concludes that the fatigue strength of concrete for the life of 10 million cycles of load and a probability of failure of 50%, regardless of whether the specimen is loaded in compression, tension, or flexure, is approximately 55% of the static strength.

In AASHTO [A5.5.3.1], the limiting tensile stress in flexure before the section is considered as cracked is $0.25\sqrt{f'_c}$, which is 40% of the static strength $f_r = 0.63\sqrt{f'_c}$. Further, because the stress range is typically the difference between a minimum stress due to permanent load and a maximum stress due to permanent load plus the transitory fatigue load, the limits on the compressive stress in Table 7.7 should keep the stress range within $0.40f'_c$. Both of these limitations are comparable to the recommendations of ACI Committee 215 (1992) for the fatigue strength of concrete.

Fatigue of Reinforcing Bars. Observations of deformed reinforcing bars subjected to repeated loads indicate that fatigue cracks start at the base of a transverse deformation where a stress concentration exists. With repeated load cycles, the crack grows until the cross-sectional area is reduced and the bar fails in tension. The higher the stress range S_r of the repeated load, the fewer the number of cycles N before the reinforcing bar fails.

Results of experimental tests on straight deformed reinforcing bars are shown by the S_r–N curves in Figure 7.22. These curves were generated by bars whose size varied from No. 15 to No. 35. The curves begin to flatten out at about 1 million cycles, indicating that reinforcing bars may have a stress endurance limit below which the fatigue life will be infinite.

The stress range S_r is the difference between the maximum stress S_{max} and the minimum stress S_{min} of the repeating load cycles. The higher the minimum stress level, the higher the average tensile stress in the reinforcing bar and the lower the fatigue strength.

The stress concentrations produced at the base of a deformation or at the intersection of deformations can also be produced by bending and welding of the reinforcing bars. Investigations reported by ACI Committee 215 (1992) indicate the fatigue strength of bars bent through an angle of 45° to be about 50% that of straight bars and the fatigue strength of bars with stirrups attached by tack welding to be about 67% that of bars with stirrups attached by tie wires.

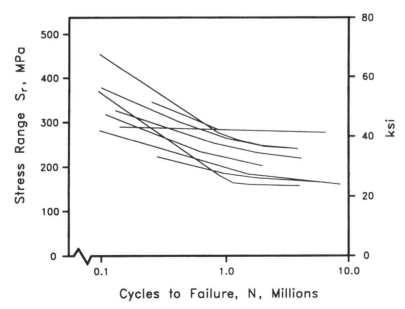

Fig. 7.22 Stress range versus fatigue life for reinforcing bars. [ACI Committee 215, 1992.] (Used with permission of American Concrete Institute).

In AASHTO [A5.5.3.2], minimum stress f_{min} and deformation geometry are considered in setting a limit on the fatigue stress range f_f for straight deformed reinforcing bars, that is,

$$f_f = 145 - 0.33f_{min} + 55 \left(\frac{r}{h}\right) \tag{7.54}$$

where r/h is the ratio of base radius to height of rolled-on transverse deformations; if the actual value is not known, 0.3 may be used. All of the values are in units of megapascals.

In the case of a single span girder, the minimum stress f_{min} produced by the fatigue truck is zero. Using the default value for r/h of 0.3, the permissible fatigue stress range f_f is 162 MPa. This value compares well with a lower bound to the curves in Figure 7.22 for 1–10 million cycles to failure.

As recommended by ACI Committee 215 (1992), Eq. 7.54 should be reduced by 50% for bent bars or bars to which auxiliary reinforcement has been tack welded. As a practical matter, primary reinforcement should not be bent in regions of high stress range and tack welding should be prohibited.

Fatigue of Prestressing Tendons. If the precompression due to prestressing is sufficient so that the concrete cross section remains uncracked and never sees tensile stresses, fatigue of prestressing tendons is seldom a problem. However, designs are allowed that result in cracked sections under service loads (see Table 7.7) and it becomes necessary to consider fatigue. The AASHTO [A5.5.3.1] states that fatigue must be considered when the fatigue truck with a load factor of 1.5 produces tension in the precompressed cross section.

Fatigue tests have been conducted on individual prestressing wires and on seven-wire strand, which are well documented in the literature cited by ACI Committee 215 (1992). However, the critical component that determines the fatigue strength of prestressing tendons is their anchorage. Even though the anchorages can develop the static strength of prestressing tendons, they develop less than 70% of the fatigue strength. Bending at an anchorage can cause high local stresses not seen by a direct tensile pull of a prestressing tendon.

The S_r–N curves shown in Figure 7.23 are for proprietary anchorages for strand and multiple wire tendons. Similar curves are also given by ACI Committee 215 (1992) for anchorages of bars. From Figure 7.23, an endurance limit for the anchorages occurs at about 2 million cycles to failure (arrows indicate specimens for which failure did not occur). A lower bound for the stress range is about $0.07 f_{pu}$, which for $f_{pu} = 1860$ MPa translates to $S_r = 130$ MPa.

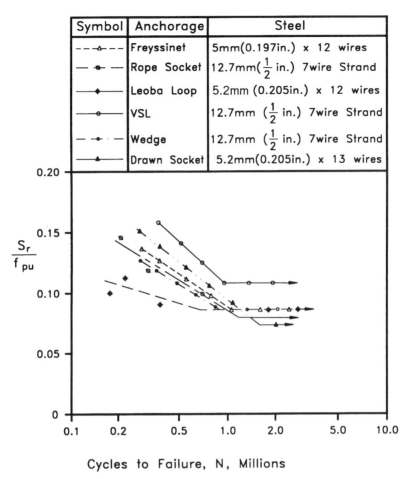

Symbol	Anchorage	Steel
– – –△– – –	Freyssinet	5mm(0.197in.) x 12 wires
— –●– —	Rope Socket	12.7mm($\frac{1}{2}$ in.) 7wire Strand
—––◆–––—	Leoba Loop	5.2mm (0.205in.) x 12 wires
———○———	VSL	12.7mm ($\frac{1}{2}$ in.) 7wire Strand
— –●– · —	Wedge	12.7mm ($\frac{1}{2}$ in.) 7wire Strand
———▲———	Drawn Socket	5.2mm(0.205in.) x 13 wires

$\frac{S_r}{f_{pu}}$

Cycles to Failure, N, Millions

Fig. 7.23 Stress range versus fatigue life for strand and multiple wire anchorages [ACI Committee 215, 1992.] (Used with permission of American Concrete Institute).

Bending of the prestressing tendons also occurs when it is held down at discrete points throughout its length. Fatigue failures can initiate when neighboring wires rub together or against plastic and metal ducts. This fretting fatigue can occur in both bonded and unbonded post-tensioning systems.

The limiting fatigue stress range given for prestressing tendons [A5.5.3.3] varies with the radius of curvature of the tendon as shown in Table 7.9 The sharper the curvature, the lower the fatigue strength (stress range). There is no distinction between bonded and unbonded tendons. This lack of distinction differs from ACI Committee 215 (1992), which considers the anchorages of unbonded tendons to be more vulnerable to fatigue and recommends a reduced stress range comparable to the 70 MPa for the sharper curvature.

TABLE 7.9   Fatigue Stress Range Limit for Prestressing Tendons[a,b]

Radius of Curvature (mm)	Stress Range Limit (MPa)
> 9000	125
< 3600	70

[a]A linear interpolation may be used for radii between 3600 and 9000 mm.
[b]In AASHTO [A5.5.3.3]. [From *AASHTO LRFD Bridge Design Specifications,* Copyright © 1994 by the American Association of State Highway and Transportation Officials, Washington, DC. Used by permission.]

7.6.3   Strength Limit State

A strength limit state is one that is governed by the static strength of the materials in a given cross section. There are five different strength load combinations specified in Table 3.1. For a particular component of a bridge structure, only one or maybe two of these load combinations will need to be investigated. The differences in the strength load combinations are associated mainly with the load factors applied to the live load. A smaller live load factor is used for a permit vehicle and in the presence of wind, which seems logical. The load combination that produces the maximum load effect is then compared to the strength or resistance provided by the cross section of a member.

In calculating the resistance to a particular factored load effect such as axial load, bending, shear, or torsion, the uncertainties are represented by an understrength or resistance factor φ. The φ factor is multiplied times the calculated nominal resistance R_n and the adequacy of the design is then determined by whether or not the inequality expressed by Eq. 7.49 is satisfied.

In the case of a reinforced concrete member there are uncertainties in the quality of the materials, cross-sectional dimensions, placement of reinforcement, and equations used to calculate the resistance.

Some modes of failure can be predicted with greater accuracy than others and the consequence of their occurrence is less costly. For example, beams in flexure are usually designed as underreinforced so that failure is precipitated by gradual yielding of the tensile reinforcement while columns in compression may fail suddenly without warning. A shear failure mode is less understood and is a combination of a tension and compression failure mode. Therefore, its φ factor should be somewhere between that of a beam in flexure and a column in compression. The consequence of a column failure is more serious than that of a beam because when a column fails it will bring down a number of beams; therefore, its margin of safety should be greater. All of these considerations, and others, are reflected in the resistance factors specified [A5.5.4.2] and presented in Table 7.10.

For the case of combined flexure and compression, the compression ϕ factor can be increased linearly from 0.75 at small axial load to the ϕ factor for pure flexure at an axial load of zero. A small axial load is defined as 0.10 $f'_c A_g$, where f'_c is the 28-day compressive strength of the concrete and A_g is the gross cross-sectional area of the compression member.

For beams with or without tension that are a mixture of nonprestressed and prestressed reinforcement, the ϕ factor depends on the partial prestressing ratio *PPR* and is given by

$$\phi = 0.90 + 0.10 PPR \tag{7.55}$$

in which

$$PPR = \frac{A_{ps} f_{py}}{A_{ps} f_{py} + A_s f_y} \tag{7.56}$$

where A_{ps} is the area of prestressing steel, f_{py} is the yield strength of prestressing steel, A_s is area of nonprestressed tensile reinforcement, and f_y is the yield strength of the reinforcing bars.

TABLE 7.10 Resistance Factors for Conventional Construction[a]

Strength Limit State	ϕ Factor
For flexure and tension	
Reinforced concrete	0.90
Prestressed concrete	1.00
For shear and torsion	
Normal weight concrete	0.90
Lightweight concrete	0.70
For axial compression with spirals or ties, except for	0.75
Seismic Zones 3 and 4	
For bearing on concrete	0.70
For compression in strut-and-tie models	0.70
For compression in anchorage zones	
Normal weight concrete	0.80
Lightweight concrete	0.65
For tension in steel in anchorage zones	1.00

[a]In AASHTO [A5.5.4.2]. [From *AASHTO LRFD Bridge Design Specifications*, Copyright © 1994 by the American Association of State Highway and Transportation Officials, Washington, DC. Used by permission.]

7.6.4 Extreme Event Limit State

Extreme event limit states are unique occurrences with return periods in excess of the design life of the bridge. Earthquakes, ice loads, vehicle collisions, and vessel collisions are considered to be extreme events and are to be used one at a time. However, these events may be combined with a major flood (recurrence interval >100 years but <500 years), or with the effects of scour of a major flood, as shown in Table 3.1. For example, it is possible that scour from a major flood may have reduced support for foundation components when the design earthquake occurs, or that ice flows are colliding with a bridge during a major flood.

The resistance factors ϕ for an extreme event limit state are to be taken as unity. This choice of ϕ may result in excessive distress and structural damage, but the bridge structure should survive and not collapse.

7.7 FLEXURAL STRENGTH OF REINFORCED CONCRETE MEMBERS

The AASHTO (1994) Bridge Specifications present unified design provisions that apply to concrete members reinforced with a combination of conventional steel bars and prestressing strands. Such members are often called partially prestressed. The expressions developed are also applicable to conventional reinforced and prestressed concrete when one reinforcement or the other is not present. Background for the development of these provisions can be found in Loov (1988) and Naaman (1992).

7.7.1 Depth to Neutral Axis for Beams with Bonded Tendons

Consider the flanged cross section of a reinforced concrete beam shown in Figure 7.24 and the accompanying linear strain diagram. For bonded tendons, the compatibility condition gives the strain in the surrounding concrete as

$$\varepsilon_{cp} = -\varepsilon_{cu}\frac{d_p - c}{c} = -\varepsilon_{cu}\left(\frac{d_p}{c} - 1\right) \tag{7.57}$$

where ε_{cu} is the limiting strain at the extreme compression fiber, d_p is the distance from the extreme compression fiber to the centroid of the prestressing tendons, and c is the distance from the extreme compression fiber to the neutral axis. Again, tensile strains are considered positive and compressive strains are negative.

From Eq. 7.44, the strain in the prestressing tendon becomes

$$\varepsilon_{ps} = -\varepsilon_{cu}\left(\frac{d_p}{c} - 1\right) + \Delta\varepsilon_{pe} \tag{7.58}$$

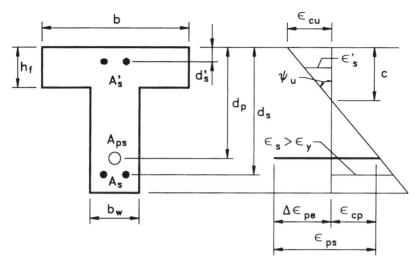

Fig. 7.24 Strains in a reinforced concrete beam. [Reproduced with permission from R. E. Loov (1988). "A General Equation for the Steel for Bonded Prestressed Concrete Members," *PCI Journal,* Vol. 33, No. 6, Nov.–Dec., pp. 108–137.]

where $\Delta\varepsilon_{pe}$ is approximately equal to f_{pe}/E_p and remains essentially constant throughout the life of the member (Collins and Mitchell, 1991). At the strength limit state, AASHTO [A5.7.2.1] defines $\varepsilon_{cu} = -0.003$ if the concrete is unconfined. For confined concrete, ε_{cu} can be an order of magnitude greater than for unconfined concrete (Mander et al., 1988). With both $\Delta\varepsilon_{pe}$ and ε_{cu} being constants depending on the prestressing operation and the lateral confining pressure, respectively, the strain in the prestressing tendon ε_{ps} and the corresponding stress f_{ps} is a function only of the ratio c/d_p.

Equilibrium of the forces in Figure 7.25 can be used to determine the depth of the neutral axis c. However, this requires the determination of f_{ps} that is a function of the ratio c/d_p. Such an equation has been presented by Loov (1988), endorsed by Naaman (1992), and adopted by AASHTO [A5.7.3.1.1] as

$$f_{ps} = f_{pu}\left(1 - k\frac{c}{d_p}\right) \tag{7.59}$$

$$k = 2\left(1.04 - \frac{f_{py}}{f_{pu}}\right) \tag{7.60}$$

For low-relaxation strands with $f_{pu} = 1860$ MPa, Table 7.4 gives $f_{py}/f_{pu} = 0.90$, which results in $k = 0.28$. By using $E_p = 197$ GPa, neglecting ε_{ce}, and assuming that $\varepsilon_{cu} = -0.003$ and $f_{pe} = 0.56f_{pu}$, Eqs. 7.58 and 7.59 become

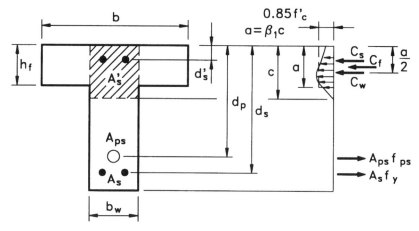

Fig. 7.25   Forces in a reinforced concrete beam.

$$\varepsilon_{ps} = 0.003\frac{d_p}{c} + 0.0023 \qquad (7.58a)$$

$$f_{ps} = 1860\left(1 - 0.28\frac{c}{d_p}\right) \qquad (7.59a)$$

Substituting values of c/d_p from 0.05 to 0.50 into Eq. 7.58a and 7.59a, the approximate stress–strain curve has been generated and compared to the Ramberg–Osgood curve of Eq. 7.41 in Figure 7.26. Also shown on Figure 7.26 is the 0.2% offset strain often used to determine the yield point of rounded stress–strain curves and its intersection with $f_{py} = 0.9f_{pu}$. The agreement with both curves is very good.

When evaluating the compressive forces in the concrete, it is convenient to use an equivalent rectangular stress block. In AASHTO [A5.7.2.2], the following familiar provisions for the stress block factors have been adopted:

- Uniform concrete compressive stress of $0.85f_c'$.
- Depth of rectangular stress block $a = \beta_1 c$.

Here,

$$\beta_1 = 0.85 \text{ for } f_c' \leq 28 \text{ MPa}$$

$$\beta_1 = 0.65 \text{ for } f_c' \geq 56 \text{ MPa} \qquad (7.61)$$

$$\beta_1 = 0.85 - 0.05\frac{(f_c' - 28)}{7} \text{ for } 28 \text{ MPa} \leq f_c' \leq 56 \text{ MPa}$$

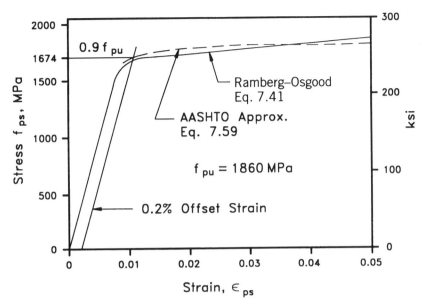

Fig. 7.26 Comparison of stress–strain curves for 1860 MPa low-relaxation. Prestressing strands. [Reproduced with permission from R. E. Loov (1988). "A General Equation for the Steel for Bonded Prestressed Concrete Members," *PCI Journal,* Vol. 33, No. 6, Nov.–Dec., pp. 108–137.]

Note that in Eq. 7.61 and in the derivations that follow, the compressive stresses f_c' and f_y' are taken as their absolute values.

Equilibrium of the forces in the beam of Figure 7.25 requires that the total nominal compressive force equal the total nominal tensile force, that is,

$$C_n = T_n \qquad (7.62)$$

in which

$$C_n = C_w + C_f + C_s \qquad (7.62a)$$

$$T_n = A_{ps}f_{ps} + A_sf_y \qquad (7.62b)$$

where

C_w = concrete compressive force in the web

C_f = concrete compressive force in the flange

C_s = compressive force in the nonprestressed steel

A_{ps} = area of prestressing steel

f_{ps} = average stress in prestressing steel at nominal bending resistance of member as given by Eq. 7.59

A_s = area of nonprestressed tension reinforcement

f_y = specified minimum yield strength of tension reinforcement

The concrete compressive force in the web C_w is over the cross-hatched area in Figure 7.25 of width equal to the web width b_w that extends through the flange to the top fibers. It is equal to

$$C_w = 0.85f'_c a b_w = 0.85\beta_1 f'_c c b_w \tag{7.63a}$$

which can be thought of as an average stress in the concrete of $0.85\beta_1 f'_c$ over the area $c b_w$. Because this average stress is over a portion of the concrete in the flange, to be consistent, the concrete area in the remaining flange should be subject to the same average concrete stress, that is,

$$C_f = 0.85\beta_1 f'_c (b - b_w) h_f \tag{7.63b}$$

as discussed in AASHTO [C5.7.3.2.2], the inclusion of β_1 in C_f allows a smooth transition for the determination of c as the value of the width of the compression flange b approaches b_w. It also allows for the realistic criterion that the beginning of T-section behavior occurs when c, not a, exceeds h_f. The compressive force in the compression steel C_s, assuming that its compressive strain ε'_s in Figure 7.24 is greater than or equal to the yield strain ε'_y, is

$$C_s = A'_s f'_y \tag{7.64}$$

where A'_s is the area of the compression reinforcement and f'_y is the absolute value of specified yield strength of the compression reinforcement. The assumption of yielding of the compression steel can be checked by calculating ε'_s from similar strain triangles in Figure 7.24 and comparing to $\varepsilon'_y = f'_y/E_s$, that is,

$$\varepsilon'_s = \varepsilon_{cu} \frac{c - d'_s}{c} = \varepsilon_{cu}\left(1 - \frac{d'_s}{c}\right) \tag{7.65}$$

where d'_s is the distance from the extreme compression fiber to the centroid of the compression reinforcement.

If we substitute f_{ps} from Eq. 7.59 into Eq. 7.62b, the total tensile force becomes

$$T_n = A_{ps}f_{pu}\left(1 - k\frac{c}{d_p}\right) + A_sf_y \tag{7.66}$$

By substituting the compressive forces from Eqs. 7.63 and 7.64 into Eq. 7.62a, the total compressive force becomes

$$C_n = 0.85\beta_1 f'_c c b_w + 0.85\beta_1 f'_c(b - b_w)h_f + A'_s f'_y \tag{7.67}$$

If we equate the total tensile and compressive forces, and solve for c, we have

$$c = \frac{A_{ps}f_{pu} + A_sf_y - A'_sf'_y - 0.85\beta_1 f'_c(b - b_w)h_f}{0.85\beta_1 f'_c b_w + kA_{ps}f_{pu}/d_p} \geq h_f \tag{7.68}$$

If c is less than h_f, the neutral axis is in the flange and c should be recalculated with $b_w = b$ in Eq. 7.68. This expression for c is completely general and can be used for prestressed beams without reinforcing bars ($A_s = A'_s = 0$) and for reinforced concrete beams without prestressing steel ($A_{ps} = 0$).

Equation 7.68 assumes that the compression reinforcement A'_s has yielded. If it has not yielded, the stress in the compression steel is calculated from $f'_s = \varepsilon'_s E_s$, where ε'_s is determined from Eq. 7.65. This expression for f'_s replaces the value of f'_y in Eq. 7.68 and results in a quadratic equation for determining c. As an alternative, one can simply neglect the contribution of the compression steel when it has not yielded and set $A'_s = 0$ in Eq. 7.68.

7.7.2 Depth to Neutral Axis for Beams with Unbonded Tendons

When the tendons are *not* bonded, strain compatibility with the surrounding concrete cannot be used to determine the strain and, therefore, the stress in the prestressing tendon. Instead the total change in length of the tendon must now equal the total change in length of the concrete over the distance between anchorage points.

This results in a uniform stress in the prestressing tendon between anchorage points that depends on the deformations of the entire structure. To determine the stress in the unbonded tendon at the limit state, Naaman and Alkhaini (1991) developed a prediction equation that is based on analysis and experimental results. Starting with Eq. 7.48, they assume that the average strain of the concrete $\overline{\varepsilon_{cp}}$ can be expressed in terms of the bonded concrete strain ε_{cp} modified by a bond reduction factor Ω_u and that $\Delta\varepsilon_{pe}$ is approximately equal to f_{pe}/E_p. Thus, for an unbonded tendon

$$\varepsilon_{ps} = \Omega_u \varepsilon_{cp} + \frac{f_{pe}}{E_p} \tag{7.69}$$

where Ω_u is a "bond" reduction coefficient at ultimate nominal resistance for an unbonded tendon given as [A5.7.3.1.2]

$$\Omega_u = \frac{3}{L/d_p} \qquad \text{for uniform and near third-point loading}$$

$$\Omega_u = \frac{1.5}{L/d_p} \qquad \text{for near midspan loading}$$

for which L is the span length in the same units as d_p (see Fig. 7.27).

If we substitute Eq. 7.57 for ε_{cp} and assume the unbonded tendon stress f_{ps} at the limit state is in the elastic range, then Eq. 7.69, when multiplied by E_p, yields

(a)

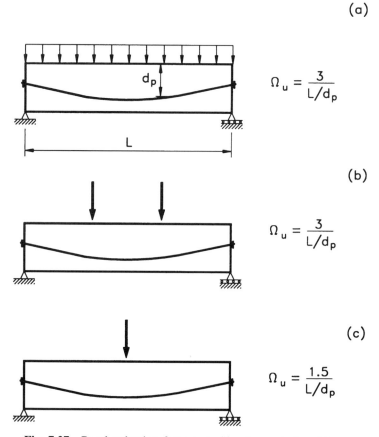

(b)

(c)

Fig. 7.27   Bond reduction factors at ultimate nominal resistance.

$$f_{ps} = f_{pe} - \Omega_u \varepsilon_{cu} E_p \left(\frac{d_p}{c} - 1 \right) \leq f_{py} \tag{7.70}$$

Equation 7.70 was developed for single spans and needs to be modified when the unbonded tendon is continuous over multiple spans of total length L_2 between anchorages when one or more spans of total length L_1 are loaded (see Fig. 7.28). The change in strain due to load in L_1 is averaged over L_2 by the ratio L_1/L_2 to give f_{ps} for an unbonded tendon as

$$f_{ps} = f_{pe} - \Omega_u \varepsilon_{cu} E_p \left(\frac{d_p}{c} - 1 \right) \frac{L_1}{L_2} \leq f_{py} \tag{7.71}$$

The expression in AASHTO [A5.7.3.1.2] is similar but does not have the

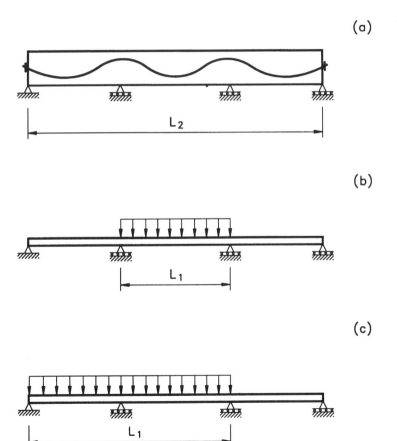

Fig. 7.28 Definition of tendon length and loaded lengths.

minus sign because they use the absolute value for ε_{cu}. Further, the AASHTO equation has a reduced upper limit of $0.94f_{py}$ to provide an increased margin for elastic behavior, that is, for an unbonded tendon

$$f_{ps} = f_{pe} + \Omega_u \varepsilon_{cu}^* E_p \left(\frac{d_p}{c} - 1 \right) \frac{L_1}{L_2} \le 0.94f_{py} \qquad (7.72)$$

where ε_{cu}^* is the absolute value of ε_{cu}.

In the AASHTO Standard Specifications (1996), the second term in Eq. 7.71 is taken simply as a constant value of 100 MPa. This simplification can be used as a first approximation for f_{ps} if an interative method is used to determine c.

Following the same procedure as for the bonded tendon in establishing force equilibrium, the expression for the distance from the extreme compression fiber to the neutral axis for an unbonded tendon is

$$c = \frac{A_{ps}f_{ps} + A_s f_y - A_s' f_y' - 0.85\beta_1 f_c'(b - b_w)h_f}{0.85\beta_1 f_c b_w} \ge h_f \qquad (7.73)$$

where f_{ps} is determined from Eq. 7.72. If c is less than h_f, the neutral axis is in the flange and c should be recalculated with $b_w = b$ in Eq. 7.73. If the strain in the compression reinforcement calculated by Eq. 7.65 is less than the yield strain ε_y', f_y' in Eq. 7.73 should be replaced by f_s' as described previously for Eq. 7.68.

Substituting Eq. 7.72 into Eq. 7.73 results in a quadratic equation for c that has the root

$$c = \frac{-B_1 + \sqrt{B_1^2 - 4A_1 C_1}}{2A_1} \qquad (7.74)$$

where

$A_1 = 0.85\beta_1 f_c' b_w$

$B_1 = A_{ps}(\Omega_u \varepsilon_{cu}^* E_p L_1/L_2 - f_{pe}) + A_s' f_y' - A_s f_y + 0.85\beta_1 f_c'(b - b_w)h_f$

$C_1 = -A_{ps}\Omega_u \varepsilon_{cu}^* E_p d_p L_1/L_2$

Alternatively, an iterative method can be used starting with a first trial value for the unbonded tendon stress of

$$f_{ps} = f_{pe} + 100 \text{ MPa} \qquad (7.75)$$

in Eq. 7.73. With c known, f_{ps} is calculated from Eq. 7.72, compared with the previous trial, and a new value chosen. These steps are repeated until convergence within an acceptable tolerance is attained.

7.7.3 Nominal Flexural Strength

With c and f_{ps} known for either bonded or unbonded tendons, it is a simple matter to determine the nominal flexural strength M_n for a reinforced concrete beam section. If we refer to Figure 7.25 and balance the moments about C_w we get

$$M_n = A_{ps}f_{ps}\left(d_p - \frac{a}{2}\right) + A_s f_y\left(d_s - \frac{a}{2}\right) + C_s\left(\frac{a}{2} - d_s'\right) + C_f\left(\frac{a}{2} - \frac{h_f}{2}\right)$$

where $a = \beta_1 c$ and c is not less than the compression flange thickness h_f. Substitution of Eqs. 7.63b and 7.64 for C_f and C_s results in

$$M_n = A_{ps}f_{ps}\left(d_p - \frac{a}{2}\right) + A_s f_y\left(d_s - \frac{a}{2}\right) + A_s' f_y'\left(\frac{a}{2} - d_s'\right)$$
$$+ 0.85\beta_1 f_c'(b - b_w)h_f\left(\frac{a}{2} - \frac{h_f}{2}\right) \qquad (7.76)$$

If the depth to the neutral axis from the extreme compression fiber c is less than the compression flange thickness h_f, or if the beam has no compression flange, the nominal flexural strength M_n for the beam section is calculated from Eq. 7.76 with b_w set equal to b.

EXAMPLE 7.7

For the beam cross section in Figure 7.29, determined the distance from the extreme compression fiber to the netural axis c, the average stress in the prestressing steel f_{ps}, and the nominal moment strength M_n for (a) bonded tendons and (b) unbonded tendons. Use normal weight concrete with $f_c' = 40$ MPa, Grade 400 mild steel reinforcement, and 12.70 mm, 1860-MPa low-relaxation prestressing tendons. The beam is uniformly loaded with a single span length of 10.67 m.

1. **Material Properties.**

$$\beta_1 = 0.85 - 0.05\left(\frac{f_c' - 28}{7}\right) = 0.85 - 0.05\left(\frac{40 - 28}{7}\right) = 0.76$$
$$E_c = 4800\sqrt{f_c'} = 4800\sqrt{40} = 30 \text{ GPa}$$

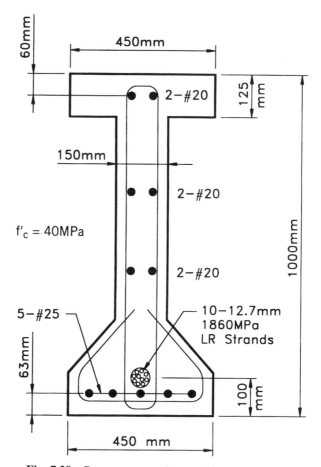

Fig. 7.29   Beam cross-section used in Example 7.7.

$$\varepsilon_{cu} = -0.003$$

$$f_y = |f_y'| = 400 \text{ MPa}$$

$$E_s = 200 \text{ GPa}$$

$$\varepsilon_y = |\varepsilon_y'| = \frac{f_y}{E_s} = \frac{400}{200\,000} = 0.002$$

$$f_{py} = 0.9 f_{pu} = 0.9(1860) = 1674 \text{ MPa}$$

$$k = 2\left(1.04 - \frac{f_{py}}{f_{pu}}\right) = 2(1.04 - 0.9) = 0.28$$

$$\text{Assume } f_{pe} = 1030 \text{ MPa} > 0.5 f_{pu} = 930 \text{ MPa}$$

$$E_p = 197 \text{ GPa}$$

2. **Section Properties.**

$$b = 450 \text{ mm}, \ b_w = 150 \text{ mm}, \ h = 1000 \text{ mm}, \ h_f = 125 \text{ mm}$$

$$d'_s = 60 \text{ mm}, \ d_s = h - 63 = 937 \text{ mm}$$

$$d_p = h - 100 = 900 \text{ mm}$$

$$A_s = 2500 \text{ mm}^2, \ A'_s = 600 \text{ mm}^2$$

$$A_{ps} = 10(98.71) = 987 \text{ mm}^2$$

3. **Depth to Neutral Axis and Stress in Prestressing Steel.**

Bonded Case. From Eq. 7.68

$$c = \frac{A_{ps}f_{pu} + A_s f_y - A'_s|f'_y| - 0.85\beta_1 f'_c(b - b_w)h_f}{0.85\beta_1 f'_c b_w + (kA_{ps}f_{pu}/d_p)}$$

$$c = \frac{987(1860) + 2500(400) - 600(400) - 0.85(0.76)(40)(450 - 150)(125)}{0.85(0.76)(40)(150) + \dfrac{0.28(987)(1860)}{900}}$$

$c = \underline{366 \text{ mm}} > h_f = 125 \text{ mm}$, neutral axis in web

From Eq. 7.65

$$\varepsilon'_s = \varepsilon_{cu}\left(1 - \frac{d'_s}{c}\right) = -0.003\left(1 - \frac{60}{366}\right) = -0.002\,51$$

$|\varepsilon'_s| = 0.002\,51 > |\varepsilon'_y| = 0.002$ ∴ compression steel has yielded

From Eq. 7.59

$$f_{ps} = f_{pu}\left(1 - k\frac{c}{d_p}\right)$$

$$f_{ps} = 1860\left(1 - 0.28\frac{366}{900}\right) = \underline{1648 \text{ MPa}}$$

Unbonded Case

$$L = L_1 = L_2 = 10.67 \text{ m}, \ \Omega_u = 3d_p/L = 3(900)/10\,670 = 0.253$$

From Eq. 7.74

$$c = \frac{-B_1 + \sqrt{B_1^2 - 4A_1 C_1}}{2A_1}$$

$$A_1 = 0.85\beta_1 f_c' b_w = 0.85(0.76)(40)(150) = 3876 \text{ N}$$

$$B_1 = A_{ps}(\Omega_u|\varepsilon_{cu}|E_p L_1/L_2 - f_{pe}) + A_s'|f_y'|$$
$$- A_s f_y + 0.85\beta_1 f_c'(b - b_w)h_f$$

$$B_1 = 987[0.253(0.003)(197\,000)(1.0) - 1030] + 600(400)$$
$$- 2500(400)$$
$$+ 0.85(0.76)(40)(450 - 150)125 = -660\,030 \text{ N}$$

$$C_1 = -A_{ps}\Omega_u|\varepsilon_{cu}|E_p d_p \frac{L_1}{L_2} =$$
$$-(987)(0.253)(0.003)(197\,000)(900)(1.0)$$
$$C_1 = -132\,820\,000 \text{ N}$$

$$c = \frac{660\,030 + \sqrt{(660\,030)^2 - 4(3876)(-132.82 \times 10^6)}}{2(3876)}$$
$$= \underline{289 \text{ mm}} > h_f = 125 \text{ mm, neutral axis in web}$$

From Eq. 7.65

$$\varepsilon_s' = \varepsilon_{cu}\left(1 - \frac{d_s'}{c}\right) = -0.003\left(1 - \frac{60}{289}\right) = -0.002\,38$$

$$|\varepsilon_s'| = 0.002\,38 > |\varepsilon_y'| = 0.002 \qquad \therefore \text{ compression steel has yielded.}$$

From Eq. 7.72

$$f_{ps} = f_{pe} + \Omega_u|\varepsilon_{cu}|E_p\left(\frac{d_p}{c} - 1\right)\frac{L_1}{L_2} \leq 0.94 f_{py}$$

$$f_{ps} = 1030 + 0.253(0.003)(197\,000)\left(\frac{900}{289} - 1\right)(1.0)$$

$$f_{ps} = \underline{1346 \text{ MPa}} < 0.94(1674)$$
$$= 1574 \text{ MPa, stress below upper bound}$$

4. **Nominal Flexural Strength.**

Bonded Case.

$$a = \beta_1 c = 0.76(366) = 278 \text{ mm}$$

From Eq. 7.76

$$M_n = A_{ps}f_{ps}\left(d_p - \frac{a}{2}\right) + A_s f_y\left(d_s - \frac{a}{2}\right) + A_s'|f_y'|\left(\frac{a}{2} - d_s'\right)$$

$$+ \ 0.85\beta_1 f_c'(b - b_w)h_f\left(\frac{a}{2} - \frac{h_f}{2}\right)$$

$$M_n = 987(1648)\left(900 - \frac{278}{2}\right) + 2500(400)\left(937 - \frac{278}{2}\right)$$

$$+ \ 600(400)\left(\frac{278}{2} - 60\right)$$

$$+ \ 0.85(0.76)(40)(450 - 150)125\left(\frac{278}{2} - \frac{125}{2}\right)$$

$$M_n = 2129 \times 10^6 \ \text{Nmm} = \underline{2129 \ \text{kNm}}$$

Unbonded Case.

$$a = \beta_1 c = 0.76(289) = 220 \ \text{mm}$$

From Eq. 7.76

$$M_n = 987(1346)\left(900 - \frac{220}{2}\right) + 2500(400)\left(937 - \frac{220}{2}\right)$$

$$+ \ 600(400)\left(\frac{220}{2} - 60\right)$$

$$+ \ 0.85(0.76)(40)(450 - 150)125\left(\frac{220}{2} - \frac{125}{2}\right)$$

$$M_n = 1935 \times 10^6 \ \text{Nmm} = \underline{1935 \ \text{kNm}}$$

For the unbonded case, with the same reinforcement as the bonded case, the nominal flexural strength is less than that of the bonded case.

7.7.4 Ductility and Maximum Tensile Reinforcement

Ductility in reinforced concrete beams is an important factor in their design because it allows large deflections and rotations to occur without collapse of the beam. Ductility also allows redistribution of load and bending moments in multibeam deck systems and in continuous beams. It is also important in seismic design for dissipation of energy under hysteretic loadings.

A ductility index u, defined as the ratio of the limit state curvature ψ_u to the yield curvature ψ_y,

$$u = \frac{\psi_u}{\psi_y} \tag{7.77}$$

has been used as a measure of the amount of ductility available in a beam. An idealized bilinear moment-curvature relationship for a reinforced concrete beam is shown in Figure 7.30, where the elastic and plastic flexural stiffnesses K_e and K_p can be determined from the two points (ψ_y, M_y) and (ψ_u, M_u). At the flexural limit state, the curvature ψ_u can be determined from the strain diagram in Figure 7.24 as

$$\psi_u = \frac{\varepsilon_{cu}}{c} \tag{7.78}$$

where ε_{cu} is the limit strain at the extreme compression fiber and c is the distance from the extreme compression fiber to the neutral axis. The yield curvature ψ_y is determined by dividing the yield moment M_y, often expressed as a fraction of M_u, by the flexural stiffness EI for the transformed elastic-cracked section. In design, a beam is considered to have sufficient ductility if the value of the ductility index u is not less than a specified value. The larger the ductility index, the greater the available curvature capacity, and the larger the deformations in the member before collapse occurs.

A better measure of ductility, as explained by Skogman et al. (1988), is the rotational capacity of the member developed at a plastic hinge. A simply supported beam with a single concentrated load at midspan is shown in Figure 7.31. The moment diagram is a triangle and the curvature diagram at the limit state is developed from the moment–curvature relationship shown in Figure

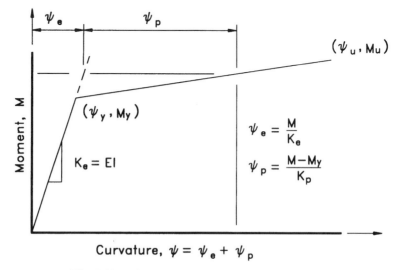

Fig. 7.30   Bilinear moment–curvature relationship.

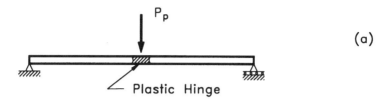

(a)

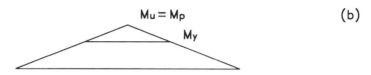

(b)

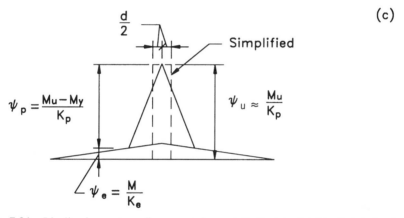

(c)

Fig. 7.31 Idealized curvature diagram at flexural limit state. (a) Limit state load, (b) moment diagram, and (c) curvature diagram.

7.30. The sharp peak in the plastic portion of the curvature diagram is not realistic because when hinging begins it spreads out due to cracking of the concrete and yielding of the steel. Sawyer (1964) recommended that this spread in plasticity be extended a distance of one-half the effective depth d at each moment concentration. Because the elastic contribution to the curvature diagram is small compared to the plastic curvature, it can be neglected. From moment-area principles, the approximate plastic rotation θ_p in the hinge is the area of the simplified curvature diagram in Figure 7.31(c). Using the relationship in Eq. 7.78, the ductility measure becomes

$$\theta_p = \psi_u d = \varepsilon_{cu}\frac{d}{c} \qquad (7.79)$$

From Eq. 7.79 it is clear that ductility can be improved by increasing the limit strain ε_{cu} or by decreasing the depth to the neutral axis c. As shown in

Figure 7.5, confining the concrete with sprials or lateral ties can substantially increase ε_{cu}. The neutral axis depth c depends on the total compressive force which, in turn, must be balanced by the total tensile force. Therefore, c can be decreased by increasing the concrete compressive strength f'_c, by increasing the area of the compression reinforcement A'_s, or by decreasing the tensile steel areas A_{ps} and A_s. The effect of these parameters on c can also be observed in Eq. 7.68.

In previous editions of the bridge specifications, the ductility control for reinforced concrete was to limit the compressive force subject to a brittle failure by specifying a maximum tensile steel reinforcement ratio ρ_{max} as 0.75 of the balanced steel ratio ρ_b, that is,

$$\rho = \frac{A_s}{bd} \le \rho_{max} = 0.75\rho_b \tag{7.80}$$

where ρ is the tensile reinforcement ratio and ρ_b is the reinforcement ratio that produces balanced strain conditions. The balanced strain conditions require that the concrete strain ε_c is at ε_{cu} when the steel strain ε_s reaches ε_y. By equating the balanced tensile and compressive forces in a rectangular singly reinforced concrete beam, and by using similar strain triangles, the balanced steel ratio, becomes

$$\rho_b = \frac{A_{sb}}{bd} = \frac{0.85\beta_1 f'_c}{f_y} \frac{|\varepsilon_{cu}|}{|\varepsilon_{cu}| + \varepsilon_y} \tag{7.81}$$

where A_{sb} is the balanced tensile steel area. Introducing the mechanical reinforcement index ω as

$$\omega = \rho\frac{f_y}{f'_c} \tag{7.82}$$

If we multiply both sides of Eq. 7.80 by f_y/f'_c and substitute Eq. 7.81, we get

$$\omega \le 0.64\beta_1 \frac{|\varepsilon_{cu}|}{|\varepsilon_{cu}| + \varepsilon_y} \tag{7.83}$$

If we substitute $\varepsilon_{cu} = -0.003$, $\varepsilon_y = 0.002$ yields

$$\omega \le 0.38\beta_1 \tag{7.84}$$

A similar limitation was also placed on the prestressed mechanical reinforcement index ω_p in previous editions of the bridge specifications as

$$\omega_p = \frac{A_{ps}f_{ps}}{bd_p f'_c} \leq 0.36\beta_1 \qquad (7.85)$$

The disadvantage of using tensile reinforcement ratios to control brittle compression failures is that they must be constantly modified, sometimes in a confusing manner, to accommodate changes in the compressive force caused by the addition of flanges, compression reinforcement, and combinations of nonprestressed and prestressed tensile reinforcement. A better approach is to control the brittle concrete compressive force by setting limits on the distance c from the extreme compressive fiber to the neutral axis. Consider the left-hand side of Eq. 7.84 defined by Eq. 7.82 and substitute the compressive force in the concrete for the tensile force in the steel, so that

$$\omega = \frac{A_s \, f_y}{bd_s \, f'_c} = \frac{0.85f'_c\beta_1 cb}{bd_s f'_c} = 0.85\beta_1 \frac{c}{d_s} \qquad (7.86)$$

where d_s is the distance from the extreme compression fiber to the centroid of the nonprestressed tensile reinforcement. Similarly, for Eq. 7.85

$$\omega_p = \frac{A_{ps}f_{ps}}{bd_p f'_c} = \frac{0.85f'_c\beta_1 cb}{bd_p f'_c} = 0.85\beta_1 \frac{c}{d_p} \qquad (7.87)$$

where d_p is the distance from the extreme compression fiber to the centroid of the prestressing tendons. Thus, putting limits on the neutral axis depth is the same as putting limits on the tensile reinforcement, only it can be much simpler. *Further, by limiting the maximum value for the ratio c/d, this assures a minimum ductility in the member as measured by the rotational capacity at the limit state of Eq. 7.79.*

All that remains is to decide on a common limiting value on the right-hand sides of Eqs. 7.84 and 7.85 and a unified definition for the effective depth to the tensile reinforcement. These topics have been presented by Skogman et al. (1988) and discussed by Naaman et al. (1990). In AASHTO [A5.7.3.1], the recommendations proposed by Naaman (1992) that the right-hand sides of Eqs. 7.84 and 7.85 be $0.36\beta_1$ and that the effective depth from the extreme compression fiber to the centroid of the tensile force in the tensile reinforcement be defined as

$$d_e = \frac{A_{ps}f_{ps}d_p + A_s f_y d_s}{A_{ps}f_{ps} + A_s f_y} \qquad (7.88)$$

where f_{ps} is calculated by either Eq. 7.59 or Eq. 7.72 or in a preliminary design can be assumed to be f_{py}. Finally, the ductility and maximum tensile reinforcement criterion becomes

$$0.85\beta_1 \frac{c}{d_e} \leq 0.36\beta_1$$

or simply [A5.7.3.3.1]

$$\frac{c}{d_e} \leq 0.42 \qquad (7.89)$$

EXAMPLE 7.8

Check the ductility requirement for the beam in Figure 7.29 with the properties given in Example 7.7.

Bonded Case

$$c = 366 \text{ mm} \qquad f_{ps} = 1648 \text{ MPa}$$

$$d_e = \frac{A_{ps}f_{ps}d_p + A_s f_y d_s}{A_{ps}f_{ps} + A_s f_y} = \frac{987(1648)(900) + 600(400)(937)}{987(1648) + 600(400)} = 905 \text{ mm}$$

$$\frac{c}{d_e} = \frac{366}{905} = \underline{0.40} < 0.42, \text{ ductility} \quad \text{OK}$$

Unbonded Case

$$c = 289 \text{ mm}, f_{ps} = 1346 \text{ MPa}$$

$$d_e = \frac{987(1346)(900) + 600(400)(937)}{987(1346) + 600(400)} = 906 \text{ mm}$$

$$\frac{c}{d_e} = \frac{289}{906} = \underline{0.32} < 0.42, \text{ ductility} \quad \text{OK}$$

7.7.5 Minimum Tensile Reinforcement

Minimum tensile reinforcement is required to guard against a possible sudden tensile failure. This sudden tensile failure could occur if the moment strength provided by the tensile reinforcement is less than the cracking moment strength of the gross concrete section. To account for the possibility that the moment resistance M_n provided by nonprestressed and prestress tensile reinforcement may be understrength while the moment resistance M_{cr} based on the concrete tensile strength may be overstrength, AASHTO [A5.7.3.3.2] gives the criterion

$$\phi M_n \geq 1.2 M_{cr} \tag{7.90}$$

for which

$$M_{cr} = \frac{f_r I_g}{y_t} \tag{7.91}$$

where f_r is the concrete tensile rupture stress given by Eq. 7.19, I_g is the gross moment of inertia of the cross section, and y_t is the distance from the neutral axis to the extreme tensile fiber. The understrength or resistance factor ϕ is calculated from Eq. 7.55. The 1996 AASHTO LRFD Interim Specifications are expected to include the lesser of 1.33 times the factored moment required for strength or 1.2 M_{cr} for the right side of Eq. 7.90.

Consider a rectangular beam of width b and overall depth h with only nonprestressed tensile reinforcement A_s. If we assume the lever arm jd is $0.9h$, the factored moment resistance is

$$\phi M_n = \phi A_s f_y (jd) = 0.9 A_s f_y (0.9h) = 0.8 A_s f_y h$$

By assuming $f_r = 0.12 f_c'$, the cracking moment strength is

$$M_{cr} = \tfrac{1}{6} f_r b h^2 = 0.02 f_c' b h^2$$

By substituting these expressions for ϕM_n and M_{cr} into Eq. 7.90 and solving for the minimum tensile reinforcement area

$$A_s \geq 0.03 b h \frac{f_c'}{f_y} \tag{7.92}$$

where f_c' is the 28-day concrete compressive strength and f_y is the tensile reinforcement yield strength. Equation 7.92 can be used to determine the minimum tensile reinforcement for rectangular beams without prestressing tendons. Here AASHTO [A5.7.3.3.2] permits Eq. 7.92 to be used for nonrectangular nonprestressed cross sections by substituting the section's gross concrete area for bh.

7.7.6   Loss of Prestress

After a reinforced concrete member is precompressed by prestressing tendons, a number of things happen that reduce the effectiveness of the prestress force. Some of the losses occur almost instantaneously while others take years before they finally dampen out. Immediate prestress losses are due to slip of the tendons in the anchorages Δf_{pA}, elastic compression (shortening) of the concrete Δf_{pES}, and friction between a tendon and its conduit Δf_{pF}. Long-time prestress losses are due to shrinkage of concrete Δf_{pSR}, creep of concrete Δf_{pCR}, and relaxation of the prestressing tendon Δf_{pR}. The total prestress loss Δf_{pT} is the accumulation of the losses that occur at the different load stages throughout the life of the member. The total prestress losses will depend on the method used to apply the prestress force.

For Pretensioned Members

$$\Delta f_{pT} = \Delta f_{pES} + \Delta f_{pSR} + \Delta f_{pCR} + \Delta f_{pR} \tag{7.93}$$

For Post-tensioned Members

$$\Delta f_{pT} = \Delta f_{pA} + \Delta f_{pES} + \Delta f_{pF} + \Delta f_{pSR} + \Delta f_{pCR} + \Delta f_{pR} \tag{7.94}$$

The evaluation of the prestress losses indicated by the terms in Eqs. 7.93 and 7.94 will be discussed in the sections that follow. The expressions developed to calculate the prestress losses should be considered as only estimates of the magnitudes of the different quantities. There are too many variables associated with the prestressing operation, the placing and curing of the concrete, and the service environment to make accurate calculations. However, the expressions are sufficiently accurate for designing members with prestressing tendons and estimating their strength.

Anchorage Set Loss. In post-tensioned construction not all of the stress developed by the jacking force is transferred to the member because the tendons slip slightly as the wedges or other mechanical devices seat themselves in the anchorage. The anchorage slip or set Δ_A is assumed to produce an average strain over the length of a tendon L, which results in an anchorage set loss of

$$\Delta f_{pA} = \frac{\Delta_A}{L} E_p \tag{7.95}$$

where E_p is the modulus of elasticity of the prestressing tendon. The range of Δ_A varies from 3 to 10 mm with a value of 6 mm often assumed. For long tendons the anchorage set loss is relatively small, but for short tendons it could become significant.

Elastic Shortening Loss. When the strands at the ends of a pretensioned member are cut, the prestress force is transferred to and produces compression in the concrete. The compressive force on the concrete causes the member to shorten. Compatibility of the strains in the concrete and in the tendon results in a reduction in the elongation of the tendon and an accompanying loss of prestress. Equating the strain in the tendon due to the change in prestress Δf_{pES} and the strain in the concrete due to the concrete stress at the centroid of the tendon f_{cgp} yields

$$\frac{\Delta f_{pES}}{E_p} = \frac{f_{cgp}}{E_{ci}}$$

By solving for the prestress loss due to elastic shortening of the concrete in a pretensioned member, we have

$$\Delta f_{pES} = \frac{E_p}{E_{ci}} f_{cgp} \qquad (7.96)$$

where E_{ci} is the modulus of elasticity of concrete at transfer of the prestressing force.

If the centroid of the prestressing force is below the centroid of the concrete member, the member will be lifted upward at transfer and the self-weight of the member will be activated. The elastic concrete stress at the centroid of the tendon is then given by the first three terms of Eq. 7.53 with $y = e$

$$f_{cgp} = -\frac{P_i}{A_g} - \frac{(P_i e)e}{I_g} + \frac{M_g e}{I_g} \qquad (7.97)$$

where P_i is the prestressing force at transfer. These linear elastic concrete stresses are shown in Figure 7.20. The force P_i will be slightly less than the transfer force based on the transfer stresses given in Table 7.8, because these stresses will be reduced by the elastic shortening of the concrete and the relaxation of tendons between the time of jacking and transfer. Thus, for low-relaxation strand, P_i can be expressed as

$$P_i = A_{ps}(0.74 f_{pu} - \Delta f_{pES} - \Delta f_{pR1}) \qquad (7.98)$$

where Δf_{pR1} is the prestress loss due to relaxation of the tendons at transfer. Realizing that P_i is changed a relatively small amount, AASHTO [A5.9.5.2.3a] allows P_i to be based on a prestressing tendon stress of $0.70 f_{pu}$ for low-relaxation strands and $0.65 f_{pu}$ for stress-relieved strands and high-strength bars.

In the case of a post-tensioned member, there will be no loss of prestress due to elastic shortening if all the tendons are tensioned simultaneously. No loss occurs because the post-tensioning force compensates for the elastic shortening as the jacking operation progresses. If the tendons are tensioned sequentially, the first tendon anchored will experience a loss due to elastic shortening given by Eq. 7.96 for a pretensioned member.

Each subsequent tendon that is post-tensioned will see a fraction of the pretensioned loss, with the last tendon anchored having no loss. The average post-tensioned loss would be one-half of the pretensioned loss if the last tendon also had a loss. Because the last tendon anchored does not have a loss, the loss of prestress due to elastic shortening for post-tensioned members is given by [A5.9.5.2.3b]

$$\Delta f_{pES} = \frac{N - 1}{2N} \frac{E_p}{E_{ci}} f_{cgp} \qquad (7.99)$$

where N is the number of identical prestressing tendons.

Friction Loss. In post-tensioned members, friction develops between the tendons and the ducts in which they are placed. If the tendon profile is curved or draped, the ducts will be placed in the member to follow the profile. When the tendons are tensioned after the concrete has hardened, they will tend to straighten out and develop friction along the wall of the duct. This friction loss is referred to as the *curvature* effect. Even if the tendon profile is straight, the duct placement may vary from side to side or up and down and again friction is developed between the tendon and the duct wall. This friction loss is referred to as the *wobble* effect.

Consider the post-tensioned member in Figure 7.32(a) with a curved tendon having an angle change α over a length x from the jacking end. A differential element of length of the curved tendon is shown in Figure 7.32(b) with tensile forces P_1 and P_2 that differ by the friction component dP_1 developed by the normal force N, that is,

$$P_1 - P_2 = dP_1 = \mu N$$

where μ is the coefficient of friction between the tendon and the duct due to the curvature effect. Assuming P_1 and P_2 are nearly equal and that $d\alpha$ is a small angle, the normal force N can be determined from the force polygon of Figure 7.32(c) as $P_1 d\alpha$ so that

$$dP_1 = \mu P_1 d\alpha$$

Wobble friction losses over the tendon length dx are expressed as $KP_1 dx$, where K is the coefficient of friction between the tendon and the surrounding concrete due to the wobble effect. Thus, the total friction loss over length dx becomes

$$dP_1 = \mu P_1 d\alpha + KP_1 dx$$

or

$$\frac{dP_1}{P_1} = \mu d\alpha + K dx \qquad (7.100)$$

The change in tendon force between two points A and B is given by integrating both sides of Eq. 7.100, that is,

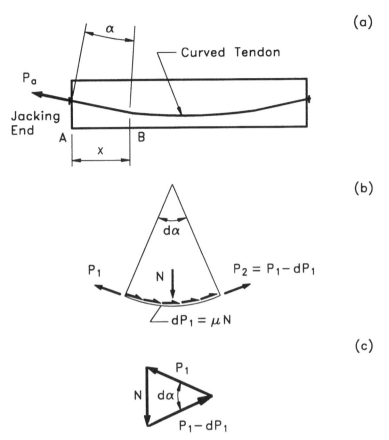

Fig. 7.32 Curvature friction loss (after Nawy, 1989). (a) Tendon profile, (b) differential length element, (c) force polygon.

$$\int_{P_B}^{P_A} \frac{dP_1}{P_1} = \mu \int_0^\alpha d\alpha + K \int_0^x dx$$

which results in

$$\log_e P_A - \log_e P_B = \mu\alpha + Kx$$

or

$$\log_e \frac{P_B}{P_A} = -\mu\alpha - Kx$$

By taking the antilogarithm of both sides and multiplying by P_A we get

$$P_B = P_A e^{-(\mu\alpha + Kx)}$$

By dividing both sides by the area of the prestressing tendon and subtracting from the stress at A, the change in stress between two points x distance apart can be expressed as

$$f_A - f_B = f_A - f_A e^{-(\mu\alpha + Kx)}$$

or

$$\Delta f_{pF} = f_{pj}[1 - e^{-(\mu\alpha + Kx)}] \qquad (7.101)$$

where Δf_{pF} is the prestress loss due to friction and f_{pj} is the stress in the tendon at the jacking end of the member.

A conservative approximation to the friction loss is obtained if it is assumed that P_1 in Eq. 7.100 is constant over the length x, so that the integration yields

$$\Delta P_1 \approx P_1(\mu\alpha + Kx)$$

or in terms of stresses

$$\Delta f_{pF} \approx f_{pj}(\mu\alpha + Kx) \qquad (7.102)$$

This approximation is comparable to using only the first two terms of the series expansion for the exponential in Eq. 7.101. The approximation should be sufficiently accurate because the quantity in parenthesis is only a fraction of unity.

The friction coefficients μ and K depend on the type of tendons, the rigidity of the sheathing, and the form of construction. Design values for these coefficients are given in AASHTO [A5.9.5.2.2b] and are reproduced in Table 7.11. Note that some of these coefficients can vary by an order of magnitude; therefore, it is important to know the characteristics of the post-tensioning system that is to be used to accurately estimate friction losses.

Shrinkage Loss. Shrinkage of concrete is a time-dependent loss that is influenced by the curing method used, the volume to surface ratio V/S of the member, the water content of the concrete mix, and the ambient relative humidity H. The total long-time shrinkage strain can range from 0.4×10^{-3} mm/mm to 0.8×10^{-3} mm/mm over the life of a member with about 80% occurring in the first year (see Fig. 7.12).

The shrinkage strain ε_{SR} for steam-cured concrete can be estimated from a modified version of Eq. 7.24 (Collins and Mitchell, 1991) as

TABLE 7.11 Friction Coefficients for Post-tensioning Tendons[a]

Type of Tendons and Sheathing	Wobble Coefficient, K (1/mm)	Curvature Coefficient, μ (1/rad)
Tendons in rigid and semirigid galvanized ducts Seven-Wire strands	6.6×10^{-7}	0.05–0.25
Pregreased tendons Wires and seven-wire strands	$9.8 \times 10^{-7} - 6.6 \times 10^{-6}$	0.05–0.15
Mastic-coated tendons Wires and seven-wire strands	$3.3 \times 10^{-6} - 6.6 \times 10^{-6}$	0.05–0.15
Rigid steel pipe deviators	6.6×10^{-7}	0.25 Lubrication probably required

[a]In AASHTO Table 5.9.5.2.2b-1. [From *AASHTO LRFD Bridge Design Specifications,* Copyright © 1994 by the American Association of State Highway and Transportation Officials, Washington, DC. Used by permission.]

$$\varepsilon_{SR} = -k_s k_h \left[\frac{t}{t + 55}\right] 0.56 \times 10^{-3} \tag{7.103}$$

where k_s is the size factor given in Figure 7.10 for different V/S ratios, k_h is the humidity factor given in Table 7.2, and t is the drying time in days. This shortening of the concrete due to shrinkage strain converts to a tensile pre-stress loss in the tendons when multiplied by E_p, that is,

$$\Delta f_{pSR} = -\varepsilon_{SR} E_p \tag{7.104}$$

Taking $t = 500$ days and $V/S = 100$ mm, then k_s from Figure 7.10 is about 0.7. By assuming that k_h can be represented by

$$k_h = 1.7 - 0.015H$$

where H is the relative humidity in percent, and using $E_p = 197 \times 10^3$ MPa, Eqs. 7.103 and 7.104 yield

$$\Delta f_{ps} = 0.7(1.7 - 0.015H) \left[\frac{500}{500 + 55}\right] 0.56 \times 197$$

which results in

$$\Delta f_{pSR} = 117 - 1.03H \tag{7.105}$$

as an estimate of the prestress loss (MPa) due to shrinkage for a pretensioned member.

For a post-tensioned member, the shrinkage loss in the tendons will be less than for the pretensioned member because the concrete has additional drying time before the prestress is applied. Here AASHTO [A5.9.5.4.2] specifies Eq. 7.105 for pretensioned members and yields

$$\Delta f_{pSR} = 93 - 0.85H \qquad (7.106)$$

for post-tensioned members. This shrinkage loss is about 80% of the loss for pretensioned members and corresponds to an additional drying time of 5 days before application of prestress (Nawy, 1989).

Creep Loss. Creep of concrete is a time-dependent phenomenon in which deformation increases under constant stress due primarily to the viscous flow of the hydrated cement paste. Creep depends on the age of the concrete, the type of cement, the stiffness of the aggregate, the proportions of the concrete mixture, and the method of curing. The additional long-time concrete strains due to creep can be more than double the initial strain ε_{ci} at the time load is applied.

The creep strain ε_{CR} is a function of the age of the concrete t and the time t_i at which the permanent loads are applied as shown in Figure 7.14 and by Eq. 7.25, repeated here for convenience,

$$\varepsilon_{CR}\,(t,t_i) = \psi(t,t_i)\varepsilon_{ci} \qquad (7.107)$$

where $\psi(t,t_i)$ is the creep coefficient given by Eq. 7.26. A time-dependent concrete modulus of elasticity $E_{CR}(t,t_i)$ can be expressed as

$$E_{CR}(t,t_i) = \frac{f_{ci}}{\varepsilon_{CR}(t,t_i)}$$

where f_{ci} is the concrete stress due to the applied load at time t_i. Substituting Eq. 7.107 and using the relationship $E_{ci} = f_{ci}/\varepsilon_{ci}$, we have

$$E_{CR}(t,t_i) = \frac{E_{ci}}{\psi(t,t_i)} \qquad (7.108)$$

where for normal weight concrete Eq. 7.2 gives $E_{ci} = 4800\sqrt{f'_{ci}}$.

A time-dependent modular ratio $n_{CR}(t,t_i)$ can be defined as

$$n_{CR}(t,t_i) = \frac{E_p}{E_{CR}(t,t_i)}$$

where E_p is assumed to be constant with time. Substituting Eq. 7.108 results in

$$n_{CR}(t,t_i) = \frac{E_p}{E_{ci}} \psi(t,t_i) \tag{7.109}$$

which can be used to estimate the stress in the prestressing tendon from the stress in the surrounding concrete.

In general, the stress f_{cp} in the concrete surrounding the prestressing tendon in a noncomposite section can be expressed as

$$f_{cp} = -\frac{P_i}{A_g} - \frac{(P_i e - M_g)e}{I_g} + \frac{(M_{DC} + M_{DW})e}{I_g} \tag{7.110}$$

where P_i is the prestressing force at transfer, A_g is the cross-sectional area, e is the eccentricity of the prestressing force, M_g is the moment due to the self-weight of the girder, I_g is the moment of inertia of the section, M_{DC} is the moment due to superimposed dead load, and M_{DW} is the moment due to a future wearing surface. The eccentricity is assumed to be below the center of gravity of the section and the sign convention is that tensile stresses are positive. If the section is composite, the third term in Eq. 7.110 may have to be split into those dead loads, such as the bridge deck, that act on the non-composite section and the remainder that act on the composite section, that is,

$$f_{cp} = -\frac{P_i}{A_g} - \frac{(P_i e - M_g)e}{I_g} + \frac{M_{DC1} e}{I_g} + \frac{(M_{DC2} + M_{DW})e}{I_c} \tag{7.111}$$

where I_c is the moment of inertia of the composite section. A general description of these concrete stresses is given in Figure 7.20. Both Eqs. 7.110 and 7.111 can be written as

$$f_{cp} = -f_{cgp} + \Delta f_{cdp} \tag{7.112}$$

where f_{cgp} is the concrete stress at the center of gravity of the prestressing tendons at transfer represented by the first two terms and Δf_{cdp} is the change in concrete stress at the center of gravity of the prestressing tendons due to permanent loads DC and DW that are applied later.

Usually f_{cgp} will be greater than Δf_{cdp} and the resulting concrete stress will be compressive. This compressive concrete stress will cause a reduction in the tensile stress of the prestressing tendon Δf_{pCR} that can be determined by multiplying by the creep modular ratio of Eq. 7.108 and reversing the signs. The time-dependent modular ratios are going to be different for the two stresses in Eq. 7.111 because the stresses occur under different conditions, that is,

$$\Delta f_{pCR} = n_{CR,TR}(t,t_{i,TR})f_{cgp} - n_{CR,LT}(t,t_{i,LT})\Delta f_{cdp} \tag{7.113}$$

where $n_{CR,TR}$ is the creep modular ratio at transfer, $t_{i,TR}$ is the age of the concrete at transfer, $n_{CR,LT}$ is the creep modular ratio for the permanent loads, and $t_{t,LT}$ is the age of the concrete when the permanent loads are applied. In writing Eq. 7.113 in this manner, it is assumed that superposition of stresses is also valid for creep stresses, which may or may not be correct.

To estimate representative values for the creep modular ratios for concrete at an age of one year, consider the following quantities to be given: $t = 365$ days, $H = 70\%$, $V/S = 100$ mm, and $E_p = 197$ GPa. Assuming conditions at transfer of prestress to be: $t_i = 5$ days and $f'_{ci} = 24$ MPa, which gives $E_{ci} = 23.5$ GPa (Eq. 7.2), $k_f = 0.94$ (Eq. 7.27), and $k_c = 0.67$ (Fig. 7.13) so that the creep coefficient (Eq. 7.26) becomes

$$\psi(365,5) = 3.5(0.67)(0.94)\left(1.58 - \frac{70}{120}\right)5^{-0.118}\frac{360^{0.6}}{10 + 360^{0.6}} = 1.41$$

The creep modular ratio at transfer is then

$$n_{CR,TR}(365,5) = \frac{E_p}{E_{ci}}\psi(365,5) = \frac{197}{23.5}(1.41) = 11.8$$

If we assume conditions when additional permanent loads are applied to be: $t_i = 30$ days and $f'_{ci} = 34.5$ MPa, which gives $E_{ci} = 28.2$ GPa, $k_f = 0.81$, and $k_c = 0.67$ so that the creep coefficient becomes

$$\psi(365,30) = 3.5(0.67)(0.81)(0.997)30^{-0.118}\frac{335^{0.6}}{10 + 335^{0.6}} = 0.97$$

The creep modular ratio for additional permanent loads is then

$$n_{CR,LT}(365,30) = \frac{E_p}{E_{ci}}\psi(365,30) = \frac{197}{28.2}(0.97) = 6.8$$

Substituting these values for the creep modular ratios into Eq. 7.113 yields the creep prestress loss as

$$\Delta f_{pCR} = 11.8f_{cgp} - 6.8\Delta f_{cdp}$$

This expression, based on assumed typical conditions, is similar to the one given in AASHTO [A5.9.5.4.3] for prestress loss due to creep, that is,

$$\Delta f_{pCR} = 12.0 f_{cgp} - 7.0 \Delta f_{cdp} \geq 0 \tag{7.114}$$

which can be used for all prestressed concrete members. The signs on the stresses in Eq. 7.114 are based on the usual case where the tendon eccentricity e in Eq. 7.110 or Eq. 7.111 is below the center of gravity of the section and opposing the dead load moments. It is possible that for some eccentricity and dead load moment combinations, the signs could be different; therefore, it is necessary to check that the prestress loss is always positive.

Relaxation Loss. Relaxation of the prestressing tendons is a time-dependent loss of prestress that occurs when the tendon is held at constant strain. The total relaxation loss Δf_{pR} is separated into two components

$$\Delta f_{pR} = \Delta f_{pR1} + \Delta f_{pR2} \tag{7.115}$$

where Δf_{pR1} is the relaxation loss at transfer of the prestressing force and Δf_{pR2} is the relaxation loss after transfer.

The prestress loss due to relaxation at transfer Δf_{pR1} for a pretensioned member with an initial prestress at transfer f_{pt} greater than $0.50\ f_{pu}$ is specified in AASHTO [A5.9.5.4.4b], for stress-relieved strand, as

$$\Delta f_{pR1} = \frac{\log 24t}{10} \left[\frac{f_{pi}}{f_{py}} - 0.55 \right] f_{pi} \tag{7.116}$$

where t is the time in days from prestressing to transfer, f_{py} is the specified yield strength of the prestressing tendon, and f_{pi} is the initial stress in the tendon at the end of prestressing, that is,

$$f_{pi} = f_{pt} - \Delta f_{pES} - \Delta f_{pR1} \tag{7.117}$$

which may require one or two iterations before converging. Equation 7.116 was first developed by Magura et al. (1964) to represent the results of their tests. The constant 24 is to convert the time into hours. The relaxation losses at transfer for low-relaxation strands, as the name implies, are one-fourth (25%) of those given by Eq. 7.116.

The prestress loss due to relaxation after transfer Δf_{pR2} for stress-relieved strands is a basic value of 138 MPa, which is reduced continually with time as the other prestress losses reduce the tendon stress. The elastic shortening loss Δf_{pES} occurs almost instantaneously so that its effect is largest. The losses due to shrinkage Δf_{pSR} and creep Δf_{pCR} take place over a period of time and have a smaller effect. The losses due to friction Δf_{pF} are somewhere between the two. Here AASHTO [A5.9.5.4.4c] approximates these effects by the expression

$$\Delta f_{pR2} = 138 - 0.4\Delta f_{pES} - 0.3\Delta f_{pF} - 0.2(\Delta f_{pSR} + \Delta f_{pCR}) \qquad (7.118)$$

where Δf_{pF} is zero for pretensioned members. The relaxation losses after transfer for low-relaxation strands are 30% of those given by Eq. 7.118.

Lump Sum Estimate of Time-Dependent Losses. It is not always necessary to make detailed calculations for the time-dependent losses of shrinkage, creep, and relaxation if the designs are routine and the conditions are average. Here AASHTO [A5.9.5.3] provides approximate lump sum estimates of the time-dependent prestress losses, which are duplicated in Table 7.12. The losses given in Table 7.12 cover shrinkage and creep in concrete and relaxation of the prestressing tendon. The instantaneous elastic shortening Δf_{pES} must be added to these time-dependent losses to obtain the total prestress loss.

The values in Table 7.12 can be used for prestressed and partially prestressed, nonsegmental, post-tensioned members and pretensioned members made with normal weight concrete under standard construction procedures and subjected to average environmental conditions. If the standard conditions given in AASHTO [A5.9.5.3] are not met then a detailed analysis is required.

In the case of wires and strands, both an upper bound estimate and an average estimate are given. A reasonable approach during preliminary design would be to use the upper bound estimate when evaluating the flexural strength and the average estimate when calculating service load effects. According to Zia et al. (1979), overestimation of prestress losses can be almost as detrimental as underestimation, since the former can result in excessive camber and horizontal movement.

7.8   SHEAR STRENGTH OF REINFORCED CONCRETE MEMBERS

Reinforced concrete members subjected to loads perpendicular to their axis must resist shear forces as well as flexural and axial forces. The shear force resistance mechanism is different for deep beams than for slender beams. The AASHTO Specifications direct a designer to use the strut-and-tie model [A5.6.3] whenever the distance from the point of zero shear to the face of a support is less than twice the effective depth of the beam, or when a load that causes at least one-half of the shear at a support is within twice the effective depth. For a beam with deep-beam proportions, plane sections no longer remain plane and a better representation of the load carrying mecha-

TABLE 7.12 Time-Dependent Prestress Losses[a]

Type of Beam Section	Level	For Wires and Strands with f_{pu} = 1620, 1725, or 1860 MPa	For Bars with f_{pu} = 1000 or 1100 MPa
Rectangular beams and solid slabs	Upper bound	$200 + 28\ PPR\ (-41)^b$	$130 + 41\ PPR$
	average	$180 + 28\ PPR\ (-41)^b$	
Box girder	Upper bound	$145 + 28\ PPR\ (-28)^b$	100
	average	$130 + 28\ PPR\ (-28)^b$	
I-girder	Average	$230\left[1.0 - 0.15\dfrac{f'_c - 41}{41}\right] + 41\ PPR\ (-41)^b$	$130 + 41\ PPR$
Single T, double T, and solid, hollow core and voided slab	Upper bound	$270\left[1.0 - 0.15\dfrac{f'_c - 41}{41}\right] + 41\ PPR\ (-55)^b$	$210\left[1.0 - 0.15\dfrac{f'_c - 41}{41}\right] + 41\ PPR$
	Average	$230\left[1.0 - 0.15\dfrac{f'_c - 41}{41}\right] + 41\ PPR\ (-55)^b$	

[a]In AASHTO Table 5.9.5.3-1. [From *AASHTO LRFD Bridge Design Specifications*, Copyright © 1994 by the American Association of State Highway and Transportation Officials, Washington, DC. Used by Permission.]
[b]Values in parentheses are subtractons for low-relaxation strands.

nism is with the concrete compression struts and steel tension ties as shown in Figure 7.33.

The proportions of typical bridge girders are slender so that plane sections before loading remain plane after loading and engineering beam theory can be used to describe the relationships between stresses, strains, cross-sectional

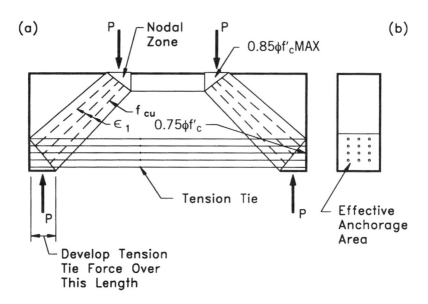

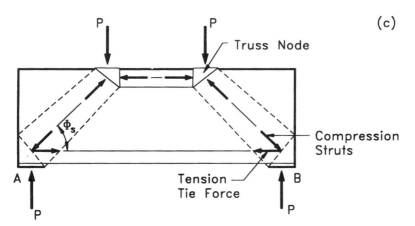

Fig. 7.33 Strut-and-tie model for a deep beam (a) Flow of forces, (b) end view and (c) truss model. (AASHTO Fig. C5.6.3.2-1). [From *AASHTO LRFD Bridge Design Specifications*. Copyright © 1994 by the American Association of State Highway and Transportation Officials, Washington, DC. Used by permission.]

properties, and the applied forces. Reinforced concrete girders are usually designed for a flexural failure mode at locations of maximum moment. However, this flexural capacity cannot be developed if there is a premature shear failure due to inadequate web dimensions and web reinforcement. To evaluate the shear resistance of typical bridge girders the sectional design model of AASHTO [A5.8.3] is used. This model satisfies force equilibrium and strain compatibility, and utilizes experimentally determined stress–strain curves for reinforcement and diagonally cracked concrete. Background and details of the sectional model can be found in the papers by Vecchio and Collins (1986, 1988) and the book by Collins and Mitchell (1991).

The nominal shear strength V_n for the sectional design model can be expressed as

$$V_n = V_c + V_s + V_p \qquad (7.119)$$

where V_c is the nominal shear strength of the concrete, V_s is the nominal shear strength of the web reinforcement, and V_p is the nominal shear strength provided by the vertical component of any inclined prestress force. In Eq. 7.119, V_p can be determined from the geometry of the tendon profile while V_c and V_s are determined by satisfying equilibrium and compatibility of a diagonally cracked reinforced concrete web. The development of expressions for V_c and V_s based on a variable-angle truss model and the modified compression field theory are given in the sections that follow.

7.8.1   Variable-Angle Truss Model

To provide a connection to the past and to introduce a model that satisfies equilibrium, the variable-angle truss model is presented. The truss analogy model is one of the earliest analytical explanations of shear in reinforced concrete beams. According to Mitchell and Collins (1991), it is about 100 years old since it was described by Ritter in 1899 and elaborated by Morsch in 1902.

An example of a variable-angle truss model of a uniformly loaded beam is given in Figure 7.34a. It is similar to one in Hsu (1993). The dotted lines represent concrete compression struts for the top chord and diagonal web members of the truss. The solid lines represent steel tension ties for the bottom chord and vertical web members. The bottom chord steel area is the longitudinal reinforcement selected to resist flexure and the vertical web members are the stirrups at spacing s required to resist shear.

The top chord concrete compression zone balances the bottom chord tensile steel and the two make up the couple that resists the moment due to the applied load. The diagonal concrete compressive struts are at an angle θ with the axis of the beam and run from the top of a stirrup to the bottom chord. The diagonal struts fan out at the centerline and at the supports to provide a

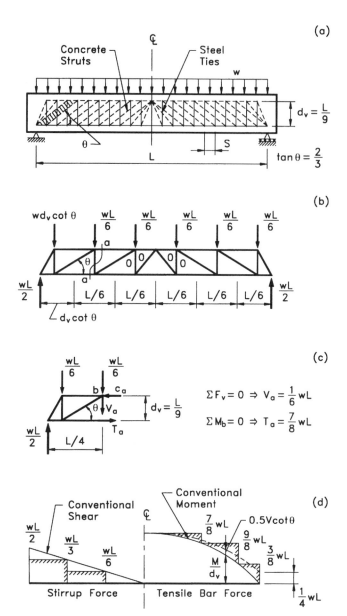

Fig. 7.34 Truss model for a uniformly loaded beam. (a) Variable angle truss model, (b) Simplified strut-and-tie design model, (c) free-body-diagram for section a–a, and (d) staggered diagrams for truss bar forces. [Reprinted with permission from T. T. C. Hsu (1993). *Unified Theory of Reinforced Concrete,* CRC Press, Boca Raton, FL. Copyright CRC Press, Boca Raton, FL © 1993.]

load path for the bottom and top of each stirrup. The fanning of the diagonals also results in a midspan chord force that matches the one obtained by dividing the conventional beam moment by the lever arm d_v.

In defining the lever arm d_v used in shear calculations, the location of the centroid of the tensile force is known a priori but not that of the compressive force. To assist the designer, AASHTO [A5.8.2.7] defines d_v as the effective shear depth taken as the distance, measured perpendicular to the neutral axis, between the resultants of the tensile and compressive forces due to flexure, but it need not be taken less than the greater of $0.9d_e$ or $0.72h$. The effective depth d_e from the extreme compression fiber to the centroid of the tensile force is given by Eq. 7.88 and h is the overall depth of the member.

It is not necessary in design to include every stirrup and diagonal strut when constructing a truss model for concrete beams. Stirrup forces can be grouped together in one vertical member over some tributary length of the beam to give the simplified truss design model of Figure 7.34(b). Obviously, there is more than one way to configure the design truss. For this example, the beam has been divided into six panels, each with a panel load of $wL/6$. Choosing the effective shear depth $d_v = L/9$, then $\tan \theta = \frac{2}{3}$. The bar forces in the members of the truss can then be determined using free-body-diagrams such as the one in Figure 7.34(c).

The variation in the stirrup force and the tensile bar force is shown in Figure 7.34(d). Because of the discrete nature of the truss panels, these force diagrams are like stair steps. The staggered stirrup force diagram is always below the conventional shear force diagram for a uniformly loaded beam. The staggered tensile bar force diagram is always above the tensile bar force diagram derived from a conventional moment diagram divided by the lever arm d_v. If the staggered compressive bar force in the top chord had also been shown, it would be below the compressive bar force derived from conventional moment. This variation can be explained by looking at equilibrium of joints at the top and bottom chords. The presence of compression from the diagonal strut reduces the tension required in a vertical stirrup, reduces the compression in the top chord, and increases the tension in the bottom chord.

To derive an expression for the shear force carried by a stirrup in the variable-angle truss, consider the equilibrium conditions in Figure 7.35 for a section of the web in pure shear (M = 0). Balancing vertical forces in Figure 7.35(a) results in

$$V = f_2 b_v d_v \cos \theta \sin \theta$$

or
$$f_2 = \frac{V}{b_v d_v \cos\theta \sin \theta} \qquad (7.120)$$

where f_2 is the principal compressive stress in the web and b_v is the minimum web width within the depth d_v. From the force polygon, $\tan \theta = V/N_v$ and

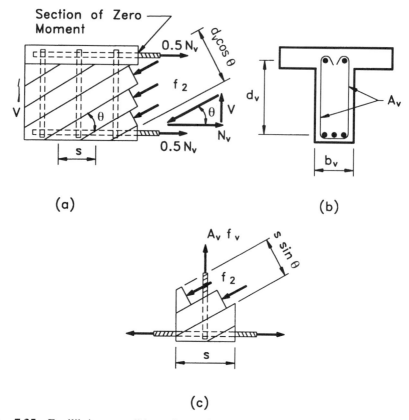

Fig. 7.35 Equilibrium conditions for variable-angle truss. (a) Diagonally cracked web, (b) cross-section, (c) tension in web reinforcement. (After Collins and Mitchell, 1991.) [Reprinted by permission of Prentice Hall, Upper Saddle River, NJ.]

$$N_v = V \cot\theta \qquad (7.121)$$

where N_v is the tensile force in the longitudinal direction required to balance the shear force V on the section. This tensile force N_v is assumed to be divided equally between the top and bottom chords of the truss model, adding to the tension in the bottom and subtracting from the compression in the top. The additional tensile force $0.5V\cot\theta$ is shown added to the tensile force M/d_v in the right half of Figure 7.34(d). The resulting dotted line is a good approximation to a smoothed-out representation of the staggered tensile bar forces.

A bottom chord joint with a tributary length equal to the stirrup spacing s is shown in Figure 7.35(c). Balancing the vertical force in the stirrup with the vertical component of the diagonal compressive force applied over the stirrup spacing s results in

$$A_v f_v = f_2 s b_v \sin\theta \sin\theta$$

where A_v is the total area of the stirrup legs resisting shear and f_v is the tensile stress in the stirrup. Substituting f_2 from Eq. 7.120 yields

$$A_v f_v = \frac{V s b_v \sin\theta \, \sin\theta}{b_v d_v \cos\theta \, \sin\theta} = \frac{V s}{d_v} \tan\theta$$

$$V = \frac{A_v f_v d_v}{s} \cot\theta \qquad\qquad (7.122)$$

It is not possible to obtain a closed-form solution for the shear capacity V from the three equilibrium equations: Eqs. 7.120–7.122; because they contain four unknowns: θ, f_v, N_v, and f_2. One design strategy is to assume $\theta = 45°$ and a value for f_v, such as a fraction of f_y for strength design. In either case, Eq. 7.122 gives a shear capacity of a reinforced concrete beam that depends on the tensile stress in the stirrups and the orientation of the principal compressive stress in the concrete. It does not include any contribution of the tensile strength in the concrete. In other words, by using a variable-angle truss model only the contribution of V_s in Eq. 7.119 is included. The contribution from the tensile strength of the concrete V_c is considered to be zero.

In summary, the variable-angle truss model clearly shows by Eq. 7.121 that a transverse shear force on a cross-section results in an axial force that increases the tension in the longitudinal reinforcement. However, it has two shortcomings: it cannot predict the orientation of the principal stresses and it ignores the contribution of the concrete tensile strength. Both of these shortcomings are overcome by the modified compression field theory, where strain compatibility gives a fourth condition permitting a rationale solution.

7.8.2 Modified Compression Field Theory

In the design of the relatively thin webs of steel plate girders, the web panels between transverse stiffeners subjected to shearing stresses, are considered to support tensile stresses only because the compression diagonal is assumed to have buckled. The postbuckling strength of the plate girder webs depends on the orientation of the principal tensile stress, stiffener spacing, girder depth, web thickness, and yield strength of the material. A tension field theory has been developed to determine the relationships between these parameters and to predict the shear strength of plate girder webs. See Chapter 8 for details.

In the webs of reinforced concrete beams subjected to shearing stresses, an analogous behavior occurs, except the tension diagonal cracks, and the compression diagonal is the dominant support in the web. Instead of a tension field theory, a compression field theory has been developed to explain the behavior of reinforced concrete beams subjected to shear. Originally, the compression field theory assumed that once web cracking occurred, the principal tensile stress vanished. The theory was later modified to include the principal tensile stress and to give a more realistic description of the shear failure mechanism.

Figure 7.36 illustrates pure shear stress fields in the web of a reinforced concrete beam before and after cracking. A Mohr stress circle for the concrete is also shown for each of the cases. Before cracking [Fig. 7.36(a)], the reinforced concrete web is assumed to be homogeneous and Mohr's circle of stress is about the origin with radius v and $2\theta = 90°$. After cracking [Fig. 7.36(b)], the web reinforcement carries the tensile stresses and the concrete struts carry the compressive stresses, as a result, the orientation of the principal stresses changes to an angle θ less than 45°. If the concrete tensile strength is not ignored and carries part of the tensile force, the stress state of the modified compression field theory [Fig. 7.36(c)] is used to describe the behavior.

The Mohr stress circle for the concrete compression strut of Figure 7.36(c) is more fully explained in Figure 7.37. A reinforced concrete element subjected to pure shear will have a Mohr stress circle of radius v about the origin [Fig. 7.37(a)]. Interaction within the element develops compression in the concrete struts [Fig. 7.37(b)] and tension in the steel reinforcement [Fig. 7.37(c)]. The concrete portion of the element is assumed to carry all of the shear, along with the compression, which results in the Mohr stress circles of Figures 7.36(c) and 7.37(b). The angle 2θ rotates, depending on the relative values of shear and compression, even though the comparable angle of the reinforced concrete element remains fixed at 90°.

There is no stress circle for the steel reinforcement because its shear resistance (dowel action) has been ignored. The tensile stresses f_s^* and f_v^* are psuedoconcrete tensile stresses, or smeared steel tensile stresses, that are equivalent to the tensile forces in the reinforcement. Using superposition and diagrams in Figures 7.37(b) and 7.37(c) yields

$$f_s^* b_v s_x = f_s A_s$$

$$f_s^* = \frac{A_s}{b_v s_x} f_s = \rho_x f_s$$

and

$$f_v^* b_v s = f_v A_v$$

$$f_v^* = \frac{A_v}{b_v s} f_v = \rho_v f_v$$

where s_x is the vertical spacing of longitudinal reinforcement including skin reinforcement, s is the horizontal spacing of stirrups,

$$\rho_x = \frac{A_s}{b_v s_x} = \text{longitudinal reinforcement ratio, and} \qquad (7.125)$$

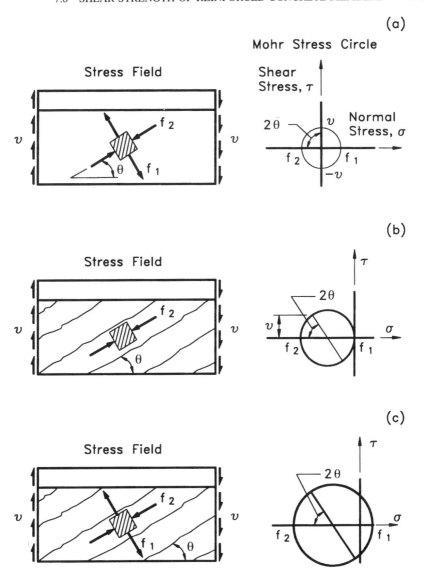

Fig. 7.36   Stress fields in web of a reinforced concrete beam subjected to pure shear (a) Before cracking, $f_1 = f_2 = v$, $\theta = 45°$, (b) compression field theory, $f_1 = 0$, $\theta < 45°$, and (c) modified compression field theory, $f_1 \neq 0$, $\theta < 45°$. (after Collins and Mitchell, 1991). [Reprinted by permission of Prentice Hall, Upper Saddle River, NJ.]

$$\rho_v = \frac{A_v}{b_v s} = \text{transverse reinforcement ratio} \qquad (7.126)$$

The stresses between the concrete and reinforcement may be dissimilar after cracking because of different material moduli, but the strains are not. In fact, the condition of strain compatibility provides the additional relationships,

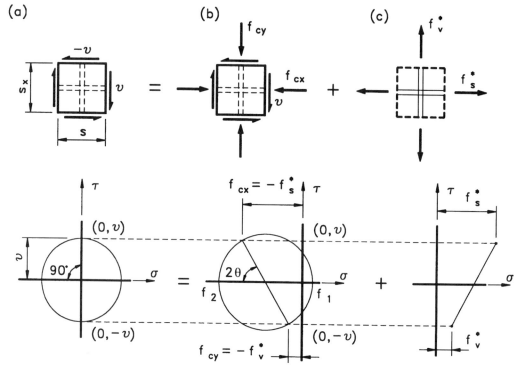

Fig. 7.37 Reinforced concrete element subjected to pure shear (a) Reinforced concrete, (b) concrete struts, and (c) reinforcement. [Reprinted with permission from T. T. C. Hsu (1993). *Unified Theory of Reinforced Concrete,* CRC Press, Boca Raton, FL. Copyright CRC Press, Boca Raton, FL © 1993.]

coupled with the equilibrium equations, to uniquely determine the angle θ and the shear strength of a reinforced concrete member. This unique determination can be done by considering the web of a reinforced concrete beam to behave like a membrane element with in-plane shearing and normal stresses and strains that can be analyzed using Mohr stress and strain circles.

Before writing the equilibrium equations for the modified compression field theory, the compatibility conditions based on a Mohr strain circle are developed. Consider the cracked reinforced concrete web element in Figure 7.38(a), which is subjected to a biaxial state of stress and has strain gages placed to record average strains in the longitudinal ε_x, transverse ε_t, and 45° ε_{45} directions. The strain gages are assumed to be long enough so that the average strain is over more than one crack. The definition of normal strains [Fig. 7.38(b)] is an elongation per unit length while shearing strains [Fig. 7.38(c)] are defined as the change in angle γ from an original right angle. Because of the assumed symmetry in the material properties, this angle is split equally between the two sides originally at right angles. The direction of the shearing

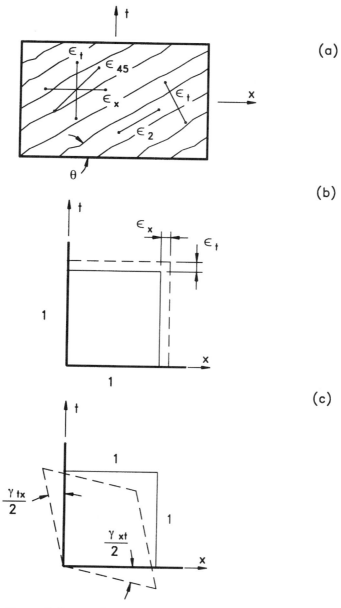

Fig. 7.38 Compatibility conditions for a cracked web element (a) Average strains in a cracked web element, (b) normal strains, (c) shearing strains. [Reprinted with permission from T. T. C. Hsu (1993). *Unified Theory of Reinforced Concrete,* CRC Press, Boca Raton, FL. Copyright CRC Press, Boca Raton, FL © 1993.]

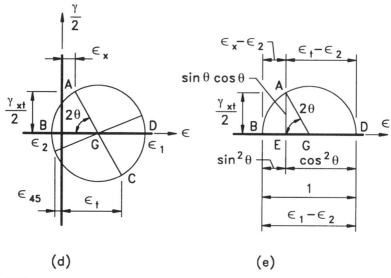

(d)   **(e)**

Fig. 7.38 Compatibility conditions for a cracked web element (d) Mohr strain circle, and (e) geometric relations. [Reprinted with permission from T. T. C. Hsu (1993). *Unified Theory of Reinforced Concrete,* CRC Press, Boca Raton, FL. Copyright CRC Press, Boca Raton, FL © 1993.]

strains corresponds to the direction assumed for positive shearing stresses in Figure 7.37.

A Mohr strain circle [Fig. 7.38(d)] can be constructed if three strains at a point and their orientation to each other are known. The three given average strains are ε_x, ε_t, and ε_{45}. The relationships between these strains and the principal average strains ε_1 and ε_2 and the angle θ, which defines the inclination of the compression struts, are required.

In deriving the compatibility conditions, consider the top one-half of the Mohr strain circle in Figure 7.38(e) to be completely in the first quadrant, that is, all the strain quantities are positive (the $\gamma/2$ axis is to the left of the figure). [With all of the strains assumed to be positive, the derivation is straightforward and does not require intuition as to which quantities are positive and which are negative. (The positive and negative signs should now take care of themselves.)] First, the center of the circle can be found by taking the average of ε_x and ε_t, or the average of ε_1 and ε_2, that is,

$$\frac{\varepsilon_x + \varepsilon_t}{2} = \frac{\varepsilon_1 + \varepsilon_2}{2}$$

so that the principal tensile strain is

$$\varepsilon_1 = \varepsilon_x + \varepsilon_t - \varepsilon_2 \tag{7.127}$$

By using a diameter of unity in Figure 7.38(e), the radius is one-half and the vertical line segment AE is

$$AE = \tfrac{1}{2}\sin 2\theta = \sin \theta \cos \theta$$

By recalling $\sin^2 \theta + \cos^2 \theta = 1$, the line segment ED is given by

$$ED = \tfrac{1}{2}\cos 2\theta + \tfrac{1}{2} = \tfrac{1}{2}(\cos^2 \theta - \sin^2 \theta) + \tfrac{1}{2}(\cos^2 \theta + \sin^2 \theta) = \cos^2 \theta$$

so that the line segment BE becomes

$$BE = 1 - \cos^2 \theta = \sin^2 \theta$$

From these relationships and similar triangles, the following three compatibility equations can be written:

$$\varepsilon_x - \varepsilon_2 = (\varepsilon_1 - \varepsilon_2)\sin^2 \theta \tag{7.128}$$

$$\varepsilon_t - \varepsilon_2 = (\varepsilon_1 - \varepsilon_2)\cos^2 \theta \tag{7.129}$$

$$\gamma_{xt} = 2(\varepsilon_1 - \varepsilon_2)\sin \theta \cos \theta \tag{7.130}$$

Dividing Eq. 7.128 by Eq. 7.129 results in an expression that does not contain ε_1, that is,

$$\tan^2 \theta = \frac{\varepsilon_x - \varepsilon_2}{\varepsilon_t - \varepsilon_2} \tag{7.131}$$

The relative magnitudes of the principal strains ε_1 and ε_2 shown in Figure 7.38d with ε_1 being an order of magnitude greater than ε_2, are to be expected because the average tensile strain ε_1 is across cracks that offer significantly less resistance than the direct compression in the concrete struts.

Equilibrium conditions for the modified compression field theory are determined by considering the free body diagrams in Figure 7.39. The cracked reinforced concrete web shown in Figure 7.39(a) is the same as the one in Figure 7.35(a) except for the addition of the average principal tensile stress f_1 in the concrete. The actual tensile stress distribution in the concrete struts is shown with a peak value within the strut, which then goes to zero at a crack. The constitutive laws developed for concrete in tension in cracked webs (Fig. 7.9) are based on stresses and strains measured over a finite length, and therefore the values for f_1 and ε_1 should be considered as average values over this length.

Equilibrium of vertical forces in Figure 7.39(a) results in

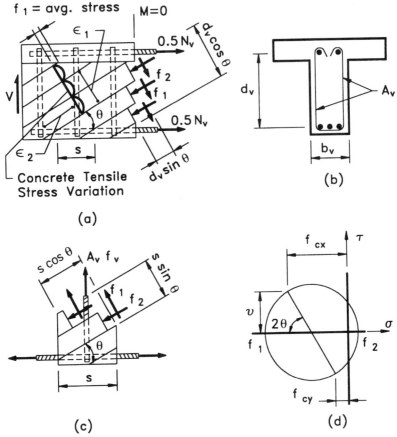

Fig. 7.39 Equilibrium conditions for modified compression field theory (a) Cracked reinforced concrete web, (b) cross-section, (c) tension in web reinforcement, and (d) Mohr stress circle for concrete. (after Collins and Mitchell, 1991). [Reprinted by permission of Prentice Hall, Upper Saddle River, NJ.]

$$V = f_2 b_v d_v \cos \theta \sin \theta + f_1 \, b_v d_v \sin \theta \cos \theta$$

from which the principal compressive stress f_2 can be expressed as

$$f_2 = \frac{v}{\sin \theta \cos \theta} - f_1 \qquad (7.132)$$

where v is the average shear stress,

$$v = \frac{V}{b_v d_v} \qquad (7.133)$$

In Eq. 7.132, f_2 is assumed to be a compressive stress in the direction shown in Figures 7.39(a) and 7.39(c).

Equilibrium of the vertical forces in Figure 7.39(c) results in

$$A_v f_v = f_2 s b_v \sin^2 \theta - f_1 s b_v \cos^2 \theta$$

Substitution of Eq. 7.132 for f_2, Eq. 7.133 for v, and rearranging terms gives

$$V = f_1 b_v d_v \cot \theta + \frac{A_v f_v d_v}{s} \cot \theta \qquad (7.134)$$

which represents the sum of the contributions to the shear resistance from the concrete and the web reinforcement tensile stresses. By comparing Eq. 7.122 with Eq. 7.134, the modified compression field theory provides the concrete tensile stress shear resistance missing from the variable-angle truss model.

Equilibrium of the longitudinal forces in Figure 7.39(a) results in

$$N_v = f_2 b_v d_v \cos^2 \theta - f_1 b_v d_v \sin^2 \theta$$

Substituting for f_2 from Eq. 7.132 and combining terms gives

$$N_v = (v \cot \theta - f_1) b_v d_v$$

If there is no axial load on the member, N_v must be resisted by the longitudinal reinforcement, that is,

$$N_v = A_{sx} f_{sx} + A_{px} f_{px}$$

where A_{sx} is the total area of longitudinal nonprestressed reinforcement, A_{px} is the total area of longitudinal prestressing tendons, f_{sx} and f_{px} are the "smeared stresses" averaged over the area $b_v d_v$ in the longitudinal nonprestressed reinforcement and longitudinal prestressing tendons, respectively. Equating the above two expressions for N_v and dividing by $b_v d_v$ results in

$$\rho_{sx} f_{sx} + \rho_{px} f_{px} = v \cot \theta - f_1 \qquad (7.135)$$

where

$$\rho_{sx} = \frac{A_{sx}}{b_v d_v} = \text{nonprestressed reinforced ratio} \qquad (7.136)$$

$$\rho_{px} = \frac{A_{px}}{b_v d_v} = \text{prestressed reinforcement ratio} \qquad (7.137)$$

With the strain compatibility conditions and stress equilibrium requirements written, only the constitutive relations linking together the stresses and strains remain to complete the definition of the modified compression field theory. The stress–strain relations for concrete in compression (Fig. 7.4), concrete in tension (Fig. 7.9), nonprestressed reinforcement (Fig. 7.17), and prestressing reinforcement (Fig. 7.19) were presented earlier and are summarized in Figure 7.40 for convenience.

A few comments on the four stress–strain curves in Figure 7.40 are appropriate. The importance of compression softening of concrete [Fig. 7.40(a)] due to tension cracking in the perpendicular direction cannot be overemphasized. The recent discovery (1972) and quantification (1981) of this phenomenon was called by Hsu (1993) "the major breakthrough in understanding the shear and torsion problem in reinforced concrete."

Prior to this discovery, the compression response of concrete was obtained from uniaxial tests on concrete cylinders and the predictions of shear strength

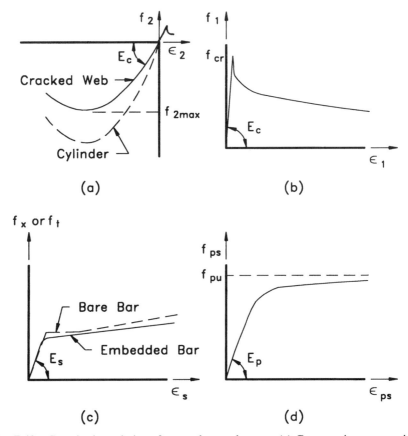

Fig. 7.40 Constitutive relations for membrane elements (a) Concrete in compression, (b) concrete in tension, (c) nonprestressed steel, and (d) prestressing tendon. [Reprinted with permission from T. T. C. Hsu (1993). *Unified Theory of Reinforced Concrete,* CRC Press, Boca Raton, FL. Copyright CRC Press, Boca Raton, FL © 1993.]

based on the truss model consistently overestimated the tested response. The current relationships given in Eqs. 7.3 and 7.4 are based on relatively thin (75 mm) membrane elements with one layer of reinforcement (Vecchio and Collins, 1986). Recent tests (Adebar and Collins, 1994) are being conducted on thicker (300 mm) elements with two layers of reinforcement. The effect of confinement provided by through-the-thickness reinforcement may change the compression relationships in the future.

The average stress–strain response for a reinforced concrete web in tension is shown in Figure 7.40(b). The curve shown is an enlargement of the upper right-hand corner (first quadrant) of Figure 7.40(a). The concrete modulus of elasticity E_c is the same in both figures but is distorted in Figure 7.40(b) because the stress scale has been expanded while the strain has not. The maximum principal tensile strain ε_1 is of the same order of magnitude as the maximum principal compressive strain ε_2, even though the tensile stresses are not. Expressions for the ascending and descending branches are given in Eqs. 7.22 and 7.23.

As shown in Figure 7.40(c), the stress–strain response of a reinforcing bar embedded in concrete is different than that of a bare bar. The embedded bar is stiffened by the concrete surrounding it and does not exhibit a flat yield plateau. At strains beyond the yield strain of the bare bar, an embedded bar develops stresses that are lower than those in a bare bar. A bilinear approximation to the average stress–strain response of a mild steel bar embedded in concrete is given by Eqs. 7.37 and 7.38.

A typical stress–strain curve for a bare prestressing tendon is shown in Figure 7.40(d). Expression for low-relaxation and stress-relieved strands are given by Eqs. 7.41 and 7.42, respectively. For bonded and unbonded prestressing strands, approximate expressions for the prestressing stress f_{ps} are given by Eqs. 7.59 and 7.72, respectively.

In the development of the concrete tensile response shown in Figure 7.40(b), two conditions have been implied: (1) average stresses and average strains across more than one crack have been used and (2) the cracks are not so wide that shear cannot be transferred across them. The first condition has been emphasized more than once, but the second condition requires further explanation.

A diagonally cracked beam web is shown in Figure 7.41(a) with a diagram of the actual tensile stress variation and the average principal stress f_1 related to a principal tensile strain ε_1 taken over a finite gage length. For the cracked web the average principal tensile strain ε_1 is due mostly to the opening of the cracks because the elastic tensile strain is relatively small, that is,

$$\varepsilon_1 \approx \frac{w}{s_{m\theta}} \qquad (7.138)$$

where w is the crack width and $s_{m\theta}$ is the mean spacing of the diagonal cracks. If the crack width w becomes too large, it will not be possible to transfer

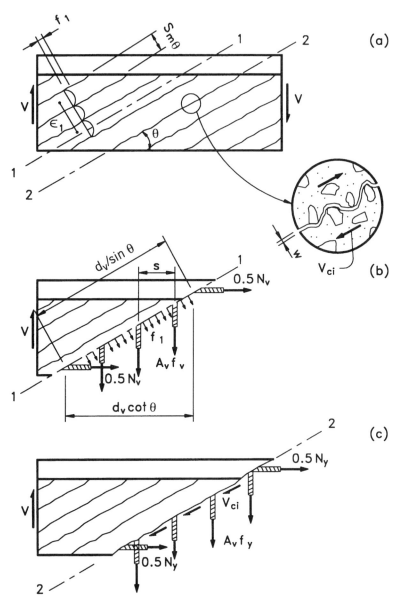

Fig. 7.41 Transmitting tensile forces across a crack (a) Beam web cracked by shear, (b) average stresses between cracks, and (c) local stresses at a crack. (after Collins and Mitchell, 1991). [Reprinted by permission of Prentice Hall, Upper Saddle River, NJ.]

shear across the crack by the aggregate interlock mechanism shown in the detail at a crack. In other words, if the cracks are too wide, shear failure occurs by slipping along the crack surface.

The aggregate interlock mechanism is modeled in the definitive work by Walraven (1981). It is based on a statistical analysis of the contact areas and the wedging action that occurs between irregular crack faces. At a crack, local shear stresses v_{ci} are developed, which enable tensile forces to be transmitted across the crack. Experimental pushoff tests on externally restrained specimens were conducted and they verified the analytical model. The variables in the tests were concrete strength, maximum aggregate size, total aggregate volume per unit volume of concrete, external restraint stiffness, and initial crack width. In fitting the experimental results to the theoretical model, the best results were obtained with a coefficient of friction of 0.4 and a matrix yielding strength that is a function of the square root of the concrete compressive strength. By using Walraven's experimental data, Vecchio and Collins (1986) derived a relationship between shear transmitted across a crack and the concrete compressive strength. Their expression was further simplified by Collins and Mitchell (1991) who dropped the effect of local compressive stresses across the crack and recommended that the limiting value of v_{ci} be taken as

$$v_{ci} \le \frac{0.18\sqrt{f_c'}}{0.3 + 24w/(a_{max} + 16)} \tag{7.139}$$

where w is the crack width (mm), a_{max} is the maximum aggregate size (mm) and f_c' is concrete compressive strength in mega pascals (MPa). By limiting the shear stress on the crack v_{ci} to the value of Eq. 7.139, crack slipping failures should not occur.

The average stresses on section 1-1 in Figure 7.41(a) within a concrete compressive strut that were used in developing the equilibrium Eqs. 7.132, 7.134, and 7.135 are repeated in Figure 7.41(b). The stresses in the transverse and longitudinal reinforcement are also average stresses because the stiffening effect of a bar embedded in concrete shown in Figure 7.40(c) applies. At a crack along section 2-2 in Figures 7.41(a) and 7.41(c), the concrete tensile stress vanishes, the aggregate interlock mechanism is active, and (as in Fig. 7.16) the reinforcement stress increases until it reaches its yield strength.

Both sets of stresses in Figures 7.41(b) and 7.41(c) must be in equilibrium with the same vertical shear force V. This vertical equilibrium can be stated as

$$A_v f_v \frac{d_v \cot \theta}{s} + f_1 \frac{b_v d_v}{\sin \theta} \cos \theta = A_v f_y \frac{d_v \cot \theta}{s} + v_{ci} \frac{b_v d_v}{\sin \theta} \sin \theta$$

and solving for the average principal tensile stress, we have

$$f_1 \leq v_{ci} \tan \theta + \frac{A_v}{b_v s}(f_y - f_v) \tag{7.140}$$

where f_1 is limited by the value of v_{ci} in Eq. 7.139.

The two sets of stresses in Figures 7.41(b) and 7.41(c) must also result in the same horizontal force, that is,

$$N_v + f_1 \frac{b_v d_v}{\sin \theta} \sin \theta = N_y + v_{ci} \frac{b_v d_v}{\sin \theta} \cos \theta$$

Substitution for v_{ci} from Eq. 7.140 and rearranging terms yields

$$N_y = N_v + f_1 b_v d_v + \left[f_1 - \frac{A_v}{b_v s}(f_y - f_v) \right] b_v d_v \cot^2 \theta \tag{7.141}$$

in which

$$N_y = A_{sx} f_y + A_{px} f_{ps} \tag{7.142}$$

$$N_v = A_{sx} f_{sx} + A_{px} f_{px} \tag{7.143}$$

where A_{sx} is the total area of longitudinal nonprestressed reinforcement, A_{px} is the total area of longitudinal prestressing tendons, f_y is the yield stress of the bare nonprestressed reinforcement, f_{ps} is the stress in the prestressing tendon from Eq. 7.59, f_{sx} and f_{px} are the smeared stresses averaged over the area $b_v d_v$ in the embedded longitudinal nonprestressed reinforcement and prestressing tendons, respectively. Equation 7.141 is a second limitation on f_1 that states that if the longitudinal reinforcement begins to yield at a crack, the maximum principal concrete tensile stress f_1 has been reached and cannot exceed

$$f_1 \leq \frac{N_y - N_v}{b_v d_v} \sin^2 \theta + \frac{A_v}{b_v s}(f_y - f_v) \cos^2 \theta$$

which can be written in terms of stresses as

$$f_1 \leq [\rho_{sx}(f_y - f_{sx}) + \rho_{px}(f_{ps} - f_{px})] \sin^2 \theta + \rho_v(f_y - f_v) \cos^2 \theta \tag{7.144}$$

where the reinforcement ratios ρ_{sx}, ρ_{px}, and ρ_v are defined in Eqs. 7.136, 7.137, and 7.126, respectively.

The response of a reinforced concrete beam subjected to shear forces can now be determined from the relationships discussed above. In Collins and Mitchell (1991), a 17 step procedure is outlined that covers the calculations and checks necessary to determine V of Eq. 7.134 as a function of the principal tensile strain ε_1. Similarly, Hsu (1993) presents a flow chart and an example problem illustrating the solution procedure for generating the shearing stress–strain curve.

Obviously these solution procedures are too cumbersome for practical design applications and design aids are needed to reduce the effort. These aids have been developed by Collins and Mitchell (1991) and are available in AASHTO [A5.8.3.4.2]. They are discussed in Section 7.8.3.

7.8.3 Shear Design Using Modified Compression Field Theory

Returning to the basic expression for nominal shear resistance given by Eq. 7.119, we have

$$V_n - V_p = V_c + V_s \tag{7.145}$$

Substituting the shear resistance from the concrete and web reinforcement determined by the modified compression field theory (Eq. 7.134), we get

$$V_n - V_p = f_1 b_v d_v \cot \theta + \frac{A_v f_v d_v}{s} \cot \theta \tag{7.146}$$

If we assume that $f_v = f_y$ when the limit state is reached, Eqs. 7.139 and 7.140 will yield an upper bound for the average principal tensile stress

$$f_1 \le v_{ci} \tan \theta \le \frac{0.18 \sqrt{f'_c}}{0.3 + \dfrac{24w}{a_{max} + 16}} \tan \theta \tag{7.147}$$

and we can express Eq. 7.146 as

$$V_n - V_p = \beta \sqrt{f'_c} b_v d_v + \frac{A_v f_y d_v}{s} \cot \theta \tag{7.148}$$

where

$$\beta \le \frac{0.18}{0.3 + \dfrac{24w}{a_{max} + 16}} \tag{7.149}$$

Now the crack width w can be expressed as the product of the average principal tensile strain ε_1 and the mean spacing of the diagonal cracks $s_{m\theta}$ to yield

$$w = \varepsilon_1 s_{m\theta} \qquad (7.150)$$

To simplify the calculations, Collins and Mitchell (1991) assume that the crack spacing $s_{m\theta}$ is 300 mm and that the maximum aggregate size a_{max} is 20 mm. This results in an upper bound for β of

$$\beta \leq \frac{0.18}{0.3 + 200\varepsilon_1} \qquad (7.151)$$

In addition to the limitation imposed on f_1 in Eq. 7.147 by the shear stress on a diagonal crack, f_1 is also assumed to follow the constitutive relationship shown in Figure 7.40(b) and given by Eq. 7.23 with $f_{cr} = 0.33\sqrt{f'_c}$, that is,

$$f_1 = \frac{\alpha_1\alpha_2 0.33\sqrt{f'_c}}{1 + \sqrt{500\varepsilon_1}} \qquad (7.152)$$

If we substitute this expression into Eq. 7.146 and relate it to Eq. 7.148, we get

$$\beta = \frac{\alpha_1\alpha_2 0.33 \cot \theta}{1 + \sqrt{500\varepsilon_1}}$$

assuming the tension stiffening or bond factors $\alpha_1\alpha_2$ are equal to unity, we have a second relationship for β that depends on the average principal tensile strain ε_1

$$\beta = \frac{0.33 \cot \theta}{1 + \sqrt{500\varepsilon_1}} \qquad (7.153)$$

At this point it is informative to compare the modified compression field theory Eq. 7.148 with the traditional expression for shear strength. From AASHTO (1996) Standard Specifications, the nominal shear strength for non-prestressed beams is (for inch-pound units)

$$V_n = 2\sqrt{f'_c}b_w d + \frac{A_v f_y d}{s} \qquad (7.154)$$

By comparing this result with Eq. 7.148, and realizing that $b_w = b_v$ and d is nearly equal to d_v, the two expressions will give the same results if $\theta = 45°$ and $\beta = 2$ (*Note:* β in inch-pound units must be divided by 12 to compare with SI units). Thus, the improvements introduced by the modified compression field theory are the ability to consider a variable orientation and a change in magnitude of the principal tensile stress across a cracked compression web. The orientation and magnitude are not fixed but vary according to the relative magnitude of the local shear stress and longitudinal strain.

In both Eqs. 7.151 and 7.153, an increase in tensile straining (represented by the average principal tensile strain ε_1) decreases β and the shear that can be resisted by the concrete tensile stresses. To determine this important parameter ε_1, the modified compression field theory introduces the compatibility conditions of a Mohr strain circle developed in Eqs. 7.127–7.131. Substituting Eq. 7.131 into Eq. 7.127 yields

$$\varepsilon_1 = \varepsilon_x + (\varepsilon_x - \varepsilon_2)\cot^2\theta \tag{7.155}$$

which shows that ε_1, depends on the longitudinal tensile strain ε_x, the principal compressive strain ε_2, and the orientation of the principal strains (or stresses) θ.

The principal compressive strain ε_2 can be obtained from the constitutive relationship shown in Figure 7.40(a) and given by Eq. 7.3. Set the strain ε'_c at peak compressive stress f'_c to -0.002, and solve the resulting quadratic equation to get

$$\varepsilon_2 = -0.002\left(1 - \sqrt{1 - \frac{f_2}{f_{2\,\text{max}}}}\right) \tag{7.156}$$

where $f_{2\,\text{max}}$ is the important reduced peak stress given by Eq. 7.4, that is,

$$f_{2\,\text{max}} = \frac{f'_c}{0.8 + 170\varepsilon_1} \tag{7.157}$$

which decreases as the tensile straining increases. Now the principal compressive stress f_2 is relatively large compared to the principal tensile stress f_1 as shown in Figure 7.40(a). Therefore, f_2 can reasonably and conservatively be estimated from Eq. 7.132 as

$$f_2 = \frac{v}{\sin\theta\,\cos\theta} \tag{7.158}$$

where the nominal shear stress on the concrete v includes the reduction provided by the vertical component V_p of an inclined prestressing tendon, that is,

$$v = \frac{V_n - V_p}{b_v d_v} = \frac{V_u - \phi V_p}{\phi b_v d_v} \tag{7.159}$$

By substituting Eqs. 7.157 and 7.158 into Eq. 7.156, and then substituting that result into Eq. 7.155, we get

$$\varepsilon_1 = \varepsilon_x + \left[\varepsilon_x + 0.002 \left(1 - \sqrt{1 - \frac{v}{f_c'} \frac{0.8 + 170\varepsilon_1}{\sin\theta\cos\theta}} \right) \right] \cot^2\theta \quad (7.160)$$

which can be solved for ε_1 once θ, v/f_c', and ε_x are known.

Before calculating the longitudinal strain ε_x in the web on the flexural tension side of the member, the relationships between some previously defined terms need to be clarified. This clarification can be done by substituting and rearranging some previously developed equilibrium equations and seeing which terms are the same and cancel out and which terms are different and remain. First, consider Eq. 7.135, which express longitudinal equilibrium in Figure 7.39(a), written as

$$N_v = (v\cot\theta - f_1)b_vd_v = V\cot\theta - f_1b_vd_v \quad (7.161)$$

where N_v is the total axial force due to all of the longitudinal reinforcement on the overall cross section, prestressed and nonprestressed, multiplied by smeared stresses averaged over the area b_vd_v. Second, consider Eq. 7.141, which expresses longitudinal equilibrium in Figure 7.41, written with $f_v = f_y$ as

$$N_y \geq N_v + f_1b_vd_v + f_1b_vd_v\cot^2\theta \quad (7.162)$$

where N_y is the total axial force due to all of the longitudinal reinforcement on the overall cross section, prestressed and nonprestressed, multiplied by the prestressing tendon stress and the yield stress of the base reinforcement, respectively. By substituting Eq. 7.161 into Eq. 7.162 and using Eq. 7.142 to express N_y, we get

$$A_{sx}f_y + A_{px}f_{ps} \geq V\cot\theta + f_1b_vd_v\cot^2\theta \quad (7.163)$$

Next, substituting Eq. 7.134, which expresses vertical equilibrium in Figure 7.39(c), into Eq. 7.163 yields

$$A_{sx}f_y + A_{px}f_{ps} \geq 2f_1b_vd_v\cot^2\theta + \frac{A_vf_yd_v}{s}\cot^2\theta = (2V_c + V_s)\cot\theta \quad (7.164)$$

where

$$V_c = f_1b_vd_v\cot\theta = \beta\sqrt{f_c'}b_vd_v \quad (7.165)$$

and

$$V_s = \frac{A_v f_y d_v}{s} \cot \theta \qquad (7.166)$$

Because the majority of the longitudinal reinforcement is on the flexural tension side of a member, Eq. 7.164 can be written in terms of the more familiar tensile steel areas A_s and A_{ps} by assuming that the shear depth d_v has been divided by two to yield

$$A_s f_y + A_{ps} f_{ps} \geq (V_c + 0.5 V_s) \cot \theta \qquad (7.167)$$

However, from Eq. 7.145, we can write

$$V_c + 0.5 V_s = V_n - 0.5 V_s - V_p$$

so that the longitudinal tensile force requirement caused by shear becomes

$$A_s f_y + A_{ps} f_{ps} \geq \left(\frac{V_u}{\phi_v} - 0.5 V_s - V_p \right) \cot \theta \qquad (7.168)$$

Thus, after all the manipulations with the previously developed equilibrium equations, it comes down to the requirement expressed in Eq. 7.168 that additional longitudinal tensile force must be developed to resist that caused by shear. This phenomenon was observed early in the study of shear resistance using truss analogies (see Fig. 7.34) where the presence of shear force was shown to add to the tensile chord force and subtract from the compressive chord force. Unfortunately, this concept was not included when the shear design procedures used in current practice were developed. This omission can be a serious shortcoming, especially in regions of high shear force, and it will undoubtedly be corrected in future editions of the other design codes.

In addition to the shear requirement given in Eq. 7.168, the longitudinal tensile reinforcement must also resist the tensile force produced by any applied moment M_u and axial load N_u as shown in Figure 7.42. This consideration leads to the following requirement for the longitudinal tensile reinforcement as given in AASHTO [A5.8.3.5]

$$A_s f_y + A_{ps} f_{ps} \geq \frac{M_u}{d_v \phi_f} + 0.5 \frac{N_u}{\phi_a} + \left(\frac{V_u}{\phi_v} - 0.5 V_s - V_p \right) \cot \theta \qquad (7.169)$$

where ϕ_f, ϕ_a, and ϕ_v are the resistance factors from Table 7.10 for flexure, axial load, and shear, respectively.

Return now to the parameter ε_x, which is used to measure the stiffness of the section when it is subjected to moment, axial load, and shear. If ε_x is small, the web deformations are small and the concrete shear strength V_c is high. If ε_x is larger, the deformations are larger and V_c decreases. If we recall

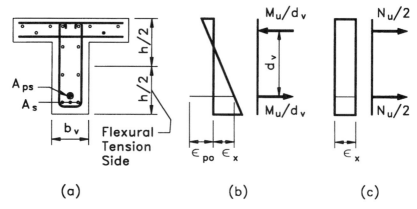

Fig. 7.42 Longitudinal strain and forces due to moment and tension. (a) Cross section, (b) strains and forces due to moment M_u, (c) strains and forces due to tension N_u. (After Collins and Mitchell, 1991). [Reprinted by permission of Prentice Hall, Upper Saddle River, NJ.]

that ε_x is the longitudinal strain in the web on the flexural tension side of the member, it can be conservatively estimated by calculating its value at the level of the flexural tensile reinforcement as shown in Figure 7.42. The longitudinal tensile force of Eq. 7.169 divided by a weighted stiffness quantity $E_s A_s + E_p A_{ps}$, and considering the precompression force $A_{ps} f_{po}$, results in the following expression given in AASHTO [A5.8.3.4.2]

$$\varepsilon_x = \frac{(M_u/d_v) + 0.5 N_u + 0.5 V_u \cot \theta - A_{ps} f_{po}}{E_s A_s + E_p A_{ps}} \le 0.002 \qquad (7.170)$$

where f_{po} is the stress in the prestressing tendon when the stress in the surrounding concrete is zero. Notice that the expression involving V_u has been simplified, the ϕ factors have not been included, and the maximum value for ε_x is 0.002.

If Eq. 7.170 gives a negative value for ε_x because of a relatively large precompression force, then the concrete area A_c on the flexural tension side will participate and increase the longitudinal stiffness. In that case the denominator of Eq. 7.170 should be changed to $E_c A_c + E_s A_s + E_p A_{ps}$.

When a designer is preparing shear envelopes and moment envelopes for combined force effects, the extreme values for shear and moment at a particular location do not usually come from the same position of live load. The commentary of AASHTO [C5.8.3.4.2] indicates that it is conservative to use the moment envelope values for M_u and V_u when calculating ε_x. In other words, it is not necessary to calculate M_u for the same live load position as was used in determining the maximum value of V_u.

Given applied forces, v calculated from Eq. 7.159, an estimated value of θ, and ε_x calculated from Eq. 7.170; ε_1 can be found by solving Eq. 7.160.

With ε_1 known, β can be determined from Eqs. 7.151 and 7.153. Then the concrete shear strength V_c can be calculated from Eq. 7.165, the required web reinforcement strength V_s from Eq. 7.145, and the required stirrup spacing s from Eq. 7.166. Thus, for an estimated value of θ, the required amount of web reinforcement to resist given force effects can be calculated directly.

To determine whether the estimated θ results in the minimum amount of web reinforcement, a designer must try a series of values for θ until the optimum is found. This possibly lengthy procedure has been shortened by the development of design aids, in the form of tables and figures, for selecting θ and β. The tables originally appeared in Collins and Mitchell (1991). They were expanded to include negative values of ε_x and put into graphical form for the AASHTO (1994) Bridge Specifications. The family of curves for members with web reinforcement taken from AASHTO [A5.8.3.4.2] is given as Figure 7.43.

When developing the curves in Figure 7.43, Collins and Mitchell (1991) were guided by the limitations that the principal compressive stress in the concrete f_2 did not exceed $f_{2\,max}$ and that the strain in the web reinforcement ε_v was at least 0.002, that is, $f_v = f_y$. Fayyaz (1994) conducted an analysis of the tables found in Collins and Mitchell (1991) using this criteria and concluded that there were exceptions. In a subsequent discussion with Collins (1994), this observation was confirmed. In cases of low relative shear stresses v/f_c', the optimum value for θ is obtained when β is at its maximum, even though $\varepsilon_v < 0.002$. Other exceptions were noted and attributed to engineering judgment acquired through use of the proposed provisions in trial designs.

The shear design of members with web reinforcement using the modified compression field theory consists of the following steps (Collins and Mitchell, 1991):

Step 1

Determine the factored shear V_u and moment M_u envelopes due to the Strength I limit state. Values are usually determined at the tenth points of each span. Interpolations can easily be made for values at critical sections such as a distance d_v from the face of a support. In the derivation of the modified compression field theory, d_v is defined as the lever arm between the resultant compressive force and the resultant tensile force in flexure. The AASHTO [A5.8.2.7] adds that d_v need not be less than $0.9d_e$ or $0.72h$, where d is the distance from the extreme compression fiber to the centroid of the tensile reinforcement and h is the overall depth of the member.

Step 2

Calculate the nominal shear stress v from Eq. 7.159 and divide by the concrete strength f_c' to obtain the shear stress ratio v/f_c'. If this ratio is higher than 0.25, a larger cross section must be chosen.

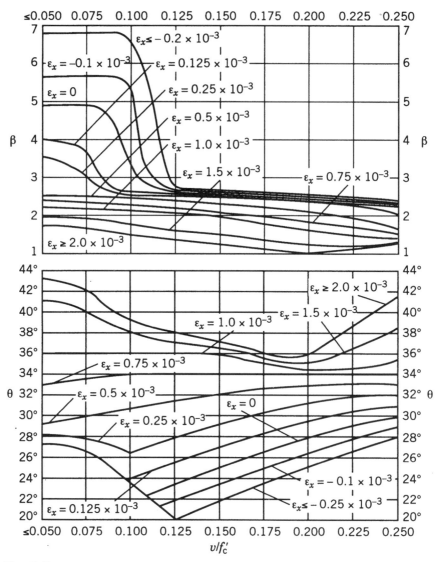

Fig. 7.43 Values of θ and β for sections with web reinforcement (AASHTO Fig. 5.8.3.4.2-1). [From *AASHTO LRFD Bridge Design Specifications*. Copyright © 1994 by the American Association of State Highway and Transportation Officials, Washington, DC. Used by permission.]

Step 3

Estimate a value of θ, say 40°, and calculate the longitudinal strain ε_x from Eq. 7.170. For a prestressed beam $f_{po} = f_{pe} + f_{pc}E_p/E_c$, where f_{pe} is the effective stress in the prestressing tendon after all losses and f_{pc} is the compressive stress in the concrete after all prestress losses at the centroid of the cross section.

Step 4

Use the calculated values of v/f'_c and ε_x, to determine θ from Figure 7.43 and compare with the value estimated in Step 3. If different, recalculate ε_x and repeat Step 4 until the estimated value of θ agrees with the value from Figure 7.43. When it does, select β from the top half of Figure 7.43. This figure was developed for inch-pound units and β must be divided by 12 for SI units.

Step 5

Calculate the required web reinforcement strength V_s from Eqs. 7.145 and 7.165 to give

$$V_s = \frac{V_u}{\phi_v} - V_p - 0.083\beta\sqrt{f'_c}b_v d_v \qquad (7.171)$$

Step 6

Calculate the required spacing of stirrups from Eq. 7.166 as

$$s \le \frac{A_v f_y d_v}{V_s}\cot\theta \qquad (7.172)$$

This spacing must not exceed the value limited by the minimum transverse reinforcement of AASHTO [A5.8.2.5] that is,

$$s \le \frac{A_v f_y}{0.083\sqrt{f'_c}b_v} \qquad (7.173)$$

It must also satisfy the maximum spacing requirements of AASHTO [A5.8.2.7]:

- If $V_u < 0.1f'_c b_v d_v$, then $s \le 0.8d_v \le 600$ mm   (7.174)
- If $V_u \ge 0.1f'_c b_v d_v$, then $s \le 0.4d_v \le 300$ mm   (7.175)

Step 7

Check the adequacy of the longitudinal reinforcement using Eq. 7.169. If the inequality is not satisfied, either add more longitudinal reinforcement or increase the amount of stirrups.

EXAMPLE 7.9

Determine the required spacing of No. 10 stirrups for the nonprestressed T-beam of Figure 7.44 at a positive moment location where V_u = 700 kN and M_u = 300 kN m. Use f'_c = 30 MPa and f_y = 400 MPa.

Step 1

Given V_u = 700 kN and M_u = 300 kN m.

$$A_s = 2000 \text{ mm}^2 \qquad b_v = 400 \text{ mm} \qquad b = 2000 \text{ mm}$$

Assume *NA* is in flange

$$a = \frac{A_s f_y}{0.85 f'_c b} = \frac{2000(400)}{0.85(30)(2000)} = 16 \text{ mm} < h_f = 200 \text{ mm, OK}$$

$$d_v = \max \begin{cases} d_e - a/2 = (1000 - 68) - 16/2 = \underline{924 \text{ mm}}, \text{ governs} \\ 0.9 d_e = 0.9(932) = 839 \text{ mm} \\ 0.72h = 0.72(1000) = 720 \text{ mm} \end{cases}$$

Step 2

Calculate $\dfrac{v}{f'_c}$ $V_p = 0$ $\phi_v = 0.9$

Eq. 7.159

$$v = \frac{V_u}{\phi_v b_v d_v} = \frac{700\,000}{0.9(400)(924)} = 2.10 \text{ N/mm}^2 = 2.10 \text{ MPa}$$

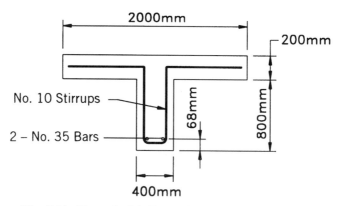

Fig. 7.44 Example 7.9. Determination of stirrup spacing.

$$\frac{v}{f'_c} = \frac{2.10}{30} = \underline{0.070} < 0.25, \text{ OK}$$

Step 3

Calculate ε_x from Eq. 7.170. $N_u = 0$. $A_{ps} = 0$.

   Estimate $\theta = 40°$ $\cot \theta = 1.192$

$$\varepsilon_x = \frac{M_u/d_v + 0.5V_u \cot \theta}{E_s A_s} = \frac{300 \times 10^6/924 + 0.5(700 \times 10^3)1.192}{200 \times 10^3(2000)}$$

$$\varepsilon_x = 1.85 \times 10^{-3}$$

Step 4

Determine θ and β from Figure 7.43. $\theta \approx 41.5°$, $\cot \theta = 1.130$

$$\varepsilon_x = \frac{300 \times 10^6/924 + 0.5(700 \times 10^3)1.130}{200 \times 10^3(2000)} = \underline{1.80 \times 10^{-3}}$$

Use $\underline{\theta = 41.5°}$ and $\underline{\beta = 1.75}$

Step 5

Calculate V_s from Eq. 7.171

$$V_s = \frac{V_u}{\phi_v} - 0.083\beta\sqrt{f'_c}b_v d_v$$

$$= 700 \times 10^3/0.9 - 0.083(1.75)\sqrt{30}(400)924$$

$$= 778\ 000 - 294\ 000 = \underline{484\ 000N}$$

Step 6

Calculate the required stirrup spacing from Eqs. 7.172–7.175 and using $A_v = 200$ mm²

$$s \le \frac{A_v f_y d_v}{V_s} \cot \theta = \frac{200(400)(924)}{484\ 000}(1.130) = 173 \text{ mm}$$

$$s \le \frac{A_v f_y}{0.083\sqrt{f'_c}b_v} = \frac{200(400)}{0.083\sqrt{30}(400)} = 440 \text{ mm}$$

$$V_u < 0.1f'_c b_v d_v = 0.1(30)(400)(924) = 1.109 \times 10^6 \text{ N}$$

$$s \le 0.8d_v = 0.8(924) = 739 \text{ or } 600 \text{ mm}$$

The stirrup spacing of $s = 173$-mm controls.

Step 7

Check the additional demand on the longitudinal reinforcement caused by shear as given by Eq. 7.169.

$$A_s f_y \geq \frac{M_u}{d_v \phi_f} + \left(\frac{V_u}{\phi_v} - 0.5 V_s\right) \cot \theta$$

$$2000(400) \ ? \ \frac{300 \times 10^6}{924(0.9)} + \left(\frac{700 \times 10^3}{0.9} - \frac{484\ 000}{2}\right) 1.130$$

$$800\ 000\ \text{N} \leq 361\ 000 + (778\ 000 - 242\ 000)1.130$$
$$= 967\ 000\ \text{N, no good}$$

Increasing V_s to satisfy the inequality

$$V_s \geq 2\left[\frac{V_u}{\phi_v} - \left(A_s f_y - \frac{M_u}{d_v \phi_f}\right) \tan \theta\right]$$

$$\geq 2[778\ 000 - (800\ 000 - 361\ 000) \tan 41.5°] = 779\ 000\ \text{N}$$

requires the stirrup spacing to be

$$s \leq \frac{200(400)924}{779\ 000}(1.130) = 107\ \text{mm}$$

which is likely not cost effective. It is better to simply increase A_s to satisfy the inequality, that is,

$$A_s \geq \frac{967\ 000}{f_y} = \frac{967\ 000}{400} = 2418\ \text{mm}^2$$

<u>Use 2-No. 35's plus 1-No. 25</u> $A_s = 2500\ \text{mm}^2$ and
<u>No. 10 U-stirrups at 170 mm</u>

7.9   CONCRETE BARRIER STRENGTH

The purpose of a concrete barrier, in the event of a collision by a vehicle, is to redirect the vehicle in a controlled manner. The vehicle shall not overturn or rebound across traffic lanes. The barrier shall have sufficient strength to survive the initial impact of the collision and to remain effective in redirecting the vehicle.

To meet the design criteria, the barrier must satisfy both geometric and strength requirements. The geometric conditions will influence the redirection of the vehicle and whether it will be controlled or not. This control must be provided for the complete mix of traffic from the largest trucks to the smallest automobiles. Geometric shapes and profiles of barriers that can control collisions have been developed over the years and have been proven by crash testing. Any variation from the proven geometry may involve risk and is not recommended. A typical solid concrete barrier cross section with sloping face on the traffic side is shown in Figure 7.45.

The strength requirements for barriers depend on the truck volume and speed of the traffic anticipated for the bridge. For given traffic conditions, a performance level for the barrier can be selected and the collision forces defined [A13.7.2]. The design forces and their location relative to the bridge deck are given for three performance levels in Table 4.5. The concrete barrier in Figure 7.45 has a height sufficient for performance level *PL-2*.

7.9.1 Strength of Uniform Thickness Barrier Wall

The lateral load carrying capacity of a uniform thickness solid concrete barrier was analyzed by Hirsh (1978). The expressions developed for the strength of the barrier are based on the formation of yield lines at the limit state. The assumed yield line pattern caused by a truck collision that produces a force F_t that is distributed over a length L_t is shown in Figure 7.46.

The fundamentals of yield line analysis are given in Section 6.3. Essentially, for an assumed yield line pattern that is consistent with the geometry and boundary conditions of a wall or slab, a solution is obtained by equating the external work due to the applied loads to the internal work done by the resisting moments along the yield lines. The applied load determined by this method is either equal to or greater than the actual load, that is, it is noncon-

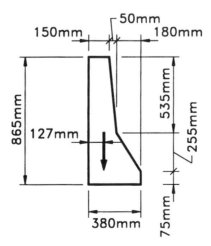

Fig. 7.45 Concrete barrier.

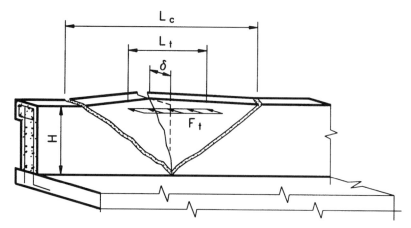

Fig. 7.46 Yield line pattern for barrier wall. [After Hirsh, 1978.]

servative. Therefore, it is important to minimize the load for a particular yield line pattern. In the case of the yield line pattern shown in Figure 7.46, the angle of the inclined yield lines can be expressed in terms of the critical length L_c. The applied force F_t is minimized with respect to L_c to get the least value of this upper bound solution.

External Work by Applied Loads. The original and deformed positions of the top of the wall are shown in Figure 7.47. The shaded area represents the integral of the deformations through which the uniformly distributed load $w_t = F_t/L_t$ acts. For a virtual displacement δ, the displacement x is

$$x = \frac{L_c - L_t}{L_c} \delta \qquad (7.176)$$

and the shaded area becomes

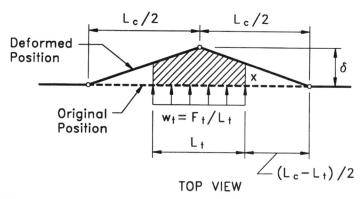

Fig. 7.47 External work by distributed load. [After Calloway, 1993.]

$$\text{area} = \frac{1}{2}(\delta + x)L_t = \frac{\delta}{2}\left(1 + \frac{L_c - L_t}{L_c}\right)L_t = \delta\,\frac{L_t}{L_c}\left(L_c - \frac{L_t}{2}\right) \quad (7.177)$$

so that the external work W done by w_t is

$$W = w_t(\text{area}) = \frac{F_t}{L_t}\,\delta\,\frac{L_t}{L_c}\left(L_c - \frac{L_t}{2}\right) = F_t\,\frac{\delta\left(L_c - \dfrac{L_t}{2}\right)}{L_c} \quad (7.178)$$

Internal Work Along Yield Lines. The internal work along the yield lines is sum of the products of the yield moments and the rotations through which they act. The segments of the wall are assumed to be rigid so that all of the rotation is concentrated at the yield lines. At the top of the wall (Fig. 7.48), the rotation θ of the wall segments for small deformations is

$$\theta \approx \tan\theta = \frac{2\delta}{L_c} \quad (7.179)$$

The barrier can be analyzed by separating it into a beam at the top and a uniform thickness wall below. At the limit state, the top beam will develop plastic moments M_b equal to its nominal bending strength M_n and form a mechanism as shown in Figure 7.48. Assuming that the negative and positive plastic moment strengths are equal, the internal work U_b done by the top beam is

$$U_b = 4M_b\theta = \frac{8M_b\delta}{L_c} \quad (7.180)$$

The wall portion of the barrier will generally be reinforced with steel in both the horizontal and vertical directions. The horizontal reinforcement in the wall develops moment resistance M_w per unit of length about a vertical

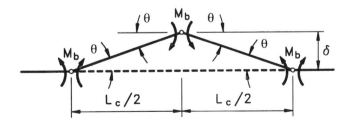

TOP VIEW

Fig. 7.48 Plastic hinge mechanism for top beam. [After Calloway, 1993.]

axis. The vertical reinforcement in the wall develops a cantilever moment resistance M_c per unit length about a horizontal axis. These two components of moment will combine to develop a moment resistance M_α about the inclined yield line as shown in Figure 7.49. When determining the internal work along inclined yield lines, it is simpler to use the projections of moment on and rotation about the vertical and horizontal axes.

If we assume that the positive and negative bending resistance M_w about the vertical axis are equal, and we realize that the projection on the horizontal plane of the rotation about the inclined yield line is θ, the internal work U_w done by the wall moment $M_w H$ is

$$U_w = 4M_w H\theta = \frac{8M_w H\delta}{L_c} \qquad (7.181)$$

The projection on the vertical plane of the rotation about the inclined yield line is δ/H, and the internal work U_c done by the cantilever moment $M_c L_c$ is

$$U_c = \frac{M_c L_c \delta}{H} \qquad (7.182)$$

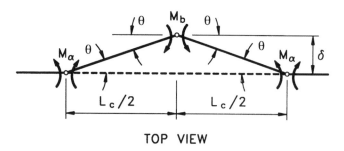

TOP VIEW

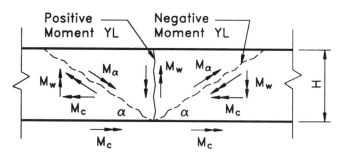

FRONT VIEW

Fig. 7.49 Internal work by barrier wall. [After Calloway, 1993.]

Nominal Railing Resistance to Transverse Load R_w. By equating the external work W to the internal work U, we have

$$W = U_b + U_w + U_c$$

If we substitute Eqs. 7.178, 7.180–7.182, we get

$$\frac{F_t\left(L_c - \dfrac{L_t}{2}\right)\delta}{L_c} = \frac{8M_b\delta}{L_c} + \frac{8M_wH\delta}{L_c} + \frac{M_cL_c\delta}{H}$$

Solve for the transverse vehicle impact force F_t

$$F_t = \frac{8M_b}{\left(L_c - \dfrac{L_t}{2}\right)} + \frac{8M_wH}{\left(L_c - \dfrac{L_t}{2}\right)} + \frac{M_cL_c^2}{H\left(L_c - \dfrac{L_t}{2}\right)} \qquad (7.183)$$

This expression depends on the critical length L_c that determines the inclination of α of the negative moment yield lines in the wall. The value for L_c that minimizes F_t can be determined by differentiating Eq. 7.183 with respect to L_c and setting the result equal to zero, that is,

$$\frac{dF_t}{dL_c} = 0 \qquad (7.184)$$

This minimization results in a quadratic equation that can be solved explicitly to give

$$L_c = \frac{L_t}{2} + \sqrt{\left(\frac{L_t}{2}\right)^2 + \frac{8H(M_b + M_wH)}{M_c}} \qquad (7.185)$$

When this value of L_c is used in Eq. 7.183, then the minimum value for F_t results, and the result is denoted as R_w, that is,

$$\min F_t = R_w \qquad (7.186)$$

where R_w is the nominal railing resistance to transverse load. By rearranging Eq. 7.183, we can write

$$R_w = \frac{2}{2L_c - L_t}\left(8M_b + 8M_wH + \frac{M_cL_c^2}{H}\right) \qquad (7.187)$$

7.9.2 Strength of Variable Thickness Barrier Wall

Most of the concrete barrier walls have sloping faces, as shown in Figure 7.45, and are not of uniform thickness. Calloway (1993) investigated the yield line approach applied to barrier walls of changing thickness. Equations were developed for R_w based on continuously varying moment resistances whose product with rotation was integrated over the height of the wall to obtain the internal work. The results of this more exact approach were compared to those obtained from the Hirsh equations (Eqs. 7.185 and 7.187) using various methods for calculating M_w and M_c. The recommended procedure is to use the Hirsh equations with average values for M_w and M_c. In the cases examined by Calloway, the R_w calculated by using average values was 4% less (conservative) than the more exact approach. This procedure is illustrated in the deck overhang design of Example Problem 7.10.1.

7.9.3 Crash Testing of Barriers

It should be emphasized that a railing system and its connection to the deck shall be approved only after they have been shown to be satisfactory through crash testing for the desired performance level [A13.7.3.1]. If minor modifications have been made to a previously tested railing system that do not affect its strength, it can be used without further crash testing. However, any new system must be verified by full-scale crash testing.

7.10 EXAMPLE PROBLEMS

In this section, a number of typical concrete beam and girder superstructure designs are given. The first example is the design of a concrete deck followed by design examples of solid slab, *T*-beam, prestressed girder, and multiple box-girder bridges.

Table 5.1 describes the notation used to indicate locations of critical sections for moments and shears. This notation is used throughout the example problems.

References to the AASHTO LRFD Specifications (1994) are enclosed in brackets and denoted by the letter A followed by the article number, for example, [A4.6.2.1.3]. If a commentary is cited the article number is preceded by the letter C. Figures and tables that are referenced are also enclosed in brackets to distinguish them from figures and tables in the text, for example, [Fig. A3.6.1.2.2-1] and [Table A4.6.2.1.3-1].

Appendix B includes tables that may be helpful to a designer when selecting bars sizes and prestressing tendons. These are referenced by the letter B followed by a number and are not enclosed in brackets.

The units chosen for the force effects in the example problems are kilonewton meter (kN m) for moments and kilonewton (kN) for shears. The resistance expression in the AASHTO specifications are in newtons (N) and millimeters (mm), which means that multipliers of 10^6 and 10^3 are added as necessary when force effects are compared to resistance.

The rationale for choosing kilonewtons and meters is to give the reader who is familiar with the kips and feet system, values for the force effects that are of the same order of magnitude as previously calculated. For example, bending moments in kilonewton meter are approximately 1.33 times the value in kip-feet. Thus, if a designer is anticipating a bending moment in a beam to be 600 kip ft, this now becomes 800 kN m in the SI system.

For shear forces and reactions, the multiplier between kilonewtons and kips is approximately 4. This multiplier is easy to apply and allows a mental adjustment of an anticipated shear force of 150 kips to a value of 600 kN.

Similarly, the integer multiplier between megapascals (or N/mm²) and ksi is 7. The familiar 60 ksi yield strength of reinforcing steel becomes 400 MPa, and a concrete compressive strength of 30 MPa is comparable to the common 4 ksi concrete.

The design examples generally follow the outline of *Appendix A—Basic Steps for Concrete Bridges* given at the end of Section 5 of the AASHTO (1994) LRFD Bridge Specifications. Care has been taken in preparing these examples, but they should not be considered as fully complete in every detail. Each designer must take the responsibility for understanding and correctly applying the provisions of the specifications. Additionally, the AASHTO LRFD Bridge Design Specifications will likely be altered each year by addendums that define interim versions. The computations outlined herein may not be current with the most recent interim.

7.10.1 Concrete Deck Design

Problem Statement. Use the approximate method of analysis [A4.6.2] to design the deck of the reinforced concrete *T*-beam bridge section of Figure E7.1-1 for a *HL*-93 live load and a *PL*-2 performance level concrete barrier (Fig. 7.45). The *T*-beams supporting the deck are 2440 mm on centers and

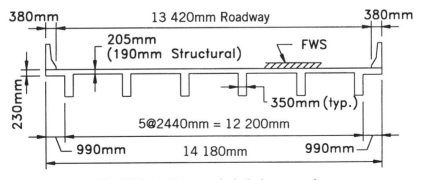

Fig. E7.1-1 Concrete deck design example.

have a stem width of 350 mm. The deck overhangs the exterior T-beam approximately 0.4 of the distance between T-beams. Allow for sacrificial wear of 15 mm of concrete surface and for a future wearing surface of 75-mm thick bituminous overlay. Use $f'_c = 30$ MPa, $f_y = 400$ MPa, and compare the selected reinforcement with that obtained by the empirical method [A9.7.2].

A.   Deck Thickness. The minimum thickness for concrete deck slabs is 175 mm [A9.7.1.1]. Traditional minimum depths of slabs are based on the deck span length S to control deflection to give [Table A2.5.2.6.3-1]

$$h_{\min} = \frac{S + 3000}{30} = \frac{2440 + 3000}{30} = 181 \text{ mm} > 175 \text{ mm}$$

Use $h_s = 190$ mm for the structural thickness of the deck. By adding the 15 mm allowance for the sacrificial surface, the dead weight of the deck slab is based on $h = 205$ mm. Because the portion of the deck that overhangs the exterior girder must be designed for a collision load on the barrier, its thickness has been increased by 25 mm to $h_o = 230$ mm.

B.   Weights of Components [Table A3.5.1-1]. For a 1 mm width of a transverse strip.

Barrier

$$P_b = 2400 \times 10^{-9} \text{ kg/mm}^3 \times 9.81 \text{ N/kg} \times 197\ 325 \text{ mm}^2 = 4.65 \text{ N/mm}$$

Future wearing surface

$$w_{DW} = 2250 \times 10^{-9} \times 9.81 \times 75 = 1.66 \times 10^{-3} \text{ N/mm}^2$$

Slab 205 mm thick

$$w_s = 2400 \times 10^{-9} \times 9.81 \times 205 = 4.83 \times 10^{-3} \text{ N/mm}^2$$

Cantilever overhang 230 mm thick

$$w_o = 2400 \times 10^{-9} \times 9.81 \times 230 = 5.42 \times 10^{-3} \text{ N/mm}^2$$

C.   Bending Moment Force Effects—General   An approximate analysis of strips perpendicular to girders is considered acceptable [A9.6.1]. The extreme positive moment in any deck panel between girders shall be taken to apply to all positive moment regions. Similarly, the extreme negative moment over any girder shall be taken to apply to all negative moment regions [A4.6.2.1.1].

The strips shall be treated as continuous beams with span lengths equal to the center-to-center distance between girders. The girders shall be assumed to be rigid [A4.6.2.1.6].

For ease in applying load factors, the bending moments will be determined separately for the deck slab, overhang, barrier, future wearing surface, and vehicle live load.

1. Deck Slab

$$h = 205 \text{ mm}, w_s = 4.83 \times 10^{-3} \text{ N/mm}^2, S = 2440 \text{ mm}$$

$$FEM = \pm\frac{w_s S^2}{12} = \pm\frac{(4.83 \times 10^{-3})(2440)^2}{12} = 2396 \text{ N mm/mm}$$

Placement of the deck slab dead load and results of a moment distribution analysis for negative and positive moments in a 1-mm wide strip is given in Figure E7.1-2.

A deck analysis design aid based on influence lines is given in Table A.1 of Appendix A. For a uniform load, the tabulated areas are multiplied by S for shears and by S^2 for moments.

$$R_{200} = w_s(\text{net area w/o cantilever})S$$

$$= 4.83 \times 10^{-3}(0.3928)2440 = 4.63 \text{ N/mm}$$

$$M_{204} = w_s(\text{net area w/o cantilever})S^2$$

$$= 4.83 \times 10^{-3}(0.0772)2440^2 = 2220 \text{ N mm/mm}$$

$$M_{300} = w_s(\text{net area w/o cantilever})S^2$$

$$= 4.83 \times 10^{-3}(-0.1071)2440^2 = -3080 \text{ N mm/mm}$$

Comparing the results from the design aid with those from moment distribution shows good agreement. In determining the remainder of the bending moment force effects, the design aid of Table A.1 will be used.

2. Overhang. The parameters $h_o = 230$ mm, $w_o = 5.42 \times 10^{-3}$ N/mm², and $L = 990$ mm. Placement of the overhang dead load is shown in Figure E7.1-3. By using the design aid Table A.1, the reaction on the exterior T-beam and the bending moments are

$$R_{200} = w_o(\text{net area cantilever})L$$

$$= 5.42 \times 10^{-3}\left(1.0 + 0.635\frac{990}{2440}\right)990 = 6.75 \text{ N/mm}$$

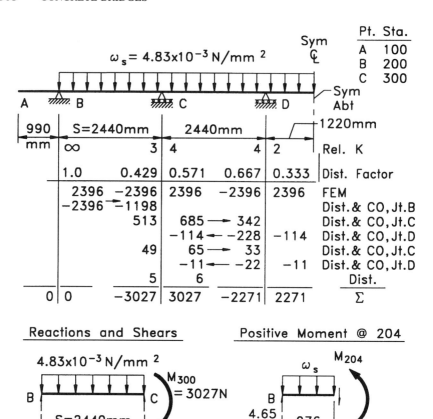

Fig. E7.1-2 Moment distribution for deck slab dead load.

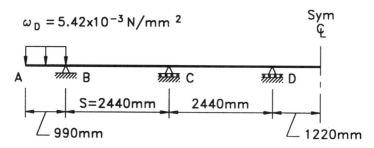

Fig. E7.1-3 Overhang dead load placement.

$$M_{200} = w_o(\text{net area cantilever})L^2$$

$$= 5.42 \times 10^{-3}(-0.5000)990^2 = -2656 \text{ N mm/mm}$$

$$M_{204} = w_o(\text{net area cantilever})L^2$$

$$= 5.42 \times 10^{-3}(-0.2460)990^2 = -1307 \text{ N mm/mm}$$

$$M_{300} = w_o(\text{net area cantilever})L^2$$

$$= 5.42 \times 10^{-3}(0.1350)990^2 = 717 \text{ N mm/mm}$$

3. Barrier. The parameters $P_b = 4.65$ N/mm and $L = 990 - 127 = 863$ mm. Placement of the center of gravity of the barrier dead load is shown in Figure E7.1-4. By using the design aid Table A.1 for the concentrated barrier load, the intensity of the load is multiplied by the influence line ordinate for shears and reactions. For bending moments, the influence line ordinate is multiplied by the cantilever length L.

$$R_{200} = P_b(\text{influence line ordinate}) = 4.65\left(1.0 + 1.270\frac{863}{2440}\right) = 6.74 \text{ N/mm}$$

$$M_{200} = P_b(\text{influence line ordinate})L = 4.65(-1.0000)(863) = -4013 \text{ N mm/mm}$$

$$M_{204} = P_b(\text{influence line ordinate})L$$

$$= 4.65(-0.4920)(863) = -1974 \text{ N mm/mm}$$

$$M_{300} = P_b(\text{influence line ordinate})L$$

$$= 4.65(0.2700)(863) = 1083 \text{ N mm/mm}$$

4. Future Wearing Surface. $FWS = w_{DW} = 1.66 \times 10^{-3}$ N/mm². The 75 mm of bituminous overlay is placed curb-to-curb as shown in Figure E7.1-5. The length of the loaded cantilever is reduced by the base width of the barrier

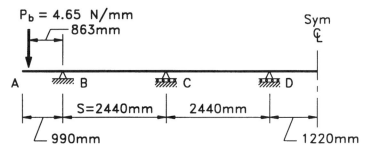

Fig. E7.1-4   Barrier dead load placement.

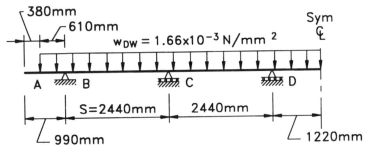

Fig. E7.1-5   Future wearing surface dead load placement.

to give $L = 990 - 380 = 610$ mm. If we use the design aid Table A.1, we have

$$R_{200} = w_{DW}[(\text{net area cantilever})L + (\text{net area w/o cantilever})S]$$

$$= 1.66 \times 10^{-3}\left[\left(1.0 + 0.635\frac{610}{2440}\right)610 + (0.3928)2440\right]$$

$$= 2.76 \text{ N/mm}$$

$$M_{200} = w_{DW}(\text{net area cantilever})L^2$$

$$= 1.66 \times 10^{-3}(-0.5000)(610)^2 = -309 \text{ N mm/mm}$$

$$M_{204} = w_{DW}[(\text{net area cantilever})L^2 + (\text{net area w/o cantilever})S^2]$$

$$= 1.66 \times 10^{-3}[(-0.2460)610^2 + (0.0772)2440^2] = 611 \text{ N mm/mm}$$

$$M_{300} = w_{DW}[(\text{net area cantilever})L^2 + (\text{net area w/o cantilever})S^2]$$

$$= 1.66 \times 10^{-3}[(0.135)610^2 + (-0.1071)2440^2] = -975 \text{ N mm/mm}$$

D. Vehicular Live Load—General   Where decks are designed using the approximate strip method [A4.6.2.1], and the strips are transverse, they shall be designed for the 145-kN axle of the design truck [A3.6.1.3.3]. Wheel loads on an axle are assumed to be equal and spaced 1800 mm apart [Fig. A3.6.1.2.2-1]. The design truck shall be positioned transversely to produce maximum force effects such that the center of any wheel load is not closer than 300 mm from the face of the curb for the design of the deck overhang and 600 mm from the edge of the 3600 mm wide design lane for the design of all other components [A3.6.1.3.1].

The width of equivalent interior transverse strips (mm) over which the wheel loads can be considered distributed longitudinally in CIP concrete decks is given as [Table A4.6.2.1.3-1]

- Overhang, $1140 + 0.833X$
- Positive moment, $660 + 0.55S$
- Negative moment, $1220 + 0.25S$

where X is the distance from the wheel load to centerline of support and S is the spacing of the T-beams. For our example, X is 310 mm (see Fig. E7.1-6) and S is 2440 mm.

Tire contact area [A3.6.1.2.5] shall be assumed as a rectangle with width of 510 mm and length ℓ given by

$$\ell = 2.28\gamma\left(1 + \frac{IM}{100}\right)P$$

where γ is the load factor, IM is the dynamic-load allowance, and P is the wheel load.

For our example,

$$\gamma = 1.75, \ IM = 33\%, \ P = 72.5 \text{ kN, and}$$

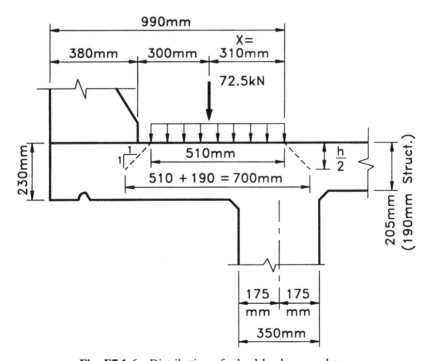

Fig. E7.1-6 Distribution of wheel load on overhang.

$$\ell = 2.28(1.75)(1 + 33/100)72.5 = 385 \text{ mm.}$$

Thus, the tire contact area is 510×385 mm with the 510 mm in the transverse direction as shown in Figure E7.1-6.

When calculating the force effects, wheel loads may be modeled as concentrated loads or as patch loads distributed transversely over a length along the deck span of 510 mm plus the slab depth [A4.6.2.1.6]. This distributed model is shown in Figure E7.1-6 and represents a 1:1 spreading of the tire loading to middepth of the beam. For our example, length of patch loading = $510 + 190 = 700$ mm. If the spans are short, the calculated bending moments in the deck using the patch loading can be significantly lower than those using the concentrated load. In the design example, force effects will be calculated conservatively by using concentrated wheel loads.

The number of design lanes N_L to be considered across a transverse strip is the integer value of the roadway width divided by 3600 mm [A3.6.1.1.1]. For our example,

$$N_L = \text{INT}\left(\frac{13\,420}{3600}\right) = 3$$

The multiple presence factor m is 1.2 for one loaded lane, 1.0 for two loaded lanes, and 0.85 for three loaded lanes. (If only one lane is loaded, we must consider the probability that this single truck can be heavier than each of the trucks traveling in parallel lanes [A3.6.1.1.2].)

1. Overhang Negative Live Load Moment. The critical placement of a single wheel load is shown in Figure E7.1-6. The equivalent width of a transverse strip is $1140 + 0.833X = 1140 + 0.833(310) = 1400$ mm and "m" $= 1.2$. Therefore,

$$M_{200} = \frac{-1.2(72.5 \times 10^3)(310)}{1400} = -19\,260 \text{ N mm/mm} = -19.26 \text{ kN m/m}$$

If the concrete barrier is structurally continuous, it will be effective in distributing the wheel loads in the overhang and the above moment can be reduced [A3.6.1.3.4]. However, as we shall see later, the overhang negative moment caused by horizontal forces from a vehicle collision [A13.7.2] will be greater than both of these moments.

2. Maximum Positive Live Load Moment. For repeating equal spans, the maximum positive bending moment occurs near the 0.4 point of the first interior span, that is, at Location 204. In Figure E7.1-7, the placement of wheel loads is given for one and two loaded lanes. For both cases, the equivalent width of a transverse strip is $660 + 0.55S = 660 + 0.55(2440) = 2000$

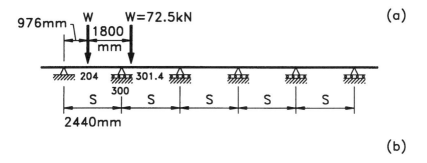

(a)

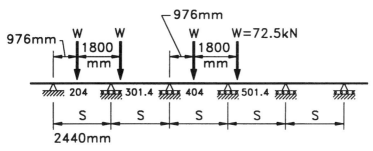

(b)

Fig. E7.1-7   Live load placement for maximum positive moment. (a) One loaded lane, "*m*" = 1.2, and (b) two loaded lanes, "*m*" = 1.0.

mm. If we use the influence line ordinates from Table A.1, the exterior girder reaction and positive bending moment with one loaded lane (m = 1.2) are

$$R_{200} = 1.2(0.5100 - 0.0486)\frac{72.5 \times 10^3}{2000} = 20.1 \text{ N/mm} = 20.1 \text{ kN/m}$$

$$M_{204} = 1.2(0.2040 - 0.0195) \times 2440\frac{72.5 \times 10^3}{2000}$$

$$= 19\,580 \text{ N mm/mm} = 19.58 \text{ kN m/m}$$

and for two loaded lanes (m = 1.0)

$$R_{200} = 1.0(0.5100 - 0.0486 + 0.0214 - 0.0039)\frac{72.5 \times 10^3}{2000}$$

$$= 17.4 \text{ N/mm} = 17.4 \text{ kN/m}$$

$$M_{204} = 1.0(0.2040 - 0.0195 + 0.0086 - 0.0016) \times 2440\frac{72.5 \times 10^3}{2000}$$

$$= 16\,900 \text{ N mm/mm} = 16.9 \text{ kN m/m}$$

Thus, the one loaded lane case governs.

3. Maximum Interior Negative Live Load Moment. The critical placement of live load for maximum negative moment is at the first interior deck support with one loaded lane ($m = 1.2$) as shown in Figure E7.1-8. The equivalent transverse strip width is $1220 + 0.25S = 1220 + 0.25(2440) = 1830$ mm. By using influence line ordinates from Table A.1, the bending moment at Location 300 becomes

$$M_{300} = 1.2(-0.1007 - 0.0781) \times 2440 \frac{72.5 \times 10^3}{1830}$$

$$= -20\,740 \text{ N mm/mm} = -20.74 \text{ kN m/m}$$

Note that the small increase due to a second truck is less than the 20% ($m = 1.0$) required to control. Therefore, only the one lane case is investigated.

4. Maximum Live Load Reaction on Exterior Girder. The exterior wheel load is placed 300 mm from the curb or 310 mm from the centerline of the support as shown in Figure E7.1-9. The width of the transverse strip is conservatively taken as that of the overhang. If we use influence line ordinates from Table A.1,

$$R_{200} = 1.2(1.1614 + 0.2869) \frac{72.5 \times 10^3}{1400} = 90.0 \text{ N/mm} = 90.0 \text{ kN/m}$$

E. Strength Limit State   The gravity load combination can be stated as [Table A3.4.1-1]

$$\eta \sum \gamma_i Q_i = \eta[\gamma_p DC + \gamma_p DW + 1.75(LL + IM)]$$

where

$$\eta = \eta_D \eta_R \eta_I \geq 0.95 \qquad \text{[A1.3.2.1-2]}$$

$$\eta_D = 0.95, \text{ reinforcement designed to yield [A1.3.3]}$$

$$\eta_R = 0.95, \text{ continuous [A1.3.4]}$$

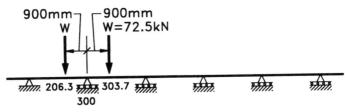

Fig. E7.1-8   Live load placement for maximum negative moment.

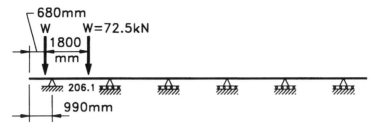

Fig. E7.1-9 Live load placement for maximum reaction at exterior girder.

$$\eta_I = 1.05, \text{ operationally important [A1.3.5]}$$

so that $\eta = 0.95(0.95)(1.05) = 0.95$.

The factor for permanent loads γ_p is taken at its maximum value if the force effects are additive and at its minimum value if it subtracts from the dominant force effect [Table A3.4.1-2]. The dead load DW is for the future wearing surface and DC represents all the other dead loads.

The dynamic load allowance IM [A3.6.2.1] is 33% of the live load force effect.

$$R_{200} = 0.95[1.25(4.63 + 6.75 + 6.74) + 1.50(2.76) + 1.75(1.33)(90.0)]$$

$$= 224.5 \text{ N/mm} = 224.5 \text{ kN/m}$$

$$M_{200} = 0.95[1.25(-2656 - 4013) + 1.50(-309) + 1.75(1.33)(-19\,260)]$$

$$= -50\,950 \text{ N mm/mm} = -50.96 \text{ kN m/m}$$

$$M_{204} = 0.95[1.25(2220) + 0.9*(-1307 - 1974) + 1.50(611)$$

$$+ 1.75(1.33)(19\,580)]$$

$$= 44\,000 \text{ N mm/mm} = 44.0 \text{ kN m/m}$$

$$M_{300} = 0.95[1.25(-3080) + 0.9*(717 + 1083) + 1.50(-975)$$

$$+ 1.75(1.33)(-20\,740)]$$

$$= -49\,370 \text{ N mm/mm} = -49.37 \text{ kN m/m}$$

The two negative bending moments are nearly equal, which confirms choosing the length of the overhang as $0.4S$. For selection of reinforcement, these moments can be reduced to their value at the face of the support [A4.6.2.1.6]. The T-beam stem width is 350 mm, so the design sections will be 175 mm

---

*The application of a minimum load factor to a term that helps reduce the total moment may be an unnecessary refinement.

on either side of the support centerline used in the analysis. The critical negative moment section is at the interior face of the exterior support as shown in the free body diagram of Figure E7.1-10.

The values for the loads in Figure E7.1-10 are for a 1-mm wide strip. The concentrated wheel load is for one loaded lane, that is, $W = 1.2(72\,500)/1400 = 62.14$ N/mm. In calculating the moment effect, the loads are kept separate so that correct R_{200} values are used.

1. Deck Slab

$$M_s = -\tfrac{1}{2}w_s x^2 + R_{200}x$$
$$= -\tfrac{1}{2}(4.83 \times 10^{-3})(175)^2 + 4.63(175) = 736 \text{ N mm/mm}$$

2. Overhang

$$M_o = -w_o L\left(\frac{L}{2} + x\right) + R_{200}x$$
$$= -(5.42 \times 10^{-3})(990)\left(\frac{990}{2} + 175\right) + 6.75(175) = -2414 \text{ N mm/mm}$$

3. Barrier

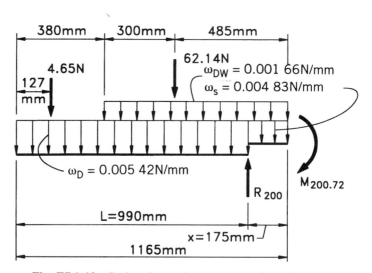

Fig. E7.1-10   Reduced negative moment at face of support.

$$M_b = -P_b(L + x - 127) + R_{200}x$$

$$= -4.65(1165 - 127) + 6.74(175) = -3647 \text{ N mm/mm}$$

4. *Future Wearing Surface*

$$M_{DW} = -\tfrac{1}{2}w_{DW}(L + x - 380)^2 + R_{200}x$$

$$= -\tfrac{1}{2}(0.001\ 66)(1165 - 380)^2 + 2.76(175) = -28 \text{ N mm/mm}$$

5. *Live Load*

$$M_{LL} = -W(485) + R_{200}x$$

$$= -62.14(485) + 90.00(175) = -14\ 388 \text{ N mm/mm}$$

6. *Strength I Limit State*

$$M_{200.72} = 0.95[0.9(736) + 1.25(-2414 - 3647) + 1.50(-28)$$

$$+ 1.75(1.33)(-14\ 388)]$$

$$= -38\ 420 \text{ N mm/mm} = -38.42 \text{ kN m/m}$$

This negative bending design moment represents a significant reduction from the value at $M_{200} = -50.95$ kN m/m. Because the extreme negative moment over any girder applies to all negative moment regions [A4.6.2.1.1], the extra effort required to calculate the reduced value is justified. Note that the moment at the outside face is smaller and can be calculated to be -27.3 kN m/m.

F.   Selection of Reinforcement—General   The material strengths are $f'_c = 30$ MPa and $f_y = 400$ MPa. Use epoxy-coated reinforcement in the deck and barrier. To keep track of units, it is convenient to recall that a MPa = N/mm^2.

The effective concrete depths for positive and negative bending will be different because of different cover requirements (see Fig. E7.1-11).

Concrete Cover [Table A5.12.3-1]

Deck surfaces subject to wear	60 mm
Bottom of CIP slabs	25 mm

Assuming a No. 15 bar, $d_b = 16$ mm, $A_b = 200$ mm^2

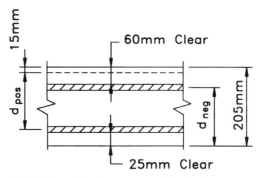

Fig. E7.1-11 Effective concrete depths for deck slabs.

$$d_{\text{pos}} = 205 - 15 - 25 - 16/2 = 157 \text{ mm}$$

$$d_{\text{neg}} = 205 - 60 - 16/2 = 137 \text{ mm}$$

A simplified expression for the required area of steel can be developed by neglecting the compressive reinforcement in the resisting moment to give [A5.7.3.2]

$$\phi M_n = \phi A_s f_y \left(d - \frac{a}{2} \right) \tag{E7.1-1}$$

where

$$a = \frac{A_s f_y}{0.85 f_c' b} \tag{E7.1-2}$$

Assuming that the lever arm $(d - a/2)$ is independent of A_s, we can replace it by jd and solve for an approximate A_s required to resist $\phi M_n = M_u$.

$$A_s \approx \frac{(M_u/\phi)}{f_y(jd)} \tag{E7.1-3}$$

Further, if we substitute $f_y = 400 \text{ N/mm}^2$, $\phi = 0.9$ [A5.5.4.2.1], and assume that for lightly reinforced sections $j \approx 0.92$, a trial steel area can be expressed as

$$\text{trial } A_s \approx \frac{M_u}{330d} \tag{E7.1-4}$$

Because it is an approximate expression, it will be necessary to verify the moment capacity of the selected reinforcement.

Maximum reinforcement [A5.7.3.3.1] is limited by the ductility requirement of $c \leq 0.42d$ or $a \leq 0.42 \, \beta_1 d$. For our example, $\beta_1 = 0.85 -$
$0.05(2/7) = 0.836$ [A5.7.2.2] to yield

$$a \leq 0.35d \qquad \text{(E7.1-5)}$$

Minimum reinforcement [A5.7.3.3.2] for components containing no prestressing steel is satisfied if

$$\rho = \frac{A_s}{(bd)} \geq 0.03\frac{f'_c}{f_y} \qquad \text{(E7.1-6)}$$

For the given material properties, the minimum area of steel per unit width
of slab is

$$\min A_s = \frac{0.03(30)}{400}(1)d = 0.002\,25\,d\frac{mm^2}{mm} \qquad \text{(E7.1-7)}$$

Maximum spacing of primary reinforcement [A5.10.3.2] for slabs is 1.5
times the thickness of the member or 450 mm. By using the structural slab
thickness of 190 mm,

$$s_{max} = 1.5(190) = 285 \text{ mm}$$

1. *Positive Moment Reinforcement*

$$M_u = 44.0 \text{ kN m/m}, \qquad d = 157 \text{ mm}$$

$$\text{trial } A_s \approx \frac{M_u}{330d} = \frac{44\,000}{330(157)} = 0.85 \text{ mm}^2/\text{mm}$$

$$\min A_s = 0.002\,25\,d = 0.002\,25(157) = 0.35 \text{ mm}^2/\text{mm, OK}$$

From Appendix B, Table B.4, try No. 15 @ 225 mm, provided $A_s = 0.889$
mm^2/mm.

$$a = \frac{A_s f_y}{0.85f'_c b} = \frac{0.889(400)}{0.85(30)(1)} = 14.0 \text{ mm}$$

Check ductility.

$$a \leq 0.35\,d = 0.35(157) = 55 \text{ mm, OK}$$

Check moment strength

$$\phi M_n = \phi A_s f_y \left(d - \frac{a}{2} \right)$$

$$= 0.9(0.889)(400)\left(157 - \frac{14.0}{2} \right) = 48\ 000 \text{ N mm/mm}$$

$$= 48.0 \text{ kN m/m} > 44.0 \text{ kN m/m, OK}$$

For transverse bottom bars,

<u>Use No. 15 @ 225 mm</u>

2. *Negative Moment Reinforcement*

$$M_u = -38.42 \text{ kN m/m} \qquad d = 137 \text{ mm}$$

$$\text{trial } A_s \approx \frac{38\ 420}{330(137)} = 0.85 \text{ mm}^2/\text{mm}$$

$$\text{min } A_s = 0.002\ 25(137) = 0.31 \text{ mm}^2/\text{mm, OK}$$

From Table B.4, try No. 15 @ 225 mm, provided $A_s = 0.889 \text{ mm}^2/\text{mm}$.

$$a = \frac{0.889(400)}{0.85(30)(1)} = 13.9 \text{ mm} < 0.35(137) = 48 \text{ mm, OK}$$

Check moment strength

$$\phi M_n = 0.9(0.889)(400)\left(137 - \frac{13.9}{2} \right)$$

$$= 41\ 620 \text{ N mm/mm} = 41.6 \text{ kN m/m} > 38.42 \text{ kN m/m, OK}$$

For transverse top bars,

<u>Use No. 15 @ 225 mm</u>

3. *Distribution Reinforcement.* Secondary reinforcement is placed in the bottom of the slab to distribute wheel loads in the longitudinal direction of the bridge to the primary reinforcement in the transverse direction. The required area is a percentage of the primary positive moment reinforcement. For primary reinforcement perpendicular to traffic [A9.7.3.2]

$$\text{percentage} = \frac{3840}{\sqrt{S_e}} \leq 67\%$$

where S_e is the effective span length [A9.7.2.3]. For monolithic T-beams, S_e is the distance face to face of stems, that is, $S_e = 2440 - 350 = 2090$ mm, and

$$\text{percentage} = \frac{3840}{\sqrt{2090}} = 84\%, \quad \text{use } 67\%$$

$$\text{dist } A_s = 0.67(\text{pos } A_s) = 0.67(0.889) = 0.60 \text{ mm}^2/\text{mm}$$

For longitudinal bottom bars,

<u>Use No. 10 @ 150 mm</u>, $A_s = 0.667 \text{ mm}^2/\text{mm}$

4.   *Shrinkage and Temperature Reinforcement.* The minimum amount of re-inforcement in each direction shall be [A5.10.8.2]

$$\text{temp } A_s \geq 0.75\frac{A_g}{f_y}$$

where A_g is the gross area of the section. For the full 205-mm thickness,

$$\text{temp } A_s \geq 0.75\frac{(205 \times 1)}{400} = 0.38 \text{ mm}^2/\text{mm}$$

The primary and secondary reinforcement already selected provide more than this amount, however, for members greater than 150 mm in thickness the shrinkage and temperature reinforcement is to be distributed equally on both faces. The maximum spacing of this reinforcement is 3.0 times the slab thickness or 450 mm. For the top face longitudinal bars,

$$\tfrac{1}{2}(\text{temp } A_s) = 0.19 \text{ mm}^2/\text{mm}$$

<u>use No. 10 @ 450 mm</u>, provided $A_s = 0.222 \text{ mm}^2/\text{mm}$

G.   *Control of Cracking—General*   Cracking is controlled by limiting the tensile stress in the reinforcement under service loads f_s to an allowable tensile stress f_{sa} [A5.7.3.4]

$$f_s \leq f_{sa} = \frac{Z}{(d_c A)^{1/3}} \leq 0.6 f_y$$

where

$$Z = 23\ 000 \text{ N/mm for severe exposure conditions}$$

d_c = depth of concrete from extreme tension fiber to center of closest bar

  ≤ 50 mm

A = effective concrete tensile area per bar having the same centroid as the reinforcement.

Service I Limit State applies to the investigation of cracking in reinforced concrete structures [A3.4.1]. In the Service I Limit State, the load modifier η is 1.0 and the load factors for dead and live load are 1.0. Therefore, the moment used to calculate the tensile stress in the reinforcement is

$$M = M_{DC} + M_{DW} + 1.33 M_{LL}$$

The calculation of service load tensile stress in the reinforcement is based on transformed elastic, cracked section properties [A5.7.1]. The modular ratio $n = E_s/E_c$ transforms the steel reinforcement into equivalent concrete. The modulus of elasticity E_s of steel bars is 200 000 MPa [A5.4.3.2]. The modulus of elasticity E_c of concrete is given by [A5.4.2.4]

$$E_c = 0.043 \gamma_c^{1.5} \sqrt{f_c'}$$

where

$$\gamma_c = \text{density of concrete} = 2400 \text{ kg/m}^3$$
$$f_c' = 30 \text{ MPa}$$

so that

$$E_c = 0.043(2400)^{1.5}\sqrt{30} = 27\,700 \text{ MPa}$$

and

$$n = \frac{200\,000}{27\,700} = 7.2, \quad \underline{\text{Use } n = 7}$$

1. Check of Positive Moment Reinforcement. Service I positive moment at Location 204 is

$M_{204} = M_{DC} + M_{DW} + 1.33 M_{LL}$

  $= (2220 - 1307 - 1974) + 611 + 1.33(19\,580) = 25\,600 \text{ N mm/mm}$

  $= 25.6 \text{ kN m/m}$

The calculation of the transformed section properties is based on a 1-mm wide doubly reinforced section as shown in Figure E7.1-12. Because of its relatively large cover, the top steel is assumed to be on the tensile side of the neutral axis. Sum of statical moments about the neutral axis yields

$$0.5bx^2 = nA_s'(d' - x) + nA_s(d - x)$$

$$0.5(1)x^2 = 7(0.889)(53 - x) + 7(0.889)(157 - x)$$

$$x^2 + 24.9x - 2614 = 0$$

Solve, $x = 40.2$ mm, which is less than 53 mm, so the assumption is correct. The moment of inertia of the transformed cracked section is

$$I_{cr} = \frac{bx^3}{3} + nA_s'(d' - x)^2 + nA_s(d - x)^2$$

$$= \frac{(1)(40.2)^3}{3} + 7(0.889)(53 - 40.2)^2 + 7(0.889)(157 - 40.2)^2$$

$$= 107\ 600\ \text{mm}^4/\text{mm}$$

and the tensile stress in the bottom steel becomes

$$f_s = n\left(\frac{My}{I_{cr}}\right) = 7\left(\frac{(25\ 600)(157 - 40.2)}{107\ 600}\right) = 195\ \text{MPa}$$

(The tensile stress was also calculated using a singly reinforced section and was found to be 200 MPa. The contribution of the top bars is small and can be safely neglected.)

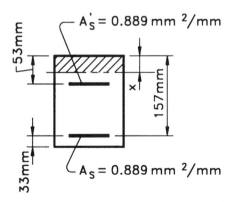

Fig. E7.1-12   Positive moment cracked section.

The positive moment tensile reinforcement of No. 15 bars at 25 mm on centers is located 33 mm from the extreme tension fiber. Therefore,

$$d_c = 33 \text{ mm} \leq 50 \text{ mm}$$

$$A = 2(33)(225) = 14\,850 \text{ mm}^2$$

and

$$f_{sa} = \frac{23\,000}{(33 \times 14\,850)^{1/3}} = 292 \text{ MPa} > 0.6 f_y$$

so use

$$f_{sa} = 0.6 f_y = 0.6(400) = 240 \text{ MPa} > f_s = 195 \text{ MPa, OK}$$

2. Check of Negative Moment Reinforcement. Service I negative moment at Location 200.72 is

$$M_{200.72} = M_{DC} + M_{DW} + 1.33\, M_{LL}$$

$$= (736 - 2414 - 3647) + (-28) + 1.33(-14\,388)$$

$$= -24\,490 \text{ N mm/mm} = -24.49 \text{ kN m/m}$$

The cross section for negative moment is shown in Figure E7.1-13 with compression in the bottom. This time x will be assumed greater than $d' = 33$ mm, so that the bottom steel will be in compression. Balancing statical moments about the neutral axis gives

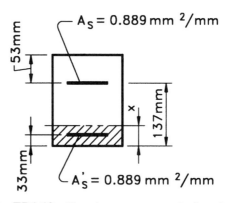

Fig. E7.1-13 Negative moment cracked section.

$$0.5bx^2 + (n - 1)A'_s(x - d') = nA_s(d - x)$$

$$0.5(1)x^2 + (6)(0.889)(x - 33) = 7(0.889)(137 - x)$$

$$x^2 + 23.1x - 2057 = 0$$

Solve, $x = 35.3$ mm, which is greater than 33 mm, so the assumption is correct. The moment of inertia of the transformed cracked section becomes

$$I_{cr} = \tfrac{1}{3}(1)(35.3)^3 + 6(0.889)(35.3 - 33)^2 + 7(0.889)(137 - 35.3)^2$$

$$= 79\,050 \text{ mm}^4/\text{mm}$$

and the tensile stress in the top steel is

$$f_s = 7\frac{(24\,490)(137 - 35.3)}{79\,050} = 221 \text{ MPa}$$

(The tensile stress was calculated to be 220 MPa by using a singly reinforced section. There really is no need to do a doubly reinforced beam analysis.)

The negative moment tensile reinforcement of No. 15 bars at 225 mm on centers is located 53 mm from the tension face. Therefore, d_c is the maximum value of 50 mm, and

$$A = 2(50)(175) = 17\,500 \text{ mm}^2$$

$$f_{sa} = \frac{23\,000}{(50 \times 17\,500)^{1/3}} = 240.5 \text{ MPa} > 0.6f_y$$

so use

$$f_{sa} = 0.6f_y = 240 \text{ MPa} > f_s = 221 \text{ MPa, OK}$$

H. Fatigue Limit State Fatigue need not be investigated for concrete decks in multigirder applications [A9.5.3].

I. Traditional Design for Interior Spans The design sketch in Figure E7.1-14 summarizes the arrangement of the transverse and longitudinal reinforcement in four layers for the interior spans of the deck. The exterior span and deck overhang have special requirements that must be dealt with separately.

J. Empirical Design of Concrete Deck Slabs Research has shown that the primary structural action of concrete decks is not flexure, but internal arching. The arching creates an internal compressive dome. Only a minimum amount

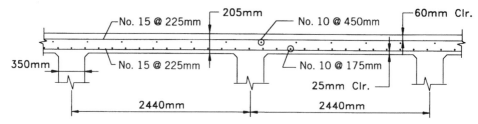

Fig. E7.1-14 Traditional design of interior deck spans.

of isotropic reinforcement is required for local flexural resistance and global arching effects [C9.7.2.1].

1. Design Conditions [A9.7.2.4]. Design depth excludes the loss due to wear, $h = 190$ mm. The following conditions must be satisfied:

• Supporting components are made of steel and/or concrete, YES
• The deck is fully CIP and water cured, YES
• $6.0 < S_e/h = 2090/190 = 11.0 < 18.0$, OK
• Core depth $= 205 - 60 - 25 = 120$ mm > 100 mm, OK
• Effective length [A9.7.2.3] $= 2090$ mm < 4100 mm, OK
• Minimum slab depth $= 175$ mm < 190 mm, OK
• Overhang $= 990$ mm $> 5h = 5 \times 190 = 950$ mm, OK
• $f'_c = 30$ MPa > 28 MPa, OK
• Deck must be made composite with girder, YES

2. Reinforcement Requirements [A9.7.2.5].

• Four layers of isotropic reinforcement, $f_y \geq 400$ MPa
• Outer layers placed in direction of effective length.
• Bottom layers: min $A_s = 0.570$ mm^2/mm, No. 15 @ 350 mm
• Top layers: min $A_s = 0.380$ mm^2/mm, No. 10 @ 250 mm
• Max spacing $= 450$ mm
• Straight bars only, hooks allowed, no truss bars.
• Only lap splices, no welded or mechanical splices permitted.
• Overhang designed for [A9.7.2.2 and A3.6.1.3.4].
  — Wheel loads using equivalent strip method if barrier discontinuous.
  — Equivalent line loads if barrier continuous.
  — Collision loads using yield line failure mechanism [A.A13.2].

3. Empirical Design Summary. With the empirical design approach there is no need for any analysis. When the design conditions have been met, the minimum reinforcement in all four layers is predetermined. The design sketch in Figure E7.1-15 summarizes the reinforcement arrangement for the interior deck spans.

K. Comparison of Reinforcement Quantities   The weight of reinforcement for the traditional and empirical design methods are compared in Table E7.1-1 for a 1-m wide transverse strip. Significant savings, in this case 74% of the traditionally designed reinforcement is required, can be made by adopting the empirical design method.

L. Deck Overhang Design   Neither the traditional method nor the empirical method for the design of deck slabs includes the design of the deck overhang. The design loads for the deck overhang [A9.7.1.5 and A3.6.1.3.4] are applied to a free-body diagram of a cantilever that is independent of the deck spans. The resulting overhang design can then be incorporated into either the traditional or empirical design by anchoring the overhang reinforcement into the first deck span.

Two limit states must be investigated: Strength I [A13.6.1] and Extreme Event II [A13.6.2]. The strength limit state considers vertical gravity forces and it seldom governs, unless the cantilever span is very long. The extreme event limit state considers horizontal forces caused by collision of a vehicle with the barrier. (These forces are given in Appendix A of Section 13 of the AASHTO LRFD Bridge Specifications and reference to articles there will be preceded by the letters AA.) The extreme event limit state will usually govern the design of the deck overhang.

I. Strength I Limit State. The design negative bending moment is taken at the exterior face of the support shown in Figure E7.1-6 for the loads given in Figure E7.1-10. Because the overhang has a single load path and is, therefore, a nonredundant member, then $\eta_R = 1.05$ and

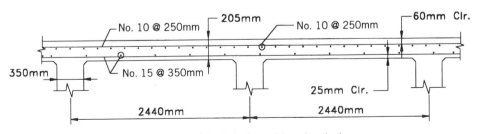

Fig. E7.1-15   Empirical design of interior deck spans.

TABLE E7.1-1 Comparison of Reinforcement Quantities[a]

Design Method	Transverse		Longitudinal		Totals	
	Top	Bottom	Top	Bottom	(kg)	(kg/m²)
Traditional	No. 15@225	No. 15@225	No. 10@450	No. 10@175		
Mass (kg)	117.0	117.0	24.3	63.6	321.9	22.7
Empirical	No. 10@250	No. 15@350	No. 10@250	No. 15@350		
Mass (kg)	44.2	75.2	44.2	75.2	238.8	16.8

[a](Area = 1 m × 14.18 m)

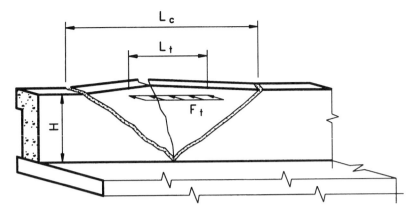

Fig. E7.1-16 Loading and yield line pattern for concrete barrier.

$$\eta = \eta_D\eta_R\eta_I = (0.95)(1.05)(1.05) = 1.05$$

The individual cantilever bending moments for a 1-mm wide design strip are

$$M_b = -P_b(990 - 175 - 127) = -4.65(688) = -3200 \text{ N mm/mm}$$
$$M_o = -w_o(990 - 175)^2/2 = -5.42 \times 10^{-3}(815)^2/2 = -1800 \text{ N mm/mm}$$
$$M_{DW} = -w_{DW}(815 - 380)^2/2$$
$$= -1.66 \times 10^{-3}(435)^2/2 = -157 \text{ N mm/mm}$$
$$M_{LL} = -W(310 - 175) = -62.14(135) = -8390 \text{ N mm/mm}$$

the factored design moment at Location 108.23 becomes

$$M_{108.23} = \eta[1.25M_{DC} + 1.50M_{DW} + 1.75(1.33M_{LL})]$$
$$= 1.05[1.25(-3200 - 1800) + 1.50(-157) + 1.75 \times 1.33(-8390)]$$
$$= -27\,300 \text{ N mm/mm} = -27.3 \text{ kN m/m}$$

When compared to the previously determined negative bending moment at the centerline of the support ($M_{200} = -50.95$ kN m/m), the reduction in negative bending to the face of the support is significant. This reduced negative bending moment is not critical in the design of the overhang.

2. Extreme Event II Limit State. The forces to be transmitted to the deck overhang due to a vehicular collision with the concrete barrier are determined from a strength analysis of the barrier. In this example, the loads applied to the barrier are for performance level *PL-2*, which is suitable for [A13.7.2] *high-speed mainline structures on freeways, expressways, highways, and areas with a mixture of heavy vehicles and maximum tolerable speeds.*

The minimum edge thickness of the deck overhang is 200 mm [A13.7.3.1.2] and the minimum height of barrier for *PL-2* is 810 mm [A13.7.3.2]. The design forces for *PL-2* that must be resisted by the barrier and its connection to the deck are given in Table E7.1-2 [AA13.2][a] and illustrated in Figure E7.1-16. The transverse and longitudinal forces are distributed over a length of barrier of 1070 mm. This length represents the approximate diameter of a truck tire, which is in contact with the wall at time of impact. The vertical force distribution length represents the contact length of a truck lying on top of the barrier after a collision. The design philosophy is that if any failures are to occur they should be in the barrier, which can be readily repaired, rather than in the deck overhang. The procedure is to calculate the barrier strength and then to design the deck overhang so that it is stronger. When calculating the resistance to extreme event limit states, the resistance factors ϕ are taken as 1.0 [A1.3.2.1] and the vehicle collision load factor is 1.0 [Tables A3.4.1-1 and A13.6.2].

M. Concrete Barrier Strength All traffic railing systems shall be proven satisfactory through crash testing for a desired performance level [A13.7.3.1]. If a previously tested system is used with only minor modifications that do not change its performance, additional crash testing is not required [A13.7.3.1.1]. The concrete barrier and its connection to the deck overhang shown in Figure E7.1-17 is similar to the profile and reinforcement arrangement of traffic barrier type T5 analyzed by Hirsch (1978) and tested by Buth et al. (1990).

As developed by a yield line approach in Section 7.9, the following expressions [AA13.3.1] can be used to check the strength of the concrete barrier away from an end or joint and to determine the magnitude of the loads that must be transferred to the deck overhang. From Eqs. 7.187 and 7.185

$$R_w = \left(\frac{2}{2L_c - L_t}\right)\left(8M_b + 8M_wH + \frac{M_cL_c^2}{H}\right) \qquad (E7.1-8)$$

[a]Reference to articles in Appendix A of AASHTO Section 13 are preceded by the letters AA.

TABLE E7.1-2 Design Forces for a *PL*-2 Barrier

Direction	Force (N)	Length (mm)
Transverse	240 000	1070
Longitudinal	80 000	1070
Vertical	80 000	5500

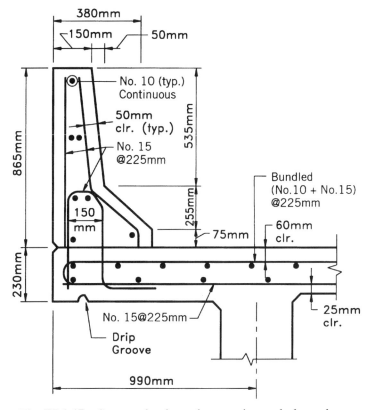

Fig. E7.1-17 Concrete barrier and connection to deck overhang.

$$L_c = \frac{L_t}{2} + \sqrt{\left(\frac{L_t}{2}\right)^2 + \frac{8H(M_b + M_w H)}{M_c}} \qquad \text{(E7.1-9)}$$

where

M_b = moment strength of beam, if any, at top of wall

M_w = moment strength of wall about vertical axis

M_c = moment strength of wall about horizontal axis

H = height of wall

L_t = longitudinal distribution length of impact force

L_c = critical wall length of yield line pattern

For the barrier wall in Figure E7.1-17, $M_b = 0$ and $H = 865$ mm.

1. Moment Strength of Wall About Vertical Axis, $M_w H$. The moment strength about the vertical axis is based on the horizontal reinforcement in the wall. Both the positive and negative moment strengths must be determined because the yield line mechanism develops both types (Fig. E7.1-16). The thickness of the barrier wall varies and it is convenient to divide it for calculation purposes into three segments as shown in Figure E7.1-18.

Neglecting the contribution of compressive reinforcement, the positive and negative bending strengths of Segment *I* are approximately equal and calculated as

$$(f'_c = 25 \text{ MPa}, f_y = 400 \text{ MPa})$$

$$A_s = 2 - \text{No. } 10's = 2(100) = 200 \text{ mm}^2$$

$$d_{ave} = \frac{(75 + 72 + 36)}{2} = 92 \text{ mm}$$

$$a = \frac{A_s f_y}{0.85 f'_c b} = \frac{(200)(400)}{0.85(25)(535)} = 7.0 \text{ mm}$$

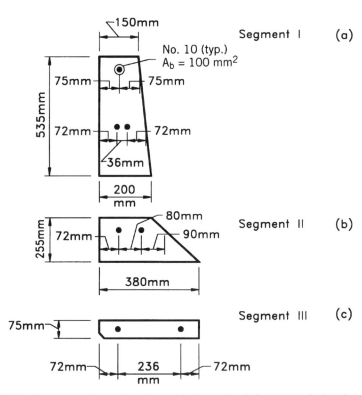

Fig. E7.1-18 Approximate location of horizontal reinforcement in barrier wall.

$$\phi M_{n_I} = \phi A_s f_y \left(d - \frac{a}{2} \right)$$

$$= 1.0(200)(400)(92 - 7.0/2) = 7.08 \times 10^6 \text{ N mm}$$

For Segment *II*, the moment strengths are slightly different. Considering the moment positive if it produces tension on the straight face, we have

$$A_s = 1 - \text{No. } 10 = 100 \text{ mm}^2$$

$$d_{\text{pos}} = 80 + 90 = 170 \text{ mm}$$

$$a = \frac{100(400)}{0.85(25)(255)} = 7.4 \text{ mm}$$

$$\phi M_{n_{\text{pos}}} = 1.0(100)(400)\left(170 - \frac{7.4}{2} \right) = 6.65 \times 10^6 \text{ N mm}$$

$$d_{\text{neg}} = 72 + 80 = 152 \text{ mm}$$

$$\phi M_{n_{\text{neg}}} = 1.0(100)(400)\left(152 - \frac{7.4}{2} \right) = 5.93 \times 10^6 \text{ N mm}$$

and the average value is

$$\phi M_{n_{II}} = \frac{(\phi M_{n_{\text{pos}}} + \phi M_{n_{\text{neg}}})}{2} = 6.29 \times 10^6 \text{ N mm}$$

For Segment III, the positive and negative bending strengths are equal and

$$A_s = 1 - \text{No. } 10 = 100 \text{ mm}^2$$

$$d = 236 + 72 = 308 \text{ mm}$$

$$a = \frac{100(400)}{0.85(25)(75)} = 25.1 \text{ mm}$$

$$\phi M_{n_{III}} = 1.0(100)(400)\left(308 - \frac{25.1}{2} \right) = 11.82 \times 10^6 \text{ N mm}$$

The total moment strength of the wall about the vertical axis is the sum of the strengths in the three segments

$$M_w H = \phi M_{n_I} + \phi M_{n_{II}} + \phi M_{n_{III}}$$

$$= (7.08 + 6.29 + 11.82) \times 10^6 = 25.2 \times 10^6 \text{ N mm}$$

$$\underline{M_w H = 25\ 200 \text{ kN mm}}$$

It is interesting to compare this value of M_wH with one determined by simply considering the wall to have uniform thickness and the same area as the actual wall, that is,

$$h_{ave} = \text{cross-sectional area/height of wall} = 196\,075/865 = 227 \text{ mm}$$

$$d_{ave} = 227 - 72 = 155 \text{ mm}$$

$$A_s = 4 - \text{No. } 10's = 4(100) = 400 \text{ mm}^2$$

$$a = \frac{400(400)}{0.85(25)(865)} = 8.7 \text{ mm}$$

$$M_wH = \phi M_n = \phi A_s f_y \left(d - \frac{a}{2} \right)$$

$$= 1.0(400)(400)\left(155 - \frac{8.7}{2} \right) = 24.1 \times 10^6 \text{ N mm}$$

$$\underline{M_wH = 24\,100 \text{ kN mm}}$$

This value is acceptably close to that calculated previously and is calculated with a lot less effort.

2. Moment Strength of Wall About Horizontal Axis, M_c. The moment strength about the horizontal axis is determined from the vertical reinforcement in the wall. The yield lines that cross the vertical reinforcement (Fig. E7.1-16) produce only tension in the sloping face of the wall, so that only the negative bending strength need be calculated.

The depth to the vertical reinforcement increases from top to bottom of the wall, therefore, the moment strength will also increase from top to bottom. Matching the spacing of the vertical bars in the barrier with the spacing of the bottom bars in the deck, the vertical bars become No. 15 bars at 225 mm ($A_s = 0.889 \text{ mm}^2/\text{mm}$) for the traditional design (Fig. E7.1-14). For Segment *I*, the average wall thickness is 175 mm and the moment strength about the horizontal axis becomes

$$d = 175 - 50 - 8 = 117 \text{ mm}$$

$$a = \frac{A_s f_y}{0.85 f'_c b} = \frac{0.889(400)}{0.85(25)(1)} = 16.7 \text{ mm}$$

$$M_{cI} = \phi A_s f_y \left(d - \frac{a}{2} \right) = 1.0(0.889)(400)\left(117 - \frac{16.7}{2} \right) = 38\,600 \text{ N mm/mm}$$

At the bottom of the wall the vertical reinforcement at the wider spread is not anchored into the deck overhang. Only the hairpin dowel at a narrower spread is anchored. The bending strength about the horizontal axis for Segments *II* and *III* may increase slightly where the vertical bars overlap, but it is reasonable to assume it is constant and determined by the hairpin dowel. The effective depth for the tension leg of the hairpin dowel is (Fig. E7.1-17)

$$d = 50 + 16 + 150 + 8 = 224 \text{ mm}$$

and

$$M_{cII+III} = 1.0(0.889)(400)\left(224 - \frac{16.7}{2}\right) = 76\,700 \text{ N mm/mm}$$

A weighted average for the moment strength about the horizontal axis is given by

$$M_c = \frac{[M_{cI}(535) + M_{cII+III}(255 + 75)]}{865} = \frac{[(38\,600)(535) + (76\,700)(330)]}{865}$$

$$M_c = 53\,100 \text{ N mm/mm} = 53.1 \text{ kN m/m}$$

3. *Critical Length of Yield Line Pattern, L_c.* With the moment strengths determined and $L_t = 1070$ mm, then Eq. E7.1-9 yields

$$L_c = \frac{L_t}{2} + \sqrt{\left(\frac{L_t}{2}\right)^2 + \frac{8H(M_b + M_w H)}{M_c}}$$

$$= \frac{1070}{2} + \sqrt{\left(\frac{1070}{2}\right)^2 + \frac{8(865)(0 + 24\,100)}{53.1}}$$

$$L_c = 2393 \text{ mm}$$

4. *Nominal Resistance to Transverse Load, R_w.* From Eq. E7.1-8, we have

$$R_w = \left(\frac{2}{2L_c - L_t}\right)\left(8M_b + 8M_w H + \frac{M_c L_c^2}{H}\right)$$

$$= \frac{2}{2(2393) - 1070}\left[0 + 8(24\,300) + \frac{53.1(2393)^2}{865}\right]$$

$$R_w = 293.8 \text{ kN} > F_t = 240 \text{ kN, OK}$$

5. *Shear Transfer Between Barrier and Deck.* The nominal resistance R_w must be transferred across a cold joint by shear friction. Free body diagrams of the forces transferred from the barrier to the deck overhang are shown in Figure E7.1-19.

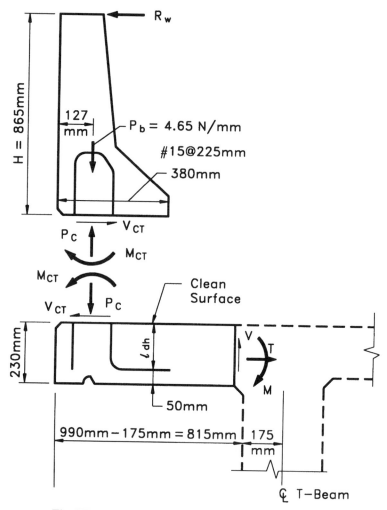

Fig. E7.1-19 Force transfer between barrier and deck.

Assuming that R_w spreads out at a 1:1 slope from L_c, the shear force at the base of the wall from the vehicle collision V_{CT}, which becomes the tensile force T per unit of length in the overhang, is given by [AA13.4.2]

$$T = V_{CT} = \frac{R_w}{(L_c + 2H)} \tag{E7.1-10}$$

$$T = \frac{293\,800}{[2393 + 2(865)]} = 71.3 \text{ N/mm} = 71.3 \text{ kN/m}$$

The nominal shear resistance V_n of the interface plane is given by [A5.8.4.1]

$$V_n = cA_{cv} + \mu(A_{vf}f_y + P_c) \qquad \text{(E7.1-11)}$$

which shall not exceed $0.2 f'_c A_{cv}$ or $5.5 A_{cv}$, where

A_{cv} = shear contact area = 380(1) = 380 mm²/mm

A_{vf} = dowel area across shear plane = 0.889 mm²/mm

f_y = yield strength of reinforcement = 400 MPa

P_c = permanent compressive force = P_b = 4.65 N/mm

f'_c = strength of weaker concrete = 25 MPa

c = cohesion factor [A5.8.4.2] = 0.52 MPa

μ = friction factor [A5.8.4.2] = 0.6.

The last two factors are for concrete placed against hardened concrete clean and free of laitance, but not intentionally roughened. Therefore, for a 1-mm wide design strip

$$V_n \leq 0.2f'_c A_{cv} = 0.2(25)(380) = 1900 \text{ N/mm}$$

$$\leq 5.5A_{cv} = 5.5(380) = 2090 \text{ N/mm}$$

$$= cA_{cv} + \mu(A_{vf}f_y + P_c) = 0.52(380) + 0.6[0.889(400) + 4.65]$$

$$= 414 \text{ N/mm}$$

$$\underline{V_n = 414 \text{ kN/m} > V_{CT} = T = 71.3 \text{ kN/m, OK}}$$

where V_{CT} is the shear force produced by a truck collision.

In the above calculations, only one leg of the hairpin is considered as a dowel, because only one leg is anchored in the overhang. The minimum cross-sectional area of dowels across the shear plane is [A5.8.4.1]

$$A_{vf} = 0.35\frac{b_v s}{f_y} \qquad \text{(E7.1-12)}$$

$$= 0.35\frac{(380)(225)}{400} = 75 \text{ mm}^2 < 200 \text{ mm}^2, \text{ OK}$$

which is satisfied by a single No. 15 bar.

The basic development length ℓ_{hb} for a hooked bar with $f_y = 400$ MPa is given by [A5.11.2.4.1]

$$\ell_{hb} = \frac{100 d_b}{\sqrt{f_c'}} \qquad \text{(E7.1-13)}$$

and shall not be less than $8d_b$ or 150 mm. For a No. 15 bar, $d_b = 16$ mm and

$$\ell_{hb} = \frac{100(16)}{\sqrt{30}} = 292 \text{ mm}$$

which is greater than $8(16) = 128$ and 150 mm. The modification factors of 0.7 for adequate cover and 1.2 for epoxy-coated bars [A5.11.2.4.2] apply, so that the development length ℓ_{dh} is changed to

$$\ell_{dh} = 0.7(1.2)\ell_{hb} = 0.84(292) = 245 \text{ mm}$$

The available development length (Fig. E7.1-19) is $230 - 50 = 180$ mm, which is not adequate, unless the required area is reduced to

$$A_s \text{ required} = (A_s \text{ provided})\left(\frac{180}{245}\right) = 0.889 \frac{180}{245} = 0.653 \text{ mm}^2$$

By using this area to recalculate M_c, L_c, and R_w, we get

$$a = \frac{0.653(400)}{0.85(25)(1)} = 12.3 \text{ mm}$$

$$M_{c_I} = 1.0(0.653)(400)\left(117 - \frac{12.3}{2}\right) = 28\,950 \text{ N mm/mm}$$

$$M_{c_{II+III}} = 1.0(0.653)(400)\left(224 - \frac{12.3}{2}\right) = 56\,900 \text{ N mm/mm}$$

$$M_c = \frac{[(28\,950)(535) + (56\,900)(330)]}{865}$$

$$= 39\,600 \text{ N mm/mm} = 39.6 \text{ kN m/m}$$

$$L_c = \frac{1070}{2} + \sqrt{\left(\frac{1070}{2}\right)^2 + \frac{8(865)(24\,300)}{39.6}} = 2664 \text{ mm}$$

$$R_w = \frac{2}{2(2664) - 1070}\left[8(24\,300) + \frac{39.6(2664)^2}{865}\right]$$

$$R_w = 244 \text{ kN} > 240 \text{ kN, OK}$$

The standard 90° hook with an extension of $12d_b = 12(16) = 192$ mm at the free end of the bar is adequate [A5.10.2.1].

6. *Top Reinforcement in Deck Overhang.* The top reinforcement must resist the negative bending moment over the exterior beam due to the collision and the dead load of the overhang. Based on the strength of the 90° hooks, the collision moment M_{CT} (Fig. E7.1-19) distributed over a wall length of $(L_c + 2H)$ is

$$M_{CT} = -\frac{R_w H}{L_c + 2H}$$

$$= -\frac{(244\ 000)(865)}{[2664 + 2(865)]} = -48\ 030 \text{ N mm/mm} = -48.03 \text{ kN m/m}$$

The dead load moments were calculated previously for Strength *I* so that for the Extreme Event *II* limit state, we have

$$M_u = \eta[1.25M_{DC} + 1.50M_{DW} + M_{CT}]$$

$$= 1.0[1.25(-3200 - 1800) + 1.50(-157) - 48\ 030]$$

$$= -54\ 500 \text{ N mm/mm} = -54.5 \text{ kN m/m}$$

Bundling a No. 10 bar with the No. 15 bar at 225 mm on centers, the negative moment strength becomes

$$A_s = \frac{(100 + 200)}{225} = 1.333 \text{ mm}^2/\text{mm}$$

$$d = 230 - 60 - \frac{16}{2} = 162 \text{ mm}$$

$$a = \frac{1.333(400)}{0.85(30)(1)} = 20.9 \text{ mm}$$

$$\phi M_n = 1.0(1.333)(400)\left(162 - \frac{20.9}{2}\right)$$

$$= 80\ 820 \text{ N mm/mm} = 80.82 \text{ kN m/m}$$

This moment strength will be reduced because of the axial tension force $T = R_w/(L_c + 2H)$.

$$T = \frac{(244\ 000)}{[2664 + 2(865)]} = 55.5 \text{ N/mm} = 55.5 \text{ kN/m}$$

By assuming the interaction curve between moment and axial tension is a straight line (Fig. E7.1-20)

$$\frac{P_u}{\phi P_n} + \frac{M_u}{\phi M_n} \leq 1.0$$

and solving for M_u, we get

$$M_u \leq \phi M_n \left(1.0 - \frac{P_u}{\phi P_n} \right) \tag{E7.1-14}$$

where $P_u = T$ and $\phi P_n = \phi A_{st} f_y$. The total longitudinal reinforcement A_{st} in the overhang is the combined area of the top and bottom bars.

$$A_{st} = \text{No. 10 @ 225, No. 15 @ 225, No. 15 @ 225}$$

$$= \frac{(100 + 200 + 200)}{225} = 2.222 \ \text{mm}^2/\text{mm}$$

$$\phi P_n = 1.0(2.222)(400) = 889 \ \text{N/mm}$$

so that

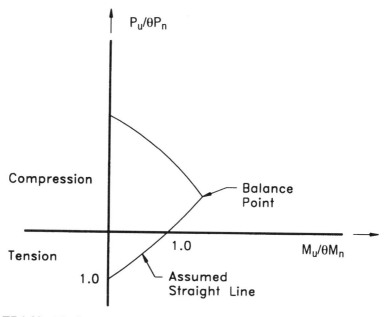

Fig. E7.1-20 Idealized interaction diagram for reinforced concrete members with combined bending and axial load.

$$M_u \leq 80.82\left(1 - \frac{55.5}{889}\right) = 75.77 \text{ kN m/m}$$

The Extreme Event *II* design moment M_u is 62.3 kN m/m, so for the top reinforcement of the overhang

<u>Use bundled (No. 10 and No. 15) @ 225 mm</u>

The top reinforcement must resist M_{CT} = 55.8 kN m/m directly below the barrier. Therefore, the free end of the No. 10 and No. 15 bars must terminate in a standard 180° hook. The development length ℓ_{dh} for a standard hook is [A5.11.2.4.1]

$$\ell_{dh} = \ell_{hb} \cdot \text{modification factors}$$

The modification factors of 0.7 for adequate cover and 1.2 for epoxy-coated bars [A5.11.2.4.2] apply and the ratio of (A_s required)/(A_s provided) can be approximated by the ratio of (M_u required)/(ϕM_n provided). Thus, the required development length for a No. 15 bar with ℓ_{hb} = 292 mm and ϕ = 1.0 becomes

$$\ell_{dh} = 292(0.7)(1.2)\left(\frac{48.03}{75.77}\right) = 155 \text{ mm}$$

The development length available (Fig. E7.1-17) for the hook in the overhang before reaching the vertical leg of the hairpin dowel is

$$\text{available } \ell_{dh} = 16 + 150 + 8 = 174 \text{ mm} > 155 \text{ mm}$$

and the connection between the barrier and the overhang shown in Figure E7.1-17 is satisfactory.

7. Length of the Additional Deck Overhang Bars. The additional No. 10 bars placed in the top of the deck overhang must extend beyond the centerline of the exterior *T*-beam into the first interior deck span. To determine the length of this extension, it is necessary to find the distance where theoretically the No. 10 bars are no longer required. This theoretical distance occurs when the collision plus dead load moments equal the negative moment strength of the continuing No. 15 bars at 225 mm. This negative moment strength was previously determined as −41 600 N mm/mm with ϕ = 0.9. For the extreme event limit state, ϕ = 1.0 and the negative moment strength increases to −46 200 N mm/mm.

Assuming a carryover factor of 0.5 and no further distribution, the collision moment diagram in the first interior deck span is shown in Figure E7.1-21.

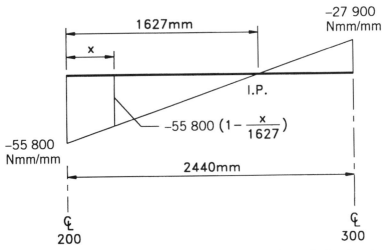

Fig. E7.1-21 Approximate moment diagram for collision forces in first interior deck span.

At a distance x from the centerline of the exterior T-beam, the collision moment is approximately

$$M_{CT}(x) = -55\ 800\left(1 - \frac{x}{1627}\right)$$

The dead load moments can be calculated as before from the loadings in Figure E7.1-10.

Barrier $-4.65(863 + x) + 6.74x$
Overhang $-(0.005\ 42)(990)(495 + x) + 6.75x$
Deck slab $-(0.004\ 83x^2/2) + 4.63x$
Future wearing surface   conservative to neglect

The distance x is found by equating the moment strength of $-46\ 200$ N mm/mm to the Extreme Event II load combination, that is,

$$-46\ 200 = M_u(x) = \eta\sum\gamma_i Q_i = 1.0[1.25M_{DC}(x) + M_{CT}(x)]$$

Solve for the resulting quadratic, $x = 416$ mm.

To account for the uncertainties in the theoretical calculation, an additional length of 15 $d_b = 15(11.3) = 170$ mm must be added to the length x before the bar can be cut off [A5.11.1.2]. This total length of $416 + 170 = 586$ mm beyond the centerline must be compared to the development length from the face of the support and the larger length selected.

The basic development length ℓ_{db} for a No. 10 bar is the larger of [A5.11.2]

$$\ell_{db} = 0.02\frac{A_b f_y}{\sqrt{f_c'}} = 0.02\frac{(100)(400)}{\sqrt{30}} = 146 \text{ mm}$$

or

$$0.06 d_b f_y = 0.06(11.3)(400) = 271 \text{ mm, controls}$$

The modification factors are

If epoxy coated [A5.11.2.1.2] = 1.2
Bundled bars [A5.11.2.3] = 1.0 (no increase for two-bar bundle)
So that the development length $\ell_d = 271(1.2)(1.0) = 325$ mm

The distance from the centerline of the 350 mm wide T-beam to the end of the development length is $325 + 175 = 500$ mm, which is less than the 586 mm calculated from the moment requirement. The length determination of the additional No. 10 bars in the deck overhang is summarized in Figure E7.1-22.

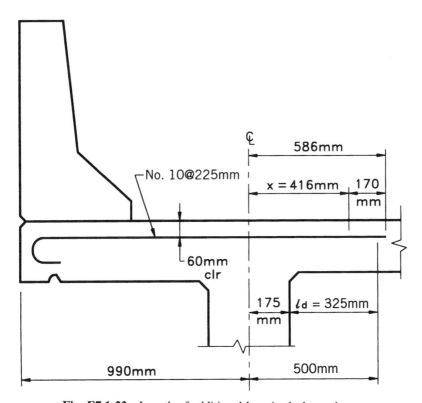

Fig. E7.1-22 Length of additional bars in deck overhang.

N. Closing Remarks   This example is general in most respects for application to decks supported by different longitudinal girders. However, the effective span length must be adjusted for the different girder flange configurations.

Designers are encouraged to use the empirical design procedure. The savings in design effort and reinforcement can be appreciable. Obviously, the details for the additional bars (Fig. E7.1-22) in the top of the deck overhang will be different for the empirical design than the traditional design.

The performance level *PL*-2 chosen for the concrete barrier in this example may have to be increased for some traffic environments. This choice of performance level is another decision that must be made when the design criteria for a project are being established.

7.10.2   Solid Slab Bridge Design

Problem Statement   Design the simply supported solid slab bridge of Figure E7.2-1 with a span length of 10 670 mm center to center of bearings for a *HL*-93 live load. The roadway width is 13 400-mm curb to curb. Allow for a future wearing surface of 75-mm thick bituminous overlay. Use $f'_c = 30$ MPa and $f_y = 400$ MPa. Follow the slab bridge outline in Appendix A5.4 and the beam and girder bridge outline in Section 5-Appendix A5.3 of the AASHTO (1994) LRFD Bridge Specifications.

A. **Check Minimum Recommended Depth [Table A2.5.2.6.3-1]**

$$h_{\min} = \frac{1.2(S + 3000)}{30} = \frac{1.2(10\ 670 + 3000)}{30} = 546.8 \text{ mm}$$

$$\underline{\text{Use } h = 550 \text{ mm}}$$

B. **Determine Live Load Strip Width [A4.6.2.3]**   *puge* A5.3

Span = 10 670 mm, primarily in the direction parallel to traffic

Span > 4600 mm, therefore the longitudinal strip method for slab-type bridges applies [A4.6.2.1.3]

1. *One-Lane Loaded*   Multiple presence factor included [C4.6.2.3]

$$E = \text{equivalent width (mm)}$$

$$E = 250 + 0.42\sqrt{L_1 W_1}$$

where $L_1 = $ modified span length

$$= \min \begin{bmatrix} 10\ 670 \text{ mm} \\ 18\ 000 \text{ mm} \end{bmatrix} = 10\ 670 \text{ mm}$$

(a)

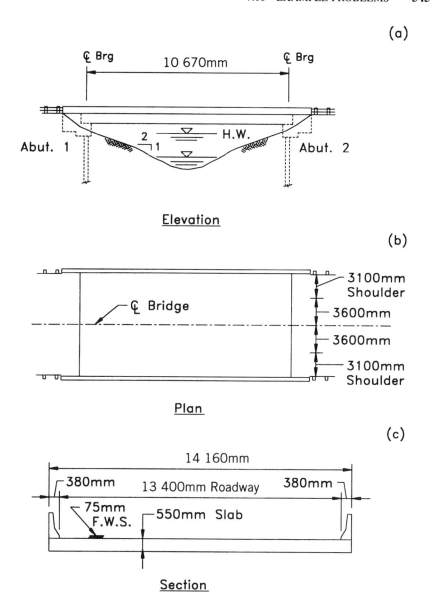

Fig. E7.2-1 Solid slab bridge design example. (a) Elevation, (b) plan, and (c) section.

W_1 = modified edge-to-edge width

$$= \min \begin{bmatrix} 14\ 160 \text{ mm} \\ 9000 \text{ mm} \end{bmatrix} = 9000 \text{ mm}$$

$$E = 250 + 0.42\sqrt{(10\ 670)(9000)} = 4370 \text{ mm}$$

2. *Multiple Lanes Loaded*

$$E = 2100 + 0.12\sqrt{L_1 W_1} \le \frac{W}{N_L}$$

where $L_1 = 10\ 670$ mm

$$W_1 = \min\begin{bmatrix} 14\ 160 \text{ mm} \\ 18\ 000 \text{ mm} \end{bmatrix} = 14\ 160 \text{ mm}$$

W = actual edge-to-edge width = 14 160 mm

N_L = number of design lanes [A3.6.1.1.1] = $INT\left(\dfrac{w}{3600}\right)$

where w = clear roadway width = 13 400 mm

$$N_L = INT\left(\frac{13\ 400}{3600}\right) = 3$$

$$E = 2100 + 0.12\sqrt{(10\ 670)(14\ 160)}$$

$$= 3580 \text{ mm} \le \frac{14\ 160}{3} = 4730 \text{ mm}$$

Use $E = 3580$ mm

C. **Applicability of Live Load for Decks and Deck Systems**   Slab-type bridges shall be designed for all of the vehicular live loads specified in AASHTO [A3.6.1.2], including the lane load [A3.6.1.3.3]
  1. *Maximum Shear Force—Axle Loads* (Fig. E7.2-2)
      Truck: [A3.6.1.2.2]

$$V_A^{Tr} = 145(1.0 + 0.60) + 35(0.19) = 238.7 \text{ kN}$$

Lane: [A3.6.1.2.4]

$$V_A^{Ln} = \tfrac{1}{2}(9.3)(10\ 670) = 49.6 \text{ kN}$$

Tandem: [A3.6.1.2.3]

$$V_A^{Ta} = 110\left(1 + \frac{9.47}{10.67}\right) = 207.6 \text{ kN not critical}$$

Impact Factor = $1 + IM/100$, where $IM = 33\%$ [A3.6.2.1]

Impact Factor = 1.33, not applied to design lane load

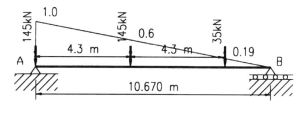

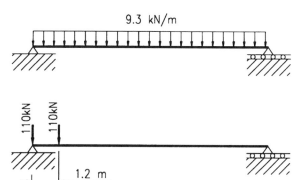

Fig. E7.2-2 Live load placement for maximum shear force.

$$V_{LL+IM} = 238.7(1.33) + 49.6 = 367.1 \text{ kN}$$

2. *Maximum Bending Moment at Midspan—Axle Loads* (Fig. E7.2-3)
   Truck:

$$M_c^{Tr} = [145(2668 + 518) + 35(518)] \times 10^{-3} = 480.1 \text{ kN m}$$

Lane:

$$M_c^{Ln} = 9.3[\tfrac{1}{2}(2668)(10\ 670)] \times 10^{-3} = 132.4 \text{ kN m}$$

Tandem:

$$M_c^{Ta} = \frac{110(2668)}{10^3}\left(1 + \frac{4135}{5335}\right) = 521.0 \text{ kN m, governs}$$

$$M_{LL+IM} = 521.0(1.33) + 132.4 = 825.3 \text{ kN m}$$

D. **Select Resistance Factors (Table 7.10) [A5.5.4.2.1]**

Strength Limit State	ϕ
Flexure and tension	0.90
Shear and torsion	0.90

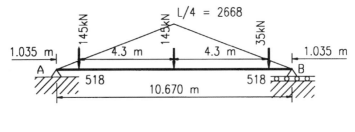

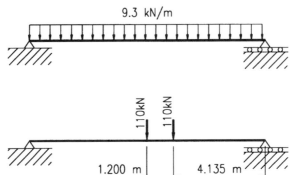

Fig. E7.2-3 Live load placement for maximum bending moment.

Axial compression	0.75
Bearing on concrete	0.70
Compression in strut-and-tie models	0.70

E. Select Load Modifiers [A1.3.2.1]

	Strength	Service	Fatigue	
1. Ductility, η_D	0.95	1.0	1.0	[A1.3.3]
2. Redundancy, η_R	1.05	1.0	1.0	[A1.3.4]
3. Importance, η_I	1.05	N/A[a]	N/A	[A1.3.5]
$\eta = \eta_D\eta_R\eta_I$	1.05	1.0	1.0	

F. Select Applicable Load Combinations (Table 3.1) [Table A3.4.1-1]
Strength I Limit State

$$U = \eta[1.25DC + 1.50DW + 1.75(LL + IM) + 1.0FR + \gamma_{TG}TG]$$

Service I Limit State

[a]N/A denotes not applicable.

$$U = 1.0(DC + DW) + 1.0(LL + IM) + 0.3(WS + WL) + 1.0FR$$

Fatigue Limit State

$$U = 0.75(LL + IM)$$

G. Calculate Live Load Force Effects

1. *Interior Strip*   Shear and moment per lane are given in Section 7.10.2, Parts C.1 and C.2. Shear and moment per *m* width of strip is critical for multiple lanes loaded because one-lane live load strip width = 4370 mm > 3580 mm

$$V_{LL+IM} = \frac{367.1}{3.580} = 102.5 \text{ kN/m}$$

$$M_{LL+IM} = \frac{825.3}{3.580} = 230.5 \text{ kN m/m}$$

2. *Edge Strip* [A4.6.2.1.4].   Longitudinal edges strip width for a line of wheels

$$= \text{distance from edge to face of barrier} + 300 \text{ mm}$$

$$+ \tfrac{1}{2} \text{ strip width} \le \text{full strip width or } 1800 \text{ mm}$$

$$= 380 + 300 + \frac{3580}{2} = 2470 \text{ mm} > 1800 \text{ mm}$$

Because strip width is limited to 1800 mm, one-lane loaded (wheel line = $\tfrac{1}{2}$ lane load) with a multiple presence factor of 1.2 will be critical (Fig. E7.2-4):

$$V_{LL+IM} = \frac{\tfrac{1}{2}(367.1)(1.2)}{1.800} = 122.4 \text{ kN/m}$$

$$M_{LL+IM} = \frac{\tfrac{1}{2}(825.3)(1.2)}{1.800} = 275.1 \text{ kN m/m}$$

H. Calculate Force Effects from Other Loads

1. *Interior Strip, 1 m wide*

$$DC \quad \rho_{conc} = 2400 \text{ kg/m}^3 \text{ [Table A3.5.1-1]}$$

$$w_{DC} = (2400)(9.81)(10^{-9})(550)$$

$$= 0.012\,949 \text{ N/mm}^2 = 12.95 \text{ kN/m}^2$$

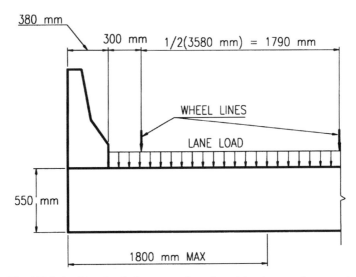

Fig. E7.2-4 Live load placement for edge strip shear and moment.

$$V_{DC} = \tfrac{1}{2}(12.95)(10.67) = 69.1 \text{ kN/m}$$

$$M_{DC} = \frac{w_{DC}L^2}{8} = \frac{(12.95)(10.67)^2}{8} = 184.3 \text{ kN m/m}$$

DW Bituminous Wearing Surface, 75 mm thick

$$\rho_{DW} = 2250 \text{ kg/m}^3 \text{ [Table A3.5.1-1]}$$

$$w_{DW} = (2250)(9.81)(10^{-9})(75)$$
$$= 0.001\ 655 \text{ N/mm}^2 = 1.66 \text{ kN/m}^2$$

$$V_{DW} = \tfrac{1}{2}(1.66)(10.67) = 8.8 \text{ kN/m}$$

$$M_{DW} = \frac{(1.66)(10.67)^2}{8} = 23.6 \text{ kN m/m}$$

2. *Edge strip, 1 m wide, barrier* = 4.65 N/mm = 4.65 kN/m Assume barrier load spread over width of one-half strip at edge of 1800 mm.

slab wt for interior strip.

$$DC:\quad w_{DC} = 12.95 + \frac{4.65 \text{ kN/m}}{1.800 \text{ m}} = 15.53 \text{ kN/m}^2$$

$$V_{DC} = \tfrac{1}{2}(15.53)(10.67) = 82.9 \text{ kN/m}$$

$$M_{DC} = \frac{(15.53)(10.67)^2}{8} = 221.0 \text{ kN m/m}$$

$$DW: \quad w_{DW} = \frac{(1.66)(1800 - 380)}{1800} = 1.31 \text{ kN/m}^2$$

$$V_{DW} = \tfrac{1}{2}(1.31)(10.67) = 7.0 \text{ kN/m}$$

$$M_{DW} = \frac{(1.31)(10.67)^2}{8} = 18.6 \text{ kN m/m}$$

I. Investigate Service Limit State

1. *Durability* [A5.12]   Cover for unprotected main reinforcing steel deck surface subject to tire wear: 60 mm

Bottom of CIP slabs: 25 mm

Effective depth for No. 25 bars:

$$d = 550 - 25 - \frac{25.2}{2} = 512 \text{ mm}$$

$$\eta_D = \eta_R = \eta_I = 1.0, \text{ therefore } \eta = 1.0 \text{ [A1.3]} \quad pg \ ^{100}$$

a. Moment—Interior Strip:

$$M_{\text{interior}} = \eta \sum \gamma_i Q_i = 1.0[1.0 M_{DC} + 1.0 M_{DW} + 1.0 M_{LL+IM}]$$

$$= 1.0[184.3 + 23.6 + 230.5)]$$

$$= 438.4 \text{ kN m/m}$$

Trial reinforcement:

$$A_s \approx \frac{M}{f_s j d} \quad \sim \quad \text{allowable or design-working stress} $$

Assume $j = 0.875$ and $f_s = 0.6 f_y = 240$ MPa $j = \frac{7}{8} = 0.875$

$f_s = 0.6 f_y$ unused

$$A_s \approx \frac{438\ 400}{240(0.875)(512)} = 4.08 \text{ mm}^2/\text{mm}$$

Try No. 30 bars @ 175 mm ($A_s = 4.00$ mm^2/mm) (Table B.4)

check stress in concrete as well

Revised $d = 550 - 25 - \tfrac{1}{2}(29.9) = 510$ mm OK

b. Moment—Edge Strip:

$$M_{\text{edge}} = \eta \sum \gamma_i Q_i = 1.0[221.0 + 18.6 + 275.1]$$

$$= 514.7 \text{ kN m/m}$$

Trial reinforcement:

$$A_s \approx \frac{M}{f_s jd} = \frac{514\ 700}{240(0.875)(510)} = 4.81 \text{ mm}^2/\text{mm}$$

Try No. 30 bars @ 140 mm ($A_s = 5.00 \text{ mm}^2/\text{mm}$)

2. *Control of Cracking* [A5.7.3.4]

$$f_s \le f_{sa} = \frac{Z}{(d_c A)^{1/3}} \le 0.6 f_y$$

a. Interior Strip—checking tensile stress against f_r [A5.4.2.6, A5.7.3.4]

$$M_{\text{interior}} = 438.4 \text{ kN m/m}$$

$$f_c = \frac{M}{\frac{1}{6}bh^2} = \frac{438\ 400}{\frac{1}{6}(1)(550)^2} = 8.70 \text{ MPa}$$

$$0.8 f_r = 0.8(0.63\sqrt{f_c'}) = 0.8(0.63)\sqrt{30} = 2.76 \text{ MPa}$$

$$f_c > 0.8 f_r, \text{ section is cracked}$$

Elastic-cracked section with No. 30 @ 175 mm ($A_s = 4000$ mm²/m) [A5.7.1] (Fig. E7.2-5)

$$n = \frac{E_s}{E_c} = 7.0, \text{ from deck design}$$

$$n A_s = 7.0(4000) = 28\ 000 \text{ mm}^2/\text{m}$$

Location of neutral axis:

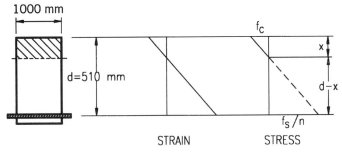

1000 mm

d=510 mm

STRAIN   STRESS

Fig. E7.2-5   Elastic-cracked section.

$$\tfrac{1}{2}bx^2 = nA_s(d - x)$$

$$\tfrac{1}{2}(10^3)x^2) = (28.0 \times 10^3)(510 - x)$$

solving, $x = 143$ mm
Moment of inertia of cracked section:

$$I_{cr} = \tfrac{1}{3}bx^3 + nA_s(d - x)^2$$

$$= \tfrac{1}{3}(10^3)(143)^3 + (28.0 \times 10^3)(510 - 143)^2$$

$$= 4.746 \times 10^9 \text{ mm}^4/\text{m}$$

Steel stress:

$$\frac{f_s}{n} = \frac{M(d - x)}{I_{cr}} = \frac{438\,400(510 - 143)10^3}{4.746 \times 10^9} = 33.9 \text{ MPa}$$

$$f_s = 7(33.9) = 237 \text{ MPa}$$

$$f_s \le 0.6f_y = 0.6(400) = 240 \text{ MPa}$$

For $Z = 23\,000$ N/mm, $d_c = 40$ mm, No. 30 @ 175 mm

$$A = \frac{2d_c b}{N} = \frac{2(40)(175)}{1} = 14\,000 \text{ mm}^2$$

$$f_{sa} = \frac{Z}{(d_c A)^{1/3}} = \frac{23\,000}{(40 \times 14\,000)^{1/3}} = 279 \text{ MPa} > 0.6f_y = 240 \text{ MPa}$$

$$f_s = 237 \text{ MPa} < f_{sa} = 240 \text{ MPa, OK}$$

Use No. 30 @ 175 mm for interior strip

   b. Edge Strip

$$M_{\text{edge}} = 514.7 \text{ kN m/m}$$

Try No. 30 @ 140 mm, $A_s = 5000$ mm^2/m

$$nA_s = 7(5000) = 35 \times 10^3 \text{ mm}^2/\text{m}$$

Location of neutral axis: (Fig. E7.2-5)

$$\tfrac{1}{2}(10^3)(x^2) = (35 \times 10^3)(510 - x)$$

$$\text{solving } x = 157 \text{ mm}$$

Moment of inertia of cracked section:

$$I_{cr} = \tfrac{1}{3}(10^3)(157)^3 + 35 \times 10^3(510 - 157)^2$$
$$= 5.651 \times 10^9 \text{ mm}^4/\text{m}$$

Steel stress:

$$\frac{f_s}{n} = \frac{514\,700(510 - 157) \times 10^3}{5.651 \times 10^9} = 32.2 \text{ MPa}$$

$$f_s = 7(32.2) = 225 \text{ MPa} < 240 \text{ MPa}$$

For $Z = 23\,000$ N/mm, $d_c = 40$ mm, No. 30 @ 140 mm

$$A = \frac{2(40)(140)}{1} = 11\,200 \text{ mm}^2$$

$$f_{sa} = \frac{Z}{(d_c A)^{1/3}} = \frac{23\,000}{(40 \times 11\,200)^{1/3}}$$
$$= 301 \text{ MPa} > 0.6 f_y = 240 \text{ MPa}$$

$$f_s = 225 \text{ MPa} < f_{sa} = 240 \text{ MPa, OK}$$

Use No. 30 @ 140 mm for edge strip

3. *Deformations* [A5.7.3.6]

a. Dead Load Camber: [A2.5.2.6]

$$w_{DC} = (12.95)(14.16) + 2(4.65) = 192.9 \text{ kN/m}$$

$$w_{DW} = (1.66)(13.40) = \underline{22.3 \text{ kN/m}}$$

$$w_{DL} = w_{DC} + w_{DW} = 215.2 \text{ kN/m}$$

$$M_{DL} = \tfrac{1}{8} w_{DL} L^2 = \frac{(215.2)(10.67)^2}{8} = 3063 \text{ kN m}$$

By using I_e: [A5.7.3.6.2]

$$\Delta_{DL} = \frac{5}{384} \frac{w_{DL} L^4}{E_c I_e}$$

$$I_e = \left(\frac{M_{cr}}{M_a}\right)^3 I_g + \left[1 - \left(\frac{M_{cr}}{M_a}\right)^3\right] I_{cr}$$

$$M_{cr} = f_r \frac{I_g}{y_t}$$

$$f_r = 0.63\sqrt{30} = 3.45 \text{ MPa}$$

$$I_g = \frac{1}{12}(14\,160)(550)^3 = 196 \times 10^9 \text{ mm}^4$$

$$M_{cr} = 3.45 \frac{196.6 \times 10^9}{(10^6)(550/2)} = 2466 \text{ kN m}$$

$$\left(\frac{M_{cr}}{M_a}\right)^3 = \left(\frac{2466}{3063}\right)^3 = 0.522$$

$$I_{cr} = (4.746 \times 10^9)(14.16) = 67.3 \times 10^9 \text{ mm}^4$$

$$I_e = (0.522)(196.6 \times 10^9) + (1 - 0.522)(67.3 \times 10^9)$$

$$I_e = 134.8 \times 10^9 \text{ mm}^4$$

$$\Delta_{DL} = \frac{5}{384(27\,700)} \frac{(215.2)(10\,670)^4}{(134.8 \times 10^9)} = 10 \text{ mm instantaneous}$$

Long-time deflection factor for $A_s' = 0$ is equal to

$$3 - 1.2\left(\frac{A_s'}{A_s}\right) = 3.0$$

Camber $= (3.0)(10) = 30$ mm upward

By using I_g: [A5.7.3.6.2]

$$\Delta_{DL} = (10)\left(\frac{134.8 \times 10^9}{196.6 \times 10^9}\right) = 7 \text{ mm}$$

Longtime deflection factor $= 4.0$

Camber $= (4.0)(7)$

$= 28$ mm upward,

comparable to the value based on I_e

b. Live Load Deflection: (Optional) [A2.5.2.6.2]

$$\Delta_{LL+IM}^{\text{allow}} = \frac{\text{span}}{800} = \frac{10\,670}{800} = 13 \text{ mm}$$

Use design truck alone or design lane load plus 25% truck load [A3.6.1.3.2]. When design truck alone, it should be placed so that the distance between its resultant and the nearest wheel is bisected by the span centerline. All design lanes should be loaded: [A2.5.2.6.2] (Fig. E7.2-6)

$$N_L = 3, m = 0.85$$

$$\sum P_{LL+IM} = 1.33(145 \times 3)(0.85) = 491.8 \text{ kN}$$

The value of I_e changes with the magnitude of the applied moment M_a. The moment associated with the live load deflection includes the dead load moment plus the truck moment from Section 7.10.2, Part C.2

$$M_{DC+DW+LL+IM} = 3063 + 3(0.85)(480.1)(1.33) = 4690 \text{ kN m}$$

so that

$$I_e = \left(\frac{2466}{4690}\right)^3 (196 \times 10^9) + \left[1 - \left(\frac{2466}{4690}\right)^3\right](67.3 \times 10^9)$$

$$= 86.0 \times 10^9 \text{ mm}^4$$

$$E_c I_e = (27\ 700)(86.0 \times 10^9) = 2.38 \times 10^{15} \text{ N mm}^2$$

From Case 8, AISC (1994) Manual (see Fig. E7.2-7),

$$\Delta_x(x < a) = \frac{Pbx}{6EIL}(L^2 - b^2 - x^2)$$

First load: $P = 491.8$ kN, $a = 8907$ mm, $b = 1763$ mm, $x =$

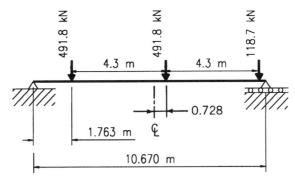

Fig. E7.2-6 Design truck placement for maximum deflection in span.

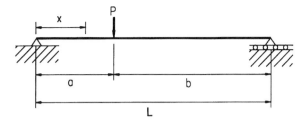

Fig. E7.2-7 Concentrated load placement for calculation of deflection.

4607 mm

$$\Delta_x = \frac{(491.8 \times 10^3)(8907)(4607)}{6(2.38 \times 10^{15})(10\ 670)}[(10\ 670)^2$$

$$- (1763)^2 - (4607)^2] = 12\ \text{mm}$$

Second load: $P = 491.8$ kN, $a = x = 6063$ mm, $b = 4607$ mm

$$\Delta_x = \frac{(491.8 \times 10^3)(4607)(6063)}{6(2.38 \times 10^{15})(10\ 670)}[(10\ 670)^2$$

$$- (4607)^2 - (6063)^2] = 5\ \text{mm}$$

Third load: $P = 118.7$ kN, $a = 10\ 363$ mm, $b = 307$ mm, $x = 6063$ mm

$$\Delta_x = \frac{(118.7 \times 10^3)(307)(6063)}{6(2.38 \times 10^{15})(10\ 670)}[(10\ 670)^2$$

$$- (307)^2 - (6063)^2]\ \text{is} = 0.1\ \text{mm}$$

$\Delta_{LL+IM} = \sum \Delta_x = 17$ mm

> 13 mm, deflection may be a problem

Design lane load:

$$w = 1.33(9.3)(3)(0.85) = 31.54\ \text{N/mm} = 31.54\ \text{kN/m}$$

$$M = \frac{1}{8}wL^2 = \frac{(31.54)(10.67)^2}{8} = 448.8\ \text{kN m}$$

$$\Delta_{\text{lane}} = \frac{5}{48}\frac{ML^2}{E_cI_e} = \frac{5(448.8 \times 10^6)(10\ 670)^2}{48(2.38 \times 10^{15})} = 2\ \text{mm}$$

$$25\%\ \text{Truck} = \tfrac{1}{4}(17) = 4\ \text{mm}$$

$$\Delta_{LL+IM} = 6 \text{ mm, not critical}$$

Single concentrated tandem load at midspan:

$$P = 1.33(220\ 000 \times 3)(0.85) = 746\ 130 \text{ N}$$

$$\Delta = \frac{PL^3}{48E_cI_e} = \frac{(746\ 130)(10\ 670)^3}{48(2.38 \times 10^{15})} = 8 \text{ mm, not critical}$$

The live load deflection estimate of 17 mm is conservative because I_e was based on the maximum moment at midspan rather than an average I_e over the entire span. Also, the additional stiffness provided by the concrete barriers (which can be significant) has been neglected, as well as the compression reinforcement in the top of the slab. Bridges typically deflect less than calculations predict and, as a result, the deflection check has been made optional.

4. *Concrete stresses* [A5.9.4.3]. No prestressing, does not apply.

5. *Fatigue* [A5.5.3]

$$U = 0.75(LL + IM) \text{ (Table 3.1) [Table A3.4.1-1]}$$

$$IM = 15\% \text{ [A3.6.2.1]}$$

Fatigue load shall be one design truck with 9000-mm axle spacing [A3.6.1.1.2]. Because of the large rear axle spacing, the maximum moment results when the two front axles are on the bridge. As shown in Figure E7.2-8, the two axle loads are placed on the bridge so that the distance between the resultant of the axle loads on the bridge and the nearest axle is divided equally by the centerline of the span (Case 42, AISC Manual, 1994). No multiple presence fac-

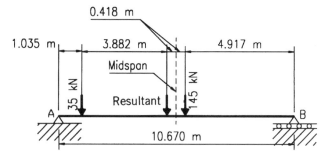

Fig. E7.2-8   Fatigue truck placement for maximum bending moment.

tor is applied ($m = 1$). From Figure E7.2-8,

$$R_B = 180\left(\frac{3882 + 1035}{10\,670}\right) = 82.9 \text{ kN}$$

$$M_C = (82.9)(4.917) = 407.6 \text{ kN m}$$

$$\eta\sum\gamma_i Q_i = 1.0(0.75)(407.6)(1.15) = 351.6 \text{ kN m/lane}$$

a. Tensile Live Load Stresses:
one lane loaded, $E = 4370$ mm

$$M_{LL+IM} = \frac{351.6}{4370}(10^3) = 80.46 \text{ kN m/m}$$

$$\frac{f_s}{n} = \frac{(80.46)(510 - 143)}{4.746 \times 10^9}(10^6) = 6.22 \text{ MPa}$$

$$f_{max} = 7(6.22) = 43.6 \text{ MPa}$$

b. Reinforcing Bars: [A5.5.3.2]
Maximum stress range:

$$f_f = 145 - 0.33f_{min} + 55(r/h)$$

where $f_{min} = 0$, because deck is treated as a simple beam; thus, there is no moment reversal

$$\frac{r}{h} = 0.3$$

$$f_f = 145 - 0.33(0) + 55(0.3) = 161.5 \text{ MPa} > f_{max}$$
$$= 43.6 \text{ MPa, OK}$$

J. Investigate Strength Limit State

1. *Flexure* [A5.7.3.2] Rectangular stress distribution [A5.7.2.2]

$$\beta_1 = 0.85 - 0.05(\tfrac{2}{7}) = 0.836$$

a. Interior Strip:
Eq. 7.73 with $A_{ps} = 0$, $b = b_w$, $A'_s = 0$
A_s = No. 30 @ 175 mm = 4.00 mm²/mm for service limit state.

$$c = \frac{A_s f_y}{0.85f'_c\beta_1 b} = \frac{(4.00)(400)}{0.85(30)(0.836)(1)} = 75 \text{ mm}$$

$$a = \beta_1 c = (0.836)(75) = 63 \text{ mm}$$

$$d_s = 550 - 25 - \tfrac{1}{2}(29.9) = 510 \text{ mm}$$

$$\frac{c}{d_s} = \frac{75}{510} = 0.147 < 0.42, \text{ OK [A5.7.3.3.1]}$$

$$\rho = \frac{A_s}{bd} = \frac{4.00}{(1)(510)} = 0.007 \, 84$$

$$\rho_{\min} = 0.03\frac{f'_c}{f_y} = 0.03\left(\frac{30}{400}\right) = 0.002 \, 25 < \rho, \text{ OK [A5.7.3.3.2]}$$

Eq. 7.76 with $A_{ps} = 0$, $b = b_w$, $A'_s = 0$, $A_s = 4.0 \text{ mm}^2/\text{mm}$

$$M_n = A_s f_y\left(d_s - \frac{a}{2}\right) = \frac{(4.0)(400)}{10^3}\left(510 - \frac{63}{2}\right)$$

$$= 765.6 \text{ kN m/m}$$

Factored resistance $= \phi M_n = 0.9(765.6) = 689.04 \text{ kN m/m}$
Strength *I*

$$M_u = \eta\sum\gamma_i Q_i = 1.05[1.25M_{DC} + 1.50M_{DW}$$
$$+ 1.75M_{LL+IM} + \gamma_{TG}M_{TG}]$$

For simple span bridges, temperature gradient effect reduces gravity load effects. Because temperature gradient may not always be there, assume $\gamma_{TG} = 0$.

$$M_u = \gamma\sum\gamma_i Q_i = 1.05[1.25(184.3) + 1.50(23.6) + 1.75(230.5)]$$

$$M_u = 702.54 \text{ kN m/m} > \phi M_n = 689.0 \text{ kN m/m, not good}$$

Try No. 30 @ 150 mm, $A_s = 4.667 \text{ mm}^2/\text{mm}$:

$$c = \frac{A_s f_y}{0.85 f'_c \beta_1 b} = \frac{(4.667)(400)}{0.85(30)(0.836)(1)} = 88 \text{ mm}$$

$$a = \beta_1 c = (0.836)(88) = 74 \text{ mm}$$

$$\frac{c}{d_s} = \frac{88}{510} = 0.173 < 0.42, \text{ OK [A5.7.3.3.1]}$$

$$\rho = \frac{A_s}{bd} = \frac{4.667}{(1)(510)} = 0.009 \, 15 > \rho_{\min} = 0.002 \, 25, \text{ OK}$$

$$M_n = \frac{(4.667)(400)}{10^3}\left(510 - \frac{74}{2}\right) = 883.0 \text{ kN m/m}$$

$$\phi M_n = 0.9(883.0) = 794.7 \text{ kN m/m} > M_u$$
$$= 702.54 \text{ kN m/m, OK}$$

Strength limit state governs.
Use No. 30 @ 150 mm for interior strip

b. Edge Strip:

$A_s = $ No. 30 @ 140 mm, A_s
$$= 5.0 \text{ mm}^2/\text{mm for service limit state}$$

$$c = \frac{A_s f_y}{0.85 f'_c \beta_1 b} = \frac{(5.0)(400)}{0.85(30)(0.836)(1)} = 94 \text{ mm}$$

$$a = \beta_1 c = (0.836)(94) = 79 \text{ mm}$$

$$\frac{c}{d_s} = \frac{94}{510} = 0.184 < 0.42, \text{ OK [A5.7.3.3.1]}$$

$$\rho = \frac{A_s}{bd} = \frac{5.0}{(1)(510)} = 0.0098 > \rho_{min} = 0.002\ 25 \text{ OK}$$

$$\phi M_n = \frac{0.9(5.0)(400)}{10^3}\left(510 - \frac{79}{2}\right) = 846.9 \text{ kN m/m}$$

Strength *I*:

$$M_u = \eta \sum \gamma_i Q_i = 1.05[1.25(221.0) + 1.50(18.6) + 1.75(275.1)]$$

$$M_u = 824.95 \text{ kN m/m} < \phi M_n = 846.9 \text{ kN m/m, OK}$$

Use No. 30 @ 140 mm for edge strip

2. *Shear* [A5.14.4.1] Slab bridges designed for moment in conformance with AASHTO [A4.6.2.3] may be considered satisfactory for shear. If longitudinal tubes are placed in the slab to create voids and reduce the cross section, the shear resistance must be checked.

K. **Distribution Reinforcement [A5.14.4.1]** The amount of bottom transverse reinforcement may be taken as a percentage of the main reinforcement required for positive moment as

$$\frac{1750}{\sqrt{L}} \leq 50\%$$

$$\frac{1750}{\sqrt{10\,670}} = 16.9\%$$

a. Interior Strip:

Positive moment reinforcement = No. 30 @ 150 mm,

$$A_s = 4.667 \text{ mm}^2/\text{mm}$$

Transverse reinforcement = 0.169(4.667)

$$= 0.789 \text{ mm}^2/\text{mm}$$

Use No. 15 @ 250 mm transverse bottom bars, $A_s = 0.800$ mm²/mm

b. Edge Strip:

Positive moment reinforcement = No. 30 @ 140 mm,

$$A_s = 5.0 \text{ mm}^2/\text{mm}$$

Transverse reinforcement = 0.169(5.0)

$$= 0.845 \text{ mm}^2/\text{mm}$$

Use No. 15 @ 200 mm, transverse bottom bars, $A_s = 1.000$ mm²/mm. For ease of placement, use No. 15 @ 200 mm across the entire width of the bridge.

L. **Shrinkage and Temperature Reinforcement** Transverse reinforcement in the top of the slab [A5.10.8]

$$\text{Temp } A_s \geq 0.11\frac{A_g}{f_y} = 0.11\left[\frac{(1)(550)}{400}\right]$$

$$= 1.375 \text{ mm}^2/\text{mm each direction, both faces}$$

$$\text{Top layer } A_s = \tfrac{1}{2}(1.375) = 0.688 \text{ mm}^2/\text{mm}$$

Use No. 15 @ 250 mm, transverse top bars, $A_s = 0.800$ mm²/mm

M. **Design Sketch** The design of the solid slab bridge is summarized in the half-section of Figure E7.2-9.

7.10.3 *T*-Beam Bridge Design

Problem Statement Design a reinforced concrete *T*-beam bridge for a 13 420 mm wide roadway and three-spans of 10 670 mm − 12 800 mm −

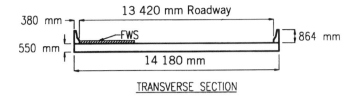

13 420 mm Roadway

380 mm

FWS

550 mm

864 mm

14 180 mm

TRANSVERSE SECTION

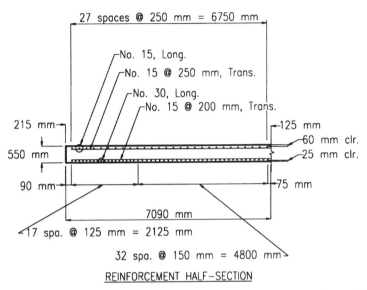

27 spaces @ 250 mm = 6750 mm

No. 15, Long.

No. 15 @ 250 mm, Trans.

No. 30, Long.

No. 15 @ 200 mm, Trans.

215 mm

125 mm

60 mm clr.

550 mm

25 mm clr.

90 mm

75 mm

7090 mm

17 spa. @ 125 mm = 2125 mm

32 spa. @ 150 mm = 4800 mm

REINFORCEMENT HALF-SECTION

Fig. E7.2-9 Design sketch for solid slab bridge. (a) Transverse section and (b) reinforcement half-section.

10 670 mm with a skew of 30° as shown in Figure E7.3-1. Use the concrete deck of Figures E7.1-14 and E7.1-17 previously designed for a *HL*-93 live load, a bituminous overlay, and a 2440-mm spacing of girders in Example Problem 7.10.1. Use $f'_c = 30$ MPa, $f_y = 400$ MPa, and follow the outline of AASHTO (1994) LRFD Bridge Specifications, Section 5-Appendix A5.3.

A. *Develop General Section*
   The bridge is to carry interstate traffic over a normally small stream that is subject to high water flows during the rainy season (Fig. E7.3-1).

B. *Develop Typical Section and Design Basis*
   1. *Top Flange Thickness* [A5.14.1.3.1a]
      • As determined in Section 9 [A9.7.1.1]

      Minimum depth of concrete deck = 175 mm

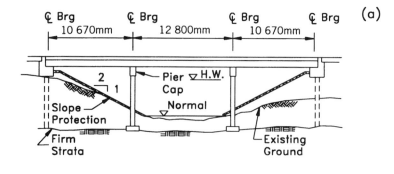

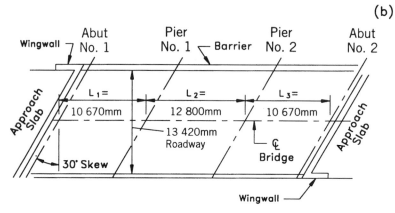

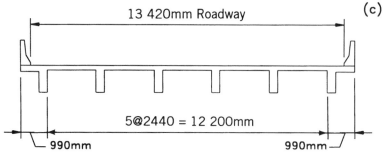

Fig. E7.3-1   T-Beam bridge design example. (a) Elevation, (b) plan, and (c) section.

From deck design, structural thickness = <u>190 mm</u>, OK

- Maximum clear span = 20(190) = 3800 mm > 2440 mm, OK
2. *Bottom Flange Thickness* (not applicable to *T*-Beam)
3. *Web Thickness* [A5.14.1.3.1c and C5.14.1.3.1c]
   - Minimum of 200 mm without prestressing ducts
   - Minimum concrete cover for main bars, exterior 50 mm [A5.12.3]

- Three No. 35 bars in one row require a beam width of [A5.10.3.1.1]

$$b_{min} \approx 2(50) + 3d_b + 2(1.5d_b) = 100 + 6(35.7) = 314 \text{ mm}$$

- To give a little extra room for bars, try $\underline{b_w = 350 \text{ mm}}$

4. *Structure Depth* (Table 2.1) [Table A2.5.2.6.3-1]
   - Minimum depth continuous spans = 0.065L

$$h_{min} = 0.065(12\ 800) = 832 \text{ mm, try } \underline{h = 990 \text{ mm}}$$

5. *Reinforcement Limits*
   - Deck overhang: at least $\frac{1}{3}$ of bottom layer of transverse reinforcement [A5.14.1.3.2a]
   - Minimum reinforcement: the lesser of $\phi M_n > 1.2 M_{cr}$ or $\phi M_n \geq$ 1.33 times the factored moment required for the Strength I limit state. [A5.7.3.3.2]

$$\rho_{min} \geq 0.03\frac{f'_c}{f_y} = 0.03\frac{30}{400} = 0.002\ 25$$

   - Crack control: $f_s \leq f_{sa} = \dfrac{Z}{(d_c A)^{1/3}} \leq 0.6 f_y$ [A5.7.3.4]
   - Flanges in tension: tension reinforcement shall be distributed over the lesser of the effective flange width or a width equal to 1/10 of the average of the adjacent spans [A4.6.2.6, A5.7.3.4]
   - Longitudinal skin reinforcement required if web depth > 900 mm [A5.7.3.4]
   - Shrinkage and temperature reinforcement [A5.10.8]

$$A_s \geq 0.75\frac{A_g}{f_y}$$

6. *Effective Flange Widths* [A4.6.2.6.1]
   - Effective span length for continuous spans

     = distance between points of permanent load inflections

   - Interior beams

$$b_i \leq \begin{cases} \frac{1}{4} \text{ effective span} \\ 12t_s + b_w \\ \text{average spacing of adjacent beams} \end{cases}$$

• Exterior beams

$$b_e - \frac{1}{2}b_i \leq \begin{cases} \frac{1}{8}\text{effective span} \\ 6t_s + \frac{1}{2}b_w \\ \text{width of overhang} \end{cases}$$

7. *Identify Strut and Tie Areas, if any* not applicable.
The trial section for the T-beam bridge is shown in Figure E7.3-2.

C. *Design Conventionally Reinforced Concrete Deck* The reinforced concrete deck for this bridge is designed in Section 7.10.1. The design sketches for the deck are given in Figures E7.1-14 and E7.1-17.

D. *Select Resistance Factors (Table 7.10)* [A5.5.4.2]

1. *Strength Limit State*	ϕ [A5.5.4.2.1]
Flexure and tension	0.90
Shear and torsion	0.90
Axial compression	0.75
Bearing	0.70
2. *Nonstrength Limit States*	1.0 [A1.3.2.1]

E. *Select Load Modifiers [A1.3.2.1]*

	Strength	Service	Fatigue	
Ductility, η_D	0.95	1.0	1.0	[A1.3.3]
Redundancy, η_R	0.95	1.0	1.0	[A1.3.4]
Importance, η_I	1.05	N/A	N/A	[A1.3.5]
$\eta = \eta_D\eta_R\eta_I$	0.95	1.0	1.0	

F. *Select Applicable Load Combinations (Table 3.1)* [Table A3.4.1-1]

Strength *I* Limit State

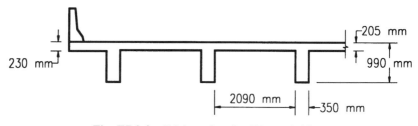

230 mm   205 mm   990 mm   2090 mm   350 mm

Fig. E7.3-2 Trial section for *T*-beam bridge.

$$U = \eta[1.25DC + 1.50DW + 1.75(LL + IM)$$
$$+ 1.0(WA + FR) + \cdots]$$

Service *I* Limit State

$$U = 1.0(DC + DW) + 1.0(LL + IM)$$
$$+ 1.0WA + 0.3(WS + WL) + \cdots$$

Fatigue Limit State

$$U = 0.75(LL + IM)$$

G. *Calculate Live Load Force Effects*

1. *Select Number of Lanes* [A3.6.1.1.1]

$$N_L = \text{INT}\left(\frac{w}{3600}\right) = \text{INT}\left(\frac{13\,420}{3000}\right) = 3$$

2. *Multiple Presence* (Table 4.6) [A3.6.1.1.2]

No. of Loaded Lanes	m
1	1.20
2	1.00
3	0.85

3. *Dynamic Load Allowance* (Table 4.7) [A3.6.2.1] Not applied to the design lane load.

Component	IM (%)
Deck joints	75
Fatigue	15
All other	33

4. *Distribution Factors for Moment* [A4.6.2.2.2] Applicability: constant deck width, at least four parallel beams of nearly same stiffness, roadway part of overhang (Fig. E7.3-3), $d_e = 990 - 380 = 610$ mm < 910 mm. OK

Cross-section type (e) (Table 2.2) [Table A4.6.2.2.1-1]

$$\text{No. of beams } N_b = 6, \ t_s = 190 \text{ mm,}$$
$$S = 2440 \text{ mm, } L_1 = L_3 = 10\,670 \text{ mm, } L_2 = 12\,800 \text{ mm}$$

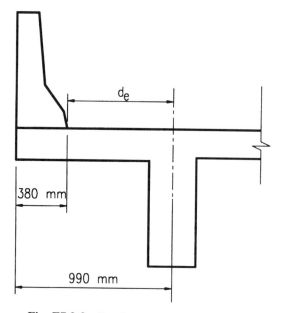

Fig. E7.3-3 Roadway part of overhang, d_e.

a. Interior beams with concrete decks (Table 6.5) [A4.6.2.2.2b and Table A4.6.2.2.2b-1]

For preliminary design $\dfrac{K_g}{Lt_s^3} = 1.0$ and $\dfrac{I}{J} = 1.0$

One Design Lane Loaded: range of applicability satisfied

$$mg_M^{SI} = 0.06 + \left(\frac{S}{4300}\right)^{0.4}\left(\frac{S}{L}\right)^{0.3}\left(\frac{K_g}{Lt_s^3}\right)^{0.1}$$

mg = girder distribution factor with multiple presence factor included

SI = single lane loaded, interior M = moment

Two or More Design Lanes Loaded

$$mg_M^{MI} = 0.075 + \left(\frac{S}{2900}\right)^{0.6}\left(\frac{S}{L}\right)^{0.2}\left(\frac{K_g}{Lt_s^3}\right)^{0.1}$$

MI = multiple lanes loaded, interior M = moment

Distribution Factor	$L_1 = 10\,670$ mm	$L_{ave} = 11\,735$ mm	$L_2 = 12\,800$ mm
mg_M^{SI}	0.572	0.558	0.545
mg_M^{MI}	0.746	0.734	0.722

For interior girders, distribution factors are governed by multiple lanes loaded.

b. Exterior beams (Table 6.5) [A4.6.2.2.2d and Table A4.6.2.2.2d-1]

One Design Lane Loaded—Lever Rule, $m = 1.2$ (Fig. E7.3-4)

$$R = 0.5P\left(\frac{2450 + 650}{2440}\right) = 0.635P$$

$$g_M^{SE} = 0.635 \qquad SE = \text{single lane, exterior}$$

$$mg_M^{SE} = 1.2(0.635) = \underline{0.762}, \text{ governs}$$

Two or More Design Lanes Loaded, $d_e = 610$ mm

$$mg_M^{ME} = emg_M^{MI} \qquad ME = \text{multiple lanes loaded, exterior}$$

$$\text{where } e = 0.77 + \frac{d_e}{2800} = 0.77 + \frac{610}{2800}$$

$$= 0.99 < 1.0 \qquad \text{Use } e = 1.0$$

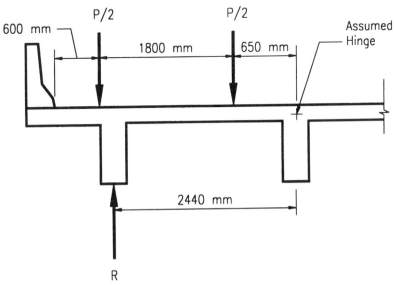

Fig. E7.3-4   Definition of level rule.

therefore, $mg_M^{ME} = mg_M^{MI} = 0.746, 0.734, 0.722$

For exterior girders, the critical distribution factor is by the lever rule with one lane loaded $= 0.762$.

c. Skewed bridges (Table 6.5) [A4.6.2.2.2e] Reduction of live load distribution factors for moment in longitudinal beam on skewed supports is permitted. $S = 2440$ mm, $\theta = 30°$

$$r_{skew} = 1 - c_1(\tan \theta)^{1.5} = 1 - 0.4387c_1$$

where
$$c_1 = 0.25\left(\frac{K_g}{Lt_s^3}\right)^{0.25}\left(\frac{S}{L}\right)^{0.5} \quad \text{[Table A4.6.2.2.2e-1]}$$

Range of applicability is satisfied.

Reduction Factor	$L_1 = 10\,670$ mm	$L_{ave} = 11\,735$ mm	$L_2 = 12\,800$ mm
c_1	0.120	0.114	0.109
r_{skew}	0.948	0.950	0.952

d. Distributed live load moments

$$M_{LL+IM} = mgr\left[(M_{Tr} \text{ or } M_{Ta})\left(1 + \frac{IM}{100}\right) + M_{Ln}\right]$$

Location 104 (Fig. E7.3-5). For relatively short spans, design tandem governs positive moment (see Table 5.8a). Influence line coefficients are from Table 5.4.

$$M_{Ta} = 110(0.207\,00 + 0.158\,07)10.67 = 428.5 \text{ kN m}$$
$$M_{Ln} = 9.3(0.102\,14)(10.67)^2 = 108.1 \text{ kN m}$$

Interior Girders:

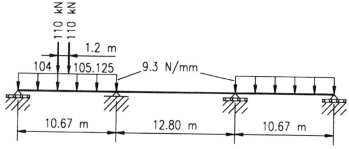

Fig. E7.3-5 Live load placement for maximum positive moment in exterior span.

$$M_{LL+IM} = 0.746(0.948)[428.5(1.33) + 108.1] = 479.5 \text{ kN m}$$

Exterior Girders:

$$M_{LL+IM} = 0.762(0.948)[428.5(1.33) + 108.1] = 489.8 \text{ kN m}$$

Location 200 (Fig. E7.3-6). For negative moment at support, a single truck governs with the second axle spacing extended to 9000 mm (see Table 5.8a). The distribution factors are based on the average length of span 1 and span 2.

$$M_{Tr} = [145(-0.094\ 29 - 0.102\ 71) + 35(-0.058\ 96)]10.67$$

$$= -326.8 \text{ kN m}$$

$$M_{Ln} = 9.3(-0.138\ 53)(10.67)^2 = -146.7 \text{ kN m}$$

$$1.33M_{Tr} + M_{Ln} = 1.33(-326.8) - 146.7 = -581.3 \text{ kN m}$$

Interior Girders:

$$M_{LL+IM} = 0.734(0.950)(-581.3) = -405.4 \text{ kN m}$$

Exterior Girders:

$$M_{LL+IM} = 0.762(0.950)(-581.3) = -420.8 \text{ kN m}$$

Location 205 (Fig. E7.3-7). Tandem governs (see Table 5.8a)

$$M_{Ta} = 110(0.203\ 57 + 0.150\ 94)10.67 = 416.1 \text{ kN m}$$

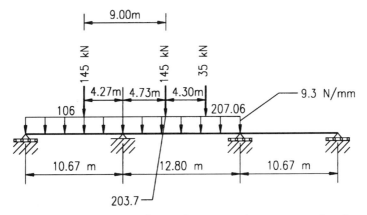

Fig. E7.3-6   Live load placement for maximum negative moment at interior support.

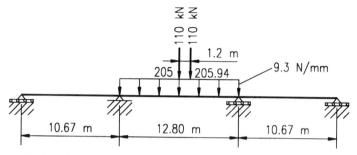

Fig. E7.3-7 Live load placement for maximum positive moment in interior span.

$$M_{Ln} = 9.3(0.102\ 86)(10.67)^2 = 108.9 \text{ kN m}$$

$$1.33M_{Ta} + M_{Ln} = 662.3 \text{ kN m}$$

Interior Girders:

$$M_{LL+IM} = 0.722(0.952)(662.3) = 455.2 \text{ kN m}$$

Exterior Girders:

$$M_{LL+IM} = 0.762(0.952)(662.3) = 480.5 \text{ kN m}$$

5. *Distribution Factors for Shear* [A4.6.2.2.3] Cross-section type (e) (Table 2.2) [Table A4.6.2.2.1-1], $S = 2440$ mm, mg is independent of span length

a. Interior beams (Table 6.5) [A4.6.2.2.3a and Table A4.6.2.2.3a-1]

$$mg_V^{SI} = 0.36 + \frac{S}{7600} = 0.36 + \frac{2440}{7600} = 0.681$$

$$mg_V^{MI} = 0.2 + \frac{S}{3600} - \left(\frac{S}{10\ 700}\right)^2$$

$$= 0.2 + \frac{2440}{3600} - \left(\frac{2440}{10\ 700}\right)^2 = 0.826, \text{ governs}$$

$$V = \text{shear}$$

b. Exterior beams (Table 6.5) [A4.6.2.2.3b and Table A4.6.2.2.3b-1]

Lever Rule: $mg_V^{SE} = 0.762$, governs

$$mg_V^{ME} = emg_V^{MI}, \text{ where } e = 0.6 + \frac{d_e}{3000} = 0.6 + \frac{610}{3000} = 0.803$$

$$mg_V^{ME} = 0.803(0.826) = 0.664$$

c. Skewed bridges (Table 6.5) [A4.6.2.2.3c and Table A4.6.2.2.3c-1] All beams treated like beam at obtuse corner.

$$\theta = 30° \qquad \left(\frac{Lt_s^3}{K_g}\right) = 1.0$$

$$r_{skew} = 1.0 + 0.20\left(\frac{Lt_s^3}{K_g}\right)^{0.3} \tan\theta$$

$$= 1.0 + 0.20(1.0)^{0.3}(0.577) = 1.115$$

d. Distributed live load shears

$$V_{LL+IM} = mgr[(V_{Tr} \text{ or } V_{Ta})1.33 + V_{Ln}]$$

Location 100 (Fig. E.7.3-8). Truck governs (see Table 5.8b).

$$V_{Tr} = 145(1.0 + 0.514\,21) + 35(0.125\,01) = 223.9 \text{ kN}$$

$$V_{Ln} = 9.3(0.455\,36)10.67 = 45.2 \text{ kN}$$

$$1.33V_{Tr} + V_{Ln} = 1.33(223.9) + 45.2 = 343 \text{ kN}$$

Interior Girders:

$$V_{LL+IM} = 0.826(1.115)(343) = 315.9 \text{ kN}$$

Exterior Girders:

$$V_{LL+IM} = 0.762(1.115)(343) = 291.4 \text{ kN}$$

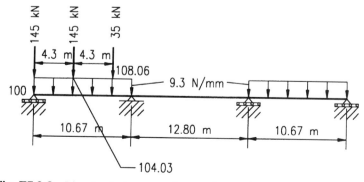

Fig. E7.3-8   Live load placement for maximum shear at exterior support.

Location 110 (Fig. E7.3-9) Truck governs (see Table 5.8b).

$$V_{Tr} = 145(-1.0 - 0.691\,22) + 35(-0.239\,77)$$

$$= -253.6 \text{ kN}$$

$$V_{Ln} = 9.3(-0.638\,53)10.67 = -63.4 \text{ kN}$$

$$1.33V_{Tr} + V_{Ln} = 1.33(-253.6) - 63.4 = -400.7 \text{ kN}$$

Interior Girders:

$$V_{LL+IM} = 0.826(1.115)(-400.7) = -369.0 \text{ kN}$$

Exterior Girders:

$$V_{LL+IM} = 0.762(1.115)(-400.7) = -340.4 \text{ kN}$$

Location 200 (Fig. E7.3-10). Truck governs (see Table 5.8b).

$$V_{Tr} = 145(1.0 + 0.690\,64) + 35(0.300\,28) = 255.7 \text{ kN}$$

$$V_{Ln} = 9.3(0.665\,10)10.67 = 66.0 \text{ kN}$$

$$1.33V_{Tr} + V_{Ln} = 1.33(255.7) + 66.0 = 406.1 \text{ kN}$$

Interior Girders:

$$V_{LL+IM} = 0.826(1.115)(406.1) = 374.0 \text{ kN}$$

Exterior Girders:

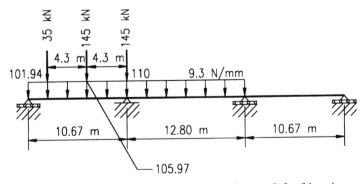

Fig. E7.3-9 Live load placement for maximum shear to left of interior support.

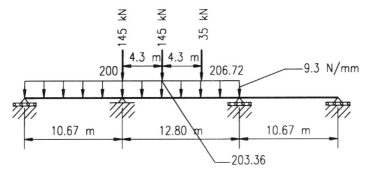

Fig. E7.3-10 Live load placement for maximum shear to right of interior support.

$$V_{LL+IM} = 0.762(1.115)(406.1) = 345.0 \text{ kN}$$

6. *Reactions to Substructure* [A3.6.1.3.1] The following reactions are per design lane without any distribution factors. The lanes shall be positioned transversely to produce extreme force effects.

Location 100

$$R_{100} = V_{100} = 1.33V_{Tr} + V_{Ln} = 343.0 \text{ kN/lane}$$

Location 200 (Fig. E7.3-11)

$$R_{200} = 1.33[145(1.0 + 0.791\,15) + 35(0.721\,37)] + 63.4 + 66.0$$

$$= 508.5 \text{ kN/lane}$$

H. **Calculate Force Effects from Other Loads** Analysis for a uniformly distributed load w (Fig. E7.3-12). See Table 5.4 for coefficients.

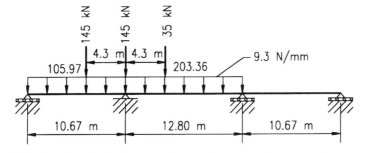

Fig. E7.3-11 Live load placement for maximum reaction at interior support.

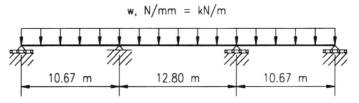

Fig. E7.3-12   Uniformly distributed dead load, w.

Moments

$$M_{104} = w(0.071\ 29)(10.67)^2 = 8.116w \text{ kN m}$$

$$M_{200} = w(-0.121\ 79)(10.67)^2 = -13.866w \text{ kN m}$$

$$M_{205} = w(0.058\ 21)(10.67)^2 = 6.627w \text{ kN m}$$

Shears

$$V_{100} = w(0.378\ 21)(10.67) = 4.036w \text{ kN}$$

$$V_{110} = w(-0.621\ 79)(10.67) = -6.634w \text{ kN}$$

$$V_{200} = w(0.600\ 00)(10.67) = 6.402w \text{ kN}$$

1. *Interior Girders*

    DC   Slab $(2400 \times 10^{-9} \times 9.81)(205)2440 = 11.78$ N/mm

    girder stem $(2400 \times 10^{-9} \times 9.81)(350)785 = \underline{6.47}$

$$w_{DC} = 18.25 \text{ N/mm}$$

  DW:FWS $w_{DW} = (2250 \times 10^{-9} \times 9.81)(75)2440 = 4.04$ N/mm

By multiplying the general expressions for uniform loads by the values of the interior girder uniform loads, the unfactored moments and shears are generated in Table E.7.3-1.

TABLE E7.3-1   Interior Girder Unfactored Moments and Shears

Load Type	w (kN/m)	Moments (kN m)			Shears (kN)		
		M_{104}	M_{200}	M_{205}	V_{100}	V_{110}	V_{200}
DC	18.25	148.3	−253.3	121.1	73.7	−121.2	117.0
DW	4.04	32.8	−56.0	26.8	16.3	−26.8	25.9
LL + IM	N/A	479.5	−405.4	455.2	315.9	−369.0	374.0

2. *Exterior Girders* By using deck design results for reaction on exterior girder from Section 7.10.1, Part C

DC	Deck Slab	4.63 N/mm
	Overhang	6.75
	Barrier	6.74
	Girder stem	$\underline{6.37}$ = 2400 × 10^{-9} × 9.81 × 175 [(990 −
		230) + (990 − 205)]

$$w_{DC} = 24.49 \text{ N/mm}$$

$$DW{:}FWS \quad w_{DW} = 2.76 \text{ N/mm}$$

By multiplying the generic expressions for uniform loads by the values of the exterior girder uniform loads, the unfactored moments and shears in Table E7.3-2 are generated.

I. **Investigate Service Limit State**

1–3. *Prestress Girders* Not applicable.

4. *Investigate Durability* [C5.12.1] It is assumed that concrete materials and construction procedures provide adequate concrete cover, nonreactive aggregates, thorough consolidation, adequate cement content, low water/cement ratio, thorough curing, and air-entrained concrete.

Concrete Cover for Unprotected Main Reinforcing Steel [Table 5.12.3-1]

Exposure to deicing salts	60 mm | cover to ties
Exterior other than above	50 mm | and stirrups
Bottom of CIP slabs, Up to No. 35	25 mm | 12 mm less

Effective Depth—assume No. 35 bar, $d_b = 35.7$ mm

TABLE E7.3-2 Exterior Girder Unfactored Moments and Shears

Load Type	w (kN/m)	Moments (kN m)			Shears (kN)		
		M_{104}	M_{200}	M_{205}	V_{100}	V_{110}	V_{200}
DC	24.49	198.8	−339.6	162.3	98.8	−162.5	156.8
DW	2.76	22.4	−38.3	18.3	11.1	−18.3	17.7
LL + IM	N/A	489.8	−420.8	480.5	291.4	−340.4	345.0

Note: Interior girder has larger shears. Exterior girder has larger moments.

Positive bending $d_{pos} = (990 - 15)$

$$-\left(50 + \frac{35.7}{2}\right) = 907 \text{ mm}$$

Negative bending $d_{neg} = 990 - \left(60 + \frac{35.7}{2}\right) = 912 \text{ mm}$

5. *Crack Control* [A5.7.3.4]

$$f_s \le f_{sa} = \frac{Z}{(d_c A)^{1/3}} \le 0.6 f_y$$

Use $Z = 23\,000$ N/mm and Service I Limit State [A3.4.1]

a. Effective flange width [A4.6.2.6.1] Depends on effective span length, which is defined as the distance between points of permanent load inflection for continuous beams (Fig. E7.3-13). Positive Bending M_{104}

$$L_{eff} = 8070 \text{ mm}$$

$$b_i \le \begin{cases} \frac{1}{4}L_{eff} = \frac{1}{4}(8070) = \underline{2018 \text{ mm}}, \text{ governs} \\ 12t_s + b_w = 12(190) + 350 = 2630 \text{ mm} \\ S = 2440 \text{ mm} \end{cases}$$

$$b_e - \tfrac{1}{2}b_i \le \begin{cases} \frac{1}{8}L_{eff} = \frac{1}{8}(8070) = 1009 \text{ mm} \\ 6t_s + \tfrac{1}{2}b_w = 6(190) + 175 = 1315 \text{ mm} \\ \text{overhang} = 990 \text{ mm, governs} \end{cases}$$

$$b_e = 990 + \tfrac{1}{2}(2018) = \underline{1999 \text{ mm}}$$

Use $b_i = 2020$ mm, $b_e = 2000$ mm

b. Positive bending reinforcement—exterior girder (Table 3.1) [Table A3.4.1-1] Service I Limit State, $\eta = 1.0$, gravity load factors $= 1.0$, moments from Table E7.3-2

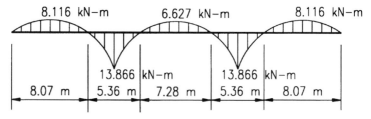

Fig. E7.3-13 Length between inflection points for permanent load.

$$M_{104} = \eta\Sigma\gamma_i Q_i = M_{DC} + M_{DW} + mgr M_{LL+IM}$$

$$= (198.8 + 22.4 + 489.8) = \underline{711.0 \text{ kN m}}$$

$$f'_c = 30 \text{ MPa}, f_y = 400 \text{ MPa}, d_{pos} = 907 \text{ mm}$$

Assume $j = 0.875$ and $f_s = 0.6f_y = 240 \text{ MPa} = 240 \text{ N/mm}^2$

$$A_s \approx \frac{M}{f_s jd} = \frac{711.0 \times 10^6}{240 \times 0.875 \times 907} = 3730 \text{ mm}^2$$

Try 6 No. 30 bars, provided $A_s = 4200 \text{ mm}^2$ (Table B.3)

Minimum beam width must consider bend diameter of tie [Table A5.10.2.3-1].

For No. 15 stirrup and No. 30 bar (Fig. E7.3-14)

$$2d_s > \tfrac{1}{2}d_b$$

$$2(16) = 32 \text{ mm} > \tfrac{1}{2}(30) = 15 \text{ mm}$$

No. 30 bar will be away from vertical leg of stirrup a distance of $32 - 15 = 17$ mm.

$$b_{min} = 2(38 + 3d_s) + 2d_b + 2(1.5d_b)$$

$$= 2(38 + 3 \times 16) + 5(30) = 322 \text{ mm}$$

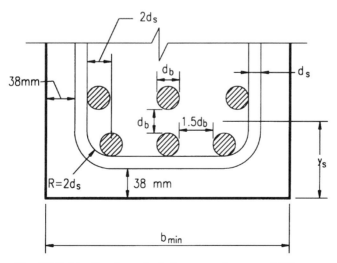

Fig. E7.3-14 Spacing of reinforcement in stem of *T*-beam.

3 No. 30 bars will fit in one layer of $b_w = 350$ mm

$$y_s = 38 + 16 + 30 + \tfrac{1}{2}(30) = 99 \text{ mm}$$

$$d_{\text{pos}} = 990 - 15 - 99 = 876 \text{ mm}$$

Elastic-Cracked transformed section analysis required to check crack control [A5.7.3.4]

$$n = \frac{E_s}{E_c} = 7 \quad \text{assume NA (neutral axis) in flange (Fig. E7.3-15)}$$

$$x = -\frac{nA_s}{b} + \sqrt{\left(\frac{nA_s}{b}\right)^2 + \frac{2nA_s d}{b}}$$

$$x = \frac{-7(4200)}{2000} + \sqrt{\left(\frac{7 \times 4200}{2000}\right)^2 + \frac{2(7)(4200)(876)}{2000}}$$

$$= 146 \text{ mm} < h_f = 190 \text{ mm}$$

NA lies in flange; therefore, assumption OK

The actual steel stress must be compared to the allowable steel stress for crack control (Fig. E7.3-16).

$$I_{cr} = \tfrac{1}{3}bx^3 + nA_s(d - x)^2$$

$$I_{cr} = \tfrac{1}{3}(200)(146)^3 + 7(4200)(876 - 146)^2 = 17.74 \times 10^9 \text{ mm}^4$$

$$f_s = \frac{nM(d - x)}{I_{cr}} = \frac{7(711 \times 10^6)(730)}{17.74 \times 10^9} = 205\frac{N}{\text{mm}^2} = 205 \text{ MPa}$$

$$A = \frac{2y_s b_w}{N} = \frac{2(99)(350)}{6} = 11\ 550 \text{ mm}^2$$

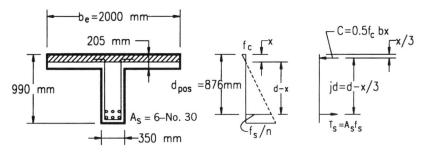

Fig. E7.3-15 Elastic-cracked transformed positive moment section at Location 104.

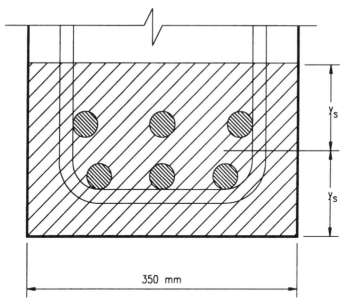

350 mm

Fig. E7.3-16 Concrete area with same centroid as principal tensile reinforcement.

$$f_{sa} = \frac{Z}{(d_c A)^{1/3}} = \frac{23\ 000}{(50 \times 11\ 550)^{1/3}}$$

$$= 276\ \text{MPa} > 0.6 f_y = 240\ \text{MPa} > f_s = 205\ \text{MPa}$$

6 No. 30 bottom bars OK

c. Negative bending reinforcement—exterior girder

Service I Limit State, $\eta = 1.0$, gravity load factors = 1.0, moments from Table E7.3-2

$$M_{200} = \eta \sum \gamma_i Q_i = M_{DC} + M_{DW} + mgrM_{LL+IM}$$

$$= (-339.6 - 38.3 - 420.8) = -798.7\ \text{kN m}$$

$d_{neg} = 912\ \text{mm}$ assume $j = 0.875$ and $f_s = 240\ \text{MPa}$

$$A_s \approx \frac{M}{f_s jd} = \frac{799 \times 10^6}{240 \times 0.875 \times 912} = 4170\ \text{mm}^2$$

Try 6 No. 30 bars provided $A_s = 4200\ \text{mm}^2$ (Table B.3)

Tension reinforcement in flange distributed over the lesser of: effective flange width or $\frac{1}{10}$ span. [A5.7.3.4]

Effective flange width [A5.7.3.4]

$$L_{\text{eff}} = 5360 \text{ mm} \qquad b_i = \tfrac{1}{4}L_{\text{eff}} = \tfrac{1}{4}(5360) = 1340 \text{ mm}$$

$$b_e = \tfrac{1}{2}b_i + \tfrac{1}{8}L_{\text{eff}} = \tfrac{1}{2}(1340) + \tfrac{1}{8}(5360) = 1340 \text{ mm}$$

$$\tfrac{1}{10} \text{ span} = \tfrac{1}{10}(11\ 735) = 1174 \text{ mm, governs.}$$

Effective flange width greater than one-tenth span, additional reinforcement required in outer portions of the flange.

Additional $A_s > 0.004$ (excess slab area)

$$> 0.004(190)(1340 - 1174) = 126 \text{ mm}^2$$

2 No. 10 bars additional reinforcement, additional $A_s = 200 \text{ mm}^2$ (Fig. E7.3-17).

Revised d_{neg} inside slab bars

$$d_{\text{neg}} = 990 - 60 - 16 - \frac{29.9}{2} = 899 \text{ mm}$$

$$b = b_w = 350 \text{ mm}$$

$$\frac{nA_s}{b} = \frac{7(4200)}{350} = 84 \text{ mm}$$

$$\frac{2nA_s d}{b} = 2(84)899 = 151\ 032 \text{ mm}^2$$

$$x = -84 + \sqrt{84^2 + 151\ 032} = 314 \text{ mm}$$

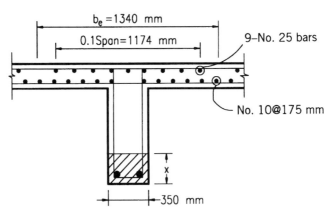

Fig. E7.3-17   Elastic-cracked transformed negative moment section at Location 200.

$$I_{cr} = \tfrac{1}{3}(350)(314)^3 + 7(4200)(899 - 314)^2 = 13.67 \times 10^9 \text{ mm}^4$$

$$f_s = \frac{nM(d - x)}{I_{cr}} = \frac{7(799 \times 10^6)(585)}{13.67 \times 10^9} = 239 \text{ MPa}$$

$$f_{sa} = \frac{Z}{(d_c A)^{1/3}} \qquad y_s = 60 + 16 + \frac{30}{2} = 91 \text{ mm}$$

$$d_c = 50 \text{ mm} \qquad A = \frac{2(91)(1174)}{6} = 35\,611 \text{ mm}^2$$

$$f_{sa} = \frac{23\,000}{(50 \times 35\,611)^{1/3}} = 190 \text{ MPa} < f_s = 239 \text{ MPa, no good}$$

Must use a larger number of smaller bars. Try 9 No. 25 bars, provided

$$A_s = 4500 \text{ mm}^2 \qquad d = 901 \text{ mm}$$

$$\frac{nA_s}{b} = \frac{7(4500)}{350} = 90 \text{ mm}$$

$$\frac{2nA_s d}{b} = 2(90)(901) = 162\,180 \text{ mm}^2$$

$$x = -90 + \sqrt{90^2 + 162\,180} = 323 \text{ mm}$$

$$I_{cr} = \tfrac{1}{3}(350)(323)^3 + 7(4500)(901 - 323)^2 = 14.46 \times 10^9 \text{ mm}^4$$

$$f_s = \frac{7(800 \times 10^6)(578)}{14.46 \times 10^9} = 224 \text{ MPa}$$

$$A = \frac{2(89)(1174)}{9} = 23\,219 \text{ mm}^2$$

$$f_{sa} = \frac{23\,000}{(50 \times 23\,219)^{1/3}} = 219 \text{ MPa} \approx f_s = 224 \text{ MPa}$$

2% difference, 9 No. 25 top bars, OK

6. *Investigate Fatigue*

Fatigue Limit State (Table 3.1) [Table A3.4.1-1]

$$U_f = \eta \Sigma \gamma_i Q_i = 0.75(LL + IM)$$

Fatigue Load
- One design truck with constant spacing of 9000 mm between 145-kN axles [A3.6.1.4].
- Dynamic load allowance: $IM = 15\%$ [A3.5.2.1].

- Distribution factor for one traffic lane shall be used [A3.6.1.4.3b].
- Multiple presence factor of 1.2 should be removed [C3.6.1.1.2].

a. Determination of need to consider fatigue [A5.5.3.1] Prestressed beams may be precompressed, but for the continuous *T*-beam without prestress, there will be regions, sometimes in the bottom of the beam, sometimes in the top of the beam, where the permanent loads do not produce compressive stress. In these regions, such as Locations 104 and 200, fatigue must be considered.

b. Allowable fatigue stress range f_f in reinforcement [A5.5.3.2]

$$f_f = 145 - 0.33 f_{min} + 55\left(\frac{r}{h}\right), \text{ MPa}$$

where

f_{min} = algebraic minimum stress level from fatigue

load given above,

positive if tension.

$\dfrac{r}{h}$ = ratio of base radius to height of rolled-on transverse

deformations; if the actual value

is not known, 0.3 may be used.

c. Location 104 (Fig. E7.3-18) [C3.6.1.1.2] Exterior girder— distribution factor

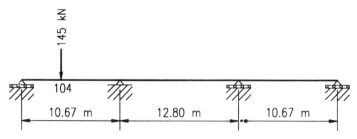

Fig. E7.3-18 Fatigue truck placement for maximum tension in positive moment reinforcement.

$$g_M^{SE}r = mg_M^{SE}\frac{r}{m} = \frac{0.762(0.948)}{1.2} = 0.602 \qquad [\text{C3.6.1.1.2}]$$

Fatigue load moment for maximum tension in reinforcement. Influence line ordinates taken from Table 5.4.

$$\text{pos } M_u = 145(0.207\ 00)10.67 = 320.3 \text{ kN m}$$

$$\text{pos } M_{104} = 0.75[g_M^{SE}rM_u(1 + IM)]$$

$$= 0.75[(0.602)(320.3)(1.15)] = 166.3 \text{ kN m}$$

Maximum tension in reinforcement using 6 No. 30 bars

$$f_{\text{max}} = \frac{nM(d - x)}{I_{cr}} = \frac{7(166.3 \times 10^6)(876 - 146)}{17.74 \times 10^9} = 48 \text{ MPa}$$

Fatigue load moment for maximum compression in reinforcement (Fig. E7.3-19)

$$\text{neg } M_{LL} = [145(-0.041\ 35 + 0.005\ 33) + 35(0.009\ 76)]10.67$$

$$= -52.08 \text{ kN m}$$

$$\text{neg } M_{104} = 0.75[0.602(-52.1)(1.15)] = -27.0 \text{ kN m}$$

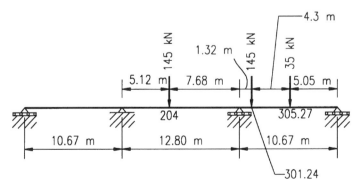

Fig. E7.3-19 Fatigue truck placement for maximum compression in positive moment reinforcement.

Section Properties for Compression in Bottom Fibers (Fig. E7.3-20)

Consider slab reinforcement as A_s

$$\text{No. 10 @ 450} \qquad A_s = 0.222 \text{ mm}^2/\text{mm}$$

$$\text{No. 10 @ 175} \qquad \underline{A_s = 0.571}$$

$$A_s = 0.793 \text{ mm}^2/\text{mm}$$

$$\text{over } L/10 \text{ width} = 1067 \text{ mm}$$

$$A_s = 0.793(1067) = 846 \text{ mm}^2$$

$$\text{plus } 2 - \text{No. } 25 = \underline{1000}$$

$$\text{Total } A_s = 1846 \text{ mm}^2$$

$$A_s' = 6 - \text{No. } 30 = 4200 \text{ mm}^2$$

By balancing statical moments about NA, $n = 7$

$$\tfrac{1}{2}b_w x^2 + (n - 1)A_s'(x - d') = nA_s(d - x)$$

$$\tfrac{1}{2}(350)x^2 + (6)(4200)(x - 99) = 7(1846)(899 - x)$$

$$\text{Solving } x = 195 \text{ mm}$$

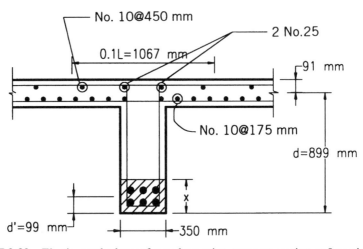

Fig. E7.3-20   Elastic-cracked transformed negative moment section at Location 104.

$$I_{cr} = \tfrac{1}{3}b_w x^3 + (n-1)A_s'(x-d')^2 + nA_s(d-x)^2$$
$$= \tfrac{1}{3}(350)(195)^3 + (6)(4200)(195-99)^2$$
$$+ (7)(1846)(899-195)^2$$
$$= 7.50 \times 10^9 \text{ mm}^4$$

$$f_{min} = f_s' = \frac{nM(x-d')}{I_{cr}}$$

$$= \frac{7(-2.70 \times 10^6)(195-99)}{7.50 \times 10^9} = -2.4 \text{ MPa}$$

Stress Range for Fatigue at Location 104

$$f_{max} - (f_{min}) = 48 - (-2.4) = 50.4 \text{ MPa}$$

Allowable Stress Range

$$f_f = 145 - 0.33f_{min} + 55\left(\frac{r}{h}\right) = 145 - 0.33(-2.4) + 55(0.3)$$
$$= 162 \text{ MPa} > 50 \text{ MPa, OK}$$

d. Location 200 Based on previous calculations, the moments due to $LL + IM$ at Location 200 are less than those at Location 104. Therefore, by inspection, the fatigue stresses will not be critical.

7. *Calculate Deflection and Camber* (Table 3.1) [Table A3.4.1-1]

Service I Limit State, $\eta = 1.0$, gravity load factors $= 1.0$

$$U = \eta\Sigma\gamma_i Q_i = DC + DW + LL + IM$$

a. Live load deflection criteria (optional) [A2.5.2.6.2]
• Distribution factor for deflection [C2.5.2.6.2]

$$mg = \frac{N_L}{N_B} = \frac{3}{6} = \underline{0.5}$$

N_L = No. design lanes N_B = No. of beams [A3.6.1.1.2]
• A right cross section may be used for skewed bridges.
• Use one design truck or lane load plus 25% design truck [A3.6.1.3.2].

• Live load deflection limit, first span [A2.5.2.6.2].

$$\Delta_{\text{allow}} = \frac{\text{span}}{800} = \frac{10\,670}{800} = 13 \text{ mm}$$

b. Section properties at Location 104 Transformed cracked section from Section 7.10.3, Part I.5b

$$d_{\text{pos}} = 876 \text{ mm} \qquad x = 146 \text{ mm} \qquad I_{cr} = 17.74 \times 10^9 \text{ mm}^4$$

Gross or uncracked section (Fig. E7.3-21)

$$A_g = 2000(190) + 350(785) = 380\,000 + 274\,750$$

$$A_g = 654\,750 \text{ mm}^2$$

$$\bar{y} = \frac{380\,000(880) + 274\,750(392.5)}{654\,750} = 675.4 \text{ mm}$$

$$I_g = \frac{1}{12}(2000)(190)^3 + 380\,000(204.6)^2 + \frac{1}{12}(350)(785)^3$$

$$+ \; 274\,750(282.9)^2 = 53.15 \times 10^9 \text{ mm}^4$$

$$f_c' = 30 \text{ MPa}$$

$$E_c = 4800\sqrt{f_c'} = 4800\sqrt{30} = 26\,290 \text{ MPa} \quad [\text{C5.4.2.4}]$$

$$f_r = 0.63\sqrt{f_c'} = 0.63\sqrt{30} = 3.45 \text{ MPa} \quad [\text{A5.4.2.6}]$$

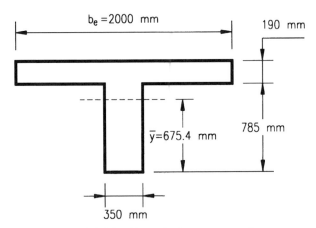

Fig. E7.3-21   Uncracked or gross section.

$$M_{cr} = f_r \frac{I_g}{y_t} = 3.45 \frac{53.15 \times 10^9}{675.4}$$

$$= 271.5 \times 10^6 \text{ N mm} = 271.5 \text{ kN m}$$

c. Estimated live load deflection at Location 104. Assume deflection is maximum where moment is maximum (Fig. E7.3-22)

$$M_{104} = 110(0.207\ 00 + 0.158\ 07)10.67$$

$$= 428.5 \text{ kN m} \quad \text{(Table 5.4)}$$

$$M_{200} = 110(-0.082\ 50 - 0.092\ 36)10.67 = -205.2 \text{ kN m}$$

Total moment at 104,

$$M_a = M_{DC} + M_{DW} + mgM_{LL}(1 + IM)$$

$$= 198.8 + 22.4 + 0.5(428.5)(1.33) = 506.2 \text{ kN m}$$

Effective moment of inertia [A5.7.3.6.2]

$$I_e = \left(\frac{M_{cr}}{M_a}\right)^3 I_g + \left[1 - \left(\frac{M_{cr}}{M_a}\right)^3\right] I_{cr} \le I_g,$$

$$\left(\frac{M_{cr}}{M_a}\right)^3 = \left(\frac{271.5}{506.2}\right)^3 = 0.154$$

$$I_e = (0.154)(53.15 \times 10^9) + (1 - 0.154)(17.74 \times 10^9)$$

$$= 23.19 \times 10^9 \text{ mm}^4$$

$$EI = E_c I_e = (26\ 290)(23.19 \times 10^9) = 610 \times 10^{12} \text{ N mm}^2$$

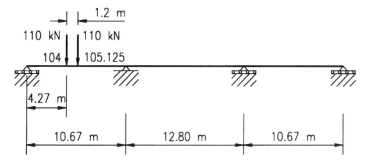

Fig. E7.3-22 Live load placement for deflection at Location 104.

Calculate deflection at Location 104 by considering first span as a simple beam with an end moment and use superposition (Fig. E7.3-23). Deflections for a design truck are (Eq. 5.19)

$$y_1 = \frac{L^2}{6EI}[M_{ij}(2\xi - 3\xi^2 + \xi^3) - M_{ji}(\xi - \xi^3)] \qquad \xi = \frac{x}{L}$$

$$M_{ij} = 0 \qquad M_{ji} = M_{200}$$

$$= -205 \times 10^6 \text{ N mm}$$

$$L = 10\,670 \text{ mm} \qquad \xi = 0.4$$

$$y_1 = \frac{(10\,670)^2}{6(610 \times 10^{12})}[-(-205 \times 10^6)(0.4 - 0.4^3)] = 2.1 \text{ mm} \uparrow$$

$$y_2 = \Delta_x(x < a) = \frac{Pbx}{6EIL}(L^2 - b^2 - x^2)$$

[AISC Manual (1986), Case 8]

For $P = 110$ kN, $x = 0.4L = 4268$ mm, $b_2 = 0.6L = 6402$ mm

$$y_2 = \frac{110\,000(6402)(4268)}{6(610 \times 10^{12})(10\,670)}(10\,670^2 - 6402^2)$$

$$- 4268^2) = 4.2 \text{ mm} \downarrow$$

For $P = 110$ kN, $x = 0.4L$, $a = 0.5125(10\,670) = 5468$ mm, $b = L - a = 5202$ mm

$$y_3 = \frac{110\,000(5202)(4268)}{6(610 \times 10^{12})(10\,670)}(10\,670^2 - 5202^2)$$

$$- 4268^2) = 4.3 \text{ mm} \downarrow$$

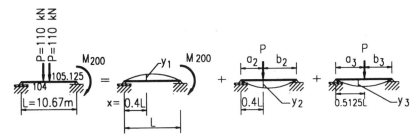

Fig. E7.3-23   Deflection estimate by superposition.

Estimated LL + IM deflection at 104.

With three lanes of traffic supported on six girders, each girder carries only a half-lane load. Including impact, the estimated live load deflection is

$$\Delta_{104}^{LL+IM} = mg(-y_1 + y_2 + y_3)(1 + IM)$$

$$= 0.5(-2.1 + 4.2 + 4.3)(1.33)$$

$$= 4 \text{ mm} < \Delta_{\text{allow}} = 13 \text{ mm, OK}$$

d. Dead load camber [A5.7.3.6.2]

The dead loads taken from Tables E7.3-1 and E7.3-2 are

Dead Loads	Interior Girder	Exterior Girder
w_{DC}	18.27 N/mm	24.49 N/mm
w_{DW}	4.04	2.76
w_{DL}	22.31 N/mm	27.25 N/mm

Unit Load Analysis (Fig. E7.3-24)

Deflection Equations

Simple beam at distance x from left end, uniform load

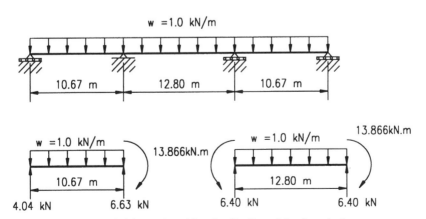

Fig. E7.3-24   Unit uniformly distributed load analysis.

$$\Delta_x = \frac{wx}{24EI}(L^3 - 2Lx^2 + x^3) \qquad \Delta_{centerline} = \frac{5}{384}\frac{wL^4}{EI}$$

[AISC Manual (1986), Case 1]

Simple beam at $\xi = x/L$ from i end, due to end moments

$$y = \frac{L^2}{6EI}[M_{ij}(2\xi - 3\xi^2 + \xi^3) - M_{ji}(\xi - \xi^3)] \qquad \xi = \frac{x}{L}$$

Flexural Rigidity EI for Longtime Deflections

The instantaneous deflection is multiplied by a creep factor λ to give a longtime deflection.

$$\Delta_{LT} = \lambda\Delta_i$$

so that

$$\Delta_{camber} = \Delta_i + \Delta_{LT} = (1 + \lambda)\Delta_i$$

If instantaneous deflection is based on I_g: $\lambda = 4.0$ [A4.5.2.2]

If instantaneous deflection is based on I_e:

$$\lambda = 3.0 - 1.2\left(\frac{A_s'}{A_s}\right) \geq 1.6$$

Location 104, x = 4268 mm

$$w = 1.0 \text{ N/mm}, M_{ij} = 0, M_{ji}$$
$$= -13.866 \times 10^6 \text{ N mm}, x = 0.4L$$
$$\Delta_i = \frac{1.0(4268)}{24 \times 610 \times 10^{12}}[(10\,670)^3$$
$$- 2(10\,670)(4268)^2 + (4268)^3]$$
$$- \frac{(10\,670)^2}{6 \times 610 \times 10^{12}}[-(-13.866$$
$$\times 10^6)(0.4 - 0.064)]$$
$$\Delta_i = 0.263 - 0.145 = 0.118 \text{ mm} \downarrow$$

Using $A_s = 4200 \text{ mm}^2$, $A_s' = 1846 \text{ mm}^2$

$$\lambda = 3.0 - 1.2\frac{1846}{4200} = 2.47$$

Exterior Girder, $w_e = 27.25$ N/mm

$$\Delta_{camber} = 27.25(1 + 2.47)(0.118)$$

$$= 11.16 \text{ mm}$$

$$(w_i = 22.31 \text{ N/mm}) = 9.14 \text{ mm, \underline{say 10 mm}, average}$$

Location 205

Assume same *EI* as at 104.

$$w = 1.0 \text{ N/mm}, M_{ij} = -M_{ji} = 13.866 \times 10^6 \text{ N mm}, x = 0.5L$$

$$\Delta_i = \frac{5}{384}\frac{1.0(12\ 800)^4}{610 \times 10^{12}} - \frac{(12\ 800)^2}{6 \times 610 \times 10^{12}}\left[13.866\right.$$

$$\left.\times 10^6\left(1 - \frac{3}{4} + \frac{1}{8} + \frac{1}{2} - \frac{1}{8}\right)\right]$$

$$= 0.573 - 0.466 = 0.107 \text{ mm} \downarrow$$

By using $\lambda = 2.47$ and $w_e = 27.25$ N/mm

$$\Delta_{camber} = 27.25(1 + 2.47)(0.107) = 10.12 \text{ mm}$$

$$(w_i = 22.31 \text{ N/mm}) = 8.28 \text{ mm, \underline{say 9 mm}, average}$$

Dead Load Deflection Diagram—All Girders (Fig. E7.3-25)

Upward camber should be placed in the formwork to offset the estimated longtime dead load deflection. The deflections are summarized in Figure E7.3-25.

J. *Investigate Strength Limit State* The previous calculations for the service limit state considered only a few critical sections at Locations 104,

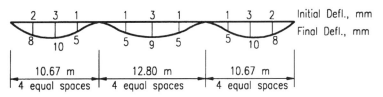

Fig. E7.3-25 Dead load deflection diagram—all girders.

200, and 205 to verify the adequacy of the trial section given in Figure E7.3-2. Before proceeding with the strength design of the girders, it is necessary to construct the factored moment and shear envelopes from values calculated at the tenth points of the spans. The procedure for generating the live load values is given in Chapter 5 and summarized in Tables 5.8a and 5.8b for spans of 10.67, 12.80, and 10.67 m.

The Strength *I* limit state can be expressed as

$$U = \eta[1.25DC + 1.50DW + 1.75(mgr)LL(1 + IM)] \quad \text{(E7.3-1)}$$

With the use of permanent loads given in Tables E7.3-1 and E7.3-2, live loads from Tables 5.8a and 5.8b, and live load distribution factors (*mgr*) determined earlier, the envelope values for moment and shear are generated for interior and exterior girders. These values are given in Tables E7.3-3 and E7.3-4 in the columns with the heading "eta*Sum," where eta = η and Sum is the quantity in brackets of Eq. E7.3-1. The envelope values for moment and shear are plotted in Figure E7.3-26. Notice how close together the curves are for the interior and exterior girders. One design will be sufficient for both.

1. *Flexure*

    a. and b. *Prestressed beams* Not applicable.

    c. *Factored flexural resistance* [A5.7.3.2, Table A3.4.1-1] Exterior girder has slightly larger moment. Neglect TG.

$$M_u = \eta\, \Sigma\gamma_i M_i = 0.95(1.25M_{DC} + 1.50M_{DW} + 1.75M_{LL+IM})$$

Location 104 Unfactored values for moment from Table E7.3-2

$$M_{104} = 0.95[1.25(198.8) + 1.50(22.4) + 1.75(489.8)]$$

$$= 1082 \text{ kN m}$$

This number is comparable to the value of 1085 kN m found in Table E7.3-3.

Check resistance provided by bars selected for crack control (Fig. E7.3-27)

Assume $a < t_s = 190$ mm

$$a = \frac{A_s f_y}{0.85 f'_c b_e} = \frac{4200(400)}{0.85(30)(2000)} = 33 \text{ mm} \quad \text{[A5.7.3.2.2]}$$

All compression is in flange.

TABLE E7.3-3 Moment Envelope for 10.67, 12.80, 10.67 m T-Beam (kN m)

Location	Unit Dead Load	Positive Moment				Negative Moment			
		Truck or Tandem	Lane	eta times Sum Int. Gir.	eta times Sum Ext. Gir.	Truck or Tandem	Lane	eta times Sum Int. Gir.	eta times Sum Ext. Gir.
100	0.0	0	0	0	0	0	0	0	0
101	3.8	200	43	468	497	-28	-8	51	71
102	6.4	327	75	775	823	-56	-16	70	103
103	7.8	400	97	954	1013	-84	-24	55	95
104	8.2	428	108	1022	1085	-112	-33	12	53
105	7.4	420	109	987	1045	-140	-40	-62	-27
106	5.4	380	99	859	904	-167	-48	-170	-147
107	2.3	304	78	630	655	-195	-57	-309	-304
108	-1.9	200	47	316	313	-223	-65	-478	-497
109	-7.2	77	22	-51	-88	-251	-90	-696	-747
110	-13.7	65	18	-255	-327	-327[a]	-145	-1048	-1150
200	-13.7	65	18	-255	-327	-327[a]	-145	-1048	-1150
201	-6.3	94	19	-9	-35	-208	-77	-577	-635
202	-0.6	230	43	381	400	-178	-48	-342	-363
203	3.5	335	79	695	748	-147	-47	-181	-177
204	5.9	399	102	885	958	-117	-47	-69	-49
205	6.8	416	110	944	1024	-87	-47	2	29

[a]Truck with 9 m rear axle spacing governs.

TABLE E7.3-4 Shear Envelope for 10.67, 12.80, 10.67 m T-Beam (kN)

Location	Unit Dead Load	Positive Shear				Negative Shear			
		Truck or Tandem	Lane	eta times Sum Int. Gir.	eta times Sum Ext. Gir.	Truck or Tandem	Lane	eta times Sum Int. Gir.	eta times Sum Ext. Gir.
100	4.0	224	46	636	617	-26	-7	45	72
101	3.0	188	36	520	503	-26	-8	16	38
102	1.9	153	28	407	390	-39	-10	-43	-25
103	0.8	125	21	309	292	-66	-13	-132	-116
104	-0.2	100	15	222	203	-92	-18	-220	-204
105	-1.3	77	10	137	117	-117	-22	-307	-294
106	-2.4	56	7	59	36	-140	-29	-395	-383
107	-3.4	37	5	-12	-37	-168	-36	-492	-480
108	-4.5	20	3	-79	-107	-198	-45	-596	-584
109	-5.6	6	2	-138	-171	-226	-54	-697	-686
110	-6.6	6	2	-166	-204	-254	-63	-794	-784
200	6.4	256	66	797	785	-24	-7	117	158
201	5.1	225	55	681	668	-24	-7	82	115
202	3.8	191	44	560	546	-27	-8	37	63
203	2.6	156	34	442	428	-48	-11	-44	-21
204	1.3	124	27	329	314	-72	-15	-134	-114
205	0.0	98	20	229	211	-98	-20	-229	-211

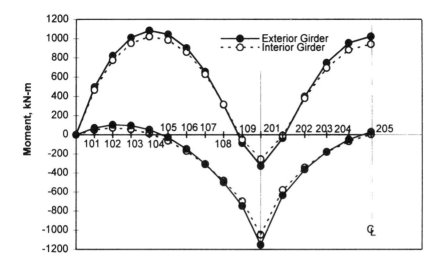

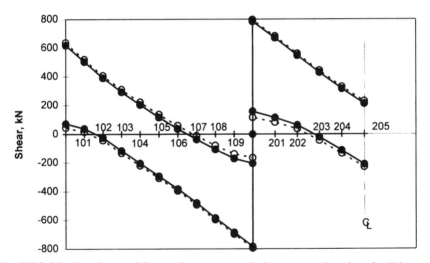

Fig. E7.3-26 Envelopes of factored moments and shears at tenth points for *T*-beams.

$$\phi M_n = \phi A_s f_y \left(d - \frac{a}{2} \right) = 0.9(4200)(400)\left(876 - \frac{33}{2} \right)$$

$$\phi M_n = 1300 \times 10^6 N\cdot mm = 1300 \text{ kN m} > M_u$$

$$= 1082 \text{ kN m, OK}$$

Use 6 No. 30 bottom bars

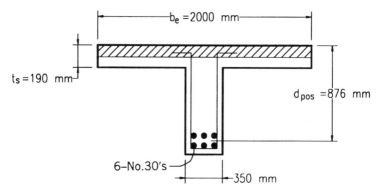

Fig. E7.3-27   Positive moment design section.

Location 200 Unfactored values for moment from Table E7.3-2

$$M_{200} = 0.95[1.25(-339.6) + 1.50(-38.3) + 1.75(-420.8)]$$
$$= -1157 \text{ kN m}$$

This number is comparable to the value of -1150 kN m found in Table E7.3-3.

Check resistance provided by bars selected for crack control (Fig. E7.3-28). Neglecting compression reinforcement

$$a = \frac{4500(400)}{0.85(30)(350)} = 202 \text{ mm}$$

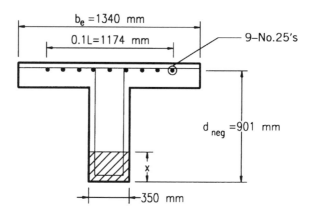

Fig. E7.3-28   Negative moment design section.

$$\phi M_n = 0.9(4500)(400)\left(901 - \frac{202}{2}\right)$$

$$\phi M_n = 1296 \times 10^6 \text{ N mm} = 1296 \text{ kN m} > M_u$$

$$= 1218 \text{ kN m, OK}$$

Use 9 No. 25, top bars

d. Limits for reinforcement

$$\beta_1 = 0.85 - 0.05 \frac{30 - 28}{7} = 0.836 \qquad [A5.7.2.2]$$

maximum reinforcement such that $\dfrac{c}{d_e} \le 0.42$ [A5.7.3.3.1]

minimum reinforcement such that $\phi M_n \ge 1.2 M_{cr}$ [A5.7.3.3.2] or

$$\rho = A_s / A_g > \rho_{min} = 0.03 f'_c / f_y$$

Location 104 $\quad \dfrac{c}{d_e} = \dfrac{a/\beta_1}{d_s} = \dfrac{33/0.836}{876} = 0.045 \le 0.42$, OK

$$\phi M_n = 1300 \text{ kN m} > 1.2 M_{cr} = 1.2(271.5) = 326 \text{ kN m}$$

Location 200 $\quad \dfrac{c}{d_e} = \dfrac{a/\beta_1}{d_s} = \dfrac{202/0.836}{901} = 0.27 \le 0.42$, OK

$$\rho_{min} = 0.03 \frac{30}{400} = 0.0023$$

$$\rho = \frac{A_s}{A_g} = \frac{4500}{190(1340) + 350(785)} = 0.0085 > \rho_{min}, \text{ OK}$$

2. *Shear (Assuming No Torsional Moment)*
   a. General requirements
      • Transverse reinforcement shall be provided where [A5.8.2.4]

$$V_u \ge 0.5\phi(V_c + V_p) \qquad \phi = \phi_v = 0.9$$

   • Minimum transverse reinforcement [A5.8.2.5]

$$A_v \ge 0.083\sqrt{f'_c}\frac{b_v s}{f_y}$$

- Maximum spacing of transverse reinforcement [A5.8.2.7]

If $V_u < 0.1f'_c b_v d_v$, then $s \le 0.8d_v \le 600$ mm

If $V_u \ge 0.1f'_c b_v d_v$, then $s \le 0.4d_v \le 300$ mm

where

b_v = minimum web width within depth d_v

d_v = effective shear depth taken as the distance between the resultants of the tensile and compressive forces due to flexure, but it need not be taken less than the greater of $0.9d_e$ or $0.72h$.

b. Sectional design model [A5.8.3]
- Based on equilibrium of forces and compatibility of strains (Collins and Mitchell, 1991).
- Where the reaction force produces compression at a support, the critical section for shear shall be taken as the larger of $0.5d_v \cot\theta$ or d_v from the internal face of the bearing (see Fig. E7.3-29) [A5.8.3.2].

Nominal Shear Resistance V_n [A5.8.3.3]
- Shall be the lesser of

$$V_n = V_c + V_s + V_p$$
$$V_n = 0.25f'_c b_v d_v + V_p$$

- Nominal concrete shear resistance

$$V_c = 0.083\beta\sqrt{f'_c}b_v d_v \qquad \text{(traditional value of } \beta = 2.0)$$

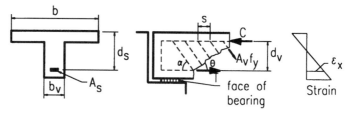

Fig. E7.3-29 Shear sectional design model.

• Nominal transverse reinforcement shear resistance

$$V_s = \frac{A_v f_y d_v (\cot \theta + \cot \alpha) \sin \alpha}{s}$$

for vertical stirrups $\alpha = 90°$ and

$$V_s = \frac{A_v f_y d_v \cot \theta}{s} \qquad \text{(traditional value of } \theta = 45°)$$

Determination of β and θ [A5.8.3.4.2]

Use tables and figures in AASHTO [A5.8.3.4.2] to determine β and θ. These tables depend on the following parameters for non-prestressed beams without axial load:

• Nominal shear stress in the concrete.

$$v = \frac{V_u}{\phi b_v d_v}$$

• Tensile strain in the longitudinal reinforcement.

$$\varepsilon_x = \frac{M_u/d_v + 0.5 V_u \cot \theta}{E_s A_s} \le 0.002$$

Longitudinal Reinforcement [A5.8.3.5]

Shear causes tension in the longitudinal reinforcement that must be added to that caused by flexure. Thus,

$$A_s f_y \ge \frac{M_u}{\phi_f d_v} + \left(\frac{V_u}{\phi_v} - 0.5 V_s \right) \cot \theta$$

If this equation is not satisfied, either the tensile reinforcement A_s must be increased or the stirrups must be placed closer together to increase V_s.

The procedure outlined in Section 7.8.3 for the shear design of members with web reinforcement is illustrated for a section at a distance d_v from an exterior support. The factored V_u and moment M_u envelopes for the Strength I limit state are plotted in Figure E7.3-26 from the values in Tables E7.3-3 and E7.3-4.

Step 1 Determine V_u and M_u at a distance d_v from an exterior support [A5.8.2.7]. From Figure E7.3-27

$$A_s = 6 \text{ No. } 30 = 4200 \text{ mm}^2, \ b_v = 350 \text{ mm}, \ b_i = 2000 \text{ mm}$$

$$a = \frac{A_s f_y}{0.85 f'_c b_i}$$

$$= \frac{(4200)(400)}{0.85(30)(2000)} = 33 \text{ mm},$$

$$d_e = d_s = 990 - 15 - 99 = 876 \text{ mm}$$

$$d_v = \max \begin{cases} d - a/2 = 876 - 33/2 = 860 \text{ mm, governs} \\ 0.9 d_e = 0.9(876) = 788 \text{ mm} \\ 0.72 h = 0.72(990) = 713 \text{ mm} \end{cases}$$

Distance from support as a percentage of the span

$$\frac{d_v}{L_1} = \frac{860}{10\,670} = 0.0806$$

Interpolating from Tables E7.3-3 and E7.3-4 for the factored shear and moment at Location 100.806 for an interior girder:

$$V_{100.806} = 636 - 0.806(636 - 520) = 543 \text{ kN}$$

$$M_{100.806} = 0.806(468) = 377 \text{ kN m}$$

These values are used to calculate the strain ε_x on the flexural tension side of the member [A5.8.3.4.2]. They are both extreme values at the section and have been determined from different positions of the live load. It is conservative to take the highest value of M_u at the section, rather than a moment coincident with V_u [C5.8.3.4.2].

Step 2 Calculate the shear stress ratio v/f'_c. From Eq. 7.159

$$v = \frac{V_u}{\phi b_v d_v} = \frac{543\,000}{0.9(350)(860)} = 2.00 \ \frac{N}{mm^2} = 2.00 \text{ MPa}$$

so that

$$\frac{v}{f'_c} = \frac{2.00}{30} = 0.0667$$

Step 3 Estimate an initial value for θ and calculate ε_x from Eq. 7.170.

First Trial $\quad \theta = 40°$, cot $\theta = 1.192$, $E_s = 200$ GPa

$$\varepsilon_x = \frac{(M_u/d_v) + 0.5V_u\cot \theta}{E_sA_s} = \frac{\dfrac{377\ 000}{860} + 0.5(543)(1.192)}{200(4200)} = 0.91 \times 10^{-3}$$

Step 4 Determine θ and β from Figure 7.43 and iterate until θ converges. Second Trial: $\theta = 35°$, cot $\theta = 1.428$

$$\varepsilon_x = \frac{(377\ 000/860) + 0.5(543)(1.428)}{200(4200)} = 0.98 \times 10^{-3}$$

Third Trial: $\theta = 36°$, cot $\theta = 1.376$

$$\varepsilon = \frac{\dfrac{377\ 000}{860} + 0.5(543)(1.376)}{200(4200)} = 0.97 \times 10^{-3}$$

Use $\theta = 36°$, $\beta = 2.2$

Step 5 Calculate the required web reinforcement strength V_s from Eq. 7.171:

$$V_s = \frac{V_u}{\phi_v} - 0.083\beta\sqrt{f_c'}b_vd_v$$

$$= \frac{543}{0.9} - 0.083(2.2)\sqrt{30}(350)(860) \times 10^{-3} = 301 \text{ kN}$$

Step 6 Calculate the required spacing of stirrups.

Eq. 7.172: No. 10 U-stirrups, $A_v = 2(100) = 200 \text{ mm}^2$

$$s \le \frac{A_vf_yd_v}{V_s}\cot \theta = \frac{200(400)(860)}{301 \times 10^3}(1.376) = 315 \text{ mm}$$

Eq. 7.173: $s \le \dfrac{A_vf_y}{0.083\sqrt{f_c'}b_v} = \dfrac{200(400)}{0.083\sqrt{30}(350)} = 500 \text{ mm}$

Eq. 7.174: $\begin{array}{l} V_u < 0.1f_c'b_vd_v = 0.1(30)(350)(860) \times 10^{-3} = 903 \text{ kN} \\ s \le 0.8d_v = 0.8(860) = 688 \text{ mm or } 600 \text{ mm} \end{array}$

Use $s = 315$ *mm*

Step 7 Check the adequacy of the longitudinal reinforcement by Eq. 7.169.

$$A_s f_y \geq \frac{M_u}{d_v \phi_f} + \left(\frac{V_u}{\phi_v} - 0.5 V_s\right) \cot \theta$$

$$V_s = \frac{200(400)(860)}{315}(1.376) \times 10^{-3} = 301 \text{ kN}$$

$$\frac{4200(400)}{10^3} \geq \frac{377 \times 10^3}{860(0.9)} + \left[\frac{543}{0.9} - 0.5(301)\right](1.376)$$

$$1680 \text{ kN} \geq 1110 \text{ kN, OK}$$

The above procedure is repeated for each of the tenth points. The results are summarized in Table E7.3-5 and plotted in Figure E7.3-30. Stirrup spacings are then selected to have values less than the calculated spacings. Starting at the left end and proceeding to midspan of the T-beam, the spacings are 6 @ 275 mm, 4 @ 380 mm, 5 @ 480 mm, 4 @ 350 mm, 7 @ 200 mm, 45 @ 100 mm, 8 @ 200 mm, 4 @ 350 mm, and $2\frac{1}{2}$ @ 480 mm. The selected stirrup spacings are shown by the dotted line in Figure E7.3-30. This completes the design of the T-beam bridge example. Tasks remaining include the determination of cut-off points for the main flexural reinforcement, anchorage requirements for the stirrups, and side reinforcement in the beam stems.

TABLE E7.3-5 Summary of Stirrup Spacing for T-Beam

Location	θ	β	s req'd (mm)	s prov'd (mm)
$100 + d_v$	35.5	2.25	315	275
101	36.5	2.1	310	275
102	39.0	2.0	475	380
103	41.0	1.9	500	480
104	41.5	1.9	500	480
105	41.5	1.9	500	480
106	40.5	1.9	450	350
107	38.5	2.0	315	200
108	37.0	2.1	200	200
109	38.0	1.8	150	100
$110 - d_v$	38.5	1.75	140	100
$200 + d_v$	39.0	1.5	110	100
201	37.5	2.0	150	100
202	36.0	2.2	290	200
203	39.0	2.0	375	350
204	40.5	2.0	500	480
205	41.0	2.0	500	480

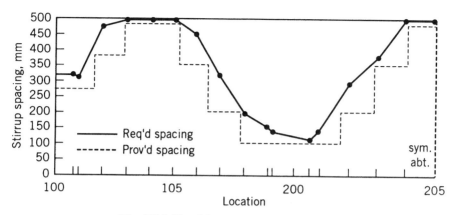

Fig. E7.3-30 Stirrup spacing for *T*-beam.

7.10.4 Prestressed Girder Bridge

Problem Statement Design the simply supported pretensioned prestressed concrete girder bridge of Figure E7.4-1 with a span length of 30 480 mm center to center of bearings for a *HL*-93 live load. The roadway width is 13 420 mm curb-to-curb. Allow for a future wearing surface of 75-mm thick bituminous overlay and use the concrete deck design of Example Problem 7.10.1 (f'_c = 30 MPa). Follow the beam and girder bridge outline in Section 5-Appendix A5.3 of the AASHTO (1994) LRFD Bridge Specifications. Use f'_{ci} = 40 MPa, f'_c = 55 MPa, f_y = 400 MPa, and 1860 MPa, low-relaxation 12.70 mm, seven-wire strands.

A. ***Develop General Section*** The bridge is to carry interstate traffic in Virginia over a single-track railroad with minimum vertical clearance of 7000 mm (Fig. E7.4-1).

B. ***Develop Typical Section*** Use a precast pretensioned Nebraska University girder (Green and Tadros, 1994) made composite with the deck. These are relatively new girder shapes and have been developed in metric to incorporate the best features of strength, efficiency, and constructability (Fig. E7.4-2). They may not be available in all regions of the United States. If other shapes are used, the procedures given in this example are still valid. The only difference is that different values are given for the section properties.

1. *Minimum Thickness* [A5.14.1.2.2]

$$\text{Top flange} \geq 50 \text{ mm, OK}$$

$$\text{Web} \geq 125 \text{ mm, OK}$$

$$\text{Bottom flange} \geq 125 \text{ mm, OK}$$

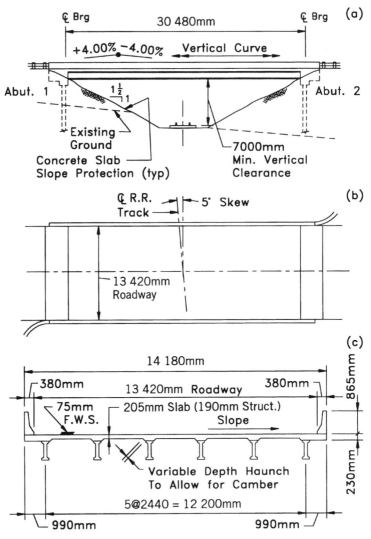

Fig. E7.4-1 Prestressed concrete girder bridge design example (a) elevation, (b) plan, and (c) section.

2. *Minimum Depth* (includes deck thickness) [A2.5.2.6.3]

$$h_{min} = 0.045L = 0.045(30\ 480) = 1372 \text{ mm}$$

$$< h = 1350 + 190 = 1540 \text{ mm, OK}$$

3. *Effective Flange Widths* [A4.6.2.6.1]

Effective span length = 30 480 mm

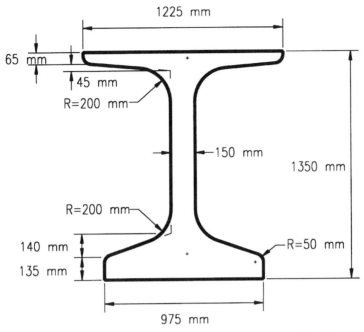

Fig. E7.4-2 Precast pretensioned NU 1350 girder. $A_g = 486\,051\ mm^2$.

Interior girders

$$b_i \leq \begin{cases} \frac{1}{4} \text{ effective span} = \frac{1}{4}(30\,480) = 7710 \text{ mm} \\ 12t_s + \frac{1}{2}b_f = 12(190) + \frac{1}{2}(1225) = 2890 \text{ mm} \\ \text{ctr. to ctr. spacing of girders} = 2440\text{mm, governs} \end{cases}$$

Exterior girders

$$b_e - \frac{b_i}{2} \leq \begin{cases} \frac{1}{8} \text{ effective span} = \frac{1}{8}(30\,480) = 3810 \text{ mm} \\ 6t_s + \frac{1}{4}b_f = 6(190) + \frac{1}{4}(1225) = 1446 \text{ mm} \\ \text{width of overhang} = 990\text{mm, governs} \end{cases}$$

$$b_e = \frac{2440}{2} + 990 = 2210\text{mm}$$

C. Design Conventionally Reinforced Concrete Deck The design section for negative moments in the deck slab is at one-third the flange width, but not more than 380 mm, from the centerline of the support for precast concrete beams [A4.6.2.1.6]. One-third of the flange width $b_f/3 = 1225/3 = 408$ mm is greater than 380 mm; therefore, the critical distance is 380 mm from the centerline of the support.

The deck design in Section 7.10.1, Part E, is for a monolithic T-beam girder and the design section is at the face of the girder or 175

mm from the centerline of the support (Fig. E7.1-10). The design negative moment for the precast *I*-beam, and resulting reinforcement, can
be reduced by using the 380-mm distance rather than 175 mm.

By following the procedures in Section 7.10.1, Part F.2, the top reinforcement at an interior support is reduced from No. 15 bars @ 225
mm to No. 10 bars @ 250 mm (Fig. E7.1-14). This latter selection is
the same as that required by the empirical method of design (Fig. E7.1-
15).

The deck overhang design remains the same as for the *T*-beam (Fig.
E7.1-17). It is governed by the truck collision and providing sufficient
moment capacity to develop the strength of the barrier. The changes in
the total design moment are small when the gravity loads are included
at different distances from the centerline of the support. The dominant
effect is the collision moment at the free end of the overhang and that
remains the same, so the overhang design remains the same.

D. *Select Resistance Factors (Table 7.10)* [A5.5.4.2]

1. *Strength Limit State*	ϕ		[A5.5.4.2.1]
Flexure and tension	1.00		
Shear and torsion	0.90		
Compression in anchorage zones	0.80		
2. *Nonstrength Limit States*	1.00		[A1.3.2.1]

E. *Select Load Modifiers* [A1.3.2.1]

	Strength	Service	Fatigue	
Ductility, η_D	0.95	1.0	1.0	[A1.3.3]
Redundancy, η_R	0.95	1.0	1.0	[A1.3.4]
Importance, η_I	1.05	N/A	N/A	[A1.3.5]
$\eta = \eta_D\eta_R\eta_I$	0.95	1.0	1.0	

F. *Select Applicable Load Combinations* (Table 3.1) [Table A3.4.1-1]

Strength I Limit State

$$U = \eta[1.25DC + 1.50DW + 1.75(LL + IM) + 1.0FR + \gamma_{TG}TG]$$

Service I Limit State

$$U = 1.0(DC + DW) + 1.0(LL + IM) + 0.3(WS + WL) + 1.0FR$$

Fatigue Limit State

$$U = 0.75(LL + IM)$$

Service III Limit State

$$U = 1.0(DC + DW) + 0.80(LL + IM) + 1.0WA + 1.0FR$$

G. *Calculate Live Load Force Effects*
  1. *Select Number of Lanes:* [A3.6.1.1.1]

$$N_L = INT\left(\frac{w}{3600}\right) = INT\left(\frac{13\ 420}{3600}\right) = 3$$

  2. *Multiple Presence Factor:* (Table 4.6) [A3.6.1.1.2]

No. of Loaded Lanes	m
1	1.20
2	1.00
3	0.85

  3. *Dynamic Load Allowance* (Table 4.7) [A3.6.2.1] Not applied to the design lane load.

Component	IM (%)
Deck joints	75
Fatigue	15
All other	33

  4. *Distribution Factors for Moment:* [A4.6.2.2.2]

Cross-Section Type (k) (Table 2.2) [Table A4.6.2.2.1-1]

$$\left.\begin{array}{ll} \text{Beam} & \text{55-MPa concrete} \\ \text{Deck} & \text{30-MPa concrete} \end{array}\right\}$$

n = modular ratio between beam and deck materials

$$= \sqrt{\frac{55}{30}} = 1.354$$

stiffness factor, K_g (See Fig. E7.4-8 for additional cross section properties.)

$$e_g = 742 + 50 + \frac{190}{2} = 887 \text{ mm}$$

$$K_g = n(I_g + Ae_g^2) = 1.354[126\ 011 + (486\ 051)(887)^2]$$

$$K_g = 517.8 \times 10^9 \text{ mm}^4$$

$$\frac{K_g}{Lt_s^3} = \frac{517.8 \times 10^9}{(30\ 480)(190)^3} = 2.477$$

$$S = 2440 \text{ mm}, \quad L = 30\ 480 \text{ mm}$$

a. Interior beams with concrete decks (Table 6.5) [A4.6.2.2.2b and Table A4.6.2.2.2b-1]

One Design Lane Loaded

$$mg_M^{SI} = 0.06 + \left(\frac{S}{4300}\right)^{0.4}\left(\frac{S}{L}\right)^{0.3}\left(\frac{K_g}{Lt_s^3}\right)^{0.1}$$

$$mg_M^{SI} = 0.06 + \left(\frac{2440}{4300}\right)^{0.4}\left(\frac{2440}{30\ 480}\right)^{0.3}(2.477)^{0.1} = 0.469$$

Two Design Lanes Loaded

$$mg_M^{MI} = 0.075 + \left(\frac{S}{2900}\right)^{0.6}\left(\frac{S}{L}\right)^{0.2}\left(\frac{K_g}{Lt_s^3}\right)^{0.1}$$

$$mg_M^{MI} = 0.075 + \left(\frac{2440}{2900}\right)^{0.6}\left(\frac{2440}{30\ 480}\right)^{0.2}(2.477)^{0.1}$$

$$= 0.671, \text{ governs}$$

b. Exterior beams with concrete decks (Table 6.5) [A4.6.2.2.2d and Table A4.6.2.2.2d-1]

One Design Lane Loaded—Lever Rule (Fig. E7.4-3)

$$R = \frac{P}{2}\left(\frac{650 + 2450}{2440}\right) = 0.635P$$

$$g_M^{SE} = 0.635$$

$$mg_M^{SE} = 1.2(0.635) = 0.762, \text{ governs}$$

Two or More Design Lanes Loaded

$$d_e = 990 - 380 = 610 \text{ mm}$$

$$e = 0.77 + \frac{d_e}{2800} = 0.77 + \frac{610}{2800} = 0.988 < 1.0$$

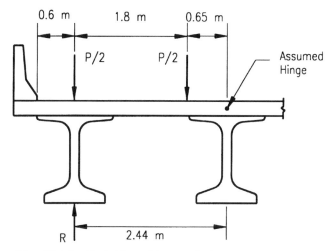

Fig. E7.4-3 Definition of lever rule for an exterior girder.

$$\text{Use } e = 1.0$$

$$mg_M^{ME} = emg_M^{MI} = 0.671$$

5. *Distribution factors for shear* [A4.6.2.2.3] Cross-Section Type (*k*) (Table 2.2) [Table A4.6.2.2.1-1]

a. Interior beams (Table 6.5) [A4.6.2.2.3a and Table A4.6.2.2.3a-1]

One Design Lane Loaded

$$mg_V^{SI} = 0.36 + \frac{S}{7600} = 0.36 + \frac{2440}{7600} = 0.68$$

Two Design Lanes Loaded

$$mg_V^{MI} = 0.2 + \frac{S}{3600} - \left(\frac{S}{10\,700}\right)^{2.0}$$

$$mg_V^{MI} = 0.2 + \frac{2440}{3600} - \left(\frac{2440}{10\,700}\right)^{2.0} = 0.826, \text{ governs}$$

b. Exterior beams (Table 6.5) [A4.6.2.2.3b and Table A4.6.2.2.3b-1]

One Design Lane Loaded—Lever Rule (Fig. E7.4-3):

$$mg_V^{SE} = 0.762, \text{ governs}$$

Two or More Design Lanes Loaded

$$d_e = 610 \text{ mm}$$

$$e = 0.6 + \frac{d_e}{3000} = 0.6 + \frac{610}{3000} = 0.803$$

$$mg_V^{ME} = e \cdot mg_V^{MI} = (0.803)(0.826) = 0.663$$

6. *Calculation of shears and moments due to live loads* The shears and moments at tenth points along the span are found next. Calculations are shown below for Locations 100, 101, and 105 only. Concentrated loads are multiplied by influence line ordinates. Uniform loads are multiplied by the area under the influence line. As discussed in Chapter 5, the influence functions are straight lines for simple spans. Shears and moments at the other locations are found in a similar manner. Results of these calculations are summarized in Tables E7.4-3 and E7.4-4.

Location 100 (Fig. E7.4-4)

Truck

$$V_{100}^{Tr} = 145\left(1 + \frac{26\,180}{30\,480}\right) + 35\left(\frac{21\,880}{30\,480}\right) = 294.7 \text{ kN}$$

$$M_{100}^{Tr} = 0$$

Lane

$$V_{100}^{Ln} = \frac{(15\,240)(9.3)}{10^3} = 141.7 \text{ kN}$$

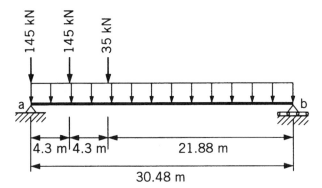

Fig. E7.4-4 Live load placement at Location 100.

$$M^{Ln}_{100} = 0$$

Location 101 (Fig. E7.4-5):

Truck

$$V^{Tr}_{101} = R_a = 145\left(\frac{23\,132 + 27\,432}{30\,480}\right) + 35\left(\frac{18\,832}{30\,480}\right) = 262.2 \text{ kN}$$

$$M^{Tr}_{101} = \frac{262.2(3048)}{10^3} = 799.2 \text{ kN m}$$

Tandem

$$V^{Ta}_{101} = 110\left(\frac{26\,232 + 27\,432}{30\,480}\right) = 193.7 \text{ kN}$$

$$M^{Ta}_{101} = \frac{193.7(3048)}{10^3} = 590.4 \text{ kN m}$$

Lane

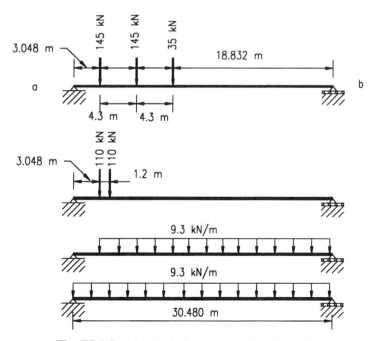

Fig. E7.4-5   Live load placement at Location 101.

$$V_{101}^{Ln} = \frac{9.3(27\,432)}{10^3}\left(\frac{13\,716}{30\,480}\right) = 114.8 \text{ kN}$$

$$M_{101}^{Ln} = \frac{1}{2}wab = \frac{1}{2}\frac{(9.3)(3048)(27\,432)}{10^6} = 388.8 \text{ kN m}$$

Location 105 (Fig. E7.4-6):

Truck

$$V_{105}^{Tr} = 145\left(\frac{15\,240 + 10\,940}{30\,480}\right) + 35\left(\frac{6640}{30\,480}\right) = 132.2 \text{ kN}$$

$$R_a = 145\left(\frac{15\,240 + 10\,940}{30\,480}\right) + 35\left(\frac{19\,540}{30\,480}\right) = 147 \text{ kN}$$

$$M_{105}^{Tr} = \frac{(147)(15\,240) - 35(4300)}{10^3} = 2090 \text{ kN m}$$

Tandem

$$V_{105}^{Ta} = 110\left(\frac{15\,240 + 14\,040}{30\,480}\right) = 105.7 \text{ kN}$$

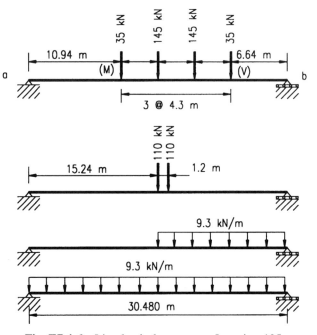

Fig. E7.4-6 Live load placement at Location 105.

$$M_{105}^{Ta} = \frac{(105.7)(15\ 240)}{10^3} = 1610 \text{ kN m}$$

Lane

$$V_{105}^{Ln} = \frac{9.3(15\ 240)}{10^3}\left(\frac{7620}{30\ 480}\right) = 35.4 \text{ kN}$$

$$M_{105}^{Ln} = \frac{1}{8}wL^2 = \frac{1}{8}\frac{(9.3)(30\ 480)^2}{10^6} = 1080 \text{ kN m}$$

H. *Calculate Force Effects from Other Loads*
   1. *Interior girders*

DC   Weight of concrete $= (2400)(9.81)(10^{-9}) = 2.3544 \times 10^{-5} \text{ N/mm}^3$

$$
\begin{aligned}
\text{Slab} \quad &(2.3544 \times 10^{-5})(205)(2440) = 11.78 \text{ N/mm} \\
\text{50-mm haunch} \quad &(2.3544 \times 10^{-5})(50)(1225) = 1.44 \text{ N/mm} \\
\text{Girder} \quad &(2.3544 \times 10^{-5})(486\ 051) = \underline{11.44 \text{ N/mm}} \\
&\phantom{(2.3544 \times 10^{-5})(486\ 051)} = 24.66 \text{ N/mm}
\end{aligned}
$$

Estimate diaphragm size   300 mm thick, 1200 mm deep

Diaphragms @ 1/3 points $(2.3544 \times 10^{-5})(300)(1200)(2290)$

$$= 19\ 410 \text{ N}$$

DW   75 mm bituminous paving $= (2250)(9.81)(10^{-9})(75)(2440)$

$$= 4.04 \text{ N/mm}$$

   2. *Exterior girders*

$$
\begin{aligned}
DC1\text{: Overhang} \quad &(5.42 \times 10^{-3})(990) &&= 5.36 \text{ N/mm} \\
\text{Slab} \quad &(4.83 \times 10^{-3})(1220) &&= 5.89 \text{ N/mm} \\
\text{Girder + Haunch} \quad & &&= \underline{12.88 \text{ N/mm}} \\
& &&= 24.13 \text{ N/mm}
\end{aligned}
$$

Diaphragms @ $\frac{1}{3}$ points $19\ 410/2 = 9705\ N = 9.705$ kN

$DC2$: Barrier $= 4.65$ N/mm

DW 75-mm bituminous paving

$$= (1.66 \times 10^{-3})(990 - 380 + 1220)$$

$$= 3.04 \text{ N/mm}$$

(*DC*2 and *DW* act on the composite section)

From Figure E7.4-7, shears and moments due to a unit uniform load are found at tenth points (Table E7.4-1), where

$$V_x = w\left(\frac{L}{2} - x\right) = wL(0.5 - \xi), \ \xi = \frac{x}{L}$$

$$M_x = \frac{w}{2}x(L - x) = 0.5wL^2(\xi - \xi^2)$$

From Figure E7.4-7, shears and moments due to the diaphragms for interior girders are found at tenth points (Table E7.4-2). Values for exterior girders are one-half the values for interior girders.

3. *Summary of force effects*

   a. Interior girders (Table E7.4-3)

$$mg_M = 0.671, \ mg_V = 0.826, \ IM^{TR} = 33\%, \ IM^{LN} = 0,$$

$$w_g = 11.44 \ \text{N/mm}$$

$$DC1 = 24.66 \ \text{N/mm, Diaphragm} = 19 \ 410 \ \text{N},$$

$$DW = 4.04 \ \text{N/mm}$$

   b. Exterior girders (Table E7.4-4)

$$mg_M = 0.762, \ mg_V = 0.762, \ IM^{TR} = 33\%, \ IM^{LN} = 0$$

$$DC1 = 24.13 \ \text{N/mm, Diaphragm} = 9705 \ \text{N},$$

$$DC2 = 4.65 \ \text{N/mm}, \ DW = 3.04 \ \text{N/mm}$$

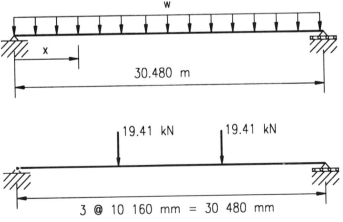

Fig. E7.4-7   Uniform dead and diaphragm loads.

TABLE E7.4-1   Shears and Moments for $w = 1.0$ N/mm $= 1.0$ kN/m

	$\xi = 0$	$\xi = 0.1$	$\xi = 0.2$	$\xi = 0.3$	$\xi = 0.4$	$\xi = 0.5$
V_x (kN)	15.240	12.192	9.144	6.096	3.048	0
M_x (kN m)	0	41.8	74.3	97.5	111.5	116.1

I. *Investigate Service Limit State*

1. *Stress limits for prestressing tendons:* (Table 7.8) [A5.9.3]

$f_{pu} = 1860$ MPa, low-relaxation 12.70 mm, seven-wire strands
$A = 98.71$ mm^2 (Table B.2), $E_p = 197\,000$ MPa [A5.4.4.2]

Pretensioning

At jacking $f_{pj} = 0.78f_{pu} = 0.78(1860) = 1451$ MPa

After transfer $f_{pt} = 0.74f_{pu} = 0.74(1860) = 1376$ MPa

$f_{py} = 0.9f_{pu} = 0.9(1860) = 1674$ MPa

After losses $f_{pe} = 0.80f_{py} = 0.80(1674) = 1339$ MPa

2. *Stress limits for concrete:* (Tables 7.6 and 7.7) [A5.9.4]

$f'_c = 55$ MPa, 28-day compressive strength
$f'_{ci} = 0.75f'_c = 40$ MPa strength at time of initial prestressing

Temporary stresses before losses—fully prestressed components:

Compressive stresses $f_{ci} = 0.6f'_{ci} = 0.6(40) = 24$ MPa

Tensile stresses $f_{ti} = 0.25\sqrt{f'_{ci}} = 0.25\sqrt{40} = 1.58$ MPa

$1.58 > 1.38$ (Use $f_{ti} = 1.38$ MPa)

Stresses at service limit state after losses—fully prestressed components [A5.9.4.2]:

TABLE E7.4-2   Shears and Moments due to Diaphragm, Interior Girders

	$\xi = 0$	$\xi = 0.1$	$\xi = 0.2$	$\xi = 0.3$	$\xi = 0.4$	$\xi = 0.5$
V_x (kN)	19.410	19.410	19.410	19.410	0	0
M_x (kN m)	0	59.2	118.3	177.5	197.2	197.2

TABLE E7.4-3   Summary of Force Effects for Interior Girder

Force Effect	Load Type	Distance from Support					
		0	0.1L	0.2L	0.3L	0.4L	0.5L
	Service *I* Loads						
M_s (kN m)	Girder self-weight	0	478	850	1115	1276	1328
	$DC1$ (incl. diaph.) on girder alone	0	1090	1951	2582	2947	3060
	DW on composite section	0	169	300	394	451	469
	$mg_M(LL + IM)$	0	974	1714	2216	2518	2590
V_s (kN)	$DC1$ (incl. diaph.) on girder alone	395	320	245	170	75	0
	DW on composite section	62	49	37	25	12	0
	$mg_V(LL + IM)$	441	383	327	274	223	175
	Strength *I* Loads						
M_u (kN m)	$\eta[1.25DC + 1.50DW + 1.75(LL + IM)]$	0	3154	5594	7312	8328	8608
V_u (kN)	$\eta[1.25DC + 1.50DW + 1.75(LL + IM)]$	1291	1087	887	693	477	290

Compressive stresses $f_c = 0.45f'_c$

$= 0.45(55) = 24.75$ MPa

Tensile stresses $f_t = 0.5\sqrt{f'_c}$

$= 0.5\sqrt{55}$

$= 3.71$ MPa   Use Service *III*

Modulus of Elasticity

$$E_{ci} = 4800\sqrt{f'_{ci}} = 4800\sqrt{40} = 30\ 360 \text{ MPa}$$
$$E_c = 4800\sqrt{f'_c} = 4800\sqrt{55} = 35\ 600 \text{ MPa}$$

3. *Preliminary choices of prestressing tendons*   Controlled either by the concrete stress limits at service loads or by the sectional strength under factored loads. For the final load condition, the composite cross section properties are needed. To transform the CIP deck into equivalent girder concrete, the modular ratio is taken as $n = \sqrt{30/55} = 0.74$.

If we assume for convenience that the haunch depth is 50 mm

TABLE E7.4-4 Summary of Force Effects for Exterior Girder

Force Effect	Load Type						
		0	0.1L	0.2L	0.3L	0.4L	0.5L
	Service *I* Loads						
M_s (kN m)	Girder self-weight	0	478	850	1115	1276	1328
	DC1 (incl. diaph.) on girder alone	0	1038	1852	2441	2789	2900
	DC2 (barrier) on composite section	0	194	346	453	519	540
	DW on composite section	0	127	226	296	339	353
	$mg_M(LL + IM)$	0	1106	1946	2516	2860	2941
V_s (kN)	DC1 (incl. diaph.) on girder alone	377	304	230	157	74	0
	DC2 (barrier) on composite section	71	57	43	28	14	0
	DW on composite section	46	37	28	19	9	0
	$mg_V(LL + IM)$	407	353	302	253	206	161
	Strength *I* Loads						
M_u (kN m)	$\eta[1.25DC + 1.50DW + 1.75(LL + IM)]$	0	3483	6167	8041	9166	9477
V_u (kN)	$\eta[1.25DC + 1.50DW + 1.75(LL + IM)]$	1274	1068	866	667	460	268

and use the effective flange width of 2210 mm for an exterior girder, the composite section dimensions are shown in Figure E7.4-8.

Section properties for the girder are as follows (Green and Tadros, 1994):

$$A_g = 486\,051 \text{ mm}^2$$

$$I_g = 126\,011 \times 10^6 \text{ mm}^4$$

$$S_{tg} = \frac{I_g}{y_{tg}} = \frac{126.0 \times 10^9}{742} = 169.8 \times 10^6 \text{ mm}^3$$

$$S_{bg} = \frac{I_g}{y_{bg}} = \frac{126.0 \times 10^9}{608} = 207.3 \times 10^6 \text{ mm}^3$$

Section properties for the composite girder are calculated below.

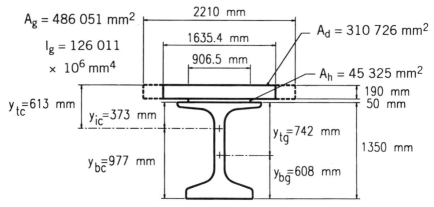

Fig. E7.4-8   Composite section properties.

The distance to the neutral axis from the top of the deck is

$$\bar{y} = \frac{(310\ 726)(95) + (45\ 325)(215) + (486\ 051)(982)}{310\ 726 + 45\ 325 + 486\ 051} = 613\ \text{mm}$$

$$I_c = (126\ 011 \times 10^6) + (486\ 051)(369)^2 + (906.5)(50)^3/12$$
$$\quad + (45\ 325)(398)^2 + (1635.4)(190)^3/12 + (310\ 726)(518)^2$$
$$\quad = 283.7 \times 10^9\ \text{mm}^4$$

$$S_{tc} = \frac{I_c}{y_{tc}} = \frac{283.7 \times 10^9}{613} = 462.8 \times 10^6\ \text{mm}^3\ \text{(top of deck)}$$

$$S_{ic} = \frac{I_c}{y_{ic}} = \frac{283.7 \times 10^9}{373} = 760.6 \times 10^6\ \text{mm}^3\ \text{(top of girder)}$$

$$S_{bc} = \frac{I_c}{y_{bc}} = \frac{283.7 \times 10^9}{977} = 290.4 \times 10^6\ \text{mm}^3\ \text{(bot of girder)}$$

Preliminary Analysis—Exterior Girder at Midspan

The minimum value of prestress force F_f to ensure that the tension in the bottom fiber of the beam does not exceed the limit of 3.71 MPa in the composite section under final conditions can be expressed as (Eq. 7.53)

$$f_{bg} = -\frac{F_f}{A_g} - \frac{F_f e_g}{S_{bg}} + \frac{M_{dg} + M_{ds}}{S_{bg}} + \frac{M_{da} + M_L}{S_{bc}} \leq 3.71\ \text{MPa}$$

where

M_{dg} = moment due to self-weight of girder = 1328 kN m

M_{ds} = moment due to dead load of wet concrete + diaphragm

= 1572 kN m

M_{da} = moment due to additional dead load after concrete hardens

= 893 kN m

M_L = moment due to live load

+ impact (Service *III*) = 0.8(2941)

= 2353 kN m

e_g = distance from *cg* of girder to centroid of pretensioned strands

= 608 − 100 = 508 mm

$$f_{bg} = -\frac{F_f}{486\ 051} - \frac{F_f(508)}{207.3 \times 10^6} + \frac{2900 \times 10^6}{207.3 \times 10^6}$$

$$+ \frac{3246 \times 10^6}{290.4 \times 10^6} \leq 3.71 \text{ MPa}$$

$$= -\ [(2.057 \times 10^{-6}) + (2.451 \times 10^{-6})]F_f + 13.989$$

$$+ 11.118 \leq 3.71$$

$$(4.508 \times 10^{-6})F_f \geq 21.397$$

$$F_f \geq \frac{21.397}{4.508 \times 10^{-6}} = 4.75 \times 10^6 \text{ N} = 4750 \text{ kN}$$

Assuming stress in strands after all losses is $0.6f_{pu}$ = 0.6(1860) = 1116 MPa = 1116 N/mm²,

$$A_{ps} \geq \frac{F_f}{0.6f_{pu}} = \frac{4.75 \times 10^6}{1116} = 4250 \text{ mm}^2$$

From Collins and Mitchell (1991), in order to satisfy strength requirements, the following approximate expression can be used

$$\phi M_n = \phi(A_{ps} \times 0.95f_{pu} + A_s f_y)\ 0.9h \geq M_u$$

where

$$\phi = 1.0$$

$$PPR = 1.0 \text{ (prestress ratio) [A5.5.4.2.1]}$$

$$h = \text{overall depth of composite section} = 1590 \text{ mm}$$

$$M_u = \text{Strength } I \text{ factored moment} = 9477 \text{ kN m}$$

$$A_{ps} \geq \frac{M_u}{\phi 0.95 f_{pu}(0.9h)} = \frac{9477 \times 10^6}{1.0(0.95)(1860)(0.9)(1590)}$$

$A_{ps} \geq 3748 \text{ mm}^2 < 4250 \text{ mm}^2$, strength limit is not critical

Try 52—12.70-mm strands $A_{ps} = 52(98.71) = 5133 \text{ mm}^2$ (Fig. E7.4-9)

(*Note:* Other strand patterns were tried. To save space, only the final iteration is given here.)

At Midspan			At End Section		
N	y	Ny	N	y	Ny
32	75	2400	28	75	2 100
10	150	1500	8	150	1 200
6	200	1200	4	200	800
4	275	1100	12	1175	14 100
52		6200	52		18 200

$$\bar{y} = \frac{6200}{52} = 119 \text{ mm} \qquad \bar{y} = \frac{18\ 200}{52} = 350 \text{ mm}$$

$$e_{CL} = 608 - 119 = 489 \text{ mm} \qquad e_{end} = 608 - 350 = 258 \text{ mm}$$

4. *Evaluate prestress losses* [A5.9.5]

$$\Delta f_{pT} = \Delta f_{pES} + \Delta f_{pSR} + \Delta f_{pCR} + \Delta f_{pR}$$

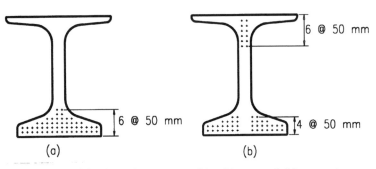

Fig. E7.4-9   Strand patterns at (a) midspan and (b) support.

a. Elastic shortening, Δf_{pES} (Eq. 7.96) [A5.9.5.2.3a]

$$\Delta f_{pES} = \frac{E_p}{E_{ci}} f_{cgp}$$

where

$$E_p = 197\ 000\ \text{MPa}$$

$$E_{ci} = 4800 \sqrt{40} = 30\ 360\ \text{MPa}$$

f_{cgp} = sum of concrete stresses at cg of A_{ps} due

to F_i and M_{dg} at centerline

$$F_i = 0.70 f_{pu} A_{ps} = 0.70(1860)(52)(98.71) \times 10^{-3}$$

$$F_i = 6683\ \text{kN}$$

$$M_{dg} = 1328\ \text{kN m}$$

$$f_{cgp} = -\frac{F_i}{A_g} - \frac{F_i e_{CL}^2}{I_g} + \frac{M_{dg} e_{CL}}{I_g}, \qquad CL = \text{centerline}$$

$$= -\frac{6.683 \times 10^6}{486\ 051} - \frac{(6.683 \times 10^6)(489)^2}{126\ 011 \times 10^6}$$

$$+ \frac{(1328 \times 10^6)(489)}{126\ 011 \times 10^6}$$

$f_{cgp} = -21.3$ MPa (Minus sign indicates elastic shortening
of concrete. This results in a positive prestress loss.)

$$\Delta f_{pES} = \frac{E_p}{E_{ci}} f_{cgp} = \frac{197\ 000}{30\ 360} (21.3) = 138.1\ \text{MPa}$$

Relaxation, Δf_{pR1} (Eq. 7.115) [A5.9.5.4.4]

$$\Delta f_{pR} = \Delta f_{pR1} + \Delta f_{pR2}$$

where

$$\Delta f_{pR1} = \text{Relaxation loss at transfer (Eq. 7.116)}$$

$$= \frac{\log(24t)}{40.0} \left(\frac{f_{pi}}{f_{py}} - 0.55 \right) f_{pi}$$

$$t = \text{Estimated time from prestressing to transfer}$$
$$= 4 \text{ days (max)}$$
$$f_{pi} = \text{Initial stress in tendon at the end of stressing}$$

1st Iteration (Eq. 7.117)

$$f_{pi} = f_{pt} - \Delta f_{pES} = 0.74(1860) - 138.1 = 1238.3 \text{ MPa}$$
$$f_{py} = 0.9 f_{pu} = 0.9(1860) = 1674 \text{ MPa}$$
$$\Delta f_{pR1} = \frac{\log[24(4)]}{40}\left(\frac{1238.3}{1674} - 0.55\right)1238.3 = 11.6 \text{ MPa}$$

Recalculating f_{pi} and Δf_{pR1}

$$f_{pi} = 1238.3 - 11.6 = 1226.7 \text{ MPa}$$
$$\Delta f_{pR1} = \frac{\log[24(4)]}{40}\left(\frac{1226.7}{1674} - 0.55\right)1226.7 = 11.1 \text{ MPa}$$

2nd Iteration (Eq. 7.117)

$$f_{pi} = 0.74(1860) - (138.1 + 11.1) = 1227.2 \text{ MPa}$$
$$F_i = (52)(98.71)(1.2272) = 6299.1 \text{ kN}$$
$$f_{cgp} = -\frac{6.2291 \times 10^6}{486\ 051} - \frac{(6.2991 \times 10^6)(489)^2}{126\ 011 \times 10^6}$$
$$+ \frac{(1328.2 \times 10^6)(489)}{126\ 011 \times 10^6} = -19.8 \text{ MPa}$$
$$\Delta f_{pES} = \frac{197\ 000}{30\ 360}(19.8) = 128.5 \text{ MPa}$$
$$f_{pi} = 0.74(1860) - (128.5 + 11.1) = 1236.8 \text{ MPa}$$
$$\Delta f_{pR1} = \frac{\log[24(4)]}{40}\left(\frac{1236.8}{1674} - 0.55\right)1236.8 = 11.57 \text{ MPa}$$

3rd Iteration

$$f_{pi} = 0.74(1860) - (128.5 + 11.57) = 1236.3 \text{ MPa}$$
$$F_i = (52)(98.71)(1.2363) = 6346 \text{ kN}$$

$$f_{cgp} = -\frac{6.346 \times 10^6}{486\ 051} - \frac{(6.346 \times 10^6)(489)^2}{126\ 011 \times 10^6}$$

$$+ \frac{(1328.2 \times 10^6)(489)}{126\ 011 \times 10^6} = -19.9 \text{ MPa}$$

$$\Delta f_{pES} = \frac{197\ 000}{30\ 360}(19.9) = 129.1 \text{ MPa}$$

$$f_{pi} = 0.74(1860) - (129.1 + 11.57) = 1235.7 \text{ MPa,} \quad \text{OK}$$

b. Shrinkage, Δf_{pSR} (Eq. 7.105) [A5.9.5.4.2]

$$\Delta f_{pSR} = 117 - 1.03H$$

where

H = relative humidity = 70% for Virginia [Fig. 5.4.2.3.3-1]
$$\Delta f_{pSR} = 117 - 1.03(70) = 44.9 \text{ MPa}$$

c. Creep, Δf_{pCR} (Eq. 7.114) [A5.9.5.4.3]

$$\Delta f_{pCR} = 12.0 f_{cgp} - 7.0 \Delta f_{cdp} \geq 0$$

where

f_{cgp} = concrete stress at cg of A_{ps} at transfer = 19.9 MPa

Δf_{cdp} = change in concrete stress at cg of A_{ps} due to permanent loads except the load acting when F_i is applied, that is subtract M_{dg}

$$\Delta f_{cdp} = -\frac{(2900 - 1328)10^6(489)}{126\ 011 \times 10^6}$$

$$- \frac{(540 + 353)10^6(977 - 119)}{283.7 \times 10^9} = -8.8 \text{ MPa}$$

(Again, a negative concrete stress results in a positive prestress loss.)

$$\Delta f_{pCR} = 12.0(19.9) - 7(8.8) = 177.2 \text{ MPa}$$

d. Relaxation, Δf_{pR} (Eq. 7.115) [A5.9.5.4.4]

$$\Delta f_{pR} = \Delta f_{pR1} + \Delta f_{pR2}$$

where Δf_{pR1} = relaxation loss at transfer = 11.57 MPa

$$\Delta f_{pR2} = 0.3[138 - 0.4\Delta f_{pES} - 0.2(\Delta f_{pSR} + \Delta f_{pCR})]$$
$$= 0.3[138 - 0.4(129.1) - 0.2(44.9 + 177.2)]$$
$$\Delta f_{pR2} = 12.6 \text{ MPa}$$

e. Total losses (Eq. 7.93)

$$\Delta f_{pT} = \text{(initial losses)} + \text{(long-term losses)}$$
$$= (\Delta f_{pES} + \Delta f_{pR1}) + (\Delta f_{pSR} + \Delta f_{pCR} + \Delta f_{pR2})$$
$$= (129.1 + 11.57) + (44.9 + 177.2 + 12.6)$$
$$\Delta f_{pT} = 375.4 \text{ MPa}$$

Approximate Lump Sum Estimate of Time-Dependent Losses

Time-dependent losses can be approximated by the following formula, given in Table 7.12 [Table A5.9.5.3-1]

$$\text{Average} = 230\left[1 - 0.15\frac{f'_c - 41}{41}\right] + 41PPR$$

where

$$PPR = 1.0 \quad [\text{A5.5.4.2.1}]$$
$$\text{Average} = 230\left[1 - 0.15\frac{55 - 41}{41}\right] + 41 = 259.2 \text{ MPa}$$

For low relaxation strand,

$$\text{Average} = 259.2 - 41 = 218.2 \text{ MPa}$$

$\Delta f_{pT} = 129.1 + 218.2 = 347.3$ MPa, compared to the losses of 375.4 MPa, calculated by the longer procedure. (This amounts to a underestimation of total stresses of 7.5%.)

5. *Calculate Girder Stresses at Transfer*

$$F_i = 6346 \text{ kN}$$

$$e_{CL} = 489 \text{ mm} \qquad e_{end} = 238 \text{ mm}$$

At Midspan

$$f_{ti} = -\frac{F_i}{A_g} + \frac{F_i e_{CL}}{S_{tg}} - \frac{M_{dg}}{S_{tg}}$$

$$= -\frac{6.346 \times 10^6}{486\ 051} + \frac{(6.346 \times 10^6)(489)}{169.8 \times 10^6} - \frac{1328 \times 10^6}{169.8 \times 10^6}$$

$$= -2.5 \text{ MPa} < 1.38 \text{ MPa}, \qquad \text{OK}$$

$$f_{bi} = -\frac{F_i}{A_g} - \frac{F_i e_{CL}}{S_{bg}} + \frac{M_{dg}}{S_{bg}}$$

$$= -\frac{6.346 \times 10^6}{486\ 051} + \frac{(6.346 \times 10^6)(489)}{207.3 \times 10^6} - \frac{1328 \times 10^6}{207.3 \times 10^6}$$

$$= -21.6 \text{ MPa} > f_{ci} = -24 \text{ MPa}, \qquad \text{OK}$$

At Beam End

$$f_{ti} = -\frac{6.346 \times 10^6}{486\ 051} - \frac{(6.346 \times 10^6)(258)}{169.8 \times 10^6}$$

$$= -3.4 \text{ MPa} < 1.38 \text{ MPa}, \qquad \text{OK}$$

$$f_{bi} = -\frac{6.346 \times 10^6}{486\ 051} - \frac{(6.346 \times 10^6)(258)}{207.3 \times 10^6}$$

$$= -21.0 \text{ MPa} > f_{ci} = -24 \text{ MPa}, \qquad \text{OK}$$

6. *Girder Stresses After Total Losses*

$$f_{pf} = 0.74f_{pu} - \Delta f_{pT} = 1376.4 - 375.4 = 1001.0 \text{ MPa}$$

$$F_f = \frac{(5133)(1001)}{10^3} = 5140 \text{ kN}$$

At Midspan

$$f_{tf} = -\frac{F_f}{A_g} + \frac{F_f e_{CL}}{S_{tg}} - \frac{M_{dg} + M_{ds}}{S_{tg}} - \frac{M_{da} + M_L}{S_{ic}}$$

$$= -\frac{5.14 \times 10^6}{486\ 051} + \frac{(5.14 \times 10^6)(489)}{169.8 \times 10^6} - \frac{(1328 + 1572) \times 10^6}{169.8 \times 10^6}$$

$$- \frac{(893 + 2353) \times 10^6}{760.6 \times 10^6}$$

$$= -17.1 \text{ MPa} > f_c = 0.45 f_c' = -24.75 \text{ MPa}, \qquad \text{OK}$$

$$f_{bf} = -\frac{5.14 \times 10^6}{486\ 051} - \frac{(5.14 \times 10^6)(489)}{207.3 \times 10^6} + \frac{2900 \times 10^6}{207.3 \times 10^6}$$

$$+ \frac{3246 \times 10^6}{290.4 \times 10^6}$$

$$= 2.44 \text{ MPa} < 3.71 \text{ MPa}, \qquad \text{OK}$$

52 − 12.70-mm low-relaxation strands satisfy Service Limit State

7. *Check Fatigue Limit State* [A5.5.3]

   a. Live load moment due to fatigue truck (*FTr*) at midspan (Fig. E7.4-10)

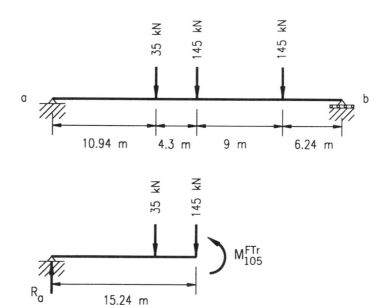

Fig. E7.4-10 Fatigue truck placement for maximum positive moment at midspan.

$$R_a = 145\left(\frac{6.240 + 15.240}{30.480}\right) + 35\left(\frac{19.540}{30.480}\right) = 124.6 \text{ kN}$$

$$M_{105}^{FTr} = [(124.6)(15.240) - (35)(4.300)] = 1748.4 \text{ kN m}$$

Exterior Girder DF—remove 1.2 multiple presence for fatigue

$$g_M^{SE} = \frac{0.762}{1.2} = 0.635$$

Factored Moment including $IM = 15\%$

$$M_{\text{fatigue}} = 0.75(0.635)(1748.4)(1.15) = 957.6 \text{ kN m}$$

b. *Dead load moments at midspan*

Exterior girder   (Table E7.4-4)

Noncomp. $M_{DC1} = 2900$ kN m

Comp. $M_{DC2} + M_{DW} = (540 + 353) = 893$ kN m

If section is in compression under *DL* and two times fatigue load, fatigue does not need to be investigated [A5.5.3.1].

$$f_b = -\frac{F_f}{A_g} - \frac{F_t e_{CL}}{S_{bg}} + \frac{M_{DC1}}{S_{bg}}$$

$$+ \frac{M_{DC2} + M_{DW} + 2M_{\text{fatigue}}}{S_{bc}}$$

$$= -\frac{5.14 \times 10^6}{486\,051} - \frac{(5.14 \times 10^6)(489)}{207.3 \times 10^6} + \frac{2900 \times 10^6}{207.3 \times 10^6}$$

$$+ \frac{[893 + 2(957.6)]10^6}{290.4 \times 10^6}$$

$$= 0.94 \text{ MPa, tension; therefore,}$$

fatigue shall be considered

Stress range due only to $M_{\text{fatigue}} = 957.6$ kN m

Section Properties: Cracked Section properties [A5.5.3.1] if

1.5M_{fatigue} and tensile stress exceeds $0.25\sqrt{f'_c}$

$$= 0.25\sqrt{55} = 1.85 \text{ MPa}$$

$$f_b = -\frac{5.14 \times 10^6}{486\ 051} - \frac{(5.14 \times 10^6)(489)}{207.3 \times 10^6} + \frac{2900 \times 10^6}{207.3 \times 10^6}$$

$$+ \frac{[893 + 1.5(957.6)]10^6}{290.4 \times 10^6}$$

$$= -0.71 \text{ MPa} < 1.85 \text{ MPa; therefore,}$$

use gross section properties

Concrete stress at cg of prestress tendons due to fatigue load

$$f_{cgp} = \frac{M_{\text{fatigue}}(977 - 119)}{283.7 \times 10^9} = \frac{957.6 \times 10^6}{330.7 \times 10^6} = 2.9 \text{ MPa}$$

Stress in tendon due to fatigue load (*FL*)

$$f_{pFL} = f_{cgp}\frac{E_p}{E_c} = (2.9)\left(\frac{197\ 000}{35\ 600}\right) = 16.0 \text{ MPa}$$

Stress range in prestressing tendons shall not exceed (Table 7.9) [A5.5.3.3]

- 125 MPa for radii of curvature greater than 9000 mm
- 70 MPa for radii of curvature less than 3600 mm

Harped Tendons (Fig. E7.4-11)

$$e_{\text{end}} = 258 \text{ mm}, \Delta y = 489 - 258 = 231 \text{ mm},$$

$$e_{0.33L} = e_{CL} = 489 \text{ mm}$$

At hold down point, the radius of curvature depends on the hold down device and is small, therefore $R < 3600$ mm.

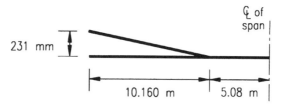

Fig. E7.4-11   Schematic of harped tendons.

$$f_{pFL} = 16.0 \text{ MPa} < 70 \text{ MPa}, \qquad \text{OK}$$

Satisfies fatigue limit state.

8. *Calculate Deflection and Camber*

   a. Immediate deflection due to live load and impact (Fig. E7.4-12)

$$\Delta(x < a)x = \frac{Pbx}{6EIL}(L^2 - b^2x^2), \; b = L - a$$

$$\Delta x\left(x = \frac{L}{2}\right) = \frac{PL^3}{48EI}$$

Use *EI* for $f'_c = 55$ MPa and composite section

$$E_c = 35\,600 \text{ MPa}, \; I_c = 283.7 \times 10^9 \text{ mm}^4$$

$$E_c I_c = 10.1 \times 10^{15} \text{ N mm}^2 = 10.1 \times 10^6 \text{ kN m}^2$$

$\underline{P = 35 \text{ kN}}, \; x = 15.240 \text{ m}, \; a = 19.540 \text{ m}, \; b = 10.940 \text{ m}$

$$\Delta_x^{35} = \frac{(35)(10.940)(15.240)}{6(EI)(30.480)}(30.480^2 - 10.940^2 - 15.240^2)$$

$$= \frac{18.4 \times 10^3}{EI} = \frac{18.4 \times 10^3}{10.1 \times 10^6} = 0.002 \text{ m} = 2 \text{ mm}$$

$\underline{P = 145 \text{ kN}}, \; x = a = b = 15.240 \text{ m}$

$$\Delta_x^{145} = \frac{(145)(30.480)^3}{48EI} = \frac{85.5 \times 10^3}{EI}$$

$$= \frac{85.5 \times 10^3}{10.1 \times 10^6} = 0.008 \text{ m} = 8 \text{ mm}$$

$\underline{P = 145 \text{ kN}}, \; x = 15.240 \text{ m}, \; a = 19.540 \text{ m}, \; b = 10.940 \text{ m}$

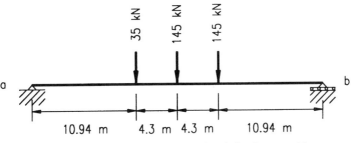

Fig. E7.4-12 Live load placement for deflection at midspan.

$$\Delta_x^{145'} = \frac{(145)(10.940)(15.240)}{6EI(30.480)} (30.480^2 - 10.940^2 - 15.240^2)$$

$$= \frac{76.3 \times 10^3}{EI} = \frac{76.3 \times 10^3}{10.1 \times 10^6} = 0.008 \text{ m} = 8 \text{ mm}$$

Total deflection due to truck

$$\Delta_{105}^{Tr} = \frac{(18.4 + 85.5 + 76.3) \times 10^3}{EI} = \frac{180.2 \times 10^3}{10.1 \times 10^6}$$

$$= 0.018 \text{ m} = 18 \text{ mm}$$

Deflection DF $= N_L/N_G = 3/6 = 0.5$, $IM = 33\%$

$$\Delta_{105}^{L+I} = 0.5(18)(1.33) = 12 \text{ mm} \downarrow$$

$$= 12 \text{ mm} \leq \frac{L}{800} = \frac{30\ 480}{800} = 38 \text{ mm}, \quad \text{OK}$$

b. Long-term deflections (Collins and Mitchell, 1991) Loads on exterior girder from Section 7.10.4, Part H.2.

• Elastic deflections due to girder self weight at release of prestress

$$E_{ci} = 30\ 360 \text{ MPa}, \qquad I_g = 126\ 011 \times 10^6 \text{ mm}^4$$

$$E_{ci}I_g = 3.83 \times 10^6 \text{ kN m}^2$$

$$\Delta_{gi} = \frac{5}{384} \frac{wL^4}{EI} = \frac{5}{384} \frac{(11.44)(30.480)^4}{3.83 \times 10^6}$$

$$= 0.034 \text{ m} = 34 \text{ mm} \downarrow$$

• Elastic camber due to prestress at time of release for double harping point with $\beta L = 0.333L$ (Collins and Mitchell, 1991)

$$\Delta_{pi} = \left[\frac{e_c}{8} - \frac{\beta^2}{6} (e_c - e_e) \right] \frac{F_i L^2}{EI}$$

$$= \left[\frac{489}{8} - \frac{(0.333)^2}{6} (489 - 258) \right] \frac{(6346)(30.480)^2}{3.83 \times 10^6}$$

$$= 88 \text{ mm} \uparrow$$

At release, net upward deflection:

$$88 - 34 = 54 \text{ mm} \uparrow$$

- Elastic deflection due to deck and diaphragms on exterior girder

$$DC1 - w_g = 24.13 - 11.44 = 12.69 \text{ N/mm} = 12.69 \text{ kN/m}$$

Diaphragm = 9.705 kN

$$E_c = 35\ 600 \text{ MPa}$$

$$E_c I_g = 4.49 \times 10^6 \text{ kN} \cdot \text{m}^2$$

$$b = \frac{L}{3} = 10.160 \text{ m}$$

$$
\begin{aligned}
\Delta_{DC} &= \frac{5}{384} \frac{wL^4}{EI} + \frac{Pb}{24EI}(3L^2 - 4b^2) \\
&= \frac{5}{384} \frac{(12.69)(30.480)^4}{4.49 \times 10^6} \\
&\quad + \frac{(9.705)(10.160)}{24(4.49 \times 10^6)}[3(30.480)^2 - 4(10.160)^2] \\
&= 0.032 + 0.002 = 0.034 \text{ m} = 34 \text{ mm} \downarrow
\end{aligned}
$$

- Elastic deflection due to additional dead load acting on composite section

$$DW + \text{Barrier} = 3.04 + 4.65 = 7.69 \text{ N/mm} = 7.69 \text{ kN/m}$$

$$\Delta_c = \frac{5}{384} \frac{wL^4}{EI} = \frac{5}{384} \frac{(7.69)(30.480)^4}{10.1 \times 10^6} = 0.009 \text{ m} = 9 \text{ mm} \downarrow$$

Note the full barrier load is conservatively applied to the exterior girder. Many designers distribute this load equally to all girders.

Long-Term Deflections

Using the multipliers in Table E7.4-5 (PCI, 1992) to approximate the creep effect, the net upward deflection at the time the deck is placed is

$$\Delta_l = 1.80(88) - 1.85(34) = 96 \text{ mm} \uparrow$$

TABLE E7.4-5   Suggested Multipliers To Be Used as a Guide in Estimating Long-Time Cambers and Deflections for Typical Members[a]

	Without Composite Topping	With Composite Topping
At erection		
1. Deflection (downward) component—apply to the elastic deflection due to the member weight at release of prestress	1.85	1.85
2. Camber (upward) component—apply to the elastic camber due to prestress at the time of release of prestress	1.80	1.80
Final		
3. Deflection (downward) component—apply to the elastic deflection due to the member weight at release of prestress	2.70	2.40
4. Camber (upward) component—apply to the elastic camber due to prestress at the time of release of prestress	2.45	2.20
5. Deflection (downward)—apply to elastic deflection due to superimposed dead load only	3.00	3.00
6. Deflection (downward)—apply to elastic deflection caused by the composite topping		2.30

[a]In PCI Table 4.6.2. [From *PCI Design Handbook: Precast and Prestressed Concrete*, 4th ed., Copyright © 1992 by the Precast/Prestressed Concrete Institute, Chicago, IL.]

The net long-term upward deflection is

$$\Delta_{LT} = 2.20(88) - 2.40(34) - 2.30(34) - 3.00(9) = 7 \text{ mm} \uparrow$$

After construction has been completed the center of the bridge is estimated to creep downward from an initial upward deflection of 96 mm to a final upward value of 7 mm.

J. *Investigate Strength Limit State*

1. *Flexure*

   a. Stress in prestressing steel bonded tendons (Eq. 7.59) [A5.7.3.1.1]

$$f_{ps} = f_{pu}\left(1 - k\frac{c}{d_p}\right)$$

where (Eq. 7.60)

$$k = 2\left(1.04 - \frac{f_{py}}{f_{pu}}\right) = 2\left(1.04 - \frac{1674}{1860}\right) = 0.28$$

By using transformed section (Fig. E7.4-8)

$$b = 1635.4 \text{ mm}, \ d_p = (1350 + 50 + 190) - 119 = 1471 \text{ mm}$$

$$f_c' = 55 \text{ MPa}, \ A_s = A_s' = 0$$

$$\beta_1 = 0.85 - \frac{55 - 28}{7}(0.05) = 0.657$$

from Eq. 7.68

$$c = \frac{A_{ps}f_{pu} + A_sf_y + A_s'f_y' - 0.85\beta_1 f_c'(b - b_w)h_f}{0.85f_c'\beta_1 b_w + kA_{ps}(f_{pu}/d_p)}$$

$$= \frac{(5133)(1860) - 0.85(0.657)(55)(1635.4 - 150)(190)}{0.85(55)(0.657)(150) + 0.28(5133)(1860/1471)}$$

$$= 136.8 \text{ mm}$$

$$f_{ps} = 1860\left[1 - 0.28\left(\frac{136.8}{1471}\right)\right] = 1812 \text{ MPa}$$

$$T_p = A_{ps}f_{ps} = \frac{(5133)(1812)}{10^3} = 9300 \text{ kN}$$

b. Factored Flexural Resistance—Flanged Sections [A5.7.3.2.2]

$$a = \beta_1 c = (0.657)(136.8 \text{ mm}) = 89.9 \text{ mm}$$

$$\phi = 1.0$$

from Eq. 7.76

$$\phi M_n = \phi\left[A_{ps}f_{ps}\left(d_p - \frac{a}{2}\right) + A_sf_y\left(d_s - \frac{a}{2}\right)\right.$$

$$\left. - A_s'f_y'\left(d_s' - \frac{a}{2}\right) + 0.85\beta_1 f_c'(b - b_w)h_f\left(\frac{a}{2} - \frac{h_f}{2}\right)\right]$$

$$= 1.0\left[5133(1812)\left(1471 - \frac{89.9}{2}\right)\right.$$

$$+ \ 0.85(0.657)(55)(1635.4$$

$$\left. - \ 150)(190)\left(\frac{89.9}{2} - \frac{190}{2}\right)\right]$$

$$\phi M_n = 12.8 \times 10^9 \text{ N mm} = 12\ 800 \text{ kN m}$$

$$M_u = 9477 \text{ kN m (Table E7.4-4)}$$

$$\phi M_n > M_u, \qquad \text{OK}$$

c. Limits for Reinforcement [A5.7.3.3]
  • Maximum reinforcement limited by (Eq. 7.89)

$$\frac{c}{d_e} \leq 0.42 \qquad \text{for} \qquad d_e = d_p$$

$$\frac{c}{d_p} = \frac{136.8}{1471} = 0.093 < 0.42, \qquad \text{OK}$$

  • Minimum reinforcement limited by (Eq. 7.90)

$$\phi M_n \geq 1.2 M_{cr}$$

$$M_{cr} \qquad \text{based on} \qquad f_r = 0.63 \sqrt{f'_c} = 0.63 \sqrt{55} = 4.67 \text{ MPa}$$

Under final conditions at midspan with Service *III* live load $M_L = 2353$ kN m, the bottom tensile stress is $f_{bf} = 2.44$ MPa. To cause cracking, an additional tensile stress of $\Delta f_b = 4.67 - 2.44 = 2.23$ MPa is required. Additional moment to cause this stress is

$$\Delta M = S_{bc} \Delta f_b = (290.4 \times 10^6)(2.23)$$

$$= 648 \times 10^6 \text{ N mm} = 648 \text{ kN m}$$

Addition of ΔM with the service dead and live load moments yields

$$M_{cr} = (2900 + 893 + 2353 + 648) = 6794 \text{ kN m}$$

$$1.2 M_{cr} = 1.2(6794) = 8153 \text{ kN m}$$

$$\phi M_n > 1.2 M_{cr}, \qquad \text{OK}$$

52 − 12.70 mm low-relaxation strands satisfy strength limit state

2. *Shear* [A5.8]
  a. General

$$\phi_v = 0.9$$
$$\eta = 0.95 \qquad \text{[A5.5.4.2.1]}$$

$$V_n = V_c + V_s + V_p \leq 0.25 f'_c b_v d_v \text{ [A5.8.3.3]}$$

$$d_e = 1471 \text{ mm}$$

$$d_v = d_e - \frac{a}{2}$$

$$\geq \max \begin{cases} 0.9d_e = 0.9(1471) = 1324 \text{ mm, governs at } CL \\ 0.72h = 0.72(1590) = 1145 \text{ mm} \end{cases}$$

At CL: $a = \beta_1 c = (0.657)(136.8) = 89.9$ mm

$$d_v = 1471 - \frac{89.9}{2} = 1426 \text{ mm [A5.8.2.7]}$$

At the end of the beam

$$d_e = 1590 - 350 = 1240 \text{ mm}$$

$$d_v = \max \begin{cases} 0.9d_e = 0.9(1240) = 1116 \text{ mm} \\ 0.72h = 0.72(1590) = 1145 \text{ mm, governs} \end{cases}$$

b_v = minimum web width within d_v = 150 mm

f'_c(girder) = 55 MPa

b. Prestress contribution to shear resistance

V_p = vertical component of prestressing force

Transfer length = 60 strand diameters = 60(12.70) = 762 mm [A5.8.2.3]

Critical section for shear $\geq 0.5d_v \cot\theta$ or d_v = 1145 mm [A5.8.3.2]

$d_v >$ transfer length, therefore, full value of V_p can be used.

cg of harped strands (12) = 1350 − 175 = 1175 mm from bottom of girder

= 1175 − 119 = 1056 mm from cg A_{ps} at

centerline (Fig. E7.4-13)

$$\gamma = \tan^{-1} \frac{1056}{10\ 160} = 5.934°$$

$$F_f = 5140 \text{ kN}$$

$$V_p = \frac{12}{52} F_f \sin\gamma = \frac{12}{52}(5140) \sin 5.934 = 122.6 \text{ kN}$$

c. Design for shear Design for shear at a distance of d_v from the support and at tenth points along the span. Calculations are

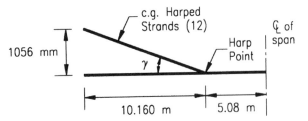

Fig. E7.4-13 Harped tendon profile.

shown below for a distance d_v from the support (Fig. E7.4-14) and Location 101. The same procedure is used for the remaining points with final results given later in Table E7.4-6 of Section 7.10.4, Part J.2.e.

$$d_v = 1145 \text{ mm}$$

$$\xi = \frac{d_v}{L} = \frac{1145}{30\ 480} = 0.0376$$

For a unit load, $w = 1.0 \text{ N/mm} = 1.0 \text{ kN/m}$

$$V_x = wL(0.5 - \xi) = w30.480(0.5 - 0.0376) = 14.094w \text{ kN}$$

$$M_x = 0.5wL^2(\xi - \xi^2) = 0.5w(30.480)^2(0.0376 - 0.0376^2)$$

$$= 16.8w \text{ kN m}$$

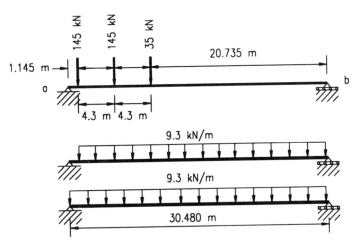

Fig. E7.4-14 Live load placement for maximum shear and moment at Location 100.376.

Exterior girders dead loads taken from Section 7.10.4, Part H.2

$$DC_1 = 24.13 \text{ N/mm}$$

$$DC_2 = 4.65 \text{ N/mm}$$

$$DW = 3.04 \text{ N/mm}$$

$$DIAPH = 9.705 \text{ kN}$$

$$IM = 0.33$$

$$V^{Tr}_{100.376} = \left[145 \left(\frac{25.035 + 29.335}{30.480} \right) + 35 \left(\frac{20.735}{30.480} \right) \right]$$

$$= 282.5 \text{ kN}$$

$$M^{Tr}_{100.376} = 1.145(282.5) = 323.5 \text{ kN m}$$

$$V^{Ln}_{100.376} = \frac{1}{2} (9.3) \left(\frac{29.335}{30.480} \right)^2 = 131.3 \text{ kN}$$

$$M^{Ln}_{100.376} = \tfrac{1}{2} (1.145)(29.335)(9.3) = 156.2 \text{ kN m}$$

$$V_u = \eta[1.25DC + 1.50DW + 1.75(LL + IM)]$$

$$= 0.95\{1.25[(24.13 + 4.65)(14.094) + 9.705]$$

$$+ 1.50[(3.04)(14.094)] + 1.75(0.762)[(282.5)(1.33)$$

$$+ 131.3]\}$$

$$= 1197 \text{ kN}$$

$$M_u = 0.95\{1.25[(24.13 + 4.65)(16.8) + 9.705(1.145)]$$

$$+ 1.50[(3.04)(16.8)] + 1.75(0.762)[(323.5)(1.33)$$

$$+ (156.2)]\}$$

$$= 1403 \text{ kN m}$$

From Eq. 7.159 [A5.8.3.4.2]

$$v = \frac{V_u - \phi V_p}{\phi b_v d_v}$$

$$= \left[\frac{(1.197 \times 10^6) - 0.9(122.6 \times 10^3)}{0.9(150)(1145)} \right]$$

$$= 7.02 \text{ MPa}$$

$$\frac{v}{f'_c} = \frac{7.02}{55} = 0.128 > 0.100$$

therefore, $s_{\max} = \min \begin{cases} 0.4d_v = 0.4(1145) = 458 \text{ mm} \\ 300 \text{ mm, governs} \end{cases}$ [A5.8.2.7]

$d_e = d_p = 1590 - 350 + \dfrac{1145}{10\ 160}(489 - 258) = 1266 \text{ mm}$

1st Iteration

Assume $\theta = 30°, f_{po} \approx f_{se} = 1001 \text{ MPa}$

$$d_e = 1266 \text{ mm}$$

$d_v = \max \begin{cases} d_e - \dfrac{a}{2} = 1266 - \dfrac{89.9}{2} = 1221 \text{ mm, governs} \\ 0.9d_e = 0.9(1266) = 1139 \text{ mm} \\ 0.72h = 0.72(1590) = 1145 \text{ mm} \end{cases}$

$$d_v = 1221 \text{ mm}$$

From Eq. 7.170 [A5.8.3.4.2]

$$\varepsilon_x = \dfrac{(M_u/d_v) + 0.5N_u + 0.5V_u \cot \theta - A_{ps}f_{po}}{E_sA_s + E_pA_{ps}}$$

$$= \dfrac{(1.402 \times 10^9/1221) + 0.5(1.197 \times 10^6) \cot 30 - (5133)(1001)}{(197\ 000)(5133)}$$

$$= -2.9 \times 10^{-3} \text{ (compression)}$$

Because ε_x is negative, it shall be reduced by the factor [A5.8.3.4.2]

$$F_\varepsilon = \dfrac{E_sA_s + E_pA_{ps}}{E_cA_c + E_sA_s + E_pA_{ps}} = \dfrac{E_pA_{ps}}{E_cA_c + E_pA_{ps}}$$

Where A_c is the area of concrete on flexural tension side of member defined as concrete below $h/2$ of member [Fig. A5.8.3.4.2-3]

$$h = 1350 + 50 + 190 = 1590 \text{ mm}, \frac{h}{2} = 795 \text{ mm}$$

$$A_c \cong (135)(975) + 2(\tfrac{1}{2})(140)(412.5) + (660)(150)$$

$$\approx 288\ 375 \text{ mm}^2$$

$$E_c = 35\ 600 \text{ MPa}$$

$$F_\varepsilon = \frac{(197\ 000)(5133)}{(35\ 600)(288\ 375) + (197\ 000)(5133)} = 0.0897$$

$$\varepsilon_x = (-0.0029)(0.0897) = -0.26 \times 10^{-3}$$

Using $v/f_c' = 0.128$ and ε_x with Figure 7.43 $\Rightarrow \theta = 20°$

$$\cot \theta = 2.747 \qquad \beta = 2.75$$

2nd Iteration

$$\theta = 20°$$

$$M_{DC1} = (24.13)(16.8 \times 10^6) + (9705)(1221)$$

$$= 416.4 \times 10^6 \text{ N mm}$$

$$e = d_p - (y_{tg} + t_s + 50)$$

$$= 1221 - (742 + 190 + 50) = 239 \text{ mm}$$

$$f_{pc} = -\frac{F_f}{A_g} + \frac{F_f e(y_{bc} - y_{bg})}{I_g} - \frac{M_{DC1}(y_{bc} - y_{bg})}{I_g}$$

$$= -\frac{5.14 \times 10^6}{486\ 051} + \frac{(5.14 \times 10^6)(239)(977 - 608)}{126\ 011 \times 10^6}$$

$$- \frac{(416.4 \times 10^6)(977 - 608)}{126\ 011 \times 10^6}$$

$$f_{pc} = -8.20 \text{ MPa}$$

$$f_{po} = f_{se} + f_{pc}\frac{E_p}{E_c} = 1001 + 8.2\frac{197\ 000}{35\ 600}$$

$$= 1046.5 \text{ MPa (precompression)}$$

$$\varepsilon_x = \frac{\dfrac{1.402 \times 10^9}{1221} + 0.5(1.196 \times 10^6)(2.747) - (5133)(1046.5)}{(197\ 000)(5133)}$$

$$= -0.0026$$

$$F_\varepsilon \varepsilon_x = 0.0897(-0.0026) = -0.23 \times 10^{-3}$$

Figure 7.43 $\Rightarrow \theta = 20°$, converged,

$$\text{use } \cot \theta = 2.747 \qquad \beta = 2.75$$

$$V_c = 0.083\beta\sqrt{f_c'}b_v d_v$$

$$= 0.083(2.75)\sqrt{55}(150)(1221) = 310 \times 10^3 \text{ N}$$

Required

$$V_s = \frac{V_u}{\phi} - V_c = \frac{1.196 \times 10^6}{0.9} - 310 \times 10^3 = 1.019 \times 10^6 \text{ N}$$

Spacing of No. 15 stirrups, (Eq. 7.172)

$$d_s = 16 \text{ mm}, \qquad A_v = 2(200 \text{ mm}^2) = 400 \text{ mm}^2$$

$$s \le \frac{A_v f_y d_v \cot \theta}{V_s} = \frac{(400)(400)(1221)(2.747)}{1.019 \times 10^6}$$

$$s \le 526 \text{ mm} > s_{\max} = 300 \text{ mm}$$

Check Longitudinal Reinforcement (Eq. 7.169) [A5.8.3.5]

$$A_s f_y + A_{ps} f_{ps} \ge \left[\frac{M_y}{d_v \phi_f} + 0.5 \frac{N_u}{\phi_a} + \left(\frac{V_u}{\phi_v} - 0.5 V_s - V_p \right) \cot \theta \right]$$

$$V_s = 1.019 \times 10^6 \left[\frac{526}{300} \right] = 1.787 \times 10^6 \text{ N}$$

$$(5133)(1812) \ge \frac{1.402 \times 10^9}{(1221)(1.0)} + \left[\frac{1.196 \times 10^6}{0.9} \right.$$

$$\left. - 0.5(1.787 \times 10^6) - 122.6 \times 10^3 \right] 2.747$$

$$9.301 \times 10^6 \text{ N} > 2.007 \times 10^6 \text{ N}, \qquad \text{OK}$$

Use s = 300 mm at Location 100.38

d. Location 101

$$d_e = 1590 - 350 + \frac{3048}{10\ 160}(489 - 258) = 1309 \text{ mm}$$

$$d_v = \max \begin{cases} d_e - \dfrac{a}{2} = 1309 - \dfrac{89.9}{2} = 1264 \text{ mm, governs} \\ 0.9 d_e = 0.9(1309) = 1178 \text{ mm} \\ 0.72 h = 0.72(1590) = 1145 \text{ mm} \end{cases}$$

$$d_v = 1264 \text{ mm}$$

$$V_u = 1067.9 \text{ kN}, \qquad M_u = 3481.4 \text{ kN m}$$

$$v = \frac{V_u - \phi V_p}{\phi b_v d_v} = \frac{[1067.9 - 0.9(122.6)]10^3}{0.9(150)(1264)} = 5.61 \text{ MPa}$$

$$\frac{v}{f'_c} = \frac{5.61}{55} = 0.102 > 0.100, \text{ therefore, } s_{max} = 300 \text{ mm}$$

1st Iteration

$$\text{Assume } \theta = 28°, \qquad f_{po} = 1050 \text{ N/mm}^2$$

$$\varepsilon_x = \frac{\dfrac{3481.4 \times 10^6}{1264} + 0.5(1067.9 \times 10^3) \cot 28 - (5133)(1050)}{(197\ 000)(5133)}$$

$$= -0.0016 \qquad \text{(compression)}$$

$$F_\varepsilon \varepsilon_x = (0.0897)(-0.0016) = -0.145 \times 10^{-3}$$

$$\text{Figure 7.43} \Rightarrow \theta = 23.5°, \qquad \cot \theta = 2.300$$

2nd Iteration

$$M_{DC1} = 1036.1 \text{ kN m}$$

$$e = 1309 - (742 + 50 + 190) = 327 \text{ mm}$$

$$f_{pc} = \frac{5.14 \times 10^6}{486\ 051} + \frac{(5.14 \times 10^6)(327)(369)}{126\ 011 \times 10^6}$$

$$- \frac{(1036.1 \times 10^6)(369)}{126\ 011 \times 10^6} = -8.69 \text{ MPa}$$

$$f_{po} = 1001 + 8.69\left(\frac{197\ 000}{35\ 600}\right) = 1049 \text{ MPa (precompression)}$$

$$\varepsilon_x = \frac{\dfrac{3481.4 \times 10^6}{1264} + 0.5(1067.9 \times 10^3)(2.300) - (5133)(1049)}{(197\ 000)(5133)}$$

$$= -1.387 \times 10^{-3}$$

$$F_\varepsilon \varepsilon_x = (0.0897)(-1.387 \times 10^{-3}) = -0.124 \times 10^{-3}$$

$$\text{Figure 7.43} \Rightarrow \theta = 23.5°, \qquad \beta = 5, \qquad \cot \theta = 2.300$$

$$V_c = 0.083\beta\sqrt{f'_c}b_v d_v$$

$$= 0.083(5)\sqrt{55}(150)(1264) = 583.5 \times 10^3 \text{ N}$$

Requires

$$V_s = \frac{V_u}{\phi} - V_c = \frac{1067.9 \times 10^3}{0.9} - 583.5 \times 10^3$$

$$= 603.1 \times 10^3 \text{ N}$$

$$s \leq \frac{(400)(400)(1264)(2.300)}{603.1 \times 10^3} = 771 \text{ mm} > s_{\max}$$

$$s_{\max} = 300 \text{ mm}$$

$$V_s = 603.1 \times 10^3 \left(\frac{771}{300}\right) = 1550 \times 10^3 \text{ N}$$

Check Longitudinal Reinforcement

$$(5133)(1812) \geq \frac{3481.4 \times 10^6}{(1264)(1.0)} + \left(\frac{1067.9 \times 10^3}{0.9}\right.$$

$$\left. - 0.5(1550 \times 10^3) - 122.6 \times 10^3 \right)2.300$$

$$9.301 \times 10^6 \text{ N} \geq 3.419 \times 10^6 \text{ N}, \quad \text{OK}$$

Use s = 300 mm at Location 101

e. Summary of shear design (Table E7.4-6)

f. Horizontal shear [A5.8.4] At interface between two concretes cast at different times the nominal shear resistance shall be taken as

$$V_{nh} = cA_{cv} + \mu(A_{vf}f_y + P_c) \begin{cases} \leq 0.2f'_c A_{cv} \\ \leq 5.5A_{cv} \end{cases}$$

where

$$A_{cv} = \text{area of concrete engaged in shear transfer}$$

$$= (1225 \text{ mm})(1 \text{ mm}) = 1225 \text{ mm}^2$$

$$A_{vf} = \text{area of shear reinforcement crossing the shear plane}$$

$$= 2(200 \text{ mm}^2) = 400 \text{ mm}^2, \quad 2 \text{ legs}$$

$$f_y = \text{yield strength of reinforcement} = 400 \text{ MPa}$$

TABLE E7.4-6 Summary of Shear Design

	Location					
	100.38	101	102	103	104	105
V_u (kN)	1196	1068	865	666	459	268
M_u (kN m)	1.4	3481	6162	8036	9159	9470
V_p (kN)	122.6	122.6	122.6	122.6	0	0
d_v (mm)	1221	1204	1334	1403	1426	1426
$\dfrac{v}{f'_c}$	0.128	0.102	0.0762	0.0533	0.043	0.0253
f_{po} (MPa)	1047	1049	1056	1060	1064	1066
$\varepsilon_x \times 10^3$	−0.23	−0.124	0.0469	0.781	1.236	1.320
θ, degrees	20	23.5	27	34	38.5	39
β	2.75	5	4.4	2.3	2.1	2.1
V_c (kN)	310	584	296	246	230	230
Required V_s (kN)	1019	603	420	442	224	209
s (mm)	526	771	999	752	1281	13 482
$A_{ps}f_{ps} = 9.301 \times 10^6$ N $\geq (\times10^6$ N)	2.007	3.419	5.580	6.233	6.764	6.719
Prov'd s (mm)	300	300	600	600	600	600
$\phi_f = 1.0, \ \phi_v = 0.9$						
$\dfrac{M_u}{(\phi_f d_v)}$ $(\times 10^6)$	1.148	2.754	4.619	5.728	6.423	6.641
$\dfrac{V_u \cot\theta}{\phi_v}$ $(\times 10^6)$	3.650	2.729	1.887	1.098	0.642	0.367
$-0.5V_s \cot\theta$ $(\times 10^6)$	−2.454	−1.783	−0.685	−0.411	−0.300	−0.290
$-V_p \cot\theta$ $(\times 10^6)$	−0.337	−0.282	−0.241	−0.182	0	0
Σ last 3 rows $(\times 10^6)$	0.859	0.665	0.961	0.505	0.342	0.077
Σ last 4 rows $(\times 10^6)$	2.007	3.419	5.580	6.233	6.764	6.719

$$c = \text{cohesion factor} = 0.70 \text{ MPa} \Bigg\}$$
$$\mu = \text{friction factor} = 1.0$$

intentionally roughened [A5.8.4.2]

P_c = permanent net compressive force normal to shear plane

= overhang + slab + haunch + barrier

= [5.37 + 5.83 + 0.61 + 4.65] = 16.46 N/mm

$$V_{nh} = 0.70(1225) + 1.0\left[\left(\frac{400}{s}\right)(400) + 16.46\right] \quad \text{(E7.4-1)}$$

$$= 873.96 + \frac{160\ 000}{s} \text{ N/mm}$$

s = spacing of shear reinforcement, mm

$$V_{nh} \leq \min\begin{cases} 0.2f'_c A_{cv} = 0.2(30)(1225) = 7350 \text{ N/mm} \\ 5.5A_{cv} = 5.5(1225) = 6740 \text{ N/mm, governs} \end{cases}$$

$$\phi_v V_{nh} \geq \eta V_{uh}$$

where

V_{uh} = horizontal shear due to barrier, *FWS* and *LL + IM*

$$= \frac{V_u Q}{I_c}$$

$$I_c = 283.7 \times 10^9 \text{ mm}^4$$

$$Q = A\bar{y} = (1635.4)(190)\left(1590 - 977 - \frac{190}{2}\right)$$

$$= 160.96 \times 10^6 \text{ mm}^3$$

$$V_u = 1.25DC2 + 1.50DW + 1.75(LL + IM)$$

Location 100 (Table E7.4-4)

$$V_u = 1.25(71) + 1.50(46) + 1.75(407) = 870 \text{ kN}$$

$$V_{uh} = \frac{(870 \times 10^3)(160.96 \times 10^6)}{283.7 \times 10^9} = 493.4 \text{ N/mm}$$

$$\frac{\eta V_{uh}}{\phi_v} = \frac{0.95(493.4)}{0.9} = 520.8 \text{ N/mm} < 6740 \text{ N/mm} \quad \text{(E7.4-2)}$$

Equating Eqs. E7.4-1 and E7.4-2

$$873.96 + \frac{160\ 000}{s} \geq 520.8$$

$$s \leq \frac{160\ 000}{873.96 - 520.8} = 453 \text{ mm}$$

$$s_{max} = 300 \text{ mm}$$

Use s = 300 mm at Location 100

At Location 100.38 Interpolating between Locations 100 and 101

$$V_u = 1.25(65.7) + 1.50(42.6) + 1.75(386.5) = 822.4 \text{ kN}$$

$$V_{uh} = \frac{(822.4 \times 10^3)(160.96 \times 10^6)}{283.7 \times 10^9} = 466.5 \text{ N/mm}$$

$$\frac{\eta V_{uh}}{\phi_v} = \frac{0.95(466.5)}{0.9} = 492.4 \text{ N/mm} < 6740 \text{ N/mm}$$

$$s \leq \frac{160\ 000}{873.96 - 486.1} = 413 \text{ mm} > s_{max} = 300 \text{ mm}$$

Use s = 300 mm at Location 100.38

At Location 101 (Table E7.4-4)

$$V_u = 1.25(57) + 1.50(37) + 1.75(353) = 745 \text{ kN}$$

$$V_{uh} = \frac{(745 \times 10^3)(160.96 \times 10^6)}{283.7 \times 10^9} = 422.7 \text{ N/mm}$$

$$\frac{\eta V_{uh}}{\phi_v} = \frac{0.95(422.7)}{0.9} = 446.2 \text{ N/mm}$$

$$s \leq \frac{160\ 000}{873.96 - 446.2} = 374 \text{ mm} > s_{max}$$

$$s_{max} = 300 \text{ mm}$$

Use s = 300 mm at Location 101

By inspection, horizontal shear does not govern strand spacing for any of these or remaining locations.

g. Check details

Anchorage Zone [A5.10.10]

The factored resistance provided by transverse reinforcement P_r shall not be less than 4% of the factored prestressing force [A3.4.3] [1.3 times jacking force = $1.3(6.34 \times 10^6) = 8.24 \times 10^6$ N]

$$P_r = \phi_{ca} f_y A_s$$

where

ϕ_{ca} = 0.80 (compression in anchorage zones) (Table 7.10)

[A5.5.4.2.1]

$$f_y = 400 \text{ MPa}$$

A_s = total area of transverse reinforcement within $d/4$

of end of beam

d = depth of precast beam = 1350 mm

$\phi_{ca} f_y A_s \geq 0.04 F_{ui} = 0.04(8.24 \times 10^6) = 329.6 \times 10^3$ N

$$A_s \geq \frac{329.6 \times 10^3}{0.8(400)} = 1030 \text{ mm}^2$$

$$\text{within } \frac{d}{4} = \frac{1350}{4} = 337.5 \text{ mm}$$

Number of No. 15 U stirrups required: $\dfrac{1030}{400} = 2.6$

Use 3 No. 15 stirrups at 130 mm

Confinement Reinforcement: [A5.10.10.2]

For a distance of $1.5d = 1.5(1350) = 2025$ mm from the end of the beam, reinforcement shall be placed to confine the prestressing steel in the bottom flange.

Use 14 No. 10 at 150 mm shaped to enclose the strands

K. *Design Sketch* The design of the prestressed concrete girder is summarized in Figure E7.4-15. The design utilized the NU 1350 girder shape developed by Nebraska University, f'_c = 55 MPa, and f'_{ci} = 40 MPa. The prestressing steel consists of $52 - 1860$ MPa, low-relaxation 12.70 mm, seven-wire strands.

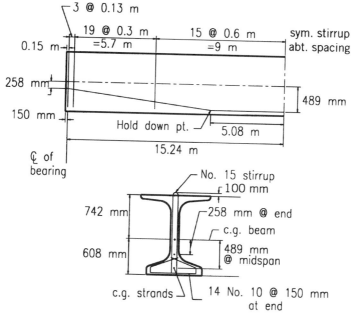

Fig. E7.4-15 Design sketch for prestressed girder.

7.10.5 Concrete Box-Girder Bridge

Problem Statement Design the continuous post-tensioned concrete box girder bridge of Figure E7.5-1 with 30-m–36-m–30-m spans for a HL-93 live load. Roadway width is 13 420 mm curb to curb. Allow for a future wearing surface of 75-mm thick bituminous overlay. Use the empirical method for deck slabs [A9.7.2] to design the top flange of the box girder. Use $f'_c = 35$ MPa, $f_y = 400$ MPa, and 1860 MPa, low-relaxation 12.70 mm, 7-wire strands. Follow the outline of AASHTO (1994) LRFD Bridge Specifications, Section 5, Appendix A5.3.

A. Develop General Section

The bridge is to carry interstate traffic over a secondary highway with a minimum clearance of 6100 mm (Fig. E7.5-1).

B. Develop Typical Section

1. *Top Flange* [A5.14.1.3]

   - Minimum deck slab thickness = 175 mm [A9.7.1.1]
   - Slab plus sacrificial surface = 188 mm
   - $h_{tf} \geq \frac{1}{20}$ (clear span between fillets)

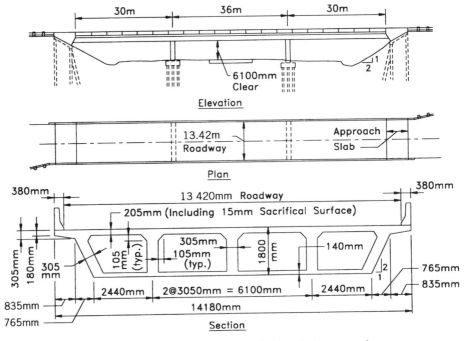

Fig. E7.5-1   Concrete box girder bridge design example.

$$h_{tf} > \tfrac{1}{20}\,(3050 - 305 - 210) = 127 \text{ mm, not critical}$$

Empirical design—effective length $= 3050 - 305 = 2745$ mm [A9.7.2.4]

$$h_{tf} \geq \tfrac{1}{18}\,(\text{effective length}) = 153 \text{ mm, not critical}$$

Try 205 mm Top Flange (including 15-mm sacrificial surface)

2. *Bottom Flange*   [A5.14.1.3]

- Minimum thickness $= 140$ mm
- $h_{bf} \geq \tfrac{1}{30}$ (clear span between webs)
- $h_{bf} > \tfrac{1}{30}\,(2745 \text{ mm}) = 92$ mm, not critical

Try 140-mm Bottom Flange

3. *Webs*

- Minimum thickness for prestressing ducts $= 300$ mm [C5.14.1.3]

Try 305-mm Web Thickness

4. *Structure Depth*   [A2.5.2.6] Minimum depth for prestressed CIP box beams = $h \geq 0.040L$

$$h \geq 0.040(36\ 000) = 1440 \text{ mm}$$

Try 1830-mm Structure Depth

5. *Reinforcement*   [A5.14.5.3]

   a. Deck reinforcement may be determined by either the traditional or empirical design method [A9.7.2 or A9.7.3]. At least one-third of the bottom layer of the transverse reinforcement shall extend into the overhang and be anchored there.

   b. Bottom slab parallel to girder span, total $A_{s\parallel}$ = 0.4% flange area

   $$A_{s\parallel} = 0.004(140 \text{ mm})(10\ 980 \text{ mm}) = 6150 \text{ mm}^2$$

   or

   $$A_{s\parallel} = 0.004(140)(1000) = 560 \text{ mm}^2/\text{m}$$

   Spacing is not to exceed 450 mm

   Try No. 15 @ 350-mm single layer

   Transverse to girder span, $A_{s\perp}$ = 0.5% slab cross-section area

   $$A_{s\perp} = (0.005)(140)(1000) = 700 \text{ mm}^2/\text{m}$$

   Spacing not to exceed 450 mm

   Try No. 15 @ 250-mm single layer   code two layers a must

   The transverse reinforcement shall be anchored in the outside face of exterior web with 90° hook.

   c. Minimum reinforcement [A5.7.3.3]

   $$\phi M_n \geq 1.2 M_{cr}, \text{ based on } f_r' = 0.63\sqrt{f_c'}$$

   d. Shrinkage and temperature reinforcement [A5.10.8]

   $$A_s \geq 0.75 A_g/f_y, \text{ each direction}$$

   Maximum spacing = 3 × slab thickness or 450 mm

   Deck slab

$$h = 205 \text{ mm} > 150 \text{ mm}$$

$$A_S = \frac{0.75(205)(1000)}{400} = 385 \text{ mm}^2/\text{m}$$

Use No. 10's @ 250 mm both faces, not critical

Bottom slab

$$h = 140 \text{ mm} < 150 \text{ mm}$$

$$A_S = \frac{0.75(140)(1000)}{400} = 263 \text{ mm}^2/\text{m}$$

Use No. 10's @ 350-mm single layer, not critical

6. *Effective flange widths* [A4.6.2.6] Based on effective span lengths between points of inflection under permanent loads as shown in Figure E7.5-2. For multicell superstructures, consider full width of deck slab or each web as a beam.
Interior web, b_w = 305 mm

b_i shall be taken as the lesser of:

$$\begin{cases} \text{one-fourth effective span (check all spans)} \\ 12t_s + b_w \\ \text{web spacing} \end{cases}$$

$$b_i \le \begin{cases} 0.25(22\ 698) = 5675 \text{ mm} \\ 0.25(20\ 469) = 5117 \text{ mm} \\ 12(190) + 305 = 2585 \text{ mm} \\ 3050 \text{ mm} \end{cases}$$

$$b_i = 2585 \text{ mm}$$

Exterior web, b_w = 305 $\times \dfrac{\sqrt{5}}{2}$ = 341 mm

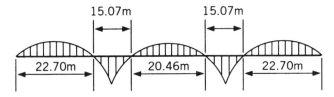

Fig. E7.5-2 Length between inflection points for permanent load.

b_e shall be taken as one-half of b_i, plus the lesser of:

$$\begin{cases} \text{one-eighth effective span} \\ 6t_s + \tfrac{1}{2}b_w \\ \text{width of overhang} \end{cases}$$

$$b_e - b_i/2 \leq \begin{cases} 0.125(22\ 698) = 2837 \text{ mm} \\ 6(190) + \tfrac{1}{2}(341) = 1310.5 \text{ mm} \\ 835 + \tfrac{1}{2}(341) = 1006 \text{ mm} \end{cases}$$

$b_e = 1006 + 2585/2 = 2298$ mm

C. Design Reinforced Concrete Deck

Empirical Design [A9.7.2]
1. *Overhang*   Use the same design as for *T*-beam bridge in Section 7.10.1, Part M (Fig. E7.10.1-17)
2. *Effective length* = face-to-face distance = 3050 − 305 = 2745 mm
3. *Design Conditions*   Supported on concrete, CIP, uniform depth

$$6.0 < \frac{\text{effective length}}{\text{design depth}} = \frac{2745}{190} = 14.4 < 18$$

Core depth = slab depth − top and bottom cover = 190 − 60 − 25 = 105 mm > 100 mm

Effective length = 2745 mm < 4100 mm

Minimum depth of slab = 190 mm > 175 mm

Overhang = 1006 mm > $5t_s$ = 5(190) = 950 mm

f'_c = 30 MPa > 28.0 MPa

Monolithic construction

All conditions satisfied

4. *Reinforcement requirements* [A9.7.2] (Fig. E7.5-3) Four layers with the outermost layers in the direction of the effective length each bottom layer, $A_s \geq 0.570$ mm²/mm

Use No. 15's @ 350 mm

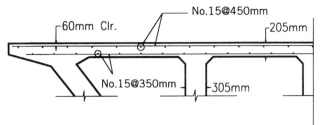

Fig. E7.5-3 Empirical deck design summary.

each top layer, $A_s \geq 0.380$ mm²/mm

Use No. 15's @ 450 mm

maximum spacing $= 450$ mm

D. Select Resistance Factors [A5.5.4.2]

1. For flexure and tension, $\phi = 0.9 + 0.1\ PPR$

   where $PPR = \dfrac{A_{ps}f_{py}}{A_{ps}f_{py} + A_s f_y}$

2. For shear and torsion, $\phi = 0.9$
3. For bearing on concrete, $\phi = 0.7$
4. For compression in anchorage zones, $\phi = 0.8$
5. For tension in steel in anchorage zones, $\phi = 1.0$

E. Select Load Modifiers [A1.3.2.1]

		Strength	Service, Fatigue	
Ductility	η_D	0.95	1.0	[A1.3.3]
Redundancy	η_R	0.95	1.0	[A1.3.4]
Importance	η_I	1.05	1.0	[A1.3.5]
$\eta = \eta_D \eta_R \eta_I$		0.95	1.0	

F. Select Load Combinations and Load Factors (Table 3.1) [Table A3.4.1-1]

Strength *I*, Service *I*, Service *III*, and Fatigue Limit States must be considered.

G. Calculate Live Load Force Effects

1. *Select Number of Lanes* [A3.6.1.1.1]

$$\text{No. of lanes} = \text{INT} \left(\frac{13\ 420}{3600} \right) = 3 \text{ lanes}$$

2. *Multiple Presence Factor* (Table 4.6) [Table A3.6.1.1.2-1]

No. Loaded Lanes	Multiple Presence Factor
1	1.20
2	1.00
3	0.85

3. *Dynamic Load Allowance* (Table 4.7) [Table A3.6.2.1-1]

$$\text{Impact} = 33\%$$

$$\text{Fatigue} = 15\%$$

4. *Distribution Factors for Moment* [A4.6.2.2.2]

Cross-section type (d) (Table 2.2) [Table A4.6.2.2.1-1]

$$2100 \text{ mm} < S = 3050 \text{ mm} < 4000 \text{ mm}$$

$$18\ 000 \text{ mm} < L = 36\ 000 \text{ mm or } 30\ 000 \text{ mm} < 73\ 000 \text{ mm}$$

$N_c = 4 > 3$, N_c = number of cells in a concrete box girder

Therefore, the range of applicability for cast-in-place concrete multicell box cross sections is satisfied.

a. Interior webs (Table 6.21b) [A4.6.2.2.2b and Table A4.6.2.2.2b-1], $S = 3050$ mm

L = 30 000 mm

$$mg_M^{SI} = \left(1.75 + \frac{S}{1100} \right) \left(\frac{300}{L} \right)^{0.35} \left(\frac{1}{N_c} \right)^{0.45}$$

$$mg_M^{SI} = \left(1.75 + \frac{3050}{1100} \right) \left(\frac{300}{30\ 000} \right)^{0.35} \left(\frac{1}{4} \right)^{0.45} = 0.484$$

$$mg_M^{MI} = \left(\frac{13}{N_c}\right)^{0.3}\left(\frac{S}{430}\right)\left(\frac{1}{L}\right)^{0.25}$$

$$mg_M^{MI} = \left(\frac{13}{4}\right)^{0.3}\left(\frac{3050}{430}\right)\left(\frac{1}{30\ 000}\right)^{0.25} = 0.768,\ \text{governs}$$

$L = 33\ 000\ mm$

$$mg_M^{SI} = \left(1.75 + \frac{3050}{1100}\right)\left(\frac{300}{33\ 000}\right)^{0.35}\left(\frac{1}{4}\right)^{0.45} = 0.468$$

$$mg_M^{MI} = \left(\frac{13}{4}\right)^{0.3}\left(\frac{3050}{430}\right)\left(\frac{1}{33\ 000}\right)^{0.25} = 0.749,\ \text{governs}$$

$L = 36\ 000\ mm$

$$mg_M^{SI} = \left(1.75 + \frac{3050}{1100}\right)\left(\frac{300}{36\ 000}\right)^{0.35}\left(\frac{1}{4}\right)^{0.45} = 0.454$$

$$mg_M^{MI} = \left(\frac{13}{4}\right)^{0.3}\left(\frac{3050}{430}\right)\left(\frac{1}{36\ 000}\right)^{0.25} = 0.733,\ \text{governs}$$

b. Exterior webs (Table 6.21b) [A4.6.2.2.2d and Table A4.6.2.2.2d-1]

W_e = one-half of web spacing + total overhang

$$= 0.5(3050) + 1006 = 2531\ mm < S = 3050\ mm$$

$$mg_M^{SE} = 1.2\left(\frac{W_e}{4300}\right) = 1.2\left(\frac{2531}{4300}\right) = 0.706,\ \text{governs}$$

$$mg_M^{ME} = 1.0\left(\frac{2531}{4300}\right) = 0.589$$

5. Distribution factors for shear [A4.6.2.2.3]

Cross-section type (d) (Table 2.2) [Tables A4.6.2.2.1-1 and A4.6.2.2.3a-1]

$$1800\ mm < S = 3050\ mm < 4000\ mm$$

$$6000\ mm < L = 36\ 000\ mm\ \text{and}\ 30\ 000\ mm < 73\ 000\ mm$$

$$890\ mm < d = 1830\ mm < 2800\ mm$$

$$N_c = 4 > 3$$

Therefore, the range of applicability for CIP concrete multicell box cross sections is satisfied.

a. Interior webs (Table 6.21b) [Table A4.6.2.2.3a-1]

$L = 30\ 000\ mm$

$$mg_V^{SI} = \left(\frac{3050}{2900}\right)^{0.6}\left(\frac{1830}{30\ 000}\right)^{0.1} = 0.779$$

$$mg_V^{MI} = \left(\frac{3050}{2200}\right)^{0.6}\left(\frac{1830}{30\ 000}\right)^{0.1} = 1.01,\ governs$$

$L = 33\ 000\ mm$

$$mg_V^{SI} = \left(\frac{3050}{2900}\right)^{0.6}\left(\frac{1830}{33\ 000}\right)^{0.1} = 0.772$$

$$mg_V^{MI} = \left(\frac{3050}{2200}\right)^{0.6}\left(\frac{1830}{33\ 000}\right)^{0.1} = 1.005,\ governs$$

$L = 36\ 000\ mm$

$$mg_V^{SI} = \left(\frac{3050}{2900}\right)^{0.6}\left(\frac{1830}{36\ 000}\right)^{0.1} = 0.765$$

$$mg_V^{MI} = \left(\frac{3050}{2200}\right)^{0.6}\left(\frac{1830}{36\ 000}\right)^{0.1} = 0.996,\ governs$$

b. Exterior webs (Table 6.21b) [A4.6.2.2.3b and Table A4.6.2.2.3b-1]

One design lane loaded, $m = 1.2$, lever rule applies (Fig. E7.5-4)

$$mg_V^{SE} = 1.2\left[0.5 \times \frac{\sqrt{5}}{2} \times \frac{(1260 + 3060)}{3034}\right] = 0.955,\ governs$$

Two or more design lanes loaded

$$-600\ mm < d_e = 1006 - 385 = 621\ mm < 1500\ mm$$

$$mg_V^{ME} = e \cdot mg_V^{MI}$$

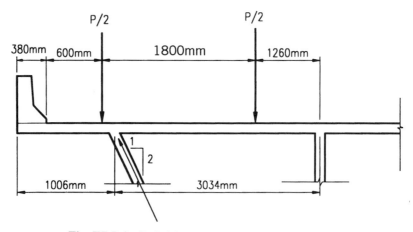

Fig. E7.5-4 Definition of lever rule for exterior web.

$$e = 0.64 + \frac{d_e}{3800} = 0.64 + \frac{621}{3800} = 0.803$$

$$mg_V^{ME} = 0.803(1.005) = 0.811$$

6. *Live Load Actions.* Interior webs govern. Influence line values are taken from Table 5.4.

$$M_{LL+IM} = mg\left[(M_{Tr} \text{ or } M_{Ta})\left(1 + \frac{IM}{100}\right) + M_{Ln}\right]$$

$$V_{LL+IM} = mg\left[(V_{Tr} \text{ or } V_{Ta})\left(1 + \frac{IM}{100}\right) + V_{Ln}\right]$$

Location 100 (Maximum shear at left support) (Fig. E7.5-5)

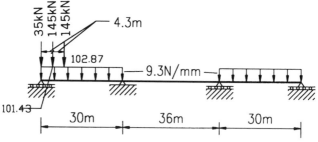

Fig. E7.5-5 Live load placement for maximum shear at left support.

$$V_{Tr} = 145(1 + 0.822\ 873) + 35(0.648\ 96) = 287.0 \text{ kN}$$

$$V_{Ln} = 9.3(30)(0.455\ 36) = 127.0 \text{ kN}$$

$$V_{100} = mg_V^{MI} V_{LL+I} = 1.01[287(1 + 0.33) + 127] = 513.8 \text{ kN}$$

Location 104 (Maximum positive moment in first span) (Fig. E7.5-6)

$$M_{Tr} = [145(0.2070 + 0.145\ 59) + 35(0.130\ 81)]30.0 = 1671 \text{ kN m}$$

$$M_{Ln} = 9.3(0.102\ 14)(30.0)^2 = 855 \text{ kN m}$$

$$M_{104} = mg_M^{MI} M_{LL+I} = 0.768[1671(1.33) + 855] = 2363 \text{ kN m}$$

Location 110 (Maximum negative shear at interior support) (Fig. E7.5-7)

$$V_{Tr} = 145(-1.000 - 0.911\ 34) + 35(-0.798\ 46) = -305.10 \text{ kN}$$

$$V_{Ln} = 9.3(30)(-0.638\ 53) = -178.15 \text{ kN}$$

$$V_{110} = mg_V^{MI} V_{LL+I} = 1.005[-305.1(1.33) - 178.15] = -586.9 \text{ kN}$$

Location 200 (Maximum positive shear at interior support) (Fig. E7.5-8)

$$V_{Tr} = 145(1.000 + 0.909\ 71) + 35(0.795\ 245) = 304.7 \text{ kN}$$

$$V_{Ln} = 9.3(0.665\ 10)(30.0) = 185.6 \text{ kN}$$

$$V_{200} = mg_V^{MI} V_{LL+I} = 1.005[304.7(1.33) + 185.6] = 593.8 \text{ kN}$$

Location 200 (Maximum negative moment at the interior support) (Fig. E7.5-9)

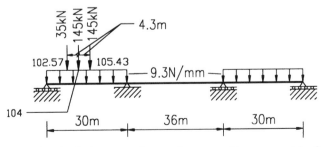

Fig. E7.5-6 Live load placement for maximum positive moment in first span.

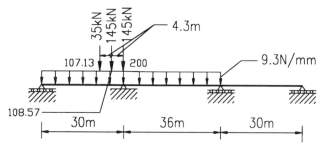

Fig. E7.5-7 Live load placement for maximum negative shear at interior support.

$$M_{Tr} = [145(-0.094\ 29 - 0.103\ 37 - 0.087\ 96 - 0.096\ 75)$$

$$+ 35(-0.080\ 37 - 0.093\ 64)] \times 30.0 = -1846 \text{ kN m}$$

$$M_{Ln} = 9.3(-0.138\ 53)(30)^2 = -1159 \text{ kN m}$$

$$M_{200} = 0.9mg_M^{MI} M_{LL+I} = 0.9\{0.749[-1846(1.33) - 1159]\}$$

$$= -2436 \text{ kN m}$$

Location 205 (Maximum position in second span) (Fig. E7.5-10)

$$M_{Tr} = [145(0.203\ 57 + 0.138\ 23) + 35(0.138\ 23)]30$$

$$= 1632 \text{ kN m}$$

$$M_{Ln} = 9.3(30)^2(0.102\ 86) = 861 \text{ kN m}$$

$$M_{205} = mg_M^{MI} M_{LL+I} = 0.733[1632(1.33) + 861] = 2222 \text{ kN m}$$

H. Calculate Force Effects of Other Loads

1. *Interior Webs* (Fig. E7.5-11)

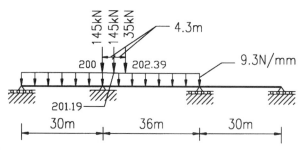

Fig. E7.5-8 Live load placement for maximum positive shear at interior support.

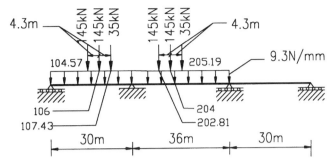

Fig. E7.5-9 Truck train placement for maximum negative moment at interior support.

DC

Slab	$2400 \times 9.81 \times 205 \times 3050/10^9$	= 14.72 N/mm
Top fillets	$2400 \times 9.81 \times 105 \times 105/10^9$	= 0.26 N/mm
Web	$2400 \times 9.81 \times 305 \times (1830 - 205 - 140)/10^9$	= 10.66 N/mm
Bottom flange	$2400 \times 9.81 \times 140 \times 3050/10^9$	= 10.05 N/mm
Barriers, one-fifth share	$(\frac{1}{5} \times 2 \times 4.68)$	= 1.87 N/mm
	w'_{DC}	= 37.56 N/mm

DW   75 mm bituminous overlay
$$w'_{DW} = 2250 \times 9.81 \times 3050 \times 75/10^9 = 5.05 \text{ N/mm}$$

2. *Exterior Webs* (Fig. E7.5-12)

DC

Slab $\dfrac{2400 \times 9.81}{10^9} \times \left[1509(205) + \dfrac{1006(305 + 180)}{2} \right]$

$= 13.03$ N/mm

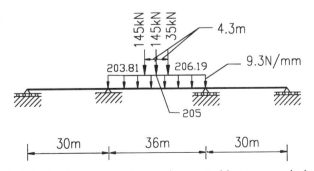

Fig. E7.5-10 Live load placement for maximum positive moment in interior span.

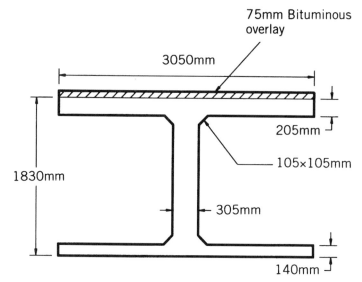

Fig. E7.5-11 Interior web cross section (b_i = 2585 mm).

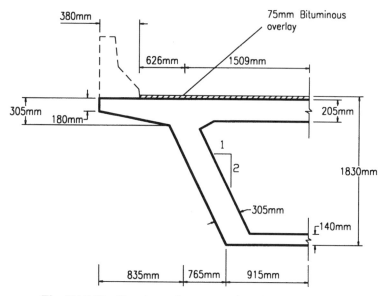

Fig. E7.5-12 Exterior web cross section (b_e = 2298 mm).

$$\text{Fillet} \quad \tfrac{1}{2}(0.26) = 0.13 \text{ N/mm}$$

$$\text{Web} \quad \frac{2400 \times 9.81}{10^9} \times \left[305 \times \frac{\sqrt{5}}{2}(1830 - 140 - 255) \right]$$

$$= 11.52 \text{ N/mm}$$

$$\text{Bottom Flange} \quad \frac{2400 \times 9.81}{10^9}(140 \times 915) = 3.02 \text{ N/mm}$$

$$\text{Barrier, one-fifth share} \quad = 1.87 \text{ N/mm}$$

$$w_{DC}^E = 29.6 \text{ N/mm}$$

$$DW \quad w_{DW}^E = 2250 \times 9.81 \times 2135 \times 75/10^9 = 3.53 \text{ N/mm}$$

3. *Analysis of Uniformly Distributed Load, w* (Fig. 7.5-13)

   a. Moments

$$M_{104} = 0.071\ 29w\ (30)^2 = 64.16w \text{ kN m}$$
$$M_{200} = -0.121\ 79w\ (30)^2 = -109.61w \text{ kN m}$$
$$M_{205} = 0.058\ 21w\ (30)^2 = 52.39w \text{ kN m}$$

   b. Shears

$$V_{100} = 0.378\ 21w\ (30) = 11.35w \text{ kN}$$
$$V_{110} = -0.621\ 79w\ (30) = -18.65w \text{ kN}$$
$$V_{200} = 0.6000w\ (30) = 18.0w \text{ kN}$$

4. *Combined Force Effects*

By multiplying the uniform load expressions for moments and shears by the values for dead load on the interior and exterior webs, and com-

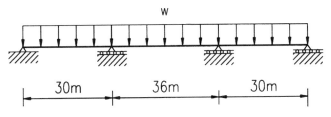

Fig. E7.5-13 Uniformly distributed load, *w*.

bining them with the live load force effects, the values in Table E7.5-1 are generated.

The procedure above can be repeated to find the effects of all loads at tenth points along the span. Results are summarized in Tables E7.5-2 and E7.5-3, where eta times Sum is the effect of all loads after multiplying them by the appropriate impact factors, distribution factors, load factors, and by the load modifier, η. The values for eta times Sum are used to graph the moment and shear envelopes in Figure E7.5-14.

I. Calculate Design Stresses

1. *Stress Limitations for Post tensioning Tendons* Using AASHTO M203 (ASTM A416) uncoated seven-wire low-relaxation strands (Grade 270)

$$\text{Diameter} = 12.70 \text{ mm}, \text{ Area} = 98.71 \text{ mm}^2/\text{strand}$$

a. Material properties

$$f_{pu} = 1860 \text{ MPa}$$
$$f_{py} = 0.9f_{pu} = 1674 \text{ MPa}$$
$$E_p = 197\ 000 \text{ MPa}$$

TABLE E7.5-1   Summary of Force Effects at Critical Locations[a]

Load Type	Value (N/mm)	Moments (kN m)			Shears (kN)		
		M_{104}	M_{200}	M_{205}	V_{100}	V_{110}	V_{200}
Uniform	1.0	64.16	−109.61	52.39	11.35	−18.65	18.0
DC^I	37.56	2410	−4117	1968	426.2	−700.6	676.1
DW^I	5.05	324	−554	265	57.3	−94.2	91.0
DC^E	29.6	1900	−3244	1551	335.9	−552.1	532.8
DW^E	3.53	226	−387	185	40.1	−65.8	63.5
$LL + IM^I$	Tr + Ln	2363	−2436	2222	513.8	−586.9	593.8
		(2364)	(−2429)	(2226)	(514.16)	(−589.19)	(593.25)
Strength *I* (Internal webs)		$U = \eta[1.25DC + 1.5DW + 1.75(LL + IM)]$					
		7252	−9728	6409	1441.96	−1941.92	1919.74
		(7276)	(−9658)	(6472)	(1445.29)	(−1942.99)	(1918.65)
Service *I* (Internal webs)		$U = \eta[DC + DW + (LL + IM)]$					
		5097	−7107	4455			
Service *III* (Internal webs)		$U = \eta[DC + DW + 0.8(LL + IM)]$					
		4624	−6620	4011			

[a]Values in parentheses were obtained from the BT Beam program.

TABLE E7.5-2 Moment Envelope for 30-, 36-, 30-m Box Girder (kN m)

Loc.	Unit Dead Load	Positive Moment				Negative Moment			
		Truck or Tandem	Lane	eta times Sum Int. Web	eta times Sum Ext. Web	Truck or Tandem	Lane	eta times Sum Int. Web	eta times Sum Ext. Web
100	0.0	0	0	0	0	0	0	0	0
101	29.6	742	339	3226	2750	−97	−64	1292	961
102	50.3	1251	595	5489	4680	−195	−127	2124	1566
103	61.9	1543	768	6807	5806	−292	−192	2484	1804
104	64.6	1670	857	7276	6219	−389	−256	2383	1685
105	58.2	1637	861	6895	5915	−487	−320	1810	1200
106	42.9	1474	782	5724	4952	−584	−383	778	358
107	18.5	1161	620	3722	3291	−681	−447	−727	−850
108	−14.8	736	374	961	998	−779	−512	−2693	−2416
109	−57.2	297	179	−2230	−1623	−876	−711	−5299	−4506
110	−108.5	239	139	−5037	−3822	−1846[a]	−1148	−9658	−8176
200	−108.5	239	139	−5063	−3822	−1846[a]	−1148	−9658	−8176
201	−50.2	351	146	−1854	−1296	−766	−612	−4631	−3937
202	−4.8	857	336	1550	1544	−654	−380	−1806	−1665
203	27.6	1285	626	4275	3857	−543	−370	70	−176
204	47.0	1557	807	5941	5275	−431	−370	1260	778
205	53.5	1631	868	6472	5724	−319	−370	1782	1214

[a]Truck train with trucks spaced 17 826 mm apart governs

TABLE E7.5-3 Shear Envelope for 30-, 36-, 30-m Box Girder (kN)

Loc.	Unit Dead Load	Positive Shear				Negative Shear			
		Truck or Tandem	Lane	eta times Sum Int. Web	eta times Sum Ext. Web	Truck or Tandem	Lane	eta times Sum Int. Web	eta times Sum Ext. Web
100	11.4	287	127	1445	1266	-32	-21	482	356
101	8.4	247	101	1157	1021	-32	-23	324	232
102	5.4	208	78	876	781	-49	-28	124	69
103	2.4	171	59	605	551	-82	-37	-121	-136
104	-0.6	136	43	344	330	-121	-48	-382	-355
105	-3.6	103	30	92	119	-158	-63	-645	-578
106	-6.6	72	20	-146	-80	-193	-82	-910	-803
107	-9.6	45	13	-374	-269	-226	-102	-1173	-1025
108	-12.6	25	8	-582	-440	-256	-126	-1435	-1246
109	-15.6	10	6	-776	-596	-282	-151	-1692	-1462
110	-18.6	8	5	-938	-723	-305	-178	-1943	-1674
200	18.0	305	185	1919	1660	-31	-18	834	630
201	14.4	277	153	1617	1407	-31	-20	644	482
202	10.8	244	124	1309	1146	-33	-23	448	328
203	7.2	208	98	999	884	-62	-31	185	110
204	3.6	171	74	690	623	-96	-41	-95	-123
205	0.0	133	56	388	369	-133	-56	-388	-369

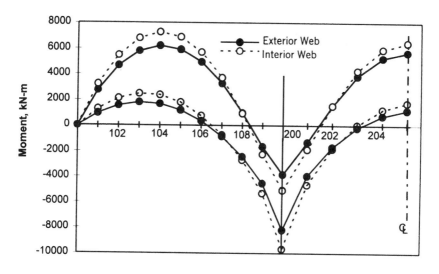

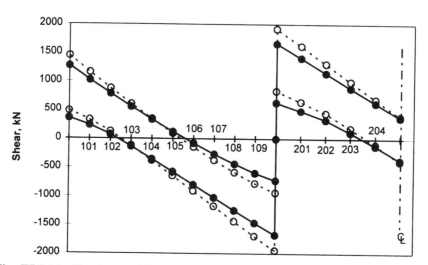

Fig. E7.5-14 Envelopes of factored moments and shears at tenth points for box girder.

b. Stress limits (Table 7.8) [Table A5.9.3-1]

	Stress Limits
At jacking (f_{pj})	$0.8f_{pu}$ = 1488 MPa
At transfer (f_{pt})—at anchors	$0.7f_{pu}$ = 1302 MPa
—general	$0.74f_{pu}$ = 1376 MPa
At service after losses (f_{pe})	$0.8f_{py}$ = 1339 MPa

2. *Stress Limitations for Concrete*
   a. At jacking and at transfer (before losses) (Table 7.6) [A5.9.4.1]

$$\text{Compressive stresses } (DC \text{ only}) = 0.55f'_{ci}$$

$$\text{Tensile stresses } (DC \text{ only}) = 0$$

   b. At service (after losses) (Table 7.7) [A5.9.4.2]

$$\text{Compressive stresses } [DC + DW + (LL + IM)] = 0.6f'_c$$

$$\text{Compressive stresses } (DC + DW) = 0.45f'_c$$

$$\text{Tensile stresses } [DC + DW + (LL + IM)]$$

$$= 0 \quad (\text{for unbonded tendons})$$

Sign convention for stresses in concrete:

Positive is compression, negative is tension

J. Develop Preliminary Tendon Profile

1. *Section Properties of Effective Interior Section* (Fig. E7.5-15)
   (Note: Bottom fillets are small compared to top fillets and are not included.)

$$A = (2585 \times 205) + (105 \times 105) + (305 \times 1485) + (140 \times 2585)$$

$$= 1\ 355\ 775 \text{ mm}^2$$

$$c_b = \frac{\begin{array}{c}(2585)(205)(1727.5) + (105)(105)(1590)\\ + (305)(1485)(882.5) + (140)(2585)(70)\end{array}}{1\ 355\ 775}$$

$$c_b = 1001.7 \text{ mm}; \ c_t = 828.3 \text{ mm}$$

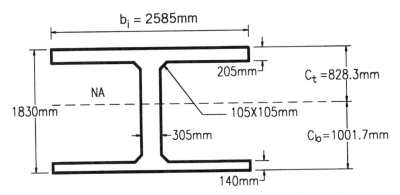

Fig. E7.5-15 Section properties of effective interior section.

$$I = (2585)(205)\left(828.3 - \frac{205}{2}\right)^2 + \frac{1}{12}(2585)(205)^3$$

$$+ \frac{1}{36}(105)(105)^3(2)$$

$$+ (105)(105)\left(828.3 - 205 - \frac{105}{3}\right)^2$$

$$+ (305)(1485)\left(1001.7 - 140 - \frac{1485}{2}\right)^2$$

$$+ \frac{1}{12}(305)(1485)^3 + (2585)(140)\left(1001.7 - \frac{140}{2}\right)^2$$

$$+ \frac{1}{12}(2585)(140)^3$$

$$= 689 \times 10^9 \text{ mm}^4$$

$$K_t = \frac{I}{Ac_b} = 507.4 \text{ mm} \qquad K_b = \frac{I}{Ac_t} = 613.5 \text{ mm}$$

$$S_t = \frac{I}{c_t} = \frac{689 \times 10^9}{828.3} = 831.8 \times 10^6 \text{ mm}^3$$

$$S_b = \frac{I}{c_b} = \frac{689 \times 10^9}{1001.7} = 687.8 \times 10^6 \text{ mm}^3$$

2. *Dead Load Moment Diagram* (for internal webs) (Fig. E7.5-16) The tendon profile is selected based on the shape of the moment diagram for the dead loads on the structure.

3. *Maximum Eccentricities* Maximum eccentricities occur at points of maximum dead load moments. A clearance of 50 mm is arbitrarily selected to allow for clearance between the tendons and the mild reinforcement of the bottom flange.

Location 104 $e_{max} = c_b - t_{flange} - \text{clearance} = 1001.7 - 140$

    $- 50 = 811.7 \text{ mm}$

Location 200 $e_{max} = c_t - t_{flange} - \frac{1}{2} \text{ fillet} = 828.3$

    $- 205 - 50 = 573.3 \text{ mm}$

Location 205 $e_{max} = c_b - t_{flange} - \text{clearance} = 1001.7$

    $- 140 - 50 = 811.7 \text{ mm}$

4. *Preliminary Prestressing Profile* (Fig. E7.5-17)

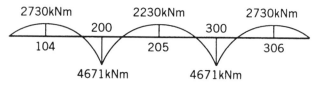

Fig. E7.5-16 Dead load moment diagram for interior webs.

Jacking at both ends
Uncoated seven-wire low-relaxation strands (Grade 270) 1860 MPa
Rigid galvanized ducts

Points B, C, and D lie on a straight line, therefore

$$\frac{h}{15} = \frac{1575 - 190}{15 + 3} \qquad h = 1154.2 \text{ mm}$$

Points D, E, and F lie on a straight line, therefore

$$\frac{m}{14.4} = \frac{1575 - 190}{14.4 + 3.6} \qquad m = 1108 \text{ mm}$$

K. Calculate Moments Due to Prestressing

1. *Prestress Losses*

   a. Anchor losses. The magnitude of prestress loss will depend on the prestressing system used. For this design, the amount of prestress

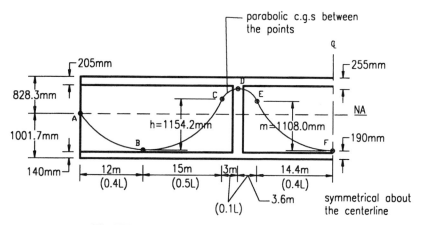

Fig. E7.5-17 Preliminary prestressing profile.

loss due to anchor setting will be such that it reduces the stress in
the tendons from $0.8f_{pu}$ to $0.7f_{pu}$, which is a loss of 186 MPa.
b. Friction losses [A5.9.5.2.2b] Losses due to friction between the prestressing tendons and the ducts (Eq. 7.101).

$$\Delta f_{pF} = f_{pj}(1 - e^{-(Kx+\mu\alpha)})$$

$$f_{pj} = \text{stress in steel at jacking (MPa)}$$

$$x = \text{length of prestressing tendon between any two points (mm)}$$

$$K = \text{Wobble coefficient (mm}^{-1})$$

$$\mu = \text{coefficient of friction}$$

$$\alpha = \text{absolute value of angular change of prestressing path between two points}$$

$$cumm.\ \alpha = \text{sum of absolute value of angular change of prestressing path from jacking end to any point}$$

For seven-wire strand tendons in rigid galvanized ducts,

$K = 6.6 \times 10^{-7}$ mm^{-1} neglect effects of Kx due to small value
    of Kx compared to $\mu\alpha$

Use $\mu = 0.2$/rad and assume jacking occurs at both ends

Segment	α	$cumm.\ \alpha$	$e^{-\mu\alpha}$
AB	0.136	0.136	0.973
BC	0.154	0.290	0.944
CD	0.154	0.444	0.915
DE	0.154	0.598	0.887
EF	0.154	0.752	0.860

The value of $e^{-\mu\alpha}$ indicates the force at any point as a fraction of
the force at the jacking end.
    By assuming a parabolic curve between points 1 and 2

$$\alpha_{12} = 2y/x$$

where y = vertical distance between points 1 and 2
      x = horizontal distance between points 1 and 2

$$\alpha_{AB} = \frac{2 \times 811.7}{12\ 000}; \ \alpha_{BC} = \frac{2 \times 1154.2}{15\ 000}; \ \alpha_{CD} = \frac{2 \times 230.8}{3000}$$

$$\alpha_{DE} = \frac{2 \times 277}{3600}; \ \alpha_{EF} = \frac{2 \times 1108}{14\,400}$$

    c. *Long-term losses.* Long-term losses due to creep, shrinkage, and relaxation will be such that they reduce the stress in the tendons to $0.8f_{py}$ at service from $0.74f_{pu}$ at transfer.

2. *Force along length* [Fig. E7.5-18(a)] Defining the parameters P_E as the force at the jacking end at service without anchor losses and P as the force at any point in the tendon at service.

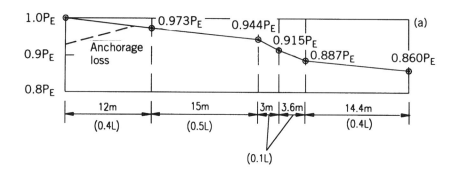

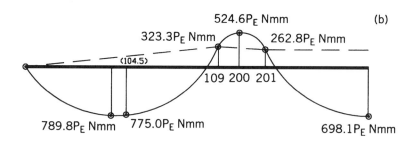

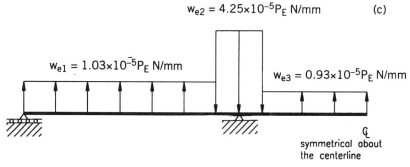

Fig. E7.5-18 Equivalent load system for calculating moments due to prestressing.

Location	P(N)	e (mm)	Pe
104	$0.973\,P_E$	811.7	$789.8\,P_E$
104.5	$0.970\,P_E$	799	$775.0\,P_E$
109	$0.944\,P_E$	342.5	$323.3\,P_E$
200	$0.915\,P_E$	573.3	$524.6\,P_E$
201	$0.887\,P_E$	296.3	$262.8\,P_E$
205	$0.860\,P_E$	811.7	$698.1\,P_E$

3. *Pe Diagram* Assume that the *Pe* graph approximates parabolic curves through the inflection points shown in Figure E7.5-18(b).
4. *Equivalent Load System* [Fig. E7.5-18(c)] Uniform load that results in the same moment produced by the tendon between the inflection points.

$$w_{e_1} = \frac{8P_E\left(\dfrac{323.3}{2} + 775.0\right)}{(0.9 \times 30\,000)^2} = 1.03 \times 10^{-5}\,P_E \text{ N/mm} \uparrow$$

$$w_{e_2} = \frac{8P_E\left(524.6 - \dfrac{323.3}{2} - \dfrac{262.8}{2}\right)}{[(0.1)(30\,000) + (0.1)(36\,000)]^2} = 4.25 \times 10^{-5}\,P_E \text{ N/mm} \downarrow$$

$$w_{e_3} = \frac{8P_E(698.1 + 262.8)}{(0.8 \times 36\,000)^2} = 0.93 \times 10^{-5}\,P_E \text{ N/mm} \uparrow$$

5. *Moment Distribution Analysis*
   a. Distribution factors

$$K_{21} = 3EI/L = 0.1EI \text{ (far end simply supported)}$$

$$K_{23} = 2EI/L = 0.056EI \text{ (symmetry)}$$

$$r_{21} = \frac{K_{21}}{K_{21} + K_{23}} = 0.64$$

$$r_{23} = \frac{K_{23}}{K_{21} + K_{23}} = 0.36$$

b. Fixed end moments (*FEM*)

$$FEM_{12} = \frac{(4.25 \times 10^{-5})P_E \times (30\ 000)^2}{12} [4 - 3(0.1)](0.1)^3$$

$$+ \frac{(-1.03 \times 10^{-5})P_E \times (30\ 000)^2}{12} [6 - 8(0.9)$$

$$+ 3(0.9)^2](0.9)^2$$

$$= -757.9\ P_E\ \text{N mm}$$

$$FEM_{21} = \frac{(-4.25 \times 10^{-5})P_E \times (30\ 000)^2}{12} [6 - 8(0.1)$$

$$+ 3(0.1)^2](0.1)^2$$

$$+ \frac{(1.03 \times 10^{-5})P_E \times (30\ 000)^2}{12} [4 - 3(0.9)](0.9)^3$$

$$= 565.4\ P_E\ \text{N mm}$$

$$FEM_{23} = \frac{(4.25 + 0.93) \times 10^{-5}\ P_E \times (36\ 000)^2}{12} [6 - 8(0.1) + 3(0.1)^2](0.1)^2$$

$$+ \frac{(4.25 + 0.93) \times 10^{-5}\ P_E \times (36\ 000)^2}{12} [4 - 3(0.1)](0.1)^3$$

$$+ \frac{(-0.93 \times 10^{-5})P_E \times (36\ 000)^2}{12}$$

$$= -691.1\ P_E\ \text{N mm}$$

Symmetry axis

DF			0.64	0.36
FEM	$-757.9P_E$		$564.4P_E$	$-691.1P_E$
D&CO	$+757.9P_E$	$\times \frac{1}{2} \rightarrow$	$379P_E$	
D	0		$-162.1P_E$	$-91.2P_E$
Σ			$782.3P_E$	$-782.3P_E$

Centerline

6. *Calculate M^P Values*

M^P = Moment due to prestressing force at service limit state, after all the losses.

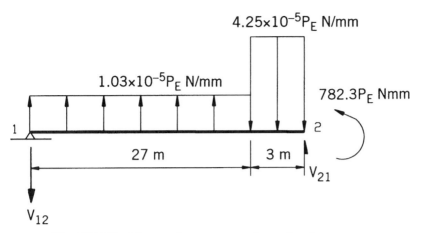

Fig. E7.5-19 Moment due to prestressing at interior support.

$$\Sigma M_2 = 0 \text{ (Fig. E7.5-19)}$$

$$V_{12} = \frac{(1.03 \times 10^{-5})P_E(27\ 000)\left(3000 + \dfrac{27\ 000}{2}\right)}{30\ 000} \\ \frac{- \dfrac{(4.25 \times 10^{-5})P_E(3000)^2}{2} - 782.3P_E}{30\ 000}$$

$$= 0.1205P_E \text{ N}$$

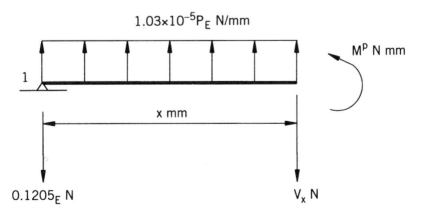

Fig. E7.5-20 Moment due to prestressing at distance x from left support.

$$\Sigma M_x = 0 \text{ (Fig. E7.5-20)}$$

$$M^P_{104} = \frac{(1.03 \times 10^{-5})P_E(12\ 000)^2}{2} - 0.1205P_E(12\ 000) = -704.4P_E \text{ N mm}$$

$$M^P_{109} = \frac{(1.03 \times 10^{-5})P_E(27\ 000)^2}{2} - 0.1205P_E(27\ 000) = 500.9P_E \text{ N mm}$$

$$M^P_{200} = 782.3P_E \text{ N mm}$$

$$\Sigma M_3 = 0 \text{ (Fig. E7.5-21)}$$

$$V_{23} = \frac{(-0.93 \times 10^{-5})P_E(28.8)\left(3.6 + \dfrac{28.8}{2}\right) + (4.25 \times 10^{-5})P_E(2 \times 3.6)\left(\dfrac{36}{2}\right)}{36}$$

$$= 0.019P_E \text{ N}$$

$$\Sigma M_{205} = 0 \text{ (Fig. E7.5-22)}$$

$$M^P_{205} = \left[0.019P_E(18\ 000) - (4.25 \times 10^{-5}\ P_E)(3600)\left(14\ 400 + \frac{3600}{2}\right)\right]$$

$$+ \frac{0.93 \times 10^{-5}\ P_E(14\ 400)^2}{2} + 783.2P_E = -389.2P_E \text{ N mm}$$

Location	M^P
104	$-704.4P_E$
109	$500.9P_E$
200	$782.3P_E$
205	$-389.2P_E$

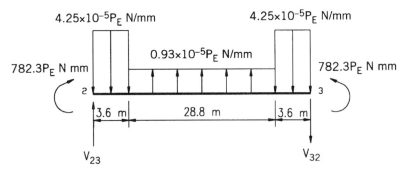

Fig. E7.5-21 Shear forces due to prestressing for interior span.

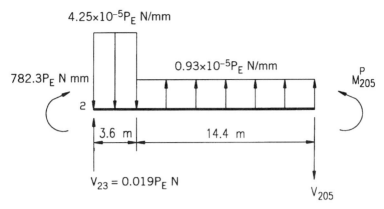

Fig. E7.5-22 Moment due to prestressing at midspan of interior span.

L. Select Jacking Force

Using the three critical locations (104, 200, and 205).

$$f = \frac{P}{A} \mp \frac{M^P}{S} \pm \frac{M^S}{S}$$

1. *Location 104*

M^{SI}_{104} = Service *I* moment = 5097 kN m (Table E7.5-1)

M^{SIII}_{104} = Service *III* moment = 4624 kN m (Table E7.5-1)

M^{P}_{104} = Moment due to force in tendons = $-704.4P_E$ N mm

Assuming that tension in the bottom fibers controls. Allowable tension stress = 0 (Section 7.10.5, Part I.2.b).

$$f_b = \frac{P_{104}}{A} + \frac{M^P_{104}}{S_b} - \frac{M^{SI}_{104}}{S_b} \text{ (unbonded tendons)}$$

$P_E = 0.9P_J$ (due to long-term losses but excluding anchor losses)

$$P_{104} = 0.973P_E \text{ (due to friction losses)}$$

where

$$P_J = \text{Jacking force}$$

P_E = Force in tendons at service at jacking ends
without anchor losses

$$f_b = 0 = \frac{0.973 \times 0.9}{1\,355\,775} P_J + \frac{(704.4 \times 0.9P_J - 5097 \times 10^6)}{687.8 \times 10^6}$$

Solving for P_J in above equation yields

$$P_J = 4.73 \times 10^6 \text{ N}$$

Check compression in the top fibers, using M_{104}^{SI}

$$f_t = \frac{(0.973)(0.9)(4.73 \times 10^6)}{1\,355\,775}$$

$$- \frac{(704.4)(0.9)(4.73 \times 10^6) - (5097 \times 10^6)}{831.8 \times 10^6}$$

$$f_t = 5.58 \text{ MPa} < 0.6f_c' = 21 \text{ MPa, OK}$$

Therefore, tension in bottom fibers controls.

2. *Location 200* (at supports)

$$M_{200}^{SI} = -7107 \text{ kN m}$$

$$M_{200}^{SIII} = -6620 \text{ kN m}$$

$$M_{200}^{P} = 782.3P_E \text{ N mm}$$

Assuming that tension in the top fibers controls,

$$f_t = \frac{P_{200}}{A} + \frac{M_{200}^{P}}{S_t} - \frac{M_{200}^{SI}}{S_t}$$

$$P_{200} = 0.915P_E \quad \text{and} \quad P_E = 0.9P_J$$

$$f_t = 0 = \frac{0.915 \times 0.9P_J}{1\,355\,775} + \frac{(782.3)(0.9)P_J - (7107 \times 10^6)}{831.8 \times 10^6}$$

Solving for P_J yields

$$P_J = 5.87 \times 10^6 \text{ N}$$

Check compression in bottom fibers, using M_{200}^{SI}

$$f_b = \frac{(0.915)(0.9)(5.87 \times 10^6)}{1\,355\,775}$$

$$- \frac{(782.3)(0.9)(5.87 \times 10^6) - (7107 \times 10^6)}{687.8 \times 10^6}$$

$$f_b = 7.89 \text{ MPa} < 0.6f'_c = 21 \text{ MPa, OK}$$

Therefore, tension in top fibers controls.

3. *Location 205*

$$M^{SI}_{205} = 4455 \text{ kN m}$$

$$M^{SIII}_{205} = 4011 \text{ kN m}$$

$$M^{P}_{205} = -389.2P_E \text{ N mm}$$

Assuming that tension in the bottom fibers controls,

$$f_b = \frac{P_{205}}{A} + \frac{M^{P}_{205}}{S_b} - \frac{M^{SI}_{205}}{S_b}$$

$$P_{205} = 0.86P_E$$

$$P_E = 0.9P_J$$

$$f_b = 0 = \frac{(0.86)(0.9)P_J}{1\ 355\ 775} + \frac{(389.2)(0.9)P_J - (4455 \times 10^6)}{687.8 \times 10^6}$$

Solving for P_J yields

$$P_J = 6.0 \times 10^6 \text{ N controls}$$

Check compression in the top fibers

$$f_t = \frac{(0.86)(0.9)(6 \times 10^6)}{1\ 355\ 775} - \frac{(389.2)(0.9)(6 \times 10^6) - (4455 \times 10^6)}{831.8 \times 10^6}$$

$$f_t = 6.25 \text{ MPa} < 0.6f'_c = 21 \text{ MPa, OK}$$

Tension in the bottom fibers controls.

Select the largest of the jacking forces

$$P_{\text{Jacking}} = 6 \times 10^6 \text{ N}$$

Amount of prestressing per interior web

$$A_{ps} = \frac{P_{\text{Jacking}}}{f_{\text{Jacking}}} = \frac{6.0 \times 10^6}{0.8f_{pu}} = \frac{6.0 \times 10^6}{(0.8)(1860 \times 10^6)} = 4032.3 \text{ mm}^2/\text{web}$$

M. Calculate Ultimate Strength Capacity

The ultimate strength capacity (ϕM_n), without any mild reinforcement, at any section should be compared with M_u and 1.2 M_{cr} [A5.7.3.3.2]

If $\phi M_n > M_u$ and 1.2 M_{cr} then no mild steel required.

If $\phi M_n < M_u$ or 1.2 M_{cr} then mild steel required.
1. *Check Limits of Reinforcement*
   a. The maximum amount of prestressed and nonprestressed reinforcement shall be such that [A5.7.3.3.1]

$$\frac{c}{d_e} \le 0.42 \quad \text{where} \quad d_e = \frac{A_{ps}f_{ps}d_p + A_s f_y d_s}{A_{ps}f_{ps} + A_s f_y}$$

$$c = a/\beta_1 \text{ where } \beta_1 = 0.8 \text{ for } f'_c = 35 \text{ MPa}$$

Summary of final design:

Section	A_{ps} (mm²)	f_{ps} (MPa)	d_p (mm)	A_s (mm²)	f_y (MPa)	d_s (mm)	c (mm)	c/d_e	Status
104	4032	1406	1640	3446	400	1760	115.0	0.07	OK
200	4032	1328	1575	7388	400	1727.5	135.0	0.08	OK
205	4032	1255	1640	739	400	1760	87.0	0.05	OK

All sections satisfy the maximum reinforcement requirements provided in the specifications [A5.7.3.3.1].
   b. Minimum amounts of prestressed and nonprestressed reinforcement should be such that each section develops factored flexural resistance M_r or at least 1.2 times the cracking strength (M_{cr}) [A5.7.3.3.2].

Cracking strengths: Using $P_J = 6 \times 10^6$ N

Location 104

Allowing a tension stress equal to $f_r = 0.63\sqrt{f'_c} = -3.73$ MPa in the bottom fibers

$$f_b = -3.73 \text{ MPa} = \frac{(0.9)(0.973)(6 \times 10^6)}{1\,355\,775}$$
$$+ \frac{(704.4)(0.9)(6 \times 10^6) - M_{cr}}{687.8 \times 10^6}$$

Solving for M_{cr} yields

$$M_{cr} = 9034 \times 10^6 \text{ N mm}$$

$$1.2 \, M_{cr} = 10\ 840 \times 10^6 \text{ N mm}$$

Location 200

Allowing a tension stress equal to $0.63\sqrt{f'_c} = -3.73$ MPa in the top fibers

$$f_t = -3.73 \text{ MPa} = \frac{(0.9)(0.915)(6 \times 10^6)}{1\ 355\ 775}$$

$$+ \frac{(782.3)(0.9)(6 \times 10^6) - M_{cr}}{831.8 \times 10^6}$$

Solving for M_{cr} yields:

$$M_{cr} = 10\ 360 \times 10^6 \text{ N mm}$$

$$1.2 \, M_{cr} = 12\ 430 \times 10^6 \text{ N mm}$$

Location 205

Allowing a tension stress of $0.63\sqrt{f'_c} = -3.73$ MPa in the bottom fibers

$$f_b = -3.73 = \frac{(0.86)(0.9)(6 \times 10^6)}{1\ 355\ 775} + \frac{(389.2)(0.9)(6 \times 10^6) - M_{cr}}{687.8 \times 10^6}$$

Solving for M_{cr} yields

$$M_{cr} = 7020 \times 10^6 \text{ N mm}$$

$$1.2 \, M_{cr} = 8440 \times 10^6 \text{ N mm}$$

2. *Ultimate Strength*
   a. Location 104 (Fig. E7.5-23)

$$M_u = 7252 \times 10^6 \text{ N mm}$$

$$1.2 \, M_{cr} = 10\ 840 \times 10^6 \text{ N mm}$$

Assuming that no mild reinforcement is required, $\phi = 1.00$

Assuming NA is in the flange,

$$\phi M_n = \phi A_{ps} f_{ps} \, [d_p - a/2]$$

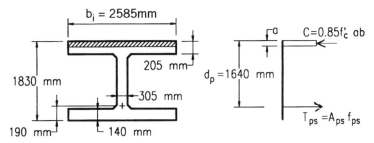

Fig. E7.5-23 Positive moment strength at Location 104.

Using the estimated value of $f_{ps} = f_{pe} + 103$, where f_{pe} is the stress in the tendons at service at Location 104,

$$f_{pe} = f_{S104} = 0.9 \times 0.973 f_J$$

where $f_J = 0.8 f_{pu}$
$$f_{pe} = 0.9 \times 0.973 \times 0.8 \times 1860 = 1303 \text{ MPa}$$
$$f_{ps} = 1303 + 103 = 1406 \text{ MPa}$$

For horizontal equilibrium, $C = T_{ps}$

$$0.85 \times 35 \times 2585 \times a = 4032 \times 1406$$

$$a = 74 \text{ mm} < 205 \text{ mm}$$

Therefore NA is in the flange

$$\phi M_n = 1.00 \times 4032 \times 1406 \times (1640 - 74/2)$$

$$\phi M_n = 9090 \times 10^6 \text{ N mm} > M_u = 7252 \times 10^6 \text{ N mm}$$

$$\text{but} < 1.2 M_{cr} = 10\ 840 \times 10^6 \text{ N mm}$$

Because $M_u < \phi M_n < 1.2 M_{cr}$, mild steel reinforcement is required for ductility purpose only.

Amount of mild steel required at Location 104 (Fig. E7.5-24)

Assuming the centroid of the mild steel to be in the middle of the flange bottom flange thickness and the NA to be in top flange:

$$C = A_{ps} f_{ps} + A_s f_y = 0.85 f'_c ab$$

Using estimate value of $f_{ps} = 1406$ MPa

$$a = \frac{(4032)(1406) + 400\ A_s}{(0.85)(35)(2585)} = 74 + 0.0052 A_s$$

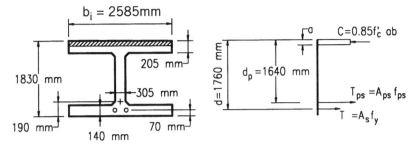

Fig. E7.5-24 Positive moment strength for partially prestressed section at Location 104.

Assume $\phi = 0.96$

$$1.2M_{cr} = \phi M_n = 0.96\left[A_{ps}f_{ps}\left(d_p - \frac{a}{2}\right) + A_s f_y\left(d - \frac{a}{2}\right)\right]$$

$$10.84 \times 10^9 = 0.96\left[\begin{array}{l} 5.67 \times 10^6\left(1640 - \dfrac{74}{2} - \dfrac{0.0052\,A_s}{2}\right) \\ + 400\,A_s\left(1760 - \dfrac{74}{2} - \dfrac{0.0052\,A_s}{2}\right) \end{array}\right]$$

$$= (87.3 \times 10^8) + (64.7 \times 10^4\,A_s) - 0.998\,A_s^2$$

Solving for A_s

$$A_s = 3277.8 \text{ mm}^2 \text{ of mild steel per 2585 mm}$$

$$A_s = 1268 \text{ mm}^2/\text{m}$$

Use No. 15 bars @ 150 mm, prov'd $A_s = 1333$ mm^2/m (Table B.4)

Check ϕ value $A_s = 1333(2.585) = 3446$ mm^2

$$\phi_{calc} = 0.9 + 0.1\left[\frac{4032 \times 1674}{(4032)(1674) + (3446)(400)}\right]$$

$$= 0.98 > \phi_{used} = 0.96$$

Therefore ϕ used was conservative.

Check if $a < t_s$

$$a = 74 + 0.0052(3446) = 91.9 \text{ mm} < 205 \text{ mm}$$

Therefore NA is in top flange.

$$C = 0.85f'_cab = 0.85(35)(91.9)(2585)10^{-3} = 7067 \text{ kN}$$

b. Location 200 (at supports) (Fig. E7.5-25)

$$M_u = -9728 \text{ kN m}$$

$$1.2 M_{cr} = -12\ 430 \text{ kN m}$$

Assuming no mild steel is required, $\phi = 1.00$, and assuming NA is in bottom flange.

Using the estimated value of $f_{ps} = f_{pe} + 103$ where

f_{pe} = stress in tendons at service at Location 200
$f_{pe} = f_{S200} = (0.9)(0.915)f_j = (0.9)(0.915)(0.8)(1860) = 1225 \text{ MPa}$
$f_{ps} = 1225 + 103 = 1328 \text{ MPa}$

For equilibrium, $T_{ps} = C$

$$A_{ps}f_{ps} = 0.85f'_cab$$

$$(4032)(1328) = (0.85)(35)(2585)(a)$$

$$a = 69.6 \text{ mm} < 140 \text{ mm}$$

Therefore NA is in the bottom flange.

$$\phi M_n = -\phi A_{ps}f_{ps}(d_p - a/2)$$
$$= -1.00 \times 4032 \times 1328 \times (1575 - 69.6/2)$$
$$= -8247 \times 10^6 \text{ N mm} < M_u = -9728 \times 10^6 \text{ N mm}$$

and

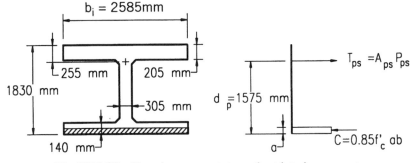

Fig. E7.5-25 Negative moment strength at interior support.

$$< 1.2\ M_{cr} = -12\ 430 \times 10^6\ \text{N mm}$$

Since $M_u > \phi M_n$ and $1.2 M_{cr} > \phi M_n$, mild steel is required for both strength and serviceability.

Amount of mild steel required at Location 200 (Fig. E7.5-26)

Assuming the centroid of the mild steel to be in the middle of the flange thickness.

For equilibrium, $C = A_{ps}f_{ps} + A_s f_y$

Assuming ϕ value of 0.97

$$1.2\ M_{cr} = \phi M_n = 0.97[A_{ps}f_{ps}(d_p - a/2) + A_s f_y(d - a/2)]$$
$$12\ 430 \times 10^6 = 0.97[(4032)(1328)(1575 - a/2) + A_s(400)(1727.5 - a/2)]$$

but
$$a = \frac{(4032)(1328) + 400\ A_s}{(0.85)(35)(2585)} = 69.6 + 0.0052\ A_s$$

Substituting for a,

$$12\ 430 \times 10^6 = 0.97\left[\begin{array}{l}(53.5 \times 10^5)\left(1575 - \dfrac{69.6}{2} - \dfrac{0.0052\ A_s}{2}\right) \\[2mm] + 400\ A_s\left(1727.5 - \dfrac{69.6}{2} - \dfrac{0.0052\ A_s}{2}\right)\end{array}\right]$$

$$= (79.9 \times 10^8) + (6.43 \times 10^5\ A_s) - 1.01 A_s^2$$
$$1.01 A_s^2 - (6.43 \times 10^5)A_s + (4.44 \times 10^9) = 0$$

Solving for A_s yields:

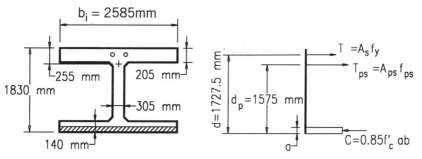

Fig. E7.5-26 Negative moment strength for partially prestressed section at interior support.

$$A_s = 6982 \text{ mm}^2 \text{ per } 2585 \text{ mm width} = 2700 \text{ mm}^2/\text{m}$$

Try two layers of No. 25 bars @ 350 mm, $A_s = 2(1429)\left(\dfrac{2585}{10^3}\right) =$ 7388 mm^2

Check ϕ value

$$\phi_{\text{calc}} = 0.9 + 0.1\left[\frac{(4032)(1674)}{(4032)(1674) + (7388)(400)}\right] = 0.97 = \phi_{\text{used}} = 0.97$$

Therefore, ϕ value used is accurate.

Check if $a < t_f$.

$$a = 69.6 + 0.0052(7388) = 108 \text{ mm} < 140 \text{ mm}$$

Therefore, NA is in bottom flange

$$C = 0.85f'_c ab = 0.85(35)(108)(2585)10^{-3} = 8306 \text{ kN}$$

Use 2 layers of No. 25 bars @ 350 mm

$$\text{prov'd } A_s = 2(1429) = 2858 \text{ mm}^2/\text{m (Table B.4)}$$

c. Location 205 (Fig. E7.5-27)

$$M_u = 6409 \text{ kN m}$$
$$1.2\, M_{cr} = 8440 \text{ kN m}$$

Assuming no mild steel is required, $\phi = 1.00$

Assuming NA to be in top flange,

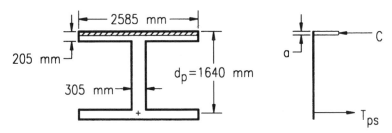

Fig. E7.5-27 Positive moment strength at Location 205.

$$0.85f'_c ab = A_{ps}f_{ps}$$

Using estimated value of $f_{ps} = f_{pe} + 103$

$$f_{pe} = f_{S205} = 0.9 \times 0.86f_J = 1152 \text{ MPa}$$

$$f_{ps} = 1152 + 103 = 1255 \text{ MPa}$$

$$a = \frac{(1255)(4032)}{0.85(35)(2585)} = 65.8 \text{ mm}$$

$a = 65.8$ mm < 205 mm, therefore NA is in top flange

$$\phi M_n = 1.00(4032)(1255)\left(1640 - \frac{65.8}{2}\right)$$

$$\phi M_n = 8132 \times 10^6 \text{ N mm} > M_u = 6409 \times 10^6 \text{ N mm}$$

$$< 1.2 \, M_{cr} = 8440 \times 10^6 \text{ N mm}$$

Because $M_u < \phi M_n < 1.2 \, M_{cr}$, mild steel reinforcement is required for ductility purposes only.

Amount of mild steel required at Location 205. (Fig. E7.5-28)
Assume centroid of mild steel to be in middle of bottom flange.

$$0.85f'_c ab = A_{ps}f_{ps} + A_s f_y$$

$$a = \frac{(4032)(1255) + A_s(400)}{0.85(35)(2585)} = 65.8 + 0.0052 \, A_s$$

$$1.2M_{cr} = \phi M_n = \phi\left[A_{ps}f_{ps}\left(d_p - \frac{a}{2}\right) + A_s f_y\left(d - \frac{a}{2}\right)\right]$$

Assume $\phi = 0.98$

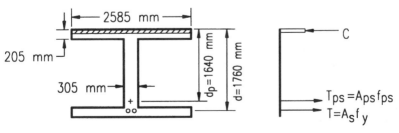

Fig. E7.5-28 Positive moment strength for partially prestressed section at Location 205.

$$8440 \times 10^6 = 0.98 \left[\begin{array}{l} (4032)(1255) \left(1640 - \dfrac{65.8}{2} - \dfrac{0.0052\,A_s}{2} \right) \\ + 400\,A_s \left(1760 - \dfrac{65.8}{2} - \dfrac{0.0052\,A_s}{2} \right) \end{array} \right]$$

$$8440 \times 10^6 = (79.7 \times 10^8) + (66.4 \times 10^4)A_s - 1.02\,A_s^2$$

Solving for A_s

$$A_s = 705.5 \text{ mm}^2 \text{ per 2585 mm width} = 273 \text{ mm}^2/\text{m}$$

Try No. 10 bars @ 350 mm, prov'd $A_s = 286$ mm^2/m (Table B.4)

Check ϕ value

$$A_s = 286(2.585) = 739 \text{ mm}^2$$

$$\phi = 0.9 + 0.1 \left[\frac{(4032)(1674)}{(4032)(1674) + (739)(400)} \right] = 0.996$$

Therefore, ϕ value used (0.98) is conservative.

Check if $a < t_s$

$$a = 65.8 + (0.0052)(739) = 69.6 \text{ mm} < 205 \text{ mm}$$

$$C = 0.85 f'_c ab = 0.85(35)(69.6)(2582)10^{-3} = 5353 \text{ kN}$$

Use No. 10 bars at 350 mm

N. Shear Design

1. *Transverse Shear Reinforcement* This reinforcement is required at any section where [A5.8.2.4]

$$V_u > 0.5\phi(V_c + V_p)$$

where

V_c = nominal shear resistance of concrete
V_p = component of prestressing force in the direction of the shear force (i.e., transverse component)

2. *Minimum Transverse Reinforcement* Where transverse reinforcement is required, the area of steel shall not be less than [A5.8.2.5]

$$A_v = 0.083\sqrt{f'_c} \frac{b_v s}{f_y}$$

where

$$A_v = \text{area of transverse reinforcement within } s \text{ (mm}^2)$$
$$s = \text{spacing of reinforcement (mm)}$$
$$b_v = \text{width of web (mm)}$$
$$f_y = \text{yield strength of transverse reinforcement (MPa)}$$

3. *Types of Transverse Reinforcement* [A5.8.2.6] U-shaped vertical stirrups are used.

4. *Maximum Spacing of Stirrups* [A5.8.2.7] If $V_u < 0.1f_c'b_vd_v$ then

$$s \leq 0.8d_v \leq 600 \text{ mm}$$

If $V_u \geq 0.1f_c'b_vd_v$ then

$$s \leq 0.4d_v \leq 300 \text{ mm}$$

5. *Amount of Transverse Reinforcement* The reinforcement at any section should be such that [A5.8.3.3]

$$V_u \leq 0.25f_c'b_vd_v + V_p$$
$$V_u \leq \phi(V_c + V_s + V_p)$$

where

$$V_c = 0.083\beta\sqrt{f_c'}b_vd_v$$
$$V_s = \frac{A_vf_yd_v(\cot\theta + \cot\alpha)\sin\alpha}{s}$$

$\alpha = 90°$ for vertical stirrups and θ and β are determined using Figure 7.43 from AASHTO [A5.8.3.4.]

6. *Critical section* This section shall be taken as the larger of $0.5d_v \cot\theta$ or d_v from internal face of support.

   a. *Location 104*

$$V_u = 382 \text{ kN (shear envelope) (Table E7.5-3)}$$

$$M_u = 7276 \text{ kN m (moment envelope) (Table E7.5-2)}$$

$$V_p = 0 \text{ since tendons are horizontal at this location}$$

$$b_v = 305 \text{ mm}$$

$$d_v = \text{moment arm} = \frac{M_n}{C} = \frac{10.84 \times 10^9}{0.96(7.067 \times 10^6)} = 1598 \text{ mm}$$

$$d_v \geq \begin{cases} 0.9d_e = 0.9(1663) = 1497 \text{ mm} \\ 0.72h = 1317.6 \text{ mm} \end{cases}$$

$$d_v = 1598 \text{ mm}$$

$\phi = 0.9$ for shear in normal weight concrete

Determine θ and β

$$v = \frac{V_u - \phi V_p}{\phi b_v d_v} = \frac{382\ 000}{0.9(305)(1598)} = 0.87 \text{ MPa}$$

$$\frac{v}{f_c'} = 0.025$$

Assuming $\theta = 27°$

$$\varepsilon_x = \frac{\dfrac{7276 \times 10^6}{1598} + 0.5(382\ 000) \cot 27 - (4032)(1303)}{(200\ 000)(3446) + (197\ 000)(4032)}$$

where

$$f_{po} \cong f_{pe} = 1303 \text{ MPa}$$

$$A_s = 3446 \text{ mm}^2$$

$$A_{ps} = 4032 \text{ mm}^2$$

$\varepsilon_x = -0.000\ 22 < 0$, therefore reduction factor F_ε should be used.

$$F_\varepsilon = \frac{E_s A_s + E_p A_{ps}}{E_c A_c + E_s A_s + E_p A_p}$$

A_c = area on tension side of member (Fig. E7.5-29)

$$F_\varepsilon = \frac{(200\ 000)(3446) + (197\ 000)(4032)}{(200\ 000)(3446) + (197\ 000)(4032) + (598\ 275)(29\ 910)} = 0.0077$$

$$\varepsilon_x = -0.000\ 22 \times 0.0077 = -1.7 \times 10^{-6}$$

From Figure 7.43, using $v/f_c' = 0.025$ and $\varepsilon_x = -1.7 \times 10^{-6}$

$$\theta = 27° \text{ and } \beta = 4.8$$

$$V_c = 0.083(4.8)\sqrt{35}(305)(1598) = 1149 \text{ kN}$$

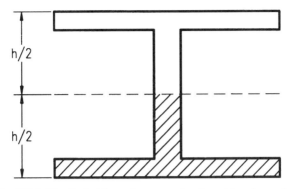

Fig. E7.5-29 Definition of area on tension side of member.

Because $V_u = 382$ kN $< 0.5\phi(V_c + V_p) = 0.5(0.9)(1149 + 0) = 517$ kN, no transverse shear reinforcement is required in the vicinity of this location.

b. *Location 200 + d_v or 200.44 (Fig. E7.5-30)*

$$d_v \text{ moment arm} = \frac{M_n}{C} = \frac{12.43 \times 10^9}{0.97(8.306 \times 10^6)} = 1543 \text{ mm}$$

$$d_v \geq \begin{cases} 0.9d_e = 0.9(1629) = 1466 \text{ mm} \\ 0.72h = 1317.6 \text{ mm} \end{cases}$$

$$d_v = 1543 \text{ mm}$$

$V_u = 1786$ kN (interpolating from shear envelope) (Table E7.5-3)

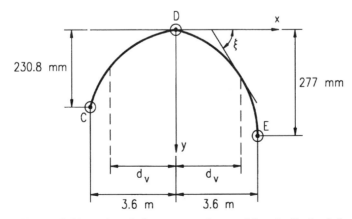

Fig. E7.5-30 Definition of angle between tendons and longitudinal reinforcement.

$M_u = -7446 \times 10^6$ N mm (interpolating from moment envelope)

(Table E7.5-2)

$$V_p = A_{ps}f_{ps} \sin\xi$$

$$f_{ps} = 0.903 \times 0.9 \times 1448 = 1176.4 \text{ MPa}$$

$$\xi = \tan^{-1}\left[\frac{2(277)(1543)}{(3600)^2}\right] = 3.8°$$

$$\left(\begin{array}{c}\text{angle in vertical plane between} \\ \text{tendons and longitudinal reinforcement}\end{array}\right)$$

$$V_p = 4032 \times 1176.4 \times \sin 3.8° = 314.4 \text{ kN}$$

$$b_v = 305 \text{ mm}$$

$$\phi = 0.9$$

Determine θ and β

$$\upsilon = \frac{V_u - \phi V_p}{\phi b_v d_v} = \frac{1\ 786\ 000 - 0.9(314\ 400)}{0.9(1543)(305)} = 3.55 \text{ MPa}$$

$$\frac{\upsilon}{f_c'} = 0.1$$

Assuming $\theta = 32°$

$$\varepsilon_x = \frac{\dfrac{7446 \times 10^6}{1543} + 0.5(1\ 786\ 000) \cot 32° - (4032)(1176.4)}{(200\ 000)(7388) + (197\ 000)(4032)}$$

where $f_{po} \cong f_{pe} = 1176.4$ MPa   (conservatively)

$$A_{ps} = 4032 \text{ mm}^2$$

$$A_s = 7388 \text{ mm}^2$$

$$\varepsilon_x = 0.67 \times 10^{-3}$$

From Figure 7.43 and using $\upsilon/f_c' = 0.1$ and $\varepsilon_x = 0.67 \times 10^{-3}$

$$\theta = 32° \text{ and } \beta = 2.3$$

$$V_c = \frac{0.083(2.3)\sqrt{35}(305)(1543)}{10^3} = 531.5 \text{ kN}$$

$$V_s \geq \left(\frac{V_u}{\phi}\right) - V_c - V_p$$

$$V_s \geq \frac{1786}{0.9} - 531.5 - 314.4 = 1138.5 \text{ kN}$$

$$V_s \geq \left(\frac{V_u}{\phi}\right) - 0.25f_c'b_vd_v - V_p$$

$$V_s \geq \frac{1786}{0.9} - 0.25(35)(305)(1543) - 314.4 = \text{neg.}$$

$$V_s = 1138.5 \text{ kN}$$

Amount of reinforcement

Using No. 10 stirrups; $A_v = 200 \text{ mm}^2$

spacing

$$s = \frac{A_v f_y d_v(\cot \theta + \cot \alpha)\sin \alpha}{V_s}$$

$$s = \frac{200(400)(1543) \cot 32°}{1\,138\,500} = 174 \text{ mm}$$

$$0.1f_c'b_vd_v = 0.1(35)(305)(1543) = 1647 \text{ kN} < V_u$$

$$s_{max} \leq \begin{cases} 0.4d_v = 617 \text{ mm} \\ 300 \text{ mm} \end{cases}$$

Use No. 10 stirrups @ 175 mm @ Location 200.44

c. *Location 110 − d_v or 109.48*

$$d_v = 1543 \text{ mm}$$

$$V_u = -1812.3 \text{ kN}$$

$$M_u = -7391 \text{ kN m}$$

$$V_p = A_{ps}f_{ps}\sin\xi$$

$$f_{ps} = 0.93 \times 0.9 \times 1448 = 1212 \text{ MPa}$$

$$\xi = \tan^{-1}\left[\frac{2(230.8)(1543)}{(3600)^2}\right] = 3.2°$$

$$V_p = 4032 \times 1212 \times \sin 3.2° = 273 \text{ kN}$$

$$b_v = 305 \text{ mm}$$

$$\phi = 0.9$$

Determine θ and β

$$v = \frac{V_u - \phi V_p}{\phi b_v d_v} = \left[\frac{1812.3 - 0.9(273)}{0.9(305)(1543)}\right]10^3 = 3.70 \text{ MPa}$$

$$\frac{v}{f_c'} = 0.106$$

Assume $\theta = 32°$

$$\varepsilon_x = \frac{\dfrac{7391 \times 10^6}{1543} + 0.5(1\,812\,300)\cot 32 - (4032)(1212)}{(200\,000)(7388) + (197\,000)(4032)}$$

$$= 0.60 \times 10^{-3}$$

$$(f_{ps} = 1212 \text{ MPa conservatively})$$

From Figure 7.43 and using $v/f_c' = 0.106$ and $\varepsilon_x = 0.60 \times 10^{-3}$

$$\theta = 32° \text{ and } \beta = 2.2$$

$$V_c = 0.083(2.2)\sqrt{35}(305)(1543) = 508 \text{ kN}$$

$$V_s \geq \left(\frac{V_u}{\phi}\right) - V_c - V_p = \frac{1812.3}{0.9} - 508 - 273.2 = 1232.5 \text{ kN}$$

$$V_s \geq \left(\frac{V_u}{\phi}\right) - 0.25 f_c'b_v d_v - V_p = \text{neg}$$

$$V_s = 1232.5 \text{ kN}$$

Amount of reinforcement

Using No. 10 stirrups; $A_v = 200 \text{ mm}^2$

spacing

$$s = \frac{200(400)(1543)\cot 32°}{1\,232\,500} = 160.3 \text{ mm}$$

$$s_{\max} \leq \begin{cases} 0.4d_v = 617 \text{ mm} \\ 300 \text{ mm} \end{cases}$$

Use No. 10 stirrups @ 160 mm @ Location 109.48

d. *Location 100 + d_v or 100.44*

$$d_v \geq \begin{cases} 0.9d_e = 0.9(1001.7) = 901.5 \text{ mm} \\ 0.72h = 0.72(1830) = 1317.6 \text{ mm} \end{cases}$$

$$d_v = 1317.6 \text{ mm}$$

$$V_u = 1318.6 \text{ kN}$$

(interpolating from shear envelope) (Table E7.5-3)

$$M_u = 1419 \text{ kN m}$$

(interpolating from moment envelope) (Table E7.5-2)

$$V_p = A_{ps} f_{ps} \sin\xi$$

where

$$\xi = \tan^{-1}\left[\frac{2(10\ 682.4)(811.7)}{(12\ 000)^2}\right] = 6.9°$$

At service,

$$f_{ps} = 0.8 f_{py} - \text{Anchor loss} = 1339 - 186 = 1153 \text{ MPa}$$

$$V_p = (4032)(1153)(\sin 6.9°) = 558.5 \text{ kN}$$

$$b_v = 305 \text{ mm and } \phi = 0.9 \quad \text{(for normal density concrete)}$$

Determine θ and β

$$v = \frac{V_u - \phi V_p}{\phi b_v d_v} = \frac{1\ 318\ 600 - 0.9(558\ 500)}{0.9(305)(1317.6)} = 2.26 \text{ MPa}$$

$$\frac{v}{f'_c} = 0.065$$

Assume $\theta = 26°$

$$f_{po} \cong f_{pe} = 1153 \text{ MPa} \quad \text{(conservatively)}$$

$$E_s = 200\ 000 \text{ MPa}$$

$$E_c = 0.043(2400)^{1.5}\sqrt{35} = 29\ 910 \text{ MPa}$$

$$E_p = 197\ 000 \text{ MPa (seven wire strands)}$$

$$A_s = 0 \text{ (no mild steel)}$$

$$\varepsilon_x = \frac{\dfrac{1419 \times 10^6}{1317.6} + 0.5(1\ 318\ 600)\cot 26° - (4032)(1153)}{(197\ 000)(4032)}$$

$$= -0.0028$$

Since $\varepsilon_x < 0$, reduction factor F_ε should be used.

$$F_\varepsilon = \frac{(197\,000)(4032)}{(197\,000)(4032) + (598\,275)(29\,910)} = 0.043$$

$$\varepsilon_x = (-0.0028)(0.043) = -0.12 \times 10^{-3}$$

From Figure 7.43, using $v/f'_c = 0.064$ and $\varepsilon_x = -0.12 \times 10^{-3}$,

$$\theta = 27° \text{ and } \beta = 4.8$$

$$V_c = 0.083(4.8)(\sqrt{35})(305)(1317.6) = 947.2 \text{ kN}$$

Since $\phi(V_c + V_p) \geq V_u \geq \frac{1}{2}\phi(V_c + V_p)$

$$1355 \text{ kN} \geq 1318.6 \text{ kN} \geq 677.5 \text{ kN}$$

minimum transverse reinforcement is required.

Using No. 10 stirrups; $A_v = 200 \text{ mm}^2$

spacing

$$s = \frac{A_v f_v}{0.083\sqrt{f'_c}b_v} = \frac{(200)(400)}{0.083\sqrt{35}(305)} = 534 \text{ mm}$$

Maximum stirrup spacing

$$V_u > 0.1f'_c b_v d_v = 1406.5 \text{ kN}$$

$$s \leq \begin{cases} 0.8d_v = 1054 \text{ mm} \\ 600 \text{ mm} \end{cases}$$

Use No. 10 stirrups @ 530 mm @ Location 100.44

e. *Location 205*

$$V_u = -388.05 \text{ kN} \quad \text{(from shear envelope) (Table E7.5-3)}$$
$$M_u = 6472 \times 10^6 \text{ N mm}$$

$$\text{(from moment envelope) (Table E7.5-2)}$$
$$V_p = 0 \text{ (tendons are horizontal at this location)}$$
$$b_v = 305 \text{ mm}$$

$$d_v = \frac{M_n}{C} = \frac{8440 \times 10^6}{0.99(5.353 \times 10^6)} = 1593 \text{ mm}$$

$$d_v \geq \begin{cases} 0.9d_e = 0.9(1647) = 1482 \text{ mm} \\ 0.72h = 0.72(1830) = 1317.6 \text{ mm} \end{cases}$$

$$d_v = 1593 \text{ mm}$$

$$\phi = 0.9$$

Determine θ *and* β

$$v = \frac{V_u - \phi V_p}{\phi b_v d_v} = \frac{388\ 050}{0.9(305)(1593)} = 0.89 \text{ MPa}$$

$$\frac{v}{f'_c} = 0.025$$

Assume $\theta = 27°$

$$\varepsilon_x = \frac{\dfrac{6472 \times 10^6}{1593} + 0.5(388\ 050) \cot 27° - (4032)(1152)}{(200\ 000)(739) + (197\ 000)(4032)}$$

$$= -0.21 \times 10^{-3}$$

where $f_{po} \cong f_{pe} = 1152$ MPa

Since $\varepsilon_x < 0$, reduction factor F_ε should be used.

$$F_\varepsilon = \frac{E_s A_s + E_p A_{ps}}{E_c A_c + E_s A_s + E_p A_p}$$

$$F_\varepsilon = \frac{(200\ 000)(739) + (197\ 000)(4032)}{(200\ 000)(739) + (197\ 000)(4032) + (29\ 910)(598\ 275)}$$

$$= 0.05$$

$$\varepsilon_x = (-0.21 \times 10^{-3})(0.05) = -1.0 \times 10^{-5}$$

From Figure 7.43, using $v/f'_c = 0.025$ and $\varepsilon_x = -1.0 \times 10^{-5}$,

$$\theta = 27.2° \text{ and } \beta = 4.8$$

$$V_c = 0.083(4.8)(\sqrt{35})(305)(1593) = 1145.2 \text{ kN}$$

Because V_u = 388.05 kN < 1/2 $\phi(V_c + V_p)$ = 515.3 kN, shear reinforcement is not required in the vicinity of this location.

O. Closing Remarks

The design of a concrete box-girder bridge for flexure and shear has been presented. Many details regarding the placement of prestressed and non-prestressed reinforcement in the webs and flanges are not discussed for lack of space. For more information, the reader is referred to the detailing recommendations of Schlaich and Sheff (1982).

REFERENCES

AASHTO (1996). *Standard Specifications for Highway Bridges,* 16th ed., American Association of State Highway and Transportation Officials, Washington, DC.

AASHTO (1994). *LRFD Bridge Design Specifications,* 1st ed., American Association of State Highway and Transportation Officials, Washington, DC.

ACI Committee 215 (1992). "Considerations for Design of Concrete Structures Subjected to Fatigue Loading," *ACI 215R-74 (Revised 1992),* American Concrete Institute, Detroit, MI, 24 pp.

Adebar, P. and M. P. Collins (1994). "Shear Design of Concrete Offshore Structures," *ACI Structural Journal,* Vol. 91, No. 3, May–June, pp. 324–335.

AISC (1994). *Load and Resistance Factor Design,* Manual of Steel Construction, Vol. I, 2nd ed., American Institute of Steel Construction, Chicago, IL.

Buth, C. E., T. J. Hirsh, and C. F. McDevitt (1990). "Performance Level 2 Bridge Railings," *Transportation Research Record 1258,* Transportation Research Board, Washington, DC.

Calloway, B. R. (1993). "Yield Line Analysis of an AASHTO New Jersey Concrete Parapet Wall," *M.S. Thesis,* Virginia Polytechnic Institute and State University, Blacksburg, VA.

Collins, M. P. and D. Mitchell (1991). *Prestressed Concrete Structures.* Prentice-Hall, Englewood Cliffs, NJ.

Collins, M. P. (1994). Personal communication.

Fayyaz, T. (1994). "A Study of Modified Compression Field Theory," *M.S. Thesis,* unpublished, Virginia Polytechnic Institute and State University, Blacksburg, VA.

Gopalaratnam, V. S. and S. P. Shah (1985). "Softening Response of Plain Concrete in Direct Tension," *ACI Journal,* Vol. 82, No. 3, May–June, pp. 310–323.

Green, K. L. and M. K. Tadros (1994). "The NU Precast/Prestressed Concrete Bridge I-Girder Series," *PCI Journal,* Vol. 39, No. 3, May–June, pp. 26–39.

Hirsch, T. J. (1978). "Analytical Evaluation of Texas Bridge Rails to Contain Buses and Trucks, *Research Report 230-2,* Texas Transportation Institute, Texas A&M University, College Station, TX, August.

Holzer, S. M., R. J. Melosh, R. M. Barker, and A. E. Somers (1975). "SINGER. A Computer Code for General Analysis of Two-Dimensional Reinforced Concrete

Structures," *AFWL-TR-74-228 Vol. 1,* Air Force Weapons Laboratory, Kirtland AFB, New Mexico.

Hsu, T. T. C. (1993). *Unified Theory of Reinforced Concrete,* CRC Press, Boca Raton, FL.

Loov, R. E. (1988). "A General Equation for the Steel Stress for Bonded Prestressed Concrete Members," *PCI Journal,* Vol. 33, No. 6, Nov.–Dec., pp. 108–137.

Magura, D., M. A. Sozen, and C. P. Siess (1964). "A Study of Stress Relaxation in Prestressing Reinforcement," *PCI Journal,* Vol. 9, No. 2, Mar.–Apr., pp. 13–57.

Mander, J. B., M. J. N. Priestley, and R. Park (1988). "Theoretical Stress–Strain Model for Confined Concrete," *Journal of Structural Engineering,* ASCE, 114(8), pp. 1804–1826.

Mörsch, E. (1908). *Concrete-Steel Construction,* English translation by E. P. Goodrich, The Engineering News Publishing Company, New York, 1909, 368 pp. [Translation from 3rd ed. (1908) of *Der Eisenbetonbau,* 1st ed., 1902.]

Naaman, A. E. et al. (1990). Discussion of Skogman et al. (1988), "Ductility of Reinforced and Prestressed Concrete Flexural Members," *PCI Journal,* Vol. 35, No. 2, Mar.–Apr., pp. 82–89.

Naaman, A. E. and F. M. Alkhairi (1991). "Stress at Ultimate in Unbonded Prestressing Tendons—Part II: Proposed Methodology," *ACI Structural Journal,* Vol. 88, No. 6, Nov.–Dec., pp. 683–692.

Naaman, A. E. (1992). "Unified Design Recommendations for Reinforced, Prestressed, and Partially Prestressed Concrete Bending and Compression Members," *ACI Structural Journal,* Vol. 89, No. 2, Mar.–Apr., pp. 200–210.

Nawy, E. G. (1989). *Prestressed Concrete: A Fundamental Approach,* Prentice Hall, Englewood Cliffs, NJ.

Precast/Prestressed Concrete Institute (1992). *PCI Design Handbook: Precast and Prestress Concrete,* 4th ed., PCI, Chicago, IL.

Richart, F. E., A. Brandtzaeg, and R. L. Brown (1928). "A Study of the Failure of Concrete Under Combined Compressive Stresses," *Bulletin 185,* University of Illinois Engineering Experiment Station, Champaign, IL, 104 pp.

Sawyer, H. A., Jr. (1964). "Design of Concrete Frames for Two Failure Stages," *Proceedings International Symposium on Flexural Mechanics of Reinforced Concrete,* Miami, FL, pp. 405–437.

Schlaich, J. and H. Sheef (1992). "Concrete Box Girder Bridges," *Structural Engineering Documents,* International Associates for Bridge and Structural Engineering, Zurich.

Scott, R. H. and P. A. T. Gill (1987). "Short-Term Distributions of Strain and Bond Stress along Tension Reinforcement," *Structural Engineer,* Vol. 658, No. 2, June, pp. 39–48.

Setunge, S., M. M. Attard, and P. LeP. Darvall (1993). "Ultimate Strength of Confined Very High-Strength Concretes," *ACI Structural Journal,* Vol. 90, No. 6, Nov.–Dec., pp. 632–641.

Sheikh, S. A. and S. M. Uzumeri (1980). "Strength and Ductility of Tied Concrete Columns," *J. Structural Division,* ASCE, Vol. 106, No. ST5, May, pp. 1079–1102.

Sheikh, S. A. (1982). "A Comparative Study of Confinement Models," *ACI Journal,* Proceedings Vol. 79, No. 4, pp. 296–306.

Skogman, B. C., M. K. Tadros, and R. Grasmick (1988). "Ductility of Reinforced and Prestressed Concrete Flexural Members," *PCI Journal,* Vol. 33, No. 6, Nov.–Dec., pp. 94–107.

Tamai, S., H. Shima, J. Izumo, and H. Okamura (1987). "Average Stress–Strain Relationship in Post Yield Range of Steel Bar in Concrete," *Concrete Library International of JSCE,* No. 11, June 1988, pp. 117–129. (Translation from *Proceedings of JSCE,* No. 378/V-6, Feb. 1987.)

Timoshenko, S. and J. N. Goodier (1951). *Theory of Elasticity,* 2nd ed., McGraw-Hill, New York.

Troxell, G. E., H. E. Davis and J. W. Kelly (1968). *Composition and Properties of Concrete,* 2nd ed., McGraw-Hill, New York.

Vecchio, F. J. and M. P. Collins (1986). "The Modified Compression Field Theory for Reinforced Concrete Elements Subjected to Shear," *ACI Journal,* Vol. 83, No. 2, Mar–Apr, pp. 219–231.

Vecchio, F. J. and M. P. Collins (1988). "Predicting the Response of Reinforced Concrete Beams Subjected to Shear Using Modified Compression Field Theory," *ACI Structural Journal,* Vol. 85, No. 3, May–June, pp. 258–268.

Walraven, J. C. (1981). "Fundamental Analysis of Aggregate Interlock," *Journal of the Structural Division,* ASCE, Vol. 107, No. ST11, Nov., pp. 2245–2270.

Zia, P., H. K. Preston, N. L. Scott, and E. B. Workman (1979). "Estimating Prestress Losses," *Concrete International: Design and Construction,* Vol. 1, No. 6, June, pp. 32–38.

8

STEEL BRIDGES

8.1 INTRODUCTION

Steel bridges have a long and proud history. Their role in the expansion of the railway system in the United States cannot be overestimated. The development of the long-span truss bridge was in response to the need of railroads to cross waterways and deep ravines without interruption. It was fortunate that analysis methods for trusses (particularly graphical statics) had been developed at the same time the steel industry was producing plates and cross sections of dependable strength. The two techniques came together and resulted in a figurative explosion of steel truss bridges as the railroads pushed westward.

Steel truss bridges continue to be built today, for example, the Greater New Orleans Bridge No. 2 of Figure 2.44. However, with advances in methods of analysis and steel-making technology, the sizes, shapes, and forms of steel bridges are almost unlimited. We now have steel bridges of every type imaginable: arches (tied and otherwise), plate girders (haunched and plain), box girders (curved and straight), rolled beams (composite and noncomposite), cable-stayed and suspension systems. More complete descriptions of these various bridge types are given in Chapter 2.

Emphasis in this book is on short (up to 15 m) to medium (up to 50 m) span bridges. For these span lengths, steel girder bridges are a logical choice. Composite rolled beams, perhaps with cover plates, for the shorter spans and built-up composite plate girders for the longer spans. These steel girder bridges are readily adapted to different terrain and alignment, and can be erected in a relatively short time with minimum interruption of traffic.

In the sections that follow, the properties of the materials are described, limit states are presented, resistance considerations are discussed, and this chapter concludes with design examples of rolled beam and plate girder bridges.

8.2 MATERIAL PROPERTIES

As we discussed at the beginning of Chapter 7, and as we showed in Figure 7.1, the material–stress–strain response is the essential element relating forces and deformations to one another. At one time, there was basically a simple stress–strain curve that described the behavior of structural steel, which is no longer true because additional steels have been developed to meet specific needs such as improved strength, better toughness, corrosion resistance, and ease of fabrication.

Before presenting the stress–strain curves of the various steels, it is important to understand what causes the curves to differ from one another. Therefore, a brief description of the manufacturing process is given. It will be shown that the different properties are a result of a combination of chemical composition and the physical treatment of the steel [Dowling et al. (1992)]. In addition to knowledge of the stress–strain behavior, a steel bridge designer must also understand how fatigue and fracture resistance are affected by the selection of material, member sizes, and connection details. These topics are discussed in Section 8.2.

In comparing the properties of different steels, the terms strength (yield and tensile), ductility, hardness, and toughness are used. Distinctions between these terms are given in the definitions that follow:

Yield strength   is the stress at which an increase in strain occurs without an increase in stress.

Tensile strength   is the maximum stress reached in a tensile test.

Ductility   is an index of the ability of the material to withstand inelastic deformations without fracture. It can be expressed as a ratio of elongation at fracture to the elongation at first yield.

Hardness   refers to the resistance to surface indentation from a standard indenter.

Toughness   is the ability of a material to absorb energy without fracture.

8.2.1 Steelmaking Process

The typical raws materials for making steel are iron ore, coke, limestone, and chemical additives. These are the basic constituents and the chemical admixtures that produce custom-designed products for specific applications, much like the process used for making concrete. However, in the case of steelmaking, it is possible to better control the process and produce a more uniformly predictable finished product.

The raw materials are placed in a ceramic-lined blast furnace and external heat is applied. The coke provides additional heat and carbon for reducing the iron ore to metallic iron. The limestone acts as a flux that combines with the impurities and accumulates on top of the liquid iron where it can be readily removed as fluid slag. The molten iron is periodically removed from the bottom of the furnace through tap holes into transfer ladles. The ladles then transfer the liquid metal to the steelmaking area.

Steel is an alloy. It is produced by combining the molten iron with other elements to give specific properties for different applications. Depending on the steel manufacturer, this can be done in a basic oxygen furnace, an open hearth furnace, or an electric-arc furnace. At this point, the molten iron from the blast furnace is combined with steel scrap and various fluxes. Oxygen is blown into the molten metal to convert the iron into steel by oxidation. The various fluxes are often other elements added to combine with the impurities and reduce the sulfur and phosphorus contents. The steel produced flows out a tap hole and into a ladle.

The ladle is used to transport the liquid steel to either ingot molds or a continuous casting machine. While the steel is in the ladle, its chemical composition is checked and adjustments to the alloying elements are made as required. Because of the importance of these alloying elements in classifying structural steels, their effect on the behavior and characteristics of carbon and alloy steels are summarized in Table 8.1.

Aluminum and silicon are identified as deoxidizers or "killers" of molten steel. They stop the production of carbon monoxide and other gases that are expelled from the molten metal as its solidifies. Killed steel products are less porous and exhibit a higher degree of uniformity than nonkilled steel products.

Carbon is the principal strengthening element in steel. However, it has a downside. Increased amounts of carbon cause a decrease in ductility, toughness, and weldability.

Chromium and copper both increase the atmospheric corrosion resistance and are used in weathering steels. When exposed to the atmosphere, they build up a tight protective oxide film that tends to resist further corrosion.

Sulfur is generally considered an undesirable element except where machinability is important. It adversely affects surface quality and decreases ductility, toughness, and weldability.

Manganese can control the harmful effects of sulfur by combining with it to form manganese sulfides. It also increases the hardness and strength of steels, but to a lesser extent than does carbon.

8.2.2 Production of Finished Products

The liquid steel from the ladle is placed in ingot molds or a continuous casting machine. Steel placed in the ingot molds is solidified as it cools. It then goes into a second process where the ingot is hot-worked into slabs (up to 230 mm thick × 1520 mm wide), blooms (up to 300 mm × 300 mm), and billets (up to 125 mm × 125 mm).

TABLE 8.1 Effects of Alloying Elements[a]

Elements	Effects
Aluminum (Al)	Deoxidizes or "kills" molten steel. Refines grain size; increases strength and toughness.
Boron (B)	Small amounts (0.0005%) increase hardenability in quenched-and-tempered steels. Used only in aluminum-killed steels. Most effective at low carbon levels.
Calcium (Ca)	Controls shape of nonmetallic inclusions.
Carbon (C)	Principal hardening element in steel. Increases strength and hardness. Decreases ductility, toughness and weldability. Moderate tendency to segregate.
Chromium (Cr)	Increases strength and atmospheric corrosion resistance.
Copper (Cu)	Increases atmospheric corrosion resistance.
Manganese (Mn)	Increases strength. Controls harmful effects of sulfur.
Molybdenum (Mo)	Increases high temperature tensile and creep strength.
Niobium (Nb)	Increases toughness and strength.
Nickel (Ni)	Increases strength and toughness.
Nitrogen (N)	Increases strength and hardness. Decreases ductility and toughness.
Phosphorus (P)	Increases strength and hardness. Decreases ductility and toughness. Increases atmospheric corrosion resistance. Strong tendency to segregate.
Silicon (Si)	Deoxidizes or "kills" molten steel.
Sulfur (S)	Considered undesirable except for machinability. Decreases ductility, toughness and weldability. Adversely affects surface quality. Strong tendency to segregate.
Titanium (Ti)	Increases creep and rupture strength and hardness.
Vanadium (V) and Columbium (Nb)	Small additions increase strength.

[a] Brockenbrough and Barsom, 1992
[b] Abbreviation symbol used for the element is in parentheses.

In the continuous casting process, gravity is utilized to directly form slabs, blooms, and billets from a reservoir of liquid steel as shown in Figure 8.1. This process is becoming the dominant production method because it yields better steel than that from ingots at reduced cost.

The slabs are reheated and squeezed between sets of horizontal rolls in a plate mill to reduce the thickness and produce finished plate products. The longitudinal edges are often flame-cut on-line to provide the desired plate

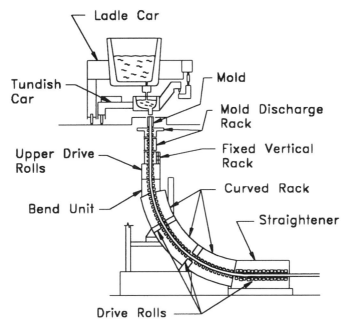

Fig. 8.1 Section schematic of a continuous caster. (Brockenbrough and Barsom, 1992.) [From *Constructional Steel Design: An International Guide,* Edited by P. J. Dowling, J. E. Harding, and R. Bjorhovde, Copyright © 1992 by Elsevier Science Ltd (now Chapman and Hall, Andover, England), with permission.]

width. After passing through leveling rolls, the plates are sheared to length. Heat treating can be done on-line or off-line.

The blooms are reheated and passed sequentially through a series of roll stands in a structural mill to produce wide-flange sections, *I*- beams, channels, angles, tees, and zees. There are four stages of roll stands, each with multiple passes that are used to reduce the bloom to a finished product. They are a breakdown stand, a roughing stand, an intermediate stand, and a finishing stand. Each stand has horizontal and vertical rolls, and in some cases edge rolls, to reduce the cross section progressively to its final shape. The structural section is cut to length, set aside to cool, and straightened by pulling or rolling.

8.2.3   Residual Stresses

Stresses that exist in a component without any applied external forces are called *residual stresses*. It is important to understand their presence, because they affect the strength of members in tension, compression, and bending. They can be induced by thermal, mechanical, or metallurgical processes. Thermally induced residual stresses are caused by nonuniform cooling under restraint. In general, tensile residual stresses develop in the metal that cools last.

Mechanically induced residual stresses are caused by nonuniform plastic deformations when a component is stretched or compressed under restraint. This nonuniform deformation can occur when a component is mechanically straightened after cooling or mechanically curved by a series of rollers.

Metallurgically induced residual stresses are caused by a change in the microstructure of the steel from fermite–pearlite to martensite (Brockenbrough and Barsom, 1992). This new material is stronger and harder than the original steel, but it is less ductile. The change to martensite results in a 4% increase in volume when the surface is heated to about 900°C and then cooled rapidly. If the volume change due to the transformation to marteniste is restrained, the residual stresses will be compressive. The tensile residual stresses induced by thermal cut edges can be partially compensated by the compressive stresses produced by the transformation.

When cross sections are fabricated by welding, complex three-dimensional (3-D) residual stresses are induced by all three processes. Heating and cooling effects take place, metallurgical changes can occur, and deformation is often restrained. High tensile residual stresses of approximately 400 MPa can be developed at a weld (Bjorhovde, 1992).

In general, rolled edges of plates and shapes are under compressive residual stress while thermally cut edges are in tension. These stresses are balanced by equivalent stresses of opposite sign elsewhere in the member. For welded members, tensile residual stresses develop near the weld and equilibrating compressive stresses elsewhere. Figure 8.2 presents simplified qualitative illustrations of the global distribution of residual stresses in as-received and welded hot-rolled steel members (Brockenbrough and Barsom, 1992). Note that the stresses represented in Figure 8.2 are lengthwise or longitudinal stresses in the members.

8.2.4 Heat Treatments

Improved properties of steel can be obtained by various heat treatments. There are slow cooling heat treatments and rapid cooling heat treatments. Slow cooling treatments are annealing, normalizing, and stress relieving. They consist of heating the steel to a given temperature, holding for a proper time at that temperature, followed by slow cooling in air. The temperature to which the steel is heated determines the type of treatment. The slow cooling treatments will improve ductility and fracture toughness, reduce hardness, and relieve residual stresses.

Rapid cooling heat treatments are indicated for the bridge steels in the AASHTO (1994) LRFD Bridge Specifications. The process is called quenching-and-tempering and consists of heating the steel to about 900°C, holding the temperature for a period of time, then rapid cooling by quenching in a bath of oil or water. After quenching, the steel is tempered by reheating to

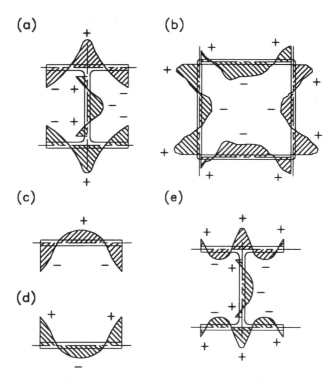

Fig. 8.2 Schematic illustration of residual stresses in as-rolled and fabricated structural components. (a) Hot-rolled shape, (b) welded box section, (c) plate with rolled edges, (d) plate with flame-cut edges, and (e) beam fabricated from flame-cut plates. (Brockenbrough and Barsom, 1992.) [From *Constructional Steel Design: An International Guide,* Edited by P. J. Dowling, J. E. Harding, and R. Bjorhovde, Copyright © 1992 by Elsevier Science Ltd (now Chapman and Hall, Andover, England), with permission.]

about 500°C, holding that temperature, then slowly cooling. Quenching-and-tempering changes the microstructure of the steel and increases its strength, hardness, and toughness.

8.2.5 Classification of Structural Steels

Mechanical properties of typical structural steels are depicted by the four stress–strain curves shown in Figure 8.3. Each of these curves represent a structural steel with specific composition to meet a particular need. It is obvious that their behavior differs from one another except for small strains near the origin. These four different steels can be identified by their chemical composition and heat treatment as (a) structural carbon steel (Grade 250), (b) high-strength low-alloy steel (Grade 345), (c) quenched-and-tempered low-alloy steel (Grade 485), and (d) high yield strength, quenched and tempered

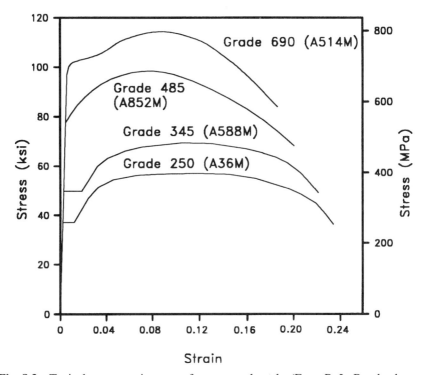

Fig. 8.3 Typical stress–strain curves for structural steels. (From R. L. Brockenbrough and B. G. Johnston, *USS Steel Design Manual,* Copyright © 1981 by R. L. Brockenbrough & Assoc, Inc., Pittsburgh, PA, with permission.)

alloy steel (Grade 690). The minimum mechanical properties of these steels are given in Table 8.2.

A unified standard specification for bridge steel is given in ASTM (1995) with the designation A709/A709M-94a (M indicates metric and 94a is the year of last revision). Six grades of steel are available in four yield strength levels as shown in Table 8.2 and Figure 8.3. Steel grades with suffix "W" indicates weathering steels that provide a substantially better atmospheric corrosion resistance than typical carbon steel and can be used unpainted for many applications.

All of the steels in Table 8.2 can be welded, but not by the same welding process. Each steel grade has specific welding requirements that must be followed.

In Figure 8.3, the number in parentheses identifying the four yield strength levels is the ASTM designation of the steel with similar tensile strength and elongation properties as the A709M steel. These numbers are given because they are familiar to designers of steel buildings and other structures. The most significant difference between these steels and the A709M steels is that the A709M steels are specifically for bridges and must meet supplementary re-

TABLE 8.2 Minimum Mechanical Properties of Structural Steel By Shape, Strength, and Thickness[a]

	Structural Steel	High-Strength Low-Alloy Steel		Quenched and Tempered Low-Alloy Steel[b]	High-Yield Strength, Quenched and Tempered Alloy Steel[b]	
AASHTO designation	M270 Grade 250	M270 Grade 345	M270 Grade 345W	M270 Grade 485W	M270 Grades 690/690W	
Equivalent ASTM designation	A709M Grade 250	A709M Grade 345	A709M Grade 345W	A709M Grade 485W	A709M Grades 690/690W	
Thickness of plates (mm)	Up to 100 incl.	Up to 100 incl.	Up to 100 incl.	Up to 100 incl.	Up–65 incl.	Over 65–100 incl.
Shapes	All groups	All groups	All groups	N/A	N/A	N/A
Minimum tensile strength, F_u, (MPa)	400	450	485	620	760	690
Minimum yield point or minimum yield strength, F_y, (MPa)	250	345	345	485	690	620

[a] In AASHTO Table 6.4.1-1. [From *AASHTO LRFD Bridge Design Specifications*, Copyright © 1994 by the American Association of State Highway and Transportation Officials. Used by permission.]
[b] Not applicable = N/A.

quirements for toughness testing. These requirements vary for nonfracture critical and fracture critical members. This concept is discussed in Section 8.2.6.

As discussed earlier, steel is an alloy and its principal component is iron. The chemical composition of the steel grades in Table 8.2 are given in Table 8.3. One component of all the structural steels is carbon which, as indicated in Table 8.1, increases strength and hardness but reduces ductility, toughness, and weldability. Other alloying elements are added to offset the negative effects and to custom design a structural steel for a particular application. Consequently, more than one type of steel is given in A709M for each yield strength level to cover the proprietary steels produced by different manufacturers. In general, a low-alloy steel has less than 1.5% total alloy elements while alloy steels have a larger percentage.

Comparing the chemical composition of bridge steels in Table 8.3 with the effects of the alloying elements in Table 8.1 shows the following relationships. Boron is added only to the quenched-and-tempered alloy steel to increase hardenability. Carbon content decreases in the higher strength steels and manganese, molybdenum, and vanadium are added to provide the increase in strength. Chromium, copper, and nickel are found in the weathering steels and contribute to their improved atmospheric corrosion resistance. Phosphorus helps strength, hardness, and corrosion but hurts ductility and toughness so its content is limited. Sulfur is considered undesirable so its maximum percentage is severely limited. Silicon is the deoxidizing agent which kills the molten steel and produces more uniform properties.

Two properties of all grades of structural steels are assumed to be constant. They are the modulus of elasticity E_s of 200 GPa and the coefficient of thermal expansion of 11.7×10^{-6} mm/mm/°C [A6.4.1].*

A brief discussion of the properties associated with each of the four levels of yield strength are given below (Brockenbrough and Johnston, 1981). To aid in the comparison between the different steels, the initial portions of their stress–strain curves and time-dependent corrosion curves are given in Figures 8.4 and 8.5, respectively.

Structural Carbon Steel The name is somewhat misleading because all structural steels contain carbon. When reference is made to carbon steel, the technical definition is usually implied. The criteria for designation as carbon steel are (AISI, 1985): (1) no minimum content is specified for chromium, cobalt, columbium, molybdenum, nickel, titanium, tungsten, vanadium, or zirconium, or any other element added to obtain a desired alloying effect; (2) the specified minimum of copper does not exceed 0.40%; or (3) the specified maximum for any of the following is not exceeded: manganese 1.65%, silicon

---

* The article numbers in the AASHTO (1994) LRFD Bridge Specifications are enclosed in brackets and preceded by the letter A if specifications and by C if commentary.

TABLE 8.3 Chemical Requirements for Bridge Steels. Heat Analysis, Percent[a,b]

Element	Carbon Steel Grade 250 Shapes	High-Strength Grade 345 Type 2	Low-Alloy Steel Grade 345W Type A	Heat-Treated, Low-Alloy Steel Grade 485	High-Strength, Heat-Treated Alloy Steel	
					Grade 690 Type C	Grade 690W Type F
Boron					0.001–0.005	0.0005–0.006
Carbon	0.26 max	0.23 max	0.19 max	0.19 max	0.12–0.20	0.10–0.20
Chromium			0.40–0.65	0.40–0.70		0.40–0.65
Copper	0.20 min	0.20 min	0.25–0.40	0.20–0.40		0.15–0.50
Manganese		1.35 max	0.80–1.25	0.80–1.35	1.10–1.50	0.60–1.00
Molybdenum					0.15–0.30	0.40–0.60
Nickel			0.40 max	0.50 max		0.70–1.00
Phosphorous	0.04 max	0.04 max	0.04 max	0.035 max	0.035 max	0.035 max
Silicon	0.40 max	0.40 max	0.30–0.65	0.20–0.65	0.15–0.30	0.15–0.35
Sulfur	0.05 max	0.05 max	0.05 max	0.04 max	0.035 max	0.035 max
Vanadium		0.01–0.15	0.02–0.10	0.02–0.10		0.03–0.08

[a]From ASTM, 1995.
[b]Note: Where a blank appears in this table there is no requirement.

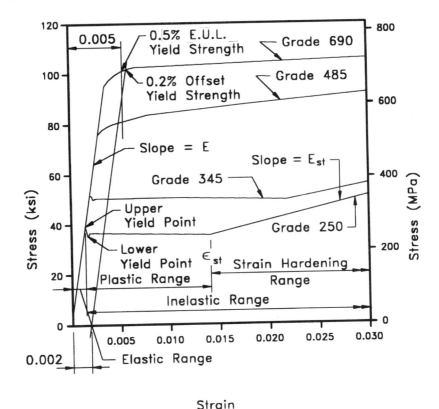

Strain

Fig. 8.4 Typical initial stress–strain curves for structural steels. (From R. L. Brockenbrough and B. G. Johnston, *USS Steel Design Manual,* Copyright © 1981 by R. L. Brockenbrough & Assoc, Inc., Pittsburgh, PA, with permission.)

0.60%, and copper 0.60%. In other words, a producer can use whatever scrap steel or junked automobiles are available to put in the furnace as long as the minimum mechanical properties of Table 8.2 are met. No exotic or fancy ingredients are necessary to make it strong. As a result, engineers often refer to it as *mild* steel.

One of the main characteristics of structural carbon steel is a well-defined yield point (F_y = 250 MPa) followed by a generous yield plateau in the plastic range. This result is shown in Figure 8.4 and indicates good ductility, which allows redistribution of local stresses without fracture. This property makes carbon steel especially well suited for connections.

Carbon steels are weldable and available as plates, bars, and structural shapes. They are intended for service at atmospheric temperature. The corrosion rate in Figure 8.5 for copper bearing carbon steel (0.20% minimum, Table 8.3) is about one-half that of plain carbon steel.

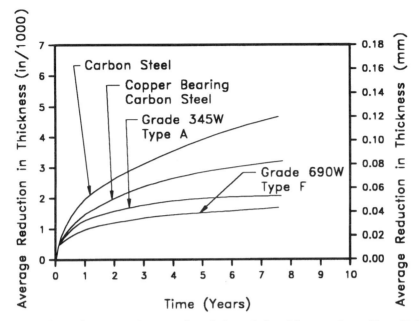

Fig. 8.5 Corrosion curves for several steels in an industrial atmosphere. (From R. L. Brockenbrough and B. G. Johnston, *USS Steel Design Manual,* Copyright © 1981 by R. L. Brockenbrough & Assoc, Inc., Pittsburgh, PA, with permission.)

High-Strength Low-Alloy Steel   These steels have controlled chemical compositions to develop yield and tensile strengths greater than carbon steel, but with alloying additions smaller than those for alloy steels (Brockenbrough, 1992). The higher yield strength (F_y = 345 MPa) is achieved in the hot-rolled condition rather than through heat treatment. As a result, they exhibit a well-defined yield point and excellent ductility as shown in Figure 8.4.

High-strength low-alloy steels are weldable and available as plates, bars, and structural shapes. These alloys also have superior atmospheric corrosion resistance as shown in Figure 8.5. Because of their desirable properties, Grade 345 steels are often the first choice of designers of small to medium span bridges.

Heat-Treated Low-Alloy Steel   High-strength low-alloy steels can be heat treated to obtain higher yield strengths (F_y = 485 MPa). The chemical composition for Grades 345W and 485W in Table 8.3 are nearly the same. The quenching-and-tempering heat treatment changes the microstructure of the steel and increases its strength, hardness, and toughness.

The heat treatment removes the well-defined yield point from the high strength steels as shown in Figure 8.4. There is a more gradual transition from elastic to inelastic behavior. The yield strength for these steels is usually determined by the 0.5% extension under load (EUL) definition or the 0.2% offset definition.

The heat-treated low-alloy steels are weldable, but are available only in plates. Their atmospheric corrosion resistance is similar to that of high-strength low-alloy steels.

High-Strength Heat-Treated Alloy Steel Alloy steels are those with chemical compositions that are not in the high-strength low-alloy classification (see Table 8.3). The quenching-and-tempering heat treatment is similar to that for the low-alloy steels, but the different composition of alloying elements develops higher strength ($F_y = 690$ MPa) and greater toughness at low temperature.

An atmospheric corrosion curve for the alloy steels (Grade 690) is given in Figure 8.5 and shows the best corrosion resistance of the four groups of steels.

Again the yield strength is determined by the 0.5% EUL definition or the 0.2% offset definition shown in Figure 8.4. By observing the complete stress–strain curves in Figure 8.3, it is obvious that the heat-treated steels reach their peak tensile strength and decrease rapidly at lower strains than the untreated steels. This lower ductility may cause problems in some structural applications and caution must be exercised when heat-treated steels are used.

The high-strength heat-treated alloy steels are weldable, but are available only in structural steel plates for bridges.

8.2.6 Effects of Repeated Stress (Fatigue)

When designing bridge structures in steel, a designer must be aware of the effect of repeated stresses. Vehicles passing any given location are repeated time and again. On a heavily traveled interstate highway with a typical mix of trucks in the traffic, the number of maximum stress repetitions can be over 1 million in a year.

These repeating stresses are produced by service loads and the maximum stresses in the base metal of the chosen cross section are less than the strength of the material. However, if there is a stress raiser due to a discontinuity in metallurgy or geometry in the base metal, the stress at the discontinuity can easily be double or triple the stress calculated from the service loads. Even though this high stress is intermittent, if it is repeated many times damage will accumulate, cracks will form, and fracture of the member can result.

This failure mechanism, which consists of the formation and growth of cracks under the action of repeated stresses, each of which is insufficient by itself to cause failure, is called fatigue (Gurney, 1992). The metal just gets tired of being subjected to moderate level stresses again and again. Fatigue is a good word to describe this phenomenon.

Determination of Fatigue Strength   Fatigue strength is not a material constant like yield strength or modulus of elasticity. It is dependent on the particular joint configuration involved and can realistically only be determined experimentally. Because most of the stress concentration problems due to discontinuities in geometry and metallurgy are associated with welded connections, most of the testing for fatigue strength has been done on welded joint configurations.

The procedure followed for each welded connection is to subject a series of identical specimens to a stress range S that is less than the yield stress of the base metal and to repeat that stress range for N cycles until the connection fails. As the stress range is reduced, the number of cycles to failure increases. The results of the tests are usually plotted as log S versus log N graphs. A typical S–N curve for a welded joint is shown in Figure 8.6. At any point on the curve, the stress value is the *fatigue strength* and the number of cycles is the *fatigue life* at that level of stress. Notice that when the stress range is reduced to a particular value, an unlimited number of stress cycles can be applied without causing failure. This limiting stress is called the fatigue limit or endurance limit of the connection.

Influence of Strength of the Base Material   The fatigue strength of unwelded components increases with the tensile strength of the base material. This fatigue strength is shown in Figure 8.7 for both solid round and notched specimens. However, if high strength steel is used in welded components, there is no apparent increase in the fatigue strength.

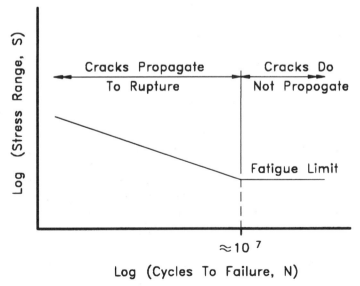

Fig. 8.6   Typical S–N curve for welded joints.

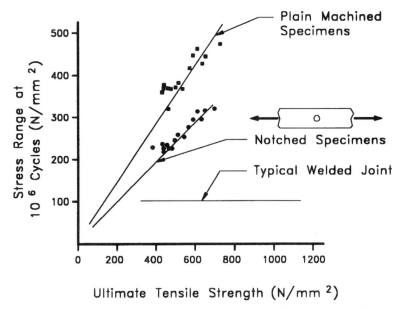

Fig. 8.7 Fatigue strength compared to static strength. (Gurney, 1992.) [From *Constructional Steel Design: An International Guide*, Edited by P. J. Dowling, J. E. Harding, and R. Bjorhovde, Copyright © 1992 by Elsevier Science Ltd (now Chapman and Hall, Andover, England), with permission.]

The explanation for the difference in behavior is that in the unwelded material cracks must first be formed before they can propagate and cause failure, while in the welded joint cracks already exist and all they need to do is propagate. Rate of crack propagation does not vary significantly with tensile strength; therefore, fatigue strength of welded joints is independent of the steel from which they are fabricated (Gurney, 1992).

Influence of Residual Stresses   In general, welded joints will not be stress relieved, so it is reasonable to assume that residual stresses σ_r will exist somewhere in the connection. If a stress cycle with range S is applied, the actual stress range will go from σ_r to $\sigma_r \pm S$, and the nominal stress range is S. Therefore, it is possible to express the fatigue behavior of a welded joint in terms of stress range alone without knowing the actual maximum and minimum values. In the AASHTO (1994) LRFD Bridge Specifications, load-induced fatigue considerations are expressed in terms of stress range and residual stresses are not considered [A6.6.1.2].

Closing Remarks on Fatigue   Fatigue is the most common cause of structural steel failure, which is largely due to the fact that the problem is not recognized at the design stage and good design is essential if a structure with high fatigue strength is to be obtained. Adequate attention to joint selection and detailing and a knowledge of service load requirements can almost eliminate the risk of failure, while ignorance of these factors can be catastrophic (Gurney, 1992).

8.2.7 Brittle Fracture Considerations

A bridge designer must understand the conditions that cause brittle fractures to occur in structural steel. Brittle fractures are to be avoided because they are nonductile and can occur at relatively low stresses. When certain conditions are present, cracks can propagate rapidly and sudden failure of a member can result.

One of the causes of a brittle fracture is a triaxial tension stress state that can be present at a notch in an element or a restrained discontinuity in a welded connection. When a ductile failure occurs, shear along slip planes is allowed to develop. This sliding between planes of the material can be seen in the necking down of the cross section during a standard tensile coupon test. The movement along the slip planes produces the observed yield plateau and increase deformation that characterizes a ductile failure. In looking at a cross section after failure, it is possible to distinguish the crystalline appearance of the brittle fracture area from the fibrous appearance of the shear plane area and its characteristic shear lip. The greater the percentage of shear area on the cross section, the greater the ductility.

In the uniaxial tension test, there is no lateral constraint to prevent the development of the shear slip planes. However, stress concentrations at a notch or stresses developed due to cooling of a restrained discontinuous weld can produce a triaxial tension state of stress in which shear cannot develop. When an impact load produces additional tensile stresses, often on the tension side caused by bending, a sudden brittle fracture occurs.

Another cause of brittle fracture is a low-temperature environment. Structural steels may exhibit ductile behavior at temperatures above 0°C, but change to brittle fracture when the temperature drops. A number of tests have been developed to measure the relative susceptibility of a steel to brittle fracture with a drop in temperature. One of these is the Charpy V-notch impact test. This test consists of a simple beam specimen with a standard size V-notch at midspan that is fractured by a blow from a swinging pendulum as shown in Figure 8.8. The amount of energy required to fracture the specimen is determined by the difference in height of the pendulum before and after striking the small beam. The fracture energy can be correlated to the percent of the cross section that fails by shear. The higher the energy the greater the percentage of shear failure. A typical plot of the results of a Charpy V-notch test with variation in temperature is given in Figure 8.9.

As seen in Figure 8.9, the energy absorbed during fracture decreases gradually as temperature is reduced until it drops dramatically at some transition temperature. The temperature at which the specimen exhibits little ductility

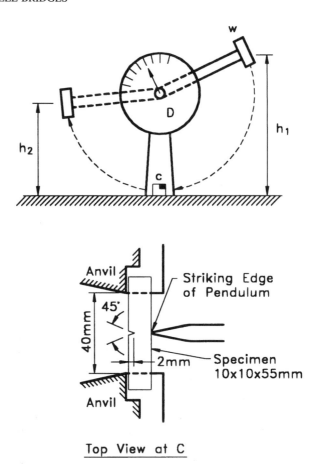

Top View at C

Fig. 8.8 Charpy V-notch impact test. (After Barsom, 1992.) [From *Constructional Steel Design: An International Guide,* Edited by P. J. Dowling, J. E. Harding, and R. Bjorhovde, Copyright © 1992 by Elsevier Science Ltd (now Chapman and Hall, Andover, England), with permission.]

is called the nil ductility transformation (NDT) temperature. The NDT temperature can be determined from the Charpy V-notch test as the temperature at a specified level of absorbed energy or the temperature at which a given percentage of the cross section fails in shear. The AASHTO (1994) LRFD Bridge Specifications give a minimum absorbed energy value of 20 N m to measure the fracture toughness of bridge steels under different temperature conditions [A6.6.2].

Welded connections must be detailed to avoid triaxial tensile stresses and the potential for brittle fracture. An example is the welded connection of intermediate stiffeners to the web of plate girders. In times past, intermediate stiffeners were full height and were often welded to both the compression and tension flanges. If the stiffener is welded to the tension flange as shown

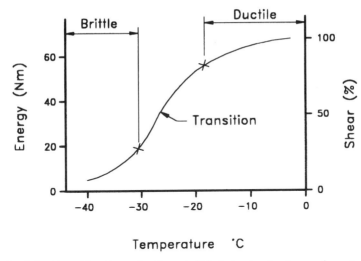

Fig. 8.9   Transition from ductile to brittle behavior for low-carbon steel.

in Figure 8.10(a), restrained cooling of welds in three directions develops triaxial tensile stresses in the web. Often a notchlike stress raiser is present in the welded connection due to material flaws, cut-outs, undercuts, or arc strikes. If principal tensile stresses due to weld residual stresses, notch stress concentrations, and flexural tension in three principal directions reach the same value, then shear stresses vanish and a brittle fracture results. If these conditions in the welded connection are accompanied by a drop in temperature, the energy required to initiate a brittle fracture drops significantly (Fig. 8.9) and the fracture can occur prematurely. Such web fractures occurred at the welded attachment of intermediate stiffeners to the tension flange of the LaFayette Street bridge during a cold winter in St. Paul, MN.

After the investigation of the web fractures in the LaFayette Street bridge, welding of intermediate stiffeners to tension flanges was no longer allowed. However, the welded connection between the stiffener and the tension flange continued to be found when inspections were made, even though not shown on the plans. Over-zealous welders had placed a fillet weld because, presumably, they thought two parts that close together should be attached.

As a result, the current specifications [A6.10.8.1.1] require that the stiffener be stopped short of the tension flange [Fig. 8.10(b)] or coped, so that they cannot be inadvertently attached.

8.3   LIMIT STATES

Structural steel bridges must be designed so that their performance under load does not go beyond the limit states prescribed by the AASHTO (1994) Bridge

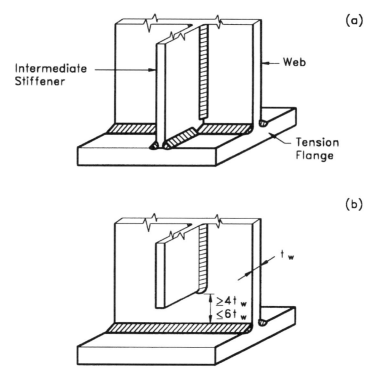

Fig. 8.10 Welded connections between an intermediate stiffener and the tension flange of a plate girder. (a) Improper detail, (b) recommended detail.

Specifications. These limit states are applicable at all stages in the life of a bridge and include service, fatigue and fracture, strength, and extreme event limit states. The condition that must be met for each of these limit states is that the factored resistance is greater than the effect of the factored load combinations, which can be expressed as

$$\phi R_n \geq \eta \sum \gamma_i Q_i \qquad (8.1)$$

where ϕ is a statistically based resistance factor for the limit state being examined; R_n is the nominal resistance; η is a load multiplier relating to ductility, redundancy and operational importance; γ_i is a statistically based load factor applied to the force effects as defined for each limit state in Table 3.1; and Q_i is a force effect. The various factors in Eq. 8.1 are discussed more fully in Chapter 3 and are repeated here only for convenience.

8.3.1 Service Limit State

Service limit states relate to the performance of a bridge subjected to the forces applied when it is put into service. In steel structures, limitations are placed on deflections and inelastic deformations under service loads. By limiting deflections, adequate stiffness is provided and vibration is reduced to an acceptable level. By controlling local yielding, permanent inelastic deformations are avoided and rideability is improved.

Because the provisions for service limit state are based on experience and engineering judgment, rather than calibrated statistically, the resistance factor ϕ, the load modifier η, and the load factors γ_i in Eq. 8.1 are taken as unity. One exception is the possibility of an overloaded vehicle that may produce excessive local stresses. For this case, the Service II limit state in Table 3.1 with a vehicle load factor of 1.30 is used.

Deflection limitations are optional. If they are required by an owner, they can be taken as the span/800 for vehicular loads [A2.5.2.6]. In calculating deflections, assumptions have to be made on load distribution to the girders, flexural stiffness of the girders in participation with the bridge deck, and stiffness contributions of attachments such as railings and concrete barriers. In general, there is more stiffness in a bridge system than is usually implied by typical engineering calculations. As a result, deflection calculations are only an estimate of the actual deflection. When this uncertainty is coupled with the subjective criteria of what constitutes an annoying vibration, the establishment of deflection limitations is not encouraged.

Inelastic deformation limitations are mandatory. Local yielding under Service *II* loads is not permitted [A6.10.3]. This local yielding will not occur for sections designed by Eq. 8.1 for a strength limit state when the maximum force effects are determined by an elastic analysis. However, if inelastic moment redistribution follows an elastic analysis [A6.10.2.2], the concept of plastic hinging is introduced and the stresses must be checked. In this case, flange stresses in positive and negative bending shall not exceed [A6.10.3.2]:

- For both steel flanges of composite sections

$$f_f \leq 0.95 R_h F_{yf} \qquad (8.2)$$

- For both steel flanges of noncomposite sections

$$f_f \leq 0.80 R_h F_{yf} \qquad (8.3)$$

where R_h is the flange-stress reduction factor for hybrid girders [A6.10.5.4.1], f_f is the elastic flange stress caused by the Service *II* loading (MPa), and F_{yf} is the yield stress of the flange (MPa). For the common case of a girder with the same steel in the web and flanges, $R_h = 1.0$. Satisfying Eq. 8.2 (or Eq. 8.3) will prevent the development of permanent deformation due to localized yielding of the flanges under an occasional service overload.

8.3.2 Fatigue and Fracture Limit State

Design for the fatigue limit state involves limiting the live load stress range produced by the fatigue truck to a value suitable for the number of stress range repetitions expected over the life of the bridge. Design for the fracture limit state involves the selection of a steel that has adequate fracture toughness (measured by Charpy V-notch test) for the expected temperature range.

Load-Induced Fatigue   When live load produces a repetitive net tensile stress at a connection detail, load-induced fatigue can occur. When cross frames or diaphragms are connected to girder webs through connection plates that restrict movement, *distortion-induced* fatigue can occur. Distortion-induced fatigue is important in many cases; however, the following discussion is only for load-induced fatigue.

As discussed in Section 8.2.6, fatigue life is determined by the tensile stress range in the connection detail. By using the stress range as the governing criteria, the values of the actual maximum and minimum tensile stresses need not be known. As a result, residual stresses are not a factor and are not to be considered [A6.6.1.2.1].

The tensile stress range is determined by considering placement of the fatigue truck live load in different spans of a bridge. If the bridge is a simple span, there is only a maximum tensile stress, the minimum stress is zero. In calculating these stresses, a linear elastic analysis is used.

In some regions along the span of a girder, the compressive stresses due to unfactored permanent loads (e.g., dead loads) are greater than the tensile live load stresses due to the fatigue truck with its load factor taken from Table 3.1. However, before fatigue can be ignored in these regions, the compressive stress must be at least twice the tensile stress, because the heaviest truck expected to cross the bridge is approximately twice the fatigue truck used in calculating the tensile live load stress [A6.6.1.2.1].

Fatigue Design Criteria   If we write Eq. 8.1 in terms of fatigue load and fatigue resistance, each connection detail must satisfy

$$\phi\,(\Delta F)_n \geq \eta\gamma(\Delta f) \qquad (8.4)$$

where $(\Delta F)_n$ is the nominal fatigue resistance (MPa) and (Δf) is the live load stress range due to the fatigue truck (MPa). For the fatigue limit state $\phi = 1.0$ and $\eta = 1.0$, so that Eq. 8.4 becomes

$$(\Delta F)_n \geq \gamma(\Delta f) \qquad (8.5)$$

where γ is the load factor in Table 3.1 for the fatigue limit state.

Fatigue Load   As discussed in Chapter 4, the fatigue load is a single design truck with a front axle spacing of 4300 mm, a rear axle spacing of 9000 mm, a dynamic load allowance of 15%, and a load factor of 0.75. Because fatigue resistance depends on the number of accumulated stress-range cycles, the frequency of application of the fatigue load must also be known. The frequency of the fatigue load shall be taken as the single-lane average daily truck traffic, ADTT_{SL} [A3.6.1.4.2]. Unless a traffic survey has been conducted, the single-lane value can be estimated from the average daily truck traffic ADTT by

$$\text{ADTT}_{\text{SL}} = p \times \text{ADTT} \qquad (8.6)$$

where p is the fraction of multiple lanes of truck traffic in a single lane taken from Table 4.3. If only the average daily traffic ADT is known, the ADTT can be determined by multiplying by the fraction of trucks in the traffic (Table 4.4). An upper bound on the total number of cars and trucks is about 20 000 vehicles per lane per day and can be used to estimate ADT.

The number N of stress-range cycles to be considered are those due to the trucks anticipated to cross the bridge in the most heavily traveled lane during its design life. For a 75-year design life, this can be expressed as

$$N = (365)(75)n(\text{ADTT}_{\text{SL}}) \qquad (8.7)$$

where n is the number of stress-range cycles per truck passage taken from Table 8.4. The values of n greater than 1.0 indicate additional cycles due to vibration that occurs after the truck has left the bridge.

EXAMPLE 8.1

Estimate the number of stress-range cycles N to be considered in the fatigue design of a two-lane, 10 670-mm simple span bridge that carries in one

TABLE 8.4 Cycles per Truck Passage, n^a

	Span Length	
Longitudinal Members	>12 000 mm	≤12 000 mm
Simple-span girders	1.0	2.0
Continuous girders		
1. Near interior support	1.5	2.0
2. Elsewhere	1.0	2.0
Cantilever girders	5.0	
Trusses	1.0	
	Spacing	
	>6000 mm	≤6000 mm
Transverse members	1.0	2.0

^{a}In AASHTO Table 6.6.1.2.5-2. [From *AASHTO LRFD Bridge Design Specifications*, Copyright © 1994 by the American Association of State Highway and Transportation Officials, Washington, DC. Used by permission.]

direction rural interstate traffic. Use an ADT of 20 000 vehicles per lane per day.

Table 4.4 gives 0.20 as the fraction of trucks in rural interstate traffic, so that

$$\text{ADTT} = 0.20(2)(20\ 000) = 8000 \text{ trucks/day}$$

Table 4.3 gives 0.85 as the fraction of truck traffic in a single lane when two lanes are available to trucks, thus Eq. 8.6 yields

$$\text{ADTT}_{\text{SL}} = 0.85(8000) = 6800 \text{ trucks/day}$$

Table 8.4 gives $n = 2.0$ as the cycles per truck passage and Eq. 8.7 results in

$$N = (365)(75)(2.0)(6800) = \underline{372 \times 10^6 \text{ cycles}}$$

Detail Categories Components and details susceptible to load-induced fatigue are grouped into eight categories according to their fatigue resistance [A6.6.1.2.3]. The categories are assigned letter grades with A being the best and E' the worst. The A and B detail categories are for plain members and well-prepared welded connections in built-up members without attachments, usually with the weld axis in the direction of the applied stress. The D and E detail categories are assigned to fillet-welded attachments and groove-welded attachments without adequate transition radius or with unequal plate thicknesses. Category C can apply to welding of attachments by providing a transition radius greater than 150 mm and proper grinding of the weld. The requirements for the various detail categories are summarized in Table 8.5 and illustrated in Figure 8.11.

Fatigue Resistance As shown in the typical *S–N* curve of Figure 8.6, fatigue resistance is divided into two types of behavior, one that gives infinite life and the other a finite life. If the tensile stress range is below the fatigue limit or threshold stress, additional loading cycles will not propagate fatigue cracks and the connection detail will have a long life. If the tensile stress range is above the threshold stress, then fatigue cracks can propagate and the detail has a finite life. This general concept of fatigue resistance is expressed for specific conditions by the following [A6.6.1.2.5]:

$$(\Delta F)_n = \left(\frac{A}{N}\right)^{\frac{1}{3}} \geq \frac{1}{2}(\Delta F)_{TH} \tag{8.8}$$

where $(\Delta F)_n$ is the nominal fatigue resistance (MPa), A is a detail category constant taken from Table 8.6 (MPa)^3, N is the number of stress-range cycles from Eq. 8.7, and $(\Delta F)_{TH}$ is the constant-amplitude fatigue threshold stress taken from Table 8.6 (MPa).

TABLE 8.5 Detail Categories for Load-Induced Fatigue[a]

General Condition	Situation	Detail Category	Illustrative Example, See Figure 8.11
Plain members	Base metal:		1, 2
	• With rolled or cleaned surfaces. Flame-cut edges with ANSI/AASHTO/AWS D5.1 (Section 3.2.2) smoothness of 0.025 mm or less	A	
	• Of unpainted weathering steel, all grades, designed and detailed in accordance with FHWA (1990)	B	
	• At net section of eyebar heads and pin plates	E	
Built-up members	Base metal and weld metal in components, without attachments, connected by		3, 4, 5, 7
	• Continuous full-penetration groove welds with backing bars removed, or	B	
	• Continuous fillet welds parallel to the direction of applied stress	B	
	• Continuous full-penetration groove welds with backing bars in place, or	B′	
	• Continuous partial-penetration groove welds parallel to the direction of applied stress	B′	
	Base metals at ends of partial-length cover plates;		
	• With bolted slip-critical end connections	B	21
	• Narrower than the flange, with or without end welds, or		7
	• Wider than the flange with end welds		
	Flange thickness ≤20 mm	E	
	Flange thickness >20 mm	E′	
	• Wider than the flange without end welds	E′	

725

TABLE 8.5 (*Continued*)

General Condition	Situation	Detail Category	Illustrative Example, See Figure 8.11
Groove-welded splice connections with weld soundness established by NDT and all required grinding in the direction of the applied stresses	Base metal and weld metal at full-penetration groove-welded splices:		
	• Of plates of similar cross sections with welds ground flush	B	8, 10
	• With 600-mm radius transitions in width with welds ground flush	B	13
	• With transitions in width or thickness with welds ground to provide slopes no steeper than 1.0–2.5		11, 12
	Grades 690/690W base metal	B′	
	Other base metal grades	B	
	• With or without transitions having slopes no greater than 1.0–2.5, when weld reinforcement is not removed	C	8, 10, 11, 12
Longitudinally loaded groove-welded attachments	Base metal at details attached by full- or partial-penetration groove welds:		
	• When the detail length in the direction of applied stress is		
	Less than 50 mm	C	6, 15
	Between 50 mm and 12 times the detail thickness, but less than 100 mm	D	15
	Greater than either 12 times the detail thickness or 100 mm		
	Detail thickness <25 mm	E	15
	Detail thickness ≥25 mm	E′	15
	• With a transition radius with the end welds ground smooth, regardless of detail length:		16
	Transition radius ≥600 mm	B	
	600 mm >transition radius ≥150 mm	C	
	150 mm > transition radius ≥50 mm	D	
	Transition radius <50 mm	E	
	• With a transition radius with end welds not ground smooth	E	16

726

Description	Category	Constant
Transversely loaded groove-welded attachments with weld soundness established by NDT and all required grinding transverse to the direction of stress		16
Base metal at detail attached by full-penetration groove welds with a transition radius:		
• With equal plate thickness and weld reinforcement removed:		
Transition radius ≥600 mm	B	
600 mm > transition radius >150 mm	C	
150 mm > transition radius >50 mm	D	
Transition radius <50 mm	E	
• With equal plate thickness and weld reinforcement not removed:		
Transition radius ≥150 mm	C	
150 mm > transition radius ≥50 mm	D	
Transition radius <50 mm	E	
• With unequal plate thickness and weld reinforcement removed:		
Transition radius ≥50 mm	D	
Transition radius <50 mm	E	
• For any transition radius with unequal plate thickness and weld reinforcement not removed	E	
Fillet-welded connections with welds normal to the direction of stress		14
Base metal:		
• At details other than transverse stiffener-to-flange or transverse stiffener-to-web connections	Lesser of C or Eq. 6.6.1.2.5-3	
• At the toe of transverse stiffener-to-flange and transfer stiffener-to-web welds	C′	6

TABLE 8.5 (*Continued*)

General Condition	Situation	Detail Category	Illustrative Example, See Figure 8.11
Fillet-welded connections with welds normal and/or parallel to the direction of stress	Shear stress on the weld throat	E	9
Longitudinally loaded fillet-welded attachments	Base metal at details attached by fillet welds:		
	• When the detail length in the direction of applied stress is		
	Less than 50 mm or stud-type shear connectors	C	15, 17, 18, 20
	Between 50 mm and 12 times the detail thickness, but less than 100 mm	D	15, 17
	Greater than either 12 times the detail thickness or 100 mm		7, 9, 15, 17
	Detail thickness <25 mm	E	
	Detail thickness ≥25 mm	E′	
	• With a transition radius with the end welds ground smooth, regardless of detail length		
	Transition radius ≥50 mm	D	16
	Transition radius <50 mm	E	
	• With a transition radius with end welds not ground smooth	E	16
Transversely loaded fillet-welded attachments with welds parallel to the direction of primary stress	Base metal at details attached by fillet welds:		
	• With a transition radius with end welds ground smooth:		
	Transition radius ≥50 mm	D	16
	Transition radius <50 mm	E	16
	• With any transition radius with end welds not ground smooth	E	
Mechanically fastened connections	Base metal:		
	• At gross section of high-strength bolted slip-critical connections, except axially loaded joints in which out-of-plane bending is induced in connected materials	B	21
	• At net section of high-strength bolted nonslip-critical connections	B	
	• At net section of riveted connections	D	

[a] In AASHTO Table 6.6.1.2.3-1. [From *AASHTO LRFD Bridge Design Specifications*, Copyright © 1994 by the American Association of State Highway and Transportation Officials, Washington DC. Used by Permission.]

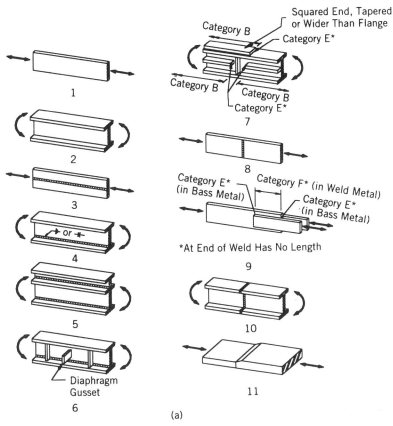

Fig. 8.11 Illustrative examples of detail categories (AASHTO Fig. 6.6.1.2.3-1). [From *AASHTO LRFD Bridge Design Specifications,* Copyright © 1994 by the American Association of State Highway and Transportation Officials, Washington, DC. Used by permission.]

The *S–N* curves for all of the detail categories are represented in Eq. 8.8. These are plotted in Figure 8.12 by taking the values from Table 8.6 for A and $(\Delta F)_{TH}$. In the finite life portion of the *S–N* curves, the effect of changes in the stress range on the number of cycles to failure can be observed by solving Eq. 8.8 for N to yield

$$N = \frac{A}{(\Delta F)_n^3} \qquad (8.9)$$

Therefore, if the stress range is cut in half, the number of cycles to failure is increased by a multiple of 8. Similarly, if the stress range is doubled, the life of the detail is divided by 8.

For the infinite life portion of the *S–N* curve given by Eq. 8.8, a factor of one-half is multiplied times the threshold stress $(\Delta F)_{TH}$, which is a conse-

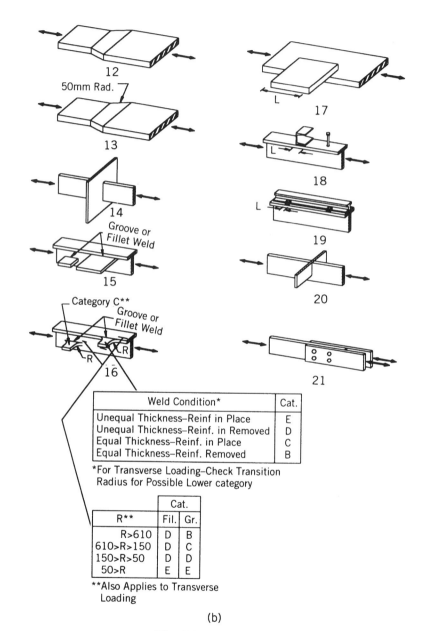

(b)

Fig. 8.11 Continued.

TABLE 8.6 Detail Category Constant, *A*, and Fatigue Thresholds[a]

Detail Category	Constant, *A* Times 10^{11} $(MPa)^3$	Fatigue Threshold (MPa)
A	82.0	165
B	39.3	110
B'	20.0	82.7
C	14.4	69.0
C'	14.4	82.7
D	7.21	48.3
E	3.61	31.0
E'	1.28	17.9
A164 (A325M) bolts in axial tension	5.61	214
M253 (A490M) bolts in axial tension	10.3	262

[a]In AASHTO Tables 6.6.1.2.5-1 and 6.6.1.2.5-3. [From *AASHTO LRFD Bridge Design Specifications,* Copyright © 1994 by the American Association of State Highway and Transportation Officials, Washington, DC. Used by permission.]

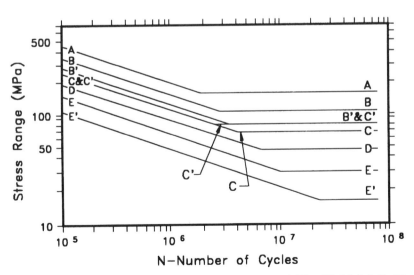

Fig. 8.12 Stress range versus number of cycles. (AASHTO Fig. C6.6.1.2.5-1). [From *AASHTO LRFD Bridge Design Specifications,* Copyright © 1994 by the American Association of State Highway and Transportation Officials, Washington, DC. Used by permission.]

quence of the possibility of the heaviest truck in 75 years being double the weight of the fatigue truck used in calculating the stress range. Logically speaking, this effect should have been applied to the load side of Eq. 8.5 instead of the resistance side. If the threshold stress controls the resistance, then Eq. 8.5 can be written

$$\tfrac{1}{2}(\Delta F)_{TH} \geq \gamma(\Delta F)$$

which is the same as

$$(\Delta F)_{TH} \geq 2\gamma(\Delta F)$$

and it is apparent that the effect of a heavier truck is considered in the infinite life portion of the fatigue resistance.

Fracture Toughness Requirements   Material in components and connections subjected to tensile stresses due to the Strength *I* limit state of Table 3.1 must satisfy supplemental impact requirements [A6.6.2]. As discussed in Section 8.2.7 on brittle fracture considerations, these supplemental impact requirements relate to minimum energy absorbed in a Charpy V-notch test at a specified temperature. The minimum service temperature at a bridge site determines the temperature zone (see Table 8.7) for the Charpy V-notch requirements.

A fracture-critical member (FCM) is defined as a member with tensile stress whose failure is expected to cause the collapse of the bridge. The material in a FCM must exhibit greater toughness and an ability to absorb more energy without fracture than a non-fracture-critical member. The Charpy V-notch fracture toughness requirements for welded components are given in Table 8.8 for different plate thicknesses and temperature zones. The FCM values for absorbed energy are approximately 50% greater than for non-FCM values at the same temperature.

TABLE 8.7   Temperature Zone Designations for Charpy V-Notch Requirements[a]

Minimum Service Temperature	Temperature Zone
$-18°C$ and above	1
$-19°C$ to $-34°C$	2
$-34°C$ to $-51°C$	3

[a]In AASHTO Table 6.6.2-1. [From *AASHTO LRFD Bridge Design Specifications,* Copyright © 1994 by the State Highway and Transportation Officials, Washington, DC. Used by permission.]

TABLE 8.8 Charpy V-Notch Fracture Toughness Requirements for Welded Components[a]

Material / Grade	Thickness (mm)	Fracture-Critical		Nonfracture-Critical	
		Zone 2 (N m @ °C)	Zone 3 (N m @ °C)	Zone 2 (N m @ °C)	Zone 3 (N m @ °C)
250	$t \leq 38$	34 @ 4.4	34 @ −12.2	20 @ 4.4	20 @ −12.2
	$38 < t \leq 100$	34 @ 4.4	34 @ −23.3	20 @ 4.4	20 @ −12.2
345/345W	$t \leq 38$	34 @ 4.4	34 @ −12.2	20 @ 4.4	20 @ −12.2
	$38 < t \leq 50$	34 @ 4.4	34 @ −23.3	20 @ 4.4	20 @ −12.2
	$50 < t \leq 100$	41 @ 4.4	41 @ −23.3	27 @ 4.4	27 @ −12.2
485W	$t \leq 38$	41 @ −6.7	41 @ −23.3	27 @ −6.7	27 @ −23.3
	$38 < t \leq 65$	41 @ −6.7	41 @ −34.4	27 @ −6.7	27 @ −23.3
	$65 < t \leq 100$	47 @ −6.7	47 @ −34.4	34 @ −6.7	34 @ −23.3
690/690W	$t \leq 65$	47 @ −17.8	47 @ −34.4	34 @ −17.8	34 @ −34.4
	$65 < t \leq 100$	61 @ −17.8	Not permitted	47 @ −17.8	47 @ −34.4

[a]In AASHTO Table 6.6.2-2. [From *AASHTO LRFD Bridge Design Specifications*, Copyright © 1994 by the American Association of State Highways and Transportation Officials, Washington, DC. Used by permission.]

8.3.3   Strength Limit States

A strength limit state is governed by the static strength of the materials or the stability of a given cross section. There are five different strength load combinations specified in Table 3.1. The differences in the strength load combinations are associated mainly with the load factors applied to the live load, for example, a smaller live load factor is used for a permit vehicle and in the presence of wind. The load combination that produces the maximum load effect is determined and then compared to the resistance or strength provided by the member.

When calculating the resistance to a particular factored load effect such as tension, compression, bending, or shear, the uncertainties are represented by an understrength or resistance factor ϕ. The ϕ factor is multiplied times the calculated nominal resistance R_n and the adequacy of the design is then determined by whether or not the inequality of Eq. 8.1 is satisfied.

In the case of structural steel members, there are uncertainties in the material properties, cross-sectional dimensions, fabrication tolerances, workmanship, and the equations used to calculate the resistance. The consequences of failure are also included in the ϕ factor. As a result, larger reductions in strength are applied to columns than beams, and to connections in general. All of these considerations are reflected in the strength limit state resistance factors given in Table 8.9 [A6.5.4.2].

8.3.4   Extreme Event Limit State

Extreme event limit states are unique occurrences with return periods in excess of the design life of the bridge. Earthquakes, ice loads, vehicle, and vessel collisions are considered to be extreme events and are to be used one at a time as shown in Table 3.1. However, these events can be combined with a major flood (recurrence interval >100 years but <500 years), or with the effects of scour of a major flood. For example, it is possible that ice floes are colliding with a bridge during a spring flood, or that scour from a major flood has reduced support for foundation components when the design earthquake occurs.

All resistance factors ϕ for an extreme event limit state are to be taken as unity, except for bolts. For bolts, the ϕ factor at the extreme event limit state shall be taken for the bearing mode of failure in Table 8.9 [A6.6.5].

8.4   GENERAL DESIGN REQUIREMENTS

Basic dimension and detail requirements are given in the AASHTO (1994) LRFD Bridge Specifications. Because these requirements can influence the design as much as force effects, a brief summary of them will be given in this section.

TABLE 8.9 Resistance Factors for the Strength Limit States[a]

Description of Mode	Resistance Factor
Flexure	$\phi_f = 1.00$
Shear	$\phi_v = 1.00$
Axial compression, steel only	$\phi_c = 0.90$
Axial compression, composite	$\phi_c = 0.90$
Tension, fracture in net section	$\phi_u = 0.80$
Tension, yielding in gross section	$\phi_y = 0.95$
Bearing on pins, in reamed, drilled or bolted holes and milled surfaces	$\phi_b = 1.00$
Bolts bearing on material	$\phi_{bb} = 0.80$
Shear connectors	$\phi_{sc} = 0.85$
A325M and A490MN bolts in tension	$\phi_t = 0.80$
A307 bolts in tension	$\phi_t = 0.67$
A325M and A490M bolts in shear	$\phi_s = 0.80$
Block shear	$\phi_{bs} = 0.80$
Weld metal in complete penetration welds:	
• Shear on effective area	$\phi_{e1} = 0.85$
• Tension or compression normal to effective area	$\phi = \text{base metal } \phi$
• Tension or compression parallel to axis of the weld	$\phi = \text{base metal } \phi$
Weld metal in partial penetration welds:	
• Shear parallel to axis of weld	$\phi_{e2} = 0.80$
• Tension or compression parallel to axis of weld	$\phi = \text{base metal } \phi$
• Tension compression normal to the effective area	$\phi = \text{base metal } \phi$
• Tension normal to the effective area	$\phi_{e1} = 0.80$
Weld metal in fillet welds:	
• Tension or compression parallel to axis of the weld	$\phi = \text{base metal } \phi$
• Shear in throat of weld metal	$\phi_{e2} = 0.80$

[a]In [A6.5.4.2]. [From *AASHTO LRFD Bridge Design Specifications,* Copyright © 1994 by the American Association of State Highway and Transportation Officials, Washington, DC. Used by permission.]

8.4.1 Effective Length of Span

The effective span length shall be taken as the center-to-center distance between bearings or supports [A6.7.1].

8.4.2 Dead Load Camber

Steel structures should be cambered during fabrication to compensate for dead load deflection and vertical curves associated with the alignment of the roadway [A6.7.2].

8.4.3 Minimum Thickness of Steel

In general, thickness of structural shall not be less than 8 mm [A6.7.3], which includes the thickness of bracing members, cross frames, and all types of gusset plates. The exceptions are webs of rolled beams or channels and of closed ribs in orthotropic decks which need be only 7 mm thick. If exposure to severe corrosion conditions is anticipated, unless a protective system is provided, an additional thickness of sacrificial metal shall be specified.

8.4.4 Diaphragms and Cross Frames

Diaphragms and cross frames are transverse bridge components that connect adjacent longitudinal beams or girders as shown in Figure 8.13. Diaphragms can be channels or beams and provide a flexural transverse connector. Cross frames are usually composed of angles and provide a truss framework transverse connector.

The function of these transverse connectors is threefold: (1) transfer of lateral wind loads to the deck and from the deck to the bearings, (2) provide stability of the beam or girder flanges during erection and placement of the deck, and (3) distribute the vertical dead load and live load to the longitudinal beams or girders [A6.7.4.1]. By transferring the wind loads on the superstructure up into the deck, the large stiffness of the deck in the horizontal plane will carry the wind loads to the supports. At the supports, diaphragms or cross frames must then transfer the wind loads down from the deck to the bearings.

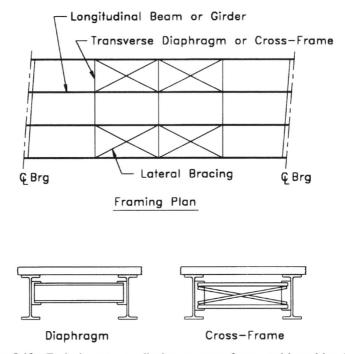

Fig. 8.13 Typical transverse diaphragm, cross frame, and lateral bracing.

To be effective, the diaphragms and cross frames shall be as deep as possible. For rolled beams they shall be at least half the beam depth [A6.7.4.2]. Intermediate diaphragms and cross frames shall be proportioned to resist the wind forces on the tributary area between lateral connections. However, end diaphragms and cross frames shall be proportioned to transmit all the accumulated wind forces to the bearings.

A rational analysis is required to determine the lateral forces in the diaphragms or cross frames. The fewer the number of transverse connectors the better, because their attachment details are prone to fatigue [C6.7.4.1].

8.4.5 Lateral Bracing

The function of lateral bracing is similar to that of the diaphragms and cross frames in transferring wind loads and providing stability, except that it acts in a horizontal plane instead of a vertical plane (Fig. 8.13). All stages of construction shall be investigated [A6.7.5.1]. Where required, lateral bracing should be placed as near the horizontal plane of the flange being braced as possible. In the first stage of composite construction, the girder must support the wet concrete, associated formwork and construction loads. It is questionable whether the formwork adequately supports the top flange in the positive moment region, and hence the unsupported length associated with the cross frames in this region must be investigated in addition to the unsupported lengths in the negative moment region. Once the concrete has hardened (stage two and subsequent stages), the top flange is adequately braced and the unsupported region of concern is in the negative moment region near the supports. Where pattern placements of concrete are used in longer bridges, the various stages and the associated unbraced length must be considered in the analysis and resistance computations. In some cases, the girder is subjected to the most critical load effects during transportation and construction. Lastly, excessive wind during construction should also be investigated. Inadequate bracing during construction has led to construction failure resulting in loss of life and significant financial resources.

Because of the favorable aspects of spreading the cross frames as far as possible, some of the more recent designs are requiring the constructor to place carefully designed temporary cross frames inside of the permanent ones to shorten unbraced lengths during construction. The temporary frames are then removed once the concrete has hardened and the top flange is adequately supported. This additional lateral support can eliminate troublesome fatigue details associated with cross frame and transverse stiffeners.

8.5 TENSILE MEMBERS

Tension members occur in the cross frames and lateral bracing of the girder bridge system shown in Figure 8.13 and are also present in truss bridges and tied-arch bridges. The cables and hangers of suspension and cable-stayed bridges are also tension members.

It is important to know how a tension member is to be used because it focuses attention on how it is to be connected to other members of the structure (Taylor, 1992). In general, it is the connection details that govern the resistance of a tension member, and they should be considered first.

8.5.1   Types of Connections

Two types of connections for tension members are considered: bolted and welded. A simple bolted connection between two plates is shown in Figure 8.14. Obviously, the bolt holes reduce the cross-sectional area of the member. A bolt hole also produces stress concentrations at the edge of the hole that can be three times the uniform stress at some distance from the hole (Fig. 8.14). The stress concentrations that exist while the material is elastic are reduced at higher load levels due to plasticity (Taylor, 1992).

A simple welded connection between two plates is shown in Figure 8.15. For the welded connection, the cross-sectional area of the member is not reduced. However, the stress in the plate is concentrated adjacent to the weld and is only uniform at some distance from the connection.

These stress concentrations adjacent to localized end connections are due to a phenomenon called *shear lag*. In the region near the hole or near the weld, shear stresses develop that cause the tensile stresses away from the hole or weld to lag behind the higher values at the edge.

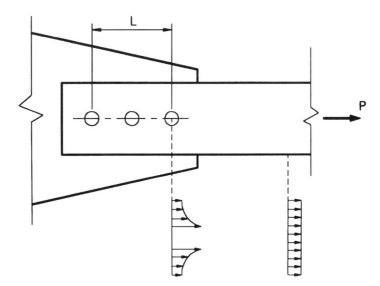

Fig. 8.14   Local stress concentration and shear lag at a bolt hole.

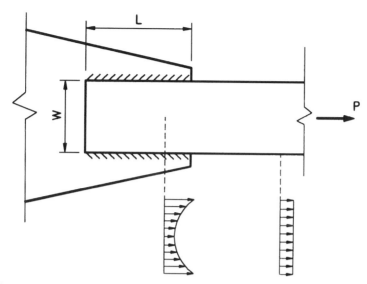

Fig. 8.15 Local stress concentration and shear lag at a welded connection.

8.5.2 Tensile Resistance

The results of typical tensile tests on bridge steels are shown by the stress–strain curves of Figure 8.4. After the yield point stress F_y is reached, plastic behavior begins. The stress remains relatively constant until strain-hardening causes the stress to increase again before decreasing and eventually failing. The peak value of stress shown for each steel in Figure 8.3 is defined as the tensile strength F_u of the steel. Numerical values for F_y and F_u are given in Table 8.2 for the various bridge steels.

When the tensile load on an end connection increases, the highest stressed point on the critical section will yield first. This point could occur at a stress concentration as shown in Figures 8.14 and 8.15 or it could occur where the tensile residual stresses (Fig. 8.2) are high. Once a portion of the critical section begins to yield and the load is increased further, a plastic redistribution of the stresses takes place. The useful tensile load-carrying limit is reached when the entire cross section becomes plastic.

The tensile resistance of an axially loaded member is governed by the lesser of [A6.8.2.1]:

- The resistance to general yielding of the gross cross section.
- The resistance to rupture on a reduced cross section at the end connection.

The factored resistance to yielding is given by

$$\phi_y P_{ny} = \phi_y F_y A_g \qquad (8.10)$$

where ϕ_y is the resistance factor for yielding of tension members taken from Table 8.9, P_{ny} is the nominal tensile resistance for yielding in the gross section (N), F_y is the yield strength (MPa), and A_g is the gross cross-sectional area of the member (mm^2).

The factored resistance to rupture is given by

$$\phi_u P_{nu} = \phi_u F_u A_e \qquad (8.11)$$

where ϕ_u is the resistance factor for fracture of tension members taken from Table 8.9, P_{nu} is the nominal tensile resistance for fracture in the net section (N), F_u is the tensile strength (MPa), and A_e is the effective net area of the member (mm^2). For bolted connections, the effective net area is

$$A_e = U A_n \qquad (8.12)$$

where A_n is the net area of the member (mm^2) and U is the reduction factor to account for shear lag. For welded connections, the effective net area is

$$A_e = U A_g \qquad (8.13)$$

The reduction factor U does not apply when checking yielding on the gross section because yielding tends to equalize the nonuniform tensile stresses caused over the cross section by shear lag [C6.8.2.1]. The resistance factor for fracture ϕ_u is smaller than the resistance factor for yielding ϕ_y because of the possibility of a brittle fracture in the strain-hardening range of the stress–strain curve.

***Reduction Factor U* [A6.8.2.2]**　When all the elements of a component (flanges, web, legs, and stem) are connected by splice or gusset plates so that force effects are transmitted uniformly, $U = 1.0$. If only a portion of the elements are connected (e.g., one leg of a single angle), the connected elements are overstressed and the unconnected ones are understressed. In the case of a partial connection, stresses are nonuniform, shear lag occurs, and $U < 1.0$.

For partial bolted connections, Munse and Chesson (1963) observed that a decrease in joint length L (Fig. 8.14) increases the shear lag effect. They proposed the following approximate expression for the reduction factor

$$U = 1 - \left(\frac{x}{L}\right) \qquad (8.14)$$

where x is the distance from the centroid of the connected area of the component to the shear plane of the connection. If a member has two symmetrically located planes of connection, x is measured from the centroid of the nearest one-half of the area. Illustrations of the distance x are given in Figure 8.16. For partial bolted connections with three or more bolts per line in the direction of load, a lower bound value of 0.85 may be taken for U in Eq. 8.14.

For partial welded connections of rolled I-shapes and tees cut from I-shapes, connected only with transverse end welds

$$U = \frac{A_{ne}}{A_{gn}} \tag{8.15}$$

where A_{ne} is the net area of the connected elements (mm^2) and A_{gn} is the net area of the rolled shape outside the connected length (mm^2).

For welded connections with longitudinal welds along both edges of the connected element (Fig. 8.15), the reduction factor may be taken as

$$\left.\begin{array}{l} U = 1.0 \quad \text{for } L > 2W \\ U = 0.87 \text{ for } 1.5W \le L < 2W \\ U = 0.75 \text{ for } W \le L < 1.5W \end{array}\right\} \tag{8.16}$$

where L is the length of the pair of welds (mm) and W is the width of the connected element (mm).

For all other members subjected to partial connection, the reduction factor may be taken as

$$U = 0.85 \tag{8.17}$$

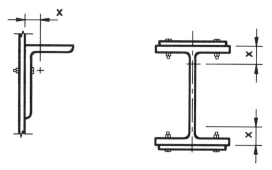

Fig. 8.16 Determination of x. (Segui, 1994.) [From *LRFD Steel Design*, by William T. Segui, Copyright © 1994 by PWS Publishing Company, Boston, MA, with permission.]

EXAMPLE 8.2

Determine the net effective area and the factored tensile resistance of a single angle $L\ 152 \times 102 \times 12.7$ tension member welded to a gusset plate as shown in Figure 8.17. Use Grade 250 structural steel.

Solution

Because only one leg of the angle is connected, the net area must be reduced by the factor U. Using Eq. 8.16 with $L = 200$ mm and $W = 152$ mm

$$L = \frac{200}{152}\ W = 1.3W \qquad U = 0.75$$

and from Eq. 8.13 with $A_g = 3060$ mm²

$$A_e = UA_g = 0.75(3060) = 2295 \text{ mm}^2$$

The factored resistance to yielding is calculated from Eq. 8.10 with $\phi_y = 0.95$ (Table 8.9) and $F_y = 250$ MPa (Table 8.2) to give

$$\phi_y P_{ng} = \phi_y F_y A_g = 0.95(250)(3060) = 727 \times 10^3 \text{ N}$$

The factored resistance to rupture is calculated from Eq. 8.11 with $\phi_u = 0.80$ (Table 8.9) and $F_u = 400$ MPa (Table 8.2) to give

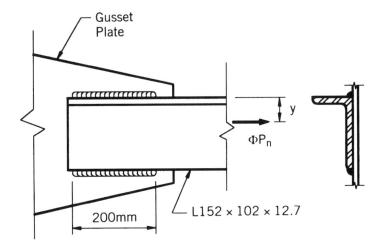

Fig. 8.17   Single-angle tension member welded to a gusset plate.

$$\phi_u P_{nu} = \phi_u F_u A_e = 0.80(400)(2295) = 734 \times 10^3 \text{ N}$$

Answer The factored tensile resistance is governed by yielding of the gross section away from the connection and is equal to 727 kN.

Net Area **[A6.8.3]** The net area A_n of a tension member is the sum of the products of thickness t and the smallest net width w_n of each element. If the connection is made with bolts, the maximum net area is with all of the bolts in a single line (Fig. 8.14). Sometimes space limitations require that more than one line be used. The reduction in cross-sectional area is minimized if a staggered bolt pattern is used (Fig. 8.18). The net width is determined for each chain of holes extending across the member along any transverse, diagonal, or zigzag line. All conceivable failure paths should be considered and the one corresponding to the smallest S_n should be used. The net width for a chain of holes is computed by subtracting from the gross width of the element the sum of the widths of all holes and adding the quantity $s^2/4g$ for each inclined line in the chain, that is,

$$w_n = w_g - \sum d + \sum \frac{s^2}{4g} \tag{8.18}$$

where w_g is the gross width of the element (mm), d is the nominal diameter of the bolt (mm) plus 3.2 mm, s is the pitch of any two consecutive holes (mm), and g is the gage of the same two holes (Fig. 8.18).

EXAMPLE 8.3

Determine the net effective area and the factored tensile resistance of a single angle L 152 × 102 × 12.7 tension member bolted to a gusset plate as shown in Figure 8.19. The holes are for 22 mm diameter bolts. Use Grade 250 structural steel.

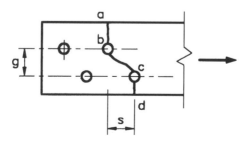

Fig. 8.18 Staggered bolt pattern.

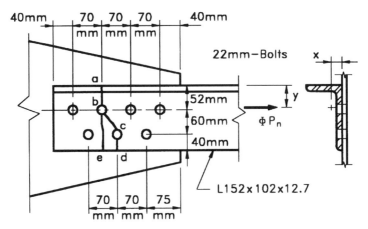

Fig. 8.19   Single-angle tension member bolted to a gusset plate.

Solution

The gross width of the cross section is the sum of the legs minus the thickness [A6.8.3].

$$w_g = 152 + 102 - 12.7 = 241.3 \text{ mm}$$

The effective hole diameter is $d = 22 + 3.2 = 25.2$ mm
Using Eq. 8.18, the net width on line *abcd* is

$$w_n = 241.3 - 2(25.2) + \frac{(35)^2}{4(60)} = 196.0 \text{ mm}$$

and on line *abe*

$$w_n = 241.3 - 1(25.2) = 216.1 \text{ mm}$$

The first case controls, so that

$$A_n = tw_n = 12.7(196.0) = 2489 \text{ mm}^2$$

Because only one leg of the cross section is connected, the net area must be reduced by the factor U. From the properties table in AISC (1992), the distance from the centroid to the outside face of the leg of the angle is $x = 25.2$ mm. Using Eq. 8.14 with $L = 3(70) = 210$ mm

$$U = 1 - \frac{x}{L} = 1 - \frac{25.2}{210} = 0.88 > 0.85$$

and from Eq. 8.12

$$A_e = UA_n = 0.88(2489) = 2190 \text{ mm}^2$$

The factored resistance to yielding is the same as in Example 8.2

$$\phi_y P_{ny} = \phi_y F_y A_g = 0.95(250)(3060) = 727 \times 10^3 \text{ N}$$

The factored resistance to rupture is calculated from Eq. 8.11 to give

$$\phi_u P_{uy} = \phi_u F_u A_e = 0.80(400)(2190) = 701 \times 10^3 \text{ N}$$

Answer

The factored tensile resistance is governed by rupture on the net section and is equal to 701 kN.

Limiting Slendernes Ratio **[A6.8.4]** Slenderness requirements are usually associated with compression members. However, it is good practice also to limit the slenderness of tension members. If the axial load in a tension member is removed and small transverse loads are applied, undesirable vibrations or deflections might occur (Segui, 1994). The slenderness requirements are given in terms of L/r, where L is the member length and r is the least radius of gyration of the cross-sectional area.

Slenderness requirements for tension members other than rods, eyebars, cables, and plates are given in Table 8.10 [A6.8.4].

TABLE 8.10 Maximum Slenderness Ratios for Tension Members

Tension Member	max (L/r)
Main members	
• Subject to stress reversals	140
• Not subject to stress reversals	200
Bracing members	240

8.5.3 Strength of Connections for Tension Members

Strength calculations for welded and bolted connections are not given in this book. The reader is referred to standard steel design textbooks and manuals that cover this topic in depth. Examples of textbooks are Gaylord et al. (1992) and Sequi (1994). Example of a manual on connections is Volume II from AISC (1994).

8.6 COMPRESSION MEMBERS

Compression members are structural elements that are subjected only to axial compressive forces that are applied along the longitudinal axis of the member and produce uniform stress over the cross section. This uniform stress is an idealized condition as there is always some eccentricity between the centroid of the section and the applied load. The resulting bending moments are usually small and of secondary importance. The most common type of compression member is a *column*. If calculated bending moments exist, due to continuity or transverse loads, they cannot be ignored and the member must be considered as a *beam column*. Compression members exist in trusses, cross frames, and lateral bracing systems where the eccentricity is small and the secondary bending can be ignored.

8.6.1 Column Stability Concepts

In structural steel, column cross sections are often slender and other limit states are reached before the material crushes. These other limit states are associated with inelastic and slender member buckling. They include lateral buckling, local buckling, and lateral-torsional buckling of the compression member. Each of the limit states must be incorporated in the design rules developed to select compression members.

The starting point for studying the buckling phenomenon is an idealized perfectly straight elastic column with pin ends. As the axial compressive load on the column increases, the column remains straight and shortens elastically until the critical load P_{cr} is reached. The critical load is defined as the lowest axial compressive load for which a small lateral displacement causes the column to bow laterally and seek a new equilibrium position. This definition of critical load is depicted schematically in the load-deflection curves of Figure 8.20.

In Figure 8.20, the point at which the behavior changes is the *bifurcation point*. The load-deflection curve is vertical until this point is reached and then the midheight of the column will move right or left depending on the direction of the lateral disturbance. Once the lateral deflection becomes nonzero, a buckling failure has occurred and small deflection theory predicts that no further increase in the axial load is possible. If large deflection theory is used,

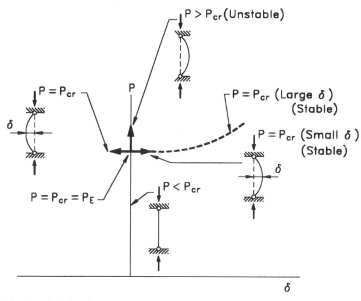

Fig. 8.20 Load-deflections curves for elastic columns. (Bjorhovde, 1992.) [From *Constructional Steel Design: An International Guide,* Edited by P. J. Dowling, J. E. Harding, and R. Bjorhovde, Copyright © 1992 by Elsevier Science Ltd (now Chapman and Hall, Andover, England), with permission.]

additional stress resultants are developed and the load-deflection response follows the dashed line in Figure 8.20.

The small deflection theory solution to the buckling problem was published by Euler in 1759. He showed that the critical buckling load P_{cr} could be given as

$$P_{cr} = \frac{\pi^2 EI}{L^2} \tag{8.19}$$

where E is the modulus of elasticity of the material, I is the moment of inertia of the column cross section about the centroidal axis perpendicular to the plane of buckling, and L is the pin-ended column length. This expression is well-known in mechanics and its derivation is not repeated here.

Equation 8.19 can also be expressed as a critical buckling stress σ_{cr} by dividing both sides by the gross area of the cross section A_s to give

$$\sigma_{cr} = \frac{P_{cr}}{A} = \frac{\pi^2 (EI/A_s)}{L^2}$$

By using the definition of the radius of the gyration r of the section as $I = Ar^2$ and rewriting the above equation, we get

$$\sigma_{cr} = \frac{\pi^2 E}{\left(\dfrac{L}{r}\right)^2} \qquad (8.20)$$

where L/r is commonly referred to as the *slenderness ratio* of the column. Buckling will occur about the centroidal axis with the least moment of inertia I (Eq. 8.19) or the least radius of gyration r (Eq. 8.20). Sometimes the critical centroidal axis is inclined, as in the case of a single-angle compression member. In any event, the maximum slenderness ratio must be found because it governs the critical stress on the cross section.

The idealized critical buckling stress given in Eq. 8.20 is influenced by three major strength parameters: end restraint, residual stresses, and initial crookedness. The first depends on how the member is connected and the last two on how it was manufactured. These parameters and their effect on the buckling strength are discussed in the following sections.

Effective Length of Columns The buckling problem solved by Euler was for an idealized column without any moment restraint at its ends. For a column of length L whose ends do not move laterally (no sidesway), end restraint provided by connections to other members will cause the location of the points of zero moment to move away from the ends of the column. The distance between the points of zero moment is the effective pinned-pinned column length KL, where in this case $K < 1$. If the end restraint is either pinned or fixed, typical values of K for the no sidesway case are shown in the first three deformed shapes of Figure 8.21.

If the ends of a column move laterally with respect to one another, the effective column length KL can be large with K considerably greater than one. This behavior is shown in the last two deformed shapes of Figure 8.21 with one end free and the other end either fixed or pinned. In general, the critical buckling stress for a column with effective length KL can be obtained by rewriting Eq. 8.20 as

$$\sigma_{cr} = \frac{\pi^2 E}{(KL/r)^2} \qquad (8.21)$$

where K is the effective length factor.

Actual column end conditions are going to be somewhere between pinned and fixed depending on the stiffness provided by the end connections. For bolted or welded connections at both ends of a compression member in which sidesway is prevented, K may be taken as 0.75 [A4.6.2.5]. Therefore, the

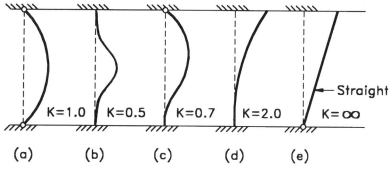

Fig. 8.21   End restraint and effective length of columns. (a) Pinned-pinned, (b) fixed-fixed, (c) fixed-pinned, (d) fixed-free, and (e) pinned-free. (Bjorhovde, 1992.) [From *Constructional Steel Design: An International Guide,* Edited by P. J. Dowling, J. E. Harding, and R. Bjorhovde, Copyright © 1992 by Elsevier Science Ltd (now Chapman and Hall, Andover, England), with permission.]

effective length of the compression members in cross frames and lateral bracing can be taken as 0.75 L, where L is the laterally unsupported length of the member.

Residual Stresses   Residual stresses have been discussed in Section 8.2.3. In general, they are caused by nonuniform cooling of the elements in a component during the manufacturing or fabrication process. The basic principle of residual stress can be summarized as follows: The fibers that cool first end up in residual compression; those that cool last will have residual tension (Bjorhovde, 1992).

The magnitude of the residual stresses can be almost equal to the yield stress of the material. Additional applied axial compressive stress can cause considerable yielding in the cross section at load levels below that predicted by $F_y A_s$. This combined stress is shown schematically in Figure 8.22, where σ_{rc} is the compressive residual stress, σ_{rt} is the tensile residual stress, and σ_a is the additional applied axial compressive stress. The outer portions of the element have gone plastic while the inner portion remains elastic.

Initial Crookedness   Residual stresses develop in an element along its length and each cross section is assumed to have a stress distribution similar to that shown in Figure 8.22. This uniform distribution of stress along the length of the element will occur only if the cooling process is uniform. What usually happens is that a member coming off the rolling line in a steel mill is cut to length and then set aside to cool. Other members are placed along side it on the cooling bed and will influence the rate of cooling that takes place.

If a hot member is on one side and a warm member is on the other side, the cooling will be nonuniform across the section. Further, the cut ends will cool faster than the sections at midlength and the cooling will be nonuniform

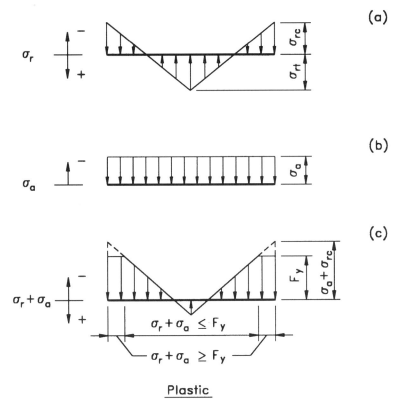

Plastic

Fig. 8.22 (a) Residual stress, (b) applied compressive stress, and (c) combined residual and applied stress. (Bjorhovde, 1992.) [From *Constructional Steel Design: An International Guide,* Edited by P. J. Dowling, J. E. Harding, and R. Bjorhovde, Copyright © 1992 by Elsevier Science Ltd (now Chapman and Hall, Andover, England), with permission.]

along the length of the member. After the member cools, the nonuniform residual stress distribution will cause the member to bow, bend, and even twist. If the member is used as a column, it can no longer be assumed to be perfectly straight, but must be considered to have initial crookedness.

A column with initial crookedness introduces bending moments when axial loads are applied. Part of the resistance of the column is used to carry these bending moments and a reduced resistance is available to support the axial load. Therefore, the imperfect column exhibits a load carrying capacity that is less than that of the ideal column.

The amount of initial crookedness in wide-flange shapes is shown in Figure 8.23 as a fraction of the member length. The mean value of the random eccentricity e_1 is $L/1500$ with a maximum value of about $L/1000$ (Bjorhovde, 1992).

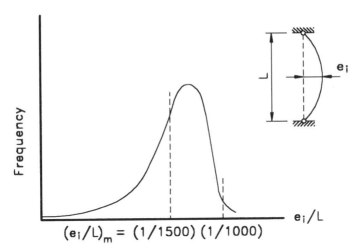

Fig. 8.23 Statistical variation of initial crookedness. (Bjorhovde, 1992.) [From *Constructional Steel Design: An International Guide,* Edited by P. J. Dowling, J. E. Harding, and R. Bjorhovde, Copyright © 1992 by Elsevier Science Ltd (now Chapman and Hall, Andover, England), with permission.]

8.6.2 Inelastic Buckling Concepts

The Euler buckling load of Eq. 8.19 was derived assuming elastic material behavior. For long, slender columns this assumption is reasonable because buckling occurs at a relatively low level and the stresses produced are below the yield strength of the material. However, for short, stubby columns buckling loads are higher and yielding of portions of the cross section takes place.

For short columns, not all fibers of the cross section reach yield at the same time. This statement is true because the locations with compressive residual stresses will yield first as illustrated in Figure 8.22. Therefore, as the axial compressive load increases the portion of the cross section that remains elastic decreases until the entire cross section becomes plastic. The transition from elastic to plastic behavior is gradual as demonstrated by the stress–strain curve in Figure 8.24 for a stub column. This stress–strain behavior is different from the relatively abrupt change from elastic to plastic usually observed in a bar or coupon test of structural steel (Fig. 8.4).

The stub column stress–strain curve of Figure 8.24 deviates from elastic behavior at the proportional limit σ_{prop} and gradually changes to plastic behavior when F_y is reached. The modulus of elasticity E represents elastic behavior until the sum of the compressive applied and maximum residual stress in Figure 8.22 equals the yield stress, that is,

$$\sigma_a + \sigma_{rc} = F_y$$

or

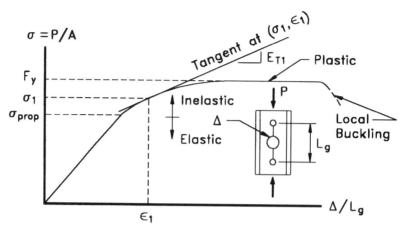

Fig. 8.24 Stub column stress–strain curve. (Bjorhovde, 1992.) [From *Constructional Steel Design: An International Guide*, Edited by P. J. Dowling, J. E. Harding, and R. Bjorhovde, Copyright © 1992 by Elsevier Science Ltd (now Chapman and Hall, Andover, England), with permission.]

$$\sigma_{\text{prop}} = F_y - \sigma_{rc} \qquad (8.22)$$

In the transition between elastic and plastic behavior, the rate of change of stress over strain is represented by the tangent modulus E_T as shown in Figure 8.24. This region of the curve where the cross section is a mixture of elastic and plastic stresses is called *inelastic*. The inelastic or tangent modulus column buckling load is defined by substituting E_T for E in Eq. 8.21 to yield

$$\sigma_T = \frac{\pi^2 E_T}{(KL/r)^2} \qquad (8.23)$$

A combined Euler (elastic) and tangent modulus (inelastic) column buckling curve is shown in Figure 8.25. The transition point that defines the change from elastic to inelastic behavior is the proportional limit stress σ_{prop} of Eq. 8.22 and the corresponding slenderness ratio $(KL/r)_{\text{prop}}$.

8.6.3 Compressive Resistance

The short or stub column resistance to axial load is a maximum when no buckling occurs and the entire cross-sectional area A_s is at the yield stress F_y. The fully plastic yield load P_y is the maximum axial load the column can support and can be used to normalize the column curves so that they are independent of structural steel grade. The axial yield load is

$$P_y = A_s F_y \qquad (8.24)$$

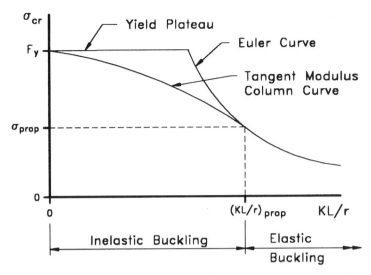

Fig. 8.25 Combined tangent modulus and Euler column curves. (After Bjorhovde, 1992.) [From *Constructional Steel Design: An International Guide*, Edited by P. J. Dowling, J. E. Harding, and R. Bjorhovde, Copyright © 1992 by Elsevier Science Ltd (now Chapman and Hall, Andover, England), with permission.]

For long columns, the critical Euler buckling load P_{cr} is obtained by multiplying Eq. 8.21 by A_s to give

$$P_{cr} = \frac{\pi^2 E A_s}{(KL/r)^2} \tag{8.25}$$

Dividing Eq. 8.25 by Eq. 8.24, we can write the normalized Euler elastic column curve as

$$\frac{P_{cr}}{P_y} = \left(\frac{\pi r}{KL}\right)^2 \frac{E}{F_y} = \frac{1}{\lambda_c^2} \tag{8.26}$$

where λ_c is the column slenderness term

$$\lambda_c = \left(\frac{KL}{\pi r}\right)\sqrt{\frac{F_y}{E}} \tag{8.27}$$

The normalized plateau and Euler column curve are shown as the top curves in Figure 8.26. The inelastic transition curve due to residual stresses is also shown. The column curve that includes the additional reduction in buckling load caused by initial crookedness is the bottom curve in Figure

8.26. This bottom curve is the column strength curve given in the specifications.

The column strength curve represents a combination of inelastic and elastic behavior. Inelastic buckling occurs for intermediate length columns from $\lambda_c = 0$ to $\lambda_c = \lambda_{prop}$, where λ_{prop} is the slenderness term for an Euler critical stress of σ_{prop} (Eq. 8.22). Elastic buckling occurs for long columns with λ_c greater than λ_{prop}. Substituting Eq. 8.22 and these definitions into Eq. 8.26, results in

$$\frac{F_y - \sigma_{rc}}{F_y}\frac{A_s}{A_s} = \frac{1}{\lambda^2_{prop}}$$

or

$$\lambda^2_{prop} = \frac{1}{1 - \dfrac{\sigma_{rc}}{F_y}} \tag{8.28}$$

The value for λ_{prop} depends on how large the residual compressive stress σ_{rc} is relative to the yield stress F_y. For example, if $F_y = 345$ MPa and $\sigma_{rc} = 190$ MPa, then Eq. 8.28 gives

$$\lambda^2_{prop} = \frac{1}{1 - \dfrac{190}{345}} = 2.23$$

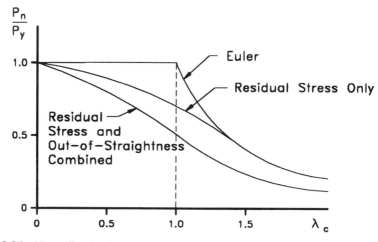

Fig. 8.26 Normalized column curves with imperfection effects. (Bjorhovde, 1992.) [From *Constructional Steel Design: An International Guide*, Edited by P. J. Dowling, J. E. Harding, and R. Bjorhovde, Copyright © 1992 by Elsevier Science Ltd (now Chapman and Hall, Andover, England), with permission.]

and λ_{prop} = 1.49. The larger the residual stress the larger the slenderness term at which the transition to elastic buckling occurs. Nearly all of the columns designed in practice behave as inelastic intermediate length columns. Seldom are columns slender enough to behave as elastic long columns that buckle at the Euler critical load.

Nominal Compressive Resistance [A6.9.4.1] To avoid the square root in Eq. 8.27, the column slenderness term λ is redefined as

$$\lambda = \lambda_c^2 = \left(\frac{KL}{\pi r}\right)^2 \frac{F_y}{E} \qquad (8.29)$$

The transition point between inelastic buckling and elastic buckling or between intermediate length columns and long columns is specified as $\lambda = 2.25$. For long columns ($\lambda \geq 2.25$), the nominal column strength P_n is given by

$$P_n = \frac{0.88 F_y A_s}{\lambda} \qquad (8.30)$$

which is the Euler critical buckling load of Eq. 8.25 reduced by a factor of 0.88 to account for initial crookedness of $L/1500$ [C6.9.4.1].

For intermediate length columns ($\lambda < 2.25$), the nominal column strength P_n is determined from a tangent modulus curve that provides a smooth transition between $P_n = P_y$ and the Euler buckling curve. The formula for the transition curve is

$$P_n = 0.66^\lambda F_y A_s \qquad (8.31)$$

The curves representing Eqs. 8.30 and 8.31 are plotted in Figure 8.27 in terms of λ_c rather than λ to preserve the shape of the curves plotted previously in Figures 8.25 and 8.26.

The final step in determining the compressive resistance of P_r of columns is to multiply the nominal resistance P_n by the resistance factor for compression ϕ_c taken from Table 8.9, that is

$$P_r = \phi_c P_n \qquad (8.32)$$

Limiting Width/Thickness Ratios [A6.9.4.2] Compressive strength of columns of intermediate length is based on the tangent modulus curve obtained from tests of stub columns. A typical stress–strain curve for a stub column is given in Figure 8.24. Because the stub column is relatively short, it will not exhibit flexural buckling. However, it could experience local buckling with a subsequent decrease in load if the width/thickness ratio of the column elements is too high. Therefore, the slenderness of plates shall satisfy

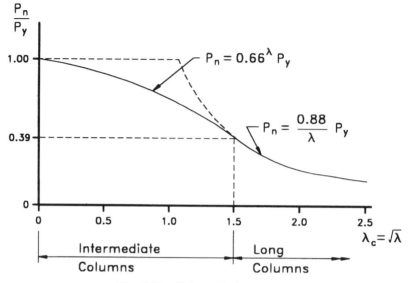

Fig. 8.27   Column design curves.

$$\frac{b}{t} \le k\sqrt{\frac{E}{F_y}} \tag{8.33}$$

where k is the plate buckling coefficient taken from Table 8.11, b is the width of the plate described in Table 8.11 (mm), and t is the plate thickness (mm). The requirements given in Table 8.11 for plates supported along one edge and plates supported along two edges are illustrated in Figure 8.28.

Limiting Slenderness Ratio [A6.9.3]   If columns become too slender they will have little strength and not be economical. The recommended limit for main members is $(KL/r) \le 120$ and for bracing member is $(KL/r) \le 140$.

EXAMPLE 8.4

Calculate the design compressive strength $\phi_c P_n$ of a W360 × 110 column with a length of 6100 mm and pinned ends. Use Grade 250 structural steel.

Properties

From AISC (1992), $A_s = 14\,100$ mm², $d = 360$ mm, $t_w = 11.4$ mm, $b_f = 256$ mm, $t_f = 19.9$ mm, $h_c/t_w = 25.3$, $r_x = 153$ mm, $r_y = 62.9$ mm.

TABLE 8.11   Limiting Width-Thickness Ratios[a]

Plates Supported Along One Edge	k	b
Flanges and projecting legs of plates	0.56	• Half-flange width of I-sections • Full-flange width of channels • Distance between free edge and first line of bolts or welds in plates • Full-width of an outstanding leg for pairs of angles in continuous contact
Stems of rolled tees	0.75	• Full-depth of tee
Other projecting elements	0.45	• Full-width of outstanding leg for single angle strut or double angle strut with separator • Full projecting width for others

Plates Supported Along Two Edges	k	b
Box flanges and cover plates	1.40	• Clear distance between webs minus inside corner radius on each side for box flanges • Distance between lines of welds or bolts for flange cover plates
Webs and other plate elements	1.49	• Clear distance between flanges minus fillet radii for webs of rolled beams • Clear distance between edge supports for all others
Perforated cover plates	1.86	• Clear distance between edge supports

[a]AASHTO Table 6.9.4.2-1. [From *AASHTO LRFD Bridge Design Specifications,* Copyright © 1994 by the American Association of State Highway and Transportation Officials, Washington, DC. Used by permission.]

Solution

slenderness ratio

$$\max \frac{KL}{r} = \frac{1.0(6100)}{62.9} = 97.0 < 120, \text{ OK}$$

$$\frac{\text{width}}{\text{thickness}}: \quad \frac{b_f}{2t_f} = \frac{256}{2(19.9)} = 6.4 < k\sqrt{\frac{E}{F_y}} = 0.56\sqrt{\frac{200\,000}{250}} = 15.8, \text{ OK}$$

$$\frac{h_c}{t_w} = 25.3 < k\sqrt{\frac{E}{F_y}} = 1.49\sqrt{\frac{200\,000}{250}} = 42.1, \text{ OK}$$

column slenderness term

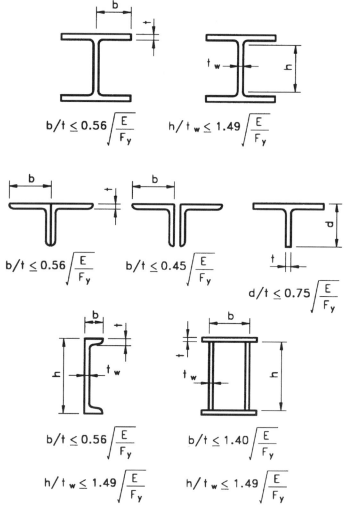

Fig. 8.28 Limiting width-thickness ratios. (After Segui, 1994.) [From *LRFD Steel Design*, by William T. Segui, Copyright © 1994 by PWS Publishing Company, Boston, MA, with permission.]

$$\lambda = \left(\frac{KL}{\pi r}\right)^2 \frac{F_y}{E} = \left(\frac{97.0}{\pi}\right)^2 \frac{250}{200\ 000} = 1.19 < 2.25$$

intermediate length column

$$P_n = 0.66^\lambda F_y A_s = (0.66)^{1.19}(250)(14\ 100) = 2.15 \times 10^6 \text{ N}$$

Answer

Design compressive strength $= \phi_c P_n = 0.90(2.15 \times 10^6)/10^3 = \underline{1935\text{ kN}}$

8.6.4 Connections for Compression Members

Strength calculations for welded and bolted connections are not given in this book. The reader is referred to standard steel design textbooks and manuals that cover this topic in depth. Examples of textbooks are Gaylord et al (1992) and Sequi (1994). Example of a manual on connections is Volume II from AISC (1994).

8.7 *I*-SECTIONS IN FLEXURE

I-Sections in flexure are structural members that carry transverse loads perpendicular to their longitudinal axis primarily in a combination of bending and shear. Axial loads are usually small in most bridge girder applications and are often neglected. If axial loads are significant, then the cross section should be considered as a *beam column*. If the transverse load is eccentric to the shear center of the cross section, then combined bending and torsion must be considered. The discussion that follows is limited to the basic behavior and design of rolled or fabricated straight steel *I*-sections that are symmetrical about a vertical axis in the plane of the web and are primarily in flexure.

8.7.1 General

The resistance of *I*-sections in flexure is largely dependent on the degree of stability provided, either locally or in an overall manner. If the section is very stable at high loads, then the *I*-section can develop a bending resistance beyond the first yield moment M_y to the full plastic moment resistance M_p. If stability is limited by either local or global buckling then the bending resistance will be less than M_p and if the buckling is severe, less than M_y.

Plastic Moment, M_p Consider the doubly symmetric *I*-section of Figure 8.29(a) that is subjected to pure bending at midspan by two equal concentrated loads. Assume stability is provided and the steel stress–strain curve is elastic-perfectly plastic. As the loads increase, plane sections remain plane, and the strains increase until the extreme fibers of the section reach $\varepsilon_y = F_y/E$ [Fig. 8.29(b)]. The bending moment at which the first fibers reach yield is defined as the yield moment M_y.

Further increase of the loads causes the strains and rotations to increase, and more of the fibers in the cross section to yield [Fig. 8.29(c)]. The limiting case is when the strains caused by the loads are so large that the entire cross section can be considered at the yield stress F_y [Fig. 8.29(d)]. When this

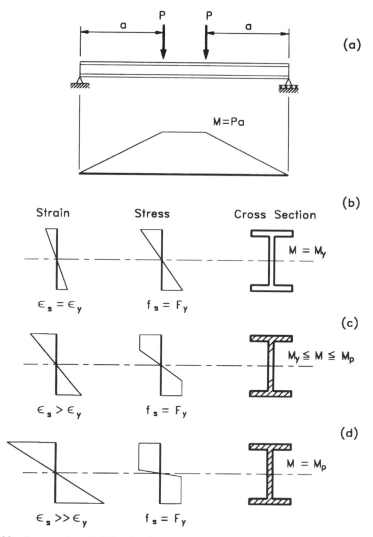

Fig. 8.29 Progressive yielding in flexure. (a) Simple beam with twin concentrated loads, (b) first yield at extreme fibers, (c) partially plastic and partially elastic, and (d) fully plastic.

occurs, the section is fully plastic and the corresponding bending moment is defined as the plastic moment M_p.

Any attempt to further increase the loads will only result in increased deformations without any increase in moment resistance. This limit of moment can be seen in the idealized moment-curvature curve in Figure 8.30. Curvature ψ is defined as the rate of change of strain or simply the slope of the strain diagram, that is,

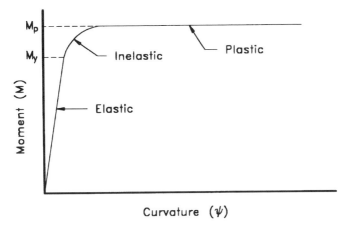

Fig. 8.30 Idealized moment-curvature response.

$$\psi = \frac{\varepsilon_c}{c} \qquad (8.34)$$

where ε_c is the strain at a distance c from the neutral axis.

The moment-curvature relation of Figure 8.30 has three parts: elastic, inelastic, and plastic. The inelastic part provides a smooth transition between elastic to plastic behavior as more and more of the fibers in the cross section yield. The length of the plastic response ψ_p relative to the elastic curvature ψ_y is a measure of ductility of the section.

Moment Redistribution When the plastic moment M_p is reached at a cross section, additional rotation will occur at the section and a hinge resisting constant moment M_p will be formed. When this plastic hinge forms in a statically determinate structure, such as the simple beam of Figure 8.29, a collapse mechanism is formed.

However, if a plastic hinge forms in a statically indeterminate structure, collapse does not occur and additional load carrying capacity remains. This increase in load can be illustrated with the propped cantilever beam of Figure 8.31(a) that is subjected to a gradually increasing concentrated load at midspan. The limit of elastic behavior is when the load causes the moment at the fixed end of the beam to reach M_y. This limiting load P_y will produce moments that are consistent with an elastic analysis as shown in Figure 8.31(b).

Further increase in the load will cause a plastic hinge to form at the fixed end. However, the structure will not collapse because a mechanism has not been formed. The beam with one fixed end has now become a simple beam with a known moment M_p at one end. A mechanism will not be formed until a second plastic hinge develops at the second highest moment location under

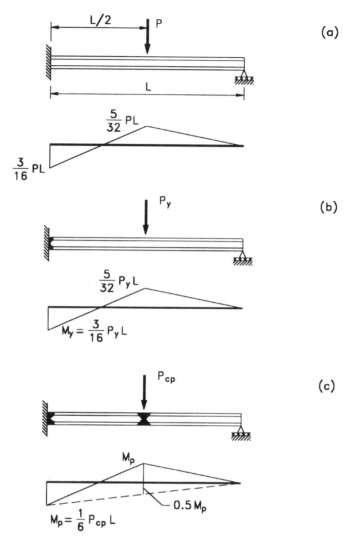

Fig. 8.31 Moment redistribution in a propped cantilever. (a) Elastic moments, (b) first yield moments, and (c) collapse mechanism moments.

the concentrated load. This condition is shown in Figure 8.31(c). This behavior for moving loads is described in detail in Chapter 6.

By assuming that $M_y = 0.9M_p$, the ratio of the collapse load P_{cp} to the yield load P_y is

$$\frac{P_{cp}}{P_y} = \frac{(6M_p/L)}{\frac{16}{3}(0.9M_p)/L} = 1.25$$

For this example, there is an approximate 25% increase in resistance to load beyond the load calculated by elastic analysis. However, for this to take place, rotation capacity had to exist in the plastic hinge at the fixed end so that moment redistribution could occur.

Another way to show that moment redistribution has taken place when plastic hinges form is to compare the ratio of positive moment to negative moment. For the elastic moment diagram in Figure 8.31(b), the ratio is

$$\left(\frac{M_{pos}}{M_{neg}}\right)_e = \frac{\frac{5}{32}PL}{\frac{3}{16}PL} = 0.833$$

while for the moment diagram at collapse [Fig. 8.31(c)]

$$\left(\frac{M_{pos}}{M_{neg}}\right)_{cp} = \frac{M_p}{M_p} = 1.0$$

Obviously, the moments have been redistributed.

If certain conditions are met, the AASHTO (1994) LRFD Bridge Specifications permit a maximum reduction of 10% of the negative moment obtained by elastic procedures [A6.10.2.2]. When a reduction in negative moment is taken, statics requires that positive moments in adjacent spans be increased. In the case of the propped cantilever of Figure 8.31, if the negative moment M_{neg} is changed by 10%, then to satisfy statics the modified positive moment M^*_{pos} at midspan must be increased by $0.05M_{neg}$, that is

$$M^*_{pos} = M_{pos} + 0.05M_{neg}$$
$$= \tfrac{5}{32}PL + 0.05(\tfrac{3}{16}PL) = 0.156PL + 0.009PL = 0.165PL$$

If both ends of the beam are continuous, the increase in positive moment could be doubled.

Moment redistribution can occur in a statically indeterminate structure where stability is provided if sufficient rotation capacity exists at the earlier formed plastic hinges. This results in a transfer of moment from higher stressed locations to locations that have reserve strength. The result is a greater load carrying capacity and a better estimate on the actual collapse load of the structure.

Stability   The key to development of the plastic moment resistance M_p is whether or not adequate stability is provided for the cross section. If global or local buckling occurs, M_p cannot be reached.

Global buckling can occur if the compression flange of a section in flexure is not laterally supported. A laterally unsupported compression flange will behave like a column and tend to buckle out-of-plane between points of lateral support. However, because the compression flange is part of a beam cross section with a tension zone that keeps the opposite flange in line, the cross-section twists when it moves laterally. This behavior is shown in Figure 8.32 and is referred to as *lateral-torsional buckling.*

Local buckling can occur if the width-thickness ratio of elements in compression become too large. Limitations on these ratios are similar to those given for columns in Figure 8.28. If the buckling occurs in the compression flange, it is called *flange local buckling.* If it occurs in the compression portion of the web, it is called *web local buckling.* Illustrations of local buckling are shown in the photographs of Figure 8.33 of a full-scale test to failure of a roof beam. Flange local buckling can be seen in the top flange of the overall view [Fig. 8.33(a)]. A close-up of the compression region of the beam [Fig. 8.33(b)] shows the buckled flange and measurement of the out-of-plane web deformation indicating web local buckling has also occurred.

Classification of Sections   Cross-sectional shapes are classified as *compact, noncompact,* or *slender* depending on the width-thickness ratios of their compression elements and bracing requirements. A *compact section* is one that can develop a fully plastic moment M_p before lateral torsional buckling or local buckling of its flange or web occurs. A *noncompact section* is one that can develop a moment equal to or greater than M_y, but less than M_p, before local buckling of any of its compression elements occurs. A *slender section*

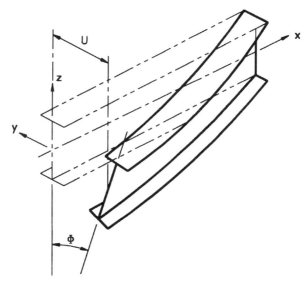

Fig. 8.32   Isometric of lateral torsional buckling. (Nethercot, 1992.) [From *Constructional Steel Design: An International Guide,* Edited by P. J. Dowling, J. E. Harding, and R. Bjorhovde, Copyright © 1992 by Elsevier Science Ltd (now Chapman and Hall, Andover, England), with permission.]

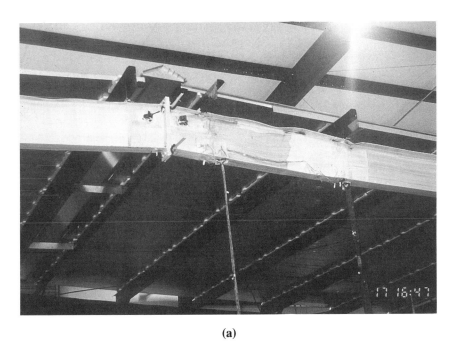

(a)

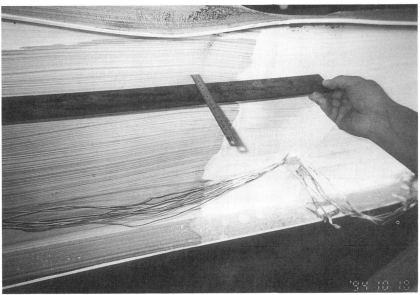

(b)

Fig. 8.33   (a) Local buckling of flange and (b) local buckling of web. (Photos courtesy of Structures/Materials Laboratory, Virginia Tech.)

is one whose compression elements are so slender that they buckle locally before the moment reaches M_y. A comparison of the moment-curvature response of these shapes in Figure 8.34 illustrates the differences in their behavior.

Sections are also classified as *composite* or *noncomposite*. A *composite section* is one where a properly designed shear connection exists between the concrete deck and the steel beam (Fig. 8.35). A section where the concrete deck is not connected to the steel beam is considered as a *noncomposite section*.

When the shear connection exists, the deck and beam act together to provide resistance to bending moment. In regions of positive moment, the concrete deck is in compression and the increase in flexural resistance can be significant. In regions of negative moment, the concrete deck is in tension and only its tensile reinforcement adds to the flexural resistance of the steel beam. The flexural resistance of the composite section is further increased because the connection of the concrete deck to the steel beam provides continuous lateral support for its compression flange and prevents lateral-torsional buckling. Because of these advantages, the AASHTO (1994) LRFD Bridge Specification recommends that, wherever technically feasible, structures shall be made composite [A6.10.6.1].

Stiffness Properties [A6.10.1.3]   In the analysis of flexural members for loads applied to a noncomposite section, only the stiffness properties of the steel beam should be used. In the analysis of flexural members for loads applied to a composite section, the transformed area of concrete used in cal-

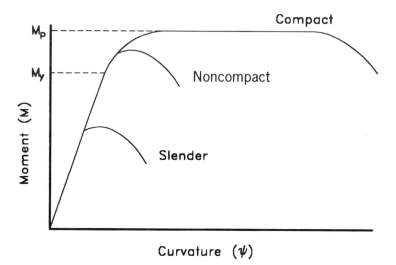

Fig. 8.34 Response of three beam classes.

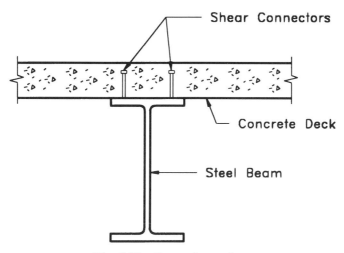

Fig. 8.35 Composite section.

culating the stiffness properties shall be based on a modular ratio of n (Table 8.12) [A6.10.5.1.1b] for transient loads and $3n$ for permanent loads. The modular ratio of $3n$ is to account for the larger increase in strain due to the creep of concrete under permanent loads. The concrete creep tends to transfer long-term stresses from the concrete to the steel, effectively increasing the relative stiffness of the steel. The multiplier on n accounts for this increase. The stiffness of the full composite section may be used over the entire bridge length, including regions of negative bending. This constant stiffness is reasonable, as well as convenient, because field tests of continuous composite bridges have shown there is considerable composite action in the negative bending regions [C6.10.1.3].

TABLE 8.12   Ratio of Modulus of Elasticity of Steel to that of Concrete. Normal Weight Concrete[a]

f_c' (MPa)	n
$16 \leq f_c' < 20$	10
$20 \leq f_c' < 25$	9
$25 \leq f_c' < 32$	8
$32 \leq f_c' < 41$	7
$41 \leq f_c'$	6

[a]From [A6.10.5.1.1b]. [From *AASHTO LRFD Bridge Design Specification,* Copyright © 1994 by the American Association of State Highway and Transportation Officials, Washington, DC. Used by permission.]

8.7.2   Limit States

I-sections in flexure must be designed to resist the load combinations for the strength, service, and fatigue limit states of Table 3.1.

Strength Limit State [A6.10.2]   For compact sections, the factored flexural resistance is in terms of moments

$$M_r = \phi_f M_n \qquad (8.35)$$

where ϕ_f is the resistance factor for flexure from Table 8.9 and $M_n = M_p$ is the nominal resistance specified for a compact section.

For noncompact sections, the factored flexural resistance is defined in terms of stress

$$F_r = \phi_f F_n \qquad (8.36)$$

where F_n is the nominal resistance specified for a noncompact section.

The factored shear resistance V_r shall be taken as

$$V_r = \phi_v V_n \qquad (8.37)$$

where ϕ_v is the resistance factor for shear from Table 8.9 and V_n is the nominal shear resistance specified for unstiffened and stiffened webs.

Service Limit State [A6.10.3]   The Service II load combination of Table 3.1 shall apply. This load combination is intended to control yielding of steel structures and to prevent objectionable permanent deflections which would impair rideability [C6.10.3.1]. When checking the flange stresses, moment redistribution may be considered if the section in the negative moment region is compact. Flange stresses in positive and negative bending for composite sections shall not exceed

$$f_f \leq 0.95 R_h F_{yf} \qquad (8.38)$$

and for noncomposite sections

$$f_f \leq 0.80 R_h F_{yf} \qquad (8.39)$$

where f_f is the elastic flange stress caused by the factored loading, R_h is the hybrid flange-stress reduction factor [A6.10.5.4.1] (for a homogeneous section, $R_h = 1.0$), and F_{yf} is the yield stress of the flange.

Fatigue Requirements for Webs [A6.10.4]   In the previous discussion on fatigue in Section 8.3.2, the concern was that the stress range from repeated loadings was not excessive. In the current discussion, the concern is to control out-of-plane flexing of the web due to repeated loadings. To control the web flexing, the maximum elastic stress in flexure or shear is limited by the web buckling stress in flexural or shear.

In calculating the maximum elastic stresses, unfactored permanent loads and double the fatigue load combination of Table 3.1 shall be used. The fatigue truck is doubled when calculating maximum stresses because the heaviest truck expected to cross the bridge is approximately twice the fatigue truck used in calculating stress ranges. Also, the distribution factor for the fatigue truck is for one lane loaded without multiple presence [A3.6.1.4] and the dynamic load allowance is 15% [A3.6.2.1].

The flexural web buckling stress is based on elastic plate buckling formulas with partially restrained edges. Besides the material constants E and F_y, the main parameter for determining web buckling capacity is the web slenderness ratio λ_w

$$\lambda_w = \frac{2D_c}{t_w} \tag{8.40}$$

where D_c is the depth of the web in compression in the elastic range and t_w is the web thickness. The compressive web depth D_c is the clear height of the web between the compression flange and the point in the web where the compression stress goes to zero [A6.10.5.1.4a]. The point of zero compressive stress can be calculated by superposition of elastic stresses from the specified load combination (see Fig. 8.36).

Theoretically, longitudinal web stiffeners can prevent flexural buckling of the web. For webs without longitudinal stiffeners, the maximum compressive elastic flexural stress in the compression flange f_{cf}, which is representative of the maximum flexural stress in the web, is limited by the following [A6.10.4.3]:

- For $\lambda_w \leq 5.76 \sqrt{\dfrac{E}{F_{yc}}}$ then $f_{cf} \leq R_h F_{yc}$ $\hspace{1cm}$ (8.41)

- For $5.76 \sqrt{\dfrac{E}{F_{yc}}} < \lambda_w \leq 6.43 \sqrt{\dfrac{E}{F_{yc}}}$ $\hspace{1cm}$ (8.42)

  then $f_{cf} \leq R_h F_{yc}\left(3.58 - 0.448\lambda_w\sqrt{\dfrac{F_{yc}}{E}}\right)$

- For $\lambda_w > 6.43 \sqrt{\dfrac{E}{F_{yc}}}$ then $f_{cf} \leq 28.9R_h \dfrac{E}{\lambda_w^2}$ $\hspace{1cm}$ (8.43)

where F_{yc} is the yield strength of the compression flange. A plot of Eqs. 8.41–8.43 is given in Figure 8.37 for $R_h = 1.0$, $E = 200$ GPa, and $F_{yc} = 345$

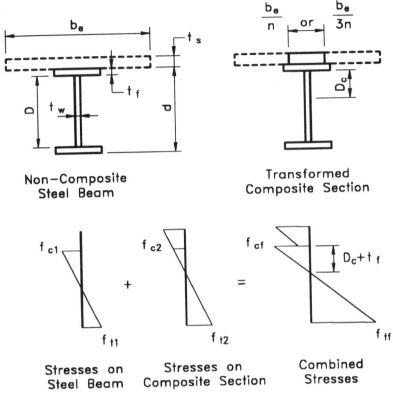

Fig. 8.36 Definition of web depth in compression.

MPa. The separation of the web flexural buckling response of Figure 8.37 into plastic, inelastic, and elastic behavior is typical of compression regions of *I*-sections in flexure. The plastic portion of the curve indicates that no web flexural buckling occurs before the yield stress is reached.

Shear buckling of the web can also occur. To increase the capacity, transverse stiffeners are provided at a spacing of d_o to subdivide the web into a series of rectangular panels with an aspect ratio of *a* of

$$a = \frac{d_o}{D} \tag{8.44}$$

where *D* is the clear depth of the web between the flanges of the beam (see Fig. 8.38).

The critical web shear buckling stress v_{cr} is dependent on the overall web slenderness ratio D/t_w and is expressed as a fraction *C* of the shear yield strength F_{yv}. The shear yield strength cannot be determined by itself, but is

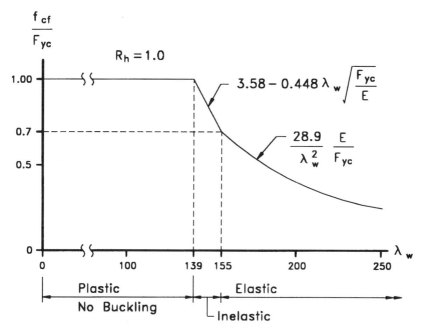

Fig. 8.37 Web flexural buckling behavior.

dependent on the shear failure criteria adopted. If the Mises shear failure criteria is used, the shear yield strength is related to the tensile yield strength by

$$F_{yv} = \frac{F_y}{\sqrt{3}} = 0.577F_y$$

so that

$$v_{cr} = CF_{yv} = 0.58CF_{yw}$$

where F_{yw} is the yield strength of the web. The maximum elastic shear stress in the web v_{cf} due to the unfactored permanent loads and double the fatigue load combination of Table 3.1 must not exceed v_{cr}, that is,

$$v_{cf} \le 0.58CF_{yw} \tag{8.45}$$

where C is defined as follows [A6.10.7.3.3a]:

- For $\dfrac{D}{t_w} < 1.10\sqrt{\dfrac{Ek}{F_{yw}}}$ then

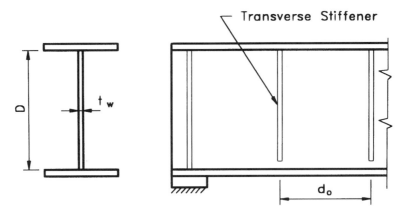

Fig. 8.38 Definition of web shear buckling terms.

$$C = 1.0 \tag{8.46}$$

- For $1.10\sqrt{\dfrac{Ek}{F_{yw}}} \le \dfrac{D}{t_w} \le 1.38\sqrt{\dfrac{Ek}{F_{yw}}}$ then

$$C = \dfrac{1.10}{\dfrac{D}{t_w}}\sqrt{\dfrac{Ek}{F_{yw}}} \tag{8.47}$$

- For $\dfrac{D}{t_w} > 1.38\sqrt{\dfrac{Ek}{F_{yw}}}$ then

$$C = \dfrac{1.52}{(D/t_w)^2}\dfrac{Ek}{F_{yw}} \tag{8.48}$$

where k is the shear buckling coefficient given by

$$k = 5 + \dfrac{5}{(d_o/D)^2} \tag{8.49}$$

A plot of Eqs. 8.46–8.48 is given in Figure 8.39 for $E = 200$ GPa, $F_{yw} = 345$ MPa and $d_o = D$. As in Figure 8.37, the plastic (no buckling), inelastic, and elastic behavior regions are also evident for web shear buckling.

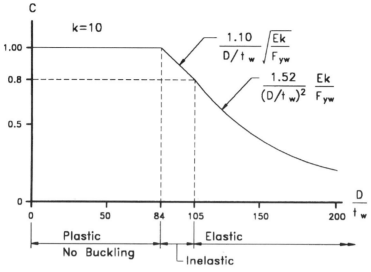

Fig. 8.39 Web shear buckling behavior.

8.7.3 Yield Moment and Plastic Moment

The bending moment capacity of *I*-sections depends primarily on the compressive force capacity of the compression flange. If the compression flange is continuously laterally supported and the web has stocky proportions, no buckling of the compression flange can occur and the cross section can develop its full plastic moment, that is, $M_n = M_p$. Cross sections that satisfy the restrictions on lateral support and on width/thickness ratios for flanges and web are called *compact sections*. These sections exhibit fully plastic behavior and their moment-curvature response is like the top curve in Figure 8.34.

If the compressed flange is laterally supported at intervals large enough to permit the compression flange to buckle locally, but not globally, then the compression flange will behave like an inelastic column. The section of the inelastic column will be *T*-shaped and part of it will have reached the yield stress and part of it will not. These cross sections are intermediate between plastic and elastic behavior and are called *noncompact sections*. They can develop the yield moment M_y, but have limited plastic response as shown in the middle curve of Figure 8.34.

If the compression flange is laterally unsupported at intervals large enough to permit lateral-torsional buckling, then the compression flange will behave as an elastic column whose capacity is an Euler-like critical buckling load reduced by the effect of torsion. The buckling of these sections with relatively high compression flange slenderness ratio occurs before the yield moment M_y can be reached and are called *slender sections*. The slender sections behavior is shown by the bottom curve in Figure 8.34.

The slender sections do not use materials effectively and most designers avoid them by providing sufficient lateral support. As a consequence, nearly all of the cross sections are designed as compact or noncompact.

Yield Moment of a Composite Section   The yield moment M_y is the moment that causes first yielding in either flange of the steel section. Because the cross section behaves elastically until first yielding, superposition of moments is valid. Therefore, M_y is the sum of the moment applied separately on the steel section, the short-term composite section, and the long-term composite section [A6.10.5.1.2].

The three stages of loading on a composite section are shown for a positive bending moment region in Figure 8.40. The moment due to factored permanent loads on the steel section before the concrete reaches 75% of its 28-day compressive strength is M_{D1} and it is resisted by the noncomposite section modulus S_{NC}. The moment due to the remainder of the factored permanent loads (wearing surface, concrete barrier) is M_{D2} and it is resisted by the long-term composite section modulus S_{LT}. The additional moment required to cause yielding in one of the steel flanges is M_{AD}. This moment is due to factored live load and is resisted by the short-term composite section modulus S_{ST}. The moment M_{AD} can be solved from the equation

$$F_y = \frac{M_{D1}}{S_{NC}} + \frac{M_{D2}}{S_{LT}} + \frac{M_{AD}}{S_{ST}} \tag{8.50}$$

and the yield moment M_y calculated from

$$M_y = M_{D1} + M_{D2} + M_{AD} \tag{8.51}$$

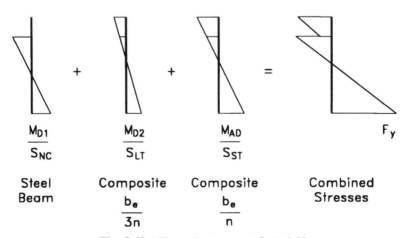

Fig. 8.40   Flexural stresses at first yield.

EXAMPLE 8.5

Determine the yield moment M_y for the composite girder cross section in Figure 8.41 subjected to factored positive moments $M_{D1} = 1180$ kN m and $M_{D2} = 419$ kN m. Use $f'_c = 30$ MPa for the concrete deck slab and Grade 345 structural steel for the girder.

Properties

The noncomposite, short-term, and long-term section properties are calculated in Tables 8.13–8.15. The modular ratio of $n = 8$ is taken from Table 8.12 for $f'_c = 30$ MPa. The transformed effective width of the slab is b_e divided by n for short-term properties and by $3n$, to account for creep, for long-term properties. The centroid of the section at each stage is calculated from the top of the steel beam and then the parallel axis theorem is used to get the moment of inertia of the components about this centroid.

$$\bar{y}_{NC} = \frac{26.784 \times 10^6}{29\,500} = 907.9 \text{ mm, below top of beam}$$

$$S^t_{NC} = \frac{10.607 \times 10^9}{907.9} = 11.68 \times 10^6 \text{ mm}^3, \text{ top of steel}$$

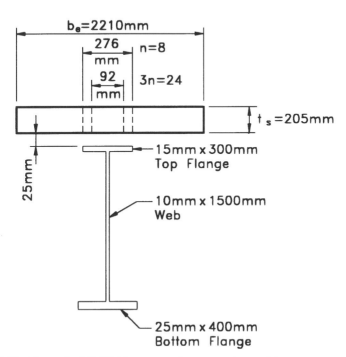

Fig. 8.41 Example 8.5. Yield moment for composite positive moment section.

TABLE 8.13 Noncomposite Section Properties

Component	A (mm^2)	y (mm)	Ay (mm^3)	$A(y - \bar{y})^2$ (mm^4)	I_o (mm^4)	I_x (mm^4)
Top flange 15 mm × 300 mm	4 500	7.5	0.034×10^6	3.649×10^9	8.44×10^4	3.649×10^9
Web 10 mm × 1500 mm	15 000	765	11.475×10^6	0.306×10^9	2.813×10^9	3.119×10^9
Bottom flange 25 mm × 400 mm	10 000	1527.5	15.275×10^6	3.839×10^9	5.21×10^5	3.839×10^9
Sum	29 500		26.784×10^6			10.607×10^9

TABLE 8.14 Short-Term Section Properties, $n = 8$

Component	A (mm^2)	y (mm)	Ay (mm^3)	$A(y - \bar{y})^2$ (mm^4)	I_o (mm^4)	I_x (mm^4)
Steel beam	29 500	907.9	26.784×10^6	13.672×10^9	10.607×10^9	24.27×10^9
Concrete slab 205 mm $\times$ (2210/8) mm	56 631	-127.5	-7.22×10^6	7.122×10^9	0.198×10^9	7.320×10^9
Sum	86 131		19.563×10^6			31.599×10^9

TABLE 8.15 Long-Term Section Properties, $3n = 24$

Component	A (mm^2)	y (mm)	Ay (mm^3)	$A(y - \bar{y})^2$ (mm^4)	I_o (mm^4)	I_x (mm^4)
Steel beam	29 500	907.9	26.78×10^6	4.815×10^9	10.607×10^9	15.422×10^9
Concrete slab 205 mm $\times$ (2210/24) mm	18 877	-127.5	-2.407×10^6	7.526×10^9	0.066×10^9	7.592×10^9
Sum	48 377		24.377×10^6			23.014×10^9

$$S_{NC}^b = \frac{10.607 \times 10^9}{1540 - 907.9} = 16.78 \times 10^6 \text{ mm}^3, \text{ bottom of steel}$$

$$\bar{y}_{ST} = \frac{19.563 \times 10^6}{86\,131} = 227.1 \text{ mm, below top of beam}$$

$$S_{ST}^t = \frac{31.599 \times 10^9}{227.1} = 139.12 \times 10^6 \text{ mm}^3, \text{ top of steel}$$

$$S_{ST}^b = \frac{31.599 \times 10^9}{1540 - 227.1} = 24.07 \times 10^6 \text{ mm}^3, \text{ bottom of steel}$$

$$\bar{y}_{LT} = \frac{24.377 \times 10^6}{48\,377} = 503.9 \text{ mm, below top of beam}$$

$$S_{LT}^t = \frac{23.014 \times 10^9}{503.9} = 45.67 \times 10^6 \text{ mm}^3, \text{ top of steel}$$

$$S_{LT}^b = \frac{23.014 \times 10^9}{1540 - 503.9} = 22.21 \times 10^6 \text{ mm}^3, \text{ bottom of steel}$$

Solution

The stress at the bottom of the girder will reach yield first. From Eq. 8.50,

$$F_y = \frac{M_{D1}}{S_{NC}} + \frac{M_{D2}}{S_{LT}} + \frac{M_{AD}}{S_{ST}}$$

$$345 = \frac{1180 \times 10^6}{16.78 \times 10^6} + \frac{419 \times 10^6}{22.21 \times 10^6} + \frac{M_{AD}}{24.07 \times 10^6}$$

$$M_{AD} = 24.07 \times 10^6 (345 - 70.3 - 18.9) = 6157 \times 10^6 \text{ N mm}$$

$$M_{AD} = 6157 \text{ kN m}$$

Answer

From Eq. 8.51, the yield moment is

$$M_y = M_{D1} + M_{D2} + M_{AD}$$

$$M_y = 1180 + 419 + 6157 = \underline{7756 \text{ kN m}}$$

Yield Moment of a Noncomposite Section   For a noncomposite section, the section moduli in Eq. 8.50 are all equal to S_{NC} and the yield moment M_y is simply

$$M_y = F_y S_{NC} \tag{8.52}$$

Plastic Neutral Axis of a Composite Section   The first step in determining the plastic moment strength of a composite section is to locate the neutral axis of the plastic forces. The plastic forces in the steel portions of the cross section is the product of the area of the flanges, web, and reinforcement times their appropriate yield strengths. The plastic forces in the concrete portions of the cross section, which are in compression are based on the equivalent rectangular stress block with uniform stress of 0.85 f'_c. Concrete in tension is neglected [A6.10.5.1.3].

The location of the plastic neutral axis (PNA) is obtained by equating the plastic forces in compression to the plastic forces in tension. If it is not obvious, it may be necessary to assume a location of the PNA and then to prove or disprove the assumption by summing plastic forces. If the assumed location does not satisfy equilibrium, then a revised expression is solved to determine the correct location of the PNA.

EXAMPLE 8.6

Determine the location of the plastic neutral axis for the composite cross section of Example 8.5 subjected to positive moment bending. Use $f'_c = 30$ MPa for the concrete and $F_y = 345$ MPa for the steel. Neglect the plastic forces in the longitudinal reinforcement of the deck slab.

Plastic Forces

The general dimensions and plastic forces are shown in Figure 8.42.

- Slab

$$P_s = 0.85f'_c b_e t_s = 0.85(30)(2210)(205) = 11.55 \times 10^6 \text{ N}$$

- Compression flange

$$P_c = F_y b_c t_c = 345(300)(15) = 1.55 \times 10^6 \text{ N}$$

- Web

$$P_w = F_y D t_w = 345(1500)(10) = 5.175 \times 10^6 \text{ N}$$

- Tension flange

$$P_t = F_y b_t t_t = 345(400)(25) = 3.45 \times 10^6 \text{ N}$$

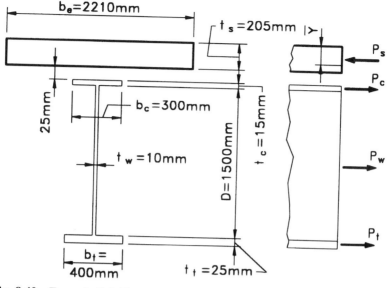

Fig. 8.42   Example 8.6. Plastic forces for composite positive moment section.

Solution

By inspection, the PNA lies in the slab because

$$P_s > P_c + P_w + P_t$$

Only a portion of the slab is required to balance the plastic forces in the steel beam, that is,

$$\frac{\overline{Y}}{t_s} P_s = P_c + P_w + P_t$$

so that the PNA is located a distance $\overline{Y}$ from the top of the slab

$$\overline{Y} = t_s \frac{P_c + P_w + P_t}{P_s} \tag{8.53}$$

Answer

By substituting the values from above

$$\overline{Y} = 205 \frac{(1.55 + 5.175 + 3.45) \times 10^6}{11.55 \times 10^6} = \underline{180.6 \text{ mm}}$$

In a region of negative bending moment where shear connectors develop composite action, the reinforcement in a concrete deck slab can be considered effective in resisting bending moments. In contrast to the positive moment region, where their lever arm is small, the contribution of the reinforcement in the negative moment region can make a difference.

EXAMPLE 8.7

Determine the location of the plastic neutral axis for the composite cross section of Figure 8.43 when subjected to negative bending moment. Use $f'_c = 30$ MPa and $F_y = 345$ MPa. Consider the plastic forces in the longitudinal reinforcement of the deck slab to be provided by two layers with 9-No. 10 bars in the top layer and 7-No. 15 bars in the bottom layer. Use $f_y = 400$ MPa.

Plastic Forces

The general dimensions and plastic forces are shown in Figure 8.43. The concrete slab is in tension and is considered to be noneffective, that is, $P_s = 0$.

- Top reinforcement

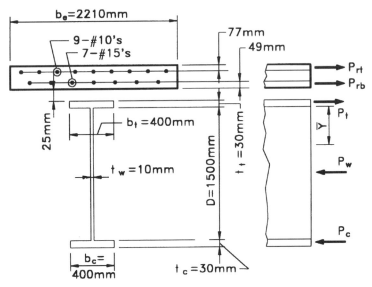

Fig. 8.43   Example 8.7. Plastic forces for composite negative moment section.

$$P_{rt} = A_{rt}f_y = 9(100)(400) = 0.36 \times 10^6 \text{ N}$$

- Bottom reinforcement

$$P_{rb} = A_{rb}f_y = 7(200)(400) = 0.56 \times 10^6 \text{ N}$$

- Tension flange

$$P_t = F_yb_tt_t = 345(400)(30) = 4.14 \times 10^6 \text{ N}$$

- Web

$$P_w = F_yDt_w = 345(1500)(10) = 5.175 \times 10^6 \text{ N}$$

- Compression flange

$$P_c = F_yb_ct_c = 345(400)(30) = 4.14 \times 10^6 \text{ N}$$

Solution

By inspection, the PNA lies in the web because

$$P_c + P_w > P_t + P_{rb} + P_{rt}$$

The plastic force in the web must be divided into tension and compression plastic forces to obtain equilibrium, that is,

$$P_c + P_w\left(1 - \frac{\overline{Y}}{D}\right) = P_w\left(\frac{\overline{Y}}{D}\right) + P_t + P_{rb} + P_{rt}$$

where $\overline{Y}$ is the distance from the top of the web to the PNA. Solving for $\overline{Y}$, we get

$$\overline{Y} = \frac{D}{2} \frac{P_c + P_w - P_t - P_{rb} - P_{rt}}{P_w} \tag{8.54}$$

Answer

By substituting the values from above

$$\overline{Y} = \frac{1500}{2} \frac{(4.14 + 5.175 - 4.14 - 0.56 - 0.36) \times 10^6}{5.175 \times 10^6} = \underline{616.7 \text{ mm}}$$

Plastic Neutral Axis of a Noncomposite Section For a noncomposite section, there is no contribution from the deck slab and the PNA is determined from Eq. 8.54 with $P_{rb} = P_{rt} = 0$. If the steel beam section is symmetric with equal top and bottom flanges, then $P_c = P_t$ and $\overline{Y} = D/2$.

Plastic Moment of a Composite Section The plastic moment M_p is the sum of the moments of the plastic forces about the PNA [A6.10.5.1.3]. It can best be described by examples. The calculations assume that global and local buckling are prevented so that plastic forces can be developed.

EXAMPLE 8.8

Determine the positive plastic moment for the composite cross section of Example 8.6 shown in Figure 8.42. The plastic forces were calculated in Example 8.6 and $\overline{Y}$ was determined to be 180.6 mm from the top of the slab.

Moment Arms

The moment arms about the PNA for each of the plastic forces can be found from the dimensions given in Figure 8.42.

- Slab

$$d_s = \frac{\overline{Y}}{2} = \frac{180.6}{2} = 90.3 \text{ mm}$$

- Compression flange

$$d_c = (t_s - \overline{Y}) + 25 + \frac{t_c}{2} = (205 - 180.6) + 25 + \frac{15}{2} = 56.9 \text{ mm}$$

- Web

$$d_w = (t_s - \overline{Y}) + 25 + t_c + \frac{D}{2}$$

$$= (205 - 180.6) + 25 + 15 + \frac{1500}{2} = 814.4 \text{ mm}$$

- Tension flange

$$d_t = (t_s - \overline{Y}) + 25 + t_c + D + \frac{t_t}{2}$$

$$= (205 - 180.6) + 25 + 15 + 1500 + \frac{25}{2} = 1576.9 \text{ mm}$$

Solution

The sum of the moments of the plastic forces about the PNA is the plastic moment.

$$M_p = \frac{\overline{Y}}{t_s} P_s d_s + P_c d_c + P_w d_w + P_t d_t \tag{8.55}$$

Answer

By substituting the values from above

$$M_p = \frac{180.6}{205} (11.55 \times 10^6)(90.3) + 1.55 \times 10^6 (56.9)$$

$$+ 5.175 \times 10^6 (814.4) + 3.45 \times 10^6 (1576.9)$$

$$M_p = 10.66 \times 10^9 \text{ N mm} = \underline{10\ 660 \text{ kN m}}$$

EXAMPLE 8.9

Determine the negative plastic moment for the composite cross section of Example 8.7 shown in Figure 8.43. The plastic forces were calculated in Example 8.7 and $\overline{Y}$ was determined to be 616.7 mm from the top of the web.

Moment Arms

The moment arms about the PNA for each of the plastic forces can be found from the dimensions given in Figure 8.43.

- Top reinforcement

$$d_{rt} = \overline{Y} + t_t + 25 + t_s - 77$$

$$= 616.7 + 30 + 25 + 205 - 77 = 799.7 \text{ mm}$$

- Bottom reinforcement

$$d_{rb} = \overline{Y} + t_t + 25 + 49 = 616.7 + 30 + 25 + 49 = 720.7 \text{ mm}$$

• Tension flange

$$d_t = \overline{Y} + \frac{t_t}{2} = 616.7 + \frac{30}{2} = 631.7 \text{ mm}$$

• Web in tension

$$d_{wt} = \frac{\overline{Y}}{2} = \frac{616.7}{2} = 308.4 \text{ mm}$$

• Web in compression

$$d_{wc} = \tfrac{1}{2}(D - \overline{Y}) = \tfrac{1}{2}(1500 - 616.7) = 441.7 \text{ mm}$$

• Compression flange

$$d_c = (D - \overline{Y}) + \frac{t_c}{2} = (1500 - 616.7) + \frac{30}{2} = 898.3 \text{ mm}$$

Solution

The plastic moment is the sum of the moments of the plastic forces about the PNA.

$$M_p = P_{rt}d_{rt} + P_{rb}d_{rb} + P_t d_t + \frac{\overline{Y}}{D} P_w d_{wt} + \frac{(D - \overline{Y})}{D} P_w d_{wc} + P_c d_c \quad (8.56)$$

Answer

By substituting the values from above

$$M_p = 0.36 \times 10^6 (799.7) + 0.56 \times 10^6 (720.7) + 4.14 \times 10^6 (631.7)$$

$$+ \frac{616.7}{1500} (5.175 \times 10^6)(308.4)$$

$$+ \frac{(1500 - 616.7)}{1500} (5.175 \times 10^6)441.7 + 4.14 \times 10^6 (898.3)$$

$$M_p = 9.028 \times 10^9 \text{ N mm} = \underline{9028 \text{ kN m}}$$

Plastic Moment of a Noncomposite Section   If no shear connectors exist between the concrete deck and the steel cross section, the concrete slab and its reinforcement do not contribute to the section properties. Consider the cross section of Figure 8.43 to be noncomposite. Then $P_{rt} = P_{rb} = 0$ and $\overline{Y} = D/2$, and Eq. 8.56 becomes

$$M_p = P_t\left(\frac{D}{2} + \frac{t_t}{2}\right) + P_w\left(\frac{D}{4}\right) + P_c\left(\frac{D}{2} + \frac{t_c}{2}\right) \tag{8.57}$$

$$= 4.14 \times 10^6\left(\frac{1500}{2} + \frac{30}{2}\right) + 5.175 \times 10^6\left(\frac{1500}{4}\right)$$

$$+ 4.14 \times 10^6\left(\frac{1500}{2} + \frac{30}{2}\right)$$

$$= 8.275 \times 10^9 \text{ N mm} = \underline{8275 \text{ kN m}}$$

Depth of Web in Compression   When evaluating the slenderness of a web as a measure of its stability, the depth of the web in compression is important. In a noncomposite cross section with a doubly symmetric steel beam, one-half of the web depth D will be in compression. For unsymmetric noncomposite cross sections and composite cross sections, the depth of web in compression will not be $D/2$ and will vary with the direction of bending in continuous girders.

When stresses due to unfactored loads remain in the elastic range, the depth of the web in compression D_c shall be the depth over which the algebraic sum of stresses due to the dead load D_1 on the steel section plus the dead load D_2 and live load $LL + IM$ on the short-term composite section are compressive [A6.10.5.1.4a].

EXAMPLE 8.10

Determine the depth of web in compression D_c for the cross section of Figure 8.41 whose elastic properties were calculated in Example 8.5. The cross section is subjected to *unfactored* positive moments $M_{D1} = 978$ kN m, $M_{D2} = 361$ kN m, and $M_{LL+IM} = 1563$ kN m.

Solution

The stress at the top of the steel for the given moments and section properties is (see Fig. 8.36)

$$f_t = \frac{M_{D1}}{S_{NC}^t} + \frac{M_{D2}}{S_{LT}^t} + \frac{M_{LL+IM}}{S_{ST}^t}$$

$$= \frac{978 \times 10^6}{11.68 \times 10^6} + \frac{361 \times 10^6}{45.67 \times 10^6} + \frac{1563 \times 10^6}{139.12 \times 10^6}$$

$$= 102.9 \text{ MPa} \qquad \text{(compression)}$$

$$f_b = \frac{M_{D1}}{S_{NC}^b} + \frac{M_{D2}}{S_{LT}^b} + \frac{M_{LL+IM}}{S_{ST}^b}$$

$$= \frac{978 \times 10^6}{16.78 \times 10^6} + \frac{361 \times 10^6}{22.21 \times 10^6} + \frac{1563 \times 10^6}{24.07 \times 10^6} = 139.5 \text{ MPa} \qquad \text{(tension)}$$

Answer

Using the proportion of the section in compression and subtracting the thickness of the compression flange with $d = 1500 + 15 + 25 = 1540$ mm

$$D_c = d \frac{f_t}{f_t + f_b} - t_c = 1540 \frac{102.9}{102.9 + 139.5} - 15 = \underline{638.7 \text{ mm}}$$

The depth of web in compression at plastic moment D_{cp} is usually determined once the PNA is located. In Example 8.6, positive bending moment is applied and the PNA is located in the deck slab. The entire web is in tension and $D_{cp} = 0$.

In Example 8.7, the cross section is subjected to negative bending moment and the PNA is located 616.7 mm from the top of the web. The bottom portion of the web is in compression, so that

$$D_{cp} = D - \bar{Y} = 1500 - 616.7 = 883.3 \text{ mm}$$

The same results are obtained using the equations in AASHTO [A6.10.5.1.4b].

8.7.4 Web Slenderness

In addition to resisting shear forces, the web has the function of supporting the flanges far enough apart so that bending is resisted effectively. When an I-section is subjected to bending, two failure mechanisms or limit states can occur in the web. The web can buckle as a vertical column that carries the compressive force which keeps the flanges apart or the web can buckle as a plate due to horizontal in-plane bending stresses. Both of these failure mechanisms require limitations on the slenderness of the web.

Vertical Buckling of the Web When bending occurs in an *I*-section, curvature produces compressive stresses between the flanges and the web of the cross section. These compressive stresses are a result of the vertical component of the flange force as shown schematically for a doubly symmetric *I*-section in Figure 8.44. To develop the yield moment of the cross section requires that the compression flange reach its yield stress F_{yc} before the web buckles. If the web is too slender, it will buckle as a column, the compression flange will lose its support and buckle vertically into the web before the yield moment is reached.

Vertical buckling of the flange into the web can be shown by considering the elemental length of web dx along the axis of the beam in Figure 8.45. It is subjected to an axial compressive stress f_{wc} from the vertical component of the compression flange force P_c. From Figure 8.44, the vertical component is $P_c d\phi$, which for a doubly symmetrical *I*-section is

$$d\phi = \frac{2\varepsilon_{fc}}{D}dx \qquad (8.58)$$

where ε_{fc} is the strain in the compression flange and D is the web depth. The axial compressive stress in the web then becomes

$$f_{wc} = \frac{P_c d\phi}{t_w dx} = \frac{2A_{fc}f_c\varepsilon_{fc}}{Dt_w} \qquad (8.59)$$

where A_{fc} is the area of the compression flange and f_c is the stress in the

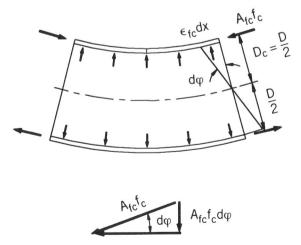

Fig. 8.44 Web compression due to curvature. (After Basler and Thürlimann, 1961.)

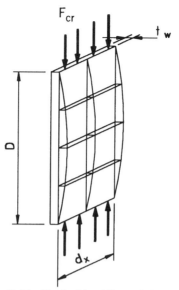

Fig. 8.45   Vertical buckling of the web.

compression flange. Equation 8.59 can be written in terms of the cross-sectional area of the web $A_w = Dt_w$ as

$$f_{wc} = \frac{2A_{fc}}{A_w} f_c \varepsilon_{fc} \qquad (8.60)$$

Thus, the vertical compressive stress in the web is proportional to the ratio of flange area to web area in the cross section, the compressive stress in the flange, and the compressive strain in the flange. The strain ε_{fc} is not simply f_c/E, but must also include the effect of residual stress f_r in the flange (Fig. 8.22), that is,

$$\varepsilon_{fc} = \frac{(f_c - f_r)}{E}$$

so that Eq. 8.60 becomes

$$f_{wc} = \frac{2A_{fc}}{EA_w} f_c(f_c + f_r) \qquad (8.61)$$

and a relationship between the compressive stress in the web and the compressive stress in the flange is determined.

By assuming the element in Figure 8.45 is from a long plate that is simply supported along the top and bottom edges, the critical elastic buckling or Euler load is

$$P_{cr} = \frac{\pi^2 EI}{D^2} \tag{8.62}$$

for which the moment of inertia I for the element plate length dx is

$$I = \frac{t_w^3 dx}{12(1 - \mu^2)} \tag{8.63}$$

where Poisson's ratio μ takes into account the stiffening effect of the two-dimensional (2-D) action of the web plate. The critical buckling stress F_{cr} is obtained by dividing Eq. 8.62 by the elemental area $t_w dx$ to yield

$$F_{cr} = \frac{\pi^2 E t_w^3 dx}{12(1 - \mu^2)D^2 t_w dx} = \frac{\pi^2 E}{12(1 - \mu^2)}\left(\frac{t_w}{D}\right)^2 \tag{8.64}$$

To prevent vertical buckling of the web, the stress in the web must be less than the critical buckling stress, that is

$$f_{wc} < F_{cr} \tag{8.65}$$

By substituting Eqs. 8.61 and 8.64 into Eq. 8.65 we get

$$\frac{2A_{fc}}{EA_w} f_c(f_c + f_r) < \frac{\pi^2 E}{12(1 - \mu^2)}\left(\frac{t_w}{D}\right)^2$$

Solving for the web slenderness ratio D/t_w results in

$$\left(\frac{D}{t_w}\right)^2 < \frac{A_w}{A_{fc}} \frac{\pi^2 E^2}{24(1 - \mu^2)} \frac{1}{f_c(f_c + f_r)} \tag{8.66}$$

To develop the yield moment M_y in the symmetric *I*-section, it is required that the compressive stress in the flange f_c reach the yield stress f_{yc} before the web buckles vertically. Assume a minimum value of 0.5 for A_w/A_{fc} and a maximum value of $0.5F_{yc}$ for f_r, then a minimum upper limit on the web slenderness ratio can be estimated from Eq. 8.66

$$\frac{D}{t_w} < \sqrt{\frac{0.5\pi^2 E^2}{24[1 - 0.3^2] F_{yc}^2(1.5)}} = 0.388 \frac{E}{F_{yc}} \tag{8.67}$$

where Poisson's ratio for steel has been taken as 0.3. Equation 8.67 is not rigorous in its derivation because of the assumptions on A_w/A_{fc} and f_r, but it can be useful as an approximate measure of web slenderness to avoid vertical buckling of the flange into the web. For example, if $E = 200$ GPa and $F_{yc} = 250$ MPa, then Eq. 8.67 requires that D/t_w be less than 310.

Bend Buckling of the Web   Because bending produces compressive stresses over a part of the web, buckling out of the plane of the web can occur as shown in Figure 8.46. The elastic critical buckling stress is given by a generalization of Eq. 8.64, that is,

$$F_{cr} = \frac{k\pi^2 E}{12(1 - \mu^2)}\left(\frac{t_w}{D}\right)^2 \tag{8.68}$$

where k is the buckling coefficient that depends on the boundary conditions of the four edges, the aspect ratio (Eq. 8.44) of the plate, and the distribution of the in-plane stresses. For all four edges simply supported and an aspect ratio greater than one, Timoshenko and Gere (1969) give the values of k for the different stress distributions shown in Figure 8.46.

Solving Eq. 8.68 for the web slenderness ratio yields

$$\left(\frac{D}{t_w}\right)^2 = \frac{k\pi^2}{12(1 - \mu^2)}\frac{E}{F_{cr}}$$

For the I-section to reach the yield moment before the web buckles, the critical buckling stress F_{cr} must be greater than F_{yc}. Therefore, setting $\mu = 0.3$, the web slenderness requirement for developing the yield moment becomes

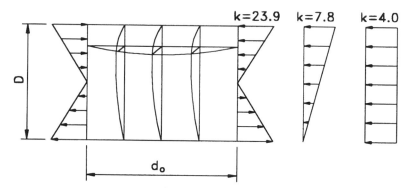

Fig. 8.46   Bending buckling of the web.

$$\frac{D}{t_w} \leq \sqrt{\frac{k(0.904)E}{F_{yc}}} = 0.95\sqrt{k}\sqrt{\frac{E}{F_{yc}}} \tag{8.69}$$

For the pure bending case of Fig. 8.46, $k = 23.9$.

$$\frac{D}{t_w} \leq 0.95\sqrt{23.9}\sqrt{\frac{E}{F_{yc}}} = 4.64\sqrt{\frac{E}{F_{yc}}} \tag{8.70}$$

Comparisons with experimental tests indicate that Eq. 8.70 is too conservative because it neglects the postbuckling strength of the web.

The AASHTO (1994) LRFD Bridge Specifications give slightly different expressions for defining the web slenderness ratio that separates elastic and inelastic buckling. To generalize the left side of Eq. 8.69 for unsymmetric *I*-sections, the depth of the web in compression D_c, defined in Figure 8.36 and calculated in Example 8.10, replaces $D/2$ for the symmetric case to yield

$$\frac{D}{t_w} = \frac{2D_c}{t_w} \tag{8.71}$$

The right side of Eq. 8.69 for unsymmetric *I*-sections is modified for the case of a stress in the compression flange f_c less than the yield stress F_{yc}. Further, to approximate the postbuckling strength and the effect of longitudinal stiffeners, the value for k is effectively taken as 50 and 150 for webs without and with longitudinal stiffeners, respectively. The AASHTO expressions are [Table A6.10.5.3.1-1].

- Without longitudinal stiffeners

$$\frac{2D_c}{t_w} \leq 6.77\sqrt{\frac{E}{f_c}} \tag{8.72}$$

- With longitudinal stiffeners

$$\frac{2D_c}{t_w} \leq 11.63\sqrt{\frac{E}{f_c}} \tag{8.73}$$

Compact Section Requirement for the Web   A compact section is one that can develop the full plastic moment M_p. Not only do the flanges go plastic, but, as shown in Figure 8.29, so does the web. Large strains must take place at the juncture of the flange and web for the plasticity to spread into the web. To prevent the web from buckling before sufficient rotation has taken place,

k is effectively taken as 16. Because the slenderness requirement is for the plastic moment, the depth of the web in compression D_{cp} based on the plastic neutral axis replaces D_c in Eq. 8.71. Substituting into Eq. 8.69, the web slenderness requirement for a compact section becomes [Table A6.10.5.2.1-1]

$$\frac{2D_{cp}}{t_w} \le 3.76\sqrt{\frac{E}{F_{yc}}} \tag{8.74}$$

Summary of Web Slenderness Effects Figure 8.47 is a generalized plot of bending moment capacity M_n as a function of a slenderness parameter λ. Once again, the three types of behavior (plastic, inelastic, and elastic) are apparent. The web slenderness parameter λ is

$$\lambda = \frac{2D_{cp}}{t_w} \quad \text{or} \quad \frac{2D_c}{t_w} \tag{8.75}$$

and the values at the transition points are

$$\lambda_p = 3.76\sqrt{\frac{E}{F_{yc}}} \tag{8.76}$$

and (for webs without longitudinal stiffeners)

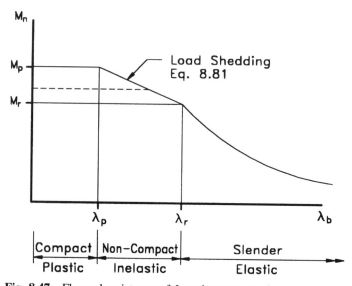

Fig. 8.47 Flexural resistance of *I*-sections versus slenderness ratio.

$$\lambda_r = 6.77\sqrt{\frac{E}{f_c}} \tag{8.77}$$

The plastic moment resistance M_p is based on F_{yc} and plastic section properties are illustrated in Examples 8.8 and 8.9. The elastic moment resistance M_r is based on the nominal flexural stress F_n and elastic section properties as illustrated in Example 8.5.

8.7.5 Load Shedding Factor

When an *I*-section is noncompact, the nominal flexural resistance is based on the nominal flexural stress F_n given by [A6.10.5.3.2a]

$$F_n = R_b R_h F_{yf} \tag{8.78}$$

where R_b is the load shedding factor, R_h is the hybrid factor, and F_{yf} is the yield strength of the flange. When the flange and web have the same yield strength, $R_h = 1.0$. A hybrid girder has a lower strength material in the web than the flange. A unit value for R_h is assumed throughout this chapter.

The load shedding factor R_b provides a transition for inelastic sections with slenderness properties between λ_p and λ_r (Fig. 8.47). From analytical and experimental studies conducted by Basler and Thürlimann (1961), the transition was given by

$$\frac{M_u}{M_y} = 1 - C(\lambda - \lambda_o) \tag{8.79}$$

in which C is the slope of the line between λ_p and λ_r, and λ_o is the value of λ when $M_u/M_y = 1$. The constant C was expressed as

$$C = \frac{A_w/A_f}{1200 + 300A_w/A_f} \tag{8.80}$$

The AASHTO (1994) LRFD Bridge Specifications use the same form as Eqs. 8.79 and 8.80 for R_b [A6.10.5.4.2a], that is,

$$R_b = 1 - \left(\frac{a_r}{1200 + 300a_r}\right)\left(\frac{2D_c}{t_w} - \lambda_b\sqrt{\frac{E}{f_c}}\right) \tag{8.81}$$

in which

$$a_r = \frac{2D_c t_w}{A_{fc}} \qquad (8.82)$$

and

$\lambda_b = 5.76$ for members with a compression flange area equal to or greater than the tension flange area

$\lambda_b = 4.64$ for members with a compression flange area less than the tension flange area

8.7.6   Compression Flange Slenderness

Because of the postbuckling strength due to increased strain capacity of the web, an *I*-section will not fail in flexure when the web-buckling load is reached. However, it will fail in flexure when one of the framing members on the edges of a web panel fails. If one of the flanges or transverse stiffeners should fail, then the web displacements would be unrestrained, the web could no longer resist its portion of the bending moment, and the *I*-section would fail.

In a doubly symmetric *I*-section subjected to bending, the compression flange will fail first in local or global buckling. Therefore, the bracing and proportioning of the compression flange are important in determining the flexural resistance of *I*-sections. To evaluate the buckling strength of the compression flange, it will be considered as an isolated column.

By assuming a hybrid connection between the web and the flange, one-half of the compression flange can be modeled as a long uniformly compressed plate (Fig. 8.48) with one longitudinal edge free and the other simply supported. Usually, the plate is long compared to its width and the boundary conditions on the loaded edges are not significant and the buckling coefficient is $k = 0.425$ for uniform compression (Maquoi, 1992).

Compact Section Requirement for the Compression Flange   To develop the plastic moment M_p resistance in the *I*-section, the critical buckling stress F_{cr} must exceed the yield stress F_{yc} of the compression flange. In a manner similar to the development of Eq. 8.69, the limit for the compression flange slenderness becomes

$$\frac{b_f}{2t_f} \leq 0.95 \sqrt{k} \sqrt{\frac{E}{F_{yc}}} \qquad (8.83)$$

For an ideally perfect plate, $k = 0.425$ and the slenderness limit can be written as

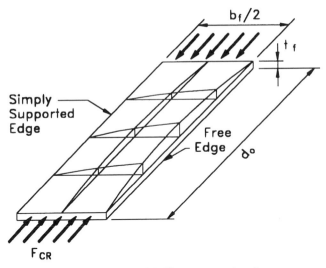

Fig. 8.48   Model of half a compression flange.

$$\frac{b_f}{2t_f} \leq 0.62\beta\sqrt{\frac{E}{F_{yc}}} \tag{8.84}$$

where β is a factor that accounts for both geometrical imperfections and residual stresses in the compression flange (Maquoi, 1992). The AASHTO (1994) LRFD Bridge Specifications take $\beta \approx 0.61$ and the compact section compression flange slenderness requirement becomes [A6.10.5.2.3c]

$$\frac{b_f}{2t_f} \leq 0.382\sqrt{\frac{E}{F_{yc}}} \tag{8.85}$$

If the steel *I*-section is composite with a concrete deck in a region of positive bending moment, the compression flange is fully supported throughout its length and the slenderness requirement does not apply.

Noncompact Section Limit for the Compression Flange   If the compression flange is too slender, elastic local buckling will occur prior to yielding. To ensure that some inelastic behavior takes place, the AASHTO (1994) LRFD Bridge Specifications require that [A6.10.5.3.3c]

$$\frac{b_f}{2t_f} \leq 1.38\sqrt{\frac{E}{f_c\sqrt{2D_c/t_w}}} \tag{8.86}$$

where f_c is the stress in the compression flange due to factored loading. Equation 8.86 is dependent on the web slenderness ratio $2D_c/t_w$ because it can vary between the values given by Eqs. 8.73 and 8.74 for noncompact sections.

As the web slenderness increases, the simply supported longitudinal edge in Figure 8.48 loses some of its vertical and transverse restraint. The effect of web slenderness on buckling of the compression flange can be shown by rewriting Eq. 8.86 as

$$\frac{b_f}{2t_f} \le C_f \sqrt{\frac{E}{f_c}} \tag{8.87}$$

in which

$$C_f = \frac{1.38}{\sqrt[4]{\dfrac{2D_c}{t_w}}} \tag{8.88}$$

where C_f is a compression flange slenderness factor that varies with $2D_c/t_w$ as shown in Figure 8.49. The value of C_f is comparable to the constant in Eq. 8.85 for compact sections. In fact, if $2D_c/t_w = 170$, they are the same. For values of $2D_c/t_w > 170$, the upper limit on $b_f/2t_f$ decreases until at $2D_c/t_w = 300$

$$\left(\frac{b_f}{2t_f}\right)_{300} = 0.332\sqrt{\frac{E}{f_c}} \tag{8.89}$$

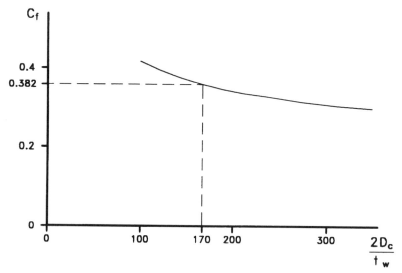

Fig. 8.49   Compression flange slenderness factor as a function of web slenderness.

Summary of Compression Flange Slenderness Effects Referring again to Figure 8.47 and the now familiar plot showing three types of behavior, the slenderness parameter λ for the compression flange is

$$\lambda = \frac{b_f}{2t_f} \tag{8.90}$$

and the values at the transition points are

$$\lambda_p = 0.382\sqrt{\frac{E}{F_{yc}}} \tag{8.91}$$

and

$$\lambda_r = 1.38\sqrt{\frac{E}{f_c\sqrt{2D_c/t_w}}} \tag{8.92}$$

The plastic moment resistance M_p is based on F_{yc} and plastic section properties, while the elastic moment resistance M_r is based on F_n of Eq. 8.78 and elastic section properties.

8.7.7 Compression Flange Bracing

Sections 8.7.5 and 8.7.6 on web slenderness and compression flange slenderness were concerned with local buckling of the compression region in *I*-sections subjected to bending. The problem of global buckling of the compression region as a column between brace points must also be addressed. As described by the stability limit state and illustrated in Figure 8.32, an unbraced compression flange will move laterally and twist in a mode known as *lateral-torsional buckling*.

If the compression flange is braced at sufficiently close intervals L_p, the compression flange material can yield before it buckles and the plastic moment M_p can be reached. If the distance between bracing points is greater than the inelastic buckling limit L_r, the compression flange will buckle elastically at a reduced moment capacity. This behavior can once again be shown by the generic moment–slenderness relationship of Figure 8.47 with the slenderness parameter given by

$$\lambda = \frac{L_b}{r_t} \tag{8.93}$$

where L_b is the distance between lateral brace points and r_t is the minimum radius of gyration of the compression flange plus one-third of the web in compression taken about the vertical axis in the plane of the web.

Because the unbraced length L_b is the primary concern in the design of I-sections for flexure, it is taken as the independent parameter rather than the slenderness ratio L_b/r_t in determining the moment resistance. Figure 8.47 is, therefore, redrawn as Figure 8.50 with L_b replacing λ. The same three characteristic regions remain: plastic (no buckling), inelastic lateral–torsional buckling, and elastic lateral-torsional buckling.

For L_b less than L_p in Figure 8.50, the compression flange is considered laterally supported and the moment resistance M_n is constant. The value of M_n depends on the classification of the cross section. If the cross section is classified as *compact*, the value of M_n is M_p. If the cross section is *noncompact* or *slender*, the value of M_n will be less than M_p. The dashed horizontal line on Figure 8.50 indicates a typical value of M_n for a section that is not compact.

For $L_b > L_r$, the compression flange will fail by elastic lateral–torsional buckling. This failure mode has a classical elasticity solution (Timoshenko and Gere, 1969) in which the moment resistance is the square root of the sum of the squares of two contributions: torsional buckling (St. Venant torsion) and lateral buckling (warping torsion), that is,

$$M_n^2 = M_{n,v}^2 + M_{n,w}^2 \tag{8.94}$$

where $M_{n,v}$ is the St. Venant torsional resistance and $M_{n,w}$ is the warping contribution. For the case of constant bending between brace points, Gaylord et al. (1992) derive the following expressions:

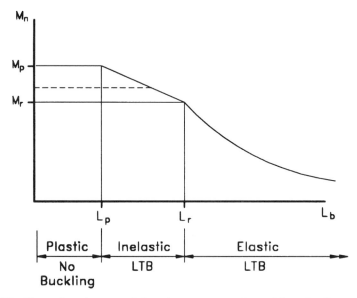

Fig. 8.50 Flexural resistance of I-sections versus unbraced length of compression flange.

$$M_{n,v}^2 = \frac{\pi^2}{L_b^2} EI_y GJ \qquad (8.95)$$

$$M_{n,w}^2 = \frac{\pi^4}{L_b^4} EI_y EC_w \qquad (8.96)$$

where I_y is the moment of inertia of the steel section about the vertical axis in the plane of the web, G is the shear modulus of elasticity, J is the St. Venant torsional stiffness constant, and C_w is the warping constant. When an *I*-section is short and stocky [Fig. 8.51(a)], pure torsional strength (St. Venant's torsion) dominates. When the section is tall and thin [Fig. 8.51(b)], warping torsional strength dominates.

For L_b between L_p and L_r, the compression flange will fail by inelastic lateral-torsional buckling. Because of its complexity the inelastic behavior is usually approximated from observations of experimental results. A straight-line estimate of the inelastic lateral-torsional buckling resistance is often used between the values at L_p and L_r.

Member Proportions   *I*-Sections in flexure shall be proportioned so that [A6.10.1.1]

$$0.1 \leq \frac{I_{yc}}{I_y} \leq 0.9 \qquad (8.97)$$

where I_{yc} is the moment of inertia of the compression flange of the steel

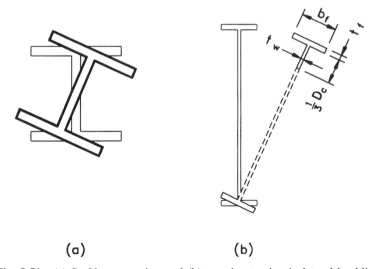

(a)   **(b)**

Fig. 8.51   (a) St. Venant torsion and (b) warping torsion in lateral buckling.

section about the vertical axis in the plane of the web and I_y is the moment of inertia of the steel section about the vertical axis in the plane of the web. If the member proportions are not within the limits given, the formulas for lateral-torsional buckling used in AASHTO (1994) LRFD Bridge Specifications are not valid.

Moment Gradient Correction Factor C_b   Equations 8.95 and 8.96 were derived for constant moment between brace points. This worst case scenario is overly conservative for the general case of varying applied moment over the unbraced length. To account for I-sections with both variable depth and variable moment, the force in the compression flange at the brace points is used to measure the effect of the moment gradient. The expression for the correction factor is given as [A6.10.5.5.2]

$$C_b = 1.75 - 1.05\left(\frac{P_1}{P_2}\right) + 0.3\left(\frac{P_1}{P_2}\right)^2 \leq 2.3 \tag{8.98}$$

where P_1 is the force in the compression flange at the brace point with the smaller force due to factored loading and P_2 is the force in the compression flange at the brace point with the larger force due to factored loading. Substituting Eqs. 8.95 and 8.96 into Eq. 8.94, solving for M_n and applying correction factor C_b, we have

$$M_n = C_b\sqrt{\frac{\pi^2}{L_b^2} EI_yGJ + \frac{\pi^4}{L_b^4} EI_yEC_w} \tag{8.99}$$

An I-section with moments M_1 and M_2 at the brace points is shown in Figure 8.52. The moment diagram between the brace points is given in Figure 8.52(a) and the corresponding compression flange forces P_1 and P_2 in Figure 8.52(b). If $P_1 = P_2$, Eq. 8.98 gives $C_b = 1.0$. As the compression flange force P_1 decreases, the lateral-torsional buckling strength increases. If $P_1 = 0$ [Fig. 8.52(c)], then $C_b = 1.75$. If P_1 goes into tension, C_b continues to increase until it reaches its maximum value of 2.3 at $P_1 = -0.46P_2$ [Fig. 8.52(d)].

In many cases the moment gradient between brace points is not linear. For example, when a uniformly distributed load is applied to an I-section between brace points, the moment variation is parabolic. Improved results are obtained for nonlinear moment gradients using the following alternative formulation for C_b [C6.10.5.5.2]

$$C_b = \frac{12.5P_{max}}{2.5P_{max} + 3P_A + 4P_B + 3P_C} \tag{8.100}$$

where P_{max} is the absolute value of the maximum compression flange force in the unbraced segment, P_A is the absolute value of the compression flange

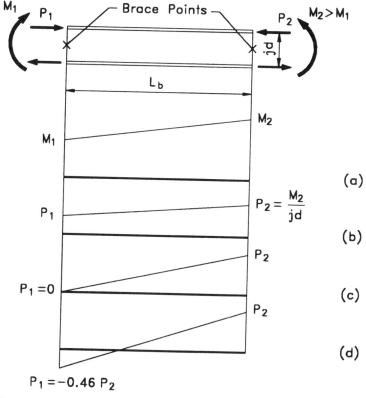

Fig. 8.52 (a) Moment gradient between brace points, (b) compression flange forces corresponding to M_1 and M_2, (c) compression flange forces when $M_1 = 0$, and (d) compression flange forces when $M_1 = -0.46M_2$.

force at the quarter point of the unbraced segment, P_B is the absolute value of the compression flange force at midspan of the unbraced segment, and P_C is the absolute value of the compression flange force at the three-quarter point of the unbraced segment. Applying Eq. 8.100 to the linear cases in Figure 8.52, the results are: for $P_1 = P_2$, $C_b = 1.0$; for $P_1 = 0$, $C_b = 1.67$; and for $P_1 = -0.46P_2$, $C_b = 2.17$. Therefore, Eq. 8.100 gives slightly conservative results for linear variations of the moment gradient compared to Eq. 8.98, and can conveniently be used to represent all cases of moment gradient.

Many of the articles in the AASHTO specification were taken directly or indirectly from specifications that address primarily stationary loads and are therefore sometimes difficult to use for bridges. The values for the compression flanges forces should be those forces that are coincidence with the forces that cause the critical load effect of the section of interest. The coincidence actions are actions at other sections when the cross section of interest is loaded for critical effect. Such actions or the load effects such as flange forces are not easily computed. The AASHTO specification addresses this issue by permitting the use of the moment envelope to estimate coincidence actions in many such cases. It is also of interest in such cases where a point of contra-flexure occurs in a composite section in the "unbraced" length, for example, with a negative moment transitioning to a positive before a brace point is encountered. Should the composite section be used at the brace point or should the moment be used at the brace point or should the composite effect be neglected? To the authors' knowledge, this is addressed both ways in practice. It should be noted that one method is not necessarily conservative for all cases.

Noncomposite Elastic I-Sections   For noncomposite *I*-sections, the compactness requirements are the same as for composite sections in negative flexure [A6.10.6.2 and A6.10.6.3]. When the unbraced length L_b exceeds the noncompact (inelastic) section requirement [A6.10.5.3.3d]

$$L_b > L_p = 1.76r_t\sqrt{\frac{E}{F_{yc}}}$$ (8.101)

then the cross section behaves elastically and has a nominal resisting moment (horizontal dashed line in Fig. 8.50) less than or equal to M_y.

If the web is relatively stocky, or if a longitudinal stiffener is provided, bend buckling of the web cannot occur and both the pure torsion and warping torsion resistances in Eq. 8.99 can be included in calculating M_n. Some simplification to Eq. 8.99 occurs if it is assumed that the *I*-section is doubly symmetric and the moment of inertia of the steel section about the weak axis I_y, neglecting the contribution of the web, is

$$I_y \approx I_{yc} + I_{yt} = 2I_{yc}$$ (8.102)

Also, the shear modulus G can be written for Poisson's ratio $\mu = 0.3$ as

$$G = \frac{E}{2(1 + \mu)} = \frac{E}{2(1 + 0.3)} = 0.385E$$ (8.103)

and the warping constant C_w for a webless *I*-section becomes (Kitipornchai and Trahair, 1980)

$$C_w \approx I_{yc}\left(\frac{d}{2}\right)^2 + I_{yt}\left(\frac{d}{2}\right)^2 = \frac{d^2}{2}I_{yc}$$ (8.104)

where d is the depth of the steel section. Substituting Eqs. 8.102–8.104 into Eq. 8.99 and factoring out the common terms results in [A6.10.6.4.1]

$$M_n = \frac{\pi EC_b}{L_b} \sqrt{(2I_{yc})(0.385)J + \frac{\pi^2}{L_b^2}(2I_{yc})\frac{d^2}{2}(I_{yc})}$$

$$M_n = \pi EC_b \frac{I_{yc}}{L_b} \sqrt{0.77\left(\frac{J}{I_{yc}}\right) + \pi^2\left(\frac{d}{L_b}\right)^2} \le M_y \qquad (8.105)$$

which is valid as long as

$$\frac{2D_c}{t_w} \le \lambda_b \sqrt{\frac{E}{F_{yc}}} \qquad (8.106)$$

where λ_b has been defined previously with Eq. 8.81 and

$$L_b < L_p = 1.76r' \sqrt{\frac{E}{F_{yc}}} \qquad (8.107)$$

where r_t of Eq. 8.101 has been replaced by r', the minimum radius of gyration of the compression flange about the vertical axis, to be consistent with the assumed webless section.

Even though Eq. 8.105 was derived for a doubly symmetric *I*-section ($I_{yc}/I_y = 0.5$), it can be used for a singly symmetric *I*-section that satisfies Eq. 8.97. For *I*-sections composed of narrow rectangular elements, the St. Venant torsional stiffness constant J can be approximated by

$$J = \frac{Dt_w^3}{3} + \sum \frac{b_f t_f^3}{3} \qquad (8.108)$$

In the development of Eq. 8.105, the hybrid factor R_h has been taken as 1.0, that is, the material in the flanges and web have the same yield strength.

For *I*-sections with webs thinner than the limit of Eq. 8.106 or without longitudinal stiffeners, cross-sectional distortion is possible and the St. Venant torsional stiffness can be neglected [C6.10.6.4.1]. Setting $J = 0$ in Eq. 8.105, the elastic lateral-torsional buckling moment for $L_b > L_r$ becomes

$$M_n = \pi^2 EC_b \frac{I_{yc}d}{L_b^2} \le M_y \qquad (8.109)$$

Reintroducing the load shedding factor R_b of Eq. 8.81 and defining L_r as the unbraced length at which $M_n = 0.5M_y$, then Eq. 8.109 can be written as

$$M_n = C_b R_b (0.5M_y)(L_r/L_b)^2 \le R_b M_y \qquad (8.110)$$

for which

$$M_y = F_{yc}S_{xc} \qquad (8.111)$$

where F_{yc} is the yield strength of the compression flange and S_{xc} is the section modulus about the horizontal axis of the I-section at the compression flange. Inserting Eq. 8.111 into Eq. 8.110, multiplying Eq. 8.109 by R_b, equating to the modified Eq. 8.110, and solving for L_r gives

$$L_r = \sqrt{\frac{2\pi^2 I_{yc} d}{S_{xc}} \frac{E}{F_{yc}}} \qquad (8.112)$$

For values of L_b between L_p and L_r a straight-line transition between $M_n = M_y$ and $M_n = 0.5My$ is given by

$$M_n = C_b R_b M_y \left[1 - 0.5 \frac{(L_b - L_p)}{(L_r - L_p)} \right] \leq R_b M_y \qquad (8.113)$$

Because the moment gradient factor C_b can be greater than 1.0 (Eq. 8.98), the elastic upper limit of M_n is given on the right side of Eq. 8.113 as $R_b M_y$.

Noncomposite Noncompact Sections Noncomposite noncompact sections in either positive or negative flexure follow the same design rules as composite noncompact sections subjected to negative flexure, except that r' replaces r_t, that is [A6.10.6.3.1]

$$L_b \leq 1.76 r' \sqrt{\frac{E}{F_{yc}}} \qquad (8.114)$$

If this bracing requirement is satisfied, the nominal flexural resistance can be based on the nominal flexural stress of each flange F_n

$$F_n = R_b R_h F_{yf} \qquad (8.115)$$

where in this book $R_h = 1$. If the bracing requirement of Eq. 8.114 is not satisfied, the nominal flexural resistance shall be based on lateral-torsional buckling of the compression flange and determined by one of the following: Eqs. 8.105, 8.110, or 8.113, whichever is applicable.

Noncomposite Compact Sections Noncomposite compact sections in either positive or negative flexure follow the same design rules as composite compact sections subjected to negative flexure [A6.10.6.2]. To qualify as compact, the compression flange shall be braced to satisfy

$$L_b \leq \left[0.124 - 0.0756\left(\frac{M_1}{M_p}\right) \right]\left(\frac{r_y E}{F_{yc}}\right) \tag{8.116}$$

where M_1 is the smaller moment due to factored loading at either end of the unbraced length. This formula was developed to provide inelastic rotation capacities of at least three times the elastic rotation corresponding to the plastic moment (Yura et al., 1978) [C6.10.5.2.3d]. If this bracing requirement is satisfied, the nominal flexural resistance M_n is equal to the plastic moment M_p. If the bracing requirement is not satisfied, the nominal flexural resistance shall be based on Eq. 8.115, as applicable.

Composite Elastic* I-*Sections   Composite sections in positive flexure have adequate lateral support for the compression flange. However, in regions of negative bending, the compression flange is not laterally supported and behaves as a column between brace points when [A6.10.5.5.2]

$$L_b > L_r = 4.44 r_t \sqrt{\frac{E}{F_{yc}}} \tag{8.117}$$

and the nominal flexural resistance in terms of compression flange stress is given by

$$F_n = C_b R_b R_h \left[\frac{\pi^2 E}{(L_b/r_t)^2} \right] \leq R_b R_h F_{yc} \tag{8.118}$$

which is the Euler critical buckling stress multiplied by the moment gradient C_b and flange reduction factors $R_b R_h$. Putting $L_b = L_r$ from Eq. 8.117 into Eq. 8.118 gives

$$F_n = C_b R_b R_h \frac{F_{yc}}{2} \tag{8.119}$$

When the unbraced length exceeds the noncompact (inelastic) section requirement [A6.10.5.3.3d]

$$L_b > L_p = 1.76 r_t \sqrt{\frac{E}{F_{yc}}} \tag{8.120}$$

then the cross section behaves elastically and has a nominal resisting moment (horizontal dashed line in Fig. 8.50) less than or equal to M_y.

For values of L_b between L_p of Eq. 8.120 and L_r of Eq. 8.117, a straight-line transition between F_{yc} and $0.5F_{yc}$ is given by

$$F_n = C_b R_b R_h F_{yc}\left[1.33 - 0.187\left(\frac{L_b}{r_t}\right)\sqrt{\frac{F_{yc}}{E}}\right] \le R_b R_h F_{yc} \qquad (8.121)$$

where in this book $R_h = 1.0$.

Composite Noncompact Sections   For composite *I*-sections in negative flexure with L_b greater than the value of Eq. 8.116 but less than the value of Eq. 8.120, then the nominal flexural resistance is based on the nominal flexural stress of the compression flange

$$F_n = R_b R_h F_{yc} \qquad (8.122)$$

Composite Compact Sections   For composite sections in negative flexure with L_b less than or equal to the value given by Eq. 8.116, the nominal flexural resistance is equal to the plastic moment, that is,

$$M_n = M_p \qquad (8.123)$$

For continuous spans with compact positive bending sections and noncompact interior sections, the nominal positive flexural resistance is limited to [A6.10.5.2.2a]

$$M_n = 1.3 R_h M_y \qquad (8.124)$$

In effect, this limits the shape factor for the compact positive bending section to 1.3. This is necessary in continuous spans because excessive yielding in the positive moment region can redistribute moments to the negative moment region that are greater than those predicted by an elastic analysis [C6.10.5.2.2a].

For compact composite sections in positive flexure, a limitation is imposed on the depth of the composite section in compression to insure that the tension flange of a steel section reaches strain hardening before the concrete slab crushes [C6.10.5.2.2b]. By assuming the concrete crushing strain to be 0.003 and the steel strain hardening strain to be 0.012, and using similar strain triangles of Figure 8.53 yields

$$\frac{D_{sh}}{d + t_s + t_h} = \frac{0.003}{0.003 + 0.012} = \frac{1}{5}$$

where D_{sh} is the compression depth of the composite section at strain hardening measured from the top of the concrete slab, d is the depth of the steel section, t_s is the thickness of the concrete slab, and t_h is the thickness of the

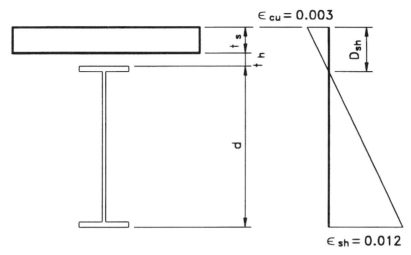

Fig. 8.53 Strain hardening depth to neutral axis.

haunch above the top flange. To provide a margin of safety for the strain in the tension flange, the depth D_{sh} is divided by 1.5 to give the requirement for the distance from the top of the slab to the neutral axis at the plastic moment D_p as [A6.10.5.2.2.b]

$$D_p \leq \frac{d + t_s + t_h}{7.5} \tag{8.125}$$

This limitation on D_p serves as the ductility requirement for composite compact sections in positive flexure.

8.7.8 Summary of *I*-Sections in Flexure

The behavior of *I*-sections in flexure is complex in details and yet simple in concept. The details are complex because of the many different conditions for which requirements must be established. Both composite and noncomposite sections subjected to positive and negative flexure must be considered for the three classes of shapes: compact, noncompact, and slender.

The concept is simple because all of the limit states follow the same pattern. Whether it be web slenderness (Fig. 8.37), compression flange slenderness (Fig. 8.47), or compression flange bracing (Fig. 8.50), the three failure modes are easily identified: no buckling, inelastic buckling, and elastic buckling. The numerous formulas describe the behavior and define the transition points for the three segments that represent the design requirements.

To organize the design requirements and present them in one place, Tables 8.16–8.18 have been developed. It is assumed in this book that the web and flange material have the same yield strength so that $R_h = 1.0$ and it does not appear in the equations. (*Note:* Due to practical and economic considerations most new designs are presently not hybrid.) The load shedding factor R_b is given by Eq. 8.81 and the moment gradient factor C_b is given by Eq. 8.98.

In Tables 8.17 and 8.18, reference is made to AASHTO [A6.10.5.6] for calculating the nominal flexural resistance when certain web and compression flange slendernesses are not satisfied. This article gives an alternative formula for flexural resistence M_n and is the result of a linear fit of experimental data between M_p and $0.7M_y$. If the following conditions are satisfied:

TABLE 8.16   Strength Limit State—Composite *I*-Sections in Positive Flexure $R_h = 1.0$

	Compact	Noncompact	Slender
Nominal flexural resistance	$M_n = M_p$ except for continuous spans with noncompact interior support sections $M_n \le 1.3M_y \le M_p$ Sections must satisfy the ductility requirement of Eq. 8.125	$F_n = R_b F_{yc}$	$F_n \le R_b F_{yc}$
Web slenderness	$\dfrac{2D_{cp}}{t_w} \le 3.76\sqrt{\dfrac{E}{F_{yc}}}$	Without longitudinal stiffeners $\dfrac{2D_c}{t_w} \le 6.77\sqrt{\dfrac{E}{f_c}}$ With longitudinal stiffeners $\dfrac{2D_c}{t_w} \le 11.63\sqrt{\dfrac{E}{f_c}}$	
Compression flange slenderness	No requirement at strength limit state		
Compression flange bracing	No requirement at strength limit state, but must satisfy $$L_b \le 1.76 r_t \sqrt{\dfrac{E}{F_{yc}}}$$ for loads applied before concrete slab hardens		

TABLE 8.17 Strength Limit State—Composite *I*-Sections in Negative Flexure $R_h = 1.0$

	Compact	Noncompact	Slender
Nominal flexural resistance	$M_n = M_p$	$F_n = R_b F_{yc}$	$F_n \leq R_b F_{yc}$ (See [A6.10.5.6])
Web slenderness	$\dfrac{2D_{cp}}{t_w} \leq 3.76\sqrt{\dfrac{E}{F_{yc}}}$	Without longitudinal stiffeners $\dfrac{2D_c}{t_w} \leq 6.77\sqrt{\dfrac{E}{f_c}}$ With longitudinal stiffeners $\dfrac{2D_c}{t_w} \leq 11.63\sqrt{\dfrac{E}{f_c}}$	(See [A6.10.5.6])
Compression flange slenderness	$\dfrac{b_f}{2t_f} \leq 0.382\sqrt{\dfrac{E}{F_{yc}}}$	$\dfrac{b_f}{2t_f} \leq 1.38\sqrt{\dfrac{E}{f_c\sqrt{\dfrac{2D_c}{t_w}}}}$	
Compression flange bracing	$L_b \leq \left[0.124 - 0.0759\left(\dfrac{M_1}{M_p}\right)\right]\left(\dfrac{r_y E}{F_{yc}}\right)$	$L_b \leq 1.76 r_t \sqrt{\dfrac{E}{F_{yc}}}$	$L_b \leq 4.44 r_t \sqrt{\dfrac{E}{F_{yc}}}$ Use Eq. 8.121 $L_b > 4.44 r_t \sqrt{\dfrac{E}{F_{yc}}}$ Use Eq. 8.118

TABLE 8.18 Strength Limit State—Noncomposite *I*-Sections in Positive and Negative Flexure $R_h = 1.0$

	Compact	Noncompact	Slender
Nominal flexural resistance	$M_n = M_p$	$F_n = R_b F_{yc}$	$F_n \leq R_b F_{yc}$
Web slenderness	$\dfrac{2D_{cp}}{t_w} \leq 3.76\sqrt{\dfrac{E}{F_{yc}}}$	Without longitudinal stiffeners $\dfrac{2D_c}{t_w} \leq 6.77\sqrt{\dfrac{E}{f_c}}$ With longitudinal stiffeners $\dfrac{2D_c}{t_w} \leq 11.63\sqrt{\dfrac{E}{f_c}}$	If $L_b > L_r$ $\dfrac{2D_c}{t_w} \leq \lambda_b\sqrt{\dfrac{E}{F_{yc}}}$ (See Eq. 8.81 for λ_b)
Compression flange slenderness	$\dfrac{b_f}{2t_f} \leq 0.382\sqrt{\dfrac{E}{F_{yc}}}$	$\dfrac{b_f}{2t_f} \leq 1.38\sqrt{\dfrac{E}{\sqrt{f_c}\sqrt{\dfrac{2D_c}{t_w}}}}$	$\dfrac{b_f}{2t_f} \leq 2.52\sqrt{\dfrac{E}{\sqrt{F_{yc}}\sqrt{\dfrac{2D_{cp}}{t_w}}}}$
Compression flange bracing	$L_b \leq \left[0.124 - 0.0759\left(\dfrac{M_1}{M_p}\right)\right]\left(\dfrac{r_y E}{F_{yc}}\right)$	$L_b \leq L_p = 1.76r'\sqrt{\dfrac{E}{F_{yc}}}$	(See [A6.10.5.6]) $L_r = \sqrt{\dfrac{2\pi^2 I_{yc} d}{S_{xc}}\dfrac{E}{F_{yc}}}$ $L_p < L_b \leq L_r$ Use Eq. 8.113 $L_b > L_r$ Use Eq. 8.110

$$\left.\begin{array}{l} \dfrac{2D_{cp}}{t_w} \le 6.77\sqrt{\dfrac{E}{F_{yc}}} \\[3ex] \dfrac{b_f}{2t_f} \le 2.52\sqrt{\dfrac{E}{F_{yc}\sqrt{\dfrac{2D_{cp}}{t_w}}}} \\[4ex] L_b \le \left[0.124 - 0.0759\left(\dfrac{M_1}{M_p}\right)\right]\left(\dfrac{r_y E}{F_{yc}}\right) \end{array}\right\} \qquad (8.126)$$

then

$$M_n = \left[1 - \left(1 - \dfrac{0.7M_y}{M_p}\right)\left(\dfrac{Q_p - Q_{fl}}{Q_p - 0.7}\right)\right]M_p \le M_p \qquad (8.127)$$

for which

$$Q_p = 5.47\dfrac{M_p}{M_y} - 3.13 \qquad \text{for unsymmetric sections}$$

$$Q_p = 3.0 \qquad \text{for symmetric sections}$$

If

$$\dfrac{b_f}{2t_f} \le 0.382\sqrt{\dfrac{E}{F_{yc}}}$$

then

$$Q_{fl} = \dfrac{30.5}{\sqrt{\dfrac{2D_{cp}}{t_w}}} \qquad (8.128)$$

Otherwise,

$$Q_{fl} = \dfrac{4.45}{(b_f/2t_f)^2\sqrt{2D_{cp}/t_w}}\dfrac{E}{F_{yc}} \qquad (8.129)$$

EXAMPLE 8.11

Determine the nominal negative flexural resistance of the composite cross section of Example 8.7 shown in Figure 8.43 if the unbraced length L_b is

6000 mm at an interior support. The PNA was determined in Example 8.7 to be 616.7 mm from the top of the web. The yield strength of the compression flange F_{yc} is 345 MPa. The negative plastic moment M_p for this section was found to be 9028 kN m in Example 8.9. The smaller factored moment M_1 at either end of the unbraced length is -2308 kN m and the larger factored moment M_2 is -6657 kN m. The algebraic sum of stresses in the steel section due to factored design moments are 290 MPa (tension) in the top flange and 316 MPa (compression) in the bottom flange.

Classification of Shape

Refer to Table 8.17.

   Web slenderness for compact section

$$\frac{2D_{cp}}{t_w} \le 3.76\sqrt{\frac{E}{F_{yc}}} = 3.76\sqrt{\frac{200\,000}{345}} = 90$$

$$D_{cp} = 1500 - 616.7 = 883.3 \text{ mm}$$

$$\frac{2D_{cp}}{t_w} = \frac{2(883.3)}{10} = 177 > 90 \qquad \text{not compact}$$

Web slenderness for noncompact section without longitudinal stiffeners

$$\frac{2D_c}{t_w} \le 6.77\sqrt{\frac{E}{f_c}} = 6.77\sqrt{\frac{200\,000}{316}} = 170$$

$$d = 1500 + 30 + 30 = 1560 \text{ mm}$$

$$D_c = d\frac{f_b}{f_b + f_t} - t_f = 1560\frac{316}{316 + 290} - 30 = 783 \text{ mm}$$

$$\frac{2D_c}{t_w} = \frac{2(783)}{10} = 157 < 170$$

longitudinal stiffeners are not required

Flange slenderness for noncompact section

$$\frac{b_f}{2t_f} \le 1.38\sqrt{\frac{E}{f_c\sqrt{\frac{2D_c}{t_w}}}} = 1.38\sqrt{\frac{200\,000}{316\sqrt{157}}} = 9.81$$

$$\frac{b_f}{2t_f} = \frac{400}{2(30)} = 6.7 < 9.81 \qquad \text{compression flange is not slender}$$

Compression flange bracing for noncompact section is

$$L_b \leq 1.76r_t\sqrt{\frac{E}{F_{yc}}} \qquad \text{(under a uniform moment)}$$

$$r_t = \sqrt{\frac{I_{yc}}{A_c + D_c t_w/3}} = \sqrt{\frac{30(400)^3/12}{30(400) + 783(10)/3}} = 104.6 \text{ mm}$$

$$L_b \leq 1.76r_t\sqrt{\frac{E}{F_{yc}}} = 1.76(104.6)\sqrt{\frac{200\,000}{345}} = 4430 \text{ mm}$$

which is less than the unbraced length required for a compact section. By considering the moment gradient, a larger unbraced length can be determined [C6.10.5.3.3d]. If we equate Eq. 8.121 to $R_b R_h F_{yc}$, we get

$$C_b\left[1.33 - 0.187\left(\frac{L_b}{r_t}\right)\sqrt{\frac{F_{yc}}{E}}\right] = 1 \qquad (8.130)$$

where C_b is the moment gradient correction factor of Eq. 8.98. For a constant section between brace points, the ratio P_1/P_2 in Eq. 8.98 can be written in terms of M_1/M_2, that is,

$$C_b = 1.75 - 1.05\left(\frac{M_1}{M_2}\right) + 0.3\left(\frac{M_1}{M_2}\right)^2 \leq 2.3$$

$$C_b = 1.75 - 1.05\left(\frac{2308}{6657}\right) + 0.3\left(\frac{2308}{6657}\right)^2 = 1.42 \leq 2.3$$

Solving Eq. 8.125 for L_b,

$$L_b = \frac{1.33 - 1/C_b}{0.187}r_t\sqrt{\frac{E}{F_{yc}}}$$

$$= \frac{1.33 - 1/1.42}{0.187}r_t\sqrt{\frac{E}{F_{yc}}} = 3.35r_t\sqrt{\frac{E}{F_{yc}}}$$

$$L_b = 3.35(104.6)\sqrt{\frac{200\,000}{345}} = 8440 \text{ mm} > 6000 \text{ mm}$$

Therefore, the cross section is not slender and is classified as noncompact.

Solution

Because the section is noncompact, the factored resistance is specified in terms of stress as

$$\phi_f F_n = \phi_f R_b R_h F_{yc} \tag{8.131}$$

where ϕ_f is the resistance factor for flexure from Table 8.9. The load shedding factor R_b is determined from Eq. 8.81

$$R_b = 1 - \left(\frac{a_r}{1200 + 300a_r}\right)\left(\frac{2D_c}{t_w} - \lambda_b\sqrt{\frac{E}{f_c}}\right)$$

where $\lambda_b = 5.76$ and

$$a_r = \frac{2D_c t_w}{A_{fc}} = \frac{2(783)(10)}{30(400)} = 1.305$$

so that

$$R_b = 1 - \left(\frac{1.305}{1200 + 300 \times 1.305}\right)\left(157 - 5.76\sqrt{\frac{200\,000}{316}}\right) = 0.990$$

For $\phi_f = 1.0$ and $R_h = 1.0$, Eq. 8.131 yields

$$\phi_f F_n = 1.0(0.990)(1.0)(345) = \underline{342\ \text{MPa}}$$

Answer

The cross section is adequate because the permissible stress of 342 MPa exceeds the maximum stress of 316 MPa produced by the applied loads.

8.7.9   Closing Remarks on *I*-Sections in Flexure

When rolled steel shapes are used as beams the web slenderness requirement does not have to be checked, because all of the webs satisfy the compact section criterion. Further, if Grade 250 steel is used, all but the W150 × 22 satisfy the flange slenderness criterion for a compact section. If Grade 345 steel is used six of the 253 W-shapes listed in AISC (1992) do not satisfy the flange slenderness criterion for a compact section. Therefore, local buckling is seldom a problem with rolled steel shapes and when they are used the emphasis is on providing adequate lateral support for the compression flange to prevent global buckling.

It should be noted that the constants associated with the slenderness limits in the AASHTO (1994) LRFD Bridge Specifications suggest more precision than that utilized during their development. For example, the web slenderness limit for compact sections is given in Tables 8.16–8.18 as

$$\frac{2D_{cp}}{t_w} \leq 3.76\sqrt{\frac{E}{F_{yc}}}$$

This requirement was adopted from the expression in the inch-pound system of AISC (1986) for webs in flexural compression

$$\frac{h_c}{t_w} \leq \frac{640}{\sqrt{F_y}} \tag{8.132}$$

where h_c is twice the distance from the neutral axis to the inside face of the compression flange less the fillet or corner radius which is effectively equal to $2D_{cp}$ and F_y is the yield strength in ksi. The constant 640 includes the square root of the elastic modulus $E = 29{,}000$ ksi. To make this hidden variable visible and to make the constant nondimensional, Eq. 8.132 is written as

$$\frac{h_c}{t_w} \leq \frac{640}{\sqrt{29\,000}}\sqrt{\frac{E}{F_y}} = 3.76\sqrt{\frac{E}{F_y}}$$

Thus, by the simple procedure of changing systems of units, the precision has gone from two significant figures to three significant figures, which is not consistent with the uncertainty considered in the development of these limits.

8.8   SHEAR RESISTANCE OF *I*-SECTIONS

When the web of an *I*-section is subjected to in-plane shear forces that are progressively increasing, small deflection beam theory can be used to predict the shear strength until the critical buckling load is reached. If the web is stiffened, additional postbuckling shear strength due to tension field action is present until web yielding occurs. Using the notation in Basler (1961a), the nominal shear resistance V_n can be expressed as

$$V_n = V_\tau + V_\sigma \tag{8.133}$$

where V_τ is the beam action shear resistance and V_σ is the tension field action shear resistance.

8.8.1   Beam Action Shear Resistance

A stress block at the neutral axis of a web of an *I*-section is shown in Figure 8.54(a). Because the flexural stresses are zero, the stress block is in a state of pure shear. A Mohr circle of stress [Fig. 8.54(b)] indicates principal stresses σ_1 and σ_2 that are equal to the shearing stress τ. These principal stresses are oriented at 45° from the horizontal. When using beam theory, it is usually assumed that the shear force V is resisted by the area of the web, that is,

$$\tau = \frac{V}{Dt_w} \tag{8.134}$$

where D is the web depth and t_w is the web thickness.

If no buckling occurs, the shear stress can reach its yield strength τ_y and the full plastic shear force V_p can be developed. Putting these values into Eq. 8.134 and rearranging, we have

$$V_p = \tau_y D t_w \tag{8.135}$$

The shear yield strength τ_y cannot be determined by itself, but is dependent on the shear failure criteria adopted. By using the Mises shear failure criterion, the shear yield strength is related to the tensile yield strength of the web σ_y by

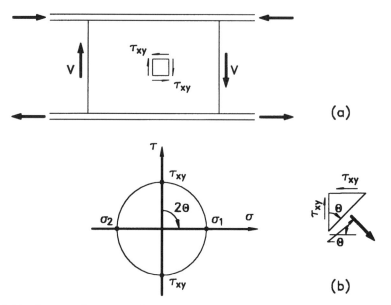

Fig. 8.54 Beam action states of stress. (a) Stress block at neutral axis and (b) Mohr circle of stress.

$$\tau_y = \frac{\sigma_y}{\sqrt{3}} \approx 0.58\sigma_y \qquad (8.136)$$

If buckling does occur, the critical shear buckling stress τ_{cr} for a rectangular panel (Fig. 8.55) is given by

$$\tau_{cr} = k \frac{\pi^2 E}{12(1 - \mu^2)} \left(\frac{t_w}{D}\right)^2 \qquad (8.137)$$

in which

$$k = 5.0 + \frac{5.0}{(d_o/D)^2} \qquad (8.138)$$

where d_o is the distance between transverse stiffeners.

By assuming that shear is carried in a beam-type manner up to τ_{cr} and then remains constant, we can express V_τ as a linear fraction of V_p, that is,

$$V_\tau = \frac{\tau_{cr}}{\tau_y} V_p \qquad (8.139)$$

8.8.2 Tension Field Action Shear Resistance

When a rectangular web panel subjected to shear is supported on four edges, tension field action on the diagonal can develop. The web panel of an *I*-section (Fig. 8.55) has two edges that are flanges and two edges that are transverse stiffeners. These two pairs of boundaries are very different. The flanges are relatively flexible in the vertical direction and cannot resist stresses from a tension field in the web. On the other hand, the transverse stiffeners can serve as an anchor for the tension stress field. As a result, the web area

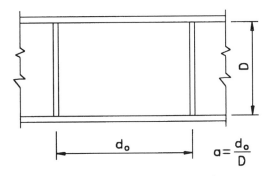

Fig. 8.55 Definition of aspect ratio α.

adjacent to the junction with the flanges is not effective and the truss-like load carrying mechanism of Figure 8.56 can be assumed. In this truss analogy, the flanges are the chords, the transverse stiffeners are compression struts, and the web is a tension diagonal.

The edges of the effective tension field in Figure 8.56 are assumed to run through the corners of the panel. The tension field width s depends on the inclination from the horizontal θ of the tensile stresses σ_t and is equal to

$$s = D \cos \theta - d_o \sin \theta \qquad (8.140)$$

The development of this partial tension field has been observed in numerous laboratory tests. An example of one from Lehigh University is shown in Figure 8.57. At early stages of loading the shear in the web is carried by beam action until the compressive principal stress σ_2 of Figure 8.54(b) reaches its critical stress and the compression diagonal of the panel buckles. At this point, no additional compressive stress can be carried, but the tensile stresses σ_t in the tension diagonal continue to increase until they reach the yield stress $\sigma_y = F_{yw}$ of the web material. The stiffened I-section in Figure 8.57 clearly shows the buckled web, the postbuckling behavior of the tension field, and the truss-like appearance of the failure mechanism.

The contribution to the shear force V_σ from the tension field action ΔV_σ is the vertical component of the diagonal tensile force (Fig. 8.56), that is,

$$\Delta V_\sigma = \sigma_t s t_w \sin \theta \qquad (8.141)$$

To determine the inclination θ of the tension field, assume that when $\sigma_t =$

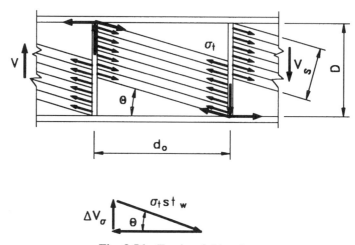

Fig. 8.56   Tension field action.

Fig. 8.57   Thin-web girder after testing (Photo courtesy of John Fisher, ATLSS Engineering Research Center, Lehigh University).

σ_y the orientation of the tension field is such that ΔV_σ is a maximum. This condition can be expressed as

$$\frac{d}{d\theta}\,(\Delta V_\sigma) = \frac{d}{d\theta}\,(\sigma_y s t_w \sin \theta) = 0$$

If we substitute Eq. 8.140 for *s*, we get

$$\sigma_y t_w \left[\frac{d}{d\theta}\,(D \cos \theta \sin \theta - d_o \sin^2 \theta)\right] = 0$$

which reduces to

$$D \tan^2 \theta + 2d_o \tan \theta - D = 0$$

Solving for tan θ

$$\tan \theta = \frac{-2d_o + \sqrt{4d_o^2 + 4D^2}}{2D} = \sqrt{1 + \alpha^2} - \alpha \qquad (8.142)$$

where α is the aspect ratio of the web panel d_o/D. Use trigonometric identities to obtain

$$\cos\theta = (\tan^2\theta + 1)^{-1/2} = [2\sqrt{1+\alpha^2}(\sqrt{1+\alpha^2} - \alpha)]^{-1/2} \quad (8.143)$$

and

$$\sin\theta = (\cot^2\theta + 1)^{-1/2} = \left[\frac{1}{2} - \frac{\alpha}{2\sqrt{1+\alpha^2}}\right]^{1/2} \quad (8.144)$$

Consider equilibrium of the free body ABDC in Figure 8.58 taken below the neutral axis of the web and between the middle of the web panels on either side of a transverse stiffener. By assuming a doubly symmetric I-section, the components of the partial tension field force on the vertical sections AC and BD are $V_\sigma/2$ vertically and F_w horizontally in the directions shown in Figure 8.58. On the horizontal section AB, the tension field stresses σ_t are inclined at an angle θ and act on a projected area $t_w d_o \sin\theta$. Equilibrium in the vertical direction gives the axial load in the stiffener F_s as

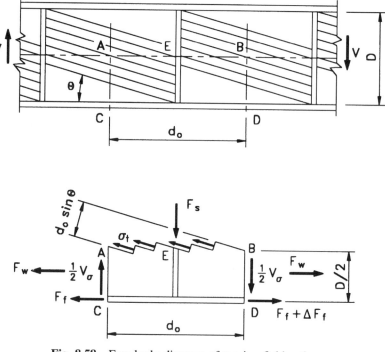

Fig. 8.58 Free body diagram of tension field action.

$$F_s = \sigma_t t_w d_o \sin\theta \sin\theta = \sigma_t t_w(\alpha D)\sin^2\theta$$

By substituting Eq. 8.144

$$F_s = \sigma_t t_w D\left(\frac{\alpha}{2} - \frac{\alpha^2}{2\sqrt{1+\sigma^2}}\right) \tag{8.145}$$

Equilibrium in the horizontal direction gives the change in the flange force ΔF_f as

$$\Delta F_f = \sigma_t t_w(\alpha D)\sin\theta\cos\theta$$

By substituting Eqs. 8.143 and 8.144 into the above expression for ΔF_f, and simplifying

$$\Delta F_f = \sigma_t t_w D\,\frac{\alpha}{2\sqrt{1+\alpha^2}} \tag{8.146}$$

Balancing moments about point E results in

$$\frac{1}{2}V_\sigma(d_o) - \Delta F_f\left(\frac{D}{2}\right) = 0$$

$$V_\sigma = \Delta F_f\frac{D}{d_o} = \frac{\Delta F_f}{\alpha}$$

So that the shear force contribution of tension field action V_σ becomes

$$V_\sigma = \sigma_t t_w D\,\frac{1}{2\sqrt{1+\alpha^2}} \tag{8.147}$$

With the use of Eqs. 8.135 and 8.136, V_σ can be written in terms of V_p as

$$V_\sigma = \frac{\sqrt{3}}{2}\frac{\sigma_t}{\sigma_y}\frac{1}{\sqrt{1+\alpha^2}}V_p \tag{8.148}$$

8.8.3   Combined Shear Resistance

If we substitute Eqs. 8.139 and 8.148 into Eq. 8.133, we obtain an expression for the combined nominal shear resistance of the web of an *I*-section

$$V_n = V_p \left[\frac{\tau_{cr}}{\tau_y} + \frac{\sqrt{3}}{2} \frac{\sigma_t}{\sigma_y} \frac{1}{\sqrt{1 + \alpha^2}} \right] \qquad (8.149)$$

where the first term in brackets is due to beam action and the second term is due to tension field action. These two actions should not be thought of as two separately occurring phenomenon where first one is observed and later the other becomes dominant. Instead they occur together and interact to give the combined shear resistance of Eq. 8.149.

Basler (1961a) develops a simple relation for the ratio σ_t/σ_y in Eq. 8.149 based on two assumptions. The first assumption is that the state of stress anywhere between pure shear and pure tension can be approximated by a straight line when using the Mises yield criterion. The second assumption is that θ is equal to the limiting case of 45°. By using these two assumptions and substituting into the stress equation representing the Mises yield criterion results in

$$\frac{\sigma_t}{\sigma_y} = 1 - \frac{\tau_{cr}}{\tau_y} \qquad (8.150)$$

Basler (1961a) conducts a numerical experiment comparing the nominal shear resistance of Eq. 8.149 with that using the approximation of Eq. 8.150. He shows that the difference is less than 10% for values of α between zero and infinity. By putting Eq. 8.150 into Eq. 8.149, the combined nominal shear resistance of the web becomes

$$V_n = V_p \left[\frac{\tau_{cr}}{\tau_y} + \frac{\sqrt{3}}{2} \frac{1 - (\tau_{cr}/\tau_y)}{\sqrt{1 + \alpha^2}} \right] \qquad (8.151)$$

In the AASHTO (1994) LRFD Bridge Specifications, Eq. 8.151 appears as [A6.10.7.3.3a]

$$V_n = V_p \left[C + \frac{0.87(1 - C)}{\sqrt{1 + (d_o/D)^2}} \right] \qquad (8.152)$$

for which

$$C = \frac{\tau_{cr}}{\tau_y} \qquad (8.153)$$

$$\alpha = \frac{d_o}{D} \qquad (8.154)$$

$$V_p = 0.58 F_{yw} D t_w \qquad (8.155)$$

8.8.4   Shear Resistance of Unstiffened Webs

The nominal shear resistance of unstiffened webs of *I*-sections can be determined from Eq. 8.152 by setting d_o equal to infinity, that is, only the beam action resistance remains

$$V_n = CV_p = 0.58CF_{yw}Dt_w \tag{8.156}$$

Substituting Eqs. 8.136 and 8.137 into Eq. 8.153 yields, for $\mu = 0.3$

$$C = \frac{\tau_{cr}}{\tau_y} = \frac{\dfrac{k\pi^2 E}{12(1-\mu^2)}\left(\dfrac{t_w}{D}\right)^2}{0.58F_{yw}} = \frac{0.90kE\left(\dfrac{t_w}{D}\right)^2}{0.58F_{yw}} \tag{8.157}$$

From Eq. 8.138 with d_o equal to infinity, $k = 5.0$, so that

$$V_n = CV_p = 0.90(5.0)E(t_w/D)^2 \, Dt_w$$

$$V_n = \frac{4.50Et_w^3}{D} \tag{8.158}$$

when shear resistance is governed by elastic shear buckling of the web.

If the web is relatively stocky, the critical shear buckling stress τ_{cr} can be greater than the shear yield stress τ_y and no buckling of the web will occur before the web material begins to yield. The limiting slenderness ratio for yielding before buckling ($V_n = V_p$) is given by

$$\tau_y \le \tau_{cr}$$

$$0.58F_{yw} \le \frac{k\pi^2 E}{12(1-\mu^2)}\left(\frac{t_w}{D}\right)^2 = 4.50E\left(\frac{t_w}{D}\right)^2$$

$$\frac{D}{t_w} \le 2.80\sqrt{\frac{E}{F_{yw}}} \tag{8.159}$$

Based on full-scale tests of welded *I*-sections, Basler (1961a) recommends that the limiting web slenderness ratio between inelastic and elastic buckling be established when

$$0.8\tau_y \le \tau_{cr}$$

or

$$\frac{D}{t_w} \le \frac{2.80}{0.8}\sqrt{\frac{E}{F_{yw}}} = 3.50\sqrt{\frac{E}{F_{yw}}} \tag{8.160}$$

The values adopted by the AASHTO (1994) LRFD Bridge Specifications are similar to, but slightly different than, those in Eqs. 8.158–8.160 for unstiffened webs. The adopted values [A6.10.7.2] are summarized in Table 8.19.

The expression for the inelastic shear buckling resistance is a straight line between the two web slenderness limits. This can be shown by writing the expression in terms of D/t_w, that is,

$$V_n = 1.48t_w^2\sqrt{EF_{yw}} = \frac{1.48t_wD}{D/t_w}\sqrt{EF_{yw}}$$

By substituting the lower limit $D/t_w = 2.46\sqrt{E/F_{yw}}$, we get

$$V_n = \frac{1.48t_wD\sqrt{EF_{yw}}}{2.46\sqrt{E/F_{yw}}} = 0.60F_{yw}Dt_w \approx V_p$$

and the upper limit $D/t_w = 3.07\sqrt{E/F_{yw}}$

$$V_n = \frac{1.48t_wD\sqrt{EF_{yw}}}{3.07\sqrt{E/F_{yw}}} = 0.48F_{yw}Dt_w \approx 0.8V_p$$

The general shape of the nominal shear resistance versus web slenderness curve is similar to Figure 8.39 for fatigue loading and Figure 8.47 for flexure. Again, the three different types of behavior—plastic, inelastic, and elastic—are shown to exist for resistance to shear just as in the other loading cases.

8.8.5 Shear Resistance of Stiffened Webs

Webs of I-sections are considered stiffened if without a longitudinal stiffener, the spacing of transverse stiffeners d_o does not exceed $3D$, or if with a longitudinal stiffener, d_o does not exceed 1.5 times the maximum subpanel depth D^* (Fig. 8.59). Otherwise, the web is considered unstiffened and the provisions in Table 8.19 apply [A6.10.7.1].

TABLE 8.19 Nominal Shear Resistance of Unstiffened Webs

	No Buckling	Inelastic Buckling	Elastic Buckling
Web slenderness	$\dfrac{D}{t_w} \le 2.46\sqrt{\dfrac{E}{F_{yw}}}$	$\dfrac{D}{t_w} \le 3.07\sqrt{\dfrac{E}{F_{yw}}}$	$\dfrac{D}{t_w} > 3.07\sqrt{\dfrac{E}{F_{yw}}}$
Nominal shear resistance	$V_n = V_p$	$V_n = 1.48t_w^2\sqrt{EF_{yw}}$	$V_n = \dfrac{4.55t_w^3E}{D}$

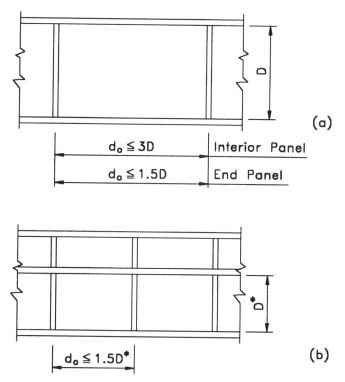

Fig. 8.59 Maximum transverse stiffener spacing.

If a longitudinal stiffener is used, its influence on the shear resistance of the web is conservatively neglected. In other words, the total depth of the web D is used to calculate the shear resistance of the web whether or not a longitudinal stiffener is present [A6.10.7.3.1].

When a web qualifies as stiffened, tension field action develops and both terms of Eq. 8.152 contribute to the shear resistance, that is,

$$V_n = V_p \left[C + \frac{0.87(1 - C)}{\sqrt{1 + (d_o/D)^2}} \right] \tag{8.161}$$

where C is the ratio of the critical shear buckling stress τ_{cr} to the shear yield stress τ_y.

Handling Requirement During fabrication and erection of *I*-sections without longitudinal stiffeners, care must be taken to prevent buckling of the web under self-weight of the steel cross sections. By using the flexural web slenderness limitation for a noncomposite doubly symmetric *I*-section before elastic buckling occurs (Table 8.18), we have, for webs without longitudinal stiffeners,

$$\frac{D}{t_w} \leq 6.77 \sqrt{\frac{E}{f_c}}$$

For

$$f_c = F_y = 250 \text{ MPa} \quad \text{and} \quad E = 200 \text{ GPa}$$

$$\frac{D}{t_w} \leq 6.77 \sqrt{\frac{200\ 000}{250}} = 191$$

For

$$f_c = F_y = 345 \text{ MPa}$$

$$\frac{D}{t_w} \leq 6.77 \sqrt{\frac{200\ 000}{345}} = 163$$

The AASHTO (1944) LRFD Bridge Specifications require that web panels without longitudinal stiffeners shall use transverse stiffeners if [A6.10.7.3.2]

$$\frac{D}{t_w} > 150 \tag{8.162}$$

Implied with this limitation is a maximum spacing of transverse stiffeners of $3D$ [C6.10.7.3.2]. If webs have $D/t_w > 150$, the maximum spacing of transverse stiffeners shall be less than $3D$ as given by the expression [A6.10.7.3.2]

$$d_o \leq D \left[\frac{260}{(D/t_w)} \right]^2 \tag{8.163}$$

whose variation with the reciprocal of $(D/t_w)^2$ is suggested by Eq. 8.137 for the critical shear buckling stress τ_{cr}. Note that with $D/t_w = 150$, d_o is $3D$.

Interior Panels of Compact Sections [A6.10.7.3.3a]   When an I-section is compact, the limiting flexural resistance (Tables 8.16–8.18) is given in terms of moments. If the moments are relatively high, the shear strength of the web is reduced, because it participates in resisting the moment. Basler (1961b) shows that the moment–shear interaction effects occur if the factored shear force V_u is greater than $0.6\phi_v V_n$ and the factored moment $M_u > 0.75\phi_v M_y$ (the resistance factors ϕ_v and ϕ_f are taken from Table 8.9).

If we assume a shape factor $M_p/M_y = 1.5$, the limiting value for the moment can be written as

$$0.75\phi_f M_y = 0.75\phi_f (M_p/1.5) = 0.50\phi_f M_p$$

When M_u is less than or equal to $0.5\phi_f M_p$, then the shear resistance for interior web panels of compact sections is given by Eq. 8.161. When M_u exceeds $0.5\phi_f M_p$, interaction between moment and shear causes a reduction in the nominal shear resistance, that is,

$$V_n = RV_p \left[C + \frac{0.87(1 - C)}{\sqrt{1 + (d_o/D)^2}} \right] \geq CV_p \qquad (8.164)$$

in which the reduction factor R is given by

$$R = \left[0.6 + 0.4 \left(\frac{M_r - M_u}{M_r - 0.75\phi_f M_y} \right) \right] \leq 1.0 \qquad (8.165)$$

where the resisting moment $M_r = \phi_f M_n$. The variation in RV_p with the moment M_u due to factor loads is shown in Figure 8.60. The nominal shear resistance from Eq. 8.164 shall be at least equal to the nominal shear resistance of an unstiffened web given by setting d_o to infinity in Eq. 8.163.

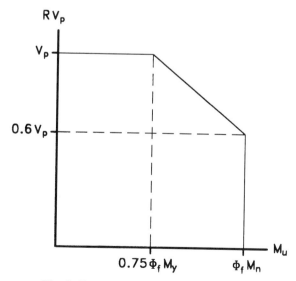

Fig. 8.60 Shear and bending interaction.

The ratio C was defined earlier in Eqs. 8.46–8.49 and is plotted as a function of D/t_w in Figure 8.39. When τ_{cr} is less than τ_y, the web panel behaves elastically and C is determined from Eq. 8.157 as

$$C = \frac{1.57}{(D/t_w)^2}\left(\frac{Ek}{F_{yw}}\right) \tag{8.166}$$

which is very close to Eq. 8.48. Basler (1961a) indicates that Eq. 8.166 is valid for τ_{cr} less than $0.8\tau_y$, so that the limiting web slenderness ratio for elastic behavior is determined by setting $C = 0.8$ in Eq. 8.166, that is,

$$\frac{D}{t_w} = \sqrt{\frac{1.57}{0.8}\frac{Ek}{F_{yw}}} = 1.40\sqrt{\frac{Ek}{F_{yw}}}$$

which is very close to the limit given for Eq. 8.48.

As in the other cases describing behavior as a function of slenderness, the inelastic response is assumed to be a straight line. Assuming the linear function of slenderness to be of the form

$$C = \frac{C_1}{(D/t_w)}\sqrt{\frac{Ek}{F_{yw}}}$$

where the constant C_1 is determined from the condition that the straight line must pass through the point: $C = 0.8$, $D/t_w = 140\sqrt{Ek/F_{yw}}$, that is,

$$0.8 = \frac{C_1}{1.40} \Rightarrow C_1 = 0.8(1.40) = 1.12$$

so that for $D/t_w < 1.40$

$$C = \frac{1.12}{(D/t_w)}\sqrt{\frac{Ek}{F_{yw}}} \leq 1.0 \tag{8.167}$$

which is very close to Eq. 8.47. The upper bound on C in Eq. 8.167 represents $\tau_{cr} = \tau_y$ where the shear buckling stress equals or exceeds the shear yield strength and full plastic behavior occurs without buckling. When $C = 1.0$ the limiting slenderness ratio is

$$\frac{D}{t_w} = 1.12\sqrt{\frac{Ek}{F_{yw}}}$$

which is very close to the limit given for Eq. 8.47.

Interior Panels of Noncompact Sections [A6.10.7.3.3b]   When an *I*-section is noncompact, the limiting flexural resistance (Tables 8.16–8.18) is given in terms of stresses, rather than moments. Therefore, the limitations on moment-shear interaction are in terms of stresses, but the expressions are the same, that is,

If

$$f_u \leq 0.75\phi_f F_y$$

then

$$V_n = V_p \left[C + \frac{0.87(1 - C)}{\sqrt{1 + (d_o/D)^2}} \right] \tag{8.168}$$

If

$$f_u > 0.75\phi_f F_y$$

then

$$V_n = RV_p \left[C + \frac{0.87(1 - C)}{\sqrt{1 + (d_o/D)^2}} \right] \geq CV_p \tag{8.169}$$

for which

$$R = \left[0.6 + 0.4 \left(\frac{F_r - f_u}{F_r - 0.75\phi_f F_y} \right) \right] \leq 1.0 \tag{8.170}$$

where f_u is the maximum stress in the compression flange in the panel under consideration due to factored loads and F_r is the factored flexural resistance of the compression flange for which f_u was determined. From Eq. 8.36 and expressions in Tables 8.16–8.18, we get

$$F_r = \phi_f F_n = \phi_f R_b F_{yc} \tag{8.171}$$

The expression for R in Eq. 8.170 is the same as in Eq. 8.165 and Figure 8.60 with moments replaced by stresses. Because the expression for R is based on stresses, the effect of strain-hardening can be utilized and the upper limit of 1.0 need not be applied to Eq. 8.170 [C6.10.7.3.3b].

End Panels [A6.10.7.3.3c]   End panels of *I*-sections have different boundary conditions than the interior panels. An end panel is at a discontinuous boundary and does not have a neighboring panel that can serve as an anchor for a tension stress field (Basler, 1961a). As a result, tension field action does not develop and only the first term of Eq. 8.152 is used for the nominal shear resistance of end panels.

Even though the end panel is considered to be stiffened, the fact that only the first term of Eq. 8.152 applies results in the same nominal shear resistance as an unstiffened web. The expression for this shear resistance is given in Eq. 8.156 and summarized in Table 8.19 for different ranges of web slenderness.

In order to eliminate premature end panel failure, Basler (1961a) recommends that a smaller stiffener spacing be used for the end panel to avoid the development of tension field action in that panel. If web buckling does not occur, then a tension field will not develop. The AASHTO (1994) LRFD Bridge Specifications take this approach for end panels and states that for webs without longitudinal stiffeners the transverse stiffener spacing shall not exceed $1.5D$ and for webs with a longitudinal stiffener the spacing shall not exceed 1.5 times the maximum subpanel depth (Fig. 8.59).

Summary of Stiffened Web Panels   The expressions for the nominal shear resistance of stiffened interior web panels are summarized in Tables 8.20 and 8.21.

EXAMPLE 8.12

Determine the web shear strength of the *I*-section of Example 8.7 shown in Figure 8.43 if the spacing of transverse stiffeners is 2000 mm for an interior web panel. In Example 8.11, for an unbraced length of the compression flange of 6000 mm in a negative moment region, the cross section is classified as noncompact. The algebraic sum of stresses in the steel section due to factored design moments are 290 MPa (tension) in the top flange and 316 MPa (compression) in the bottom flange. The yield strength of the web F_{yw} is 345 MPa.

TABLE 8.20   Nominal Shear Resistance of Stiffened Webs

	Compact	Noncompact
Nominal shear resistance	If $M_u \leq 0.5\phi_f M_p$	If $f_u \leq 0.75\phi_f F_y$
	$V_n = V_p \left[C + \dfrac{0.87(1 - C)}{\sqrt{1 + (d_o/D)^2}} \right]$	
	If $M_u > 0.5\phi_f M_p$	If $f_u > 0.75\phi_f F_y$
	$V_n = RV_p \left[C + \dfrac{0.87(1 - C)}{\sqrt{1 + (d_o/D)^2}} \right] \geq CV_p$	
Reduction factor	$R = 0.6 + 0.4 \dfrac{(M_r - M_u)}{(M_r - 0.75\phi_f M_y)} \leq 1.0$	$R = 0.6 + 0.4 \dfrac{(F_r - f_u)}{(F_r - 0.75\phi_f F_y)} \leq$

TABLE 8.21   Ratio of Shear Buckling Stress to Shear Yield Strength

	No Buckling	Inelastic Buckling	Elastic Buckling
Web Slenderness	$\dfrac{D}{t_w} \le 1.10 \sqrt{\dfrac{Ek}{F_{yw}}}$	$\dfrac{D}{t_w} \le 1.38 \sqrt{\dfrac{Ek}{F_{yw}}}$	$\dfrac{D}{t_w} > 1.38 \sqrt{\dfrac{Ek}{F_{yw}}}$
$C = \dfrac{\tau_{cr}}{\tau_y}$	$C = 1.0$	$C = \dfrac{1.10}{D/t_w} \sqrt{\dfrac{Ek}{F_{yw}}}$	$C = \dfrac{1.52}{(D/t_w)^2} \dfrac{Ek}{F_{yw}}$

Solution

By referring to Table 8.20, for a noncompact section, the amount of moment-shear interaction depends on the maximum stress f_u in the compression flange due to the factored loading. For this example, $f_u = 316$ MPa, which is greater than

$$0.75\phi_f F_y = 0.75(1.0)(345) = 259 \text{ MPa}$$

so that

$$V_n = RV_p\left[C + \frac{0.87(1 - C)}{\sqrt{1 + (d_o/D)^2}}\right] \ge CV_p$$

where

$$R = 0.6 + 0.4 \frac{(F_r - f_u)}{(F_r - 0.75\phi_f F_y)} \le 1.0$$

and from Example 8.11

$$F_r = \phi_f F_n = \phi_f R_b R_h F_{yc} = (1.0)(0.990)(1.0)(345) = 342 \text{ MPa}$$

Hence,

$$R = 0.6 + 0.4 \frac{342 - 316}{342 - 259} = 0.725$$

From Eqs. 8.154 and 8.155

$$\alpha = d_o/D = 2000/1500 = 1.33$$

and

$$V_p = 0.58 F_{yw} D t_w$$

$$= 0.58(345)(1500)(10) = 3\ 001\ 500\ \text{N} = 3002\ \text{kN}$$

By referring to Table 8.21, and calculating k from Eq. 8.138

$$k = 5.0 + \frac{5.0}{\alpha^2} = 5.0 + \frac{5.0}{(1.33)^2} = 7.81$$

so that

$$1.38 \sqrt{\frac{Ek}{F_{yw}}} = 1.38 \sqrt{\frac{(200\ 000)(7.81)}{345}} = 93$$

and

$$\frac{D}{t_w} = \frac{1500}{10} = 150 > 1.38 \sqrt{\frac{Ek}{F_{yw}}} = 93$$

Thus,

$$C = \frac{1.52}{(D/t_w)^2} \frac{Ek}{F_{yw}} = \frac{1.52}{(150)^2} \frac{(200\ 000)(7.81)}{345} = 0.306$$

and

$$CV_p = 0.306(3002) = 918\ \text{kN}$$

Answer

The nominal shear strength of the web is

$$V_n = RV_p \left[C + \frac{0.87(1 - C)}{\sqrt{1 + \alpha^2}} \right] \geq CV_p = 918\ \text{kN}$$

$$= 0.725(3002) \left[0.306 + \frac{0.87(1 - 0.306)}{\sqrt{1 + (1.33)^2}} \right]$$

$$= 2176(0.306 + 0.362) = 1454\ \text{kN}$$

and the factored web shear strength is

$$V_r = \phi_v V_n = 1.0(1454) = \underline{1454 \text{ kN}}$$

where ϕ_v is taken from Table 8.9.

8.9 SHEAR CONNECTORS

To develop the full flexural strength of a composite member, horizontal shear must be resisted at the interface between the steel section and the concrete deck slab. To resist the horizontal shear at the interface, connectors are welded to the top flange of the steel section that will become embedded in the deck slab when the concrete is placed. These shear connectors come in various types: headed studs, channels, spirals, inclined stirrups, and bent bars. Only the welded headed studs (Fig. 8.35) are discussed in this section.

In simple span composite bridges, shear connectors shall be provided throughout the length of the span [A6.10.7.4.1]. In continuous composite bridges, shear connectors are often provided throughout the length of the bridge. Placing shear connectors in the negative moment regions prevents the sudden transition from composite to noncomposite section and assists in maintaining flexural compatibility throughout the length of the bridge (Slutter and Fisher, 1967).

The larger diameter head of the stud shear connector enables it to resist uplift as well as horizontal slip. Calculations are not made to check the uplift resistance. Experimental tests (Ollgaard et al. 1971) indicate failure modes associated with shearing of the stud or failure of the concrete (Fig. 8.61). The headed studs did not pull out of the concrete and can be considered adequate to resist uplift.

Data from experimental tests are used to develop empirical formulas for resistance of welded headed studs. Tests have shown that to develop the full capacity of the connector, the height of the stud must be at least four times the diameter of its shank. Therefore, this condition becomes a design requirement [A6.10.7.4.1a].

Two limit states must be considered when determining the resistance of stud shear connectors: fatigue and strength. The fatigue limit state is examined at stress levels in the elastic range. The strength limit state depends on plastic behavior and the redistribution of horizontal shear forces among connectors.

8.9.1 Fatigue Limit State for Stud Connectors

In the experimental tests conducted by Slutter and Fisher (1967), the shear stress range was found to be the governing factor affecting the fatigue life of shear connectors. Concrete strength, concrete age, orientation of connectors, size effect, and minimum stress did not significantly influence the fatigue strength. As a result, the fatigue resistance of stud connectors can be expressed by the relationship between allowable shear stress range S_r and the

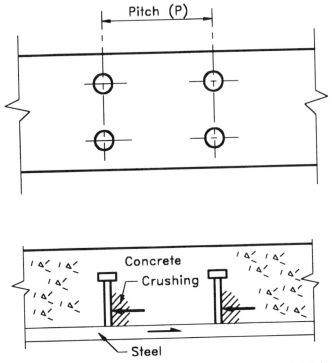

Fig. 8.61   Forces acting on a shear connector in a solid slab.

number of load cycles to failure N. The log–log plot of the S–N data for both 19 mm and 22 mm diameter studs is given in Figure 8.62. The shear stress was calculated as the average stress on the nominal diameter of the stud. The mean curve resulting from a regression analysis is given by (Slutter and Fisher, 1967)

$$S_r = 1065N^{-0.19} \tag{8.172}$$

where S_r is the shear stress range in megapascal and N is the number of loading cycles given by Eq. 8.7.

The data fits very nicely within the 90% confidence limits shown in Figure 8.62. No endurance limit was found within 10 million cycles of loading.

In AASHTO (1994) LRFD Bridge Specifications, the shear stress range S_r (MPa) becomes an allowable shear force Z_r (N) for a specific life of N loading cycles by multiplying S_r by the cross-sectional area of the stud, that is,

$$Z_r = \frac{\pi}{4} d^2 S_r = (836N^{-0.19})d^2 \tag{8.173}$$

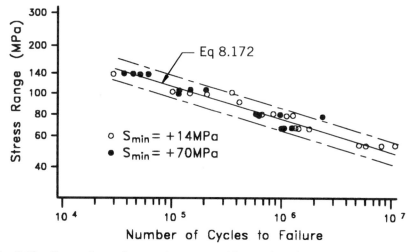

Fig. 8.62   Comparison of regression curve with test data for stud shear connectors (Slutter and Fisher, 1967.)

where d is the nominal diameter of the stud connector in millimeters. The AASHTO (1994) LRFD Bridge Specifications represent Eq. 8.173 as [A6.10.7.4.2]

$$Z_r = \alpha d^2 \geq 19.0 d^2 \qquad (8.174)$$

for which

$$\alpha = 238 - 29.5 \log N \qquad (8.175)$$

Values for α are compared in Table 8.22 with those for the quantity in parenthesis in Eq. 8.173 over the test data range of N. The expression for α in Eq. 8.175 is a reasonable approximation to the test data. (Note: The constant on the right side of Eq. 8.174 is the value of 38.0 MPa in Table 8.22 at $N = 6 \times 10^6$ divided by two.)

TABLE 8.22   Comparison of α with Regression Equation

N	$238 - 29.5 \log N$	$836 N^{0.19}$
2×10^4	111 MPa	127 MPa
1×10^5	90.5 MPa	93.8 MPa
5×10^5	69.9 MPa	69.1 MPa
2×10^6	52.1 MPa	53.1 MPa
6×10^6	38.0 MPa	43.1 MPa

Equations 8.174 and 8.175 can be used to determine the fatigue shear resistance of a single stud connector with diameter d for a specified life N. The spacing or pitch of these connectors along the length of the bridge depends on how many connectors n are at a transverse section and how large the shear force range V_{sr} (N) due to the fatigue truck is at the section.

Because fatigue is critical under repetitions of working loads, the design criteria is based on elastic conditions. If complete interaction is assumed, the horizontal shear per unit of length v_h (N/mm) can be obtained from the familiar elastic relationship

$$v_h = \frac{V_{sr}Q}{I} \tag{8.176}$$

where Q (mm³) is the first moment of the transformed deck area about the neutral axis of the short-term composite section and I (mm⁴) is the moment of inertia of the short-term composite section. The shear force per unit length that can be resisted by n connectors at a cross section with a distance p (mm) between groups (Fig. 8.61) is

$$v_h = \frac{nZ_r}{p} \tag{8.177}$$

Equating Eqs. 8.176 and 8.177 yields the pitch p in millimeters as

$$p = \frac{nZ_r I}{V_{sr}Q} \tag{8.178}$$

The center-to-center pitch of shear connectors shall not exceed 600 mm and shall not be less than 6 stud diameters [A6.10.7.4.1b].

Stud shear connectors shall not be closer than four stud diameters center-to-center transverse to the longitudinal axis of the supporting member. The clear distance between the edge of the top flange of the steel section and the edge of the nearest shear connector shall not be less than 25 mm [A6.10.7.4.1c].

The clear depth of cover over the tops of the shear connectors should not be less than 50 mm. In regions where the haunch between the top of the steel section and the bottom of the deck is large, the shear connectors should penetrate at least 50 mm into the deck [A6.10.7.4.1d].

8.9.2   Strength Limit State for Stud Connectors

Experimental tests were conducted by Ollgaard et al. (1971) to determine the shear strength of stud connectors embedded in solid concrete slabs. Variables considered in the experiments were the stud diameter, number of stud connectors per slab, type of aggregate in the concrete (lightweight and normal weight), and the concrete properties. Four concrete properties were evaluated: compressive strength, split cylinder tensile strength, modulus of elasticity, and density.

Two failure modes were observed. Either the studs sheared off the steel beam and remained embedded in the concrete slab or the concrete failed and the connectors were pulled out of the slab together with a wedge of concrete. Sometimes both of these failure modes were observed in the same test.

An examination of the data indicated that the nominal shear strength of a stud connector Q_n is proportional to its cross-sectional area A_{sc}. Multiple regression analyses of the concrete variables indicate that the concrete compressive strength f'_c and modulus of elasticity E_c are the dominant properties in determining connector shear strength. The empirical expression for the concrete modulus of elasticity (Eq. 7.2) includes the concrete density γ_c and, therefore, the effect of the aggregate type (normal or lightweight), that is,

$$E_c = 0.043\gamma_c^{1.5} \sqrt{f'_c} \qquad (7.2)$$

where γ_c is the density of concrete (kg/m³) and f'_c is the concrete compressive strength (MPa). Including the split cylinder tensile strength in the regression analyses did not significantly improve the correlation with the test results and it was dropped from the final prediction equation.

After rounding off the exponents from the regression analysis to convenient design values, the prediction equation for the nominal shear resistance Q_n (N) for a single shear stud connector embedded in a solid concrete slab is [A6.10.7.4.4c]

$$Q_n = 0.5A_{sc}\sqrt{f'_c E_c} \le A_{sc}F_u \qquad (8.179)$$

where A_{sc} is the cross-sectional area of a stud shear connector (mm²), f'_c is the specified 28-day concrete compressive strength (MPa), E_c is the concrete modulus of elasticity (MPa), and F_u is the specified minimum tensile strength of a stud shear connector. The upper bound on the nominal stud shear strength is taken as its ultimate tensile force.

When Eq. 8.179 is compared with the test data from which it was derived (Fig. 8.63), it shows that it provides a reasonable estimate to the nominal strength of a stud shear connector. The factored resistance of one shear connector Q_r must take into account the uncertainty in the ability of Eq. 8.179 to predict the resistance at the strength limit state, that is [A6.10.7.4.4a]

$$Q_r = \phi_{sc}Q_n \qquad (8.180)$$

where ϕ_{sc} is the resistance factor for shear connectors taken from Table 8.9 as 0.85.

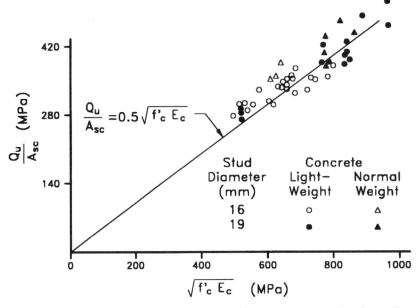

Fig. 8.63 Comparison of connector strength with concrete strength and moudlus of elasticity. (Ollgaard et al. 1971.)

Number of Shear Connectors Required   If sufficient shear connectors are provided, the maximum possible flexural strength of a composite section can be developed. The shear connectors placed between a point of zero moment and a point of maximum positive moment must resist the compression force in the slab at the location of maximum moment. This resistance is illustrated by the free-body diagrams at the bottom of Figure 8.64 for two different loading conditions. From either of these free-body diagrams, equilibrium requires that

$$n_s Q_r = V_h$$

or

$$n_s = \frac{V_h}{Q_r} \tag{8.181}$$

where n_s is the total number of shear connectors between the points of zero and maximum positive moment, V_h is the nominal horizontal shear force at the interface that must be resisted, and Q_r is the factored resistance of a single shear connector as given by Eqs. 8.179 and 8.180.

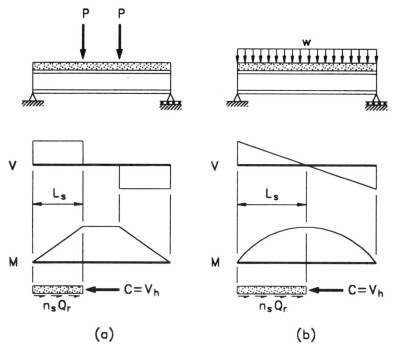

Fig. 8.64 Total number of shear connectors required. (a) Concentrated loading and (b) uniformly distributed loading.

Spacing of the Shear Connectors   Spacing of the shear connectors along the length L_s needs to be examined. For the concentrated loading of Figure 8.64(a), the vertical shear force is constant. Therefore, the horizontal shear per unit of length calculated from the elastic relationship of Eq. 8.176 is constant and spacing becomes uniform. For the uniformly distributed loading of Figure 8.64(b), the elastic horizontal shear per unit of length is variable and indicates that the connectors be closer together near the support than near midspan. These are the conditions predicted by elastic theory. At the strength limit state, conditions will be different if ductile behavior permits redistribution of the horizontal shear forces.

To test the hypothesis that stud shear connectors have sufficient ductility to redistribute horizontal shear forces at the strength limit state, Slutter and Driscoll (1965) tested three uniformly loaded simple composite beams with different connector spacings. They designed the beams with about 90% of the connectors required by Eq. 8.181 so that the connectors would control the flexural resistance. The normalized moment versus deflection response for the three beams is shown in Figure 8.65. Considerable ductility is observed and for all practical purposes the response is the same for the three beams. The conclusion is that spacing of the shear connectors along the length of the beam is not critical and can be taken as uniform [C6.10.7.4.4b].

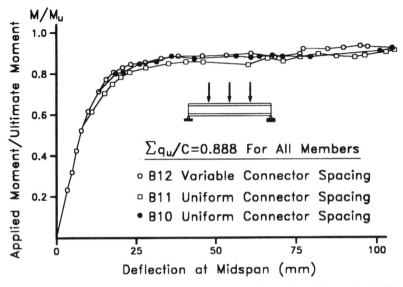

Fig. 8.65 Experimental moment-deflection curves. [Reproduced from R. G. Slutter and G. C. Driscoll (1965). "Flexural Strength of Steel-Concrete Composite Beams," *Journal of Structural Division,* ASCE, 91(ST2), pp. 71–99. With permission.]

Nominal Horizontal Shear Force $\mathbf{V_h}$ At the flexural strength limit state of a composite, the two stress distributions in Figure 8.66 are possible. A haunch is shown to indicate a gap where the shear connectors must transfer the horizontal shear from the concrete slab to the steel section.

For the first case, the plastic neutral axis is in the slab and the compressive force C is less than the full strength of the slab. However, equilibrium requires that C equal the tensile force in the steel section, so that

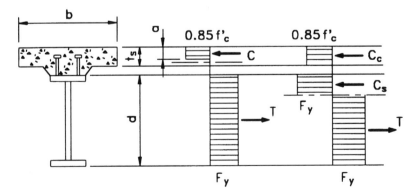

Fig. 8.66 Nominal horizontal shear force.

$$C = V_h = F_{yw}Dt_w + F_{yt}b_t t_t + F_{yc}b_c t_c \qquad (8.182)$$

where V_h is the nominal horizontal shear force shown in Figure 8.64; F_{yw}, F_{yt}, and F_{yc} are the yield strengths of the web, tension flange, and compression flange; D and t_w are the depth and thickness of the web; b_t, t_t, and b_c, t_c are the width and thickness of the tension and compression flanges. For the homogeneous steel sections in this book, this simplifies to

$$V_h = F_y A_s \qquad (8.183)$$

where F_y is the yield strength (MPa) and A_s is the total area (mm²) of the steel section.

For the second case, the plastic neutral axis is in the steel section and the compressive force $C = V_h$ is the full strength of the slab given by

$$V_h = 0.85 f'_c b t_s \qquad (8.184)$$

where f'_c is the 28-day compressive strength of the concrete (MPa), b is the effective width of the slab (mm), and t_s is the slab thickness (mm).

Techniques for locating the plastic neutral axis in positive moment regions were illustrated in Example 8.6 and Figure 8.42. In calculating V_h, this procedure can be bypassed by simply selecting the smaller value of V_h obtained from Eqs. 8.182 and 8.184 [A6.10.7.4.4b].

Continuous Composite Sections   When negative moment regions in continuous beams are made composite, the nominal horizontal shear force V_h to be transferred between the point of zero moment and maximum moment at an interior support shall be

$$V_h = A_r F_{yr} \qquad (8.185)$$

where A_r is the total area of longitudinal reinforcement (mm²) over the interior support within the effective slab width and F_{yr} is the yield strength (MPa) of the longitudinal reinforcement [A6.10.7.4.4b]. Figure 8.43 shows the forces acting on a composite section in a negative moment region. The number of shear connectors required for this region is given by Eq. 8.181.

EXAMPLE 8.13

Design stud shear connectors for the positive moment composite section of Example 8.5 shown in Figure 8.42. Assume that the shear range V_{sr} for the fatigue loading is nearly constant and equal to 230 kN in the positive moment region. Use 19-mm diameter studs 100 mm high, $F_u = 400$ MPa for the studs, $f'_c = 30$ MPa for the concrete deck, and Grade 345 for the steel beam.

General

The haunch depth is 25 mm, so the connectors project $100 - 25 = 75$ mm into the concrete deck. This projection is greater than the minimum of 50 mm. The ratio of stud height to stud diameter is

$$\frac{h}{d} = \frac{100}{19} = 5.26 > 4, \qquad \text{OK}$$

The minimum center-to-center transverse spacing of studs is 4 stud diameters and the minimum clear edge distance is 25 mm. The minimum top flange width for three 19-mm studs side by side is

$$b_{f,\min} = 2 \times 25 + 19 + 2(4)(19) = 221 \text{ mm}$$

which is less than the 300 mm provided. Therefore, use three 19-mm stud connectors at each transverse section.

Fatigue Limit State

The center to center pitch of shear connectors in the longitudinal direction shall not exceed 600 mm and shall not be less than 6 stud diameters ($6 \times 19 = 114$ mm) [A6.10.7.4.1b].

The pitch is also controlled by the fatigue strength of the studs as given by Eq. 8.178

$$p = \frac{nZ_r I}{V_{sr} Q}$$

where I and Q are elastic properties of the short-term composite section and from Eq. 8.174

$$Z_r = \alpha d^2 \geq 19 d^2$$

for which Eq. 8.175 gives

$$\alpha = 238 - 29.5 \log N$$

In Example 8.1, the number of cycles N was estimated for a 75-year life of a rural interstate bridge as 372×10^6 cycles. This value for N gives

$$\alpha = 238 - 29.5(8.57) = -15 \text{ MPa} < 19 \text{ MPa}$$

so that

$$Z_r = 19 d^2 = 19(19)^2 = 6860 \text{ N} = 6.86 \text{ kN}$$

The values of I and Q for the short term composite section are taken from Table 8.14 as

$$I = 31.6 \times 10^9 \text{ mm}^4$$

$$Q = Ay = (56\ 631)(227.1 + 25 + 205/2) = 20.1 \times 10^6 \text{ mm}^3$$

For 3 stud connectors at a transverse section and $V_{sr} = 230$ kN, the pitch is calculated as

$$p = \frac{nZ_rI}{V_{sr}Q} = \frac{3(6.86)31.6 \times 10^9}{230 \times 20.1 \times 10^6} = 140 \text{ mm}$$

This pitch is between the limits of 114 and 600 mm given earlier. By assuming that the distance from the maximum positive moment to the point of zero moment is 12 000 mm and that V_{sr} is relatively unchanged (see Section 8.11.2, Table E8.2-13), the total number of 19-mm stud connectors over this distance is

$$n = 3\left(\frac{12\ 000}{140}\right) = 257 \qquad \text{connectors}$$

Strength Limit State

The total number of shear connectors required to satisfy the strength limit state between the maximum positive moment and the point of zero moment is given by substituting Eq. 8.180 into Eq. 8.181

$$n_s = \frac{V_h}{Q_r} = \frac{V_h}{\phi_{sc}Q_n}$$

where $\phi_{sc} = 0.85$, Q_n is given by Eq. 8.179, and V_h is given by either Eq. 8.183 or Eq. 8.184. From Eq. 8.179

$$Q_n = 0.5A_{sc}\sqrt{f'_cE_c} \leq A_{sc}F_u$$

For 19-mm stud connectors

$$A_{sc} = \frac{\pi}{4}(19)^2 = 284 \text{ mm}^2$$

and for $f'_c = 30$ MPa with $\gamma_c = 2320$ kg/m³, Eq. 7.2 yields

$$E_c = 0.043\gamma_c^{1.5}\sqrt{f_c'} = 0.043(2320)^{1.5}\sqrt{30} = 26\ 320\ \text{MPa}$$

so that

$$Q_n = 0.5(284)\sqrt{30(26\ 320)} = 126\ 180\ \text{N} = 126.2\ \text{kN}$$

which is greater than the upper bound of

$$A_{sc}F_u = 284(400) = 113\ 600\ \text{N} = 113.6\ \text{kN}$$

Therefore, $Q_n = 113.6$ kN.

The nominal horizontal shear force is the lesser of the values given by Eq. 8.183 or Eq. 8.184. From Eq. 8.183 with A_s taken from Table 8.13

$$V_h = F_y A_s = 345(29\ 500) = 10.18 \times 10^6\ \text{N} = 10\ 180\ \text{kN}$$

From Eq. 8.184 with $b = 2210$ mm and $t_s = 205$ mm taken from Figure 8.42

$$V_h = 0.85f_c'bt_s = 0.85(30)(2210)(205) = 11.55 \times 10^6\ \text{N} = 11\ 550\ \text{kN}$$

Therefore, $V_h = 10\ 180$ kN and the total number of connectors required in the distance from maximum moment to zero moment is

$$n_s = \frac{V_h}{\phi_{sc}Q_n} = \frac{10\ 180}{0.85(113.6)} = 106 \quad \text{connectors}$$

Answer

The required number of shear connectors is governed by the fatigue limit state (as it often is). For the assumptions made in this example, the 19-mm diameter stud connectors placed in groups of three are spaced at a pitch of 140 mm throughout the positive moment region.

8.10 STIFFENERS

Webs of standard rolled sections have proportions such that they can reach the bending yield stress and the shear yield stress without buckling. These proportions are not the case with many built-up *I*-sections and to prevent buckling their webs must be stiffened. Both transverse and longitudinal stiffeners can be used to improve the strength of webs. In general, transverse stiffeners increase the resistance to shear while longitudinal stiffeners increase the resistance to flexural buckling of the web. The requirements for selecting the sizes of these stiffeners are discussed in the following sections.

8.10.1 Transverse Intermediate Stiffeners

Transverse intermediate stiffeners do not prevent shear buckling of web panels, but they do define the boundaries of the web panels within which the buckling occurs. These stiffeners serve as anchors for the tension field forces so that postbuckling shear resistance can develop (Fig. 8.56). The design of transverse intermediate stiffeners includes consideration of slenderness, stiffness, and strength.

Slenderness   When selecting the thickness and width of a transverse intermediate stiffener, the slenderness of projecting elements must be limited to prevent local buckling. For projecting elements in compression, Eq. 8.33 yields

$$\frac{b_t}{t_p} \leq k \sqrt{\frac{E}{F_{ys}}} \tag{8.186}$$

where b_t is the width of the projecting stiffener element, t_p is the thickness of the projecting element, k is the plate buckling coefficient taken from Table 8.11, and F_{ys} is the yield strength of the stiffener. For plates supported along one edge, Table 8.11 gives $k = 0.45$ for projecting elements not a part of rolled shapes.

Other design rules are more empirical, but are nevertheless important for the satisfactory performance of transverse intermediate stiffeners. These are the width of the stiffener b_t must be not less than 50 mm plus one-thirtieth of the depth d of the steel section and not less than one-fourth of the full-width b_f of the steel flange. Further, the slenderness ratio b_t/t_p must be less than 16 (see Fig. 8.67).

All of these slenderness requirements for transverse intermediate stiffeners are given by two expressions in the AASHTO (1994) LRFD Bridge Specifications as limits on the width b_t of each projecting stiffener element [A6.10.8.1.2]

$$50 + \frac{d}{30} \leq b_t \leq 0.48 t_p \sqrt{\frac{E}{F_{ys}}} \tag{8.187}$$

and

$$16t_p \geq b_t \geq 0.25 b_f \tag{8.188}$$

Stiffness   Transverse intermediate stiffeners define the vertical boundaries of the web panel. They must have sufficient stiffness so that they remain relatively straight and permit the web to develop its postbuckling strength.

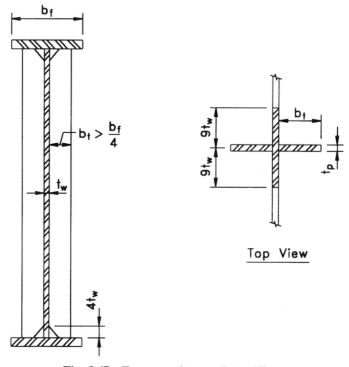

Fig. 8.67 Transverse intermediate stiffener.

A theoretical relationship can be developed by considering the relative stiffness between a transverse intermediate stiffener and a web plate. This relationship can be expressed by the nondimensional parameter (Bleich, 1952)

$$\gamma_t = \frac{(EI)_{\text{stiffener}}}{(EI)_{\text{web}}}$$

for which

$$(EI)_{\text{web}} = \frac{EDt_w^3}{12(1 - \mu^2)}$$

so that

$$\gamma_t = \frac{12(1 - \mu^2)I_t}{Dt_w^3} \qquad (8.189)$$

where μ is Poisson's ratio, D is the web depth, t_w is the web thickness, and

I_t is the moment of inertia of the transverse intermediate stiffener taken about the edge in contact with the web for single stiffeners and about the mid-thickness of the web for stiffener pairs. With $\mu = 0.3$, Eq. 8.189 can be rearranged to give

$$I_t = \frac{Dt_w^3}{10.92} \gamma_t \tag{8.190}$$

For a web without longitudinal stiffeners, the value of γ_t to ensure that the critical shear buckling stress τ_{cr} is sustained is approximately (Maquoi, 1992)

$$\gamma_t = m_t \left(\frac{21}{\alpha} - 15\alpha \right) \geq 6 \tag{8.191}$$

where α is the aspect ratio d_0/D and m_t is a magnification factor that allows for postbuckling behavior and the detrimental effect of imperfections. Taking $m_t = 1.3$ and then substituting Eq. 8.191 into Eq. 8.190, we get

$$I_t = 2.5 Dt_w^2 \left(\frac{1}{\alpha} - 0.7\alpha \right) \geq 0.55 Dt_w^2 \tag{8.192}$$

The AASHTO (1994) LRFD Bridge Specifications give the requirement for the moment of inertia of any transverse stiffener by two equations [A6.10.8.1.3]

$$I_t \geq d_0 t_w^2 J \tag{8.193}$$

and

$$J = 2.5 \left(\frac{D_p}{d_0} \right)^2 - 2.0 \geq 0.5 \tag{8.194}$$

where d_0 is the spacing of transverse intermediate stiffeners and D_p is the web depth D for webs without longitudinal stiffeners or the maximum sub-panel depth D^* for webs with longitudinal stiffeners (Fig. 8.59). Substituting Eq. 8.194 with $D_p = D$ into Eq. 8.193, and using the definition of $\alpha = d_0/D$, we can write

$$I_t \geq 2.5 Dt_w^3 \left(\frac{1}{\alpha} - 0.8\alpha \right) \geq 0.5 d_0 t_w^3 \tag{8.195}$$

By comparing Eq. 8.195 with Eq. 8.192, the code expression is very similar to the theoretically derived one.

Strength   The cross-sectional area of the transverse intermediate stiffener must be large enough to resist the vertical components of the diagonal stresses in the web. The following derivation of the required cross-sectional area is based on the work of Basler (1961a). The axial load in the transverse stiffener was derived earlier and is given by Eq. 8.145. By substituting the simple relation for σ_t from Eq. 8.150 into Eq. 8.145, and using the definition of $C = \tau_{cr}/\tau_y$, the compressive force in the transverse intermediate stiffener becomes

$$F_s = Dt_w\sigma_y(1 - C)\frac{\alpha}{2}\left(1 - \frac{\alpha}{\sqrt{1 + \alpha^2}}\right) \tag{8.196}$$

where σ_y is the yield strength of the web panel. This equation can be put in nondimensional form by dividing by $D^2\sigma_y$ to give

$$F(\alpha, \beta) = \frac{F_s}{D^2\sigma_y} = \frac{1}{2\beta}(1 - C)\left(\alpha - \frac{\alpha^2}{\sqrt{1 + \alpha^2}}\right) \tag{8.197}$$

where β is the web slenderness ratio D/t_w. In the elastic range C is given by Eq. 8.166. Defining $\varepsilon_y = F_{yw}/E$ and taking k as

$$k = 5.34 + \frac{4}{\alpha^2} \tag{8.198}$$

the expression for C becomes

$$C = \frac{1.57}{(D/t_w)^2}\left(\frac{Ek}{F_{yw}}\right) = \frac{1.57}{\varepsilon_y\beta^2}\left(5.34 + \frac{4}{\alpha^2}\right) \tag{8.199}$$

Substituting Eq. 8.199 into Eq. 8.197 yields

$$F(\alpha, \beta) = \left[\frac{1}{2\beta} - \left(4.2 + \frac{3.1}{\alpha^2}\right)\frac{1}{\varepsilon_y\beta^3}\right]\left(\alpha - \frac{\alpha^2}{\sqrt{1 + \alpha^2}}\right) \tag{8.200}$$

The maximum transverse intermediate stiffener force can be found by partial differentiation of Eq. 8.200 with respect to α and β, setting the results to zero, and solving two simultaneous equations. This gives values of $\alpha = 1.18$ and $\beta = 6.22/\sqrt{\varepsilon_y}$. Substituting $\alpha = 1.18$ into Eq. 8.196, the maximum transverse intermediate stiffener force becomes

$$\max F_s = 0.14Dt_w\sigma_y(1 - C) \tag{8.201}$$

which is the axial load in the stiffener if the maximum shear resistance of the web panel is utilized, that is, $V_u = \phi V_n$. For $V_n \leq \phi V_n$, the stiffener force will be reduced proportionately, thus

$$F_s = 0.14 Dt_w F_{yw} (1 - C) \frac{V_u}{\phi V_n} \qquad (8.202)$$

where $F_{yw} = \sigma_y$, the yield strength of the web panel.

Equation 8.202 was derived for a pair of transverse intermediate stiffeners placed symmetrically on either side of the web (Fig. 8.67). Another stiffener arrangement consists of a single stiffener on one side of the web. Basler (1961a) shows that for stiffeners made of rectangular plates, the one-sided stiffener requires at least 2.4 times the total area of stiffeners made in pairs. He also shows that an equal leg angle used as a one-sided stiffener requires 1.8 times the area of a pair of stiffeners. These variations can be incorporated into Eq. 8.202 by writing

$$F_s = 0.14 BDt_w F_{yw} (1 - C) \frac{V_u}{\phi V_n} \qquad (8.203)$$

where B is defined in Figure 8.68.

A portion of the web can be assumed to participate in resisting the vertical axial load. The AASHTO (1994) LRFD Bridge Specifications assume an effective length of web equal to $18t_w$ acting in combination with the stiffener. The force resisted by the web can be subtracted from the stiffener force in Eq. 8.203 to give

$$F_s = 0.14 BDt_w F_{yw} (1 - C) \frac{V_u}{\phi V_n} - 18t_w^2 F_{yw} \qquad (8.204)$$

The area A_s of transverse intermediate stiffeners required to carry the tension-field action of the web is obtained by dividing Eq. 8.204 by the yield strength of the stiffener F_{ys} to give [A6.10.8.1.4]

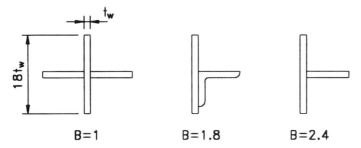

Fig. 8.68 Transverse intermediate stiffener constant B.

$$A_s \geq \left[0.15 BD t_w (1 - C) \frac{V_u}{V_r} - 18 t_w^2 \right] \left(\frac{F_{yw}}{F_{ys}} \right) \tag{8.205}$$

where $V_r = \phi V_n$ and the constant 0.14 has been rounded up to 0.15.

EXAMPLE 8.14

Select a one-sided transverse intermediate stiffener for the I-section used in Example 8.12 and shown in Figure 8.43. Use Grade 250 structural steel for the stiffener. The steel in the web is Grade 345. Assume $V_u = 1000$ kN at the section.

Slenderness

The size of the stiffener will be selected to meet slenderness requirements and then checked for stiffness and strength. From Eq. 8.188, the width of the projecting element of the stiffener must satisfy

$$b_t \geq 0.25 b_f = 0.25(400) = 100 \text{ mm}$$

and the thickness of the projecting element must satisfy

$$t_p \geq \frac{b_t}{16} = \frac{100}{16} = 6.25 \text{ mm}$$

The minimum thickness of steel elements is 8 mm [A6.7.3], so try a 8-mm $\times$ 100-mm transverse intermediate stiffener (Fig. 8.69).
From Eq. 8.187, the width b_t of the stiffener must also satisfy

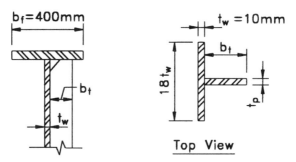

Fig. 8.69   One-sided transverse stiffener Example 8.14.

$$b_t \leq 0.48 t_p \sqrt{\frac{E}{F_{ys}}} = 0.48(8)\sqrt{\frac{200\ 000}{250}} = 109 \text{ mm}, \quad \text{OK}$$

and

$$b_t \geq 50 + \frac{d}{30} = 50 + \frac{1500 + 30 + 30}{30} = 102 \text{ mm}, \quad \text{NG (no good)}$$

Change trial size of the stiffener to 10 mm × 110 mm

$$b_t \leq 0.48 t_p \sqrt{\frac{E}{F_{ys}}} = 0.48(10)\sqrt{\frac{200\ 000}{250}} = 136 \text{ mm}, \quad \text{OK}$$

Stiffness

The moment of inertia of the one-sided stiffener is to be taken about the edge in contact with the web. For a rectangular plate, the moment of inertia taken about its base is

$$I_t = \frac{1}{3} t_p b_t^3 = \frac{1}{3}(10)(110)^3 = 4.44 \times 10^6 \text{ mm}^4$$

From Eqs. 8.193 and 8.194, the moment of inertia must satisfy

$$I_t \geq d_o t_w^3 J$$

where

$$J = 2.5\left(\frac{D_p}{d_o}\right)^2 - 2.0 \geq 0.5$$

There are no longitudinal stiffeners, so that $D_p = D = 1500$ mm. From Example 8.12, $d_o = 2000$ mm and $t_w = 10$ mm. Hence,

$$J = 2.5\left(\frac{1500}{2000}\right)^2 - 2.0 = -0.59, \quad \text{Use } J = 0.5$$

Therefore,

$$I_t \geq d_o t_w^3 J = (2000)(10)^3(0.5) = 1.0 \times 10^6 \text{ mm}^4$$

which is satisfied by the 10-mm × 110-mm stiffener ($I_t = 4.44 \times 10^6 \text{ mm}^4$).

Strength

The cross-sectional area of the stiffener

$$A_s = 10 \times 110 = 1100 \text{ mm}^2$$

must satisfy Eq. 8.205

$$A_s \geq \left[0.15 BDt_w (1 - C) \frac{V_u}{V_r} - 18t_w^2 \right]\left(\frac{F_{yw}}{F_{ys}}\right)$$

where $B = 2.4$ (Fig. 8.68) and from Example 8.12, $C = 0.306$ and $V_r = 1454$ kN. Therefore,

$$A_s \geq \left[0.15(2.4)(1500)(10)(1 - 0.306)\frac{1000}{1454} - 18(10)^2 \right]\left(\frac{345}{250}\right)$$

$$= 1073 \text{ mm}^2, \quad \text{OK}$$

Answer

Use a one-sided transverse intermediate stiffener with a thickness of $t_p = 10$ mm and a width $b_t = 110$ mm.

8.10.2 Bearing Stiffeners

Bearing stiffeners are transverse stiffeners placed at locations of support reactions and other concentrated loads. The concentrated loads are transferred through the flanges and supported by bearing on the ends of the stiffeners. The bearing stiffeners are connected to the web and provide a vertical boundary for anchoring shear forces from tension field action.

Rolled Beam Shapes   Bearing stiffeners are required on webs of rolled beams at points of concentrated forces whenever the factored shear force V_u exceeds [A6.10.8.2.1]

$$V_u > 0.75\phi_b V_n \tag{8.206}$$

where ϕ_b is the resistance factor for bearing taken from Table 8.9 and V_n is the nominal shear resistance determined in Section 8.8.

Slenderness   Bearing stiffeners are designed as compression members to resist the vertical concentrated forces. They are usually comprised of one or more pairs of rectangular plates placed symmetrically on either side of the web (Fig. 8.70). They extend the full depth of the web and are as close as practical to the outer edges of the flanges. The projecting elements of the bearing stiffener must satisfy the slenderness requirements of [A6.10.8.2.2]

$$\frac{b_t}{t_p} \le 0.48\sqrt{\frac{E}{F_{ys}}}$$ (8.207)

where b_t is the width of the projecting stiffener element, t_p is the thickness of the projecting element, and F_{ys} is the yield strength of the stiffener.

Bearing Resistance The ends of bearing stiffeners are to be milled for a tight fit against the flange from which is receives its reaction, the bottom flange at supports and the top flange for interior concentrated loads. If they are not milled, they are to be attached to the loaded flange by a full-penetration groove weld [A6.10.8.2.1].

The effective bearing area is less than the gross area of the stiffener because the end of the stiffener must be notched to clear the fillet weld between the flange and the web (Section A-A, Fig. 8.70). The bearing resistance is based

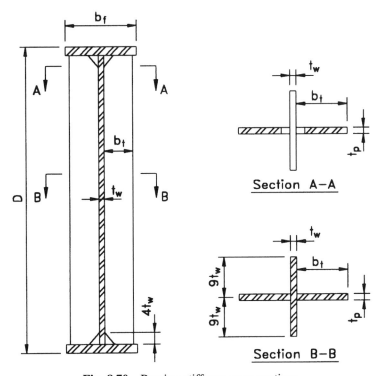

Fig. 8.70 Bearing stiffener cross sections.

on this reduced bearing area and the yield strength F_{ys} of the stiffener to give [A6.10.8.2.3]

$$B_r = \phi_b A_{pn} F_{ys} \qquad (8.208)$$

where B_r is the factored bearing resistance, ϕ_b is the bearing resistance factor taken from Table 8.9, and A_{pn} is the net area of the projecting elements of the stiffener.

Axial Resistance   The bearing stiffeners plus a portion of the web combine to act as a column to resist an axial compressive force (Section B-B, Fig. 8.70). The effective area of the column section is taken as the area of all stiffener elements, plus a centrally located strip of web extending not more than $9t_w$ on each side of the outer projecting elements of the stiffener group [A6.10.8.2.4b].

Because the bearing stiffeners fit tightly against the flanges, rotational restraint is provided at the ends and the effective pin-ended column length KL can be taken as $0.75D$, where D is the web depth [A6.10.8.2.4a]. The moment of inertia of the column section used in the calculation of the radius of gyration is taken about the centerline of the web. Designers often conservatively ignore the contribution of the web when calculating the moment of inertia and simply take the sum of the moments of inertia of the stiffeners about their edge in contact with the web.

The factored axial resistance P_r is calculated from

$$P_r = \phi_c P_n \qquad (8.209)$$

where ϕ_c is the resistance factor for compression taken from Table 8.9 and P_n is the nominal compressive resistance determined in Section 8.6.

EXAMPLE 8.15

Select bearing stiffeners for the *I*-section used in Example 8.14 and shown in Figure 8.43 to support a factored concentrated reaction $R_u = 1750$ kN. Use Grade 250 structural steel for the stiffener.

Slenderness

Selecting the width b_t of the bearing stiffener as 180 mm to support as much of the 400-mm flange width as practical, the minimum thickness for t_p is obtained from Eq. 8.207

$$\frac{b_t}{t_p} \le 0.48\sqrt{\frac{E}{F_{ys}}} = 0.48\sqrt{\frac{200\,000}{250}} = 13.6$$

$$t_p \geq \frac{b_t}{13.6} = \frac{180}{13.6} = 13.3 \text{ mm}$$

Try a 15 mm × 180 mm bearing stiffener element.

Bearing Resistance

The required area of all the bearing stiffener elements can be calculated from Eq. 8.208 for $B_r = 1750$ kN, $\phi_b = 1.0$ (milled surface), and $F_{ys} = 250$ MPa.

$$B_r = \phi_b A_{pn} F_{ys} = (1.0)A_{pn}(250)$$

$$A_{pn} = \frac{1750 \times 10^3}{250} = 7000 \text{ mm}^2$$

By using two pairs of 15-mm × 180-mm stiffener elements on either side of the web (Fig. 8.70), and allowing 40 mm to clear the web to flange fillet weld, the provided bearing area is

$$4(15)(180 - 40) = 8400 \text{ mm}^2 > 7000 \text{ mm}^2, \qquad \text{OK}$$

Try a bearing stiffener composed of four 15-mm × 180-mm elements placed in pairs on either side of the web. (Note that the 45 degree notch with $4t_w$ sides prevents the development of the unwanted triaxial tensile stress in the welds at the junction of the web, stiffener, and flange.)

Axial Resistance

By spacing the pairs of stiffeners 200 mm apart as shown in Figure 8.71, the effective area of the column cross section is

$$A = 4A_s + t_w(18t_w + 200)$$
$$A = 4(15)(180) + 10(180 + 200) = 14\ 600 \text{ mm}^2$$

and the moment of inertia of the stiffener elements about the centerline of the web is

$$I = 4I_0 + 4A_s y^2$$
$$= 4\left[\frac{1}{12}(15)(180)^3\right] + 4(15)(180)\left(\frac{180}{2} + 5\right)^2$$
$$= 126.6 \times 10^6 \text{ mm}^4$$

so that the radius of gyration for the column cross section becomes

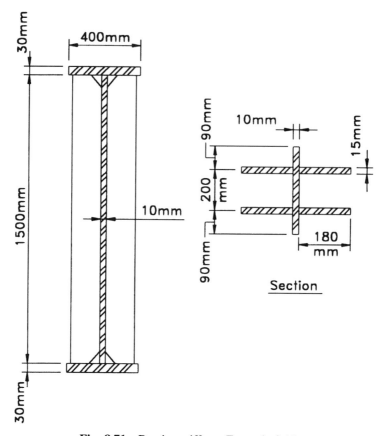

Fig. 8.71   Bearing stiffener Example 8.15.

$$r = \sqrt{\frac{I}{A}} = \sqrt{\frac{126.6 \times 10^6}{14\ 600}} = 93 \text{ mm}$$

Therefore,

$$\frac{KL}{r} = \frac{0.75D}{r} = \frac{0.75(1500)}{93} = 12.1 < 120, \qquad \text{OK}$$

and Eq. 8.29 gives

$$\lambda = \left(\frac{KL}{\pi r}\right)^2 \frac{F_y}{E} = \left(\frac{12.1}{\pi}\right)^2 \frac{250}{200\ 000} = 0.0185 < 2.25$$

so that the nominal column strength is given by Eq. 8.31

$$P_n = 0.66^\lambda F_y A_s = (0.66)^{0.0185}(250)(14\ 600) = 3.622 \times 10^6 \text{ N}$$

The factored axial resistance is calculated from Eq. 8.209 with $\phi_c = 0.90$

$$P_r = \phi_c P_n = 0.90(3622) = 3260 \text{ kN} > 1750 \text{ kN}, \quad \text{OK}$$

Answer

Use a bearing stiffener composed of two pairs of 15-mm × 180-mm stiffener elements arranged as shown in Figure 8.71.

8.11 EXAMPLE PROBLEMS

In this section, three typical steel beam and girder superstructure designs are given. The first two examples are simple span rolled steel beam bridges: one noncomposite and the other composite. The third example is a three-span continuous composite plate girder bridge.

References to the AASHTO (1994) LRFD Bridge Specifications in the examples are enclosed in brackets and denoted by a letter A followed by the article number, for example [A4.6.2.1.3]. If a commentary is cited the article number is preceded by the letter C. Referenced figures and tables are enclosed in brackets to distinguish them from figures and tables in the text, for example [Fig A3.6.1.2.2-1] and [Table A4.6.2.1.3-1]. Section properties for structural shapes are taken from AISC (1992).

Throughout the examples, comparisons are made to a bridge software program, BT Beam. The program and its availability are presented in Appendix C.

The units chosen for the force effects in the example problems are kilonewton meter (kN m) for moments and kilonewton (kN) for shears. The resistance expressions in the AASHTO specifications are in newtons (N) and millimeters (mm), which means that multipliers of 10^6 and 10^3 are added as necessary when force effects are compared to resistance.

The rationale for choosing kilonewton and meters is to give the reader, who is familiar with the kip and foot system, values for the force effects that are of the same order of magnitude as previously calculated. For example, bending moments in kilonewton meters are approximately 1.33 times the value in kip-feet. Thus, if a designer is anticipating a bending moment in a beam to be 600 kip-ft, this now becomes 800 kN m in the SI system.

For shear forces and reactions, the multiplier between kilonewton and kip is approximately 4, which is easy to apply and allows a mental adjustment of an anticipated shear force of 150 kips to a value of 600 kN.

Similarly, the integer multiplier between megapascals (or N/mm²) and ksi is seven. The familiar 36-ksi yield strength of structural steel becomes 250

MPa, a 50 ksi yield strength becomes 345 MPa, and the common concrete compressive strength of 4 ksi is comparable to 30-MPa concrete.

The design examples generally follow the outline of *Appendix B—Basic Steps for Steel Bridge Superstructures* given at the end of Section 6 of the AASHTO (1994) LRFD Bridge Specifications. Care has been taken in preparing these examples, but they should not be considered as fully complete in every detail. Each designer must take responsibility for understanding and correctly applying the provisions of the specifications. Additionally, the AASHTO LRFD Bridge Design Specifications will likely be altered each year by addendums that define interim versions. The computations outlined herein may not be current with the most recent interim.

8.11.1 Noncomposite Rolled Steel Beam Bridge

Problem Statement   Design the simple span noncomposite rolled steel beam bridge of Figure E8.1-1 with 10.5-mm span for a *HL*-93 live load. Roadway width is 13 420-mm curb to curb. Allow for a future wearing surface of 75-mm thick bituminous overlay. Use f'_c = 30 MPa and M270 Gr345 steel. Follow the outline of AASHTO (1994) LRFD Bridge Specifications, Section 6, Appendix B6.3.

A. **Develop General Section**

The bridge is to carry interstate traffic over a normally small stream that is subject to high water flows during the rainy season (Fig. E8.1-1).

1. *Roadway width:* 13 420 mm curb to curb
2. *Span arrangements:* Simple span, 10.5 m
3. *Select bridge type: I*-Girder

B. **Develop Typical Section and Design Basis**
1. I-*Girder*
   a. General [A6.10.1]. Design flexural members for
      • Strength limit state
      • Service limit state for control of permanent deflections
      • Fatigue and fracture limit state for details
      • Fatigue requirement for webs
      • Constructibility
   b. Member proportions of flexural components

$$0.1 \leq \frac{I_{yc}}{I_y} \leq 0.9$$

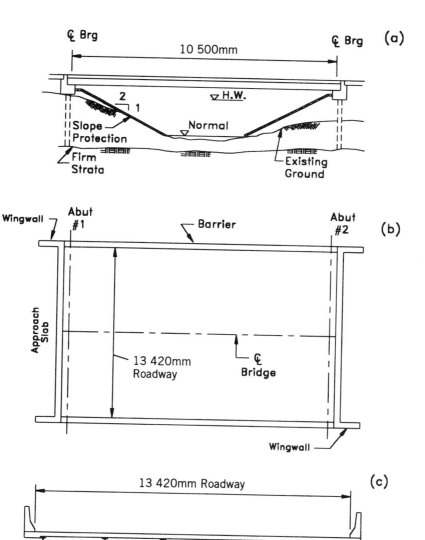

Fig. E8.1-1 Noncomposite rolled steel beam bridge design example. (a) General elevation, (b) plan view, and (c) cross section.

2. *Elastic Analysis or Inelastic Analysis* [A6.10.2.2]   Elastic analysis will be performed. The span is simply supported; thus, moment redistribution is not used.

3. *Homogeneous or Hybrid* [A6.10.5.4]   Rolled beams are homogeneous (the flanges and web are the same material and strength). For homogeneous sections, the hybrid factor, R_h shall be taken as 1.0. For compression flanges, if either a longitudinal stiffener is provided or Eq. E8.1-1 is satisfied, then the load shedding factor R_b shall be taken as 1.0.

$$\frac{2D_c}{t_w} \le \lambda_b \sqrt{\frac{E}{f_c}} \qquad (E8.1\text{-}1)$$

Otherwise,

$$R_b = 1 - \left(\frac{a_r}{1200 + 300a_r}\right)\left(\frac{2D_c}{t_w} - \lambda_b\sqrt{\frac{E}{f_c}}\right) \qquad (8.81)$$

for which

$$a_r = \frac{2D_c t_w}{A_{fc}} \qquad (8.82)$$

For tension flanges, R_b shall be taken as 1.0.

C. **Design Conventionally Reinforced Concrete Deck**

The deck was designed in Example Problem 7.10.1.

D. **Select Resistance Factors**

1. *Strength Limit State* ϕ [A6.5.4.2]
   Flexure   1.00
   Shear   1.00
2. *Nonstrength Limit States*   1.00 [A1.3.2.1]

E. **Select Load Modifiers**

		Strength	Service	Fatigue
1. Ductility, η_D	[A1.3.3]	0.95	1.0	1.0
2. Redundancy, η_R	[A1.3.4]	0.95	1.0	1.0
3. Importance, η_I	[A1.3.5]	1.05	N/A	N/A
$\eta = \eta_D\eta_R\eta_I$	[A1.3.2.1]	0.95	1.0	1.0

F. Select Applicable Load Combinations (Table 3.1)

1. *Strength I Limit State*

$$U = \eta[1.25DC + 1.50DW + 1.75(LL + IM) + 1.0FR + \gamma_{TG}TG]$$

2. *Service I Limit State*

$$U = 1.0(DC + DW) + 1.0(LL + IM) + 0.3(WS + WL) + 1.0FR$$

3. *Fatigue and Fracture Limit State*

$$U = 0.75(LL + IM)$$

G. Calculate Live Load Force Effects [A3.6.1.1.1]

1. *Select Number of Lanes:* [A3.6.1.1.1]

$$N_L = INT\left(\frac{w}{3600}\right) = INT\left(\frac{13\ 420}{3600}\right) = 3$$

2. *Multiple Presence Factor:* (Table 4.6) [A3.6.1.1.2]

No. of Loaded Lanes	m
1	1.20
2	1.00
3	0.85

3. *Dynamic Load Allowance:* (Table 4.7) [A3.6.2.1]

Component	IM (%)
Deck joints	75
Fatigue	15
All other	33

Not applied to the design lane load.

4. *Distribution Factors for Moment:* [A4.6.2.2.2]   Cross-section type (a) (Table 2.2), $S = 2440$ mm, $L = 10\ 500$ mm. Assume for preliminary design, $K_g/Lt_s^3 = 1.0$.

a. Interior beams (Table 6.5) [Table A4.6.2.2.2b-1] One design lane loaded:

$$mg_M^{SI} = 0.06 + \left(\frac{S}{4300}\right)^{0.4}\left(\frac{S}{L}\right)^{0.3}\left(\frac{K_g}{Lt_s^3}\right)^{0.1}$$

$$mg_M^{SI} = 0.06 + \left(\frac{2440}{4300}\right)^{0.4}\left(\frac{2440}{10\ 500}\right)^{0.3}(1.0)^{0.1} = 0.575$$

Two design lanes loaded:

$$mg_M^{MI} = 0.075 + \left(\frac{S}{2900}\right)^{0.6}\left(\frac{S}{L}\right)^{0.2}\left(\frac{K_g}{Lt_s^3}\right)^{0.1}$$

$$mg_M^{MI} = 0.075 + \left(\frac{2440}{2900}\right)^{0.6}\left(\frac{2440}{10\ 500}\right)^{0.2}(1.0)^{0.1}$$

$$= 0.748,\ \text{governs}$$

b. Exterior beams (Table 6.5) [Table A4.6.2.2.2d-1] One design lane loaded—lever rule (Fig. E8.1-2):

$$R = \frac{P}{2}\left(\frac{650 + 2450}{2440}\right) = 0.635P$$

$$g_M^{SE} = 0.635$$

$$mg_M^{SE} = 1.2(0.635) = 0.762,\ \text{governs}$$

Two or more design lanes loaded:

$$d_e = 990 - 380 = 610\ \text{mm}$$

$$e = 0.77 + \frac{d_e}{2800} = 0.77 + \frac{610}{2800} = 0.988 < 1.0$$

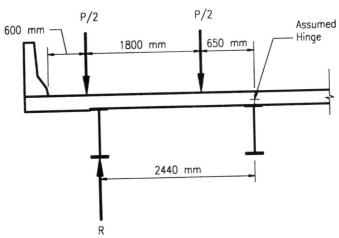

Fig. E8.1-2 Lever rule for determination of distribution factor for moment in exterior beam, one lane loaded.

$$Use \ e = 1.0$$

$$mg_M^{ME} = e \ mg_M^{MI} = 0.748$$

c. Skewed bridges—no skew.
d. Distributed live load moments (Fig. E8.1-3)

$$M_{LL+IM} = mg\left[(M_{Tr} \text{ or } M_{Ta})\left(1 + \frac{IM}{100}\right) + M_{Ln}\right]$$

$$M_{Tr} = 145(2.625) + (145 + 35)(0.475) = 466 \text{ kN m}$$

$$M_{Ta} = 110(2.625 + 2.025) = 512 \text{ kN m, governs}$$

The absolute moment due to the tandem actually occurs under the wheel closest to the resultant when the *cg* of the wheels on the span and the critical wheel are equidistant from the centerline of the span. For this span, the absolute maximum moment is 513 kN m. However, the value of 512 kN m is used because the moments due to other loads are maximum at the centerline and thus can be added to the tandem load moment.

$$M_{Ln} = \frac{9.3(10.5)^2}{8} = 128 \text{ kN m}$$

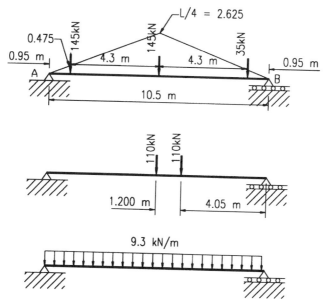

Fig. E8.1-3 Truck, tandem, and lane load placement for maximum moment at Location 105.

Interior Beams

$$M_{LL+IM} = 0.748[512(1.33) + 128] = 605 \text{ kN m}$$

Exterior Beams

$$M_{LL+IM} = 0.762[512(1.33) + 128] = 616 \text{ kN m}$$

5. *Distribution Factors for Shear* [A4.6.2.2.3]   Cross-section type (a) (Table 2.2).
   a. Interior beams (Table 6.5) [Table A4.6.2.2.3a-1] One design lane loaded:

$$mg_V^{SI} = 0.36 + \frac{S}{7600} = 0.36 + \frac{2440}{7600} = 0.681$$

Two design lanes loaded:

$$mg_V^{MI} = 0.2 + \frac{S}{3600} - \left(\frac{S}{10\ 500}\right)^{2.0}$$

$$mg_V^{MI} = 0.2 + \frac{2440}{3600} - \left(\frac{2440}{10\ 500}\right)^{2.0} = 0.826, \text{ governs}$$

   b. Exterior beams (Table 6.5) [Table A4.6.2.2.3b-1] One design lane loaded—level rule (Fig. E8.1-2)

$$mg_V^{SE} = 0.762, \text{ governs}$$

Two or more design lanes loaded

$$d_e = 610 \text{ mm}$$

$$e = 0.6 + \frac{d_e}{3000} = 0.6 + \frac{610}{3000} = 0.803$$

$$mg_V^{ME} = e \cdot mg_V^{MI} = (0.803)(0.826) = 0.663$$

   c. Skewed bridges—no skew.
   d. Distributed live load shears (Fig. E8.1-4)

$$V_{LL+IM} = mg\left[(V_{Tr} \text{ or } V_{Ta})\left(1 + \frac{IM}{100}\right) + V_{Ln}\right]$$

$$V_{Tr} = 145(1 + 0.590) + 35(0.181) = 237 \text{ kN, governs}$$

$$V_{Ta} = 110(1 + 0.886) = 207 \text{ kN}$$

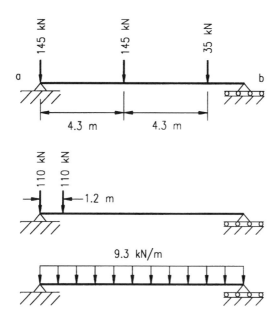

Fig. E8.1-4 Truck, tandem, and lane load placement for maximum shear at Location 100.

$$V_{Ln} = \frac{9.3(10.5)}{2} = 49 \text{ kN}$$

Interior Beams

$$V_{LL+IM} = 0.826[237(1.33) + 49] = 301 \text{ kN}$$

Exterior Beams

$$V_{LL+IM} = 0.762[237(1.33) + 49] = 278 \text{ kN}$$

6. *Reactions to Substructure* The following reactions are per design lane without any distribution factors

$$R_{100} = V_{100} = 1.33V_{Tr} + V_{Ln} = 1.33(237) + 49 = 364 \text{ kN/lane}$$

H. Calculate Force Effects from Other Loads

Analysis for a uniformly distributed load w (Fig. E8.1-5)

$$M_{\max} = M_{105} = \frac{wL^2}{8} = \frac{w(10.5)^2}{8} = 13.78 \times w$$

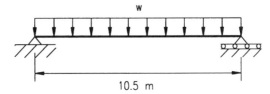

Fig. E8.1-5 Uniformly distributed load.

$$V_{max} = V_{100} = \frac{wL}{2} = \frac{w(10.5)}{2} = 5.25 \times w$$

Assume a beam weight of 1.5 kN/m.
1. *Interior Girders*

DC	Deck slab	$(2400)(10^{-9})(9.81)(205)(2440)$ = 11.78 kN/m
	Girder	= 1.5 kN/m
		w_{DC} = 13.28 kN/m

DW   75-mm bituminous paving = $(2250)(9.81)(10^{-9})(75)(2440)$

$$w_{DW} = 4.04 \text{ kN/m}$$

Unfactored moments and shears for an interior girder are summarized in Table E8.1-1.
2. *Exterior Girders.* Using deck design results for reaction on exterior girder,

DC	Deck slab	4.60 kN/m
	Overhang	6.75 kN/m
	Barrier	6.74 kN/m
	Girder	1.5 kN/m
	w_{DC} =	19.59 kN/m

TABLE E8.1-1   Interior Girder Unfactored Moments and Shears

Load Type	w (kN/m)	Moment (kN m) M_{105}	Shear (kN) V_{100}
DC	13.28	183	70
DW	4.04	56	21
LL + IM	N/A	605	301

DW   75-mm bituminous paving

$$w_{DW} = (2250)(10^{-9})(9.81)(75)\left(610 + \frac{2440}{2}\right)$$

$$= 2.77 \text{ kN/m}$$

Unfactored moments and shears for an exterior girder are summarized in Table E8.1-2.

I. Design Required Sections

1. *Strength I Limit State*

   a. Interior beam

Factored shear and moment

$$U = \eta[1.25DC + 1.50DW + 1.75(LL + IM)]$$

$$V_u = 0.95[1.25(70) + 1.50(21) + 1.75(301)] = 613 \text{ kN}$$

$$M_u = 0.95[1.25(183) + 1.50(56) + 1.75(605)] = 1302 \text{ kN m}$$

   b. Exterior beam

Factored shear and moment

$$V_u = 0.95[1.25(103) + 1.50(15) + 1.75(278)] = 606 \text{ kN}$$

$$M_u = 0.95[1.25(270) + 1.50(38) + 1.75(616)] = 1399 \text{ kN m}$$

2. *Required Plastic Section Modulus, Z*

$$\phi_f M_n \geq M_u, \; \phi_f = 1.0, \; M_n = M_p = Z F_y$$

$$Z F_y \geq M_u$$

Assuming compression flange is fully braced and section is compact,

$$\text{req'd } Z \geq \frac{M_u}{F_y} = \frac{1399 \times 10^6}{345} = 4.055 \times 10^6 \text{ mm}^3$$

TABLE E8.1-2   Exterior Girder Unfactored Moments and Shears

Load Type	*w* (kN/m)	Moment (kN m) M_{105}	Shear (kN) V_{100}
DC	19.59	270	103
DW	2.77	38	15
LL + IM	N/A	616	278

Try W760 × 134, $Z = 4.63 \times 10^6$ mm^3, $S = 4000 \times 10^3$ mm^3, $I = 1500 \times 10^6$ mm^4,

$$w_g = (134)(9.81)(10^{-3})$$

$$= 1.31 \text{ kN/m}$$

3. *Fatigue Induced by Web Flexure or Shear* [A6.10.4]

$$D_c = \tfrac{1}{2}(d - 2t_f) = \tfrac{1}{2}[750 - 2(15.5)] = 359.5 \text{ mm}$$

$$\frac{2D_c}{t_w} = \frac{2(359.5)}{11.90} = 60.42 < 5.76\sqrt{\frac{E}{F_{yc}}} = 5.76\sqrt{\frac{200\ 000}{345}} = 138.7$$

therefore, f_{cf} must be less than $R_h F_{yc}$

where

f_{cf} = maximum compressive elastic flexural stress in the compression flange due to the unfactored permanent load and twice the fatigue loading [A6.10.4.2].

The maximum flexural stress due to fatigue loading is found from Figure E8.1-6.

$$R_2 = \frac{145(5.25) + 35(5.25 + 4.3)}{10.5} = 104.3 \text{ kN}$$

$$M_{105} = 104.3(5.25) - 35(4.3) = 397 \text{ kN m}$$

$$M_{cf} = \text{Unfactored permanent load + Fatigue loading}$$

(exterior girder governs)

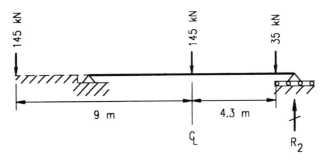

Fig. E8.1-6 Fatigue truck placement for maximum moment.

$$M_{cf} = M_{DC} + M_{DW} + 2(0.75)(LL + IM)_{Fat}/1.2$$

(no multiple presence) [A3.6.1.1.2]

$$M_{cf} = 270 + 38 + 2(0.75)(1.15)(0.762)(397)/1.2 = 743 \text{ kN m}$$

$$f_{cf} = \frac{M_{cf}}{S} = \frac{743 \times 10^6}{4000 \times 10^3}$$

$$= 186 \text{ MPa} < R_h F_{yc} = 1.0(345) = 345 \text{ MPa}, \qquad \text{OK}$$

4. *For Noncomposite Sections*

(For most rolled sections, this one included, only bracing is an issue regarding compactness. All checks are illustrated here for completeness.)

a. Compact sections shall satisfy: (Table 8.18) [A6.10.6.2]
   • Web slenderness [A6.10.5.2.3b]

$$\frac{2D_{cp}}{t_w} = \frac{2[\frac{1}{2}(d - 2t_f)]}{t_w} = \frac{2\{\frac{1}{2}[750 - 2(15.5)]\}}{11.90} = 60.42$$

$$3.76\sqrt{\frac{E}{F_{yc}}} = 3.76\sqrt{\frac{200\ 000}{345}} = 90.53 > 60.42, \qquad \text{OK}$$

   • Compression flange slenderness [A6.10.5.2.3c]

$$\frac{b_f}{2t_f} = \frac{264}{2(15.5)} = 8.52 < 0.382\sqrt{\frac{E}{F_{yc}}}$$

$$= 0.382\sqrt{\frac{200\ 000}{345}} = 9.20, \qquad \text{OK}$$

   • Compression flange bracing requirements [A6.10.5.2.3d]

   Assume compression flange is fully braced by the concrete haunch

   ∴ All requirements are satisfied; therefore, section is classified as compact.

b. Lateral-torsional buckling flexural resistance [A6.10.6.4]
   • Compression flanges [A6.10.6.4.1]
   Does not apply because section is fully braced
   • Tension flanges [A6.10.6.4.2]
   Does not apply because section is fully braced

5. *Shear Design*
   a. For beams with unstiffened webs (Table 8.19) [A6.10.7.2],

$$V_r = \phi_v V_n = 1.00 V_n$$

$$2.46 \sqrt{\frac{E}{F_{yw}}} = 2.46 \sqrt{\frac{200\ 000}{345}} = 59.23$$

$$3.07 \sqrt{\frac{E}{F_{yw}}} = 3.07 \sqrt{\frac{200\ 000}{345}} = 73.9$$

$$59.23 < \frac{D}{t_w} = \frac{d - 2t_f}{t_w} = \frac{750 - 2(15.5)}{11.90} = 60.42 < 73.9$$

$$V_r = V_n = 1.48 t_w^2 \sqrt{EF_{yw}} = 1.48(11.90)^2 \sqrt{(200\ 000)(345)}(10^{-3})$$

$$= 1741\ \text{kN}$$

$$V_r > V_u = 613\ \text{kN (Sec. 8.11.1, Part I.1.a)}, \qquad \text{OK}$$

   b. Bearing stiffener design [A6.10.8.2]

$$0.75\phi_b V_n = 0.75(1.0)(1741) = 1306\ \text{kN} > V_u = 613\ \text{kN}$$

   Therefore bearing stiffeners are not required
6. *Constructibility*
   (Lateral bracing requirements are checked later in this example.)
   a. General proportions [A6.10.1.1]

$$0.1 \leq \frac{I_{yc}}{I_y} \leq 0.9$$

$$I_y = 47.7 \times 10^6\ \text{mm}^4$$

$$I_{yc} = \frac{1}{12} t_f b_f^3 = \frac{1}{12} (15.5)(264.0)^3 = 23.77 \times 10^6\ \text{mm}^4$$

$$0.1 \leq \frac{I_{yc}}{I_y} = \frac{23.77 \times 10^6}{47.7 \times 10^6} = 0.50 < 0.9, \qquad \text{OK}$$

   b. Flexural resistance [A6.10.10.2]

$$M_n = R_b R_h M_y$$

   R_b is found from Eq. 8.81 [A6.10.5.4.2] if Eq. E8.1-1 is not satisfied.

f_c = stress in compression flange due to factored construction loading, interior girder (Table E8.1-1)

$$M = \eta[1.25\ M_{DC}] = 0.95[1.25(183)] = 217\ \text{kN m}$$

$$f_c = \frac{Mc}{I} = \frac{(217 \times 10^6)\left(\dfrac{750}{2}\right)}{1500 \times 10^6} = 54.3\ \text{MPa}$$

$$\frac{2D_c}{t_w} = 60.42 < \lambda_b \sqrt{\frac{E}{f_c}} = 5.76\sqrt{\frac{200\ 000}{54.3}} = 350$$

Therefore $R_b = 1.0$

For homogeneous sections, $R_h = 1.0$

$$M_n = (1.0)(1.0)M_y = M_y = S_x F_y = (4000 \times 10^3)(345)(10^{-6})$$
$$= 1380\ \text{kN m} > M = 217\ \text{kN m}, \quad \text{OK}$$

c. Shear resistance of unstiffened web (Table 8.19) [C6.10.10.3]

$$D = d - 2t_f = 750 - 2(15.5) = 719\ \text{mm}$$

$$\frac{D}{t_w} = \frac{719}{11.90} = 60.42$$

$$2.46\sqrt{\frac{E}{F_{yw}}} = 2.46\sqrt{\frac{200\ 000}{345}} = 59.22$$

$$3.07\sqrt{\frac{200\ 000}{345}} = 73.9$$

$$59.22 < \frac{D}{t_w} < 73.9$$

Thus, $V_n = 1.48t_w^2\sqrt{EF_{yw}} = 1.48(10)^2\sqrt{(200\ 000)(345)}$
$$= 1229\ \text{kN}$$

V = shear due to factored construction loading (Table E8.1-1)

$$= \eta[1.25\ V_{DC}] = 0.95[1.25(70)]$$

$$= 83\ \text{kN} < V_n = 1229\ \text{kN}, \quad \text{OK}$$

J. Dimension and Detail Requirements

1. *Material Thickness* [A6.7.3]   Bracing and cross frames shall not be less than 8 mm in thickness. Web thickness of rolled beams shall not be less than 7 mm.

$$t_w = 11.90 \text{ mm} > 7 \text{ mm}, \quad \text{OK}$$

2. *Optional Deflection Control* [A2.5.2.6.2]   Allowable service load deflection $\leq \dfrac{1}{800}$ span $= \dfrac{10\,500}{800} = 13$ mm

From [A3.6.1.3.2], deflection is taken as the larger of
- That resulting from the design truck alone.
- That resulting from 25% of the design truck taken together with the design lane load.

The distribution factor for deflection may be taken as the number of lanes divided by the number of beams [C2.5.2.6.2], because all design lanes should be loaded, and all supporting components should be assumed to deflect equally.

$$DF = \frac{\text{No. lanes}}{\text{No. beams}} = \frac{3}{6} = 0.5$$

a. Deflection resulting from design truck alone (Fig. E8.1-7)

$$P_1 = P_2 = 0.5(145)\left(1 + \frac{IM}{100}\right) = 0.5(145)(1.33) = 96.4 \text{ kN}$$

$$P_3 = 0.5(35)(1.33) = 23.3 \text{ kN}$$

The deflection at any point, Δ_x, due to a point load P can be found from [AISC manual (1986), Case 8] (Fig. E8.1-8) for $x \leq a$

$$\Delta_x = \frac{Pbx}{6EIL}(L^2 - b^2 - x^2)$$

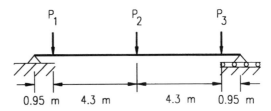

0.95 m   4.3 m   4.3 m   0.95 m

Fig. E8.1-7   Truck placement for maximum deflection.

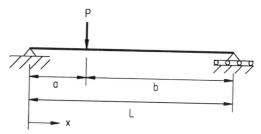

Fig. E8.1-8 General placement of point load P.

The maximum deflection (located at the center) of a simply supported span, due to a concentrated load at the center of the span, can be found from [AISC manual (1986), Case 7] (Fig. E8.1-9)

$$\Delta_{CL} = \frac{PL^3}{48EI}$$

$$\Delta_{CLTr} = \Delta_{P_1} + \Delta_{P_2} + \Delta_{P_3}$$

$$= \frac{(P_1 + P_3)(0.95)\left(\dfrac{10.5}{2}\right)(10^9)}{6(200\ 000)(1500 \times 10^6)(10.5)}$$

$$\times \left[(10.5)^2 - (0.95)^2 - \left(\frac{10.5}{2}\right)^2 \right]$$

$$+ \frac{P_2(10.5)^3(10)^9}{48(200\ 000)(1500 \times 10^6)}$$

$$= (P_1 + P_3)(215.8 \times 10^{-3}) + P_2(804 \times 10^{-3}) = 10 \text{ mm}$$

b. Deflection resulting from 25% of design truck together with the design lane load

$$\Delta_{CL25\%Tr} = 0.25(10) = 3 \text{ mm}$$

The deflection due to lane load can be found from [AISC manual (1986), Case 1] (Fig. E8.1-10)

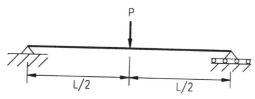

Fig. E8.1-9 Point load P at center of the span.

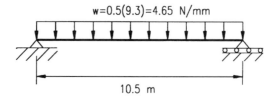

Fig. E8.1-10 Uniform lane load on the span.

$$\Delta_{\max} = \frac{5wL^4}{384EI}$$

$$\Delta_{CL\,Ln} = \frac{5(0.5)(9.3)(10\ 500)^4}{384(200\ 000)(1500 \times 10^6)} = 2 \text{ mm}$$

$$\Delta_{CL} = \Delta_{CL25\%Tr} + \Delta_{CLLn} = 3 + 2 = 5 \text{ mm}$$

$$\therefore \Delta_{CL\,Tr} = 10 \text{ mm controls}$$

$$\Delta_{CL} = 10 \text{ mm} < \Delta_{all} = 13 \text{ mm}, \qquad \text{OK}$$

3. *Service Limit State Control of Permanent Deflections* [A6.10.3] This limit state is checked to prevent permanent deflection that would impair rideability.

For both flanges of noncomposite sections:

$$f_f \le 0.80R_h F_{yf}$$

$$R_h = 1.0 \text{ for homogeneous sections [A6.10.5.4.1a]}$$

$$F_{yf} = 345 \text{ MPa}$$

The maximum Service *II* moment, which occurs at Location 105 in the exterior beam, is

$$M = 1.0(DC + DW) + 1.3(LL + IM)$$

$$= 1.0(270 + 38) + 1.3(616)$$

$$= 1109 \text{ kN m}$$

$$f_f = \frac{M}{S_x} = \frac{1109 \times 10^6}{4000 \times 10^3}$$

$$= 277 \text{ MPa} \approx 0.8(1.0)(345) = 276 \text{ MPa}, \qquad \text{OK}$$

4. *Check Construction Requirements*
a. Dead load camber

Exterior beam, $w_D = w_{DC} + w_{DW} = 19.59 + 2.77 = 22.36 \text{ N/mm}$

Interior beam, $w_D = 13.28 + 4.04 = 17.32$ N/mm

$$\Delta_{CL} = \frac{5}{384} \frac{(w_D L^4)}{EI} = \frac{5}{384} \frac{(22.36)(10\ 500)^4}{(200\ 000)(1500 \times 10^6)} = 12 \text{ mm}$$

Use 12-mm camber on all beams

b. Lateral support for compression flange is not available when fresh concrete is being placed [A6.10.5.2.3d]

$$L_b \leq \left[0.124 - 0.0759 \left(\frac{M_l}{M_p} \right) \right] \left(\frac{r_y E}{F_{yc}} \right)$$

M_l = smallest end moment = 0

$$L_b \leq 0.124 \left[\frac{(53.0)(200\ 000)}{345} \right] = 3810 \text{ mm}$$

Until the concrete cures enough to laterally support the compression flange of the steel beam, the compression flange must be braced no more than 3810 mm apart. Because AASHTO [A6.10.11.1.1c] requires a brace at the point where plastic moment is reached, provide braces at quarter points along the beam, for $L_b = 2625$ mm.

5. *Diaphragms* [A6.7.4]   Diaphragms provide:
   - Lateral support of compression flange prior to curing of the deck.
   - Transfer of wind load on exterior girder to all girders.
   - Distribution of vertical dead and live loads applied to the structure.
   - Stability of the bottom flange for all loads when it is in compression.

   For straight *I*-sections, cross frames shall be at least half the beam depth.

   a. Intermediate diaphragms (Fig. E8.1-11)

   Try C380 × 50 intermediate diaphragms at midspan, for A_s =

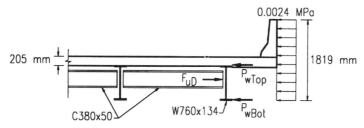

Fig. E8.1-11 Wind load acting on exterior elevation.

6430 mm^2 and $L_b = 5.25$ m

The wind load acting on the bottom half of the beam goes to the bottom flange

$$w_{\text{Bot}} = \frac{\gamma P_D d}{2} = \frac{1.4(0.0024)(750)}{2} = 1.26 \text{ N/mm}$$

$$P_{w\text{Bot}} = w_{\text{Bot}}L_b = (1.26)(5.25) = 6.6 \text{ kN}$$

The remaining wind load is transmitted to the abutment region by the deck diaphragm. The end reaction must be transferred to the bearings equally by all six girders. The resultant force is F_{uD}.

$$P_{w\text{Top}} = [1.4(0.0024)(1819) - 1.26]\left[\frac{5.25}{6 \text{ girders}}\right] = 4.2 \text{ kN}$$

$$F_{uD} = P_{w\text{Bot}} + P_{w\text{Top}} = 6.6 + 4.2 = 10.8 \text{ kN}$$

The axial resistance is [A6.9.3, A6.9.4]

$$\frac{kL}{r_y} = \frac{1.0(2440)}{22.9} = 107 < 140$$

$$\lambda = \left(\frac{kL}{r_s\pi}\right)^2 \frac{F_y}{E} = \left(\frac{1.0(2440)}{22.9\pi}\right)^2 \frac{250}{200\ 000} = 1.44 < 2.25$$

$$P_n = 0.66^\lambda F_y A_s = 0.66^{1.44}(250)(6430)(10^{-3}) = 884 \text{ kN}$$

$$P_r = \phi_c P_n = 0.9(884) = 796 \text{ kN} > P_{w\text{Bot}} = 6.6 \text{ kN}, \qquad \text{OK}$$

b. End diaphragms

Must adequately transmit all the forces to the bearings.

$$P_r = 796 \text{ kN} > F_{uD} = 10.8 \text{ kN}, \qquad \text{OK}$$

Use the same section as intermediate diaphragm.

Use C380 × 50, M270 Gr250, for all diaphragms

6. *Lateral Bracing* [A6.7.5]   Lateral bracing shall be provided at quarter points, as determined in Section 8.11.1, Part J.4.a. Use same section as diaphragms.

Use C380 × 50, M270 Gr250, for all lateral braces

7. *Check Fatigue and Fracture* [A6.5.3] Allowable fatigue stress range depends on load cycles and connection details. Fracture depends on material grade and temperature.

a. Stress cycles: Assuming a rural interstate highway with 20 000 vehicles per lane per day,

Fraction of trucks in traffic = 0.20 (Table 4.4) [Table C3.6.1.4.2-1]

$$ADTT = 0.20 \times ADT = 0.20(20\ 000)(2\ \text{lanes})$$

$$= 8000\ \text{trucks/day}$$

$$p = 0.85\ \text{(Table 4.3) [Table A3.6.1.4.2-1]}$$

$$ADTT_{SL} = p \times ADTT = 0.85(8000) = 6800\ \text{trucks/day}$$

From (Table 8.4) [Table A6.6.1.2.5-2], cycles per truck passage, for a simple-span girder of span 10 500 mm, is equal to

$$n = 2.0$$

$$N = (365)(75)(2.0)(6800) = 372 \times 10^6\ \text{cycles}$$

b. Allowable Fatigue Stress Range—Category A

$$(\Delta F)_n = \left(\frac{A}{N}\right)^{1/3} = \left(\frac{8.2 \times 10^{12}}{372 \times 10^6}\right)^{1/3} = 28.0\ \text{MPa}$$

$$\tfrac{1}{2}(\Delta F)_{TH} = \tfrac{1}{2}(165) = 82.5\ \text{MPa} > 28.0\ \text{MPa}$$

Therefore $(\Delta F)_n = 82.5$ MPa

c. The maximum stress range [C6.6.1.2.5] is assumed to be twice the live load stress range due to the passage of the fatigue load. However, the stress range need not be multiplied by 2 because the fatigue resistance is divided by 2.

For fatigue, $U = 0.75(LL + IM)$

Dynamic load allowance for fatigue is $IM = 15\%$

M_{LL+IM} is maximum in the exterior girder, no multiple presence:

$$M_{LL+IM} = 0.762[397(1.15)]/1.2 = 290\ \text{kN m}$$

$$M_{\text{fatigue}} = 0.75(290) = 217\ \text{kN m}$$

$$f = \frac{M}{S} = \frac{217 \times 10^6}{4000 \times 10^3} = 54\ \text{MPa} < 82.5\ \text{MPa}, \qquad \text{OK}$$

8. *Check Assumptions Made in Design*   All the requirements are satisfied, using a W760 × 134. This beam has a self-weight of
$(134)(9.81 \times 10^{-3}) = 1.31$ kN/m; thus, our assumed beam weight
of 1.5 kN/m is conservative. Also, for preliminary design, the value
for K_g/Lt_s^3 was taken as 1.0 in calculating the distribution factors
for moment. The actual value is calculated below.

$$K_g = n(I + Ae_g^2)$$

$$n = \frac{E_s}{E_c} = \frac{200\ 000}{4800\sqrt{30}} = 7.6, \text{ use } 8$$

$$I = 1500 \times 10^6 \text{ mm}^4$$

Because section is noncomposite $e_g = 0$

$$K_g = 8[(1500 \times 10^6) = 12.0 \times 10^9$$

$$\frac{K_g}{Lt_s^3} = \frac{12.0 \times 10^9}{(10\ 500)(205)^3} = 0.133$$

$$\left(\frac{K_g}{Lt_s^3}\right)^{0.1} = 0.817$$

This means that the live loads calculated in the preliminary design
are 18% higher than actual, which is conservative. However, the
distribution factor is not applied to the dead load so that when the
live and dead load effects are combined, the preliminary design
loads are less conservative, which is more acceptable.

K. **Design Sketch**

The design of the noncomposite, simple span, rolled steel beam bridge
is summarized in Figure E8.1-12.

8.11.2   Composite Rolled Steel Beam Bridge

Problem Statement   Design the simple span composite rolled steel beam
bridge of Figure E8.2-1 with 10.5-m span for a *HL*-93 live load. Roadway

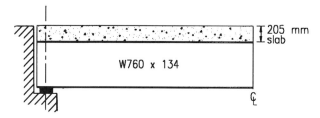

Fig. E8.1-12   Design sketch of noncomposite rolled steel girder.

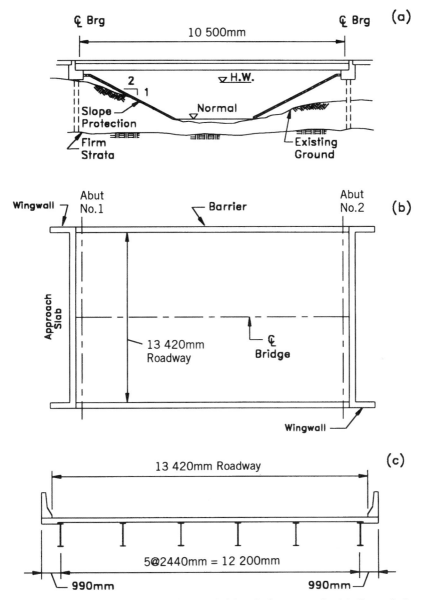

Fig. E8.2-1 Composite rolled steel beam bridge design example. (a) General elevation, (b) plan view, and (c) cross section.

width is 13 420 mm curb to curb. Allow for a future wearing surface of 75-mm thick bituminous overlay. Use f'_c = 30 MPa and M270 Gr250 steel.

A–G Same as Example Problem 8.11.1

H. Calculate Force Effects From Other Loads

$D1$ = dead load of structural components and their attachments, acting on the noncomposite section

$D2$ = future wearing surface

$D3$ = barriers that have a cross-sectional area of 197 312 mm²

A 50-mm × 300-mm average concrete haunch at each girder is used to account for camber and unshored construction.
Assume a beam weight of 1.5 kN/m.
From Section 8.11.1, Part H, for a uniformly distributed load, w,

$$M_{105} = 13.78w$$

$$V_{100} = 5.25w$$

1. *Interior Girders*

$D1$ Deck slab $(2400)(10^{-9})(9.81)(205)(2440) = 11.78$ kN/m

Girder   = 1.5 kN/m

Haunch $(2400)(10^{-9})(9.81)(300)(50) = $ 0.35 kN/m

$w'_{D1} = 13.63$ kN/m

$D2$ 75 mm bituminous paving

$$w'_{D2} = (2250)(9.81)(10^{-9})(75)(2440) = 4.04 \text{ kN/m}$$

$D3$ Barriers, one-sixth share

$$w'_{D3} = \frac{2(197\ 312)(2400)(9.81)}{6(10^9)} = 1.55 \text{ kN/m}$$

Table E8.2-1 summarizes the unfactored moments and shears at critical sections for interior girders. The values for LL + IM were determined in Section 8.11.1, Parts G.4.d and G.5.d.

TABLE E8.2-1   Interior Girder Unfactored Moments and Shears

Load Type	w (kN/m)	Moment (kN m) M_{105}	Shear (kN) V_{100}
D1	13.63	188	72
D2	4.04	56	21
D3	1.55	21	8
LL + IM	N/A	605	301

2. *Exterior Girders*

D1   Deck slab $(2400)(10^{-9})(9.81)$
$\times [(230)(990) + (205)(1220)] = 11.25$ kN/m

Girder $= 1.5$ kN/m

Haunch $= 0.35$ kN/m

$w_{D1}^E = 13.10$ kN/m

D2   75-mm bituminous paving

$w_{D2}^E = (2250)(9.81)(10^{-9})(75)(1220 + 990 - 380) = 3.03$ kN m

D3   Barriers, one-sixth share

$w_{D3}^E = 1.55$ kN/m

Table E8.2-2 summarizes the unfactored moments and shears at critical sections for exterior girders. The values for LL + IM were determined in Section 8.11.1, Parts G.4.d and G.5.d.

I. **Design Required Sections**
   1. *Strength Limit State*
      a. Interior beam—factored shear and moment

TABLE E8.2-2.   Exterior Girder Unfactored Moments and Shears

Load Type	w (kN/m)	Moment (kN m) M_{105}	Shear (kN) V_{100}
D1	13.10	181	69
D2	3.03	42	16
D3	1.55	21	8
LL + IM	N/A	616	278

$$U = \eta[1.25D1 + 1.50D2 + 1.25D3 + 1.75(LL + IM)]$$

$$V_u = 0.95[1.25(72) + 1.50(21) +$$

$$1.25(8) + 1.75(301)] = 625 \text{ kN}$$

$$M_u = 0.95[1.25(188) + 1.50(56) + 1.25(21) +$$

$$1.75(605)] = 1334 \text{ kN m}$$

b. Exterior beam—factored shear and moment

$$V_u = 0.95[1.25(69) + 1.50(16) + 1.25(8) +$$

$$1.75(278)] = 576 \text{ kN}$$

$$M_u = 0.95[1.25(181) + 1.50(42) +$$

$$1.25(21) + 1.75(616)] = 1324 \text{ kN m}$$

2. *Consider Loading and Concrete Placement Sequence* [A6.10.5.1.1a]

Case 1 Weight of girder and slab (D1). Supported by steel girder alone.
Case 2 Superimposed dead load (FWS, curbs, and railings) (D2 and D3). Supported by long-term composite section.
Case 3 Live load plus impact (LL + IM). Supported by short-term composite section.

3. *Determine Effective Flange Width* [A4.6.2.6] For interior girders the effective flange width is the least of
   a. One-quarter of the average span length
   b. Twelve times the average thickness of the slab, plus the greater of the web thickness or one-half the width of the top flange of the girder
   c. Average spacing of adjacent girders
   Assume the girder top flange is 200 mm wide

$$b_i = \min \begin{vmatrix} (0.25)(10\ 500) = 2625 \text{ mm} \\ (12)(190) + \dfrac{200}{2} = 2380 \text{ mm} \\ 2440 \text{ mm} \end{vmatrix}$$

Therefore $b_i = 2380$ mm

For exterior girders the effective flange width is one-half the effective flange width of the adjacent interior girder, plus the least of

a. One-eighth of the effective span length
b. Six times the average thickness of the slab, plus the greater of one-half of the web thickness or one-quarter of the width of the top flange of the girder
c. The width of the overhang

$$b_e = \frac{b_i}{2} + \min \begin{vmatrix} (0.125)(10\,500) = 1313 \text{ mm} \\ (6)(190) + \dfrac{200}{4} = 1190 \text{ mm} \\ 990 \text{ mm} \end{vmatrix}$$

Therefore $b_e = \dfrac{b_i}{2} + 990 = \dfrac{2380}{2} + 990 = 2180$ mm

4. *Modular Ratio*
   For $\qquad f'_c = 30$ MPa, $n = 8$ [A6.10.5.1.1b]

5. *Cover Plates*
   For economy, the lightest and shallowest beam with the largest cover plate possible gives the best design. The length of the cover plate, L_{cp}, must satisfy [A6.10.9]

$$L_{cp} \geq 2d_s + 900$$

   where d_s = depth of the steel section (mm)

6. *Trial Section Properties*
   a. Steel section at midspan

   Try W610 × 101 with 10 mm × 200-mm cover plate
   Properties of W610 × 101 are taken from AISC (1992). The calculations for the steel section properties are summarized in Table E8.2-3 and shown in Figure E8.2-2

TABLE E8.2-3   Steel Section Properties

Component	A	y	Ay	$y - \bar{y}$	$A(y - \bar{y})^2$	I_o
Beam	12 900	$\dfrac{603}{2} = 301.5$	3.889×10^6	−41.4	21.8×10^6	764×10^6
Cover plate	2 000	$603 + 5 = 608$	1.216×10^6	265.4	140.9×10^6	17×10^3
Σ	14 900		5.105×10^6		162.7×10^6	764×10^6

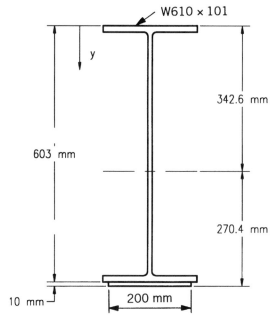

Fig. E8.2-2   Steel section at midspan.

$$I_x = (162.7 \times 10^6) + (764 \times 10^6) = 926.7 \times 10^6 \text{ mm}^4$$

$$\bar{y} = \frac{\Sigma \, Ay}{\Sigma \, A} = \frac{5.105 \times 10^6}{14 \, 900} = 342.6 \text{ mm}, \ y_t = -342.6 \text{ mm}$$

$$y_b = 613 - 342.6 = 270.4 \text{ mm}$$

$$S_t = \frac{I_x}{y_t} = \frac{926.7 \times 10^6}{-342.6} = -2.705 \times 10^6 \text{ mm}^3$$

$$S_b = \frac{I_x}{y_b} = \frac{926.7 \times 10^6}{270.4} = 3.427 \times 10^6 \text{ mm}^3$$

(Note: positive y is downward from centroid of section.)

b. Composite section, $n = 8$, at midspan. Figure E8.2-3 shows the composite section with a haunch of 25 mm, a net slab thickness (without 15 mm sacrificial wearing surface) of 190 mm, and an effective width of 2380 mm. The composite section properties calculations are summarized in Table E8.2-4.

$$I_x = (2523 \times 10^6) + (1097 \times 10^6) = 3620 \times 10^6 \text{ mm}^4$$

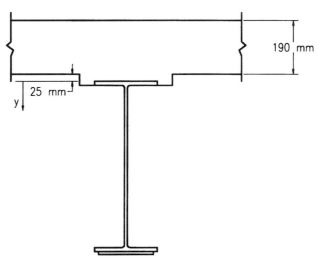

Fig. E8.2-3 Composite section at midspan.

$$\bar{y} = \frac{\Sigma\,Ay}{\Sigma\,A} = \frac{-1.678 \times 10^6}{71\,425} = -23.5 \text{ mm}, \, y_t = 23.5 \text{ mm}$$

$$y_b = 613 + 23.5 = 636.5 \text{ mm}$$

$$S_t = \frac{3620 \times 10^6}{23.5} = 154.0 \times 10^6 \text{ mm}^3$$

$$S_b = \frac{3620 \times 10^6}{636.5} = 5.69 \times 10^6 \text{ mm}^3$$

c. Composite section, $3n = 24$, at midspan. The composite section properties calculations, reduced for the effect of creep in the concrete slab, are summarized in Table E8.2-5.

$$I_x = (1780 \times 10^6) + (983.4 \times 10^6) = 2763.4 \times 10^6 \text{ mm}^4$$

$$\bar{y} = \frac{\Sigma\,Ay}{\Sigma\,A} = \frac{2.844 \times 10^6}{33\,742} = 84.3 \text{ mm}, \, y_t = -84.3 \text{ mm}$$

$$y_b = 613 - 84.3 = 528.7 \text{ mm}$$

$$S_t = \frac{2763.4 \times 10^6}{-84.3} = -32.78 \times 10^6 \text{ mm}^3$$

$$S_b = \frac{2763.4 \times 10^6}{528.7} = 5.227 \times 10^6 \text{ mm}^3$$

TABLE E8.2-4 Short-Term Composite Section Properties, $n = 8$, $b_i = 2380$ mm

Component	A	y	Ay	$y - \bar{y}$	$A(y - \bar{y})^2$	I_o
Concrete $(b_i \times t_s/n)^a$	56 525	$-25 - \dfrac{190}{2} = -120$	-6.783×10^6	-96.5	526×10^6	170×10^6
Steel	14 900	342.6	5.105×10^6	366.1	1997×10^6	927×10^6
Σ	71 425		-1.678×10^6		2523×10^6	1097×10^6

^{a}The parameter b_i is used because interior girders control the moment design

7. *Member Proportions* [A6.10.1.1]

$$0.1 \le \frac{I_{yc}}{I_y} \le 0.9$$

$$I_y = (29.5 \times 10^6) + \tfrac{1}{12}(10)(200)^3 = 36.17 \times 10^6 \text{ mm}^4$$

$$I_{yc} = \frac{1}{12}(14.9)(228)^3 = 14.72 \times 10^6 \text{ mm}^4$$

$$0.1 < \frac{I_{yc}}{I_y} = \frac{14.72 \times 10^6}{36.17 \times 10^6} = 0.407 < 0.9, \text{ OK}$$

8. *Fatigue Induced by Web Flexure* [A6.10.4.3]

$$\frac{2D_c}{t_w} = \frac{2(342.6)}{10.5} = 65.3 < 5.76\sqrt{\frac{E}{F_{yc}}} = 5.76\sqrt{\frac{200\,000}{250}} = 162.9$$

therefore, f_{cf} must be less than $R_h F_{yc}$
where

f_{cf} = maximum compressive elastic flexural stress in the compression flange due to the unfactored permanent load and twice the fatigue loading [A6.10.4.2].

The unfactored moment at Location 105 due to the fatigue loading was calculated in Section 8.11.1, Part I.3 and was found to be 397 kN m. The value given in Table E8.2-6 for M_{LL+IM} is twice the positive moment value with load factor and without multiple presence.

$$M_{LL+IM} = 2(0.75)(1.15)(0.748)(397)/1.2 = 427 \text{ kN m}$$

TABLE E8.2-5 Long-Term Composite Section Properties, $3n = 24$

Component	A	y	Ay	$y - \bar{y}$	$A(y - \bar{y})^2$	I_o
Concrete $(b_i \times t_s/3n)^a$	18 842	−120	−2.261 × 10⁶	−204.3	786 × 10⁶	56.7 × 10⁶
Steel	14 900	342.6	5.105 × 10⁶	258.3	994 × 10⁶	926.7 × 10⁶
Σ	33 742		2.844 × 10⁶		1780 × 10⁶	983.4 × 10⁶

[a]The parameter b_i is used because interior girders control the moment design

TABLE E8.2-6   Maximum Flexural Stress in the Web for Positive Flexure (Interior Girder)

Load	M_{D1}	M_{D2}	M_{D3}	M_{LL+IM}	S, Steel	S, Composite	Stress (MPa)
D1	188				-2.705×10^6		-69.5
D2		56				-32.78×10^6	-1.7
D3			21			-32.78×10^6	-0.6
LL + IM				427		154.0×10^6	2.8
Total							-69.0

$$f_{cf} = 69.0 \text{ MPa} < 250 \text{ MPa, OK}$$

9. *Stresses.* Stresses in top and bottom of girder for strength limit state are given in Tables E8.2-7 and E8.2-8. Yielding has occurred in the bottom flange.

10. *Determine If Section Is Compact*
   a. Web slenderness (Table 8.16) [A6.10.5.2.2c]

$$\frac{2D_{cp}}{t_w} \leq 3.76 \sqrt{\frac{E}{F_{yc}}}$$

Spacing of reinforcement in concrete deck is given in Figure E7.1-15. Number of bars in top of slab in effective width =

$$\frac{b_i}{\text{bar spacing}} = \frac{2290}{250} = 10 \text{ No. 10 bars}$$

TABLE E8.2-7   Compressive Stresses in Top of Steel Beam due to Factored Loading (Interior Girder)

Load	M_{D1}	M_{D2}	M_{D3}	M_{LL+IM}	S, Steel	S, Composite	Stress (MPa)
D1	235				-2.705×10^6		-86.9
D2		84				-32.78×10^6	-2.6
D3			26			-32.78×10^6	-0.8
LL + IM				1059		154.0×10^6	6.9
Total							-83.4
						$\eta = 0.95$	-79.2

TABLE E8.2-8   Tensile Stresses in Bottom of Steel Beam due to Factored Loading

Load	M_{D1}	M_{D2}	M_{D3}	M_{LL+IM}	S_b Steel	S_b Composite	Stress (MPa)
D1	235				3.427×10^6		68.6
D2		84				5.227×10^6	16.1
D3			26			5.227×10^6	5.0
LL + IM				1059		$5.69 \ \times 10^6$	186.1
Total							275.8
						$\eta = 0.95$	262.0

$$\text{Number of bars in bottom of slab} = \frac{2290}{350} = 7 \text{ No. 15 bars}$$

Plastic Forces Assume PNA is in slab above bottom bars ($c <$ 156 mm)
Top Reinf, $P_{rt} = F_{yr}A_{rt} = (400)(10)(100)(10^{-3}) = 400$ kN
Concrete Slab, $P_s = 0.85f'_c ab_i = 0.85(30)(2290)(10^{-3})a = 58a$
Bottom Reinf, $P_{rb} = F_{yr}A_{rb} = (400)(7)(200)(10^{-3}) = 560$ kN
Beam, $T_{bm} = (AF_y)_{bm} = (12\,900)(250)(10^{-3}) = 3225$ kN
Cover Plate, $T_{cp} = (AF_y)_{cp} = (10)(200)(250)(10^{-3}) = 500$ kN

Plastic Neutral Axis PNA

$$C = T$$
$$400 + 58a = 560 + 3225 + 500$$
$$a = 67.0 \text{ mm}, \quad \beta_1 = 0.85 - \frac{2}{7}(0.05) = 0.836$$
$$c = \frac{a}{\beta_1} = \frac{67.0}{0.836} = 80.1 \text{ mm} < 156 \text{ mm, OK}$$

Because the PNA is not in the web, D_{cp} shall be taken equal to 0, the web slenderness requirement is satisfied [A6.10.5.1.4b], the section is classified as compact, and $M_n = M_p$.

Calculate Plastic Moment M_p by summing moments about PNA:

Top reinforcement: $400(c - d') = 400(80.1 - 76.95)$
$(10^{-3}) = 1.3$ kN m

Slab: $58a\left(c - \dfrac{a}{2}\right) = 58(91.4)\left(80.1 - \dfrac{91.4}{2}\right)(10^{-3})$

$= 182$ kN m

Bottom reinforcement: $560(d - c) = 560(156 - 80.1)$
$(10^{-3}) = 43$ kN m

Beam: $3225(205 + 25 + \dfrac{603}{2} - 80.1)(10^{-3})$

$= 1456$ kN m

Cover plate: $500(205 + 25 + 603 + 5 - 80.1)(10^{-3})$

$= 379$ kN m

$M_p = 1.3 + 182 + 43 + 1456 + 379 = 2061$ kN m

$\phi M_p = 1.0(2061) = 2061$ kN m $> M_u = 1334$ kN m, OK

b. Compression flange slenderness [A6.10.5.2.2c]   No requirement at strength limit state for compact composite I-sections in positive flexure (Table 8.16) [Table A6.10.5.2.1-1]
c. Compression flange bracing   No requirement at strength limit state for compact composite I-sections in positive flexure (Table 8.16) [Table A6.10.5.2.1-1]
d. Calculate flexural resistance [A6.10.5.2.2a]   For simple spans, the nominal flexural resistance is taken as

$$M_n = M_p = 2061 \text{ kN m}$$

$$M_r = \phi_f M_n = 1.0(2061)$$

$$= 2061 \text{ kN m} > M_u = 1334 \text{ kN m, OK}$$

e. Check positive flexure ductility [A6.10.5.2.2b]   As calculated in Tables E8.2-7 and E8.2-8, the moment due to the factored loads results in stresses of 79.2 and 262.0 MPa in the top and bottom flanges, respectively. Because the bottom flange elastic stress exceeds the yield strength of the flange (250 MPa), the section must satisfy

$$D_p \le \frac{d + t_s + t_h}{7.5}$$

where

D_p = distance from the top of the slab to the neutral axis at the plastic moment = 80.1 mm [from Section 8.11.2, Part I.10.a]

d = depth of the steel section = 603 mm

t_s = thickness of the concrete slab = 205 mm

t_h = thickness of the concrete haunch = 25 mm

$$\frac{d + t_s + t_h}{7.5} = \frac{603 + 205 + 25}{7.5}$$

$$= 111 \text{ mm} > D_p = 80.1 \text{ mm, OK}$$

∴ All the requirements for flexure have been satisfied.

11. *Shear Design*
   a. For beams with unstiffened web (Table 8.19) [A6.10.7.2]

$$V_r = \phi_v V_n = 1.00 V_n$$

$$\frac{D}{t_w} = \frac{d - 2t_f}{t_w} = \frac{603 - 2(14.9)}{10.50} = 54.6$$

$$2.46 \sqrt{\frac{E}{F_{yw}}} = 2.46 \sqrt{\frac{200\,000}{250}} = 69.6 > 54.6$$

Therefore,

$$V_n = V_p = 0.58 F_{yw} D t_w = 0.58(250)(573.2)(10.50) = 873 \text{ kN}$$

$$V_r = 1.00 V_n = 873 \text{ kN} > V_u = 625 \text{ kN}$$

(factored shear in interior beam), OK

   b. Bearing stiffener design [A6.10.8.2]

$$0.75 \phi_b V_n = 0.75(1.0)(873) = 655 \text{ kN} > V_u = 625 \text{ kN}$$

The bearing stiffeners are not required.

12. *Constructability*

  a. General proportions [A6.10.1.1]

$$0.1 \leq \frac{I_{yc}}{I_y} \leq 0.9$$

This was found to be OK in Section 8.11.2, Part I.7.

  b. Flexure [A6.10.10.2] The section must satisfy compression flange slenderness, web slenderness, and compression flange bracing requirements. The nominal flexural resistance shall be

$$M_n = R_b R_h M_y \quad [\text{A6.10.6.1.1}]$$

Web slenderness [A6.10.10.2.2]
For webs without longitudinal stiffeners [A6.10.5.3.2b],

$$\frac{2D_c}{t_w} \leq 6.77 \sqrt{\frac{E}{f_c}}$$

where

D_c = depth of the web in compression in elastic range (mm)

f_c = stress in the compression flange due to the factored loading (MPa)

During deck placement, the section is not yet composite. The steel beam is carrying all dead loads due to structural components (i.e., slab and beam). The depth of the web in compression in the elastic range, D_c, can be found from the steel section properties calculated in Section 8.11.2, Part I.6a.

$$\bar{y}_t = 342.6 \text{ mm}$$

$$D_c = \bar{y}_t - t_f = 342.6 - 14.9 = 327.7 \text{ mm}$$

The stress due to factored loading, f_c, is from the construction loads $D1$, calculated in Section 8.11.2, Part H. The interior girder governs, with the maximum factored moment due to $D1$ equal to (Table E8.2-1)

$$M_{105} = 0.95(1.25)(188) = 223 \text{ kN m}$$

$$f_c = \frac{M_{105}}{S_t} = \frac{223 \times 10^6}{2.705 \times 10^6} = 82.4 \text{ MPa}$$

$$\frac{2D_c}{t_w} = \frac{2(327.7)}{10.50} = 62.4 < 6.77 \sqrt{\frac{E}{f_c}}$$

$$= 6.77 \sqrt{\frac{200\ 000}{82.4}} = 334, \text{ OK}$$

Compression flange slenderness [A6.10.10.2.3]

This requirement is to prevent local buckling of the top flange before the concrete deck hardens [A6.10.5.3.3c].

$$\frac{b_f}{2t_f} \leq 1.38 \sqrt{\frac{E}{f_c \sqrt{2D_c/t_w}}}$$

$$\frac{b_f}{2t_f} = \frac{228}{2(14.9)} = 7.65$$

$$1.38 \sqrt{\frac{E}{f_c \sqrt{2D_c/t_w}}} = 1.38 \sqrt{\frac{200\ 000}{82.4\sqrt{62.4}}} = 24.2 > 7.65, \text{ OK}$$

Compression flange bracing [A6.10.10.2.4]

Check required distance between braces [A6.10.5.3.3d]

$$L_b \leq 1.76 r_t \sqrt{\frac{E}{F_{yc}}}$$

where

r_t = minimum radius of gyration of the compression flange of the steel section (without one-third of the web in compression) taken about the vertical axis

$$I_t = \frac{1}{12}(14.9)(228)^3 = 14.72 \times 10^6 \text{ mm}^4$$

$$A_t = (14.9)(228) = 3397 \text{ mm}^2$$

$$r_t = \sqrt{\frac{I_t}{A_t}} = \sqrt{\frac{14.72 \times 10^6}{3397}} = 65.8 \text{ mm}$$

$$1.76r_t\sqrt{\frac{E}{F_{yc}}} = 1.76(65.8)\sqrt{\frac{200\ 000}{250}} = 3276 \text{ mm}$$

Provide braces for the compression flange at quarter points, so that

$$L_b = \frac{10\ 500}{4} = 2625 \text{ mm}$$

These braces need not be permanent because the slab will provide compression flange bracing once it is cured.
Nominal flexural resistance [A6.10.10.2.1]

$$M_n = R_b R_h M_y$$

$R_h = 1.0$ for homogeneous sections [A6.10.5.4.1a]

$$\frac{2D_c}{t_w} \le \lambda_b \sqrt{\frac{E}{f_c}} \quad \text{[A6.10.5.4.2a]}$$

$$\frac{2D_c}{t_w} = 62.4 < \lambda_b \sqrt{\frac{E}{f_c}} = 4.64\sqrt{\frac{200\ 000}{69.5}} = 249$$

Therefore $R_b = 1.0$

$$M_y = S_t F_y = (2.705 \times 10^6)(250)(10^{-6}) = 676 \text{ kN m}$$

for yielding in the top flange where S_t is from Section 8.11.2, Part I.6a.

$$M_y = S_b F_y = (3.427 \times 10^6)(250)(10^{-6}) = 857 \text{ kN m}$$

for yielding in the bottom flange.
The limit for the yield moment is when the top flange yields, so that

$$M_y = 676 \text{ kN m}$$

$$M_n = R_b R_h M_y = 1.0(1.0)(676) = 676 \text{ kN m}$$

$$\phi_f M_n = 1.0(676) = 676 \text{ kN m} > M_u = 223 \text{ kN m}$$

for interior girder, OK

Shear [A6.10.10.3] This article does not apply to sections with unstiffened webs because the shear force is limited to the shear yield or shear buckling force at the strength limit state.

J. **Dimension and Detail Requirements**

1. *Material Thickness* [A6.7.3] Bracing and cross frames shall not be less than 8 mm in thickness. Web thickness of rolled beams shall not be less than 7 mm.

$$t_w = 10.5 \text{ mm} > 7 \text{ mm, OK}$$

2. *Optional Deflection Control* [A2.5.2.6.2] This requirement was met in Example Problem 8.11.1. The only difference between this example and Example 8.11.1 is the moment of inertia, I, of the section. In Section 8.11.1, Part I.2, I was equal to 1500×10^6 mm^4. In this example, I is equal to 3592×10^6 mm^4, found in Section 8.11.2, Part I.6b. Because I is considerably greater than I in Example Problem 8.11.1, the deflections will be less, and the optional deflection control requirement is met.

3. *Service Limit State Control of Permanent Deflections* [A6.10.3] For both flanges of composite sections

$$f_f \leq 0.95 R_h F_{yf} = 0.95(1.0)(250) = 238 \text{ MPa}$$

where

$$f_f = \text{elastic flange stress caused by the factored loading}$$

The maximum Service II moment, which occurs at Location 105 in the interior beam, is due to unfactored dead loads $D1$, $D2$, and $D3$, and factored live load, $1.3(LL + IM)$ taken from Table E8.2-1. The stresses calculated from these moments are given in Tables E8.2-9 and E8.2-10.

$$\max f_f = 207.9 \text{ MPa} < 238 \text{ MPa, OK}$$

TABLE E8.2-9   Stresses in Top Flange of Steel Beam due to Service *II* Moments

Load	M_{D1}	M_{D2}	M_{D3}	M_{LL+IM}	S_t Steel	S_t Composite	Stress (MPa)
D1	188				-2.705×10^6		-69.5
D2		56				-32.78×10^6	-1.7
D3			21			-32.78×10^6	-0.6
LL + IM				787		154.0×10^6	5.1
Total							-66.7

4. *Check Construction Requirements* (Dead Load Camber)   The centerline deflection due to a uniform load on a simply support span is

$$\Delta_{CL} = \frac{5}{384} \frac{w_D L^4}{EI} = \frac{5}{384} \frac{w_D (10\,500)^4}{200\,000I} = (791.3 \times 10^6)\frac{w_D}{I}$$

By substituting the dead loads from Tables E8.2-1 and E8.2-2, and using the I values determined in Part I.6.c for long-term loads, the centerline deflections are calculated in Tables E8.2-11 and E8.2-12.
   Use a 12-mm camber on all beams
5. *Diaphragms* [A6.7.4]   Diaphragms were designed for the noncomposite bridge of Example Problem 8.11.1. The same diaphragms are adequate for this bridge.
   Use C380 × 50, M270 Gr 250 for all diaphragms
6. *Lateral Bracing* [A6.7.5]   Lateral bracing shall be provided at quarter points, as determined in Section 8.11.2, Part I.12. Use the same section as diaphragms, C380 × 50, M270 Gr250. The two braces other than the brace at midspan may be removed after the concrete cures.
7. *Check Fatigue and Fracture* [A6.5.3]   From Example Problem 8.11.1, Part J.7.b,

TABLE E8.2-10   Stresses in Bottom Flange of Steel Beam due to Service *II* Moments

Load	M_{D1}	M_{D2}	M_{D3}	M_{LL+IM}	S_b Steel	S_b Composite	Stress (MPa)
D1	188				3.427×10^6		54.9
D2		56				5.227×10^6	10.7
D3			21			5.227×10^6	4.0
LL + IM				787		5.69×10^6	138.3
Total							207.9

TABLE E8.2-11   Exterior Beam Deflection Due to Dead Loads

Load Type	Load, w (N/mm)	I (mm⁴)	Δ_{CL} (mm)
D1	13.10	926.7×10^6	11
D2	3.03	2763.4×10^6	0.9
D3	1.55	2763.4×10^6	0.5
Total			12.4

$$(\Delta F)_n = \frac{1}{2} (\Delta F)_{TH} = 82.5 \text{ MPa}$$

The maximum stress range [C6.6.1.2.5] is assumed to be twice the live load stress range due to the passage of the fatigue load. However, the stress range need not be multiplied by 2 because the fatigue resistance is divided by 2.

From Example Problem 8.11.1, Part J.7.c,

$$M_{LL+IM} = 290 \text{ kN m}$$

$$M_{\text{fatigue}} = 217 \text{ kN m}$$

$$f = \frac{M}{S_b} = \frac{217 \times 10^6}{5.69 \times 10^6} = 38 \text{ MPa} < 82.5 \text{ MPa, OK}$$

where S_b is the section modulus for the short-term composite section, calculated in Section 8.11.2, Part I.6.b.

8. *Shear Connectors* [A6.10.7.4]   Shear connectors must be provided throughout the length of the span for simple-span composite bridges.

Use 19-mm diameter studs, 100 mm high. The ratio of height to diameter is

$$\frac{100}{19} = 5.26 > 4, \text{ OK [A6.10.7.4.1a]}$$

TABLE E8.2-12   Interior Beam Deflection Due to Dead Loads

Load Type	Load, w (N/mm)	I (mm⁴)	Δ_{CL} (mm)
D1	13.63	926.7×10^6	12
D2	4.04	2763.4×10^6	1.2
D3	1.55	2763.4×10^6	0.5
Total			13.7

a. Transverse spacing: [A6.10.7.4.1c]   The center-to-center spacing of the connectors cannot be closer than 4 stud diameters, or 76 mm. The clear distance between the edge of the top flange and the edge of the nearest connector must be at least 25 mm.

b. Cover and penetration: [A6.10.7.4.1d]   Penetration into the deck should be at least 50 mm. Clear cover should be at least 50 mm.

c. Fatigue resistance: [A6.10.7.4.2]

$$Z_r = \alpha d^2 \geq 19.0 d^2$$

for which

$$\alpha = 238 - 29.5 \log N$$

where

$$N = 365(75)n(\text{ADTT})_{SL}$$

N was found to be 372×10^6 cycles in Example Problem 8.11.1, Part J.7.a.

$\alpha = 238 - 29.5 \log(372 \times 10^6) = -14.8$
$Z_r = 19.0 d^2$ [A6.10.7.4.2]
$Z_r = 19.0(19)^2 = 6859 \ N$

$$p = \frac{n Z_r I}{V_{sr} Q}$$

I = moment of inertia of short-term composite section
    $= 3620 \times 10^6 \ \text{mm}^4$
n = 3 shear connectors in a cross section
Q = First moment of the transformed area about the
    neutral axis of the short-term composite section

$$= t_s \left(\frac{b_i}{n} \right) \left(y_t + t_h + \frac{t_s}{2} \right)$$

$$= (190) \left(\frac{2290}{8} \right) \left(-20.5 + 25 + \frac{190}{2} \right)$$

$$= 5.41 \times 10^6 \ \text{mm}^3$$

V_{sr} = Shear force range under $LL + IM$ determined
    for the fatigue limit state

Shear ranges at tenth points, with required pitches, can be found in Table E8.2-13. The shear range is computed by finding the difference in the positive and negative shears at that point due to the fatigue truck, multiplied by the dynamic load allowance for fatigue (1.15), the maximum distribution factor for one design lane

TABLE E8.2-13 Shear Range for Fatigue Loading and Required Shear Connector Spacing

Location	Unfactored Maximum Positive Shear (kN)	Unfactored Maximum Negative Shear (kN)	Shear Range (kN)	Pitch (mm)
100	166	0	90.9	150
101	148	−15	89.3	153
102	130	−29	87.1	157
103	112	−44	85.4	160
104	94	−58	83.2	164
105	76	−76	83.2	164

loaded without multiple presence (0.762/1.2 for the exterior beam), and by the load factor for the fatigue limit state (0.75). Values are symmetric about the center of the bridge, Location 105.

An example calculation of the pitch is performed below, for the shear range at Location 101

$$V_{sr} = [148 - (-15)](1.15)(0.762/1.2)(0.75) = 89.3 \text{ kN}$$

$$p = \frac{3(6859)(3620 \times 10^6)}{89.3(5.41 \times 10^6)} = 153 \text{ mm}$$

$$6d_s = 114 \text{ mm} < p < 600 \text{ mm, OK}$$

Because the required pitch from Table E8.2-13 does not vary much between tenth points, use a pitch of 150 mm along the entire span.

d. *Strength Limit State* [A6.10.7.4.4]

$$Q_r = \phi_{sc} Q_n$$

$$\phi_{sc} = 0.85$$

$$Q_n = 0.5 A_{sc} \sqrt{f_c' E_c} \leq A_{sc} F_u$$

$$A_{sc} = \frac{\pi}{4}(19)^2 = 284 \text{ mm}^2$$

$$E_c = 4800\sqrt{f_c'} = 4800\sqrt{30} = 26\,291 \text{ MPa}$$

$$Q_n = 0.5(284)\sqrt{30(26\,291)}(10^{-3}) = 126 \text{ kN}$$

$$A_{sc} F_u = (284)(400)(10^{-3}) = 114 \text{ kN} > 126 \text{ kN}$$

$$Q_n = 114 \text{ kN}$$

$$Q_r = 0.85(114) = 97 \text{ kN}$$

Between sections of maximum positive moment and points of zero moment, the number of shear connectors required is

$$n = \frac{V_h}{Q_r}$$

for which

$$V_h = \min \begin{vmatrix} 0.85f'_c bt_s = 0.85(30)(2290)(190)(10^{-3}) = 11\ 095\ \text{kN} \\ A_s F_y = (12\ 900)(250)(10^{-3}) = 3225\ \text{kN} \end{vmatrix}$$

Therefore use a nominal horizontal shear force, V_h, of 3225 kN

$$n = \frac{V_h}{Q_r} = \frac{3225}{97} = 33$$

Therefore a minimum of 33 shear connectors are required at the strength limit state in half the span. This requirement is more than satisfied by the 150 mm pitch of the three shear connector group required for fatigue resistance.

K. **Design Sketch**

The design of the composite, simple span, rolled steel beam bridge is summarized in Figure E8.2-4.

8.11.3 Steel Plate Girder Bridge

Problem Statement. Design the continuous steel plate girder bridge of Figure E8.3-1 with 30 m − 36 m − 30 m spans for a *HL*-93 live load. Roadway width is 13 420 mm curb to curb. Allow for a future wearing surface of 75-mm thick bituminous overlay. Use $f'_c = 30$ MPa and M270 Gr345 steel.

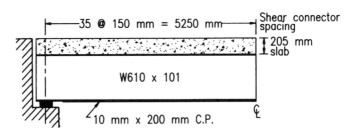

Fig. E8.2-4 Design sketch of composite rolled steel girder.

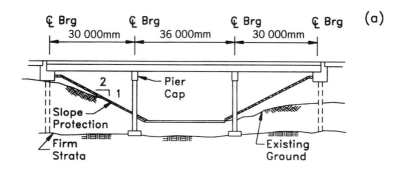

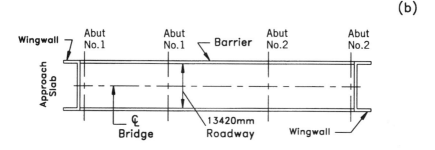

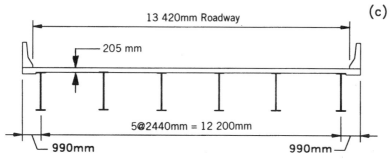

Fig. E8.3-1 Steel plate girder bridge design example. (a) General elevation, (b) plan view, and (c) cross section.

A. Develop General Section

The general elevation and plan of the three-span continuous steel plate girder bridge is shown in Figure E8.3-1. The bridge will carry two lanes of urban interstate traffic over a secondary road.

B. Develop Typical Section

A section of the bridge is shown in Figure E8.3-1(c). Six equally spaced girders are composite with the 205 mm thick concrete deck. The flanges and web of the plate girder are of the same material, so that $R_h = 1.0$.

C. Design Reinforced Concrete Deck

Use same design as concrete box girder bridge, Example Problem 7.10.5.

D. Select Resistance Factors [A6.5.4.2]

1. For flexure $\phi_f = 1.00$
2. For shear $\phi_v = 1.00$
3. For axial compression $\phi_c = 0.90$
4. For shear connectors $\phi_{sc} = 0.85$

E. Select Load Modifiers

The welded plate girder is considered to be ductile. The multiple girders and continuity of the bridge provide redundancy. The carrying of interstate traffic makes it important.

		Strength	Service, Fatigue	
1. Ductility	η_D	0.95	1.0	[A1.3.3]
2. Redundancy	η_R	0.95	1.0	[A1.3.4]
3. Importance	η_I	1.05	1.0	[A1.3.5]
$\eta = \eta_D\eta_R\eta_I$		0.95	1.0	

F. Select Load Combinations and Load Factors [A3.4.1]

Strength *I*, Service *I*, Service *II*, and Fatigue Limit States must be considered.

G. Calculate Live Load Force Effects

1. *Select Number of Lanes* [A3.6.1.1]

$$N_L = \text{INT}\left(\frac{w}{3600}\right) = \text{INT}\left(\frac{13\ 420}{3600}\right) = 3 \text{ lanes}$$

2. *Multiple Presence Factor* (Table 4.6) [Table A3.6.1.1.2-1]

No. Loaded Lanes	Multiple Presence Factor
1	1.20
2	1.00
3	0.85

3. *Dynamic Load Allowance* (Table 4.7) [Table A3.6.2.1-1]

Impact = 33%

Fatigue and fracture = 15%

4. *Distribution Factors for Moment*
a. Interior girders (Table 6.5) [Table A4.6.2.2.2b-1]. Checking that the design parameters are within the range of applicability.

1100 mm $< S =$ 2440 mm $<$ 4900 mm

6000 mm $< L =$ 36 000 mm or 30 000 mm $<$ 73 000 mm

$N_b = 6 > 4$

Therefore, the design parameters are within range, and the approximate method is applicable for concrete deck on steel beams cross section. (Note: The following properties are based on $t_s = 205$ mm. A more conservative approach would be to use a deck thickness reduced by the sacrificial wear thickness to give $t_s = 205 - 15 = 190$ mm.)
    The longitudinal stiffness parameter is taken as

$$K_g = n(I + Ae_g^2)$$

For preliminary design (K_g/Lt_s^3) is taken as unity. However, after designing the superstructure this value was changed and reevaluated for the given span. The value for K_g is different in the positive and negative moment regions since the section properties are different.
    For the negative flexural section,

$$K_g = 8[16.86 \times 10^9 + (39\,000)(907.5)^2] = 391.8 \times 10^9$$

where

$$A = 39\,000 \text{ mm}^2 \text{ from Table E8.3-4}$$

$$I = 16.86 \times 10^9 \text{ mm}^4 \text{ from Section 8.11.3, Part I.3.}$$

$$e_g = y_{\text{Top of steel}} + \text{Haunch} + t_s/2$$

$$e_g = 780 + 25 + 205/2 = 907.5 \text{ mm from Section 8.11.3,}$$

Part I.3.

$$n = 8 \text{ for } f'_c \text{ equal to 30 MPa}$$

For the positive, flexural section,

$$K_g = 8[10.607 \times 10^9 + (29\,500)(1035.4)^2] = 337.86 \times 10^9$$

where

$A = 29\,500 \text{ mm}^2$ from Table E8.3-11

$I = 10.607 \times 10^9 \text{ mm}^4$ from Section 8.11.3, Part J.1.

$e_g = y_{\text{top of steel}} + \text{Haunch} + t_s/2$

$e_g = 907.9 + 25 + 205/2 = 1035.4 \text{ mm}$ from Section 8.11.3, Part J.1

$n = 8$ for f'_c equal to 30 MPa

$$mg_M^{SI} = 0.06 + \left(\frac{S}{4300}\right)^{0.4} \left(\frac{S}{L}\right)^{0.3} \left(\frac{K_g}{Lt_s^3}\right)^{0.1}$$

$$mg_M^{MI} = 0.075 + \left(\frac{S}{2900}\right)^{0.6} \left(\frac{S}{L}\right)^{0.2} \left(\frac{K_g}{Lt_s^3}\right)^{0.1}$$

$L = 30\,000$ mm, positive flexure

$$mg_M^{SI} = 0.06 + \left(\frac{2440}{4300}\right)^{0.4} \left(\frac{2440}{30\,000}\right)^{0.3} \left(\frac{337.86 \times 10^9}{(30\,000)(205)^3}\right)^{0.1} = 0.4457$$

$$mg_M^{MI} = 0.075 + \left(\frac{2440}{2900}\right)^{0.6} \left(\frac{2440}{30\,000}\right)^{0.2} \left(\frac{337.86 \times 10^9}{(30\,000)(205)^3}\right)^{0.1} = 0.6356$$

$L_{\text{ave}} = 33\,000$ mm, negative flexure

$$mg_M^{SI} = 0.06 + \left(\frac{2440}{4300}\right)^{0.4}\left(\frac{2440}{33\,000}\right)^{0.3}\left(\frac{391.8 \times 10^9}{(33\,000)(205)^3}\right)^{0.1} = 0.4368$$

$$mg_M^{MI} = 0.075 + \left(\frac{2440}{2900}\right)^{0.6}\left(\frac{2440}{33\,000}\right)^{0.2}\left(\frac{391.8 \times 10^9}{(33\,000)(205)^3}\right)^{0.1} = 0.6280$$

L = 36 000 mm, positive flexure

$$mg_M^{SI} = 0.06 + \left(\frac{2440}{4300}\right)^{0.4}\left(\frac{2440}{36\,000}\right)^{0.3}\left(\frac{337.86 \times 10^9}{(36\,000)(205)^3}\right)^{0.1} = 0.4186$$

$$mg_M^{MI} = 0.075 + \left(\frac{2440}{2900}\right)^{0.6}\left(\frac{2440}{36\,000}\right)^{0.2}\left(\frac{337.86 \times 10^9}{(36\,000)(205)^3}\right)^{0.1} = 0.6058$$

Two or more lanes loaded controls. Because there is little difference between the maximum values, use a distribution factor of 0.636 for moment for all interior girders.

b. Exterior girders (Table 6.5) [Table A4.6.2.2.2d-1]   One design lane loaded Use the Lever Rule to determine mg_M^{SE}, where $m = 1.2$, from Figure E8.3-2.

$$\sum M_{\text{hinge}} = 0$$

$$R = 0.5P\left(\frac{2450 + 650}{2440}\right) = 0.635P$$

$$mg_M^{SE} = 1.2(0.635) = 0.762$$

Two or more design lanes loaded

$$mg_M^{ME} = e\,mg_M^{MI}$$

$$e = 0.77 + \frac{d_e}{2800} \geq 1.0$$

$$d_e = 990 - 380 = 610 \text{ mm}$$

$$e = 0.77 + \frac{610}{2800} = 0.99 < 1.0 \therefore \text{use } 1.0$$

$$mg_M^{ME} = (1.0)(0.636) = 0.636$$

Use distribution factor of 0.762 for moment for all exterior girders.

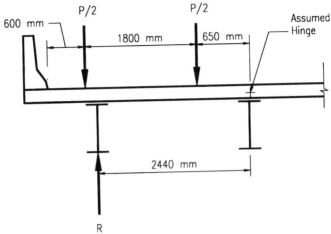

Fig. E8.3-2 Lever rule for determination of distribution factor for moment in exterior beam, one lane loaded.

5. *Distribution Factors for Shear*

    a. Interior girders (Table 6.5) [Table A4.6.2.2.3a-1] checking that the design parameters are within the range of applicability.

$$1100 \text{ mm} < S = 2440 \text{ mm} < 4900 \text{ mm}$$

$$6000 \text{ mm} < L = 36\,000 \text{ mm and } 30\,000 \text{ mm} < 73\,000 \text{ mm}$$

$$110 \text{ mm} < t_s = 205 \text{ mm} < 300 \text{ mm}$$

$$4 \times 10^9 \leq K_g \leq 3 \times 10^{12}$$

$$N_b = 6 > 4$$

Therefore, the approximate method is applicable for concrete deck on steel beam cross sections.

$$mg_V^{SI} = 0.36 + \frac{S}{7600}$$

$$mg_V^{SI} = 0.36 + \frac{2440}{7600} = 0.681$$

$$mg_V^{MI} = 0.2 + \frac{S}{3600} - \left(\frac{S}{10\,700}\right)^{2.0}$$

$$mg_V^{MI} = 0.2 + \frac{2440}{3600} - \left(\frac{2440}{10\,700}\right)^{2.0} = 0.826$$

Use distribution factor of 0.826 for shear for interior girders.

b. Exterior girders (Table 6.5) [Table 4.6.2.2.3b-1] For one design lane loaded, use the Lever Rule as before, therefore

$$mg_V^{SE} = 0.762$$

For two or more design lanes loaded,

$$mg_V^{ME} = e\, mg_V^{MI}$$

$$e = 0.6 + \frac{d_e}{3000} = 0.6 + \frac{610}{3000} = 0.803$$

$$mg_V^{ME} = 0.803(0.826) = 0.664$$

Use distribution factor of 0.762 for shear for exterior girders.
6. Live Load Actions

$$M_{LL+IM} = mg\left[(M_{Tr}\text{ or }M_{Ta})\left(1 + \frac{IM}{100}\right) + M_{Ln}\right]$$

$$V_{LL+IM} = mg\left[(V_{Tr}\text{ or }V_{Ta})\left(1 + \frac{IM}{100}\right) + V_{Ln}\right]$$

a. *Location 205* (Maximum moment at midspan in exterior girder) Influence line has the general shape shown in Figure E8.3-3 and ordinates are taken from Table 5.4. The placement of truck, tandem, and lane live loads are shown in Figures E8.3-4–E8.3-6.

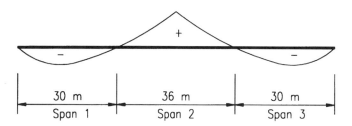

Fig. E8.3-3 Influence line for maximum moment at Location 205.

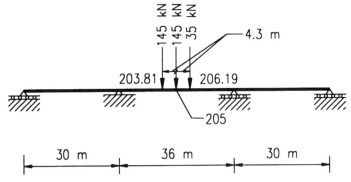

Fig. E8.3-4   Truck placement for maximum moment at Location 205.

$M_{Tr} = [145(0.138\ 23 + 0.203\ 57) + 35(0.138\ 23)](30)$

$\quad\quad = 1632$ kN m

$M_{Ta} = [110(0.203\ 57 + 0.185\ 04)](30) = 1282$ kN m

$M_{Ln} = 9.3(0.102\ 86)(30)^2 = 861$ kN m

$M_{LL+IM} = (0.762)[1632(1.33) + 861] = 2310$ kN m, which compares

well with the BT Beam result, 2341 kN m

    b. *Location 205* (Shear at midspan in interior girder)   Influence
line has the general shape shown in Figure E8.3-7 and ordinates
are taken from Table 5.4. The placement of truck, tandem, and
lane live loads are shown in Figures E8.3-8–E8.3-10.

$V_{Tr} = [145(0.5 + 0.360\ 44) + 35(0.227\ 5)] = 133$ kN

$V_{Ta} = [110(0.5 + 0.461\ 06)] = 106$ kN

$V_{Ln} = 9.3(0.065\ 10 + 0.136\ 50)(30) = 56$ kN

$V_{LL+IM} = (0.826)[133(1.33) + 56] = 192$ kN

    c. *Location 200* (Truck Train in negative moment region)
[A3.6.1.3.1]   Influence line ordinates and areas are taken from
Table 5.4.

    The truck train is applicable in the negative moment regions
and for the reactions at the interior supports of continuous superstructures. The truck train (Fig. E8.3-11) is composed of 90%
of the effect of two design trucks spaced a minimum of 15 000
mm between the rear axle of one and the front axle of the other,
combined with 90% of the design lane loading. The spacing be-

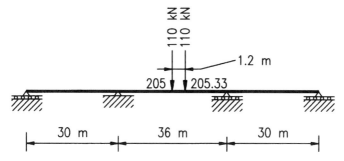

Fig. E8.3-5 Tandem placement for maximum moment at Location 205.

tween the 145-kN axles on each truck is taken as 4.3 m. The lane load is placed on spans 1 and 2 for maximum negative moment at location 200 (Fig. E8.3-12). It should be noted that the impact factor of 33% is only applied to the combined truck load.

$$M_{Tr} = \begin{bmatrix} 35(0.080\ 372 + 0.093\ 59) + 145(0.094\ 29 \\ + 0.087\ 932 + 0.103\ 37 + 0.096\ 681) \end{bmatrix} \quad (30)$$

$$= 1846\ kN\ m$$

Area span 1 = (0.061 38)(30)² = 55.242 m²
Area span 2 = (0.077 14)(30)² = 69.426 m²
M_{Ln} = (9.3)(55.242 + 69.426) = 1159.4 kN m
M_{train} = 0.9(1.33M_{Tr}) + 0.9M_{Ln} = 0.9(1.33)(1846) + 0.9(1159)
= 3253 kN m, which compares well with the BT Beam result, 3243 kNm.

H. Calculate Force Effects of Other Loads

1. *Interior Girders* Three separate dead loads must be calculated. The first is the dead load of the structural components and their attachments, $D1$, acting on the noncomposite section. The second type of dead load is $D2$, which represents the future wearing surface. The third load, $D3$, is caused by the barriers, which each have a cross-

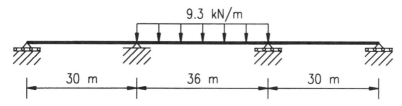

Fig. E8.3-6 Lane load placement for maximum moment at Location 205.

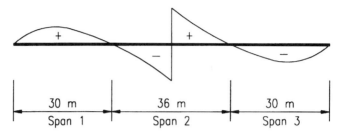

Fig. E8.3-7   Influence line for shear at Location 205.

sectional area of 197 312 mm². For this design it was assumed that
the barrier loads were distributed equally among the interior and
exterior girders. The initial cross section consists of a 10-mm ×
1500-mm web and 30-mm × 400-mm top and bottom flanges. The
girder spacing is 2440 mm, and a 50-mm × 305-mm average concrete haunch at each girder is used to accommodate camber and
unshored construction. The density of the concrete and steel are
taken as 2400 kg/m³ and 7850 kg/m³, respectively. The density of
the 75-mm bituminous future wearing surface (*FWS*) is taken as
2250 kg/m³.

*D*1
Slab $2400 \times 9.81 \times 205 \times 2440/10^9 = 11.78$ N/mm
Haunch $2400 \times 9.81 \times 305 \times 50/10^9 = 0.36$ N/mm
Web $7850 \times 9.81 \times 1500 \times 10/10^9 = 1.16$ N/mm
Flanges $2 \times 7850 \times 9.81 \times 400 \times 30/10^9 = \underline{1.85}$ N/mm
$$w'_{D1} = 15.15 \text{ N/mm}$$
*D*2 75-mm bituminous overlay
$$w'_{D2} = 2250 \times 9.81 \times 2440 \times 75/10^9 = 4.04 \text{ N/mm}$$

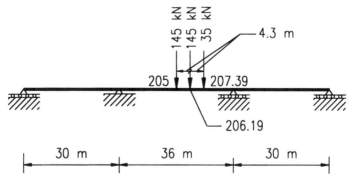

Fig. E8.3-8   Truck placement for maximum shear at Location 205.

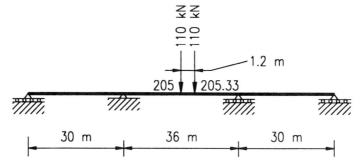

Fig. E8.3-9 Tandem placement for maximum shear at Location 205..

*D*3 Barriers, one-sixth share

$$w_{D3}^I = \frac{2(197\ 312)(2400)(9.81)}{6(10^9)} = 1.55 \text{ N/mm}$$

2. *Exterior Girders* Computing the loads for the exterior girders is based on tributary areas. This approach gives smaller loads on an exterior girder than from a consideration of the deck as a continuous beam with an overhang and finding the reaction at the exterior support.

*D*1
Slab $(2400 \times 9.81)[(230 \times 990) + (205 \times 1220)]/10^9$
$= 11.25 \text{ N/mm}$
Haunch $2400 \times 9.81 \times 305 \times 50/10^9 = 0.36 \text{ N/mm}$
Web $7850 \times 9.81 \times 1500 \times 10/10^9 = 1.16 \text{ N/mm}$
Flanges $2 \times 7850 \times 9.81 \times 400 \times 30/10^9 = \underline{1.85 \text{ N/mm}}$
$w_{D1}^E = 14.62 \text{ N/mm}$

*D*2 75-mm bituminous overlay

$w_{D2}^E = 2250 \times 9.81 \times (1220 + 990 - 380) \times 75/10^9 = 3.03 \text{ N/mm}$

*D*3 Barriers, one-sixth share

$$w_{D3}^E = \frac{2(197\ 312)(2400)(9.81)}{6(10^9)} = 1.55 \text{ N/mm}$$

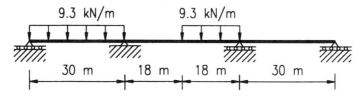

Fig. E8.3-10 Lane load placement for maximum shear at Location 205.

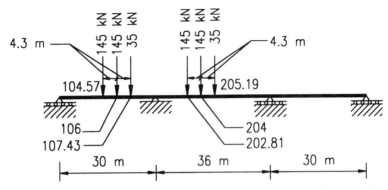

Fig. E8.3-11   Truck train placement for maximum moment at Location 200.

3. *Analysis of Uniformly Distributed Load w* (Fig. E8.3-13)
   a. Moments

$$M_{104} = 0.071\ 29w(30)^2$$

$$= 64.16w \text{ kN m (BT Beam: } 64.6w \text{ kN m)}$$

$$M_{200} = -0.121\ 79w(30)^2$$

$$= -109.61w \text{ kN m (BT Beam: } -108.5w \text{ kN m)}$$

$$M_{205} = 0.058\ 21w(30)^2$$

$$= 52.39w \text{ kN m (BT Beam: } 53.5w \text{ kN m)}$$

   b. Shears:

$$V_{100} = 0.378\ 21w(30) = 11.35w \text{ kN (BT Beam: } 11.4w \text{ kN)}$$

$$V_{110} = -0.621\ 79w\ (30) = -18.65w \text{ kN}$$

$$\text{(BT Beam: } -18.6w \text{ kN)}$$

$$V_{200} = 0.6000w\ (30) = 18.0w \text{ kN (BT Beam: } 18.0w \text{ kN)}$$

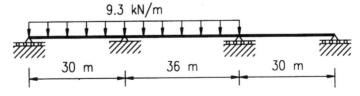

Fig. E8.3-12   Lane load placement for maximum moment at Location 200.

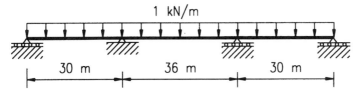

Fig. E8.3-13   Uniformly distributed load.

By substituting the values determined for dead load into the BT Beam equations for moments and shears, the values at critical locations are generated in Table E8.3-1. The LL + IM values listed in Table E8.3-1 include the girder distribution factors as illustrated in Section 8.11.3, Part G.6.

c. Effective span length

The effective span length is defined as the distance between points of permanent load inflection for continuous spans:

Span 1 (Fig. E8.3-14)

$$M = 0 = 11.35x - \frac{1}{2}(1.0)x^2$$

$$x = L_{eff} = 22.7 \text{ m}$$

Span 2 (Fig. E8.3-15)

$$M = 0 = 18x - \frac{1}{2}(1.0)x^2 - 109.61$$

$$x = 7.8 \text{ m}, 28.2 \text{ m}$$

$$L_{eff} = 28.2 - 7.8 = 20.4 \text{ m}$$

Points of inflection are points where zero moment occurs. The points of inflection due to dead load are important because it is at these locations that the flange plate transitions occur.

d. Maximum dead load moment

The maximum moment occurs where the shear is equal to zero (Fig. E8.3-16).

$$M = (11.35)^2 - \frac{(11.35)^2}{2}(1.0) = 64 \text{ kN m}$$

The shears and moments, due to dead and live loads, at the tenth points are calculated. The procedures are the same as those il-

TABLE E8.3-1 Moments and Shears at Critical Locations

Load Type	Value (N/mm)	Moments (kN m)[a]			Shears (kN)[b]		
		M_{104}	M_{200}	M_{205}	V_{100}	V_{110}	V_{200}
Uniform	1.0	64.6	−108.5	53.5	11.4	−18.6	18.0
$D1^I$	15.14	978	−1643	810	173	−282	273
$D2^I$	4.04	261	−438	216	46	−75	73
$D3^I$	1.55	100	−168	83	18	−29	28
$D1^E$	14.61	944	−1585	782	167	−272	263
$D2^E$	3.03	196	−329	162	35	−56	55
$D3^E$	1.55	100	−168	83	18	−29	28
$LL + IM$	$Tr + Ln$	2346	−2470	2314	420	−482	488
Strength I $\eta = 0.95$	$U = \eta[1.25D1 + 1.50D2 + 1.25D3 + 1.75\,(LL + IM)]$	5418	−6657	5104	991	−1277	1271
Service I $\eta = 1.0$	$U = \eta[DC + DW + (LL + IM)]$	3586	−4552	3341			
Service III $\eta = 1.0$	$U = \eta[DC + DW + 0.8(LL + IM)]$	3117	−4058	2878			

[a]Exterior girders govern for moments.

[b]Interior girders govern for shears.

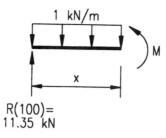

Fig. E8.3-14 Uniform load inflection point for span 1.

lustrated for the critical locations, only with different placement of the live load. The results are summarized in Tables E8.3-2 and E8.3-3 for the Strength *I* limit state. The second column in these tables gives either the moment or the shear due to a unit distributed load at each tenth point. These are multiplied by the actual distributed load given in Table E8.3-1, combined with appropriate load factors, and added to the product of the distribution factor times the factored live load plus impact. This sum is then multipied by the load modifier η and tabulated as η [sum]. The shear and moment envelopes are plotted in Figure E8.3-17.

I. Design Section for Negative Flexure

The bridge is composite in both the positive and negative regions and continuous throughout. Homogeneous sections are used and the depth of the web is constant. Only one flange plate transition is used. Longitudinal stiffeners are not used. Minimum thickness of steel is 8 mm [A6.7.3]. Deflection is not a concern.

The plate girder is initially designed for flexural requirements. The negative moment region is designed first to set the overall controlling proportions for the girder section. Following this step, the section is designed for maximum positive moment. An initial section is chosen based on similar designs. The final section for both the negative and positive moment regions is arrived at through iterations. Although a number of sections are investigated, only the final design of the section is illustrated herein.

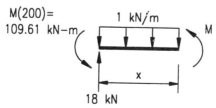

Fig. E8.3-15 Uniform load inflection point for span 2.

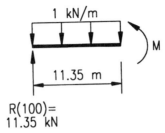

Fig. E8.3-16   Maximum moment due to uniform load in span 1.

As stated before in Section 8.11.3, Part H.1, the cross section for the maximum negative moment region consists of a 10-mm × 1500-mm web and 30-mm × 400-mm top and bottom flanges. Cross-sectional properties are computed for the steel girder alone and for the composite section. In the negative moment regions of continuous spans, the composite section is composed of the steel girder and the longitudinal reinforcement within an effective width of the slab. The concrete is neglected since it is considered cracked under tensile stress. At the interior support, stresses are checked at the top and bottom of the steel girder and in the reinforcing bars using factored moments. The steel girder alone resists moment due to *D1*. The composite section resists the moments due to *D2*, *D3*, and *LL* + *IM*.

1. *Sequence of Loading*   Consider the sequence of loading as specified in AASHTO [A6.10.5.1.1]. This article states that at any location on the composite section the elastic stress due to the applied loads shall be the sum of the stresses caused by the loads applied separately to

   a. Steel girder

   b. Short-term composite section

   c. Long-term composite section

   Permanent load that is applied before the slab reaches 75% of f'_c shall be carried by the steel girder alone. Any permanent load and live load applied after the slab reaches 75% of f'_c shall be carried by the composite section.

2. *Effective Flange Width*   Determine the effective flange width which is specified in AASHTO [A4.6.2.6]. For interior girders the effective flange width is the least of

   a. One-quarter of the effective span length (Sections 8.11.3, Part H.3.c)

   b. Twelve times the average thickness of the slab, plus the greater of the web thickness or one-half the width of the top flange of the girder

TABLE E8.3-2 Moment Envelope for 30-, 36-, and 30-m Plate Girder (kN m)

Location	Unit Dead Load	Positive Moment Truck or Tandem	Positive Moment Lane	Positive Moment η[Sum] Int. Gir.	Positive Moment η[Sum] Ext. Gir.	Negative Moment Truck or Tandem	Negative Moment Lane	Negative Moment η[Sum] Int. Gir.	Negative Moment η[Sum] Ext. Gir.
100	0.0	0	0	0	0	0	0	0	0
101	29.6	742	339	2177	2376	-97	-64	550	450
102	50.3	1251	595	3705	4044	-195	-127	873	693
103	61.9	1543	768	4602	5028	-292	-192	962	720
104	64.6	1670	857	4948	5418	-389	-256	824	538
105	58.2	1637	861	4742	5218	-487	-320	453	143
106	42.9	1474	782	4034	4483	-584	-383	-145	-461
107	18.5	1161	620	2791	3177	-681	-447	-976	-1280
108	-14.8	736	374	1070	1366	-779	-512	-2035	-2308
109	-57.2	297	179	-848	-617	-876	-711	-3471	-3721
110	-108.5	239	139	-2286	-1972	-1846[a]	-1148	-6246	-6657
200	-108.5	239	139	-2286	-1972	-1846[a]	-1148	-6246	-6657
201	-50.2	351	146	-629	-405	-766	-612	-3030	-3246
202	-4.8	857	336	1457	1757	-654	-380	-1462	-1697
203	27.6	1285	626	3206	3607	-543	-370	-463	-734
204	47.0	1557	807	4283	4750	-431	-370	192	-90
205	53.5	1631	868	4620	5104	-319	-370	518	251

[a]Truck train with trucks spaced 17 826 mm apart governs.

TABLE E8.3-3 Shear Envelope for 30-, 36-, and 30-m Plate Girder (kN)

		Positive Shear				Negative Shear			
Location	Unit Dead Load	Truck or Tandem	Lane	η[Sum] Int. Gir.	η[Sum] Ext. Gir.	Truck or Tandem	Lane	η[Sum] Int. Gir.	η[Sum] Ext. Gir.
100	11.4	287	127	991	913	−32	−21	203	186
101	8.4	247	101	806	742	−32	−23	124	113
102	5.4	208	78	626	577	−49	−28	10	9
103	2.4	171	59	454	419	−82	−37	−139	−129
104	−0.6	136	43	291	269	−121	−48	−302	−279
105	−3.6	103	30	136	126	−158	−63	−467	−431
106	−6.6	72	20	−9	−7	−193	−82	−634	−584
107	−9.6	45	13	−145	−133	−226	−102	−799	−736
108	−12.6	25	8	−265	−243	−256	−126	−962	−886
109	−15.6	10	6	−373	−342	−282	−151	−1122	−1033
110	−18.6	8	5	−455	−418	−305	−178	−1277	−1176
200	18.0	305	185	1271	1171	−31	−18	379	348
201	14.4	277	153	1084	999	−31	−20	285	261
202	10.8	244	124	892	822	−33	−23	185	169
203	7.2	208	98	699	644	−62	−31	29	26
204	3.6	171	74	506	466	−96	−41	−139	−129
205	0.0	133	56	319	294	−133	−56	−319	−294

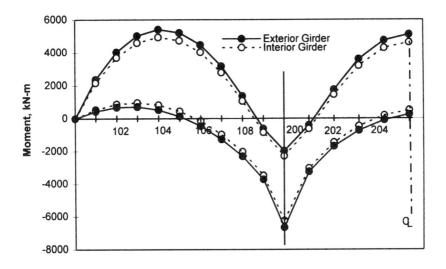

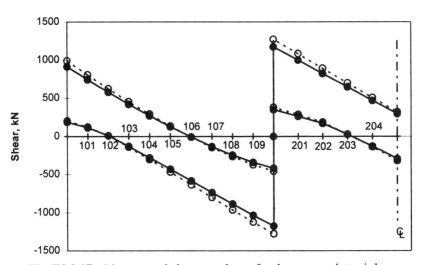

Fig. E8.3-17 Moment and shear envelopes for three-span plate girder.

c. Average spacing of adjacent girders

$$b_i = \min \begin{vmatrix} (0.25)(22\ 700) = 5675 \text{ mm} \\ (12)(190) + \dfrac{400}{2} = 2480 \text{ mm} \\ 2440 \text{ mm, governs} \end{vmatrix}$$

Therefore $b_i = 2440$ mm

For exterior girders the effective flange width may be taken as one half the effective width of the adjacent interior girder, plus the least of

a. One-eighth the effective span length

b. Six times the average thickness of the slab (using 190-mm structural), plus the greater of half the web thickness or one quarter of the width of the top flange of the basic girder

c. The width of the overhang

$$b_e = b_i/2 + \min \begin{vmatrix} (0.125)(22\ 700) = 2838 \text{ mm} \\ (6)(190) + (0.25)(400) = 1240 \text{ mm} \\ 990 \text{ mm, governs} \end{vmatrix}$$

Therefore $b_e = b_i/2 + 990 = 2440/2 + 990 = 2210$ mm

3. *Section Properties* Calculate the section properties for the steel girder alone and the composite section. Figure E8.3-18 illustrates the dimensions of the section. From Table E8.3-4 the following section properties were calculated for the steel section alone.

$$y_c = \frac{\Sigma Ay}{\Sigma A} = 0$$

$$I_{NA} = I - (y_c \times \Sigma\ Ay) = 16.86 \times 10^9 \text{ mm}^4$$

$$y_{\text{top of steel}} = \frac{D}{2} + t_f - y_c = \frac{1500}{2} + 30 - 0 = 780 \text{ mm}$$

$$y_{\text{bottom of steel}} = \frac{D}{2} + t_f + y_c = \frac{1500}{2} + 30 + 0 = 780 \text{ mm}$$

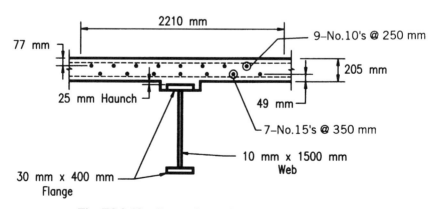

Fig. E8.3-18 Composite section for negative moment.

TABLE E8.3-4 Steel Section Properties (Negative Flexure)[a]

Component	A (mm²)	y (mm)	Ay	Ay^2	I_0 (mm⁴)	I (mm⁴)
Top flange 30 mm × 400 mm	12 000	765.0	9 180 000	7.0227×10^9	900 000	7.0236×10^9
Web 10 mm × 1500 mm	15 000	0	0	0	2.813×10^9	2.813×10^9
Bottom flange 30 mm × 400 mm	12 000	−765.0	−9 180 000	7.0227×10^9	900 000	7.0236×10^9
Total	39 000		0			16.86×10^9

[a]*Note:* y is the distance from the neutral axis of the web to the neutral axis of the component.

$$S_{\text{top of steel}} = \frac{I_{NA}}{y_t} = \frac{16.86 \times 10^9}{780} = 21.62 \times 10^6 \text{ mm}^3$$

$$S_{\text{bottom of steel}} = \frac{I_{NA}}{y_b} = \frac{16.86 \times 10^9}{780} = 21.62 \times 10^6 \text{ mm}^3$$

From Table E8.3-5 the following section properties were calculated for the composite section in negative flexure.

$$y_c = \frac{\Sigma Ay}{\Sigma A} = \frac{2\ 035\ 345}{41\ 300} = 49.282 \text{ mm}$$

$$I_{NA} = I - (y_c \times \Sigma\ Ay) = 18.56 \times 10^9 \text{ mm}^4$$

$$y_{\text{top reinf.}} = \frac{D}{2} + t_f + \text{haunch} + \text{cover} - y_c$$

$$y_{\text{top reinf.}} = \frac{1500}{2} + 30 + 25 + 128.05 - 49.282 = 883.77 \text{ mm}$$

$$y_{\text{bottom reinf.}} = \frac{1500}{2} + 30 + 25 + 49 - 49.282 = 804.72 \text{ mm}$$

$$S_{\text{top reinf.}} = \frac{I_{NA}}{y_{rt}} = \frac{18.56 \times 10^9}{883.77} = 21.001 \times 10^6 \text{ mm}^3$$

$$S_{\text{bottom reinf.}} = \frac{I_{NA}}{y_{rb}} = \frac{18.56 \times 10^9}{804.72} = 23.064 \times 10^6 \text{ mm}^3$$

$$y_{\text{top of steel}} = \frac{D}{2} + t_f - y_c = \frac{1500}{2} + 30 - 49.282 = 730.718 \text{ mm}$$

$$y_{\text{bottom of steel}} = \frac{D}{2} + t_f + y_c = \frac{1500}{2} + 30 + 49.282 = 829.282 \text{ mm}$$

$$S_{\text{top of steel}} = \frac{I_{NA}}{y_t} = \frac{18.56 \times 10^9}{730.718} = 25.40 \times 10^6 \text{ mm}^3$$

$$S_{\text{bottom of steel}} = \frac{I_{NA}}{y_b} = \frac{18.56 \times 10^9}{829.282} = 22.38 \times 10^6 \text{ mm}^3$$

4. *Member Proportions* Check the member proportions [A6.10.1.1]. This article states that flexural components are to be proportioned to meet the following requirement

$$0.1 \le \frac{I_{yc}}{I_y} \le 0.9$$

where

TABLE E8.3-5 Composite Section Properties (Negative Flexure)

Component	A (mm^2)	y (mm)	Ay	Ay^2	I_0 (mm^4)	I (mm^4)
Top flange 30 mm × 400 mm	12 000	765.0	9 180 000	7.0227×10^9	900 000	7.0236×10^9
Web 10 mm × 1500 mm	15 000	0	0	0	2.813×10^9	2.813×10^9
Bottom flange 30 mm × 400 mm	12 000	−765.0	−9 180 000	7.0227×10^9	900 000	7.0236×10^9
Top reinforcement (9 No. 10's)	900	933.0	839 745	0.784×10^9		0.784×10^9
Bottom reinforcement (7 No. 15's)	1 400	854	1 195 600	1.021×10^9		1.021×10^9
Total	41 300		2 035 345			18.66×10^9

I_y = Moment of inertia of the steel section about the vertical axis of the plane of the web (mm⁴)

I_{yc} = Moment of inertia of the compression flange of the steel section about the vertical axis in the plane of the web (mm⁴)

$$I_y = \tfrac{1}{12}(30)(400)^3 + \tfrac{1}{12}(30)(400)^3 + \tfrac{1}{12}(1500)(10)^3$$

$$= 320.1 \times 10^6 \text{ mm}^4$$

$$I_{yc} = \tfrac{1}{12}(30)(400)^3 = 160 \times 10^6 \text{ mm}^4$$

$$\frac{I_{yc}}{I_y} = 0.5 \quad \text{therefore the section is adequate}$$

5. *Check of* D_c/t_w Check D_c/t_w for fatigue induced by web flexure [A6.10.4.3] for webs without longitudinal stiffeners. The purpose of this article is to control out-of-plane flexing of the web due to flexure. The requirement involves calculating the maximum compressive elastic flexural stress in the compression flange, f_{cf}. This value is indicative of the maximum flexural stress in the web.

$$\frac{2D_c}{t_w} = \frac{2(1500/2 + 49.282)}{10} = 159.9$$

$$6.43\sqrt{\frac{E}{F_{yc}}} = 6.43\sqrt{\frac{200\ 000}{345}} = 154.82$$

If

$$\frac{2D_c}{t_w} > 6.43\sqrt{\frac{E}{F_{yc}}}$$

then (Eq. 8.43)

$$f_{cf} \leq 28.9 R_h E \left(\frac{t_w}{2D_c}\right)^2$$

where

f_{cf} = Maximum compressive elastic flexural stress in the compression flange due to unfactored permanent load from Table E8.3-1 and the fatigue loading for the exterior girder, which controls. The fatigue loading is taken as twice that calculated using the fatigue load combination in Table 3.1 [Table A3.4.1-1]. The unfactored values for the fatigue loading at the critical points are summarized in Table E8.3-6. The factored values with impact applied for the fatigue limit state are tabulated in Table E8.3-7. The distribution factor DF is $0.762/1.2 = 0.635$, because there is no multiple presence for the passage of a single fatigue truck [A3.6.1.1.2]. This stress is also indicative of the maximum flexural stress in the web (MPa)

$$D_c = \text{Depth of web in compression (mm)}$$

$$D_c = D/2 + y_c = 1500/2 + 49.282 = 799.282 \text{ mm}$$

$$R_h = \text{Hybrid section reduction factor, taken as } 1.0$$

$$f_{cf} \leq 28.9(1.0)(200\ 000)\left(\frac{1}{159.9}\right)^2 = 226.1 \text{ MPa}$$

The allowable stress of 226.1 MPa is to be compared to the maximum calculated flexural fatigue stress in the web to determine whether the section is adequate. The values for unfactored shear and moment due to fatigue loads at critical points, in Table E8.3-6, were generated by the computer program BT Beam. It is shown below how these values are calculated by hand. The values for the influence line ordinates are taken from Table 5.4.

a. $V(100)$

Positive shear (Fig. E8.3-19)

TABLE E8.3-6 Unfactored Shear and Moment Due to Fatigue Loads at Critical Points

SHEAR (kN)	V(100)	V(104)	V(110)	V(200)	V(205)
Positive	253	109	7	281	107
Negative	-30	-88	-282	-28	-107

MOMENT (kN·m)	M(100)	M(104)	M(110)	M(200)	M(205)
Positive	0	1407	219	219	1373
Negative	0	-361	-903	-903	-291

TABLE E8.3-7   Factored Moments for Fatigue Limit State for Exterior Girder

Location	Positive Moment	(kN·m)		Negative Moment	(kN·m)	
	LL+IM w/ IM=0.15	LL+IM w/ DF=0.635	LL+IM w/ LF=0.75	LL+IM w/ IM=0.15	LL+IM w/ DF=0.635	LL+IM w/ LF=0.75
0	0	0	0	0	0	0
104	1618	1027	771	-415	-264	-198
110	251	159	120	-1038	-659	-494
205	1578	1002	752	-335	-213	-160

$$V(100) = 145(1.0 + 0.632\ 97) + 35(0.470\ 38) = 253 \text{ kN}$$

Negative shear (Fig. E8.3-20)

$$V(100) = 35(-0.096\ 75) + 145(-0.103\ 37 - 0.071\ 94)$$
$$= -29 \text{ kN}$$

b. $V(104)$

Positive shear (Fig. E8.3-21)

$$V(104) = 145(0.517\ 50 + 0.212\ 34) + 35(0.098\ 64) = 109 \text{ kN}$$

Negative shear (Fig. E8.3-22)

$$V(104) = 145[-0.124\ 31 - (1 - 0.517\ 50)] = -88 \text{ kN}$$

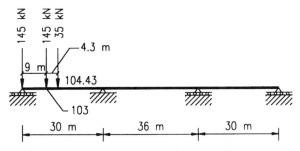

Fig. E8.3-19 Fatigue truck placement for maximum positive shear at Location 100.

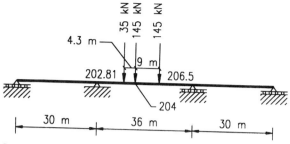

Fig. E8.3-20 Fatigue truck placement for maximum negative shear at Location 100.

c. $V(110)$

Positive shear (Fig. E8.3-23)

$$V(110) = 35(0.021\ 92) + 145(0.025\ 71 + 0.018\ 28) = 7\ \text{kN}$$

Negative shear (Fig. E8.3-24)

$$V(110) = 35(-0.650\ 34) + 145(-0.787\ 66 - 1.0) = -282\ \text{kN}$$

d. $V(200)$

Positive shear (Fig. E8.3-25)

$$V(200) = 145(1.0 + 0.783\ 75) + 35(0.653\ 185) = 282\ \text{kN}$$

Negative shear (Fig. E8.3-26)

$$V(200) = 145(-0.10 - 0.071\ 09) + 35(-0.039\ 585)$$
$$= -26\ \text{kN}$$

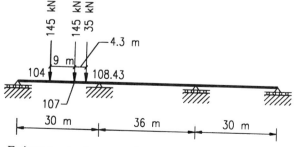

Fig. E8.3-21 Fatigue truck placement for maximum positive shear at Location 104.

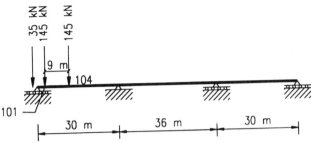

Fig. E8.3-22 Fatigue truck placement for maximum negative shear at Location 104.

e. $V(205)$

Positive shear (Fig. E8.3-27)

$$V(205) = 145(0.5 + 0.216\ 25) + 35(0.101\ 21) = 107\ \text{kN}$$

Negative shear (Fig. E8.3-28)

$$V(205) = 35(-0.101\ 21) + 145(-0.216\ 25 - 0.5) = -107\ \text{kN}$$

f. $M(100)$

Positive moment

$$M(100) = 0\ \text{kN m}$$

Negative moment

$$M(100) = 0\ \text{kN m}$$

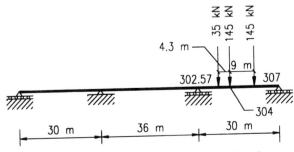

Fig. E8.3-23 Fatigue truck placement for maximum positive shear at Location 110.

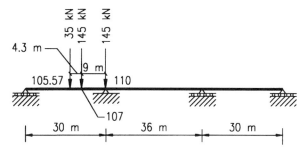

Fig. E8.3-24   Fatigue truck placement for maximum negative shear at Location 110.

g. $M(104)$

Positive moment (Fig. E8.3-29)

$$M(104) = [145(0.207 + 0.084\ 94) + 35(0.130\ 652)]30$$
$$= 1407\ \text{kN m}$$

Negative moment (Fig. E8.3-30)

$$M(104) = [145(-0.037\ 105 - 0.038\ 3025)$$
$$+ 35(-0.029\ 213)]30$$
$$= -359\ \text{kN m}$$

h. $M(110)$

Positive moment (Fig. E8.3-31)

$$M(110) = [145(0.025\ 71 + 0.018\ 28) + 35(0.021\ 91)]30$$
$$= 214\ \text{kN m}$$

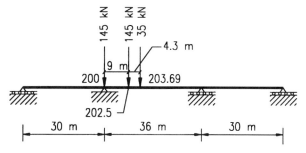

Fig. E8.3-25   Fatigue truck placement for maximum positive shear at Location 200.

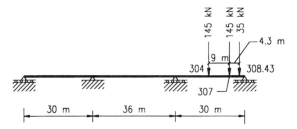

Fig. E8.3-26 Fatigue truck placement for maximum negative shear at Location 200.

Negative moment (Fig. E8.3-32)

$$M(110) = [145(-0.095\ 7525 - 0.092\ 765)$$
$$+ 35(-0.073\ 028)]30$$
$$= -897\ \text{kN m}$$

i. $M(200)$

Positive moment Same as $M(110)$

Negative moment Same as $M(110)$

j. $M(205)$

Positive moment (Fig. E8.3-33)

$$M(205) = [145(0.203\ 57 + 0.078\ 645) + 35(0.138\ 035)]30$$
$$= 1373\ \text{kN m}$$

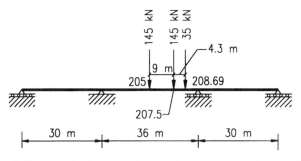

Fig. E8.3-27 Fatigue truck placement for maximum positive shear at Location 205.

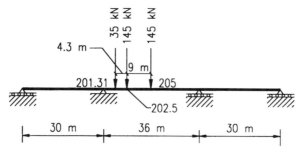

Fig. E8.3-28 Fatigue truck placement for maximum negative shear at Location 205.

Negative moment (Fig. E8.3-34)

$$M(205) = [35(-0.029\ 208) + 145(-0.034\ 29 - 0.024\ 37)]30$$
$$= -286\ \text{kN m}$$

The maximum flexural fatigue stress in the web at Location 200 is calculated in Table E8.3-8.

Maximum negative moment $= -903$ kN m from Table E8.3-6

Critical $LL + IM$ ($IM = 0.15$) $= -1038$ kN m

$LL + IM$ ($w/DF = 0.635$) $= -659$ kN m

$LL + IM$ ($w/LF = 0.75$) $= -495$ kN m

For checking fatigue in web, the moment is doubled [A6.10.4.2], that is,

$$-495 \times 2 = -990\ \text{kN m}$$

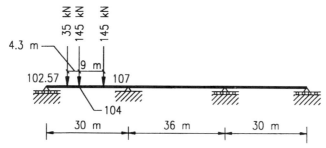

Fig. E8.3-29 Fatigue truck placement for maximum positive moment at Location 104.

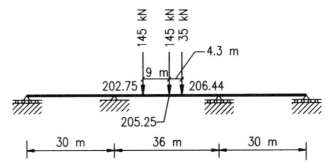

Fig. E8.3-30 Fatigue truck placement for maximum negative moment at Location 104.

M_{D1}, M_{D2}, and M_{D3} in Table E8.3-8 are the moments at Location 200 due to unfactored dead loads on the exterior girder, from Table E8.3-1. The maximum calculated stress of 139.7 MPa is less than the allowable flexural fatigue stress of 226.1 MPa calculated earlier; therefore, the section is adequate.

6. *Stresses* This step involves calculating the stresses in the top and bottom of the girder for the strength limit state. The exterior girder moments govern the flexural design. The stresses are based on factored moments from Table E8.3-1, multiplied by their load factors, at Location 200. Tables E8.3-9 and E8.3-10 summarize the maximum stresses in the tension flange (275 MPa) and compression flange (300 MPa).

7. *Determination if Compact or Noncompact* Initially, the section is assumed to be compact and is evaluated to determine if the assumption is correct.

   a. Check flexural resistance for compact sections [A6.10.5.2.3] in negative flexure.

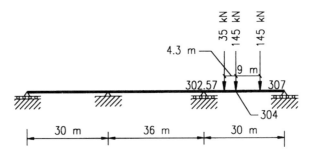

Fig. E8.3-31 Fatigue truck placement for maximum positive moment at Location 110.

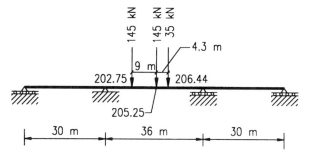

Fig. E8.3-32   Fatigue truck placement for maximum negative moment at Location 110.

Check web slenderness (Table 8.17) [A6.10.5.2.3b].

$$\frac{2D_{cp}}{t_w} \le 3.76\sqrt{\frac{E}{F_{yc}}}$$

where

D_{cp} = depth of web in compression [A6.10.5.1.4b] (mm)

F_{yc} = minimum yield strength of the compression flange (MPa)

If the web slenderness exceeds the limit then the nominal flexural resistance shall be determined as specified in AASHTO [A6.10.5.3.3] for noncompact sections.

From AASHTO [A6.10.5.1.4b] for sections in negative flexure, where the plastic neutral axis is in the web

$$D_{cp} = \frac{D}{2A_w F_{yw}}[F_{yt}A_t + F_{yw}A_w + F_{yr}A_r - F_{yc}A_c]$$

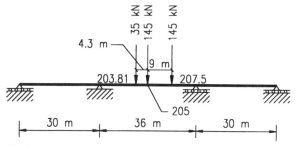

Fig. E8.3-33   Fatigue truck placement for maximum positive moment at Location 205.

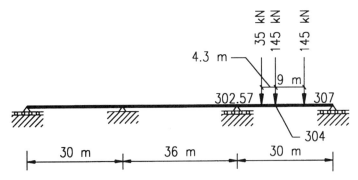

Fig. E8.3-34 Fatigue truck placement for maximum negative moment at Location 205.

where

A_w = Area of the web (mm²)

A_t = area of the tension flange (mm²)

A_r = area of the longitudinal reinforcement in the section (mm²)

A_c = area of the compression flange (mm²)

F_{yt} = minimum yield strength of the tension flange (MPa)

F_{yw} = minimum yield strength of the web (MPa)

F_{yr} = minimum yield strength of the longitudinal reinforcement (MPa)

F_{yc} = minimum yield strength of the compression flange (MPa)

For all other sections in negative flexure, D_{cp} shall be taken as

TABLE E8.3-8 Maximum Flexural Fatigue Stress in the Web for Negative Flexure at Location 200

Load	M_{D1} (kN m)	M_{D2} (kN m)	M_{D3} (kN m)	M_{LL+IM} (kN m)	S_b Steel (mm³)	S_b Composite (mm³)	Stress (MPa)
D1	1585				21.62×10^6		73.3
D2		329				22.38×10^6	14.7
D3			168			22.38×10^6	7.5
LL + IM				990		22.38×10^6	44.2
Total							139.7
						$\eta = 1.0$	139.7

TABLE E8.3-9   Stress in Top of Steel Girder (Tension) for Negative Flexure Due to Factored Loading

Load	M_{D1} (kN m)	M_{D2} (kN m)	M_{D3} (kN m)	M_{MLL+IM} (kN m)	S_t Steel (mm³)	S_t Composite (mm³)	Stress (MPa)
D1	1981				21.62×10^6		91.7
D2		493				25.40×10^6	19.4
D3			210			25.40×10^6	8.3
LL + IM				4323		25.40×10^6	170.2
Total							289.5
						$\eta = 0.95$	275.0

equal to D. Find the location of the plastic neutral axis using AASHTO (1994) LRFD Bridge Specifications, Appendix A of Section 6. The diagrams in Figure E8.3-35 illustrate the dimensions of the section and the plastic forces. The diagrams are taken from Section 6, Appendix A of AASHTO (1994) LRFD Bridge Specifications.

Plastic forces

$$\text{Top reinforcement}\quad P_{rt} = F_{yr}A_{rt} = (400)(9)(100) = 360 \text{ kN}$$

$$\text{Bottom reinforcement}\quad P_{rb} = F_{yr}A_{rb} = (400)(7)(200)$$

$$= 560 \text{ kN}$$

$$\text{Tension flange} = P_t = F_{yt}b_t t_t = (345)(400)(30) = 4140 \text{ kN}$$

$$\text{Compression flange} = P_c = F_{yc}b_t t_c = (345)(400)(30)$$

$$= 4140 \text{ kN}$$

$$\text{Web} = P_w = F_{yw}Dt_w = (345)(1500)(10) = 5175 \text{ kN}$$

TABLE E8.3-10   Stress in Bottom of Steel Girder (Compression) for Negative Flexure Due to Factored Loading

Load	M_{D1} (kN m)	M_{D2} (kN m)	M_{D3} (kN m)	M_{LL+IM} (kN m)	S_b Steel (mm³)	S_b Composite (mm³)	Stress (MPa)
D1	1981				21.62×10^6		91.7
D2		493				22.38×10^6	22.0
D3			210			22.38×10^6	9.4
LL + IM				4323		22.38×10^6	193.2
Total							316.2
						$\eta = 0.95$	300.4

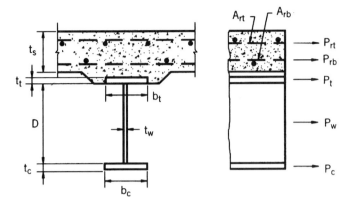

Fig. E8.3-35   Plastic neutral axis for negative moment section.

Plastic Neutral Axis: $C = T$

Check if PNA is in the web.

$$P_c + P_w \geq P_t + P_{rb} + P_{rt}$$

$$4140 + 5175 = 9315 \geq 4140 + 560 + 360 = 5060$$

Therefore the PNA is in the web.

$$D_{cp} = \frac{1500}{2(5175)} [4140 + 5175 + 360 + 560 - 4140]$$

$$= 883.3 \text{ mm}$$

$$\frac{2D_{cp}}{t_w} = \frac{2(883.3)}{10} = 176.7$$

$$3.76\sqrt{\frac{E}{F_{yc}}} = 3.76\sqrt{\frac{200\ 000}{345}} = 90.5$$

176.7 > 90.5 therefore the section is noncompact

b. Flexural resistance for noncompact sections

Check flexural resistance for noncompact sections [A6.10.5.3.3] in negative flexure (Table 8.17).
Web slenderness: The web slenderness limit state for noncompact sections is [A6.10.5.3.3b]

$$\frac{2D_c}{t_w} \le 6.77\sqrt{\frac{E}{f_c}}$$

where

f_c = Stress in the compression flange due to the factored loading

f_c = 300.4 MPa from Table E8.3-10

D_C = 1500/2 + 49.282 = 799.28 mm

$$\frac{2D_c}{t_w} = \frac{2(799.28)}{10} = 159.86$$

$$6.77\sqrt{\frac{E}{f_c}} = 6.77\sqrt{\frac{200\,000}{300.4}} = 174.7 > 159.86 \qquad \text{OK}$$

Compression flange slenderness: Check compression flange slenderness [A6.10.5.3.3c].

$$\frac{b_f}{2t_f} \le 1.38\sqrt{\frac{E}{f_c\sqrt{2D_c/t_w}}}$$

$$\frac{b_f}{2t_f} = \frac{400}{2(30)} = 6.667$$

$$1.38\sqrt{\frac{E}{f_c\sqrt{2D_c/t_w}}} = 1.38\sqrt{\frac{200\,000}{300.4\sqrt{159.86}}}$$

$$= 10.01 > 6.667$$

Therefore the section is adequate.
Check compression flange bracing [A6.10.5.3.3d].

$$L_b \le 1.76r_t\sqrt{\frac{E}{F_{yc}}}$$

where

L_b = distance between brace points along the compression flange (mm)

r_t = minimum radius of gyration of the compression flange of the steel section, plus one-third of the web in compression taken about the vertical axis (mm)

$$I_y = \frac{(D_c/3)(t_w)^3}{12} + \frac{(t_f)(b_f)^3}{12}$$

$$I_y = \frac{(799.282/3)(10)^3}{12} + \frac{(30)(400)^3}{12} = 160.0 \times 10^6 \text{ mm}^4$$

$$A = (799.282/3)(10) + (30)(400) = 14\ 664 \text{ mm}^2$$

$$r_t = \sqrt{\frac{I_y}{A}} = \sqrt{\frac{160.0 \times 10^6}{14\ 664}} = 104.46 \text{ mm}$$

Minimum $L_b \leq 1.76(104.46)\sqrt{\dfrac{200\ 000}{345}} = 4427$ mm

If the diaphragms were to be designed in this example, we would consider the section in the positive moment region before specifying diaphragm spacing. As long as the compression flange is braced at a distance less than 4427 mm, lateral torsional buckling will not be a problem.

8. *Load shedding factor*   Determine the load shedding factor, R_b, for the compression and tension flanges [A6.10.5.4.2]. This factor accounts for the nonlinear variation of stresses caused by local buckling of slender webs subjected to flexural stresses. For the tension flange, the load shedding factor is taken as 1.0. The requirement for the compression flange is as follows:

$$\frac{2D_c}{t_w} \leq \lambda_b\sqrt{\frac{E}{f_c}}$$

where

λ_b = 5.76 for members with a compression flange area equal to or greater than the tension flange area

$$\frac{2D_c}{t_w} = 159.86$$

$$\lambda_b\sqrt{\frac{E}{f_c}} = 5.76\sqrt{\frac{200\ 000}{300.4}} = 148.6 < 159.86$$

The requirement is not satisfied therefore R_b is taken as

$$R_b = 1 - \left(\frac{a_r}{1200 + 300a_r}\right)\left(\frac{2D_c}{t_w} - \lambda_b\sqrt{\frac{E}{f_c}}\right)$$

where

$$a_r = \frac{2D_c t_w}{A_{fc}}$$

A_{fc} = compression flange area (mm²)

$$a_r = \frac{2(799.282)(10)}{(30)(400)} = 1.332$$

$$R_b = 1 - \left[\frac{1.332}{1200 + 300(1.332)}\right](159.86 - 148.6)$$
$$= 0.991$$

Therefore the load shedding factor for the compression flange is 0.991.

9. *Flexural Resistance*   The final step in designing the section for the negative moment region is to calculate the resistance of the section and compare the values of the maximum stresses from Section 8.11.3, Part I.6. For a noncompact section the resistance of each flange is in terms of stress [A6.10.5.3.3a]. The nominal flexural resistance is as follows:

$$F_n = R_b R_h F_{yf}$$

where

R_b and R_h = flange stress reduction factors

R_b = 0.991 for compression flanges

R_b = 1.0 for tension flanges

R_h = 1.0 for homogeneous sections

F_{yf} = Specified minimum yield strength of the flange

For the tension flange

$$F_n = (1.0)(1.0)(345) = 345 \text{ MPa}$$
$$F_r = \phi_f F_n = (1.0)(345) = 345 \text{ MPa}$$

This value is greater than the maximum stress of 275.0 MPa from Table E8.3-9, therefore the section is adequate.

For the compression flange

$$F_n = (0.991)(1.0)(345) = 341.9 \text{ MPa}$$

$$F_r = \phi_f F_n = (1.0)(341.9) = 341.9 \text{ MPa}$$

This value is greater than the maximum stress of 300.4 MPa from Table E8.3-10, therefore the section is adequate.

All the requirements for flexure have been satisfied. Therefore use a section consisting of a 10-mm × 1500-mm web, and 30-mm × 400-mm flanges for the negative moment region.

J. Design Section for Positive Flexure

For the positive moment region, a steel section consisting of a 15-mm × 300-mm top flange, 10-mm × 1500-mm web, and 25-mm × 400-mm bottom flange is used. As stated earlier, the cross section of the web remained constant. The top flange is smaller than the bottom flange due to the additional strength provided by the concrete. Section properties are computed for the steel section alone, the short-term composite section with n equal to 8, and the long-term composite section with $3n$ equal to 24, where n is the modular ratio. The composite section in positive flexure consists of the steel section and a transformed area of an effective width of concrete slab [A6.10.5.1.1b]. For normal weight concrete, the modular ratio n for 25 MPa $\leq f'_c \leq$ 32 MPa is taken as 8, where f'_c is the 28-day compressive strength of the concrete [A6.10.5.1.1b]. Stresses are computed at the top and bottom of the steel girder and in the concrete using factored moments. The steel girder alone resists moments due to $D1$. The short term composite section resists moments due to $LL + IM$, and the long term composite section resists moments due to $D2$ and $D3$. The sequence of loading and the effective flange width are identical to that determined for the negative moment region in Section 8.11.3, Parts I.1 and I.2, respectively. The moments for the exterior girders control the positive moment region also. Therefore, the effective width used is b_e equal to 2210 mm.

1. *Section Properties* Calculate the Section Properties for the steel girder alone and the short term and long term composite sections. Figure E8.3-36 illustrates the dimensions of the section. (These are the same dimensions as the cross section of Example 8.5 where the section properties were calculated in Tables 8.13–8.15. Many designers prefer a slightly different procedure, but the results are the same as given herein.) From Table E8.3-11 the following section properties are calculated for the steel section alone.

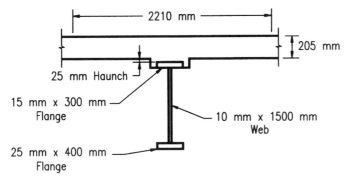

Fig. E8.3-36   Composite section for positive moment.

$$y_c = \frac{\Sigma Ay}{\Sigma A} = \frac{-4.216 \times 10^6}{29\ 500} = -142.9 \text{ mm}$$

$$I_{NA} = I - (y_c \times \Sigma\ Ay) = 10.607 \times 10^9 \text{ mm}^4$$

$$y_{\text{top of steel}} = \frac{D}{2} + t_f - y_c = \frac{1500}{2} + 15 + 142.9 = 907.9 \text{ mm}$$

$$y_{\text{bottom of steel}} = \frac{D}{2} + t_f + y_c = \frac{1500}{2} + 25 - 142.9 = 632.1 \text{ mm}$$

$$S_{\text{top of steel}} = \frac{I_{NA}}{y_t} = \frac{10.607 \times 10^9}{907.9} = 11.683 \times 10^6 \text{ mm}^3$$

$$S_{\text{bottom of steel}} = \frac{I_{NA}}{y_b} = \frac{10.607 \times 10^9}{632.1} = 16.780 \times 10^6 \text{ mm}^3$$

From Table E8.3-12, the following section properties are calculated for the short term composite section, where $n = 8$, $b_e = 2210$ mm, and $t_s = 205$ mm.

$$y_c = \frac{\Sigma Ay}{\Sigma A} = \frac{46.327 \times 10^6}{86\ 131} = 537.9 \text{ mm}$$

$$I_{NA} = I - (y_c \times \Sigma\ Ay) = 31.60 \times 10^9 \text{ mm}^4$$

$$y_{\text{top of steel}} = \frac{D}{2} + t_f - y_c = \frac{1500}{2} + 15 - 537.9 = 227.1 \text{ mm}$$

$$y_{\text{bottom of steel}} = \frac{D}{2} + t_f + y_c = \frac{1500}{2} + 25 + 537.9 = 1312.9 \text{ mm}$$

$$S_{\text{top of steel}} = \frac{I_{NA}}{y_t} = \frac{31.60 \times 10^9}{227.1} = 139.1 \times 10^6 \text{ mm}^3$$

TABLE E8.3-11 Steel Section Properties (Positive Flexure)

Component	A (mm^2)	y (mm)	Ay	Ay^2	I_0 (mm^4)	I (mm^4)
Top flange 15 mm × 300 mm	4 500	757.5	3.409×10^6	2.582×10^9	84 375	2.582×10^9
Web 10 mm × 1500 mm	15 000	0	0	0	2.813×10^9	2.813×10^9
Bottom flange 25 mm × 400 mm	10 000	−762.5	-7.625×10^6	5.814×10^9	520 833	5.815×10^9
Total	29 500		-4.216×10^6			11.209×10^9

TABLE E8.3-12 Short-Term Composite Section Properties, $n = 8$ (Positive Flexure)

Component	A (mm²)	y (mm)	Ay	Ay^2	I_0 (mm⁴)	I (mm⁴)
Top flange 15 mm × 300 mm	4 500	757.5	3.409×10^6	2.582×10^9	84 375	2.582×10^9
Web 10 mm × 1500 mm	15 000	0	0	0	2.813×10^9	2.813×10^9
Bottom flange 25 mm × 400 mm	10 000	−762.5	-7.625×10^6	5.814×10^9	520 833	5.815×10^9
Concrete $b_e \times t_s/n$	56 631	892.5	50.543×10^6	45.11×10^9	198.33×10^6	45.308×10^9
Total	86 131		46.327×10^6			56.517×10^9

$$S_{\text{bottom of steel}} = \frac{I_{NA}}{y_b} = \frac{31.60 \times 10^9}{1312.9} = 24.069 \times 10^6 \text{ mm}^3$$

$$y_{\text{top of concrete}} = \frac{D}{2} + t_f + \text{haunch} + t_s - y_c$$

$$y_{\text{top of concrete}} = \frac{1500}{2} + 15 + 25 + 205 - 537.9 = 457.1 \text{ mm}$$

$$S_{\text{top of concrete}} = \frac{I_{NA}}{y_{tc}} = \frac{31.60 \times 10^9}{457.1} = 69.13 \times 10^6 \text{ mm}^3$$

From Table E8.3-13 the following section properties are calculated for the long-term composite section, where $3n = 24$.

$$y_c = \frac{\Sigma Ay}{\Sigma A} = \frac{12.632 \times 10^6}{48\ 377} = 261.1 \text{ mm}$$

$$I_{NA} = I - (y_c \times \Sigma Ay) = 23.014 \times 10^9 \text{ mm}^4$$

$$y_{\text{top of steel}} = \frac{D}{2} + t_f - y_c = \frac{1500}{2} + 15 - 261.1 = 503.9 \text{ mm}$$

$$y_{\text{bottom of steel}} = \frac{D}{2} + t_f + y_c = \frac{1500}{2} + 25 + 261.1 = 1036.1 \text{ mm}$$

$$S_{\text{top of steel}} = \frac{I_{NA}}{y_t} = \frac{23.014 \times 10^9}{503.9} = 45.672 \times 10^6 \text{ mm}^3$$

$$S_{\text{bottom of steel}} = \frac{I_{NA}}{y_b} = \frac{23.014 \times 10^9}{1036.1} = 22.212 \times 10^6 \text{ mm}^3$$

2. *Member Proportions* Check the member proportions as specified in AASHTO [A6.10.1.1]. This article states that flexural components are to be proportioned to meet the following requirement:

$$0.1 \leq \frac{I_{yc}}{I_y} \leq 0.9$$

$$I_y = \tfrac{1}{12}(15)(300)^3 + \tfrac{1}{12}(25)(400)^3 + \tfrac{1}{12}(1500)(10)^3$$
$$= 167.2 \times 10^6 \text{ mm}^4$$

$$I_{yc} = \tfrac{1}{12}(15)(300)^3 = 33.75 \times 10^6 \text{ mm}^4$$

TABLE E8.3-13 Long-Term Composite Section Properties, $3n = 24$ (Positive Flexure)

Component	A (mm^2)	y (mm)	Ay	Ay^2	I_0 (mm^4)	I (mm^4)
Top flange 15 mm × 300 mm	4 500	757.5	3.409×10^6	2.582×10^9	84 375	2.582×10^9
Web 10 mm × 1500 mm	15 000	0	0	0	2.813×10^9	2.813×10^9
Bottom flange 25 mm × 400 mm	10 000	−762.5	-7.625×10^6	5.814×10^9	520 833	5.815×10^9
Concrete $b_e \times t_s/n$	18 877	892.5	16.848×10^6	15.037×10^9	66.11×10^6	15.103×10^9
Total	48 377		12.632×10^6			26.312×10^9

$$\frac{I_{yc}}{I_y} = 0.202 \qquad \text{therefore the section is adequate}$$

3. *Check of* D_c/t_w Check D_c/t_w for fatigue induced by web flexure [A6.10.4.3] for webs without longitudinal stiffeners. D_c is based on the superposition of stresses [C6.10.4.3].

$$\frac{2D_c}{t_w} = \frac{2(1500/2 + 142.9)}{10} = 178.58$$

$$6.43\sqrt{\frac{E}{F_{yc}}} = 6.43\sqrt{\frac{200\,000}{345}} = 154.82$$

If

$$\frac{2D_c}{t_w} > 6.43\sqrt{\frac{E}{F_{yc}}}$$

then

$$f_{cf} \le 28.9R_hE\left(\frac{t_w}{2D_c}\right)^2$$

$$D_c = D/2 + y_c = 1500/2 + 142.9 = 892.9 \text{ mm}$$

$$f_{cf} \le 28.9(1.0)(200\,000)\left(\frac{10}{2(892.9)}\right)^2 = 181.2 \text{ MPa}$$

The allowable stress of 181.2 MPa is greater than the maximum calculated compressive stress of 100.9 MPa from Table E8.3-14, therefore the section is adequate. The stresses calculated in Table E8.3-14 are based on the unfactored permanent loads and the fatigue loading described in Section 8.11.3, Part I.5. M_{LL+IM} is twice the positive moment value with load factor, at Location 104, in Table E8.3-7, that is,

$$M_{LL+IM} = 2(\text{LF})(\text{DF})M_{104}(1 + IM)$$

$$= 2(0.75)(0.636/1.2)(1407)(1.15) = 1286 \text{ kN m}$$

M_{D1}, M_{D2}, and M_{D3} are found at Location 104 in Table E8.3-1 for an interior girder. Section moduli were calculated in Section 8.11.3, Part J.1. For the exterior girder, a total stress of 98.4 MPa results. The interior girder, with a stress of 100.9 MPa, from Table E8.3-14, barely controls.

TABLE E8.3-14 Maximum Flexural Fatigue Stress in the Web for Positive Flexure, Interior Girder

Load	M_{D1} (kN m)	M_{D2} (kN m)	M_{D3} (kN m)	M_{LL+IM} (kN m)	S_t Steel (mm³)	S_t Composite (mm³)	Stress (MPa)
D1	978				11.683×10^6		83.72
D2		261				45.672×10^6	5.71
D3			100			45.672×10^6	2.19
LL + IM				1286		139.145×10^6	9.24
Total							100.9
						$\eta = 1.0$	100.9

4. *Stresses* This step involves calculating the stresses in the top and bottom of the girder for the strength limit state. The exterior girder moments govern the flexural design. The stresses are based on factoring the loads from Table E8.3-1 at Location 104. The section moduli were calculated in Section 8.11.3, Part J.1. Tables E8.3-15 and E8.3-16 summarize the maximum stresses in the tension and compression flanges.

5. *Determination if Compact or Noncompact* Initially, the section is assumed to be compact and is evaluated to determine if the assumption is correct.

Check flexural resistance for compact sections (Table 8.16) [A6.10.5.2.2] for positive flexure.

  a. Web slenderness
    Check web slenderness [A6.10.5.2.2c].

TABLE E8.3-15 Stress in Top of Exterior Steel Girder (Compression) for Positive Flexure Due to Factored Loading

Load	M_{D1} (kN m)	M_{D2} (kN m)	M_{D3} (kN m)	M_{LL+IM} (kN m)	S_t Steel (mm³)	S_t Composite (mm³)	Stress (MPa)
D1	1180				11.683×10^6		101.0
D2		294				45.672×10^6	6.4
D3			125			45.672×10^6	2.7
LL + IM				4105		139.145×10^6	29.5
Total							139.6
						$\eta = 0.95$	132.6

TABLE E8.3-16   Stress in Bottom of Exterior Steel Girder (Tension) for Positive Flexure Due to Factored Loading

Load	M_{D1} (kN m)	M_{D2} (kN m)	M_{D3} (kN m)	M_{LL+IM} (kN m)	S_b Steel (mm³)	S_b Composite (mm³)	Stress (MPa)
D1	1180				16.780×10^6		70.3
D2		294				22.212×10^6	13.2
D3			125			22.212×10^6	5.6
LL + IM				4105		24.069×10^6	170.5
Total							259.6
						$\eta = 0.95$	246.6

$$\frac{2 D_{cp}}{t_w} \le 3.76 \sqrt{\frac{E}{F_{yc}}}$$

From AASHTO [A6.10.5.1.4b] for sections in positive flexure, where the plastic neutral axis is in the web

$$D_{cp} = \frac{D}{2} \left(\frac{F_{yt} A_t - F_{yc} A_c - 0.85 f'_c A_s - F_{yr} A_r}{F_{yw} A_w} + 1 \right)$$

For all other sections in positive flexure, D_{cp} shall be taken equal to 0 and the web slenderness requirement is considered satisfied.

Find the location of the plastic neutral axis using AASHTO (1994) LRFD Appendix A of Section 6. Figure E8.3-37 illustrates the dimensions of the section and the plastic forces. The diagrams are taken from Appendix A of Section 6 of AASHTO (1994).

b. Plastic forces

Top reinforcement $P_{rt} = F_{yr} A_{rt} = (400)(9)(100) = 360$ kN

Concrete slab $P_s = 0.85 f'_c a b_e = 0.85(30)(2210)a$

$P_s = 56\ 355a$ (N) $= 56.4a$ (kN)

Bottom reinforcement $P_{rb} = F_{yr} A_{rb} = (400)(7)(200)$

$= 560$ kN

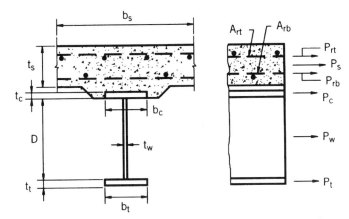

Fig. E8.3-37   Plastic neutral axis for positive moment section.

$$\text{Tension flange} = P_t = F_{yt}b_t t_t = (345)(400)(25) = 3450 \text{ kN}$$

$$\text{Compression flange} = P_c = F_{yc}b_c t_c = (345)(300)(15)$$
$$= 1552.5 \text{ kN}$$

$$\text{Web} = P_w = F_{yw}Dt_w = (345)(1500)(10) = 5175 \text{ kN}$$

c. Plastic neutral axis (PNA): $C = T$
Assume plastic neutral axis is in the slab between the reinforcement.

$$56.4a + 360 = 560 + 3450 + 1552.5 + 5175$$

Therefore $a = 184.0$ mm, $\beta_1 = 0.85 - \frac{2}{7}(0.05) = 0.836$

$$c = \frac{a}{\beta_1} = \frac{184}{0.836} = 220.2 \text{ mm} > c_{rb} = 156 \text{ mm}$$

c_{rb} is the distance from the top of the concrete slab to the bottom reinforcement. Therefore recalculate with plastic neutral axis below bottom reinforcement.

$$360 + 56.4a + 560 = 3450 + 1552.5 + 5175$$

Therefore $a = 164.1$ mm

$$c = \frac{a}{\beta_1} = \frac{164.1}{0.836} = 196.4 \text{ mm} > 156 \text{ mm} \qquad \text{OK}$$

$$P_s = \frac{1}{1000}(0.85)(30)(2210)(164.1) = 9248 \text{ kN}$$

For the case of PNA in slab below P_{rb}

$$P_t + P_w + P_c \geq \left(\frac{c_{rb}}{t_s}\right)P_s + P_{rb} + P_{rt}$$

$$3450 + 5175 + 1552.5 \geq \left(\frac{156}{205}\right)(9248) + 560 + 360$$

$$10\ 177.5 \geq 7957.5, \text{ OK}$$

The plastic neutral axis is in the slab below the bottom reinforcement, therefore D_{cp} is equal to zero. The web slenderness requirement is satisfied and the section is compact.

6. *Calculate* M$_p$

$$\bar{y} = c = 196.4 \text{ mm}$$

$$M_p = \left[\frac{\bar{y}P_s}{t_s}\right]\left(\frac{\bar{y}}{2}\right) + [P_{rt}d_{rt} + P_cd_c + P_wd_w + P_td_t + P_{rb}d_{rb}]$$

where

d_{rt} = distance from PNA to centroid of top reinforcement

$$d_{rt} = 196.4 - 76.95 = 119.45 \text{ mm}$$

d_{rb} = distance from PNA to centroid of bottom reinforcement

$$d_{rb} = 196.4 - 156 = 40.4 \text{ mm}$$

d_w = distance from PNA to centroid of web

$$d_w = 1500/2 + 15 + 25 + 205 - 196.4 = 798.6 \text{ mm}$$

d_t = distance from PNA to centroid of tension flange

$$d_t = 25/2 + 1500 + 15 + 25 + 205 - 196.4 = 1561.1 \text{ mm}$$

d_c = distance from PNA to centroid of compression flange

$$d_c = 15/2 + 25 + 205 - 196.4 = 41.1 \text{ mm}$$

$$M_p = \left[\frac{(196.4)(9248)}{205}\right]\left(\frac{196.4}{2}\right)$$

$$+ \begin{bmatrix}(360)(119.45) + (1552.5)(41.1) \\ + (5175)(798.6) + (3450)(1561.1) + (560)(40.4)\end{bmatrix}$$

$$M_p = 10.518 \times 10^6 \text{ kN mm} = 10\,518 \text{ kN m} > 5418 \text{ kN m}$$
from Table E8.3-2 (at Location 104)

Therefore, the section is adequate.

a. Compression flange slenderness
   Check compression flange slenderness [A6.10.5.2.3c]. This article applies to sections in negative flexure. There is no strength limit state requirement for compact sections in positive flexure [A6.10.5.2.2a].

b. Compression flange bracing
   Check compression flange bracing [A6.10.5.2.3d]. This article applies to sections in negative flexure. There is no strength limit state requirement for compact sections in positive flexure. However, the engineer should understand that during the placement of deck concrete the section will be noncomposite and unbraced.

7. *Flexural Resistance* Calculate the flexural resistance provided by the section [A6.10.5.2.2a]. For continuous spans with noncompact interior support sections, the nominal flexural resistance of the section is computed by the following equation:

$$M_n = 1.3R_hM_y \le M_p$$

where

$$R_h = 1.0$$

M_y = yield moment of the critical flange, tension flange in this case

If we refer to Appendix A6.2 in AASHTO (1994), the yield moment of a composite section is calculated below.

First calculate the additional stress required to cause yielding in the tension flange:

$$f_{AD} = F_y - (f_{D1} + f_{D2} + f_{D3})$$

where

F_y = the minimum yield strength of the tension flange

f_{D1} = the stress caused by the permanent load before the concrete attains 75% of its 28-day strength applied to the steel section alone, calculated in Table E8.3-16.

f_{D2} and f_{D3} = Stresses obtained from applying the remaining permanent loads to the long term composite section, calculated in Table E8.3-16.

$$f_{AD} = 345 - (70.3 + 13.2 + 5.6) = 255.9 \text{ MPa}$$

which corresponds to an additional moment:

$$M_{AD} = f_{AD} \times S_{ST}$$

where

S_{ST} = The section modulus for the short term composite section, where $n = 8$, from Section 8.11.3, Part J.1.

$$M_{AD} = (255.9)(24.069) = 6158 \text{ kN m}$$

$$M_y = M_{D1} + M_{D2} + M_{D3} + M_{AD}$$

where M_{D1}, M_{D2}, and M_{D3} are factored moments from Table E8.3-1 at Location 104

$$M_y = 1180 + 294 + 125 + 6158 = 7757 \text{ kN m}$$

$$M_n = 1.3R_h M_y = 1.3(1.0)(7757) = 10\,083 \text{ kN m}$$

$$M_n < M_p = 10\,518 \text{ kN m} \quad \text{therefore } M_n = 10\,083 \text{ kN m}$$

$$M_r = \phi_f M_n = 1.0(10\,083) = 10\,083 \text{ kN m}$$

$M_r > M_u = 5418$ kN m at Location 104 from Table E8.3-1, therefore, the section provides adequate flexural strength.

8. *Positive Flexure Ductility* The final step is to check the ductility requirement for compact composite sections in positive flexure [A6.10.5.2.2b]. The purpose of this requirement is to make sure that the tension flange of the steel section will reach strain hardening before the concrete in the slab crushes. This article only applies if the moment due to the factored loads results in a flange stress which exceeds the yield strength of the flange. If the stress due to the moments does not exceed the yield strength, then the section is considered adequate. The reason being that there will not be enough strain in the steel at or below the yield strength for crushing of the concrete to occur in the slab. From Table E8.3-16 it is shown that the flange stress in the tension flange is 246.6 MPa, which is less than the yield strength of the flange, 345 MPa, therefore the section is adequate.

All the requirements for flexure have been satisfied. Therefore use a section consisting of a 15-mm × 300-mm top flange, a 10-

mm × 1500-mm web, and a 25-mm × 400-mm bottom flange for the positive moment region.

9. *Transition Points* Transition points from the sections in positive moment regions to the sections in negative moment regions shall take place at the dead load inflection points. These locations are chosen because composite action is not considered to be developed. The permanent load inflection points were determined when the effective lengths were calculated in Section 8.11.3, Part H.3.c. Going from left to the right of the bridge, inflection points exist at the following locations:

$$x = 22.7 \text{ m}, 37.8 \text{ m}, 58.2 \text{ m}, 73.3 \text{ m}$$

The web size is a constant throughout the bridge. Only the flange size changes at the inflection points.

K. Shear Design

In general, the factored shear resistance of a girder, V_r, is taken as follows:

$$V_r = \phi_v V_n$$

where

V_n = nominal shear resistance for the stiffened web

ϕ_v = resistance factor for shear = 1.0

1. *Stiffened Web* Interior web panels of homogeneous girders without longitudinal stiffeners and with a transverse stiffener spacing not exceeding $3D$, are considered stiffened.

$$3D = 3(1500) = 4500 \text{ mm}$$

The nominal resistance of stiffened webs is given in AASHTO [A6.10.7.3.2].

a. Handling requirements
For web panels without longitudinal stiffeners, transverse stiffeners are required if

$$\frac{D}{t_w} > 150$$

$$\frac{1500}{10} = 150$$

therefore transverse stiffeners are not required for handling; however, this example will demonstrate the design of transverse stiffeners as if they were required.

Maximum spacing of the transverse stiffeners is

$$d_o \leq D \left[\frac{260}{(D/t_w)} \right]^2 = 1500 \left(\frac{260}{150} \right)^2 = 4506.7 \text{ mm}$$

b. Homogeneous sections
The requirements for homogeneous sections are in AASHTO [A6.10.7.3.3]. The purpose of this section is to determine the maximum spacing of the stiffeners while maintaining adequate shear strength within the panel. Three separate sections must be examined.

   a. End panels
   b. Interior panels for the composite section in the positive moment region
   c. Interior panels for the non-composite section in the negative moment region

From analysis, the interior girders receive the largest shear force values (Table E8.3-1).

2. *End Panels* In end panels tension field action is not permitted to be used. Only buckling strength is available in end panels. The nominal shear resistance of an end panel is confined to either the shear yield or shear buckling force.

$$V_n = CV_p$$

for which

$$V_p = 0.58 F_{yw} D t_w = 0.58(345)(1500)(10)/1000 = 3002 \text{ kN}$$

where

C = ratio of the shear buckling stress to the shear yield strength

k = shear buckling coefficient

The ratio, C, is determined [A6.10.7.3.3a] as follows:

If $\dfrac{D}{t_w} < 1.10 \sqrt{\dfrac{Ek}{F_{yw}}}$ then $C = 1.0$

If
$$1.10\sqrt{\frac{Ek}{F_{yw}}} \leq \frac{D}{t_w} \leq 1.38\sqrt{\frac{Ek}{F_{yw}}} \quad \text{then}$$

$$C = \frac{1.10}{\frac{D}{t_w}}\sqrt{\frac{Ek}{F_{yw}}}$$

If
$$\frac{D}{t_w} > 1.38\sqrt{\frac{Ek}{F_{yw}}} \quad \text{then}$$

$$C = \frac{1.52}{(D/t_w)^2}\left(\frac{Ek}{F_{yw}}\right) \leq 0.8$$

for which

$$k = 5 + \frac{5}{(d_o/D)^2}$$

For the end panels V_u equals 991 kN at Location 100, taken from Table E8.3-3. Assume

$$\frac{D}{t_w} > 1.38\sqrt{\frac{Ek}{F_{yw}}}$$

therefore,

$$C = \frac{1.52}{(D/t_w)^2}\left(\frac{Ek}{F_{yw}}\right)$$

$$V_n = \frac{1.52}{(D/t_w)^2}\left(\frac{Ek}{F_{yw}}\right)V_p$$

Solving for k in the equation above

$$k_{min} = \frac{V_n F_{yw}}{1.52 V_p E}\left(\frac{D}{t_w}\right)^2 \quad \text{use} \quad V_n = \frac{V_u}{\phi} = \frac{991}{1.0} = 991 \text{ kN}$$

$$k_{min} = \frac{(991)(345)}{1.52(3002)(200\ 000)}\left(\frac{1500}{10}\right)^2 = 8.428$$

$$k = 5 + \frac{5}{\left(\dfrac{d_o}{D}\right)^2}$$

therefore, maximum $d_o = 1811.6$ mm

Checking assumption on D/t_w for $d_o = 1500$ mm

$$k = 5 + \frac{5}{(1500/1500)^2} = 10$$

$$1.38 = \sqrt{\frac{Ek}{F_{yw}}} = 1.38\sqrt{\frac{(200\ 000)10}{345}} = 105$$

$$\frac{D}{t_w} = \frac{1500}{10} = 150 > 105, \text{ assumption OK}$$

Place stiffeners 1500 mm apart.

3. *Interior Panels of Compact Sections*   For this design, this section applies to the positive moment region only. The region considered shall be the effective lengths for the uniform unit load. These effective lengths were determined in Section 8.11.3, Part H.3.c. First, consider the 30-m span. The nominal shear resistance from (Table 8.20) [A6.10.7.3.3a] is taken as

If $M_u \leq 0.5\phi_f M_p$ then

$$V_n = V_p\left[C + \frac{0.87(1 - C)}{\sqrt{1 + \left(\dfrac{d_o}{D}\right)^2}}\right]$$

If $M_u > 0.5\phi_f M_p$ then

$$V_n = RV_p\left[C + \frac{0.87(1 - C)}{\sqrt{1 + \left(\dfrac{d_o}{D}\right)^2}}\right] \geq CV_p$$

for which

$$R = \left[0.6 + 0.4\left(\frac{M_r - M_u}{M_r - 0.75\phi_f M_y}\right)\right] \leq 1.0$$

$$V_p = 0.58F_{yw}Dt_w = 3002 \text{ kN}$$

where

M_u = maximum moment in the panel under consideration due to the factored loads, from Table E8.3-2 at Location 104 = 5418 kN m

V_n = nominal shear resistance (kN)

V_p = plastic shear force (kN) = 3002 kN

M_r = factored flexural resistance (kN m)

ϕ_f = resistance factor for flexure = 1.0

M_y = yield moment (kN m)

D = web depth (mm)

d_o = stiffener spacing (mm)

C = ratio of the shear buckling stress to the shear yield strength

$0.50\phi_f M_p = 0.50(1.0)(10\,518) = 5259$ kN m $< M_u = 5418$ kN m

therefore,

$$V_n = RV_p\left[C + \frac{0.87(1 - C)}{\sqrt{1 + (d_o/D)^2}}\right] \geq CV_p$$

Calculate the minimum spacing for the panels for the compact section using the maximum shear values. For the 30-m span the compact section exists for the first 22.66 m of the bridge, from Section 8.11.3, Part H.3.c. For design, the length of compact section will be taken as 22 m. Because the first interior stiffener is 1.5 m from the end of the bridge and the second is 1.5 m from the first, from Section 8.11.3, Part K.1, stiffeners must be spaced equally across the remaining 19 m.

The second stiffener is 3 m from the end of the bridge, which corresponds to Location 101. From Table E8.3-3,

V_u = 806 kN at 3 m from the ends of the bridge

M_y = 7757 kN m from Section 8.11.3, Part J.5

M_r = 10 083 kN m from Section 8.11.3, Part J.5

$$R = \left[0.6 + 0.4\left(\frac{10\,083 - 5418}{10\,083 - 0.75(1.0)(7757)}\right)\right] = 1.0375 > 1.0$$

therefore use $R = 1.0$

Assume

$$\frac{D}{t_w} > 1.38\sqrt{\frac{Ek}{F_{yw}}}$$

therefore

$$C = \frac{1.52}{(D/t_w)^2}\left(\frac{Ek}{F_{yw}}\right)$$

Try six equal spacings of 3250 mm over a length of $6(3.25) = 19.5$ m

$$k = 5 + \frac{5}{\left(\dfrac{3250}{1500}\right)^2} = 6.07$$

$$1.38\sqrt{\frac{(200\,000)(6.07)}{345}} = 82 < \frac{D}{t_w} = 150, \text{ assumption OK}$$

so that

$$C = \frac{1.52}{\left(\dfrac{1500}{10}\right)^2}\left[\frac{(200\,000)(6.07)}{345}\right] = 0.238$$

$$V_n = 1.0(3002)\left[0.238 + \frac{0.87(1 - 0.238)}{\sqrt{1 + \left(\dfrac{3250}{1500}\right)^2}}\right] = 1548 \text{ kN}$$

$$\phi_v V_n = 1.0(1548) = 1548 \text{ kN} > V_u = 806 \text{ kN, OK}$$

Space transverse intermediate stiffeners at 3250 mm from $x = 3$ m to $x = 22.5$ m along the 30-m span.

Consider the 36-m span for which a compact section exists from points $x = 37.8$ m to $x = 58.2$ m along the bridge. Therefore space transverse intermediate stiffeners equally along the 20.4-m distance.

$$0.5\phi_f M_p = 0.5(1.0)(10\,518) = 5259 \text{ kN m} > M_u$$

$$= 5104 \text{ kN m (from Table E8.3-2 at Location 205)}$$

therefore

$$V_n = V_p \left[C + \frac{0.87(1 - C)}{\sqrt{1 + (d_o/D)^2}} \right]$$

At $x = 37.8$ m, which corresponds to Location 202.17, from Table E8.3-3,

$$V_u = -\frac{(892 - 699)}{(203 - 202)} (202.17 - 202) + 892 = 859 \text{ kN}$$

Assume

$$\frac{D}{t_w} > 1.38 \sqrt{\frac{Ek}{F_{yw}}}$$

therefore

$$C = \frac{1.52}{\left(\dfrac{D}{t_w}\right)^2} \left(\frac{Ek}{F_{yw}}\right)$$

Try eight stiffener spacings of 2.55 m over a length of $8(2.55) = 20.4$ m

$$k = 5 + \frac{5}{\left(\dfrac{2550}{1500}\right)^2} = 6.73$$

$$1.38 \sqrt{\frac{(200\ 000)(6.73)}{345}} = 86 < \frac{D}{t_w} = 150, \text{ assumption OK}$$

$$C = \frac{1.52}{\left(\dfrac{1500}{10}\right)^2} \left(\frac{(200\ 000)(6.73)}{345}\right) = 0.264$$

$$V_n = (3002) \left[0.264 + \frac{0.87(1 - 0.264)}{\sqrt{1 + \left(\dfrac{2550}{1500}\right)^2}} \right] = 1777 \text{ kN}$$

$$\phi_v V_n = 1.0(1777) = 1777 \text{ kN} > V_u = 859 \text{ kN OK}$$

Space transverse intermediate stiffeners at 2.55 m from $x = 37.8$ m– 58.2 m.

4. *Interior Panels of Noncompact Sections*   This section applies to the negative moment region only. The region considered shall be the effective lengths for the uniform unit load. These effective lengths were determined in Section 8.11.3, Part H.3.c. The noncompact section exists between $x = 22$ m to $x = 37.8$ m and $x = 58.2$ m–74 m. Therefore, there is 8 m on the 30-m span and 7.8 m on the 36-m span from the bearing stiffener over the support to the assumed inflection point. The nominal shear resistance is taken as (Table 8.20)

If $f_u \leq 0.75\phi_f F_y$ then

$$V_n = V_p\left[C + \frac{0.87(1 - C)}{\sqrt{1 + (d_o/D)^2}}\right]$$

If $f_u > 0.75\phi_f F_y$ then

$$V_n = RV_p\left[C + \frac{0.87(1 - C)}{\sqrt{1 + (d_o/D)^2}}\right] \geq CV_p$$

for which

$$R = \left[0.6 + 0.4\left(\frac{F_r - f_u}{F_r - 0.75\phi_f F_y}\right)\right] \leq 1.0$$

$$V_p = 0.58F_{yw}Dt_w = 0.58(345)(1500)(10) = 3002 \text{ kN}$$

where

f_u = maximum stress in the compression flange in the panel under consideration due to the factored loading (MPa)

f_u = 300.4 MPa from Table E8.3-10

F_r = Factored flexural resistance of the compression flange for which f_u was determined

F_r = 341.9 MPa from Section 8.11.3, Part I.7

C = Ratio of shear buckling stress to the shear yield strength

f_u = 300.4 > $0.75\phi_f F_y$ = 0.75(1.0)(345) = 258.8 MPa

therefore,

$$V_n = RV_p\left[C + \frac{0.87(1 - C)}{\sqrt{1 + (d_o/D)^2}}\right] \geq CV_p$$

The stiffener spacing along the 30-m span is determined first. A spacing of 1500 mm will be used for all stiffeners. The spacing is checked for adequacy below. The maximum shear in the panel from $x = 22.5$ m–$x = 30$ m is at $x = 30$ m, or Location 110. $V_u = 1277$ kN from Table E8.3-3

Assume

$$\frac{D}{t_w} > 1.38\sqrt{\frac{Ek}{F_{yw}}}$$

therefore,

$$C = \frac{1.52}{(D/t_w)^2}\left(\frac{Ek}{F_{yw}}\right)$$

$$k = 5 + \frac{5}{\left(\dfrac{1500}{1500}\right)^2} = 10$$

$$1.38\sqrt{\frac{(200\ 000)(10)}{345}} = 105 < \frac{D}{t_w} = 150,\ \text{assumption OK}$$

$$C = \frac{1.52}{\left(\dfrac{1500}{10}\right)^2}\left[\frac{(200\ 000)(10)}{345}\right] = 0.392$$

$$R = \left[0.6 + 0.4\left(\frac{340.9 - 316.2}{340.9 - 258.8}\right)\right] = 0.720$$

$$V_n = (0.720)(3002)\left[0.392 + \frac{0.87(1 - 0.392)}{\sqrt{1 + \left(\dfrac{1500}{1500}\right)^2}}\right] = 1656\ \text{kN}$$

$$\phi_v V_n = 1.0(1656) = 1656\ \text{kN} > V_u = 1277\ \text{kN, OK}$$

Space transverse intermediate stiffeners at 1500 mm from $x = 22.5$ m–$x = 30$ m along the 30-m span.

Determine stiffener spacing along 36-m span. Space stiffeners equally along the 7.8 m length from $x = 30$ m–37.8 m. The maximum shear in this panel is at $x = 30$ m, or Location 200.

$V_u = 1271$ kN from Table E8.3-3

Assume

$$\frac{D}{t_w} > 1.38\sqrt{\frac{Ek}{F_{yw}}}$$

therefore,

$$C = \frac{1.52}{(D/t_w)^2}\left(\frac{Ek}{F_{yw}}\right)$$

Try four stiffener spacings of 1950 mm over a length of $4(1.95) = 7.8$ m.

$$k = 5 + \frac{5}{\left(\frac{1950}{1500}\right)^2} = 7.96$$

$$1.38\sqrt{\frac{(200\ 000)(7.96)}{345}} = 93.7 < \frac{D}{t_w} = 150,\ \text{OK}$$

$$C = \frac{1.52}{\left(\frac{1500}{10}\right)^2}\left[\frac{(200\ 000)(7.96)}{345}\right] = 0.312$$

$$V_n = (0.720)(3002)\left[0.312 + \frac{0.87(1 - 0.312)}{\sqrt{1 + \left(\frac{1950}{1500}\right)^2}}\right] = 1463\ \text{kN}$$

$$\phi_v V_n = 1.0(1463) = 1463\ \text{kN} > V_u = 1271\ \text{kN, OK}$$

Space transverse intermediate stiffeners at 1950 mm from $x = 30$–37.8 m along the 36-m span.

5. *Check* D_c/t_w *for Shear* The requirement involves calculating the maximum elastic shear stress in the web [A6.10.4.4]. The check on the requirement is performed at this time since the stiffener spacing is incorporated in calculating the shear stress. Because the same web section is used throughout the design, the shear stress is checked for the web over the supports only. The reason being that the support is the location of maximum shear and the stiffener spacing at the ends of the bridge is small enough to provide a considerable amount of strength at those locations.

$$v_{cf} \leq 0.58 F_{yw}$$

where

F_{yw} = yield strength of the web (MPa)

C = ratio of shear buckling stress to shear yield strength [A6.10.7.3.3]

v_{cf} = maximum elastic shear stress in the web from Table E8.3-17 (at Location 110) due to the unfactored permanent loads of Table E8.3-1 and the fatigue loading. The fatigue loading is taken as twice that calculated using fatigue load combination in Table 3.1 [Table A3.4.1-1]. The unfactored values for the fatigue loading at the critical points are summarized in Table E8.3-6. The factored values with impact applied are tabulated in Table E8.3-18. It should be noted that the distribution factor used for the fatigue loading is that for single lane loaded without the multiple presence factor. From Section 8.11.3, Part G.5a the distribution factor for single lane loaded was determined to be 0.681 for the interior girder which now becomes $0.681/1.2 = 0.568$.

The maximum stiffener spacing at the supports is 1950 mm determined in Section 8.11.3, Part K.1.

$$k = 5 + \frac{5}{\left(\dfrac{1950}{1500}\right)^2} = 7.96$$

$$\frac{D}{t_w} = \frac{1500}{10} = 150 > 1.38\sqrt{\frac{(200\ 000)(7.96)}{345}} = 93.7$$

TABLE E8.3-17 **Maximum Shear Stress in the Web at Location 110**

Load	V_{D1} (kN)	V_{D2} (kN)	V_{D3} (kN)	V_{LL+IM} (kN)	A_{web} (mm²)	Stress (MPa)
D1	282				15 000	18.8
D2		75			15 000	5.0
D3			29		15 000	1.9
LL + IM				276	15 000	18.4
Total						44.1

$$C = \frac{1.52}{\left(\dfrac{1500}{10}\right)^2}\left[\frac{(200\ 000)(7.96)}{345}\right] = 0.312$$

$v_{cf} \leq 0.58(0.312)(345) = 62.4$ MPa, which is greater than the maximum calculated stress of 47.8 MPa from Table E8.3-17, therefore, the section is adequate. The dead loads in Table E8.3-17 were taken from Table E8.3-1 at Location 110 for the interior girder. The value for V_{LL+IM} is twice the value for negative shear at Location 110 from Table E8.3-18 because the maximum truck crossing the bridge could be twice the size of the fatigue truck [A6.10.4.2].

6. *Transverse Intermediate Stiffener Design* The LRFD specifications for stiffener design are located in AASHTO [A6.10.8]. Transverse intermediate stiffeners are composed of plates welded to either one or both sides of the web depending on the additional shear resistance the web needs. Transverse intermediate stiffeners used as connecting elements for diaphragms must extend the full depth of the web. If the stiffeners are not to be used as connecting elements, they must be welded against the compression flange but are not to be welded to the tension flange. The allowable distance between the end of the stiffener and the tension flange is between $4t_w$ and $6t_w$. Therefore, either cut or cope the transverse stiffeners $4t_w$ or 40 mm from the tension flange.

For this design, M270 Gr250 steel is used for the stiffeners. In locations where diaphragms are to be used, a stiffener is used on each side of the web as a connecting element. For the other locations a single plate will be welded to one side of the web only. The stiffeners are designed as columns made up of either one or two plates and a centrally located strip of web.

a. Single-plate transverse stiffeners

Single-plate transverse intermediate stiffenes are used at locations where there are no connecting elements. They shall be designed based on the maximum shear for the positive and negative moment regions. This use of maximum shear is a conservative approach. The fact that the stiffeners will have more than the required strength in some areas is negligible because the amount of steel saved by changing them would be small. The stiffener size chosen is 20 mm × 140 mm for both regions. The following requirements demonstrate the adequacy of this section.

b. Projecting width

The projecting width requirement is checked to prevent local buckling of the transverse stiffeners. The width of each projecting stiffener must meet the following requirements [A6.10.8.1.2].

TABLE E8.3-18   Factored Shears for Fatigue Limit State

Location	Positive Shear (kN) (Int. Girder)			Negative Shear (kN) (Int. Girder)		
	$LL + IM$ w/ $IM = 0.15$	$LL + IM$ w/ $DF = 0.568$	$LL + IM$ w/ $LF = 0.75$	$LL + IM$ w/ $IM = 0.15$	$LL + IM$ w/ $DF = 0.568$	$LL + IM$ w/ $LF = 0.75$
100	291	165	124	−35	−20	−15
104	126	72	54	−101	−57	−43
110	8	5	3	−324	−184	−138
200	324	184	138	−33	−19	−14
205	123	70	52	−124	−70	−53

$$50 + \frac{d}{30} \le b_t \le 0.48t_p\sqrt{\frac{E}{F_{ys}}}$$

and

$$16.0t_p \ge b_t \ge 0.25b_f$$

where

d = steel section depth (mm)

t_p = thickness of projecting element (mm)

F_{ys} = minimum yield strength of the stiffener (MPa)

b_f = full width of steel flange (mm)

For the positive moment regions

$$d = 1500 + 15 + 25 = 1540 \text{ mm}$$

$$t_p = 20 \text{ mm}$$

$$F_{ys} = 250 \text{ MPa}$$

$$b_f = 300 \text{ mm (compression flange)}$$

$$50 + \frac{1540}{30} = 101 \text{ mm} \le b_t = 140 \text{ mm} \le 0.48(20)\sqrt{\frac{200\ 000}{250}}$$

$$= 272 \text{ mm OK}$$

and

$$16.0(20) = 320 \text{ mm} \ge b_t = 140 \text{ mm} \ge 0.25(300)$$

$$= 75 \text{ mm, OK}$$

For the negative moment regions

$$d = 1500 + 30 + 30 = 1560 \text{ mm}$$

$$t_p = 20 \text{ mm}$$

$$F_{ys} = 250 \text{ MPa}$$

$$b_f = 400 \text{ mm}$$

$$50 + \frac{1560}{30} = 102 \text{ mm} \leq b_t = 140 \text{ mm} \leq 0.48(20)\sqrt{\frac{200\,000}{250}}$$

$$= 272 \text{ mm, OK}$$

and

$$16(20) = 320 \text{ mm} \geq b_t = 140 \text{ mm} \geq 0.25(400)$$

$$= 100 \text{ mm, OK}$$

c. Moment of inertia
The moment of inertia of all transverse stiffeners must meet the following requirement [A6.10.8.1.3]:

$$I_t \geq d_o t_w^3 J$$

for which

$$J = 2.5\left(\frac{D_p}{d_o}\right)^2 - 2.0 \geq 0.5$$

where

I_t = moment of inertia of the transverse stiffener taken about the edge in contact with the web for single stiffeners and about the mid thickness of the web for stiffener pairs (mm⁴)

d_o = transverse stiffener spacing (mm)

D_p = web depth for webs without longitudinal stiffeners (mm)

For the positive moment regions use d_o equal to 3250 mm and for the negative moment regions use d_o equal to 1950 mm.
The moment of inertia for a single stiffener, 20 mm × 140 mm, is shown in Figure E8.3-38.

$$I = I_o + Ad^2$$

$$I = \frac{(20)(140)^3}{12} + (20)(140)(70)^2$$

$$= 18.293 \times 10^6 \text{ mm}^4$$

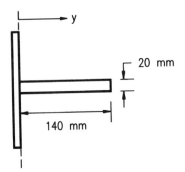

Fig. E8.3-38   Single plate transverse intermediate stiffener.

For positive moment regions:

$$J = 2.5\left(\frac{1500}{3250}\right)^2 - 2.0$$

$$= -1.47 < 0.5 \quad \text{therefore, use } J = 0.5$$

$$I_t = 18.293 \times 10^6 \geq (3250)(10)^3(0.5)$$
$$= 1.63 \times 10^6 \text{ mm}^4$$
$$\text{therefore, section is adequate}$$

For negative moment regions:

$$J = 2.5\left(\frac{1500}{1950}\right)^2 - 2.0$$

$$= -0.52 < 0.5 \quad \text{therefore, use } J = 0.5$$

$$I_t = 18.293 \times 10^6 \geq (1950)(10)^3(0.5)$$
$$= 0.975 \times 10^6 \text{ mm}^4$$
$$\text{therefore, section is adequate}$$

d. Area

The transverse stiffeners need to have enough area to resist the vertical component of the tension field. The following requirement applies [A6.10.8.1.4]:

$$A_s \geq \left[0.15 BDt_w(1 - C)\frac{V_u}{V_r} - 18.0t_w^2\right]\left(\frac{F_{yw}}{F_{ys}}\right)$$

where

V_r = factored shear resistance calculated in Section 8.11.3, Part K.1 for the compact sections and for the noncompact sections (kN)

V_u = shear due to factored loads at the strength limit state taken from Table E8.3-3 (kN)

A_s = stiffener area, total area for both stiffeners for pairs (mm²)

$$B = 1.0 \text{ for stiffener pairs}$$

$$B = 2.4 \text{ for single plate stiffeners}$$

C = ratio of shear buckling stress to the shear yield strength from Section 8.11.3, Parts K.3 and K.4 for the composite sections and for the noncomposite sections

F_{yw} = minimum yield strength of the web (MPa)

F_{ys} = minimum yield strength of the stiffener (MPa)

The following values were determined in Section 8.11.3, Part K.3 for compact sections in the positive moment regions, checking both the 30-m and 36-m spans.

For the 30-m span,

$$C = 0.238$$

$$V_u = 806 \text{ kN}$$

$$V_r = 1420 \text{ kN}$$

$$A_s = (20)(140) = 2800 \text{ mm}^2$$

$$A_s \geq \left[0.15(2.4)(1500)(10)(1 - 0.238)\left(\frac{806}{1420}\right) \right.$$

$$\left. - 18.0(10)^2 \right] \frac{345}{250} = 739 \text{ mm}^2$$

$A_s = 2800 \text{ mm}^2 > 739 \text{ mm}^2$ therefore, section is adequate

For the 36-m span,

$$C = 0.264$$

$$V_u = 859 \text{ kN}$$

$$V_r = 1777 \text{ kN}$$

$$A_s = (20)(140) = 2800 \text{ mm}^2$$

$$A_s \geq \left[0.15(2.4)(1500)(10)(1 - 0.264)\left(\frac{859}{1777}\right) \right.$$

$$\left. - 18.0(10)^2 \right] \frac{345}{250} = 167.7 \text{ mm}^2$$

$A_s = 2800 \text{ mm}^2 > 167.7 \text{ mm}^2$ therefore, section is adequate

The following values were determined in Section 8.11.3, Part K.4 for noncompact sections in the negative moment regions, checking both the 30-m and 36-m spans.

For the 30-m span,

$$C = 0.392$$

$$V_u = 1277 \text{ kN}$$

$$V_r = 1608 \text{ kN}$$

$$A_s = (20)(140) = 2800 \text{ mm}^2$$

$$A_s \geq \left[0.15(2.4)(1500)(10)(1 - 0.392)\left(\frac{1277}{1608}\right) \right.$$

$$\left. - 18(10)^2 \right] \frac{345}{250} = 1114 \text{ mm}^2$$

$A_s = 2800 \text{ mm}^2 > 1114 \text{ mm}^2$ therefore, section is adequate

For the 36-m span,

$$C = 0.312$$

$$V_u = 1271 \text{ kN}$$

$$V_r = 1463 \text{ kN}$$

$$A_s = (20)(140) = 2800 \text{ mm}^2$$

$$A_s \geq \left[0.15(2.4)(1500)(10)(1 - 0.312)\left(\frac{1271}{1463}\right) \right.$$

$$\left. - 18(10)^2 \right] \frac{345}{250} = 1970 \text{ mm}^2$$

$A_s = 2800$ mm$^2 > 1970$ mm^2 therefore, section is adequate

Therefore, use 20 mm × 140 mm single-plate transverse intermediate stiffeners where no connecting elements are present.

7. *Double-plate Transverse Stiffener Design*   Double-plate transverse intermediate stiffeners are used at locations where connecting elements such as diaphragms are used. For this design, they shall be based on the maximum shear for the positive and negative moment regions respectively. This approach is conservative. The fact that the stiffeners will have more than the required strength in some areas is negligible because the amount of steel saved by changing them would be small. A pair of stiffeners, 12 mm × 100 mm is chosen for both regions. The following requirements demonstrate the adequacy of this section.

a. Projecting width

The projecting width requirement is checked to prevent local buckling of the transverse stiffeners. The width of each projecting stiffener must meet the following requirements [A6.10.8.1.2].

$$50 + \frac{d}{30} \leq b_t \leq 0.48t_p\sqrt{\frac{E}{F_{ys}}}$$

and

$$16.0t_p \geq b_t \geq 0.25b_f$$

For the positive moment regions

$$d = 1500 + 15 + 25 = 1540 \text{ mm}$$

$$t_p = 12 \text{ mm}$$

$$F_{ys} = 250 \text{ MPa}$$

$$b_f = 300 \text{ mm}$$

$$50 + \frac{1540}{30} = 101 \text{ mm} \approx b_t = 100 \text{ mm} \leq 0.48(12)\sqrt{\frac{200\,000}{250}}$$

$$= 163 \text{ mm, OK}$$

and

$$16.0(12) = 192 \text{ mm} \geq b_t = 100 \text{ mm} \geq 0.25(300)$$

$$= 75 \text{ mm, OK}$$

For the negative moment regions

$$d = 1500 + 30 + 30 = 1560 \text{ mm}$$

$$t_p = 12 \text{ mm}$$

$$F_{ys} = 250 \text{ MPa}$$

$$b_f = 400 \text{ mm}$$

$$50 + \frac{1560}{30} = 102 \text{ mm} \approx b_t = 100 \text{ mm} \leq 0.48(12)\sqrt{\frac{200\,000}{250}}$$

$$= 163 \text{ mm, OK}$$

and

$$16.0(12) = 192 \text{ mm} \geq b_t = 100 \text{ mm} \geq 0.25(400)$$

$$= 100 \text{ mm, OK}$$

b. Moment of inertia
The moment of inertia of all transverse stiffeners must meet the following requirement [A6.10.8.1.3]:

$$I_t \geq d_o t_w^3 J$$

for which

$$J = 2.5\left(\frac{D_p}{d_o}\right)^2 - 2.0 \geq 0.5$$

For the positive moment regions use d_o equal to 3250 mm and for the negative moment regions use d_o equal to 1950 mm.

The moment of inertia for a pair of stiffeners, 12 mm × 100 mm, taken about the middle of the web, is shown in Figure E8.3-39.

$$I = I_o + Ad^2$$

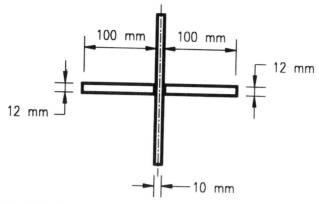

Fig. E8.3-39 Double plate transverse intermediate stiffener.

$$I = \frac{2(12)(100)^3}{12} + 2(12)(100)(55)^2$$

$$= 9.260 \times 10^6 \text{ mm}^4$$

For positive moment regions

$$J = 2.5\left(\frac{1500}{3250}\right)^2 - 2.0$$

$$= -1.47 < 0.5 \quad \text{therefore, use } J = 0.5$$

$$I_t = 9.260 \times 10^6 \geq (3250)(10)^3(0.5)$$
$$= 1.63 \times 10^6 \text{ mm}^4$$
therefore, section is adequate

For negative moment regions:

$$J = 2.5\left(\frac{1500}{1950}\right)^2 - 2.0$$

$$= -0.52 < 0.5 \quad \text{therefore, use } J = 0.5$$

$$I_t = 9.260 \times 10^6 \geq (1950)(10)^3(0.5)$$
$$= 0.975 \times 10^6 \text{ mm}^4$$
therefore, section is adequate

c. Area

The transverse stiffeners need to have enough area to resist the vertical component of the tension field. The following requirement applies [A6.10.8.1.4]:

$$A_s \geq \left[0.15BDt_w(1 - C)\frac{V_u}{V_r} - 18.0t_w^2\right]\left(\frac{F_{yw}}{F_{ys}}\right)$$

The following values were determined in Section 8.11.3, Part K.3 for compact sections in the positive moment regions, checking both spans.

For the 30-m span,

$$C = 0.238$$
$$V_u = 806 \text{ kN}$$
$$V_r = 1420 \text{ kN}$$
$$A_s = 2(12)(100) = 2400 \text{ mm}^2$$
$$A_s \geq \left[0.15(1.0)(1500)(10)(1 - 0.238)\left(\frac{806}{1420}\right) - 18.0(10)^2\right]$$
$$\times \frac{345}{250} = -1141 \text{ mm}^2$$

For the 36-m span

$$C = 0.264$$
$$V_u = 859 \text{ kN}$$
$$V_r = 1777 \text{ kN}$$
$$A_s = 2(12)(100) = 2400 \text{ mm}^2$$
$$A_s \geq \left[0.15(1.0)(1500)(10)(1 - 0.264)\left(\frac{859}{1777}\right) - 18.0(10)^2\right]$$
$$\times \frac{345}{250} = -1379 \text{ mm}^2$$

Because the values are negative, the required area is zero. This implies that the web can resist the vertical component of the tension field by itself, for both spans. Therefore the stiffener only has to meet the requirements for projecting width and moment of inertia, which have been satisfied.

The following values were determined in Section 8.11.3, Part K.4 for noncompact sections in the negative moment regions,

checking both spans.
For the 30-m span

$$C = 0.392$$
$$V_u = 1277 \text{ kN}$$
$$V_r = 1608 \text{ kN}$$
$$A_s = 2(12)(100) = 2400 \text{ mm}^2$$

$$A_s \geq \left[0.15(1.0)(1500)(10)(1 - 0.392)\left(\frac{1277}{1608}\right) - 18(10)^2 \right]$$
$$\times \frac{345}{250} = -985 \text{ mm}^2$$

For the 36-m span

$$C = 0.312$$
$$V_u = 1271 \text{ kN}$$
$$V_r = 1463 \text{ kN}$$
$$A_s = 2(12)(100) = 2400 \text{ mm}^2$$

$$A_s \geq \left[0.15(1.0)(1500)(10)(1 - 0.312)\left(\frac{1271}{1463}\right) - 18(10)^2 \right]$$
$$\times \frac{345}{250} = -628 \text{ mm}^2$$

Because the values are negative, the required area is zero. This implies that the web can resist the vertical component of the tension field by itself, for both spans. Therefore the stiffener only has to meet the requirements for projecting width and moment of inertia, which have been satisfied.

Therefore, use two plates, 12 mm × 100 mm, for transverse intermediate stiffeners where connecting elements are present.

8. *Bearing Stiffeners at Abutments*   The requirements for bearing stiffeners are taken from the LRFD specifications [A6.10.8.2]. Bearing stiffeners shall be placed on the webs of plate girders at all bearing locations and at locations of concentrated loads.

The purpose of bearing stiffeners is to transmit the full bearing force from the factored loads. They consist of one or more plates welded to each side of the web and extend the full length of the web. It is also desirable to extend them to the outer edges of the flanges. At the abutments the bearing stiffeners chosen consist of one 20 mm × 150 mm plate on each side of the web. The following requirements demonstrate the adequacy of this section.

a. Projecting width

In order to prevent local buckling of the bearing stiffener plates, the width of each projecting element has to satisfy the following [A6.10.8.2.2]:

$$b_t \le 0.48 t_p \sqrt{\frac{E}{F_{ys}}}$$

$$b_t = 150 \text{ mm} \le 0.48(20) \sqrt{\frac{200\,000}{250}} = 272 \text{ mm} \quad \text{therefore, OK}$$

b. Bearing resistance

In order to get the bearing stiffener tight against the flanges, a portion in the corner must be clipped. This clipping of the stiffener is so the fillet welding of the flange and web plates can be done. By clipping the stiffener the bearing area of the stiffener is reduced (see Fig. E8.3-40). When determining the bearing resistance, this reduced bearing area must be used. The factored bearing resistance is calculated below [A6.10.8.2.3].

$$B_r = \phi_b A_{pn} F_{ys}$$

where

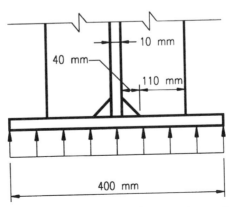

Fig. E8.3-40   Bearing stiffener at abutment.

A_{pn} = the contact area of the stiffener on the flange

$A_{pn} = 2(150 - 40)(20) = 4400$ mm²

$B_r = 1.0(4400)(250) = 1.100 \times 10^6$ N

$= 1100$ kN $> V_u = 991$ kN   therefore, OK

The value for V_u, which is equal to the reaction at the abutments, is taken from Table E8.3-3 at Location 100 (interior girder controls).

c. Axial resistance of bearing stiffeners
The axial resistance of bearing stiffeners is determined from AASHTO [A6.10.8.2.4]. The factored axial resistance P_r for components in compression is taken as [A6.9.2.1]

$$P_r = \phi_c P_n$$

where

ϕ_c = resistance factor for compression = 0.9

P_n = nominal compressive resistance (kN)

In order to calculate the nominal compressive resistance, the section properties must be determined. The radius of gyration is computed about the center of the web and the effective length is considered to be 0.75D, where D is the web depth. The reason the effective length is reduced is because of the end restraint provided by the flanges against column buckling.

d. Effective section
For stiffeners welded to the web (Fig. E8.3-41), the effective column section consists of the 20-mm × 150-mm stiffeners

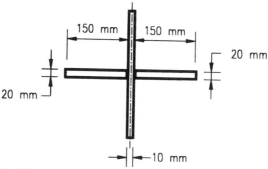

Fig. E8.3-41   Section of bearing stiffener at abutment.

[A6.10.8.2.4b].
The radius of gyration, r_s, is computed from the values in Table E8.3-19.

$$I = I_o + Ad^2 = (11.250 \times 10^6) + (38.400 \times 10^6)$$

$$= 49.650 \times 10^6 \text{ mm}^4$$

$$r_s = \sqrt{\frac{I}{A}} = \sqrt{\frac{49.650 \times 10^6}{6000}} = 90.97 \text{ mm}$$

e. Slenderness
The limiting width to thickness ratio for axial compression must be checked [A6.9.4.2]. The limiting value is as follows:

$$\frac{b}{t} \leq k\sqrt{\frac{E}{F_y}}$$

where

k = plate buckling coefficient from Table 8.11 = 0.45

b = width of plate as specified in Table 8.11 = 150 mm

t = plate thickness = 20 mm

$$\frac{b}{t} = \frac{150}{20} = 7.5 \leq 0.45\sqrt{\frac{200\,000}{250}} = 12.7 \text{ therefore, OK}$$

f. Nominal compressive resistance
The nominal compressive resistance is taken from AASHTO [A6.9.4.1] since the stiffeners are noncomposite members. The value of P_n is dependent on λ as follows:

If

$$\lambda \leq 2.25$$

TABLE E8.3-19   Effective Section for Bearing Stiffeners Over the Abutments

Part	A	y	Ay	Ay^2	I_o
Stiffener	3000	80	240 000	19.200×10^6	5.625×10^6
Stiffener	3000	−80	−240 000	19.200×10^6	5.625×10^6
Total	6000			38.400×10^6	11.250×10^6

then

$$P_n = 0.66^\lambda F_y A_s$$

If

$$\lambda > 2.25$$

then

$$P_n = \frac{0.88 F_y A_s}{\lambda}$$

$$\lambda = \left(\frac{KL}{r_s \pi}\right)^2 \frac{F_y}{E}$$

where

A_s = gross cross-sectional area (mm²)

K = effective length factor = 0.75

L = unbraced length (mm)

r_s = radius of gyration about the plane of buckling (mm)

$$\lambda = \left[\frac{0.75(1500)}{90.97\pi}\right]^2 \frac{250}{200\,000} = 0.0194 < 2.25$$

$$P_n = 0.66^\lambda F_y A_s = 0.66^{0.0194}(250)(20)(150)(2)$$

$$= 1.488 \times 10^6 \text{ N} = 1488 \text{ kN}$$

$$P_r = 0.9(1488) = 1339 \text{ kN} > V_u = 991 \text{ kN}$$

therefore, section is adequate

For the bearing stiffeners at abutments, use a pair of plates, 20 mm × 150 mm.

9. *Bearing Stiffeners at the Interior Supports*

The requirements that were specified for the bearing stiffeners at the abutments apply also to this section. At the interior supports the bearing stiffeners consist of two 20-mm × 150-mm plates on each side of the web. The two plates are spaced 150 mm apart to allow for welding. The following requirements demonstrate the adequacy of this section.

a. Projecting width

In order to prevent local buckling of the bearing stiffener plates,

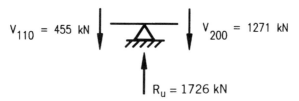

Fig. E8.3-42   Reaction at interior support.

the width of each projecting element has to satisfy the following [A6.10.8.2.2]

$$b_t \leq 0.48t_p \sqrt{\frac{E}{F_{ys}}}$$

$$b_t = 150 \text{ mm} \leq 0.48(20) \sqrt{\frac{200\,000}{250}} = 271.5 \text{ mm therefore, OK}$$

b. Bearing resistance
The factored bearing resistance is calculated below (see Fig. E8.3-40) [A6.10.8.2.3].

$$B_r = \phi_b A_{pn} F_{ys}$$

$$A_{pn} = 4(110)(20) = 8800 \text{ mm}^2$$

$$B_r = (1.0)(8800)(250) = 2.200 \times 10^6 \text{ N} = 2200 \text{ kN}$$

$$> R_u = 1726 \text{ kN, OK}$$

The value for R_u, which is equal to the reaction at either interior support (Fig. E8.3-42), is found by adding V_{110} and V_{200} from Table E8.3-3 (interior girder controls). This is a very conservative approximation because maximum values for shear due to the design truck are used. This approach was taken here for simplicity.

c. Axial resistance of bearing stiffeners
For components in compression, P_r is taken as [A6.9.2.1]

$$P_r = \phi_c P_n$$

The effective unbraced length is $0.75D = 0.75(1500) = 1125$ mm.

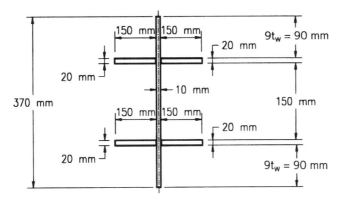

Fig. E8.3-43 Section of bearing stiffener at interior support.

d. Effective section

The effective section criteria is found in AASHTO [A6.10.8.2.4b]. For stiffeners welded to the web, the effective column section consists of the stiffeners plus a centrally located strip of web extending $9t_w$ to each side of the stiffeners as shown in Figure E8.3-43. The spacing of the stiffeners is 150 mm.

The radius of gyration r_s is computed from the values in Table E8.3-20.

$$I = I_o + Ad^2 = 22.5 \times 10^6 + 76.8 \times 10^6 = 99.3 \times 10^6 \text{ mm}^4$$

$$r_s = \sqrt{\frac{I}{A}} = \sqrt{\frac{99.3 \times 10^6}{15\,700}} = 79.53 \text{ mm}$$

e. Slenderness

The limiting width to thickness ratio for axial compression must be checked [A6.9.4.2]. The limiting value is as follows:

TABLE E8.3-20 Effective Section for Bearing Stiffeners Over the Interior Supports

Part	A	y	Ay	Ay^2	I_o
Web	3700	0	0	0	0.031×10^6
Stiffener	3000	80	240 000	19.2×10^6	5.625×10^6
Stiffener	3000	80	240 000	19.2×10^6	5.625×10^6
Stiffener	3000	−80	−240 000	19.2×10^6	5.625×10^6
Stiffener	3000	−80	−240 000	19.2×10^6	5.625×10^6
Total	15 700			76.8×10^6	22.5×10^6

$$\frac{b}{t} \le k\sqrt{\frac{E}{F_y}}$$

$$\frac{b}{t} = \frac{150}{20} = 7.5 < 0.45\sqrt{\frac{200\ 000}{250}} = 12.7; \text{ therefore, OK}$$

f. Nominal compressive resistance

The nominal compressive resistance is taken from AASHTO [A6.9.4.1] since the stiffeners are noncomposite members. The value of P_n is dependent on λ as follows:

If

$$\lambda \le 2.25$$

then

$$P_n = 0.66^{\lambda}F_y A_s$$

If

$$\lambda > 2.25$$

then

$$P_n = \frac{0.88 F_y A_s}{\lambda}$$

$$\lambda = \left(\frac{KL}{r_s \pi}\right)^2 \frac{F_y}{E}$$

$$\lambda = \left[\frac{0.75(1500)}{79.53\pi}\right]^2 \frac{250}{200\ 000} = 0.0253 < 2.25$$

$$P_n = 0.66^{\lambda}F_y A_s = 0.66^{0.0253}(250)(20)(150)(4)$$

$$= 2.969 \times 10^6 \text{ N} = 2969 \text{ kN}$$

$$P_r = 0.9(2969) = 2672 \text{ kN} > V_u$$

$$= 1726 \text{ kN; therefore, section is adequate.}$$

For the bearing stiffeners at the supports, use two pairs of stiffener plates, 20 mm $\times$ 150 mm.

A summary of the stiffener design is found in Figure E8.3-44.

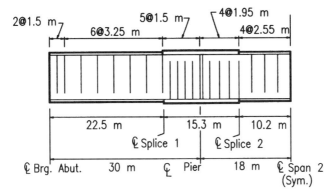

Fig. E8.3-44 Summary of stiffener design.

10. *Shear Connectors* Design of shear connectors is specified in AASHTO [A6.10.7.4]. Stud shear connectors are to be provided at the interface between the concrete slab and the steel section. The purpose of the connectors is to resist the interface shear. In continuous composite bridges, shear connectors are recommended throughout the length of the bridge including negative moment regions. Before designing, the designer must consider some general information including types of shear connectors, pitch, transverse spacing, cover and penetration.

a. Types of shear connectors

   The two primary types of connectors used are the stud and channel shear connectors. The connectors should be chosen so that the entire surface of the connector is in contact with the concrete so that it may resist any horizontal or vertical movements between the concrete and the steel section. For this design, stud shear connectors are used to provide a composite section. The ratio of the height to stud diameter is to be greater than 4.0. [A6.10.7.4.1a]. Consider 19-mm diameter studs, 100 mm high, for this design.

$$\frac{100}{19} = 5.26 > 4, \text{ OK}$$

b. Transverse spacing

   Transverse spacing of the shear connectors is discussed in AASHTO [A6.10.7.4.1c]. Shear connectors are placed transversely along the top flange of the steel section. The center to center spacing of the connectors can not be closer than 4 stud diameters, or 76 mm. The clear distance between the edge of the top flange and the edge of the nearest connector must be at least 25 mm.

c. Cover and penetration

Cover and penetration requirements are in AASHTO [A6.10.7.4.1d]. Shear connectors should penetrate at least 50 mm into the concrete deck. Also, the clear cover over the tops of the connectors should be at least 50 mm. Consider a height of 100 mm for the shear studs.

d. Fatigue resistance

Consider the fatigue resistance of shear connectors in composite sections [A6.10.7.4.2]. The fatigue resistance of an individual shear connector is as follows:

$$Z_r = \alpha d^2 \geq 19.0 d^2$$

for which

$$\alpha = 238 - 29.5 \log N$$

where

$$d = \text{stud diameter (mm)}$$
$$N = \text{number of cycles [A6.6.1.2.5]}$$

For a design life of 75 years

$$N = (365)(75)n(\text{ADTT})_{SL}$$

where

$$n = \text{number of stress range cycles per truck passage taken from Table 8.4}$$
$$(\text{ADTT})_{SL} = \text{Single lane daily truck traffic averaged over the design life [A3.6.1.4.2]}$$

$$(\text{ADTT})_{SL} = p \times \text{ADTT}$$

where

$$\text{ADTT} = \text{the number of trucks per day in one direction averaged over the design life}$$
$$p = \text{fraction of truck traffic in a single lane taken from Table 4.3} = 0.85$$

When exact data is not provided the ADTT is determined using a fraction of the average daily traffic volume. The average daily traffic includes cars and trucks. If we use the recommendations of AASHTO [C3.6.1.4.2], the ADT can be considered to be 20 000 vehicles per lane per day. The ADTT is determined by applying the appropriate fraction from Table 4.4 to the ADT. By assuming urban interstate traffic, the fraction of trucks is 15%.

$$(\text{ADTT})_{\text{SL}} = 0.85(0.15)(20\ 000)(2\ \text{lanes}) = 5100\ \text{trucks/day}$$

The number of stress range cycles per truck passage n is taken from Table 8.4. For continuous girders, a distance of one-tenth the span length of each side of an interior support is considered to be near the interior support. For the positive moment region, n equals 1.0, so that

$$N = (365)(75)(1.0)(5100) = 140 \times 10^6 \text{ cycles}$$

therefore

$$\alpha = 238 - 29.5 \log(140 \times 10^6) = -2.31$$
$$Z_r = 19(19)^2 = 6859\ \text{N}$$

Therefore, the fatigue resistance Z_r of an individual shear stud is 6.86 kN.

e. Pitch

The pitch of the shear connectors is specified in AASHTO [A6.10.7.4.1b]. The pitch is to be determined to satisfy the fatigue limit state. Furthermore, the resulting number of shear connectors must not be less than the number required for the strength limit state. The minimum center to center pitch of the shear connectors is determined as follows:

$$p = \frac{nZ_r I}{V_{sr} Q}$$

where

p = pitch of shear connectors along the longitudinal axis (mm)

n = number of shear connectors in a cross section

I = moment of inertia of the short-term composite section (mm^4)

Q = first moment of the transformed area about the neutral axis of the short-term composite section (mm^3)

V_{sr} = shear force range under *LL* + *IM* determined for the fatigue limit state

Z_r = shear fatigue resistance of an individual shear connector

d_s = shear stud diameter

and

$$6d_s = 114 \text{ mm} \leq p \leq 600 \text{ mm}$$

For the short-term composite section, the moment of inertia is 31.6×10^9 mm⁴, from Section 8.11.3, Part J.1. The first moment of the transformed area about the neutral axis for the short-term composite section is determined from Figure E8.3-45

$$Q = Ay = (190)(276.25)(347.1) = 18.218 \times 10^6 \text{ mm}^3$$

For this design three shear connectors are used in a cross section as shown in Figure E8.3-46:

Stud spacing = 100 mm > 4(19) = 76 mm
Clear distance = 40.5 mm > 25 mm

Therefore, the transverse spacing requirements [A6.10.7.4.1c] are satisfied. The required shear connector pitch is computed at the tenth points along the spans using the shear range for the fatigue

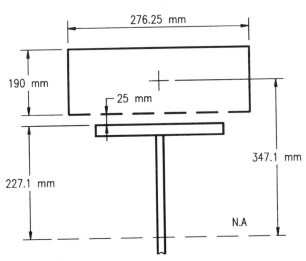

Fig. E8.3-45 Composite section properties.

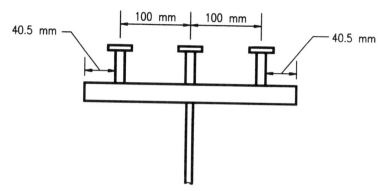

Fig. E8.3-46 Group of three shear connectors.

truck. The shear range V_{sr} is the maximum difference in shear at a specific point. It is computed by finding the difference in the positive and negative shears at that point due to the fatigue truck, multiplied by the dynamic load allowance for fatigue (1.15), the maximum distribution factor for one design lane loaded without multiple presence (0.762/1.2 for the exterior girder), and by the load factor for the fatigue limit state (0.75). The shear range and the pitch at the tenth points are tabulated in Table E8.3-21. The values in the pitch column are the maximum allowable spacing at a particular location. The required spacing is plotted on Figure E8.3-47. The spacing to be used is determined from this graph. An example calculation of the pitch is performed below, for the shear range at the end of the bridge.

$$p = \frac{3(6.86)(31.6 \times 10^9)}{(155)(18.218 \times 10^6)} = 230 \text{ mm}$$

f. Strength limit state

The strength limit state for shear connectors is taken from AASHTO [A6.10.7.4.4] The factored shear resistance of an individual shear connector is as follows:

$$Q_r = \phi_{sc} Q_n$$

where

Q_n = nominal resistance of a shear connector (kN)

TABLE E8.3-21 Shear Range for Fatigue Loading and Maximum Shear Connector Spacing

Location		Unfactored Max. Pos. Shear (kN)	Unfactored Max. Neg. Shear (kN)	Shear Range (kN)	Pitch (mm)
Sta	(m)				
100	0	253	−30	155	230
101	3	215	−30	134	266
102	6	178	−36	117	304
103	9	142	−53	107	334
104	12	109	−88	108	331
105	15	79	−124	111	321
106	18	52	−161	117	306
107	21	29	−195	123	291
108	24	10	−227	130	600[a]
109	27	7	−256	144	600[a]
110	30	7	−282	158	600[a]
200	30	281	−28	169	600[a]
201	33.6	251	−28	153	600[a]
202	37.2	217	−28	134	266
203	40.8	181	−43	123	291
204	44.4	144	−73	119	300
205	48	107	−107	117	305

[a]The maximum shear stud spacing is 600 mm [A6.10.7.4.1b]. This spacing is used in negative moment regions, assumed to be between the dead load inflection points at 22.7 m and 37.8 m.

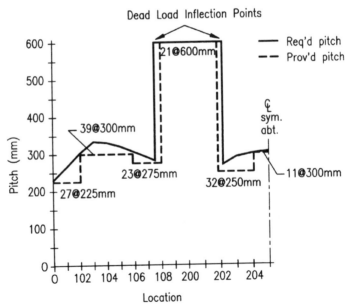

Fig. E8.3-47 Summary of shear connector spacings.

$$\phi_{sc} = \text{resistance factor for shear connectors} = 0.85$$

$$Q_n = 0.5A_{sc}\sqrt{f'_cE_c} \leq A_{sc}F_u$$

where

A_{sc} = shear connector cross-sectional area (mm²)

E_c = modulus of elasticity of concrete (MPa)

F_u = minimum tensile strength of a stud shear connector

$$F_u = 400 \text{ MPa [A6.4.4]}$$

$$A_{sc} = \frac{\pi}{4}(19)^2 = 284 \text{ mm}^2$$

$$E_c = 4800\sqrt{f'_c} = 4800\sqrt{30} = 26\,291 \text{ MPa}$$
$$Q_n = 0.5(284)\sqrt{30(26\,291)}(10^{-3}) = 126 \text{ kN}$$
$$A_{sc}F_u = (284)(400)(10^{-3}) = 114 \text{ kN} > 126 \text{ kN}$$

$$\text{Use } Q_n = 114 \text{ kN}$$

therefore

$$Q_r = (0.85)(114) = 97 \text{ kN}$$

The number of shear connectors required depends on the section. Between sections of maximum positive moment and points of 0 moment to either side, the number of shear connectors required is as follows:

$$n = \frac{V_h}{Q_r}$$

for which the nominal horizontal shear force is the lesser of the following:

$$V_h = 0.85f'_c bt_s$$

or

$$V_h = F_{yw}Dt_w + F_{yt}b_tt_t + F_{yc}b_ct_c$$

where

V_h = nominal horizontal shear force

b = effective slab width = 2210 mm

D = web depth = 1500 mm

t_s = slab thickness = 190 mm

b_c = width of compression flange = 300 mm

b_t = width of tension flange = 400 mm

t_c = thickness of compression flange = 15 mm

t_t = thickness of tension flange = 25 mm

F_y = minimum yield strengths of the respective sections
= 345 MPa

V_h = min 0.85(30)(2210)(190)(10^{-3}) = 10 707 kN

= [345(1500)(10) + 345(400)(25) + 345(300)(15)]
× 10^{-3} = 10 178 kN

Therefore, use a nominal horizontal shear force V_h of 10 178 kN and the required number of shear connectors for this region is calculated below.

$$n = \frac{10\ 178}{97} = 105$$

Therefore a minimum of 105 shear connectors are required between points of maximum positive moment and points of 0 moment. From examination of Figure E8.3-47, the number of shear connectors required by the fatigue limit state exceeds the amount required from the strength limit state.

For composite sections which are continuous, the horizontal shear force between the centerline of a support and points of 0 moment, is determined by the reinforcement in the slab. The following calculation determines the horizontal shear force.

$$V_h = A_r F_{yr}$$

where

A_r = total area of longitudinal reinforcement in the effective width over the interior support (mm^2) = 900 + 1400 = 2300 mm^2

F_{yr} = minimum yield strength of the longitudinal reinforcement = 400 MPa

V_h = (2300)(400)(10^{-3}) = 920 kN

Therefore the number of shear connectors required in this region is

$$n = \frac{920}{97} = 10$$

Ten studs are required between the interior pier and the points of 0 moment. The 600-mm maximum allowable spacing provides considerably more than this number of connectors.

Therefore, use the shear connector spacing specified on Figure E8.3-47.

L. Cross-Frame Design

In this section, intermediate and end cross frames are designed. Cross frames serve three primary purposes:

   a. Lateral support of the compression flange during placement of the deck

   b. Transfer of wind load on the exterior girder to all girders

   c. Lateral distribution of wheel load

The requirements for cross-frame design are found in AASHTO [A6.7.4]. The end cross frames must transmit all the lateral forces to the bearings. All the cross frames must satisfy acceptable slenderness requirements.

1. *Spacing*   The allowable unbraced length in the negative moment region is calculated in Section 8.11.3, Part I.7b. This value is 4427 mm. There was no compression flange requirement for sections in positive flexure. However, during concrete placement before the section becomes composite, the compression flange requires bracing. Therefore, an allowable unbraced length must be calculated for that section. The allowable unbraced length for the section in positive flexure is determined by treating the section as a noncomposite section. For the compression flange bracing of noncomposite members in positive and negative flexure regions, AASHTO [A6.10.6.3] refers to the requirements of the composite noncompact sections under negative flexure. The allowable unbraced length is calculated below

$$L_b \leq 1.76r_y\sqrt{\frac{E}{F_{yc}}}$$

where

   L_b = distance between compression flange brace points (mm)

   r_y = radius of gyration of the compression flange about the vertical axis

$$r_y = \sqrt{\frac{I_y}{A}} = \sqrt{\frac{(15)(300)^3}{12(15)(300)}} = 87 \text{ mm}$$

$$L_b = 1.76(87)\sqrt{\frac{200\ 000}{345}} = 3690 \text{ mm} = 3.69 \text{ m}$$

Therefore the maximum spacing allowed between cross frames is 3690 mm in the positive moment regions and 4427 mm in the negative moment regions. Based on the stiffener spacing in Figure E8.3-44 use a cross-frame spacing that is compatible. The cross frames are to be placed at all stiffener locations in the positive moment regions, so that the maximum unbraced length is 3250 mm. In the negative moment regions, the cross frames are to be placed at every other stiffener so that the maximum unbraced length is twice the stiffener spacing in the 36-m span negative moment region, or 3900 mm. Cross frame locations are given in Figure E8.3-48.

Alternatively, temporary framing may be used to support the positive moment region permitting the designer to increase the spacing of lateral bracing significantly.

A second alternative is to base the lateral framing on the lateral torsional buckling requirements for dead load (concrete, formwork) only. This analysis will likely yield a more economical design.

2. *Wind Load* The wind load acts primarily on the exterior girders. In bridges with composite decks, the wind force acting on the upper half of the girder, deck, barrier, and vehicle is assumed to be transmitted directly to the deck. These forces are transferred to the supports through the deck acting as a horizontal diaphragm. The wind

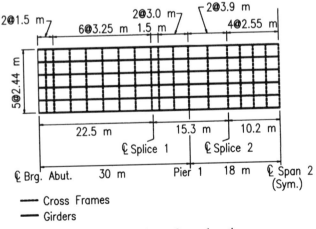

Fig. E8.3-48 Cross-frame locations.

force acting on the lower half of the girder is transmitted directly to the bottom flange. For this design the wind force, W, is applied to the bottom flange only since the top flange acts compositely with the deck. The wind force is calculated below.

$$W = \frac{\gamma p_B d}{2}$$

where

W = wind force per unit length applied to the flange

p_B = base horizontal wind pressure (MPa)

p_B = 0.0024 MPa [A3.8.1.2]

d = depth of the member (mm)

γ = load factor for the particular group loading combination from Table 3.1 [Table A3.4.1-1], for this case Strength *III* applies
   = 1.4

Consider the negative moment region first since it provides a larger value for d. The calculated wind load will be conservative for the positive moment region.

$$d = 1500 + 2(30) = 1560 \text{ mm} = 1.56 \text{ m}$$

therefore

$$W = \frac{1.4(0.0024)(1560)}{2} = 2.62 \text{ kN/m}$$

The load path taken by these forces is as follows:

a. The forces in the bottom flange are transmitted to points where cross frames exist.
b. The cross frames transfer the forces into the deck.
c. The forces acting on the top half of the girder, deck, barriers, and vehicles is transmitted directly into the deck.
d. The deck acts as a diaphragm transmitting the forces to the supports.

For this load path the maximum moment in the flange due to the wind load is as follows:

$$M_W = \frac{WL_b^2}{10}$$

where

$$L_b = \text{bracing point spacing}$$

Using the maximum unbraced length for the negative moment region of 3900 mm is conservative. Therefore the maximum lateral moment in the flange of the exterior girder due to the factored loading is

$$M_w = \frac{(2.62)(3.9)^2}{10} = 3.99 \text{ kN m}$$

The section modulus for the flange is

$$S_f = \frac{1}{6}(30)(400)^2 = 0.80 \times 10^6 \text{ mm}^3$$

and the maximum bending stress in the flange is

$$f = \frac{M_w}{S_f} = \frac{3.99 \times 10^6}{0.80 \times 10^6} = 5.0 \text{ MPa}$$

which is a small stress and any interaction with gravity loads can be neglected.

The maximum horizontal wind force applied to each brace point is also determined using maximum spacing. Therefore, the values will be conservative in most sections of the bridge.

$$P_w = WL_b = (2.62)(3.9) = 10.2 \text{ kN}$$

The cross frames must be designed to transfer all the lateral forces to the bearings. Figure E8.3-49 illustrates the transmittal of forces. As stated before, the forces acting on the deck, barrier, and upper half of the girder, $F1$, are directly transmitted into the deck. These forces are transferred to all of the girders. The forces acting on the bottom half of the girder, $F2$, are transferred to the bottom flange.

The wind force, W, was previous calculated for the bottom flange to be 2.62 kN/m. This force is referred to as $F2$ in Figure E8.3-49. F1 is calculated as

$$F1 = 1.4(0.0024)(2629) - 2.62 = 6.21 \text{ kN/m}$$

3. *Intermediate Cross-Frames*   The intermediate cross frames are designed using X-bracing along with a strut across the bottom flanges as shown in Figure E8.3-50. Single angles, M270 Grade 250 steel,

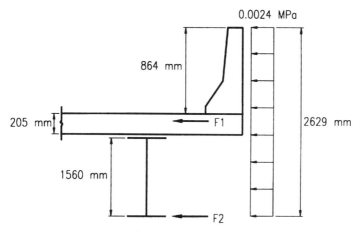

Fig. E8.3-49 Wind load acting on bridge exterior.

are used for the braces. For the cross braces acting in tension, 76 × 76 × 7.9 angles are used. For the compression strut, a 102 × 102 × 7.9 angle is used. These sections are considered as practical minimums. Section properties are given in AISC (1992).

The maximum force on the bottom flange at the brace point is

$$P_{wb} = 2.62(3.9) = 10.2 \text{ kN}$$

In order to find forces acting in the cross brace and the compression strut, the section is treated like a truss with tension diagonals only (counters) and solved using statics. From this analysis it is determined that the cross braces should be designed for a tensile force of 12.0 kN. The strut across the bottom flanges should be designed

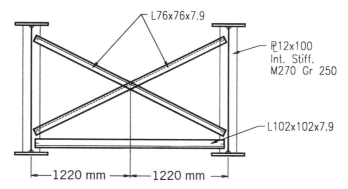

Fig. E8.3-50 Typical intermediate cross frame.

for a compressive force of 10.2 kN.

Check the 76 × 76 × 7.9 cross brace for tensile resistance [A6.8.2].

$$P_r = \phi_y P_{ny} = \phi_y F_y A_g = (0.95)(250)(1150) = 273 \text{ kN}$$

The tensile force in the cross brace is only 12.0 kN. Therefore, the cross brace has adequate strength.

Check the limiting slenderness ratio of the cross brace for tension members [A6.8.4]. For bracing members the limiting slenderness ratio is

$$\frac{L}{r} \le 240$$

where

L = the unbraced length of the cross brace (mm)

r = minimum radius of gyration of the cross brace (mm)

$$\frac{L}{r} = \frac{2860}{15.0} = 190.7 \le 240$$

Therefore use 76 × 76 × 7.9 angles for the intermediate cross braces.

Check the 102 × 102 × 7.9 strut for compressive resistance [A6.9.2.1]. However, first check the member for the limiting width/ thickness ratios for axial compression [A6.9.4.2].

$$\frac{b}{t} \le k \sqrt{\frac{E}{F_y}}$$

where

k = plate buckling coefficient from Table 8.11 = 0.45

b = full width of outstanding leg for single angle (mm) = 102 mm

t = plate thickness (mm) = 7.9 mm

$$\frac{b}{t} = \frac{102}{7.9} = 12.8 \le 0.45 \sqrt{\frac{200\,000}{250}} = 12.8$$

The nominal compressive resistance P_r is as follows:

$$P_r = \phi_c P_n$$

where

ϕ_c = resistance factor for axial compression, steel only = 0.90

P_n = nominal compressive resistance for noncomposite members

P_n is dependent on λ where

$$\lambda = \left(\frac{KL}{r_s \pi}\right)^2 \frac{F_y}{E}$$

where

K = effective length factor from [A4.6.2.5]

K = 0.75 for welded of bolted connections at both ends for bracing

L = unbraced length of the strut (mm) = 2440 mm

r_s = radius of gyration about the plane of buckling (mm)

= 20.1 mm

$$\lambda = \left[\frac{0.75(2440)}{20.1\pi}\right]^2 \frac{250}{200\,000} = 1.04$$

If $\lambda < 2.25$; then, $P_n = 0.66^\lambda F_y A_s$

$$P_n = 0.66^{(1.04)}(250)(1550) = 252 \text{ kN}$$

therefore

$$P_r = (0.9)(252) = 227 \text{ kN}$$

The compressive force in the strut is only 10.2 kN. Therefore, the strut has adequate strength.

Check the limiting slenderness ratio of the strut for compressive members [A6.9.3]. For bracing members, the limiting slenderness ratio is

$$\frac{KL}{r} \le 140$$

where

$$K = 0.75$$

$$L = 2440 \text{ mm}$$

$$r_s = 20.1 \text{ mm}$$

$$\frac{KL}{r} = \frac{0.75(2440)}{20.1} = 91.0 \leq 140$$

Therefore use $102 \times 102 \times 7.9$ struts for the intermediate cross frames.

4. *Cross Frames over Supports* The cross frames over the supports are designed using X-bracing along with a strut across the bottom flanges as shown in Figure E8.3-51. Single angles, M270 Grade 250 steel, are used for the braces. The sections used for the intermediate cross frames are also used for these sections. For the cross braces acting in tension, $76 \times 76 \times 7.9$ angles are used. For compression strut, a $102 \times 102 \times 7.9$ angle is used.

The intermediate cross frames, through their tension diagonals, transfer all of the wind load between supports into the deck diaphragm. At the supports, the cross frames transfer the total tributary wind load in the deck diaphragm down to the bearings.

The force taken from the deck diaphragm by each cross frame at the supports is approximately

$$P_{\text{frame}} = \frac{(F_1 + F_2)L_{\text{ave}}}{5 \text{ frames}} = \frac{(6.21 + 2.62)33.0}{5} = 58.3 \text{ kN}$$

The force on the bottom flange of each girder that must be transmitted through the bearings to the support is approximately

$$P_{\text{girder}} = \frac{(F_1 + F_2)L_{\text{ave}}}{6 \text{ girders}} = \frac{(6.21 + 2.62)33.0}{6} = 48.6 \text{ kN}$$

In order to find the forces acting in the cross brace and the compression strut, the section is treated like a truss with counters and solved using statics. From this analysis it is determined that the cross braces need to be designed for a tensile force of 68.3 kN. The strut across the bottom flanges needs to be designed for a compressive force of 58.3 kN. Because the forces above are less than the capacities of the members of the intermediate cross frames, and the same members are used, the chosen members are adequate. Therefore, use $76 \times 76 \times 7.9$ angles for the cross-bracing, and a $102 \times 102 \times 7.9$ strut for the cross frames over the supports.

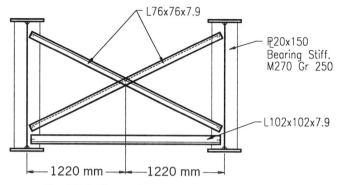

Fig. E8.3-51 Typical cross frame over supports.

5. *Cross Frames over Abutments* The cross frames over the abutments
   are designed using an inverted *V*-bracing (*K*-bracing) along with a
   diaphragm across the top flange and a strut across the bottom flange
   as shown in Figure E8.3-52. Single angles, M270 Grade 250 steel,
   are used for the braces. For the cross braces acting in tension, 76 ×
   76 × 7.9 angles are used. For the compression diaphragm across the
   top flange a W310 × 60 is used to provide additional stiffness at
   the discontinuous end of the bridge. For the strut across the bottom
   flange, a 102 × 102 × 7.9 angle is used.

   Because the tributary wind load length for the abutments is 15 m,
   the forces taken by the cross frames and girders can be determined
   from the values at the support as

$$P_{\text{frame}} = \frac{15}{33}\,(58.3) = 26.5 \text{ kN}$$

$$P_{\text{girder}} = \frac{15}{33}\,(48.6) = 22.1 \text{ kN}$$

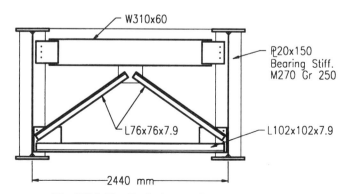

Fig. E8.3-52 Typical cross frame at abutments.

In order to find the forces acting in the cross brace and the compression diaphragm, the section is treated like a truss with counters and solved using statics. From this analysis it is determined that the cross bracing carries 35.6 kN and the strut across the bottom flanges carries 26.5 kN. Both of these loads are less than the capacities of the members. The diaphragm across the top flanges should be designed for a compressive force of 26.5 kN.

Check the W310 × 60 diaphragm for compressive resistance [A6.9.2.1]. However, first check the member for limiting width/thickness ratios for axial compression [A6.9.4.2]. The W310 × 60 dimensions are from AISC (1992).

$$\frac{b}{t} \le k\sqrt{\frac{E}{F_y}}$$

where

k = plate buckling coefficient from Table 8.11 = 1.49

b = width of plate specified in Table 8.11 (mm)

t = plate thickness (mm)

$$\frac{b}{t} = \frac{239}{7.49} = 31.9 \le 1.49\sqrt{\frac{200\,000}{250}} = 42.1$$

The nominal compressive resistance P_r is as follows:

$$P_r = \phi_c P_n$$

where

ϕ_c = resistance factor for axial compression, steel only = 0.90

P_n = nominal compressive resistance for noncomposite members

P_n is dependent on L where

$$\lambda = \left(\frac{KL}{r_s\pi}\right)^2\frac{F_y}{E}$$

where

$$K = 0.75$$

$$L = 2440 \text{ mm}$$

$$r_s = 49.1 \text{ mm}$$

$$\lambda = \left[\frac{0.75(2440)}{49.1\pi}\right]^2 \frac{250}{200\,000} = 0.176$$

If $\lambda < 2.25$; then, $P_n = 0.66^\lambda F_y A_s$

$$P_n = 0.66^{(0.176)}(250)(7600) = 1766 \text{ kN}$$

therefore

$$P_r = (0.9)(1766) = 1589 \text{ kN}$$

The compressive force in the diaphragm is only 26.5 kN. Therefore, the diaphragm has adequate strength.

Check the limiting slenderness ratio of the diaphragm for the compressive members [A6.9.3]. For bracing members, the limiting slenderness ratio is

$$\frac{KL}{r_s} \leq 140$$

where

$$K = 0.75$$

$$L = 2440 \text{ mm}$$

$$r_s = 49.1 \text{ mm}$$

$$\frac{KL}{r} = \frac{0.75(2440)}{49.1} = 37.1 \leq 140$$

Therefore use W310 × 60 diaphragm for the cross frames over the abutments.

M. Design Sketch

The design of the steel plate girder bridge is shown in Figure E8.3-53. Because the many details cannot be provided in a single drawing, the reader is referred to the figures already provided. For the cross section of the plate girder and slab, refer back to Figures E8.3-18 and E8.3-36. For stiffener spacing, refer to Figure E8.3-44. For shear stud pitch, refer back to Figure E8.3-47. For cross-frame locations, refer to Figure E8.3-48, and for cross-frame design, refer to Figures E8.3-50–E8.3-53.

The engineer must also design welds, splices, and bolted connections. These topics were not covered in this example for lack of space.

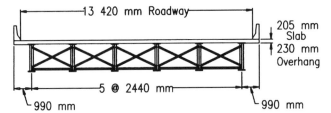

Fig. E8.3-53 Cross section of plate girder bridge showing girders and cross frames.

REFERENCES

AASHTO (1994). *LRFD Bridge Design Specifications,* 1st ed., American Association of State Highway and Transportation Officials, Washington, DC.

AISC (1986). *LRFD Manual of Steel Construction,* 1st ed., American Institute of Steel Construction, Chicago, IL.

AISC (1992). *Metric Properties of Structural Shapes,* American Institute of Steel Construction, Chicago, IL.

AISC (1994). *LRFD Manual of Steel Construction,* 2nd ed., Volume II—Connections, American Institute of Steel Construction, Chicago, IL.

AISI (1985). *Steel Product Manual—Plates,* American Iron and Steel Institute, Washington, DC.

ASTM (1995). "Standard Specifications for Carbon and High-Strength Low-Alloy Structural Steel Shapes, Plates, and Bars and Quenched-and-Tempered Alloy Structural Steel Plates for Bridges," *Annual Book of ASTM Standards,* Vol. 01.04, American Society for Testing and Materials, Philadelphia, PA.

Barsom, J. M. (1992). "Fracture Design," Chapter 5.5, *Constructional Steel Design,* Dowling et al. Eds., Elsevier Science Publishers (Chapman & Hall, Andover, England).

Basler, K. (1961a). "Strength of Plate Girders in Shear," *Journal of Structural Division,* ASCE, Vol. 87, No. ST7, October, pp. 151–180.

Basler, K. (1961b). "Strength of Plate Girders Under Combined Bending and Shear," *Journal of Structural Division,* ASCE, Vol. 87, No. ST7, October, pp. 181–197.

Basler, K. and B. Thürlimann (1961). "Strength of Plate Girders in Bending," *Journal of Structural Divisoin,* ASCE, Vol. 87, No. ST6, August, pp. 153–181.

Bjorhovde, R. (1992). "Compression Members," Chapter 2.3, *Constructional Steel Design,* Dowling et al., Eds., Elsevier Science Publishers (Chapman & Hall, Andover, England).

Bleich, F. (1952). *Buckling Strength of Metal Structures,* McGraw-Hill, New York.

Brockenbrough, R. L. (1992). "Material Properties," Chapter 1.2, *Constructional Steel Design,* Dowling et al., Eds., Elsevier Science Publishers (Chapman & Hall, Andover, England).

Brockenbrough, R. L. and J. M. Barsom (1992). "Metallurgy," Chapter 1.1, *Constructional Steel Design,* Dowling et al., Eds., Elsevier Science Publishers (Chapman & Hall, Andover, England)

Brockenbrough, R. L. and B. G. Johnston (1981). *USS Steel Design Manual,* R. L. Brockenbrough & Assoc., Pittsburgh, PA.

Dowling, P. J., J. E. Hardin, and R. Bjorhovde, Eds. (1992). *Constructional Steel Design—An International Guide,* Elsevier Science Publishers (Chapman & Hall, Andover, England).

Gaylord, E. H., Jr., C. N. Gaylord, and J. E. Stallmeyer (1992). *Design of Steel Structures,* 3rd ed., McGraw-Hill, New York.

Gurney, T. R. (1992). "Fatigue Design," Chapter 5.4, *Constructional Steel Design,* Dowling et al., Eds., Elsevier Science Publishers (Chapman & Hall, Andover, England).

Kitipornchai, S. and N. S. Trahair (1980). "Buckling Properties of Monosymmetric I-Beams," *Journal of Structural Division,* ASCE, Vol. 106, No. ST5, May, pp. 941–957.

Maquoi, R. (1992). "Plate Girders," Chapter 2.6, *Constructional Steel Design,* Dowling et al., Eds., Elsevier Science Publishers (Chapman & Hall, Andover, England).

Munse, W. H. and E. Chesson, Jr. (1963). "Riveted and Bolted Joints: Net Section Design," *Journal of Structural Division,* ASCE, Vol. 89, No. ST1, February, pp. 107–126.

Nethercot, D. A. (1992). "Beams," Chapter 2.2, *Constructional Steel Design,* Dowling, et al., Eds., Elsevier Science Publishers (Chapman & Hall, Andover, England).

Ollgaard, J. G., R. G. Slutter, and J. W. Fisher (1971). "Shear Strength of Stud Connectors in Lightweight and Normal-Weight Concrete," *AISC Engineering Journal,* Vol. 8, No. 2, April, pp. 55–64.

Segui, W. T. (1994). *LRFD Steel Design,* PWS Publishing Company, Boston, MA.

Slutter, R. G. and G. C. Driscoll, Jr. (1965). "Flexural Strength of Steel-Concrete Composite Beams," *Journal of Structural Division,* ASCE, Vol. 91, No. ST2, April, pp. 71–99.

Slutter, R. G. and J. W. Fisher (1967). "Fatigue Strength of Shear Connectors," *Steel Research for Construction,* Bulletin No. 5, American Iron and Steel Institute, Washington, DC.

Taylor, J. C. (1992). "Tension Members," Chapter 2.1, *Constructional Steel Design,* Dowling et al., Eds, Elsevier Science Publishers (Chapman & Hall, Andover, England).

Timoshenko, S. and J. M. Gere (1969). *Theory of Elastic Stability,* 3rd ed., McGraw-Hill, New York.

Yura, J. A., T. V. Galambos, and M. K. Ravindra (1978). "The Bending Resistance of Steel Beams," *Journal of Structural Division,* ASCE, Vol. 104, No. ST9, September, pp. 1355–1370.

9

WOOD BRIDGES

9.1 INTRODUCTION

The use of wood as a building material spans the course of human history. Wood and stone have been obvious materials for construction since prehistoric times and likely the first bridge was constructed of timber. The frequent backpacker can readily attest to the convenience of a fallen timber for a stream crossing. In fact, the log bridge remains a viable temporary structure in areas where large trees are readily available near the bridge site. Wood is the one bridge engineering material that is derived from a renewable resource. As one-third of the area of the United States is forest land, forest resources, if properly managed, offer an important and sustainable supply of building materials for generations (Ritter, 1990). The old growth timber can be harvested and trees replanted to renew supplies. Much of the discussion in the press involving the timber harvest is related to this process and how well environmental issues are addressed. It is beyond the scope of this book to address this issue, but it is important to mention that if timber harvesting is not thoughtfully implemented, then certainly environmental problems arise. Environmental issues are also associated with cement plants, ready mix facilities, and steel plants.

Near the middle of the nineteenth century, iron emerged as an important building material, and engineers used it to span major rivers in the United States. In the early twentieth century, reinforced concrete became an important alternative to steel, and concrete bridge construction began shortly after its use in buildings. Prior to this time, timber and masonry were used extensively for bridges, but since then, steel and concrete have become the most commonly used bridge construction materials. This choice does not obviate the need or application of wood for bridge engineering. In fact, approximately 12% (71 200) of the bridges in the United States are wood bridges. In the

U.S. Forest Service, approximately 7500 bridges are timber and the railroad industry has approximately 1500 miles of timber bridges in their inventory (Ritter, 1990).

If 12% of the U.S. bridges are wood, then why do we not see many of them? The vast majority of wood bridges are on secondary, county, and other local roads. Such roads typically have traffic volumes much lower than the primary and interstate roads. Moreover, such bridges are often located in rural areas, many in the sparsely populated mountain west, which has the majority of public lands. This fact does not denigrate the importance of such bridges. Every roadway has its function whether carrying 30 000 commuters daily or 300 people on a farm-to-market byway. It is also important to note that a significant percentage of bridges considered functionally obsolete or structurally deficient are located on non-federal-aid roadways. Such bridges are often shorter spans with small traffic volumes controlled by managers with very restricted funds. In such situations, a manager is interested in obtaining the most cost efficient (both short and long term) solution possible. In short, resources in the non-federal-aid system are very restricted and managers such as county commissioners, road superintendents, and city managers wish to provide the safest system at the lowest possible cost.

Wood has several advantages over competing materials. These advantages are listed below. (Note that some of these advantages are distinct, while others are not):

- Convenient shipping to the job site.
- Wood is relatively light, lowering transportation and initial construction costs.
- Wood is relatively light, and can be handled with smaller construction equipment, which lowers initial construction costs.
- Most job-site details can be handled without special equipment and associated highly skilled labor. For example, welding is seldom required, most connections are bolted, and field drilling can be accommodated with hand-held tools.
- Wood bridges likely use wood decks, which is advantageous in remote locations that are far from a ready mix plant. The wood deck may be placed without formwork and reinforcement placement, a significant first-cost saving. Some precast concrete systems also have this advantage.
- Wood is not susceptible to deicing agents that can cause significant problems with steel and concrete systems.
- Wood typically does not require painting or postconstruction surface treatment during its service life. Concrete and weathering steel also have this advantage over conventionally painted steel construction.
- Lastly, wood is a very aesthetic material. When properly designed, it blends very well with the surroundings.

Wood design is not frequently taught in civil engineering schools and in those departments that offer a course, it is elective. Therefore, bridge designers are often inexperienced with wood as a general construction material. In short, the bridge engineering profession has little experience with wood as compared to steel and concrete. Because of this, several misconceptions are common.

Wood is limited to short-span bridges because of limited availability of longer solid sawn timbers. While this is true for sawn sections, glued laminated (glulam) and prestressed laminated (stress-laminated) systems permit spans of any length to be shipped to the job site. Although a similar limitation exists with steel and concrete, steel can be more easily spliced in the field to create longer span systems. Prestressed concrete can also be spliced but this is not common practice. Practically, wood bridges are commonly used for spans of 6–25 m (20–80 ft).

Wood is susceptible to decay and will not last. Wood is susceptible to decay and will not last long if untreated. Fortunately, pressurized treatment is available and required for all bridge applications. The service life for a properly treated bridge is approximately 40 years (Ritter, 1990). Inventory managers with bridges in service for well over 40 years may not be impressed with this service life. Only with better bridge management methodologies can the life-cycle economy of alternative designs, material types, and maintenance and replacement strategies be properly evaluated. Such systems will consider the cost of maintenance, repair, and replacement and estimate the cost on a life-cycle basis. For example, it may be more economical to replace a wood bridge after 40 years than to maintain a steel bridge for its 75-year service life.

The primary purpose of this chapter is to introduce the reader to the design of wood bridges. In the space available, it is impossible to address all of the concepts and design methods involved with timber design. For more complete generalized treatment, the reader is referred to texts on wood engineering (Bryer, 1988; Hoyle and Woeste, 1989). For engineering bridges with wood, the reference by Ritter (1990) contains a wealth of information including the history of wood bridges, the basic properties, preservative treatments, design procedures with the working stress method, and examples of bridge designs. Inspection and maintenance are addressed in the final chapters. It is a useful reference for the experienced bridge engineer and the novice alike.

The discussion herein is limited to the design of glued laminated (glulam) and prestressed laminated (stress-laminated) systems. Although solid sawn beam systems are used for very short spans, they are not common for new construction. Furthermore, engineers familiar with the design of glulam systems can apply their knowledge to solid sawn timbers. Hence, only the most commonly used and generally applicable systems are addressed.

The topics and issues that are relevant for the selection of the deck and beams are included. These topics include a brief introduction of woods physical and mechanical properties, a discussion of the strength and serviceability limit states, and design procedures. Example designs of glulam and stress-laminated bridges are included. For other topics relevant to the complete

design and service such as connections, treatments, inspection, and maintenance, refer to Ritter (1990).

9.2   WOOD MATERIAL PROPERTIES

Wood differs from other materials because it is a natural material produced
from a living tree. Unlike the process of steel manufacturing or concrete mix
proportioning, the material suppliers do not have tight controls over the raw
product that is used. Being a natural entity (like ourselves), each tree is different and contains many factors that affect the ultimate performance of the
wood that is produced. A tree contains knots, sloped grain, compression
wood, and other strength reducing characteristics. Therefore, it is important
to implement methods to detect such characteristics and grade the material
accordingly. In short, the strength reducing characteristics of wood are inherent, so the material must be graded or enhanced through remanufacture so
these characteristics are removed or their effects minimized.

As with the design of any material, it is important for the designer to have
a basic understanding of wood and its properties. Analogous to the disciplines
of metallurgy and concrete technology, wood science is a distinct discipline
that addresses the science of wood and wood fiber. For more information on
the science of wood, specifically the science related to structural design, refer
to Bodig and Jayne (1982). Herein the scope of coverage is limited to those
properties directly related to bridge building.

9.2.1.   Structure of Wood

The primary building block of wood is the wood cell. A microscopic representation of a magnified wood structure of a softwood is shown in Figure 9.1.
Note the parallel structure that is formed by the cells. Typically, wood cells
are long, narrow, and are grouped similarly to straws as shown in Figure 9.2
(Ritter, 1990). If subjected to tension or compression along the length of the
cell structures, the cells function similarly to columns, which is the strongest
and stiffest direction and is referred to as the longitudinal direction. If a
compressive stress is applied perpendicular to the cell axis, the strength and
stiffness are significantly reduced. The three directions of interest are denoted
as longitudinal, radial, and tangential. These directions are referenced in Figure 9.3, where a piece of lumber is spatially related to the timber log. Because
the tree (log) is circular these directions are denoted by a cylindrical coordinate system. As the piece of lumber is small relative to the size of the log,
a Cartesian system is often used. Further, simplicity is used in design, where
the mechanical properties are referred to as directions parallel and perpendicular to the grain or longitudinal axis. The distinction between radial and
tangential is dropped because the mechanical properties, although not the
same, are quite similar and because the orientation of sawn lumber is some-

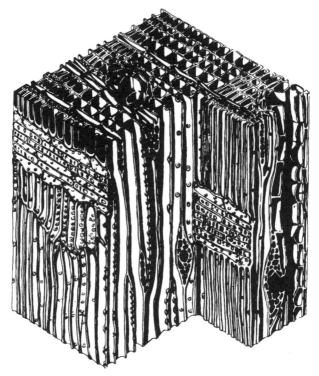

Fig. 9.1   Cell structure of wood. [After Ritter, 1990.]

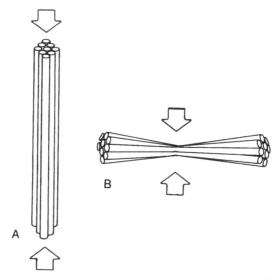

Fig. 9.2   Mechanics of wood fiber. [After Ritter, 1990.]

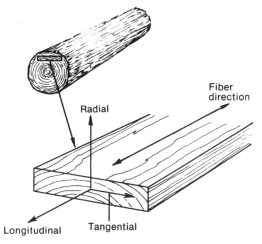

Fig. 9.3   Sawn timber related to the wood log. [After Ritter, 1990.]

what random. The more conservative, between the radial and tangential, values are used.

9.2.2   Physical Properties of Wood

Moisture Content. The moisture content (MC) of wood is defined as the weight of the water in the wood given as a percentage of the ovendry weight.

$$MC = \left(\frac{\text{moist wt.} - \text{ovendry wt.}}{\text{ovendry wt.}}\right) 100\% \qquad (9.1)$$

Water exists in wood in two forms: free water and bound water. Free water is located in the cell cavities. Bound water is bonded to the cells at the molecular level. Free water is the first removed when wood dries and escapes at a faster rate than bound water. The state at which all of the free water is removed and the bound water remains is the fiber saturation point (FSP). For most species, this is about 30%.

Wood is a hygroscopic material that absorbs and releases water with changing environmental conditions. The amount of absorption is a function of the ambient relative humidity and temperature of the final placement. For bridge applications, these conditions change continuously. Fortunately, since it takes time for the wood to respond to changed conditions, there is a lag between an environmental change and the subsequent internal moisture content. Ultimately, most of the wood cross section obtains an equilibrium state where the moisture content is essentially constant. This state is termed the equilibrium moisture content (EMC) and is important because the mechanical properties of wood are affected by the amount of moisture present. The EMC may be estimated with Table 9.1 where EMC is related to the average ambient conditions.

Studies and experience have shown that wood decks should be assumed to be used in a wet-use condition (Gutkowski and McCuthcheon, 1987) and that the wood beams supporting these decks with asphalt wearing surfaces are expected to be at the EMC, usually considered dry-use. Because of leakage of deck joints, wood located in the area of the bearing should likely be considered wet for design purposes. Moisture content factors are provided for both strength and stiffness properties in AASHTO [A4.4.4.3],* which is described in detail later.

Dimensional Stability Wood is dimensionally stable when the MC is above the FSP. Below the FSP, wood shrinks with decreasing moisture and swells with increasing moisture. In the longitudinal direction, as wood dries from green to ovendry (wood fiber) conditions, the average wood shrinkage strains are on the order of 0.001%, which is of no practical significance in wood design and is typically neglected. In the radial and tangential directions, the shrinkage is much greater. As wood dries from green to ovendry, the shrinkage strains range from 0.05 to 0.07, or 50–75 times the longitudinal values. Although not explicitly addressed in the AASHTO (1994) LRFD Bridge Specification, such strains must be considered, notably in detailing, if the wood is expected to change moisture content significantly in service. Additionally, it is important to properly specify the MC of the construction material to be approximately the same as that expected in service.

Density The unit weight is a function of species and moisture content, but acceptable weight estimates are 800 and 960 kg/m³ for softwood and hardwood, respectively. These values correspond to unit weights of 7.84×10^{-6} and 9.41×10^{-6} N/mm³. See Table 4.1 or AASHTO [Table A3.5.1-1].

Thermal expansion The coefficient of thermal expansion is generally independent of species and is estimated as 0.000 0031 − 0.000 0045/°C (0.000 0017–0.000 0025/°F).

9.2.3 Mechanical Properties of Wood

Elastic Properties Because of wood's inherent physical structure, the strength and stiffness properties are different in the radial, tangential, and longitudinal directions. Wood is an orthotropic material. Recall that isotropic materials, such as steel and concrete, have mechanical properties that are the same in all directions. A complete discussion of the elastic properties of wood may be found in Bodig and Jayne (1982) and a table of the relative elastic ratios may be found in Ritter (1990). Unless rigorous analyses are performed

*The article numbers in the AASHTO (1994) LRFD Bridge Specifications are enclosed in brackets and preceded by the letter A if specifications and by the letter C if commentary.

TABLE 9.1 Moisture Content of Wood in Equilibrium with State Temperature and Relative Humidity[a]

Temperature (°C)	Relative Humidity (%)										
	10	20	30	40	50	60	70	80	90	95	98
− 1.1	2.6	4.6	6.3	7.9	9.5	11.3	13.5	16.5	21.0	24.3	26.9
4.4	2.6	4.6	6.3	7.9	9.5	11.3	13.5	16.5	21.0	24.3	26.9
10.0	2.6	4.6	6.3	7.9	9.5	11.2	13.5	16.5	20.9	24.3	26.9
15.6	2.5	4.6	6.2	7.8	9.4	11.1	13.3	16.2	20.7	24.1	26.8
21.1	2.5	4.5	6.2	7.7	9.2	11.0	13.1	16.0	20.5	23.9	26.6
26.7	2.4	4.4	6.1	7.6	9.1	10.8	12.9	15.7	20.2	23.6	26.3
32.2	2.3	4.3	5.9	7.4	8.9	10.5	12.6	15.4	19.8	23.3	26.0
37.8	2.3	4.2	5.8	7.2	8.7	10.3	12.3	15.1	19.5	22.9	25.6
43.3	2.2	4.0	5.6	7.0	8.4	10.0	12.0	14.7	19.1	22.4	25.2
48.9	2.1	3.9	5.4	6.8	8.2	9.7	11.7	14.4	18.6	22.0	24.7

[a] Ritter (1990) Converted to SI. [From *AASHTO LRFD Bridge Design Specifications*, Copyright © 1994 by the American Association of State Highway and Transportation Officials, Washington, DC. Used by permission.]

on the complete bridge system and orthotropic constants are required, the only constant that is required by the AASHTO (1994) LRFD Bridge Specification is the modulus of elasticity in the longitudinal direction. These constants vary with species and are tabulated in AASHTO [Table A8.4.1.1.4-1] for sawn lumber and in AASHTO [Tables A8.4.1.2.3-1, A8.4.1.2.3-2, and A8.4.1.3-1] for glued laminated elements.

Strength Properties   The stress versus strain relationship for wood does not contain a yield plateau as does a mild steel. Testing results usually indicate an approximately linear relationship with a nonductile failure for shear, bending, and tension. Compression tests can exhibit some ductility. As a rule, ductility is not counted on in the design procedure. The design equations for tensile, compressive, flexural, and shear stress are based on the usual assumptions, involving linear elasticity and linear strain profile. The strength properties are generated using these assumptions, hence any nonlinear behavior or other deviation from the assumed behavior is somewhat reflected in the nominal resistance values.

As previously mentioned, the strength of wood is a function of the defects present. Thus, the grading agency must classify wood according to presence, or lack of, defective characteristics that affect strength. One classification method is to visually grade the lumber. In this process, human graders classify lumber according to the type and number of defects and their location within the piece. This process is illustrated in Figure 9.4. The rules account for how the piece is to be used, such as a joist, beam, or post. For example, a critical knot on the outside of a piece to be used as a beam would be downgraded, but that same effect would not be as critical for a column where the entire cross section is uniformly stressed in compression. This method is termed visual grading and the resulting product is referred to as visually graded lumber.

Fig. 9.4   Visually inspecting lumber. [After Ritter, 1990.]

The strength properties required for bridge design are flexural, tension, shear, compression perpendicular to the grain, and compression parallel to the grain. The basic resistance for each species and grade are given in AASHTO [Table A8.4.1.1.4-1]. The properties for Douglas Fir-Larch are given in Table 9.2. The complete AASHTO table is approximately four pages long and includes many commonly used species. For species not given by AASHTO, the strength values from National Design Specification (NDS, 1991) may be used. The NDS is based on the working stress design method, thus the NDS values are allowable stresses. These values must be adjusted upward to obtain the basic resistance values required in the LRFD procedures. The applicable adjustment factors are given in Table 9.3.

The second method of grading is to recognize that a relationship exists (empirically established) between the stiffness of lumber and its strength. Hence, the stiffness (modulus of elasticity) can be determined by mechanically deflecting the piece and measuring the resulting force. This nondestructive test is fast and generally more precise than visually grading. A schematic illustration is shown in Figure 9.5. The basic resistance values for mechanically graded lumber are given in AASHTO [Table A8.4.1.1.4-3].

The resistances (ultimate stresses) are based on the 5% exclusion values. This means that these values are derived from test data and that 95% of all the test data lie above the nominal values given in the Specification, which is depicted in Figure 9.6(a). The nominal values for the modulus of elasticity are based upon the mean or average test data and this is depicted in Figure 9.6(b).

Load Duration and Dynamic Load Allowance   Wood, as with most materials, exhibits greater strength if tested (loaded or strained) rapidly. Conversely, a long-term test exhibits lower strengths. This relationship is illustrated in Figure 9.7. A 2-month load duration is used in establishing the AASHTO nominal strength. This load duration is a conservative approach because live load effects remain on the bridge for a much shorter time, even for the case where a train of trucks are parked on the bridge. Recall in Chapter 4 that the live load impact factor for timber structures is one-half that of a bridge constructed of other materials (also see AASHTO [A3.6.2.3]). For example, a beam design is based on 16.5% rather than 33% dynamic load allowance. The justification for decreased impact is first based somewhat on the fact that wood structures have inherent damping that does not exist in steel and concrete structures, and secondly, on the fact that wood is stronger when loaded rapidly. In the previous AASHTO specifications, the design of wood bridges did not include the dynamic load effects, that is, $IM = 0$, and the 10-year load duration resistance values were used in the working stress methodology. Because the safety factors for the working stress method are the same for all load effects, it made no difference which side of the equation the adjustment for load duration was made. In the case of impact, AASHTO chose to lower the dynamic effects while also lowering allowable stress. It is our opinion that it is unlikely that damping can play a large role in the

TABLE 9.2 Base Resistance, (MPa) and Modulus of Elasticity, (MPa) for Visually Graded Sawn Lumber for Douglas Fir-Larch[a]

Grade	Size (mm)	F_{bo}	F_{to}	F_{vo}	F_{cpo}	F_{co}	E_o
Select Structural	$b = 50$–100	24	20	2.1	7.6	21	11 700
No. 1 and Better	$d \geq 50$	22	16	2.1	7.6	20	11 000
No. 1		19	14	2.1	7.6	20	10 700
No. 2		17	12	2.1	7.6	19	10 000

TABLE 9.3 **Factors to Calculate Base Resistance and Modulus of Elasticity from NDS Allowable Stresses[a]**

	F_{bo}	F_{to}	F_{vo}	F_{cpo}	F_{co}	E_o
Dimension Lumber	2.35	2.95	3.05	1.75	1.90	0.90
Beams, stringers, posts, and timbers	2.80	2.95	3.15	1.75	2.40	1.00

[a] In AASHTO 8.4.1.1.4–2 [From *AASHTO LRFD Bridge Design Specifications,* Copyright © 1994 by the American Association of State Highway and Transportation Officials, Washington, DC. Used by permission.]

dynamic response in a transient load such as vehicle impact. This behavior is especially true on short span bridges that have larger dynamic effects and the load is on and off the bridge before damping can influence the response.

In the AASHTO (1994) LRFD Bridge Specification, the resistance is based on the 2-month load duration. Note that most of the load effect is due to vehicle live load that has a very short load duration, so the 2-month duration value is rather conservative. Preservative treatment adversely affects the strength for rapid loading rates and the increases illustrated in Figure 9.7 are not as pronounced. Because all of the wood in a bridge should be treated, it is reasonable not to use a significant increase for load rate effects. In summary, the effect of treatment and the effect of increased resistance due to rapid strain rates approximately cancel.

9.2.4 Properties of Glued Laminated Wood

Glued laminated (glulam) beams and decks are a very common form of wood bridge construction. Glulams consist of dimension lumber glued together to form the component used in the final placement. Glulam has been successfully used in the United States since approximately 1935 and in bridges since the mid-1940s (Ritter, 1990). Because these components are manufactured, the quality control and engineering can be more precisely controlled than that of solid sawn elements. Glulam can be manufactured to almost any size and is not limited to the length of available lumber. This product is made by jointing the lamina to achieve longer lengths. Further, laminating disperses the strength-reducing characteristics throughout the element producing a favorable situation where one defect will not cause failure because neighboring material is stronger (Figure 9.8). Lastly, defects and weaker elements can be placed away from critical regions, and this is described next.

In bridge engineering, the most commonly used elements are beams and decks as illustrated in Figures 9.9(a) and 9.9(b), respectively. The lamina lay-up is different for these elements. In the beam (or bending) lay-up high quality

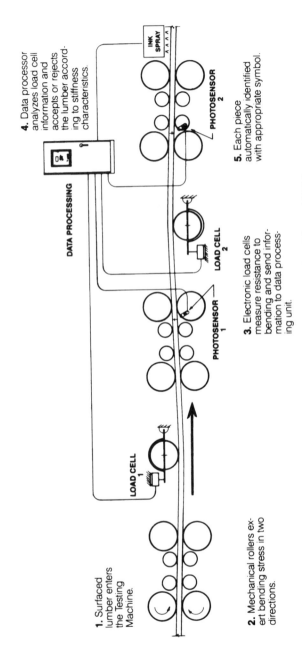

1. Surfaced lumber enters the Testing Machine.

2. Mechanical rollers exert bending stress in two directions.

3. Electronic load cells measure resistance to bending and send information to data processing unit.

4. Data processor analyzes load cell information and accepts or rejects the lumber according to stiffness characteristics.

5. Each piece automatically identified with appropriate symbol.

LOAD CELL 1

LOAD CELL 2

PHOTOSENSOR 1

PHOTOSENSOR 2

DATA PROCESSING

INK SPRAY

Fig. 9.5 Mechanical grading process. [After Ritter, 1990.]

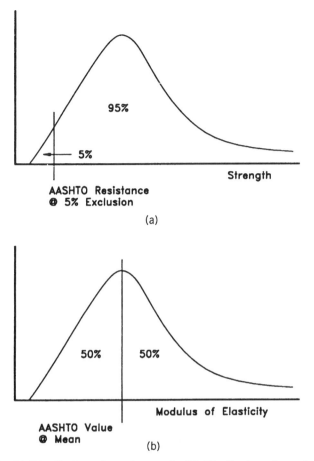

Fig. 9.6   (a) Distribution of wood strength, (b) Distribution of wood stiffness.

laminae are placed in the area of high flexural stress. If the beam is to be used in a simply-supported configuration, then high quality laminae are placed at the bottom of the beam. Lower grade materials are always placed in the interior portion to resist the shear stresses. If the beam is to be used in a continuous beam and high flexural stresses are expected on both faces, then high-grade lamina are placed in both the top and the bottom. Common lay-up combinations for such scenarios are given in Table 9.4 with their mechanical properties. Although the permissible stresses vary throughout the section, the critical values are given with the assumption that these basic stresses (resistances) are to be used with the basic mechanics equations. For example, combination symbol 24F-V8 is constructed of Douglas Fir (DF) outer and inner lamina (see Table 9.4). The section is symmetrically reinforced and the bending resistance is the same for positive and negative bending about the

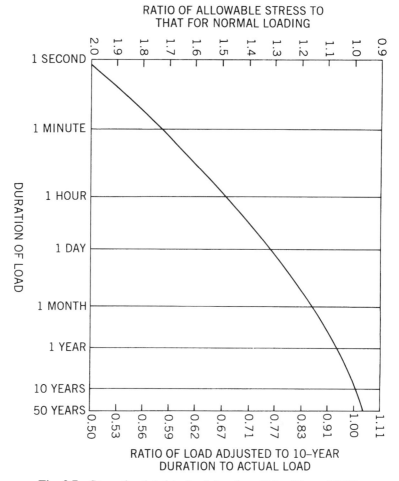

RATIO OF ALLOWABLE STRESS TO
THAT FOR NORMAL LOADING

Fig. 9.7   Strength related to load duration. [After Ritter, 1990.]

X–X axis, F_{bo} = 37 MPa (refer to the tables for axis orientation). This section would be suitable for continuous beams. Consider the combination symbol 24F-V4 where: the *24* is the allowable stress (2400 psi) used in the working stress methodology; the *F* means the section is principally a flexural section; the *V* (or *E*) means the laminae are visually or mechanically *E*-rated, and finally, the *4* denotes a lay-up combination. For this combination, the positive bending resistance is 37 MPa and the negative bending resistance is 18 MPa. This section is nonsymmetric about the X–X axis and is intended to be used for a simply supported beam. Of course, it is imperative that the tension lamina be placed on the tension face during construction. To facilitate placement, the beams are marked "top" on the compression side.

Consider bending about the Y–Y axis, for example, wind load on the exterior beam. The bending resistance for this action is significantly lower than

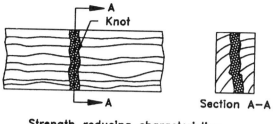

Knot

Section A–A

Strength reducing characteristics
in sawn lumber can occupy
much of the cross section
and substantially reduce strength

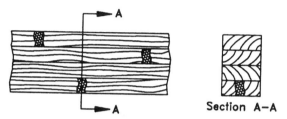

Section A–A

Laminating disperses strength
reducing characteristics, reducing
their effect on strength

Fig. 9.8 Defects in wood and laminates.

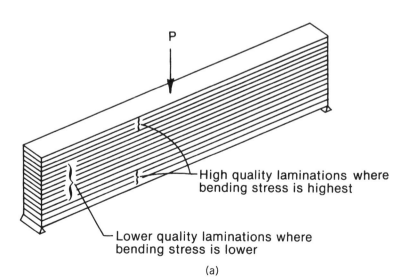

P

High quality laminations where
bending stress is highest

Lower quality laminations where
bending stress is lower

(a)

Fig. 9.9(a) Glulam beam layup. [After Ritter, 1990.]

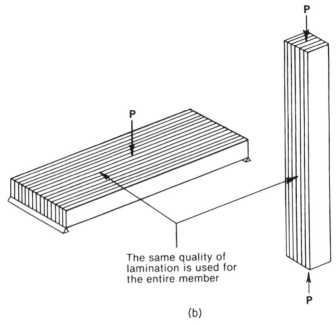

The same quality of
lamination is used for
the entire member

(b)

Fig. 9.9(b)   Glulam column and deck layup. [After Ritter, 1990.]

that for bending about the X–X axis, because the lower strength interior laminae are significantly stressed in flexure. Also note that the modulus of elasticity is higher for the bending about the X–X axis than the Y–Y axis.

The modulus of elasticity E_0 is not associated with any one lamina group
but is derived from a transformed section analysis that considers the different
stiffness of the inner and outer lamina. In short, the modulus of elasticity is
a nominal value that is used with the gross cross-section moment of inertia
and used in classical beam analysis (flexural deformations only) to estimate
the deflection. The glulam combination symbols for elements that are principally loaded in axial compression and decking are given in Table 9.5. The
laminae of the same grade are used throughout the section.

The tables of glulam properties also give the basic shear F_{vo} and compression perpendicular to the grain F_{co} resistances.

Joints   The length of glulam elements typically exceed the length of the
lumber from which they are manufactured. Thus, joints are required in the
laminae. The most common types of joints are the scarf and finger joints,
which are illustrated in Figure 9.10. The manufacturing specifications require
that these joints be staggered so that the joints in neighboring laminae do not
occur in the same region. Failure seldom occurs in the joints for beam tests
conducted to failure. The manufacturing is controlled by AASHTO 168
(AASHTO, 1993).

TABLE 9.4 Base Resistance and Modulus of Elasticity, (MPa), for Glued Laminated Timber Used Primarily in Bending

| | | Bending About X-X Axis (Loaded Perpendicular to Wide Face of Laminations) | | | | | | Bending About X-X Axis (Loaded Parallel to Wide Faces of Laminations) | | | | | Axially Loaded | | |
| | | Extreme Tensile Fiber in Bending, F_{bo} | | Compression Perpendicular to Grain, F_{cpo} | | Horizontal Shear, F_{vo} | Modulus of Elasticity, E_x / E_o | Extreme Fiber in Bending, F_{bo} | Compression Perpendicular to Grain, F_{cpo} | Horizontal Shear, F_{vo} | For members with multi-piece lams which are not edge glued — Horizontal Shear F_{vo} | Modulus of Elasticity, E_y / E_o | Combination not recommended as primary application | | Modulus of Elasticity E_o |
Combination Symbol	Species Outer Laminations / Core Laminations	Positive Bending	Negative Bending	Tension Face	Compression Face								Tension Parallel to Grain, F_{to}	Compression Parallel to Grain, F_{co}	
Visually Graded Western Species															
20F V3	DF/DF	30	15	6.2	5.2	3.1	9300	22.1	5.2	2.8	1.4	8600	16.2	20.3	8620
20F V7	DF/DF	30	30	6.2	6.2	3.1	9300	22.1	5.2	2.8	1.4	9300	16.2	21.0	9310
20F V12	AC/AC	30	15	5.2	5.2	3.8	8600	18.2	4.5	3.1	1.4	7900	14.5	19.7	7930
24F V1	DF/WW	37	18	6.2	6.2	2.8	9700	19.0	2.4	2.4	1.4	7900	16.2	17.2	7930
24F V2	HF/HF	37	18	4.8	4.8	2.8	8600	19.0	3.5	2.4	1.4	7900	15.5	17.2	7930
24F V4	DF/DF	37	18	6.2	6.2	3.1	10 300	22.8	5.2	2.8	1.4	9300	18.6	21.7	9310
24F V8	DF/DF	37	37	6.2	6.2	3.1	10 300	22.1	5.2	2.8	1.4	9300	17.9	21.7	9310
24F V9	HF/HF	37	37	4.8	4.8	2.8	8600	22.8	3.5	2.4	1.4	7900	16.2	19.0	7930
24F V10	DF/DF	37	37	6.2	6.2	2.8	10 300	21.4	3.5	2.8	1.4	9300	18.6	21.0	9310

TABLE 9.4 (Continued)

Combination Symbol	Species Outer Laminations/Core Laminations	Bending About X-X Axis — Loaded Perpendicular to Wide Face of Laminations						Bending About X-X Axis — Loaded Parallel to Wide Faces of Laminations					Axially Loaded		
		Extreme Tensile Fiber in Bending, F_{bo}		Compression Perpendicular to Grain, F_{cpo}		Horizontal Shear, F_{vo}	Modulus of Elasticity, E_x / E_o	Extreme Fiber in Bending, F_{bo}	Compression Perpendicular to Grain, F_{cpo}	Horizontal Shear, F_{vo}	For members with multi-piece lams which are not edge glued Horizontal Shear F_{vo}	Modulus of Elasticity, E_y / E_o	Combination not recommended as primary application — Tension Parallel to Grain, F_{to}	Combination not recommended as primary application — Compression Parallel to Grain, F_{co}	Modulus of Elasticity E_o
		Positive Bending	Negative Bending	Tension Face	Compression Face										
Visually Graded Southern Pine															
20F V2	SP/SP	30	15	6.2	5.2	3.8	9310	22.1	5.17	3.4	1.7	7930	17.0	20.5	7930
20F V5	SP/SP	30	30	6.2	6.2	3.8	9310	22.1	5.17	3.4	1.7	7930	17.0	20.5	7930
24F V2	SP/SP	37	18.5	6.2	6.2	3.8	9650	24.5	5.17	3.4	1.7	8620	18.0	21.0	8620
24F V3	SP/SP	37	18.5	6.2	6.2	3.8	10 300	24.5	5.17	3.4	1.7	9310	18.5	22.5	9310
24F V4	SP/SP	37	18.5	6.2	5.2	1.7	9650	19.0	4.48	2.8	1.4	7930	14.0	13.0	7930
24F V5	SP/SP	37	18.5	6.2	6.2	3.8	9650	24.5	5.17	3.4	1.7	8620	18.5	22.5	8620
26F V2	SP/SP	40	20	6.2	6.2	3.8	11 000	15.9	6.21	3.4	1.7	9650	19.5	21.5	9650
26F V4	SP/SP	40	40	6.2	6.2	3.8	11 000	15.9	5.17	3.4	1.7	9650	18.5	21.0	9650

TABLE 9.5 Base Resistance and Modulus of Elasticity, MPa, for Glued Laminated Timber Used Primarily in Axial Loading

Bending About X-X Axis — Not recommended as primary application

Loaded Perpendicular to Wide Faces of Laminations

Bending About X-X Axis — Loaded parallel to Wide Faces of Laminations

Axially Loaded

Combination Symbol	Species	Grade	Modulus of Elasticity E_o	Compression Perpendicular to Grain, F_{co}	Tension Parallel to Grain, F_{co} 2 or more Lams.	Compression Parallel to Grain, F_{co} 4 or more Lams.	Compression Parallel to Grain, F_{co} 2 or 3 Lams.	Extreme Fiber in Bending, F_{bo} 4 or more Lams.	3 Lams.	2 Lams.	4 or more lams, (for components with multiple piece lams.)	Horizontal Shear, F_{vo} 4 or more Lams.	3 Lams.	2 Lams.	Extreme Fiber in Bending, F_{bo} 2 Lams. to 375 mm Deep	4 or more Lams.	Horizontal Shear F_{vo} 2 or more Lams.
1	DF	L3	8620	5.17	14.5	20.3	15.9	22.1	19.0	15.2	1.5	2.80	2.4	2.4	19.0	22.8	3.1
2	DF	L2	9650	5.17	20.3	25.2	21.0	27.6	24.5	19.7	1.5	2.80	2.4	2.4	25.9	30.3	3.1
3	DF	L2D	10 300	6.21	23.4	30.3	24.5	32.1	28.3	23.8	1.5	2.80	2.4	2.4	30.3	35.2	3.1
5	DF	L1CL	11 400	6.21	25.9	31.7	27.6	36.5	32.1	27.6	1.5	2.80	2.4	2.4	33.4	36.5	3.1
46	SP	L1	7580	5.17	14.5	20.0	8.96	22.1	19.0	15.2	1.7	3.50	3.1	2.8	15.2		3.8
47	SP	N3C	7930	5.17	19.3	25.2	15.2	26.6	23.8	19.7	1.7	3.50	3.1	2.8	21.4	24.5	3.8
48	SP	N3M	9650	6.21	22.8	29.0	17.9	30.3	27.6	22.8	1.7	3.50	3.1	2.8	24.5	29.0	3.8
49	SP	N2	9650	5.17	21.7	27.6	19.0	29.7	26.6	22.8	1.7	3.50	3.1	2.8	27.6	32.1	3.8
50	SP	N2D	11 000	6.21	25.2	30.3	22.4	35.2	32.1	26.8	1.7	3.50	3.1	2.8	32.1	36.5	3.8
52	SP	N1	11 000	6.21	24.5	29.0	24.5	36.5	35.2	29.7	1.7	3.50	3.1	2.8	32.1	36.5	3.8
69	AC	N1D	7580	4.48	11.4	15.2	15.2	15.2	13.5	10.7	1.5	3.1	2.8	2.8	15.2	17.6	3.5
70	AC	SS	7930	4.48	16.2	19.0	20.3	19.0	16.9	14.1	1.5	3.1	2.8	2.8	20.7	23.8	3.5
71	AC	SSD	9650	5.17	20.3	25.2	26.9	25.2	22.8	19.0	1.5	3.1	2.8	2.8	25.9	30.3	3.5

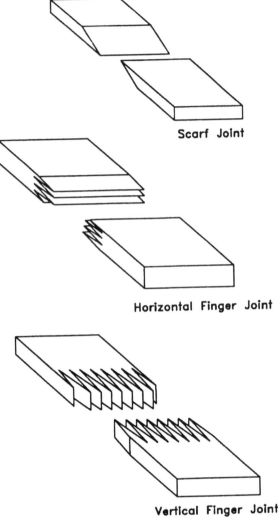

Scarf Joint

Horizontal Finger Joint

Vertical Finger Joint

Fig. 9.10 Joint types.

Appearance Glulam can be obtained in three appearance grades: industrial, architectural, and premium. The appearance grades include only surface considerations including growth characteristics, void filling, and sanding operations. Industrial grade is recommended for bridge application unless there are reasons for exceptional surface quality to satisfy special aesthetic requirements.

Standards Glulam elements must be manufactured per AASHTO 168 (AASHTO, 1993) and must be treated per the requirements of AASHTO

M133 (AASHTO, 1993). These standards outline the specific details required to meet performance requirements for strength, stiffness, and resistance to decay. The adhesives should be for wet-use and should comply with AASHTO D2559 (AASHTO, 1993).

Specification of Glulam   The dimensions used for the design and specification of glulam elements are the actual dimensions. The net dimensions of glulams are given in Table 9.6. The depth is calculated by multiplying the number of lamina times the net thickness of the dimension lumber, typically 1.5 in (38 mm).

9.2.5   Modification Factors for Resistance and Modulus of Elasticity

The basic resistance and modulus of elasticity values must be adjusted for several effects. These adjustments are

$$F = F_o C_F C_M C_D \tag{9.2}$$

and

$$E = E_o C_M \tag{9.3}$$

where

F is the applicable nominal resistance for F_b (bending), F_v (shear), F_t (tension), F_c (compression), or F_{cp} (compression perpendicular to grain) (MPA).

F_o is the basic resistance for F_{bo} (bending), F_{vo} (shear), F_{to} (tension), F_{co} (compression), or F_{cpo} (compression perpendicular to grain) (MPa).

TABLE 9.6   Net Dimensions of Glued Laminated Elements

Nominal Dimension (mm)	Western Species Net Finished Dimension (mm)	Southern Pine Net Finished Dimension (mm)
100	80	75
150	130	125
200	171	171
250	222	216
300	273	267
350	311	305
400	362	356

[a] AASHTO Table 8.4.1.2.2–1 [From *AASHTO LRFD Bridge Design Specifications*, Copyright © 1994 by the American Association of State Highway and Transportation Officials, Washington, DC. Used by Permission.]

E is the applicable modulus of elasticity (MPa).

E_o is the basic modulus of elasticity (MPa).

C_F is the size effect factor.

C_M is the moisture content factor.

C_D is the deck factor.

See Tables 9.4 and 9.5 for nominal resistance values.

Size Effects The chance of having a defect in a critical region increases as the size of the component increases. Hence, larger pieces are more likely to have strength reducing characteristics than smaller pieces. The converse is also true. The nominal strength values are based on lumber that is 12 in. (300 mm) deep. Thus, pieces that are smaller than 300 mm are typically stronger than the basic nominal values indicate and the required adjustment is upward, and vice versa for larger elements. Similarly, glulam elements require adjustment for dimensions of depth, width, and length. The adjustment factors for dimension lumber are given in AASHTO [Table A8.4.4.2-1] and are not repeated here. Adjustment for solid sawn elements is determined by

$$\text{if} \quad d \leq 300 \text{ mm, then } C_F = 1.0 \tag{9.4a}$$

and

$$\text{if} \quad d > 300 \text{ in., then} \quad C_F = \left(\frac{300}{d}\right)^{1/9} \tag{9.4b}$$

where d is the net depth (mm).

For horizontally glued laminated components (beams, not decks), when the depth, width, or length exceed 300 mm, 130 mm, or 6400 mm, then

$$C_F = \left[\left(\frac{300}{d}\right)\left(\frac{130}{b}\right)\left(\frac{6400}{L}\right)\right]^a \leq 1.0 \tag{9.5}$$

where

d is the depth of the component (mm).

b is the width of the component (mm).

L is the length of the component, measured between points of contraflexure (mm).

a is 0.05 for Southern Pine and 0.10 for all other species.

Moisture Content The basic resistances are based on 5% exclusion and 2-month load duration and the modulus of elasticity is based on mean values.

These values are based on assumed wet-use where the EMC is above 16%. If the EMC is below this threshold, and the system is glue laminated, then the basic resistance is adjusted upward by the moisture content factor C_M. For solid sawn lumber, no adjustment is required by AASHTO [A8.4.4.3], that is, $C_M = 1.0$. For glued laminated systems, the adjustments are given in Table 9.7. The regional climatic conditions should be investigated to estimate the in-service moisture conditions.

Deck Factors In laminated decks, it is unlikely that a localized failure on one lamina can cause a catastrophic failure or lead to undue serviceability problems. Therefore, it is reasonable to adjust the nominal strength upward for such a system. The adjustment factors are given in Table 9.8 [A8.4.4.4].

9.3 SERVICE LIMIT STATES

Experience has shown that wood structures that deflect excessively create problems with asphalt wearing surfaces and loosen connections between the beam and the deck (Ritter, 1990). To avoid such problems, AASHTO [A2.5.2.6.2] outlines optional deflection limits. In the design of wood bridges two are applicable and recommended:

$$\text{Live load deflection} \le \frac{\text{Span}}{425} \qquad (9.6)$$

For live load on wood planks and panels:

$$\begin{array}{l}\text{Relative live load deflection} \\ \text{between adjacent components}\end{array} \le 2.5 \text{ mm } (0.10 \text{ in.}) \qquad (9.7)$$

The limits defined in Eqs. 9.6 and 9.7 are applicable for live load with impact and are distributed to the beams assuming that all lanes are loaded and that all beams carry this load equally. This behavior implies that the distribution factor for this limit state (Service I) is equal to the number of lanes loaded, divided by the number of beams, or

$$g_{\text{deflection}} = \frac{\text{No. lanes loaded}}{\text{No. of beams}} = \frac{N_L}{N_b} \qquad (9.8)$$

TABLE 9.7 Moisture Content Factor, C_M for Glued Laminated Timber

	Property					
	F_{bo}	F_{to}	F_{co}	F_{cpo}	F_{vo}	E_o
C_M	1.25	1.25	1.35	1.90	1.15	1.20

TABLE 9.8   Deck Factor, C_D for Laminated Wood Decks

	F_{bo}	Property F_{to}	All Others
Select Structural	1.30	1.30	1.0
Grades 1 and 2	1.50	1.50	1.0

The load factors for Service Limit State I are given in Tables 3.1 and 3.2. The multiple presence factors should be applied (see Table 4.5). Typically two or more lanes loaded controls.

9.4   STRENGTH LIMIT STATES

In general, the strength limit state is defined by Eq. 3.3, restated here for convenience

$$\phi R_n \geq \eta_R \eta_I \eta_D \sum \gamma_i q_i \qquad (9.9)$$

where

ϕR_n is the factored resistance.

η_R is the redundancy factor.

η_I is the importance factor.

η_D is the ductility factor.

γ_i are load factors.

q_i is the load effect of load i.

The limit states associated with the selected design of glulam systems are described. Other limit states as described in AASHTO and/or other design specifications may certainly be applicable to other systems. The load modifier is to account for redundancy, ductility, and operational importance. The latter is often applicable for wood bridges. As these structures are mostly located on low-volume roads, the operational importance can be considered less than on the more well-traveled routes. The necessity of the bridge for emergency traffic also is a consideration. This provision is outlined in AASHTO [A1.3.5]. The importance factor may be varied from 0.95 to 1.05.

The resistance factors ϕ are defined in Table 9.9. As indicated in Eq. 9.9, these factors reduce the nominal resistance to the factored resistance level. The resistance factors for the strength combination associated with long-term sustained loads, AASHTO Strength Limit State IV, must be multiplied by 0.75 [A8.5.2.3]. This adjustment is required because the basic resistance val-

TABLE 9.9   Resistance Factors

Action	Resistance Factor
Flexure	0.85
Shear	0.75
Compression parallel to grain	0.90
Compression perpendicular to grain	0.90
Tension parallel to grain	0.80

ues are based on a load duration of 2 months. Sustained loads have a duration longer than 2 months, and therefore the resistance should be reduced (see Fig. 9.7).

Components in Flexure   In bridge engineering, flexure is always of concern and frequently controls the selection of components. The flexural action is illustrated in Figure 9.11(a). The factored resistance $M_r = \phi M_n$ is based on the familiar equation for relating flexural bending stress to bending moment and is given as

$$\phi M_n = \phi F_b S C_S \qquad (9.10)$$

where

F_b is the adjusted nominal flexural strength (MPa).
S is the elastic section modulus (mm^3).

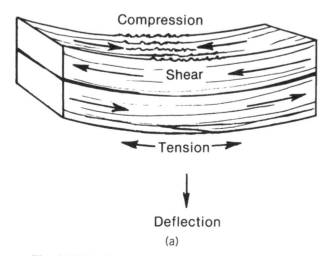

Fig. 9.11(a)   Wood in bending. [After Ritter, 1990.]

C_S is the stability factor, which is 1.0 for systems that are adequately braced to achieve a strength (no instabilities) failure mode. The spacing of the lateral bracing affects C_S. The parameter C_S is defined as

$$C_S = \frac{1 + A}{1.9} - \sqrt{\frac{(1 + A)^2}{3.61} - \frac{A}{0.95}} \le C_F \qquad (9.11a)$$

For visually graded sawn lumber,

$$A = \frac{0.438Eb^2}{L_e dF_b^*} \qquad (9.11b)$$

and for a glulam element and mechanically graded lumber,

$$A = \frac{0.609Eb^2}{L_e dF_b^*} \qquad (9.11c)$$

where L_e is the effective unbraced length, defined by

If $L_u/d < 7$, then $L_e = 2.06\ L_u$.
If $7 \le L_u/d \le 14.3$, then $L_e = L_u + 3d$.
If $L_u/d > 14.3$, then $L_e = 1.84\ L_u$.
$F_b^* = F_b$ without adjustments for size C_F and orientation C_D.

The parameter L_u is defined as the distance between points of lateral and rotational support (mm). The remaining terms have been previously defined in other equations. The coefficients in Eqs. 9.11(b) and 9.11(c) for solid sawn and glulam beam are different because of the inherent variability of the modulus of elasticity. The stability equations are based on the 5% exclusion values, but the mean modulus of elasticity is used in these equations. The 0.438 value in Eq. 9.11(b) includes a reduction of (divided by) 2.74 from the tabulated modulus of elasticity design values. For visually graded sawn lumber, this reduction represents an approximate 5% exclusion value with an additional factor of safety of 1.66. For glulam and machine graded lumber, the 2.74 reduction represents less than a 0.01% lower exclusion with a 1.66 safety factor. The AASHTO equations are a direct conversion from the NDS (1991) equations used in the working stress method.

If the component is circular, as in the case of a log bridge, then the resistance is

$$\phi M_n = \phi 1.18F_b\ S \qquad (9.12)$$

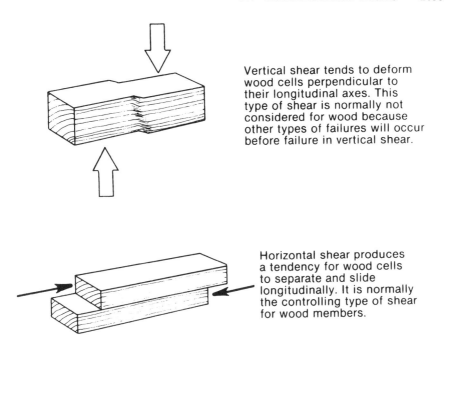

Vertical shear tends to deform wood cells perpendicular to their longitudinal axes. This type of shear is normally not considered for wood because other types of failures will occur before failure in vertical shear.

Horizontal shear produces a tendency for wood cells to separate and slide longitudinally. It is normally the controlling type of shear for wood members.

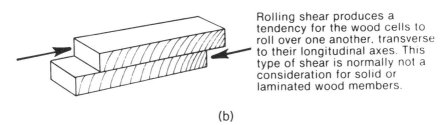

Rolling shear produces a tendency for the wood cells to roll over one another, transverse to their longitudinal axes. This type of shear is normally not a consideration for solid or laminated wood members.

(b)

Fig. 9.11(b)   Wood in shear. [After Ritter, 1990.]

The coefficient 1.18 is a form factor that accounts for the shape of the cross section. With strength parameters equal, a solid round section tests higher than a solid rectangular section. The form factor for other shapes may be found in the NDS (1991).

Components in Shear   Horizontal shear must be checked for components other than decks. The term "horizontal" shear is typically used in wood

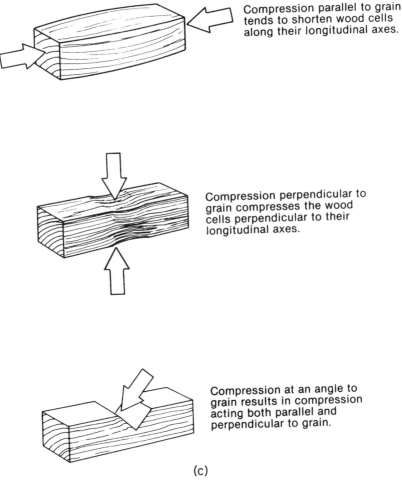

Compression parallel to grain tends to shorten wood cells along their longitudinal axes.

Compression perpendicular to grain compresses the wood cells perpendicular to their longitudinal axes.

Compression at an angle to grain results in compression acting both parallel and perpendicular to grain.

(c)

Fig. 9.11(c)   Wood in compression. [After Ritter, 1990.]

design because a shear failure initiates along the grain as shown in Figure 9.11(b). This shear failure is typically along the horizontal. The shear stress is equal in magnitude in the vertical direction but the inherent resistance is greater, and is not of concern. The AASHTO requirement is that the critical shear be investigated at a distance d away from the face of the support and that the live loads be placed for critical effect at a distance from the support equal to the lessor of either three times the depth ($3d$) or one-quarter of the span. The rationale for this provision is described with the aid of Figure 9.12 (NDS, 1991). Because of the possible presence of checking, some slippage occurs at the checks. This slippage creates a situation where the shear may be modeled by determining the resistance of the split cross section and the

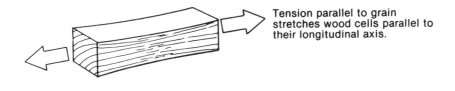

Tension parallel to grain
stretches wood cells parallel to
their longitudinal axis.

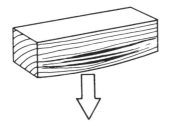

Tension perpendicular to grain
tends to separate wood cells
perpendicular to their axes,
where resistance is low.
Situations that induce this type
of stress should be avoided
in design.

(d)

Fig. 9.11(d)   Wood in tension. [After Ritter, 1990.]

resistance of the unsplit section. The solid line represents the reaction required by statics for any location of the concentrated load, that is, the influence function for the left reaction. The ordinate a represents the shear resisted by the unsplit section and the ordinate b represents the shear resisted by the split section. The resistance equation is based on the unsplit section and this shear is a maximum at approximately $3d$ away from the support. The factored shear resistance is based on the maximum shear stress, which occurs at the neutral axis,

$$\phi V_n = \frac{\phi F_v bd}{1.5} \tag{9.13}$$

Components in Compression

Compression parallel to the grain   A column is typically loaded with the compression parallel to the grain, which is limited by the nominal compressive stress (with adjustment) unless the stability is of concern. Compressive action parallel to the grain is illustrated in Figure 9.11(c). The resistance is given by

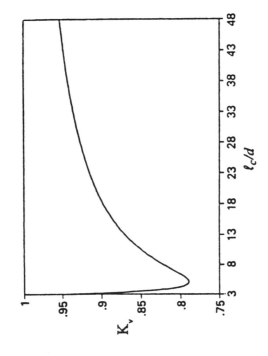

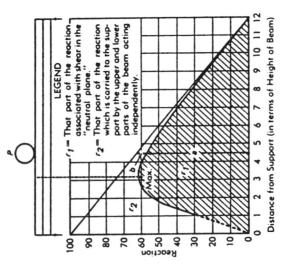

Fig. 9.12 Single beam in shear. [After National Design Specification, 1991. Used by permission.]

$$\phi P_n = \phi F_c A C_p \qquad (9.14)$$

where A is the area of the cross section (mm²) and C_p is the stability factor defined below.

For sawn lumber,

$$C_p = \frac{1 + B}{1.6} - \sqrt{\frac{(1 + B)^2}{2.56} - \frac{B}{0.80}} \le C_F \qquad (9.15)$$

for round piles,

$$C_p = \frac{1 + B}{1.7} - \sqrt{\frac{(1 + B)^2}{2.89} - \frac{B}{0.85}} \qquad (9.16)$$

for glulam and mechanically graded lumber,

$$C_p = \frac{1 + B}{1.8} - \sqrt{\frac{(1 + B)^2}{3.24} - \frac{B}{0.9}} \qquad (9.17)$$

But in the case for visually graded lumber,

$$B = \frac{4.32 \, Ed^2}{L_e F_b} \qquad (9.18)$$

for glulam and mechanically graded lumber,

$$B = \frac{60.2 \, Ed^2}{L_e F_b} \qquad (9.19)$$

where L_e is the effective column length (Euler buckling length, kL).

The rationale for the different coefficients in Eqs. 9.18 and 9.19 is attributed to different variabilities for solid sawn and glulam wood.

Compression perpendicular to the grain In the location of the supports or bearings, the wood must support compression perpendicular to the grain. Compressive action perpendicular to the grain is illustrated in Figure 9.11(c). The resistance is

$$\phi P_n = \phi F_{cpo} A_b C_b \qquad (9.20)$$

where A_b is the bearing area (mm²) and C_b is the bearing modification factor defined in Table 9.10.

TABLE 9.10   Modification Factors for Bearing

Length of bearing measured along the grain (mm)	13	25	38	50	75	100	≥ 150
C_b	1.75	1.38	1.25	1.19	1.13	1.10	1.00

Components in Tension   Tension elements are typically used in wood trusses and other discrete members, but herein it is of concern in the resistance of vehicle collision loads that impose tension in the deck. Examples of tensile actions are illustrated in Figure 9.11(d). The tensile resistance is

$$\phi P_n = \phi F_t A_n \qquad (9.21)$$

where F_t is the factored tensile resistance (MPa) and A_n is the smallest net area (area minus bolt hole areas) of the component (mm^2). Tensile actions can also be applied perpendicular to the grain as shown in Figure 9.11(d). Wood is extremely weak in tension transverse to the longitudinal direction of the cell orientation, or perpendicular to the grain. This situation should be avoided in design, and only under unusual circumstances should a detail be used that involves this type of loading. Generally, a reworking of details to avoid this situation provides a better design (NDS, 1991).

Components in Combined Flexure and Axial Loading   Components subjected to axial and flexural actions are combined with the interaction equation

$$\frac{P_u}{\phi P_n} + \frac{M_u}{\phi M_n} \leq 1.0 \qquad (9.22)$$

where

P_u is the factored axial force (N).
ϕP_n is the axial tensile resistance (N).
M_u is the factored bending moment (N mm).
ϕM_n is the flexural resistance (N mm).

Note that the resistance values include the length effects per Eqs. 9.10 and 9.14.

Bracing Requirements   Flexural and compression elements must be adequately braced to insure that instability does not occur. The resistance equations in such cases require that unbraced lengths be defined. The lateral support in the structure must be consistent with that assumed in the application of the resistance equations. This problem is discussed in more detail in the design procedures for glulam and in the design examples. Lateral support requirements are addressed in AASHTO [A8.11].

Camber Requirements   Glued and stress laminated systems may be constructed such that the components have an upward bow or camber. Here AASHTO requires that the camber be twice the dead load deflection [A8.12.1]. The upward camber deflection is decreased by the initial dead load of the system and the deflection due to long-term creep. The intent is to provide a camber such that the long-term deflection of the system is zero, providing a level bridge.

9.5   GLUED-LAMINATED BEAM SYSTEMS

A glulam beam bridge is constructed with a series of glulam beams oriented parallel to the direction of travel. Typically, three or more beams are uniformly spaced and covered with glulam deck panels oriented transversely. The deck panel may be either doweled together or placed adjacently with minimal structural connection. A common simply supported beam bridge is illustrated in Figure 9.13. The plan view is shown in Figure 9.13(a). Five beams are supporting a glulam deck. The beams are laterally supported by transverse bracing. These braces, also shown in Figure 9.13(c), are steel cross frames and do not attach directly to the deck panels. Glulam blocks may also be used as bracing. The traffic rail and its attachment to the curb and exterior beam are also illustrated. The center-to-center span length is measured from bearing to bearing and is shown in the elevation view [Fig. 9.13(b)]. The out-to-out span length dimension is also illustrated and is the total length of the beam, including the beam length plus the bearing length. A typical glulam beam bridge is shown in Figure 9.14.

9.5.1   Beam Design Procedures

The procedures for the design of beams are outlined in the following sections. In each section, the procedure is described with the applicable AASHTO Specification references and resistance equations. Where helpful, aids are given to facilitate design. The issues related to economy and constructability are also noted.

Determine the Basic Configuration   The project requirements are defined with respect to the functionality of the bridge, the expected loads, and discussions with the owner.
   Define the following:

1. Span length measured center-to-center of bearings: The span length is determined from the geometry of the crossing such as roadway geometrics or stream width and hydraulic requirements. The geology and

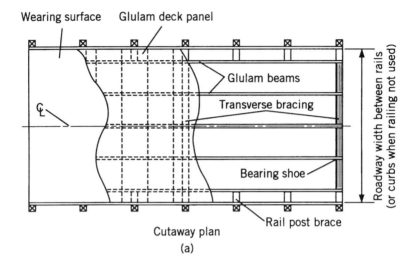

Wearing surface   Glulam deck panel

Glulam beams

Transverse bracing

Roadway width between rails
(or curbs when railing not used)

Bearing shoe

Rail post brace

Cutaway plan

(a)

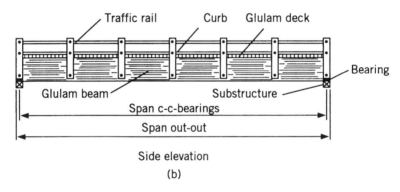

Traffic rail   Curb   Glulam deck

Bearing

Glulam beam   Substructure

Span c-c-bearings

Span out-out

Side elevation

(b)

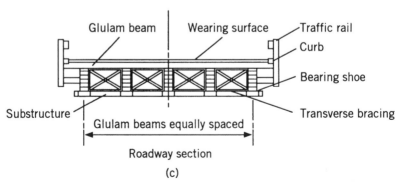

Glulam beam   Wearing surface   Traffic rail

Curb

Bearing shoe

Transverse bracing

Substructure

Glulam beams equally spaced

Roadway section

(c)

Fig. 9.13   Typical glulam beam bridge configuration. [After Ritter, 1990.]

Fig. 9.14   Wood bridge superstructure. [After Ritter, 1990.]

substructure requirements may dictate the location of the bents and abutments that affect the span length.

2. Roadway width measured to the face of the curb: The roadway and shoulder widths should meet the requirements of the bridge owner.

3. Depth limitations: The beam depth may be limited to accommodate an under pass clearance or freeboard between the bottom of the beam and the stream at flood stage.

4. Number of design lanes: The design lane is the lane designation used by the bridge engineer for live load placement. The design lane width and location may or may not be the same as the traffic lane. Here AASHTO uses a 3000 mm (10-ft) design lane and the vehicle is to be positioned within that lane for extreme effect (see Chapter 4, and AASHTO [A3.6.1.1]).

5. Number and spacing of beams: Economical spacing of beams is usually between 1500 and 1800 mm, inclusive. The number of beams is commonly three or greater with four or five typically used for two traffic lanes. The number and spacing of the beam may be estimated with the aid of Table 9.11.

6. The design loads and associated parameters are

Importance factor.

Type and depth of wearing surface.

TABLE 9.11 Recommended Beam Spacing for Glulam Beams with Transverse Glulam Decks

Roadway width (mm)[a]	Number of Beams	Beam spacing (mm)	Deck overhang (mm)[b]
One-Lane Bridge			
4 200	3	1650	450
4 800	3	1800	600
Two-Lane Bridge			
7 200	5	1500	600
7 800	5	1650	600
8 400	5	1800	600
10 200	6	1800	600

[a] Measured face-to-face of railings, or of curbs when a railing is not used.
[b] Measured from centerline of outside beam.

   Vehicles (AASHTO Design Vehicles and/or others).

   Basic wind speed.

   Seismic requirements.

   Pedestrian loads.

7. Live load deflection criteria, if applicable: The live load deflection requirement is optional but strongly recommended for wood systems with asphalt wearing surfaces. The loads for deflection are defined in AASHTO [A3.6.1.3.2] and deflection limit states are outlined in AASHTO [A2.5.2.6.2]. The load effect is the larger deflection resulting from the design truck alone and that resulting from 25% of the design truck taken together with the design lane load. The deflection should be limited to the span/425 and adjacent deck panel deflections to 2.5 mm (0.10 in.).

8. Deck and rail/curb configuration: Standard barrier rail designs are often used.

9. In-service moisture conditions: The EMC should be estimated for each component in the system.

Select the Beam Combination Symbol An initial combination symbol is selected and the basic resistance values and modulus of elasticity are determined. The combination symbol is a function of the species and the laminae lay-up. Commonly used combination symbols are shown in Table 9.12.

   The values for F_{bo}, F_{to}, F_{co}, F_{cpo}, F_{vo}, E_o are determined from Tables 9.4 and 9.5. It is convenient to establish the adjustment factors for the dry-use when referencing the basic resistance tables.

Determine Gravity Load Effects The deck depth and beam size are almost always controlled by the Strength Limit State *I* or Serviceability Limit State

TABLE 9.12 Commonly Used Glulam Combination Symbols

Beam Configuration	Western Species Combination Symbols	Southern Pine Combination Symbols
Single span	24F-V2	24F-V2
	24F-V4	24F-V3
		24F-V6
Continuous spans	24F-V8	24F-V5

1. The lateral loads seldom control the selection of these elements but may control the design of some details. The guard rail design is likely controlled by the extreme-event limit state, vehicle collision. The usual procedure is to first size the beams based on bending and then check for the remaining limit states. The order of resistance checks presented here is the order that most often controls the design, that is, bending, deflection, shear, and lateral loads. The deck design procedures are outlined later in this chapter.

The gravity load actions are established with the procedures outlined in Chapters 5 and 6. The shear and bending moments are established at the critical locations. Influence lines facilitate the analysis. Wood beams are often simply supported, which simplifies the analysis. Unless otherwise required by the owner, the AASHTO design loads should be used. The distribution factors for wood beam systems are given in Table 9.13. Alternatively, more rigorous methods such as grillage, finite element, and finite strip may be used. The

TABLE 9.13 Distribution Factors for Shear and Moment in Wood Beam Systems[a]

Location	Interior		Exterior	
Number of Design Lanes	One[b]	Two or More	One[b]	Two or More
Type of Deck[c]				
Plank	S/2000	S/2300	Lever rule	Lever rule
Stressed laminated	S/2800	S/2700	Lever rule	Lever rule
Spike laminated	S/2500	S/2600	Lever rule	Lever rule
Glued laminated panels on glued laminated beam	S/3000	S/3000	Lever rule	Lever rule
Glued laminated panels on steel beams	S/2670	S/2700	Lever rule	Lever rule

[a] In AASHTO Tables 4.6.2.2.2a–1, 4.6.2.2.3a–1, and 4.6.2.2.3b–1. [From *AASHTO LRFD Bridge Design Specifications,* Copyright © 1994 by the American Association of State Highway and Transportation. Officials, Washington, DC. Used by permission.]
[b] The distribution factors for one-lane loaded from AASHTO formulas incude $m = 1.2$ to account for the multiple presence. When the lever rule is used for one-lane loaded, the multiple presence factor of $m = 1.2$ must be used, that is, multiply the results by 1.2. When a lower multiple presence factor is used multiply the interior beam distribution by $m/1.2$ and the lever rule results by m.
[c] The range of application for planks is $S \leq 1500$ mm, for all other decks $S \leq 1800$ mm.

deck is orthotropic and should be modeled accordingly. If deflection controls the design, then there is little merit in pursuing the more rigorous methods because the distribution factor for deflection is prescribed in Eq. 9.8.

The dead load bending actions in the beams are based on the estimated beam, deck, and barrier rail weights. The final weights are not known until the design is complete. Hence, the procedure is iterative and the self-weight estimates should be revised as the procedure progresses.

The dead load actions are added to the live load actions after scaling each by the appropriate load factors given in Tables 3.1 and 3.2 for Strength *I*. This operation is defined by the right-hand side of Eq. 9.9.

Determine the Size Based on Bending   The size based on bending is controlled by Eq. 9.10 and is repeated here for convenience.

$$\text{Required} \quad M_u = \phi M_n = \phi F_b S C_S \tag{9.23}$$

Initially assume that the stability will not control the design and set C_S equal to 1.0. The factored resistance is determined by Eq. 9.23. Substitution of the deck factor $C_D = 1.0$, the moisture factor C_M, and the size factor C_F, defined by Eq. 9.5, into Eq. 9.2 yields ($C_D = 1.0$)

$$F_b = F_{bo} C_M \left(\frac{130}{b}\right)^{0.1} \left(\frac{300}{d}\right)^{0.1} \left(\frac{6400}{L}\right)^{0.1} \tag{9.24}$$

At this point in the design process, the center to center of bearing span length L, is known. A beam width b is tried. Recall that the section modulus for a rectangular section is

$$S = \frac{I}{c} = \frac{\frac{1}{12}bd^3}{\frac{1}{2}d} = \frac{bd^2}{6} \tag{9.25}$$

Substitution of Eqs. 9.4 and 9.25 into Eq. 9.23 yields

$$\phi M_n = \left[\phi F_{bo} C_M \left(\frac{130}{b}\right)^{0.1} \left(\frac{6400}{L}\right)^{0.1}\right]\left[\left(\frac{300}{d}\right)^{0.1}\left(\frac{bd^2}{6}\right)\right] C_S \tag{9.26}$$

Note that $300^{0.1} = 1.769$ and, upon substitution, Eq. 9.26 simplifies to

$$\phi M_n = \left[\phi F_{bo} C_M \left(\frac{130}{b*}\right)^{0.1} \left(\frac{6400}{L}\right)^{0.1}\right]\left[\frac{1.769\, bd^{1.9}}{6}\right] C_S \tag{9.27}$$

where the term in the second set of square brackets is a modified section modulus. All the known terms are grouped in the first set of square brackets and the unknown depth is found in the second set. Note that the width term in the size factor, denoted with the superscript*, is limited to 270 mm, hence the widths in Eq. 9.27 cannot be further combined. Equation 9.27 may be further "simplified" to ($C_S = 1.0$)

$$\phi M_n = \left[\phi F_{bo} C_M b \left(\frac{130}{b*} \right)^{0.1} \left(\frac{6400}{L} \right)^{0.1} \right] \left[\frac{1.769 \, d^{1.9}}{6} \right] \quad (9.28)$$

Collection of constants into a common constant yields

$$\phi M_n = \left[\phi \, 6.914 \, F_{bo} C_M b \left(\frac{1}{b*} \right)^{0.1} \left(\frac{1}{L} \right)^{0.1} \right] [d^{1.9}] \quad (9.29)$$

Equation 9.29 or Eq. 9.28 is used to determine the required depth d. We prefer Eq. 9.28 because the origin of each term is somewhat retained.

The required d is rounded upward to the nearest multiple of lamination depth. The actual depth is used in subsequent calculations. The stability factor C_S is initially assumed to be 1.0 in Eq. 9.28. Now that a trial section is known, the C_S factor may by determined by Eq. 9.11. The parameter C_S is limited by the size factor C_F. If C_F is lower than C_S then strength controls the design and C_S may be taken as 1.0. If C_S is lower, then stability controls the design and the calculated C_S must be used in the flexural resistance equation. From a practical and economical perspective, the lateral brace spacing may be based on the criterion that strength, not stability, controls the design. Although complex algebraic equations could be developed to establish the required lateral brace spacing, a few trials using Eqs. 9.11 and 9.28 will quickly yield a practical and constructable solution.

Check Live Load Deflection The live load deflection limit state may be checked either of two ways: (1) Determine the deflection for the trial section and compare this deflection to the limiting deflection. (2) Alternatively, set the deflection equal to the limiting deflection and solve for the depth required. Then use the larger of the depths required for deflection and deflection limit states. The deflection limit state is optional at the discretion of the owner, but is strongly encouraged for wood systems in order to maintain the wearing surface. The recommended deflection limit is the span/425.

The placement of the AASHTO design truck or design tandem, in combination with the design lanes for the absolute maximum deflection in the span, is an algebraically complex operation and is probably not warranted. The midspan deflection is sufficiently close to the absolute maximum and is used here. If the beam is continuous, then the calculation is further complicated. Influence functions must be generated for deflection at specific points assumed to be near the maximum deflection, likely between locations 104 and 105 (see Chapter 4 for span point notation). This process has been automated in several computer programs, for example, see BRASS (WYDOT,

1996) and BT Beam (BridgeTech, 1996). Alternatively, the midspan deflection of a simply supported beam may be determined and scaled by a continuity factor to account for the increase in stiffness and is discussed in more detail later.

The midspan deflection in a simply supported beam due to a concentrated load P at position $a = \beta L$ is

$$
\begin{aligned}
\Delta &= \frac{P\beta \, L^3(\frac{3}{4} - \beta^2)}{12EI} & 0 \le \beta \le \tfrac{1}{2} \\[2mm]
\Delta &= \frac{P(1 - \beta)L^3[\frac{3}{4} - (1 - \beta)^2]}{12 \, EI} & \tfrac{1}{2} \le \beta \le 1
\end{aligned}
\tag{9.30}
$$

where the terms are graphically depicted in Figure 9.15(a). In the likely case that more than one load is on the span simultaneously, then Eq. 9.30 may be used as the influence function for the midspan deflection where βL is the load position and Δ is the influence ordinate ($P = 1$). The critical position of the vehicle may be established by trial.

For continuous beams, continuity moments exist at the ends of the beam that stiffen the beam as compared to the simply supported case. The deflection

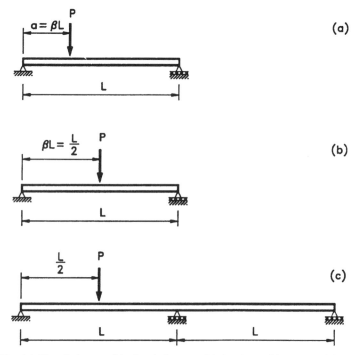

Fig. 9.15   (a) Simple beam, (b) simple beam with load at midspan, and (c) two-span beam.

as a function of position is given in Eq. 5.19 and is repeated here for convenience

$$y = \Delta = \frac{L^2}{6EI} [M_{ij}(2\varepsilon - 3\varepsilon^2 + \varepsilon^3) - M_{ji}(\varepsilon - \varepsilon^3)] \tag{9.31}$$

where x is the location where the deflection is determined, $\varepsilon = x/L$ and $y = \Delta$ is the upward translation.

The assumed maximum deflection is at midspan, therefore $\varepsilon = \frac{1}{2}$. Substitution of $\varepsilon = \frac{1}{2}$ into Eq. 5.19 yields

$$\Delta = \frac{L^2}{16EI} [M_{ij} - M_{ji}] \tag{9.32}$$

Recall that the moments are considered counterclockwise in Eqs. 5.19 and 9.31. The influence functions for the end moments for the continuous beams are available in AISC (1986) for various span ratios. For the case of equal span lengths, influence functions are available in Appendix A. As illustrated in the deck design example, Appendix A is quite helpful in estimating the deflections in a continuous system.

If influence coefficients or computer-based procedures are not available or desired, perhaps in the case of preliminary design, a simpler method is helpful. The simple-span deflection may be multiplied by a factor that accounts for continuity or increased stiffness. Consider the simple beam with a single concentrated load at midspan, as illustrated in Figure 9.15(b). By using Eq. 9.30 with $\beta = \frac{1}{2}$, the deflection at midspan determined with

$$\Delta(\text{simply supported}) = \frac{PL^3}{48EI} \tag{9.33}$$

Next, consider the midspan deflection of the two-span beam shown in Figure 9.15(c). This system conservatively models a continuous system with two or more spans of equal length. The deflection is

$$\Delta(\text{two spans}) = \frac{PL^3}{66.67 \ EI} \tag{9.34}$$

The ratio of the deflection for one- and two-span systems yields

$$\text{Deflection ratio} = \frac{48}{66.67} = 0.72 \tag{9.35}$$

Based on this ratio, the continuity factor required to convert the deflection from a simple beam analysis to that of a two-span system is 0.72. Systems with three or more spans would offer slightly more stiffness. For a five span beam, the midspan translation of the end span is

$$\Delta(\text{five spans}) = \frac{PL^3}{68.6\,EI} \tag{9.36}$$

which corresponds to a continuity factor of 48/68.6 = 0.70. Furthermore, in the case where the spans are short and a load is applied in adjacent spans the deflections are further reduced, albeit slightly. In summary, although not theoretically exact, the continuity factor approach offers simplicity and reasonable accuracy for deflection computations.

The distribution factor for deflection is not the same as for either shear or moment. As prescribed in AASHTO [A2.5.2.6.2], the deflection is assumed to be resisted by all beams equally. This assumption is somewhat inconsistent with theory and with the procedures for the distribution for internal actions. The reason for this provision is previous experience and it seems to work. The deflection distribution factor is given in Eq. 9.8.

Check Horizontal Shear The horizontal shear seldom controls the design except in very short spans where the ratio of the bending moment to shear force is low. The shear resistance check is best performed by using the section required by bending or deflection rather than solving for a required section depth (or size) directly. This suggestion is because the location of critical shear is considered to be at d away from the support (conservatively considered d away from the center of bearing). Furthermore, the live load shall be placed to create the most critical shear at a location of the smaller of $3d$ or $L/4$ [A8.7]. In short, the analysis is a function of the section depth, therefore use the best estimate for section depth at this point in the process. The analysis for shear is illustrated in the examples.

The shear resistance is checked per Eq. 9.13. If the check fails, then the required section is determined by using the estimated shear based on the depth

$$d(\text{required for shear}) = \frac{1.5\,V_u}{\phi\,F_v\,b} \tag{9.37}$$

Determine Camber The required camber is twice the dead load deflection [A8.12.1]. As with the distribution of live load for deflection, it is assumed that all beams resist deformation equally. This assumption is quite reasonable for a deflection calculation involving sustained load, because as one beam deflects more than its neighbors, perhaps due to creep or a lower modulus of elasticity, then the others tend to attract more load and deflect accordingly. Therefore, it is unlikely that one beam will deflect excessively with neighboring beams that do not. A simple way to perform this computation is to estimate the total load per foot of bridge length. This estimate includes beams, deck, wearing surface, rails, and utilities. By using the usual structural mechanics equations to predict the deflection, a simply supported system, for example, the deflection at midspan is

$$\Delta = \frac{5 w_{\text{total}} L^4}{384 \ EIN_b} \qquad (9.38)$$

where N_b is the number of beams, w_{total} is the weight per unit length of bridge, and the remaining terms were defined previously.

Design Bearing Length Component selection is not controlled by bearing stresses but rather the selected cross section is used to determine the required bearing length. Equation 9.20 is rewritten in the design format required

$$\text{Required bearing area} = A_b = \frac{P_u}{\phi \ F_{cp} C_b} \qquad (9.39)$$

Because the section width b is known, the length is established by

$$\text{Required bearing length } l_b = \frac{P_u}{\phi \ F_{cp} C_b b} \qquad (9.40)$$

This length is typically rounded upward to the nearest inch. The bearing pad is usually elastomeric and is designed independent of the type of beam supported. The design of the bearing pad is addressed later.

Check Lateral and Longitudinal Loads The lateral loads seldom control the component selection, but may control the details of the connections. The distribution and analysis of lateral loads are described in Chapter 6 and is not reiterated here. The first example provides a lateral load check of wind.

Design Transverse Bracing The transverse bracing must be sufficient to provide lateral support for the beams and support the wind loads. Often this is a very small force and the design is based on good detailing practice rather than controlled by forces. It is important to provide details that offer good connectivity with the beams and that shrinkage due to drying does not adversely loosen the system.

9.5.2 Design Examples

The normal procedure for the design of a beam bridge superstructure is to first design the barrier rail, next the deck and the beams, and finally details such as the connections and bearings. Although quite rational for design, it is easier from a pedagogic perspective to describe and design the beams first (knowing or assuming the deck design for dead load calculations). This approach is taken because almost all the design limit states and the associated resistance checks may be addressed in the beam design. The design of the deck and rail systems use only a subset of these. Hence, once the more comprehensive procedures are understood, the design of the other components can be illustrated by example without repeating a discussion of procedure. In short, the procedure is illustrated by example.

The limit states for wood design are presented in a previous section and the example procedures use this outline. After the beams are designed, the deck and rails are designed without a detailed discussion of procedure. Differences between the design of the beams and the other components are noted.

Example 9.1—Example Beam Design Design a glulam beam superstructure to span a 10 668 mm (35 ft) center to center of bearing. It must carry two traffic lanes and have a roadway width of 7300 mm (24 ft). Assume the following:

- The wearing surface is 76 mm (3 in.) asphalt (including allowance for a future overlay).
- AASHTO Design Vehicular Live Loads.
- A standard curb and rail are used and is estimated to weigh 0.73 N/mm (0.050 kips/ft).
- The deck is transversely glue laminated and is 130 mm (5.125 in) thick (the deck is designed in Example 9.2).
- The bridge is located on a secondary roadway and the operational importance factor required by the owner is 0.95 [A1.3.5].
- The bridge is located on a low-volume roadway with ADTT of approximately 50 trucks per day. The multiple presence factor may be decreased accordingly.
- The bridge is located near Laramie, WY and the pedestrian usage is minimal. Seismic requirements may be neglected for this example. The bridge is positioned at an elevation of 6100 mm (20 ft) above the surrounding terrain.

Step 1 Select the Basic Configuration The center to center (bearing) span length is given as $L = 10\,668$ mm. A roadway width of 7300 mm accommodates two traffic lanes, and 150 mm on each side is required for the barriers. No depth limitations were given in the problem statement and none are assumed. From Table 9.11 five beams are recommended for a 7600-mm bridge width and AASHTO loads are used as required. The bridge requires an operational importance $\eta_I = 0.95$. The deck and wearing surface are defined so dead weight computation may be made. The bridge site is known so the wind loads may be established. Laramie, WY is a very dry environment receiving approximately 300 mm precipitation annually, mostly in snow. The relative humidity may be conservatively estimated to be 30% and the temperature ranges between 27°C (80°F) and −34°C (−30°F). For this situation, based on Table 9.1, the EMC is estimated to be approximately 6%, well below the 16% required for wet-use. Therefore, dry-use is assumed in all resistances except for the bearing design. A standard barrier detail is used. The cross section is shown in Figure 9.16.

Step 2 Select the Beam Combination Symbol Because of the location of the bridge, western species are most appropriate. The span is a single simple span and Table 9.12 may be used to select a recommended symbol. The two symbols 24F-V2 and 24F-V4 reference the same flexural strengths, one is made of Douglas Fir (DF/DF) and the other is made of Hem Fir (HF/HF), and the principal differences are in the modulus of elasticity and compression perpendicular to the grain values. The engineer should determine the availability and make the selection accordingly. Here the Douglas Fir is used and the basic resistance properties are determined from Table 9.4. The moisture adjustment factors are found in Table 9.7 and the deck factors are 1.0 (Table 9.8). A summary of the resistance data and the associated adjustments are given in Table 9.14. No increase is taken for dry-use in the bearing area, that is, $C_M = 1.0$.

Step 3 Determine Gravity Load Effects The gravity load effects are composed of the self-weight of the components, the weight of the wearing surface, and the vehicle live load with the associated dynamic load allowance. Actions due to these loads are determined separately, then scaled by the appropriate load factors, and added. Four load cases must be considered for most resistance checks for the strength limit state: Interior beams with one-lane loaded, interior beams with multiple-lanes loaded, exterior beams with one-lane loaded, and exterior beams with multiple-lanes loaded. In general, both the dead loads and the live load distributions are different for each of these cases.

Component weights

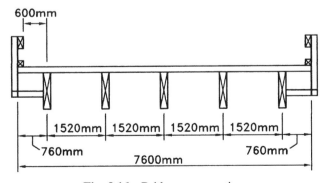

Fig. 9.16 Bridge cross section.

TABLE 9.14   Resistance Summary—Beam, 24F-V4

Action	Resistance (MPa)	C_M	C_D
F_{bo}	37	1.25	1.00
F_{cpo}	6.2	1.00	1.00
F_{vo}	3.1	1.15	1.00
E_o	10 300	1.20	1.00

$$\text{Deck } DC = (130 \text{ mm})(800 \text{ kg/m}^3)\left(\frac{\text{m}}{1000 \text{ mm}}\right)^3\left(\frac{9.81 \text{ N}}{\text{kg}}\right)$$

$$= 1.02 \times 10^{-3} \text{ N/mm}^2$$

$$\text{Deck } DW = (76 \text{ mm})(2250 \text{ kg/m}^3)\left(\frac{\text{m}}{1000 \text{ mm}}\right)^3\left(\frac{9.81 \text{ N}}{\text{kg}}\right)$$

$$= 1.68 \times 10^{-3} \text{ N/mm}^2$$

The beam weight is estimated to be 1.9 N/mm and the barrier rail is estimated to be 0.73 N/mm. The dead loads for the interior beams are based on the tributary length of 1520 mm (5 ft).

Deck $w_{DC}^I = (1520 \text{ mm})(1.02 \times 10^{-3} \text{ N/mm}^2) = 1.55 \text{ N/mm}$

Beam $w_{DC} = 1.90 \text{ N/mm}$

Total $w_{DC}^I = 3.45 \text{ N/mm}$

$$M_{DC}^I = \frac{w_{DC}^I L^2}{8} = \frac{(3.45 \text{ N/mm})(10 \text{ 668 mm})^2}{8} = 49.1 \times 10^{-6}$$

Asphalt $w_{DW}^I = (1520 \text{ mm})(1.68 \times 10^{-3} \text{ N/mm}^2) = 2.55 \text{ N/mm}$

$$M_{DW}^I = \frac{w_{DW}^I L^2}{8} = \frac{(2.55 \text{ N/mm})(10 \text{ 668 mm})^2}{8} = 36.3 \times 10^6 \text{ N mm}$$

where the superscript I indicates the load/action is for an interior beam. A superscript of E is used to denote exterior beams.

The dead load to the exterior beam is slightly greater than that obtained on a tributary basis. The influence functions introduced in Chapter 6 for the transverse analysis of decks may be used to establish the dead load to the exterior beam. Although the coefficients and associated areas are for a five beam system, they are sufficiently accurate to also model a four beam system. Refer to Appendix A and find the area under the influence function for the reaction at the exterior, that is, R_{200}. With the use of the cantilever length, $L = 760$ mm and the beam spacing $S = 1520$ mm, the net area is

$0.3928(S) = 0.3928(1520 \text{ mm}) = 597$ mm for the interior portion of the system and the area for the cantilever is

$$[1.0 + 0.635L/S]L = [1.0 + 0.635(760/1520)]760 = 1000 \text{ mm}$$

The total influence area (tributary length) is $597 + 1000 = 1597$, say 1600 mm.

$$\text{Deck } w_{DC}^E = (1600 \text{ mm})(1.02 \times 10^{-3} \text{ N/mm}^2) = 1.63 \text{ N/mm}$$

$$\text{Rail } w_{DC} = 0.73 \text{ N/mm}$$

$$\text{Rail load to beam} = 0.73 \left[1 + 1.27 \left(\frac{L = 760}{S = 1520} \right) \right] = 1.19 \text{ N/mm}$$

$$\text{Beam } w_{DC} = 1.90 \text{ N/mm}$$

$$\text{Total } w_{DC}^E = 1.63 + 1.19 + 1.90 = 4.72 \text{ N/mm}$$

$$M_{DC}^E = \frac{w_{DC}^E L^2}{8} = \frac{(4.72 \text{ N/mm})(10\ 668 \text{ mm})^2}{8} =$$

$$67.1 \times 10^6 \text{ N mm}$$

$$\text{Asphalt } w_{DW}^E = (1600 \text{ mm})(1.68 \times 10^{-3} \text{ N/mm}^2) =$$

$$2.69 \text{ N/mm}$$

$$M_{DW}^E = \frac{w_{DW}^E L^2}{8} = \frac{(2.69 \text{ N/mm})(10\ 668 \text{ mm})^2}{8}$$

$$= 38.3 \times 10^6 \text{ N mm}$$

The live load actions have been previously established in Example 5.10 and are summarized in Table 5.5. The critical moment is 654.8 kN m and is controlled by the combination of the design tandem and design lane loads. The design tandem and the design lane are $M_{\text{tandem}} = 522.5$ kN m and $M_{\text{lane}} = 132.9$ kN m, respectively. The moment is for the entire bridge and must be distributed to the individual beams. Per Table 9.13, the distribution factors for shear and moment for the interior beam are $S/3000 = 1520/3000 \cong 0.5$. The distribution factors for shear and moment for the exterior beam are established with the lever rule. The axle is placed as far out on the cantilever as permitted, that is, 600 mm from the curb or barrier rail. The curb/rail system is assumed to be 160 mm, therefore the outer wheel is placed at 600 mm from the rail as shown in Figure 9.17. Because the outer wheel is directly over the girder and the inner wheel is outside the exterior bay, the distribution factor is simply 0.5. It is convenient to include both the distribution factor and the multiple presence factor at the

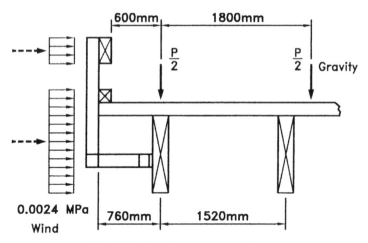

Fig. 9.17   Load over exterior girder.

same time.

The multiple presence factor for one-lane loaded is 1.2 assuming an average daily truck traffic count (ADTT) = 5000 [A3.6.1.1.2]. For bridges with ADTT = 100, this value may be factored by 0.90 or 0.90(1.2) = 1.08. This adjusted value is used to account for the reduced probability of attaining the design event with an ADTT of only 50 vehicles.

The combined factor η is

$$\eta = \eta_D \eta_I \eta_R$$

$$\eta_D = 1.05$$

$$\eta_I = 0.95$$

$$\eta_R = 0.95$$

$$\eta = 1.05(0.95)(0.95) = 0.95$$

The rationale for these assignments are that wood is nonductile, the owner specified that this bridge is of lesser importance, and the closely space stringers will permit some redundancy for capacity against total collapse if one should fail.

The load effects must be denoted by the load type, whether for an interior or exterior girder, and whether for single- or multiple-lanes loaded. Superscripts denote with Single or Multiple lanes loaded and Interior or Exterior beam and subscripts denote the load type. The factored moments are determined using the load factors from Table 3.1 and 3.2 and used to scale the moment from the dead and live loads. The dynamic load allowance for the design truck and tandem loads are taken as 0.165. Note that this allowance is neglected for the lane load. By proceeding in this manner, the vehicle and lane loads may be combined and then scaled.

$$M_u = \eta[\gamma_{DC}M_{DC} + \gamma_{DW}M_{DW}$$
$$+ mg\gamma_{LL} [M_{LL}^{\text{tandem}}(1 + IM) + M_{LL}^{\text{lane}}]]$$

$$M_u^{SI} = 0.95[1.25(49.1) + 1.5(36.3)$$
$$+ (1.08)(0.5)(1.75)[(522.5)(1.165) + (132.9)]] = 776 \text{ kN m}$$

$$M_u^{MI} = 0.95[1.25(49.1) + 1.5(36.3)$$
$$+ (1.00)(0.5)(1.75)[(522.5)(1.165) + (132.9)]] = 727 \text{ kN m}$$

$$M_u^{SE} = 0.95[1.25(67.1) + 1.5(38.3)$$
$$+ (1.08)(0.5)(1.75)[(522.5)(1.165) + (132.9)]] = 800 \text{ kN m}$$

$$M_u^{ME} = 0.95[1.25(67.1) + 1.5(38.3)$$
$$+ (1.00)(0.5)(1.75) [(522.5)(1.165) + (132.9)]] = 751 \text{ kN m}$$

The critical factored moment is 800×10^6 N mm $= 800$ kN m. The next step is to estimate the beam size based upon the critical moment.

Step 4 Determine the Beam Size Based on Flexure The flexural resistance is determined using Eq. 9.28 with $C_S = 1.0$

$$\phi M_n = \left[\phi F_{bo}C_M b \left(\frac{130}{b*}\right)^{0.1} \left(\frac{6400}{L}\right)^{0.1}\right]\left[\frac{1.769 \, d^{1.9}}{6}\right]$$

Assume a beam width of $b = 222$ mm and substitute the other known values to obtain

$$\phi M_n = \left[0.85(37)(1.25)(222) \left(\frac{130}{222}\right)^{0.1} \left(\frac{6400}{10\,668}\right)^{0.1}\right]\left[\frac{1.769 \, d^{1.9}}{6}\right]$$

which may be simplified to

$$\phi M_n = 2318d^{1.9}$$

Equating the resistance to the required factored moment yields

$$800 \times 10^6 = 2318 \, d^{1.9}$$

and the required depth is $d = 822$ mm. Divide this dimension by the lamina thickness of 38 mm to obtain the number of laminae required,

No. of laminae $= 822/38 = 21.6$, use 22 laminae or $d = 836$ mm

The beam weight is

Beam $w_{DC} = (222 \text{ mm})(836 \text{ mm})(7.84 \times 10^{-6} \text{ N/mm}^3) = 1.46$ N/mm

which is less than the 1.90 N/mm that was assumed in the analysis. Because the beam weight is a small fraction of the total load effect, it is not necessary to revise the analysis and iterate here. The size factor C_F is required in the stability check.

$$C_F = \left(\frac{130}{222}\right)^{0.1} \left(\frac{6400}{10\ 668}\right)^{0.1} \left(\frac{300}{836}\right)^{0.1} = 0.81$$

Finally, the flexural resistance is calculated as a check on the design calculations and to insure the flexural resistance is satisfied. For now, set $C_S = 1.0$. The result is

$$\phi\, M_n = \phi\, F_{bo} C_M C_F C_D C_S S$$

$$\phi\, M_n = (0.85)(37 \text{ MPa})(1.25)(0.81)(1.0)(1.0)[\tfrac{1}{6}(222 \text{ mm})(836 \text{ mm})^2]$$

$$\phi\, M_n = 823 \times 10^6 \text{ N mm} \geq 800 \times 10^6 \text{ N mm} \qquad \therefore \text{ OK}$$

Check Stability: The lateral bracing is assumed to be positioned at the quarter points or

$$L_u = \frac{10\ 668 \text{ mm}}{4} = 2667 \text{ mm}$$

$$\frac{L_u}{d} = \frac{2667 \text{ mm}}{836 \text{ mm}} = 3.2 \leq 7 \therefore \text{ use } L_e = 2.06\, L_u$$

$$L_e = 2.06\, L_u = 2.06(2667 \text{ mm}) = 5494 \text{ mm}$$

$$F_b^* = F_{bo} C_M = 37(1.25) = 46.3 \text{ MPa}$$

$$E = E_o C_M = 10\ 300(1.2) = 12\ 360 \text{ MPa}$$

$$A = \frac{0.609\ E\ b^2}{L_e d F_b^*} = \frac{0.609(12\ 360 \text{ MPa})(222 \text{ mm})^2}{5494 \text{ mm}(836 \text{ mm})(46.3 \text{ Pa})} = 1.74$$

$$C_S = \frac{1 + A}{1.9} - \sqrt{\frac{(1 + A)^2}{3.61} - \frac{A}{0.95}} \leq C_F$$

$$C_S = \frac{1 + 1.74}{1.9} - \sqrt{\frac{(1 + 1.74)^2}{3.61} - \frac{1.74}{0.95}}$$

$$C_S = 0.94 \geq C_F = 0.81 \qquad \therefore \text{ Strength Controls}$$

Because C_S is greater than C_F strength controls and the flexural limit state is satisfied. The trial section remains 836 mm (or 22 − 38 mm lams).

Step 5 Check Service Limit State—Live Load Deflection The live load deflection is determined with the aid of Eq. 9.30. This equation is the influence function for midspan deflection. The deflection limit should be checked for two load cases: (1) the design truck and, (2) 0.25 times the design truck plus the design lane load. The moment of inertia is

$$I = \tfrac{1}{12}bd^2 = \tfrac{1}{12}(222 \text{ mm})(836 \text{ mm})^3 = 10.8 \times 10^9 \text{ mm}^4$$

The design truck is positioned as shown in Figure 9.18(a) and the midspan deflection is

$$\beta = 0.5 \text{ for load 1 and } \beta = 0.097 \text{ for load 2,}$$

$$\Delta_{\text{Design truck}} = \sum_{i=1}^{2} \frac{P_i \beta_i L^3 (\tfrac{3}{4} - \beta_i^2)}{12 \ EI} =$$

$$\frac{145\,000 \text{ N}(\tfrac{1}{2})(10\,688 \text{ mm})^3(0.75 - 0.5^2)}{12(12\,360 \text{ MPa})(10.8 \times 10^9 \text{ mm}^4)}$$

$$+ \ \frac{145\,000 \text{ N}(0.097)(10\,668 \text{ mm})^3 \ [0.75 - (0.097)^2]}{12(12\,360 \text{ MPa})(10.8 \times 10^9 \text{ mm}^4)}$$

$$\Delta_{\text{Design truck}} = 28 + 8 = 36 \text{ mm}$$

The load is assumed to be shared equally between all girders, therefore the distribution factor is

$$\frac{N_L}{N_g} = \frac{2}{5} = 0.40$$

Applying the distribution factor yields

$$\Delta_{\text{Design truck}} = (0.40)(36) = 14 \text{ mm}$$

The midspan deflection due to the design lane load is

$$\Delta_{\text{Design lane}} = \frac{5 w_{\text{lane}} \ L^4}{384 \ E \ I} = \frac{5(9.3 \text{ N/mm})(10\,668 \text{ mm})^4}{384(12\,360 \text{ MPa})(10.8 \times 10^9 \text{ mm}^4)} = 12 \text{ mm}$$

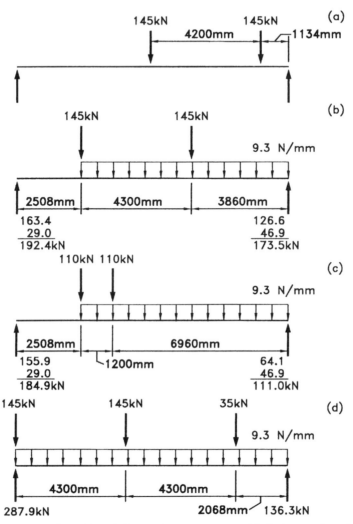

Fig. 9.18 (a) Load positioned for maximum moment, (b) design truck and design lane load positioned for maximum shear, (c) design tandem and design lane load positioned for maximum shear, and (d) design truck and design lane load positioned for maximum reaction.

The deflection due to the design lane is added to 25% of that attributed to the design truck,

$$\Delta = 12 \text{ mm} + 0.25(36 \text{ mm}) = 21 \text{ mm}$$

Again, application of the distribution factor yields

$$\Delta = 0.40(21 \text{ mm}) = 8 \text{ mm}$$

The critical deflection is 14 mm. The acceptable deflection is

$$\Delta_{\text{Limit}} = \frac{L}{425} = \frac{10\ 688 \text{ mm}}{425} = 25 \text{ mm, OK}$$

Step 6 Check Shear Limit State As previously outlined, the shear must be checked at a distance of d away from the face of the support and the live load is to be positioned at a distance of $3d$ or $L/4$, whichever is less. Using $d = 836$ mm, $3d = 3(836 \text{ mm}) = 2508$ mm, and $L/4 = 10\ 668/4 = 2667$ mm. Thus, the design truck should be positioned with the 145-kN axle at 2508 mm from the support. Because the bearing width is unknown at this point in the design, the bearing length is conservatively neglected in positioning the load and determining the critical shear at d away. The load position and reaction forces are illustrated in Figures 9.18(b) and 9.18(c). Note that the shear at d away from the support is the same as the reaction for the live load. The design truck shear of 163.4 kN is combined with that from the design lane of 29.0 kN to produce a total shear of 192.4 kN. Similarly, the design tandem combines with the lane to create a shear of 184.9 kN. The design truck shear is critical and is used in subsequent calculations.

The critical section is a d away from the support and the dead load shears must be determined at that location. The calculations follow:

$$V_{DC}^{I} = w_{DC}^{I}\left(\frac{L}{2} - d\right) = 3.45 \text{ N/mm} \left(\frac{10\ 668 \text{ mm}}{2} - 836 \text{ mm}\right) = 15.5 \text{ kN}$$

$$V_{DW}^{I} = w_{DW}^{I}\left(\frac{L}{2} - d\right) = 2.55 \text{ N/mm} \left(\frac{10\ 668 \text{ mm}}{2} - 836 \text{ mm}\right) = 11.5 \text{ kN}$$

$$V_{DC}^{E} = w_{DC}^{E}\left(\frac{L}{2} - d\right) = 4.72 \text{ N/mm} \left(\frac{10\ 668 \text{ mm}}{2} - 836 \text{ mm}\right) = 21.2 \text{ kN}$$

$$V_{DW}^{E} = w_{DW}^{E}\left(\frac{L}{2} - d\right) = 2.69 \text{ N/mm} \left(\frac{10\ 668 \text{ mm}}{2} - 836 \text{ mm}\right) = 12.1 \text{ kN}$$

The factored shears are determined in a manner similar to that used for moments.

$$V_u = \eta\{\gamma_{DC}V_{DC} + \gamma_{DW}V_{DW} + mg\gamma_{LL}\,[V_{LL}^{\text{truck}}((1 + IM) + V_{LL}^{\text{lane}}]\}$$

$$V_u^{SI} = 0.95\{1.25(15.5) + 1.5(11.5)$$

$$+ (1.08)(0.5)(1.75)[(163.4)(1.165) + (29.0)]\} = 231.7 \text{ kN}$$

$$V_u^{MI} = 0.95\{1.25(15.5) + 1.5(11.5)$$

$$+ (1.00)(0.5)(1.75)[(163.4)(1.165) + (29.0)]\} = 216.9 \text{ kN}$$

$$V_u^{SE} = 0.95\{1.25(21.2) + 1.5(12.1)$$

$$+ (1.08)(0.5)(1.75)[(163.4)(1.165) + (29.0)]\} = 239.2 \text{ kN}$$

$$V_u^{ME} = 0.95\{1.25(21.2) + 1.5(12.1)$$

$$+ (1.00(0.5)(1.75)[(163.4)(1.165) + (29.0)]\} = 224.8 \text{ kN}$$

The critical shear is $V_u = 239$ kN. The resistance is determined by using Eq. 9.13 and the appropriate adjustments for dry-use.

$$F_v = F_{vo}C_M = 3.1 \text{ MPa}(1.15) = 3.57 \text{ MPa}$$

$$\phi V_n = \frac{\phi F_v b\, d}{1.5} = \frac{0.75(3.57 \text{ MPa})(222 \text{ mm})(836 \text{ mm})}{1.5} = 331 \text{ kN}$$

Because the resistance of 331 kN exceeds the required factored shear of 239.2 kN, the section is adequate for shear.

Step 7 Determine the Bearing Length The required bearing area is proportional to the beam reaction. The dead load reactions are determined by

$$P_{DC}^I = \frac{w_{DC}^I L}{2} = 3.45 \text{ N/mm} \left(\frac{10\,668 \text{ mm}}{2}\right) = 18.4 \text{ kN}$$

$$P_{DW}^I = \frac{w_{DW}^I L}{2} = 2.55 \text{ N/mm} \left(\frac{10\,668 \text{ mm}}{2}\right) = 13.6 \text{ kN}$$

$$P_{DC}^E = \frac{w_{DC}^E L}{2} = 4.72 \text{ N/mm} \left(\frac{10\,668 \text{ mm}}{2}\right) = 25.2 \text{ kN}$$

$$P_{Dw}^E = \frac{w_{Dw}^E L}{2} = 2.69 \text{ N/mm} \left(\frac{10\,688 \text{ mm}}{2}\right) = 14.3 \text{ kN}$$

The live load reactions have been determined previously in Example 5.10 and are reported in Table 5.5. The critical reactions are 238.3 for the design truck 207.6 kN for the design tandem, and 49.6 kN for the design lane load [Fig. 9.18(d)]. Similar to the calculation for the live load moment and shear, the total reaction is distributed to the beam and at the same time the

multiple presence factor is applied.
The factor loads are determined below.

$$P_u = \eta\{\gamma_{DC}P_{DC} + \gamma_{DW}P_{DW} + mg\gamma_{LL}[P_{LL}^{truck}(1 + IM) + P_{LL}^{lane}]\}$$

$$P_u^{SI} = 0.95\{1.25(18.4) + 1.5(13.6)$$

$$+ (1.08)(0.5)(1.75)[(238.3)(1.165) + 49.6]\} = 335.0 \text{ kN}$$

$$P_u^{MI} = 0.95\{1.25(18.4) + 1.5(13.6)$$

$$+ (1.00)(0.5)(1.75)[(238.3)(1.165) + 49.6]\} = 313.2 \text{ kN}$$

$$P_u^{SE} = 0.95\{1.25(25.2) + 1.5(14.3)$$

$$+ (1.08)(0.5)(1.75)[(238.3)(1.165) + 49.6]\} = 344.1 \text{ kN}$$

$$P_u^{ME} = 0.95\{1.25(25.2) + 1.5(14.3)$$

$$+ (1.00)(0.5)(1.75)[(238.3)(1.165) + 49.6]\} = 322.2 \text{ kN}$$

The single-lane loading on the exterior girder creates the critical reaction
of 344.1 kN. Equation 9.20 is used to establish the required bearing length

$$l_b = \frac{P_u}{\phi F_{cpo}C_b b} = \frac{344\,100\ N}{0.9(6.2\ \text{MPa})(1.0)(222\ \text{mm})} = 278 \text{ mm, use } 300 \text{ mm}$$

The parameter C_b is defined in Table 9.10. Because the bearing length is
greater than 150 mm, then C_b is 1.0 as used above. Otherwise, an iterative
process is required. The required beam length is the center of bearing to
the center of bearing plus the length of the bearing, one-half on each side,
or 10 668 + 300 = 10 968 mm = 11 m.

Step 8 Determine Camber The total dead load per unit length of the bridge
is determined and applied on all five of the supporting beams. The loads
are

$$w_{DC} = 1.02 \times 10^{-3}\text{N/mm}^2 (7600 \text{ mm}) = 7.75 \text{ N/mm}$$

$$w_{DW} = 1.68 \times 10^{-3} \text{ N/mm}^2 (7600 \text{ mm}) = 12.8 \text{ N/mm}$$

$$w_{rails} = 0.73 \text{ N/mm (2 rails)} = 1.46 \text{ N/mm}$$

$$w_{beams} = 1.90 \text{ N/mm (5 beams)} = 9.5 \text{ N/mm}$$

$$w_{total} = 31.5 \text{ N/mm}$$

Recall that $E = 12\,360$ MPa and I is 12.4×10^9 mm⁴. Use these properties
to determine the midspan deflection

$$\Delta_{dead} = \frac{5(wL)L^3}{384\ N_b EI} = \frac{5(31.5\ \text{N/mm})(10\ 668\ \text{mm})(10\ 668\ \text{mm})^3}{384(5\ \text{beams})(12\ 360\ \text{MPa})(10.8\ \times\ 10^9\ \text{mm}^4)} = 8\ \text{mm}$$

The required camber is twice the immediate dead load deflection, or

$$\Delta_{camber} = 2\ \Delta_{dead} = 16\ \text{mm}$$

The camber should be specified on the contract plans.

Step 9 Check Lateral Load due to Wind The bridge is located in Laramie, WY where the basic design speed of 100 mph is conservative and reasonable. Per Table 4.9, the basic wind pressure associated with this velocity (including gusting) is 0.0024 MPa (50 psf). The bridge is located at an elevation of less than 30 ft above the surrounding terrain, so no adjustment is required to account for the velocity profile. Therefore, $P_B = P_D = 0.0024$ MPa. This uniform pressure is shown in Figure 9.17. Note that the deck and rail have wind applied directly, but the associated areas are small, and the design of these elements is not controlled by wind. The rail is controlled by vehicle collision and the deck is controlled by gravity load effects. Thus, the areas associated with these elements are combined with that of the beam, and the total load is conservatively assigned to the beam. Assume the rail has a depth of 380 mm and the deck and curb has a depth of 270 mm. These sizes are the results of the designs given in the following examples. The beam depth is 836 mm. The total area per unit length is $380 + 270 + 836 = 1486$ mm. Application of the design wind pressure to this area yields

$$w_{WS} = (0.0024\ \text{MPa})(1486\ \text{mm}) = 3.57\ \text{N/mm}$$

AASHTO 3.8.1.2 states that wind on beam components should not be less than 4.4 N/m, therefore use $w_{WS} = 4.4$ N/m.

Strength Limit State III is applicable for wind on the structure, or for bending moment,

$$M_u = \delta_{DC}M_{DC} + \delta_{DW}M_{DW} + 1.4\ M_{WS}$$

The beam moment is conservatively approximated with

$$M_{WS} = \frac{1}{10}\ w_{WS}L_B^2 = \frac{1}{10}\ (4.4\ \text{N/mm})(2667\ \text{mm})^2$$
$$= 3.13\ \times\ 10^6\ \text{N mm} = 3.13\ \text{kN m}$$

where L_B is the distance between bracing. The bending of the beam is about its minor axis, and therefore the resistance and section modulus must be based upon minor axis values. Per Table 9.4, $F_{bo} = 22.8$ MPa, and as before, the C_M is 1.25. The minor axis section modulus is also used. The resistance calculations follow:

$$F_b = F_{bo}C_M = 22.8 \text{ MPa}(1.25) = 28.5 \text{ MPa}$$

$$S = \tfrac{1}{6} bd^2 = \tfrac{1}{6}(836 \text{ mm})(222 \text{ mm})^2 = 6.87 \times 10^6$$

$$\phi M_n = \phi F_b S = 0.85(28.5 \text{ MPa})(6.87 \times 10^6)$$

$$= 166 \times 10^6 \text{ N mm} = 166 \text{ kN m}$$

The resistance is 166 kN m and the required wind load moment without dead load is $M_u = 1.4(3.13) = 4.38$ kN m. It is obvious that the design and the requirements of Strength Limit State III are met even with the addition of the dead load effects, which the wind does not control. In short,

$$\phi M_n \gg M_u$$

The total wind load to the bearings is

$$R_{ws} = 1.4(4.4 \text{ N/mm}) \left(\frac{10\ 668 \text{ mm}}{2} \right) = 32.9 \text{ kN}$$

to each side.

Step 10 Bearing Reaction to Braking The bearings pad is designed on the basis of the forces and deformations expected. Even though bearing design is not discussed in this chapter, the forces required from braking are determined here for completeness. Per Eq. 4.9,

$$F_{BR} = bW = 0.25(325) = 81.3 \text{ kN}$$

is likely to be resisted by one side of the bridge at the location of the fixed bearings.

Step 11 Summarize Design Results The beam is 222 mm × 836 mm using the Combination Symbol 24F-V4. Design results are summarized in Table 9.15.

Deck Design The design limit states and other design requirements for the deck are a subset of those for the beam design. Therefore the procedures are outlined in Example 9.2 and are not outlined explicitly in a general manner as for beam design. The steps required are listed:

- Select the combination symbol.
- Determine the basic resistance.
- Determine the gravity load bending moments.

TABLE 9.15 Beam Design Summary

Limit State	Provided	Required[a]
Flexure	823 kN m	800 kN m
Shear	331 kN	239 kN
Bearing length	300 mm	278 mm
Live load deflection	21 mm	25 mm
Camber	16 mm	N/A
Braking force	Not calculated	81.3 kN
Lateral bracing spacing	2667 mm	N/A
Wind to bearings	Not calculated	32.9 kN

[a]not applicable = N/A

- Determine the cantilever bending moment due to collision.
- Design depth based on flexure.
- Check flexure and tension interaction.
- Check shear.
- Check deflection limit(s).
- Investigate other combination symbols and deck thicknesses (optimize).
- Design the panel connections (dowels or stiffening beams).
- Summarize the design.

Example 9.2   Deck Design Example. Design a glulam deck for the bridge described in Example 9.1.

Step 1 Select the Combination Symbol Combination Symbol No. 2 gives a reasonable trial selection. As the design process proceeds, other symbols are considered. The basic resistance properties with the associated adjustments are given in Table 9.16.

Note that the wet-use moisture adjustment factor is used for the deck and the deck factor is 1.5 for bending and tension.

TABLE 9.16 Resistance Summary—Combination Symbol No. 2

Action	Resistance, (MPa)	C_M	C_D
F_{bo}	27.6	1.00	1.50
F_{cpo}	5.17	1.00	1.00
F_{vo}	1.50	1.00	1.00
F_{to}	20.3	1.00	1.50
E_o	9650	1.00	1.00

Step 2 Determine Gravity Bending Load Moments The dead load consists of the deck self-weight, rail, and wearing surface, which are all uniformly distributed along the bridge length, and therefore actions associated with these loads may be established on a unit length basis. The controlling live load is the design truck, which is comprised of a series of concentrated loads. The actions created by concentrated loads are greatest near the point of load applications and decrease with increasing distance from that point. As described in Chapter 6, it is necessary to model the deck as a continuous transverse beam supported by beams that provide rigid support. The width of the continuous beam is established by using Table 9.17.

Note that the systems that inherently have more continuity, such as the interconnected decks, distribute the load over a larger length and have larger strip widths.

The example deck is interconnected with dowels. At this point in the design, a trial deck thickness must be assumed so that the equivalent strip width may be determined for the live load and the deck self-weight may be established for the dead load. Use a trial deck thickness of 130 mm.

The equivalent strip width is

$$SW = 4h + 760 = 4(130 \text{ mm}) + 760 = 1280 \text{ mm}$$

TABLE 9.17   Equivalent Strip for Wood Decks[a]

System Type	Direction of Design	Strip Width[b] (mm)
Noninterconnected glulam	Parallel	$2.0h + 760$
	Perpendicular	$2.0h + 1020$
Interconnected glulam	Parallel	$2280 + 0.07L$
	Perpendicular	$4.0h + 760$
Stress-laminated	Parallel	$0.066S + 2740$
	Perpendicular	$0.84S + 610$
Spike-laminated Continuous or interconnected panels	Parallel	$2.0h + 760$
	Perpendicular	$2.0h + 1020$
Spike-laminated Non-interconnected panels	Parallel	$2h + 760$
	Perpendicular	$2h + 1020$
Planks	Parallel	Plank width
	Perpendicular	

[a]In AASHTO Table 4.6.2.1.3-1. [From *AASHTO LRFD Bridge Design Specifications*, Copyright © 1994 by the American Association of State Highway and Transportation Officials, Washington, DC. Used by permission.]
[b]The deck thickness = h (mm), the interior bay width = S (mm), and the span length of the deck = L (mm).

Step 3 Determine the Live Load Moments The design truck axle must be positioned transversely on the deck for critical effect as shown in Figure 9.19(a). The influence functions in Appendix A are used. Several positions are checked and the critical moment is used. First position the axle as far left on the overhang as possible. For the deck design, this is at 300 mm from the curb, which is 360 mm from the edge of the deck or 300 mm from the center of bearing of the outside beam. The continuous beam model is illustrated in Figure 9.19(b). The moment over the first support, that is, at Location 200 is

$$M_{LL}^{200} = 72.5 \text{ kN}(-1.0)(L = 0.3 \text{ m}) + 72.5 \text{ kN}(0)(S = 1.520 \text{ m})$$

$$= -21.8 \text{ kN m}$$

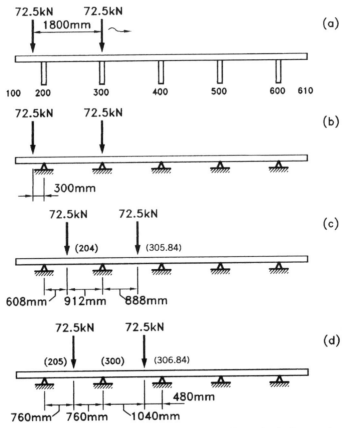

Fig. 9.19 (a) Design truck positioned for maximum effect, (b) design truck positioned for maximum cantilever moment, (c) design truck positioned for maximum moment at Location 204, and (d) design truck positioned for maximum moment at Locations 205 and 300.

Next, position the truck with the left wheel 40% of the way across the first bay (at 204) as illustrated in Figure 9.19(c). The moment at Location 204 is

$$M_{LL}^{204} = 72.5 \text{ kN}(0.2040)(S = 1.520 \text{ m}) + 72.5 \text{ kN}(-0.0257)(S = 1.520 \text{ m})$$

$$= 19.7 \text{ kN m}$$

Next move the axle slightly right so the left wheel is at Location 205 as illustrated in Figure 9.19(d). The moment at Location 205 is

$$M_{LL}^{205} = 72.5 \text{ kN}(0.1998)(S = 1.520 \text{ m}) + 72.5 \text{ kN}(-0.0250)(S = 1.520 \text{ m})$$

$$= 19.3 \text{ kN m}$$

Finally, position the left wheel at Location 205 [same as Fig. 9.19(d)] and determine the moment over the second beam, M_{300} is

$$M_{LL}^{300} = 72.5 \text{ kN}(-0.1004)(S = 1.520 \text{ m})$$

$$+ 72.5 \text{ kN}(-0.0500)(S = 1.520 \text{ m})$$

$$= -16.6 \text{ kN m}$$

These moments are assumed to be distributed uniformly over the strip width of 1280 mm. The moment due to the one-lane loaded case controls this deck design and multiple-lane cases are not investigated. Recall the multiple presence factor for one-lane loaded case is 1.08.

The uniformly distributed dead loads are the deck self-weight (*DC*) and the wearing surface (*DW*). These calculations are shown below.

Component weights:

$$\text{Deck } DC = (130 \text{ mm})(800 \text{ kg/m}^3)(\text{m/1000 mm})^3 \left(\frac{9.81 \text{ N}}{\text{kg}}\right)$$

$$= 1.02 \times 10^{-3} \text{ N/mm}^2$$

$$\text{Deck } DW = (76 \text{ mm})(2250 \text{ kg/m}^3)\left(\frac{\text{m}}{1000 \text{ mm}}\right)^3 \left(\frac{9.81 \text{ N}}{\text{kg}}\right)$$

$$= 1.68 \times 10^{-3} \text{ N/mm}^2$$

$$\text{Rail } DC = 0.073 \text{ N/mm}$$

The influence functions in Appendix A are again used. The dead load moments are established on the basis of a design strip of unit length, that is, 1.0 mm. The uniformly distributed loads (N/mm²) are applied to this strip and become a distributed line load on the continuous beam model (in N/mm). The rail load becomes a concentrated load on the design strip. The area under the functions is used with the distributed load and the ordinate at the end of the cantilever is used for the rail load.

The lower case letters denote that the actions are determined on a unit length basis. The calculations follow.

Component Moments:

$$m_{DC}^{200} = (1.02 \times 10^{-3} \text{ N/mm})(-0.500)(760 \text{ mm})^2 +$$

$$(-1.000)(0.73 \text{ N})(760 \text{ mm}) = -849 \text{ N mm/mm}$$

$$m_{DC}^{204} = (1.02 \times 10^{-3} \text{ N/mm})[(-0.2460)(760 \text{ mm})^2 +$$

$$(0.0772)(1520 \text{ mm})^2] + (-0.4920)(0.73 \text{ N})(760 \text{ mm}) =$$

$$- 236 \text{ N mm/mm}$$

$$m_{DC}^{205} = (1.02 \times 10^{-3} \text{N/mm})[(-0.1825)(760 \text{ mm})^2 +$$

$$(0.0714)(1520 \text{ mm})^2] + (-0.3650)(0.73 \text{ N})(760 \text{ mm})$$

$$= -142 \text{ N mm/mm}$$

$$m_{DC}^{300} = (1.02 \times 10^{-3} \text{ N/mm})[(0.135)(760 \text{ mm})^2 +$$

$$(-0.1071)(1520 \text{ mm})^2] + (0.2700)(0.73 \text{ N})(760 \text{ mm}) =$$

$$-23 \text{ N mm/mm}$$

Wearing Surface Moments:

$$m_{DW}^{200} = (1.68 \times 10^{-3} \text{ N/mm})(-0.500)(760 \text{ mm})^2 = - 485 \text{ N mm/mm}$$

$$m_{DW}^{204} = (1.68 \times 10^{-3} \text{ N/mm})[(-0.2460)(760 \text{ mm})^2 +$$

$$(0.0772)(1520 \text{ mm})^2] = 60.9 \text{ N mm/mm}$$

$$m_{DW}^{205} = (1.68 \times 10^{-3} \text{ N/mm})[(-0.1825)(760 \text{ mm})^2 +$$

$$(0.0714)(1520 \text{ mm})^2] = 100 \text{ N mm/mm}$$

$$m_{DW}^{300} = (1.68 \times 10^{-3} \text{ N/mm})[(0.135)(760 \text{ mm})^2 +$$

$$(-0.1071)(1520 \text{ mm})^2] = -285 \text{ N mm/mm}$$

The factored moments are determined in a manner similar to the one presented earlier for the beam design. The calculations are shown below.

$$m_u = \eta\{\gamma_{DC}m_{DC} + \gamma_{DW}m_{DW} + mg\gamma_{LL}$$
$$[m_{LL}^{truck}(1 + IM)]\}$$

$$m_u^{200} = 0.95\{1.25(-849) + 1.5(-485) + 1.08\left(\frac{1}{1280}\right)(1.75)$$

$$[-21.8 \times 10^6(1.165)]\} = -37\ 300\ \text{N mm/mm (governs)}$$

$$m_u^{204} = 0.95\{1.25(-236) + 1.5(60.9) + 1.08\left(\frac{1}{1280}\right)(1.75)$$

$$[19.7 \times 10^6(1.165)]\} = 32\ 000\ \text{N mm/mm}$$

$$m_u^{205} = 0.95\{1.25(-142) + 1.5(100) + 1.08\left(\frac{1}{1280}\right)(1.75)$$

$$[19.3 \times 10^6(1.165)]\} = 31\ 500\ \text{N mm/mm}$$

$$m_u^{300} = 0.95\{1.25(-23) + 1.5(-285) + 1.08\left(\frac{1}{1280}\right)(1.75)$$

$$[-16.6 \times 10^6(1.165)]\} = -27\ 600\ \text{N mm/mm}$$

The moment over the first beam is the critical moment for gravity load. Next the moment at this same location is determined for collision.

Step 4 Determine the Cantilever Bending Moment Due to Collision The collision load is defined by the Performance Limit—1 loading (*PL-1*). These loads are defined in Table 4.5 and are illustrated in Figure 9.20(a). The height to the middle of the rail is assumed to be 600 mm, or 90 mm above the minimum required. The deck thickness is assumed to be 130 mm, so the 120-kN load is applied at a distance above the centroid of the deck of $600 + 0.5(130) = 665$ mm. The 20-kN downward load, also illustrated in Figure 9.20(a), must also be applied, but not simultaneously with the lateral collision load. The lateral load is due to the impact and the downward load is due to the vehicle sitting on top of the rail after impact. As illustrated in Figure 9.20(b), the lever arm for the 20-kN load is 760 mm minus one-half of the railwidth, or conservatively estimated as 760 mm with an associated moment of

$$M_{CT}^{200} = 20\ 000\ \text{N}(760\ \text{mm}) = 15.2 \times 10^6\ \text{N mm}$$

The forces in the bolts are denoted as F_{curb} and are determined by balancing moments about point A [Fig. 9.20(b)].

$$F_{curb} = \frac{(690\ \text{mm} + 530\ \text{mm})(120\ \text{kN})}{690\ \text{mm}} = 212\ \text{kN}$$

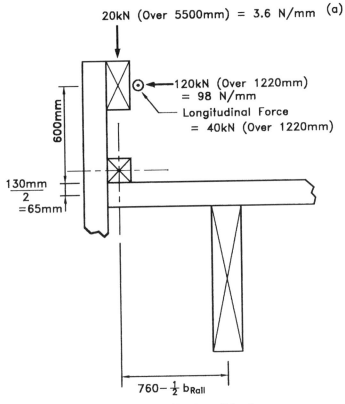

20kN (Over 5500mm) = 3.6 N/mm (a)

120kN (Over 1220mm) = 98 N/mm

Longitudinal Force = 40kN (Over 1220mm)

600mm

$\frac{130mm}{2}$ =65mm

760− $\frac{1}{2}$ b$_{Rall}$

Fig. 9.20 (a) Barrier rail loads.

This force is applied to the curb at the bolt line located at the middepth of the curb. The curb is assumed to be a nominal 6 × 6 and the required lever arm 132 mm, which results in a moment at the face of the facia beam of

$$M_{CT}^{200} = 212 \text{ kN}(132 \text{ mm}) = -28.0 \times 10^6 \text{ N mm}$$

The larger collision moment controls. This moment is distributed over the same strip width as the gravity load. The collision moment for the design strip is

$$m_{CT}^{200} = \frac{-28.0 \times 10^6}{1280 \text{ mm}} = -21\,900 \text{ N mm/mm}$$

The load combination for the collision load is controlled by the Extreme

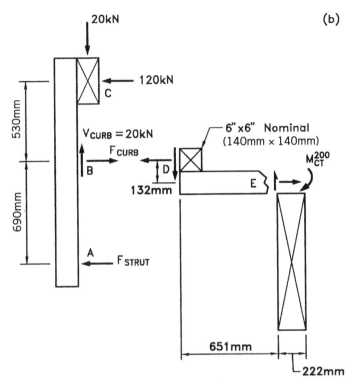

Fig. 9.20 (b) Barrier rail load analysis.

Event Limit State II. The load factors for this limit state are given in Tables 3.1 and 3.2. The factored moment calculation follows:

$$m_u^{200} = 0.95[1.25(-849) + 1.5(-485) + 1.0(-21\ 900)]$$

$$= -22\ 500\ \text{N mm/mm}$$

This moment is compared to the gravity moment of 37 300 N mm/mm, which is greater, and therefore Strength Limit State I controls. There is a slight possibility that the collision load creates enough tension stress in the deck, so that when it is combined with the collision moment, it could control. This combination is checked later. The section depth is established on the basis of 37 300 N mm/mm.

Step 5 Determine the Depth Based on Flexure The required deck thickness is determined on the basis of flexure. The basic resistance is

$$F_b = F_{bo}C_MC_FC_D = 27.6(1.0)(1.0)(1.50) = 41.4 \text{ MPa}$$

The section modulus for a unit width (1 mm) is

$$S = \tfrac{1}{6}bd^2 = \tfrac{1}{6}(1 \text{ mm})d^2 = \frac{d^2}{6} \text{ mm}^3$$

and the flexural resistance is

$$\phi\, m_n = \phi\, F_b S = 0.85(41.4)\left(\frac{d^2}{6}\right) = 5.865d^2$$

Equating the required factored moment to the resistance yields

$$m_u = \phi\, m_n$$

$$37\ 300 \text{ N mm/mm} = 5.865d^2$$

$$d = 80 \text{ mm}$$

Based solely on flexural limit state, the deck could be sized at 80 mm. Further, other limiting states such as deflection or collision with tensile action should be checked prior to resizing. The design will be continued on the basis of the 130 mm thickness.

Step 6 Check Flexure Combined with Tension for the Collision Load By previous calculations, the tensile action in the deck due to the 120-kN collision load is 197 kN. This load is distributed over a length of 1280 mm, and the resulting action is

$$p_u = \frac{197\ 000 \text{ kN}}{1280 \text{ mm}} = 154 \text{ N/mm}$$

The basic tensile resistance is

$$F_t = F_{to}C_MC_FC_D = 20.3(1.0)(1.0)(1.50) = 30.5 \text{ MPa}$$

and the tensile resistance is

$$\phi \, p_n = \phi F_t \, A_n = 0.80(30.5 \text{ MPa})(130 \text{ mm}) = 3170 \text{ N/mm}$$

The flexural resistance is

$$\phi \, m_n = \phi \, F_b \, S = 0.85(41.4 \text{ MPa})[\tfrac{1}{6}(1 \text{ mm})(130 \text{ mm})^2]$$
$$= 99.1 \times 10^3 \text{ N mm/mm}$$

By using Eq. 9.22, the interaction of tensile and flexural action is checked below.

$$\frac{P_u}{\phi \, P_n} + \frac{M_u}{\phi \, M_n} = \frac{154}{3170} + \frac{21 \, 000}{99 \, 100} = 0.049 + 0.212 = 0.26 \leq 1.0$$

Note that the tensile action has little effect on the deck in the region of the cantilever but must certainly be considered in the design of the connection of the rail to the deck.

Step 7 Check Shear Shear need not be considered in the deck design [A8.7].

Step 8 Check Live Load Deflection The critical live load deflection occurs when the left wheel is near the midspan of the first interior bay, that is, at Location 205. If we use Eq. 9.30, the deflection for the simply supported system is established. Next, by using Eq. 9.32, the deflection due to the continuity moments are determined. The two deflections are added. The continuity moments are calculated with the aid of the influence coefficients in Appendix A.
    The simple beam deflection is

$$\Delta = \frac{PS^3}{48 \, EI} = \frac{\left(\frac{72 \, 500}{1280}\right)(1.08)(1520^3)}{48(9650)[(\frac{1}{12})(130^3)]} = 2.5 \text{ mm}$$

The moments at Location 200 is zero and the moment at Location 300 from the previous analysis for strength is

$$m_{300} = -\frac{16.3 \times 10^6 \text{ N mm}}{1280 \text{ mm}} = 12 \, 730 \text{ N mm/mm}$$

Note the multiple presence factor of 1.08 is included.
    Per Eq. 9.32, the deflection due to the continuity moments is

$$\Delta = \frac{S^2}{16EI}[m_{200} - m_{300}](\text{analysis sign convention})$$

$$\Delta = \frac{(1520 \text{ mm})^2}{16(9650 \text{ MPa})[(\frac{1}{12})(1 \text{ mm})(130 \text{ mm})^3]}[0 - (-12\ 730 \text{ N mm/mm})]$$

$$= 1.0 \text{ mm (upward)}$$

The net deflection is

$$\Delta = 2.5 - 1.0 = 1.5 \text{ mm}$$

Alternatively, the simple beam deflection could be multiplied by the continuity factor of 0.70 to obtain an estimate to the more rigorously determined value. For example, $\Delta = (0.7)(2.5) = 1.8$ mm. The maximum allowable deflection is

$$\frac{S}{425} = \frac{1520 \text{ mm}}{425} = 3.6 \text{ mm}$$

Note that the continuity factor approach is conservative, and in most situations it may be used to compare with the allowable value. In cases where the deflection is unacceptable, the more rigorous approach may be used. It was presented here to illustrate both methods. Also note that if the deck panels are not made longitudinally continuous with dowels, then the deflection limit is the lesser of the span/425, or 2.5 mm. In this case, the deck stiffness would just satisfy both deflection limits.

All applicable limit states have been checked and are satisfied with the 130 mm thickness using Combination Symbol No. 2. A lower grade deck might be possible. First, use the deflection limit to establish the required modulus of elasticity for just meeting this requirement. For example,

$$E(\text{required}) = E(\text{original})\left(\frac{1.5}{3.6}\right) = 9650\left(\frac{1.5}{3.6}\right) = 4020 \text{ MPa}$$

A Combination Symbol No. 1 has an E of 8620 MPa, which satisfies the deflection limit. Next, the required resistance for flexure is 37 300 N mm/mm and the provided resistance is 99 100 N mm/mm. The required flexural resistance is

$$F_{bo}(\text{required}) = F_{bo}(\text{original})\left(\frac{37\ 300}{99\ 100}\right) = 27.6\left(\frac{37\ 300}{99\ 100}\right) = 13.6 \text{ MPa}$$

Again, Combination Symbol No. 1 meets this requirement with $F_{bo} = 22.1$

MPa. The shear resistance is the same for No. 1 and No. 2, and the resistance of Combination No. 2 is adequate.

In conclusion, the lower grade can be used and meets all applicable limit states.

Step 9 Design Panel Interconnections The interconnection of panels may be achieved by longitudinal stiffening beams, mechanical fasteners, splines, or dowels. The doweling of the deck system is intended to prevent relative deflection of adjacent glulam deck panel. With proper fabrication and construction, dowel systems have been proven effective. As a practical matter, hole alignments and the necessary field modifications reduce the effectiveness of the dowels and may lead to serviceability problems such as wearing surface cracking (Ritter, 1990). Using one stiffener beam in each bay between the beams has proven to be effective with a high degree of constructability. The requirements for longitudinal stiffener beams (used with transverse deck panels) are outlined in AASHTO [A9.9.4.3]. This article requires that the distance between stiffening beams not exceed 2400 mm, and the rigidity of the beam be $EI = 2.30 \times 10^{11}$ N mm² or greater. The stiffening beam shall be attached to the deck at intervals not exceeding 375 mm and the beam should run the full length of the deck system.

For the design example, try using the same combination number as the deck, that is, No. 1, which has a modulus of elasticity of 8620 MPa. If the EI is 2.30×10^{11} N mm², then the required moment of inertia is

$$I_{\text{required}} = \frac{2.3 \times 10^{11} \text{ N mm}^2}{8620 \text{ MPa}} = 26.7 \times 10^6 \text{ mm}^4$$

which is a relatively small section. With a section depth of 130 mm, the resulting required width is

$$I_{\text{required}} = 26.7 \times 10^6 \text{ mm}^4 = \tfrac{1}{12} bd^3 = \tfrac{1}{12}b(130 \text{ mm})^3$$

$$b = 146 \text{ mm}$$

Four laminae provide 152 mm, which also provide ample area to bolt the beam to the deck. The curb provides stiffening for the cantilever. This section is typically solid sawn and square. Given the required moment of inertia is about 26.7×10^6 mm⁴, a nominal 150×150 is adequate.

The dowel design procedure is not outlined in the AASHTO specification but the commentary refers to Ritter (1990) for the procedure. The Ritter procedure is based on the working stress method, and therefore all loads are not factored and the stresses are considered allowable. The procedure is based on the secondary moment and shears, that is, longitudinal moment and associated shear. The moment and shear may be determined by the following equations (Ritter, 1990). Note these equations have been converted to SI units. For deck spans less than or equal to 1270 mm

$$M_l = \frac{PS}{40.6}(S - 250) \tag{9.41}$$

$$R_l = \frac{6PS}{25\,400}$$

For deck spans greater than 1270 mm

$$M_l = \frac{PS}{20}\left(\frac{S - 760}{S - 250}\right) \tag{9.42}$$

$$R_l = \frac{P}{2S}(S - 500)$$

The number of dowels required for each deck span is based on the dowel diameter and properties given in Table 9.18

$$n = \frac{6.9\ \text{MPa}}{\sigma_{PL}}\left(\frac{R_l}{R_D} + \frac{M_l}{M_D}\right) \tag{9.43}$$

where

n is the number of dowels.
R_D is the shear capacity from Table 9.18.
M_D is the moment capacity from Table 9.18.

TABLE 9.18 Properties and Required Lengths of Steel Dowels for Doweled Glulam Deck Panels[a]

Dowel Diameter (mm)	Shear Capacity (kN) R_D	Moment Capacity (kN) M_D	Steel Stress Coefficients		Required Dowel Length, (mm) L_{dowel}
			C_1 $10^{-3}\left(\frac{1}{\text{mm}^3}\right)$	C_2 $10^{-3}\left(\frac{1}{\text{mm}^3}\right)$	
13	2.7	1.15	57.2	4.97	220
16	3.6	1.82	34.6	2.54	250
19	4.5	2.66	22.9	1.47	300
22	5.6	3.69	16.3	0.93	330
26	6.8	4.92	12.0	0.62	370
29	8.0	6.34	9.21	0.44	400
32	9.3	8.07	7.27	0.32	430
35	10.8	9.98	5.86	0.24	460
38	12.3	12.2	4.82	0.18	500

[a]In Ritter, 1990 (converted for SI units). The design example can be established for the secondary shear and moment by using Eq. 9.41.

σ_{PL} is the proportional limit stress for timber, perpendicular to grain, use 6.9 MPa for Douglas Fir-Larch and Southern Pine.

The steel stress must also be checked against yielding. Stress in the bar is determined and compared with the allowable value. The necessary equations follow:

$$\sigma_A = 0.80\,F_y \qquad\qquad (9.44)$$

$$\sigma = \frac{1}{n}\,(C_1 R_y + C_2 M_y)$$

where

σ_A is the allowable steel stress (MPa).
σ is the stress in the dowel due to the applied load (MPa).
F_y is the yield stress of the steel dowel (MPa).
C_1 and C_2 are reciprocals of the shear area and section modulus, respectively.

$$M_l = \frac{PS}{20}\left(\frac{S-760}{S-250}\right) = \frac{72\,500\ \text{N}(1520\ \text{mm})}{20}\left(\frac{1520-760}{1520-250}\right) =$$
$$3.3 \times 10^6\ \text{N mm}$$

$$R_l = \frac{P}{2S}\,(S-500) = \frac{72\,500\ \text{N}}{2(1520\ \text{mm})}\,(1520\ \text{mm} - 500\ \text{mm}) = 24.3 \times 10^3\ \text{N}$$

Try a dowel diameter of 29 mm. Based on the wood strength, the number of dowels per deck span is

$$n = \frac{6.9}{\sigma_{PL}}\left(\frac{R_l}{R_D} + \frac{M_l}{M_D}\right) = \frac{6.9}{6.9}\left(\frac{24.3}{8.0} + \frac{33.0}{6.34}\right) = 8.2\ \text{dowels}$$

Based on the steel strength of $F_y = 250$ MPa, the number of dowels is

$$n = \frac{1}{0.80\,F_y}\,(C_R R_y + C_M M_y)$$

$$= \frac{1}{0.80(250\ \text{MPa})}\left(\begin{array}{l}9.21 \times 10^{-3}\ 1/\text{mm}^2(24.3 \times 10^3\ N)\ + \\ 0.000\,44\ 1/\text{mm}^3(3.3 \times 10^6)\end{array}\right) = 8.4\ \text{dowels}$$

The steel strength controls the dowel design and 9 dowels 400 mm long

(150 mm on center) are required per deck span. The first dowel should be placed 75 mm from the center of the beam and 150 mm spacing thereafter, repeat for each deck span. The same arrangement is used in the cantilevers.

Note that either the stiffening beam or the dowels are required for longitudinal load distribution, but not both.

Step 10 Design Summary The provided flexural capacity and deflection were calculated by proportioning the resistances and modulus of elasticities. The deck summary is shown in Table 9.19.

Provide either a 130-mm × 150-mm stiffener beam located at the middle of each deck panel, or 29-mm dowels located at 150 mm on center. The dowels are 400 mm long.

9.6 LONGITUDINALLY LAMINATED DECKS

Longitudinally laminated decks consist of only the deck and have no beams similar to concrete slab bridges. The design is a combination of the methods used in the design of the transversely laminated decks and that used to design the beams. The process for beam design is followed with the exception that the live load is distributed to the deck on the basis of a distribution width, that is, the width of the deck that is assumed to be supporting one design lane. This width is defined in Eq. 6.15 and an example is given in Example 6.8. The following example is annotated to illustrate several salient points of deck design.

9.6.1 Deck Design Example

Example 9.3 Longitudinally Laminated Deck Design Design a longitudinally laminated superstructure to span 6100-mm center-to-center of bearing. It must carry two traffic lanes and have a roadway width of 7300 mm. Assume the following:

- The wearing surface is 76-mm asphalt (including allowance for a future overlay).
- AASHTO Design Vehicular Live Loads.
- The bridge is located on a secondary roadway and the operational importance factor required by the owner is 0.95 [A1.3.5].

TABLE 9.19 Deck Summary

Item	Provided	Required[a]
Combination Symbol No.	1	N/A
m_u (N mm/mm)	99 100	37 300
Deflection (mm)	1.5	3.6

[a]not applicable = N/A

- The bridge is located on a low-volume roadway with an average daily truck traffic (ADTT) of approximately 50 trucks per day. The multiple presence factor may be decreased accordingly.
- Western wood is available and a lamination plant is available that can lay-up decks with multiple layers of laminae.

Design the deck for the gravity loads. The wind, seismic, and collision loads need to be checked in this example.

Step 1 Select the Basic Configuration The required cross section is shown in Figure 9.21(a). The roadway surface width is 7300 mm, and 150 mm is added to each side to accommodate the barrier rail. Stiffening beams are

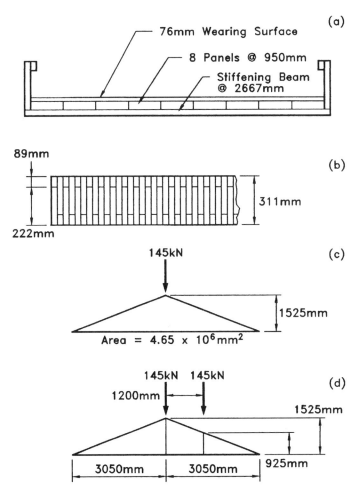

Fig. 9.21 (a) Longitudinal deck lay-up, (b) cross section, (c) design truck positioned for maximum moment, and (d) design tandem positioned for maximum moment.

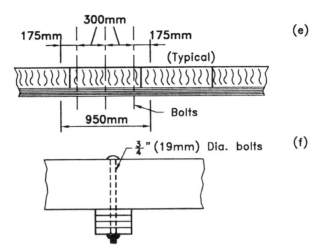

Fig. 9.21   (e) Panel lay-up and (f) stiffener attachment.

positioned transversely below the deck. These beams aid in the load distribution by helping the panels to deflect as a unit. Without these beams,
only the loaded panel would deflect creating high loading on individual
panels and significant relative deflection between the panels, which leads
to cracking of the wearing surface. A 6100-mm span often requires a depth
of 311 mm or greater and to achieve this depth in a nonprestressed system,
laminas are laid in two layers as shown in Figure 9.21(b). Here a 311-mm
deck is created with laminae of 222 and 89 mm. The 7600-mm deck is
conveniently constructed with eight panels of width 950 mm.

Step 2 Select the Combination Symbol Several combination symbols are
available for the system. Initially, No. 2 is selected and adjustment may be
made based on the closeness of the required resistance to that provided by
this lay-up.

Step 3 Determine the Gravity Load Actions For the purpose of the dead
load calculation, assume that the deck thickness is 311 mm. The dead load
calculations follow:

Deck $DC = (311 \text{ mm})(7.848 \times 10^{-6} \text{ N/mm}^3) = 2.44 \times 10^{-3} \text{ N/mm}^2$

Wearing surface $WS = (76 \text{ mm})(22.1 \times 10^{-6} \text{ N/mm}^3) =$

$1.68 \times 10^{-3} \text{ N/mm}^2$

The actions are established on the basis of a one unit design strip oriented
longitudinally. Although other widths may be used in wood design, this

approach is consistent with that used for concrete decks presented in Chapter 7. The dead load moment per unit length is

$$m_{DC} = \frac{2.44 \times 10^{-3} \text{ N/mm}^2 (6100 \text{ mm})^2}{8} = 11\,350 \text{ N mm/mm}$$

$$m_{DW} = \frac{1.68 \times 10^{-3} \text{ N/mm}^2 (6100 \text{ mm})^2}{8} = 7810 \text{ N mm/mm}$$

The design truck moment is established with the aid of the influence function shown in Figure 9.21(c). Note that the axle spacings for the design truck is 4300 mm, thus only one axle may be positioned on the bridge for critical effect. The design truck midspan moment is

$$\text{Design truck } M_{LL} = 145\,000 \text{ N}(1525 \text{ mm}) = 221 \times 10^6 \text{ N mm}$$

The design tandem has an axle spacing of 1200 mm. The critical positioning for midspan moment is shown in Figure 9.21(d). The resulting moment is

$$\text{Design tandem } M_{LL} = (110\,000)(1525 \text{ mm}) + (110\,000)(925 \text{ mm})$$

$$= 270 \times 10^6 \text{ N mm} = 270 \text{ kN m}$$

The design lane load is also required,

$$\text{Design lane } M_{LL} = 9.3 \text{ N/mm}(4.65 \times 10^6 \text{ mm}^2) = 43.3 \times 10^6 \text{ N mm}$$

The live load actions are distributed over the length prescribed in Eq. 6.15 where the widths for single- and multiple-lanes loaded are

$$E^S = 250 + 0.42\sqrt{L_1 W_1} = 250 + 0.42\sqrt{6100(7600)} = 3110 \text{ mm}$$

$$E^M = 2100 + 0.12\sqrt{L_1 W_1} = 2100 + 0.12\sqrt{6100(7600)} = 2920 \text{ mm} \le$$

$$\frac{W}{N_L} = \frac{7600 \text{ mm}}{2} = 3800 \text{ mm}$$

The dead and live load moments are factored and combined in the usual manner,

$$m_u = \eta\{\gamma_{DC}m_{DC} + \gamma_{DW}m_{DW} + mg\gamma_{LL}[m_{LL}^{\text{tandem}}(1 + IM) + m_{LL}^{\text{lane}}]\}$$

$$m_u^S = 0.95\{1.25(11\,350) + 1.5(7810) + (1.08)(\tfrac{1}{3110})(1.75)$$

$$[270 \times 10^6(1.165) + 43.3 \times 10^6]\} = 231\,000 \text{ N mm/mm}$$

$$m_u^M = 0.95\{1.25(11\,350) + 1.5(7810) + (1.00)(\tfrac{1}{2920})(1.75)$$

$$[270 \times 10^6(1.165) + 43.3 \times 10^6]\} = 228\,300 \text{ N mm/mm}$$

Step 4 Determine the Section Depth Required for Flexure The basic resistances for Combination Symbol No. 2 are given in Table 9.16.

The factored resistance is

$$F_b = F_{bo}C_MC_FC_D = 27.6 \text{ MPa}(1.0)(1.0)(1.5) = 41.4 \text{ MPa}$$

The required section depth is

$$d = \sqrt{\frac{6(231\,000(\text{N mm/mm})}{0.85(41.4 \text{ MPa})}} = 198 \text{ mm}$$

The depth indicates that a shallower deck than originally assumed is required. It is likely that a deeper section is required to satisfy the deflection limit. This requirement is checked next.

Step 5 Determine the Depth Required to Satisfy Deflection Limit As with beam design, the midspan deflections are estimated with the design truck and with 25% of the design truck applied with the design lane load. Initially, the two deflections are calculated for the deck depth of 311 mm. The associated moment of inertia for the distribution width of 2920 mm is

$$I = \tfrac{1}{12}bd^3 = \tfrac{1}{12}(2920 \text{ mm})(311 \text{ mm})^3 = 7.32 \times 10^9 \text{ mm}$$

and the deflections are

$$\Delta_{\text{design truck}} = \frac{PL^3}{48EI} = \frac{145\,000 \text{ N}(6100 \text{ mm})^3}{48(9650 \text{ MPa})(7.32 \times 10^9 \text{ mm}^4)} = 9.7 \text{ mm}$$

and

$$\Delta_{\text{design truck}} = \frac{5wL^4}{384EI} = \frac{5(9.3 \text{ N/mm})(6100 \text{ mm})^4}{384(9650 \text{ MPa})(7.32 \times 10^9 \text{ mm}^4)} = 2.4 \text{ mm}$$

Combining the deflection according to the prescribed weights yields

$$0.25(\Delta_{\text{design truck}}) + \Delta_{\text{design lane}} = 0.25(9.7 \text{ mm}) + 2.4 \text{ mm} = 4.8 \text{ mm}$$

The design truck alone controls with a deflection of 9.7 mm. The deflection limit is

$$\Delta_{\text{max}} = \frac{L}{425} = \frac{6100}{425} = 14 \text{ mm}$$

so the 311-mm depth is acceptable. By proportioning, determine the moment of inertia that is required to precisely satisfy the deflection limit.

$$I_{\text{required}} = I_{311 \text{ mm depth}} \left(\frac{\Delta_{\text{limit}}}{\Delta_{311 \text{ mm}}} \right) = 7.32 \times 10^9 \left(\frac{9.7}{14} \right) = 5.07 \times 10^9 \text{ mm}^4$$

By using a strip width of 1 mm, the required depth is readily determined.

$$d_{\text{required}} = \sqrt[3]{\frac{12 I_{\text{required}}}{b}} = \sqrt[3]{\frac{12(5.07 \times 10^9 \text{ mm}^4)}{2920 \text{ mm}}} = 275 \text{ mm}$$

Although very close to a 273 mm depth, we could consider a shallower deck. Alternatively, a different grade combination could be used. Combination No. 1 has a modulus of elasticity of 8620 MPa, which is a 11% decrease over the present 9650 MPa and easily satisfies the deflection limit. The final selection is a practical matter of availability and economy. The engineer should check with the local glulam plants. Other sources may also be helpful. Here the section depth is set at 311 mm using Combination No. 2.

Step 6 Check Shear Shear need not be considered in the design of wood decks [A8.7].

Step 7 Determine the Camber Requirement The midspan camber is twice the dead load deflection. The dead load is available per unit width and is conveniently used on that basis. The moment of inertia is

$$I = \frac{bd^3}{12} = \frac{1 \text{ mm}(311 \text{ mm})^3}{12} = 2.50 \times 10^6 \text{ mm}^4$$

The total dead load is the component load plus the wearing surface load, or $2.44 \times 10^{-3} + 1.68 \times 10^{-3} = 4.12 \times 10^{-3}$ N/mm. The midspan deflection is

$$\Delta_{dead} = \frac{5w_{dead}L^4}{384\ EI} = \frac{5(4.12 \times 10^{-3})(6100^4)}{384(9650)(2.5 \times 10^6)} = 3.1\ mm$$

The midspan camber deflection is

$$\Delta_{camber} = 2(3.1\ mm) = 6\ mm$$

Step 8 Stiffener Spacing and Requirements The transverse stiffeners are beams that are located below the deck. The required flexural rigidity is 2.3×10^{11} N mm² and the spacing should not exceed 2400 mm [A9.9.4.3]. As a practical matter, the spacing is often controlled by the requirements for the barrier rail spacing. The transverse beam is required for an effective attachment of the barrier rail to the deck. Here the Combination Symbol No. 2 is used to be consistent with the deck material. This consistency is not a requirement, in fact, material other than wood (e.g., steel) can be used effectively. The required moment of inertia is

$$I_{required} = \frac{2.30 \times 10^{11}\ N\ mm^2}{9650\ MPa} = 23.8 \times 10^6\ mm^4$$

Try a section width of 170 mm and determine the depth

$$d = \sqrt[3]{\frac{12(23.8 \times 10^6)}{170}} = 119\ mm$$

which requires four laminae or a section size of 170 mm × 152 mm. The maximum spacing of the bolts that attach the beam to the deck is 375 mm and the end bolt should be placed near the edge of each panel. Each panel is 950 mm wide and a convenient spacing is to place the first bolt at 175 mm from the edge, the next two at 300 mm, and finally, the last bolt at 175 mm from the other edge, or $175 + 300 + 300 + 175 = 950$ mm. The bolt spacing is the same for all panels. A 19-mm diameter bolt is adequate for this application. The details are illustrated in Figure 9.21(e) and (f).

Step 9 Determine the Bearing Length The load placement for the maximum support reaction is illustrated in Figure 9.22. The reaction is $P_u = 227\ 000$ N. The live load reaction is distributed over the width 2920 mm.

The reaction due the deck self-weight is

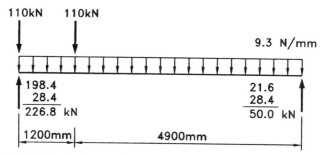

Fig. 9.22 Design tandem positioned for maximum reaction.

$$p_{DC} = \frac{2.44 \times 10^{-3} \text{ N/mm}^2(6100 \text{ mm})}{2} = 7.44 \text{ N/mm}$$

and the reaction due to the wearing surface is

$$p_{DW} = \frac{1.68 \times 10^{-3} \text{ N/mm}^2(6100 \text{ mm})}{2} = 5.12 \text{ N/mm}$$

The reactions are factored and added in the usual manner.

$$p_u^S = 0.95\{1.25(7.44) + 1.5(5.12) + (1.08)(\tfrac{1}{3110})(1.75)$$

$$[198\ 400(1.165) + 28\ 400]\} = 166 \text{ N/mm}$$

$$p_u^M = 0.95\{1.25(7.44) + 1.5(5.12) + (1.00)\ (\tfrac{1}{2920})\ (1.75)$$

$$[198\ 400(1.165) + 28\ 400]\} = 164 \text{ N/mm}$$

A 300-mm sill plate is convenient for construction and is tried here. The bearing resistance is

$$\phi p_n = \phi F_b b l_b = 0.90(5.17 \text{ MPa})(1 \text{ mm})(300 \text{ mm}) = 1400 \text{ N/mm}$$

which significantly exceeds the required force of 166 N/mm.

Step 10 Summary The deck width is 7600 mm and the thickness is 311 mm. It is to be constructed with Combination No. 2, with panel widths of 950 mm or 25–38 mm laminae each. The depth is achieved by using laminae with depths of 222 mm combined with 89 mm. Stiffening beams are spaced as required by the barrier rail design. The beam is 170 mm × 152 mm and is attached to the deck with 19-mm bolts spaced at 175, 300, 300, and 175 mm. The midspan camber is 6 mm. The bearing length is 300 mm.

A smaller section could be used with increased laminae grade. The availability and economy of each should be investigated. The barrier rail is not designed here.

Additionally, seismic and wind forces were not considered but should be in an actual design situation.

9.7 SUMMARY

A brief treatment of the design of a wood bridge is given in this chapter. An overview of the material properties is presented and AASHTO specifications are related to the design procedures of several typical systems. The design procedures are related to the analysis in the examples. Due to space limitation and the fact that the design specifications are in revision and review at the time of this writing, post-tensioned laminated deck systems are not addressed. These are an important bridge type and should be considered by the designer.

REFERENCES

AASHTO (1993). "Standard Specification for Transportation Materials and Methods of Sampling and Testing, Parts I and II," American Association of State Highway and Transportation Officials, Washington, DC.

AASHTO (1994). LRFD Bridge Design Specification, 1st Ed., American Association of State Highway and Transportation Officials, Washington, DC.

AISC (1986). *Moments, Shears, and Reactions for Continuous Bridges,* American Institute of Steel Construction, Chicago, IL.

Bodig, J. and B. A. Jayne, (1982). *Mechanics of Wood and Wood Composites,* Van Nostrand Reinhold, New York.

BridgeTech (1996). BT Beam—LRFD Analysis, User Manual, Laramie, WY.

Breyer, D. E. (1988). *Design of Wood Structures,* 2nd ed. McGraw-Hill, New York.

Gukowski, R. M and W. J. McCuthcheon, (1987), "Composite Performance of Timber Bridges," *Journal of Structural Engineering,* ASCE, Vol. 13, No. 7, July, pp. 1468–1486.

Hoyle, R. J. and F. E. Woeste (1988). *Wood Technology in the Design of Structures,* 5th ed., Iowa State University Press, Ames, IA.

National Design Specification for Wood Construction (1991). National Forest Products Association, Washington, DC.

Ritter, M. (1990). "Timber Bridges—Design, Construction, Inspection, and Maintenance," United State Department of Agriculture, Forest Service, EM 7700-8, Washington, DC.

Wyoming Department of Transportation (1996). "Bridge Rating and Structural Analysis (BRASS)" Users Manual, Cheyenne, WY.

10

SUBSTRUCTURE DESIGN*

10.1 INTRODUCTION

The design of the bridge substructure is significantly important for two reasons. First, the layout of the substructure directly influences the configuration of the superstructure, and the quality of substructure design controls the performance of the bridge. For example, the location of abutments determines the total bridge length, and the number of piers will control the depth of beams or girders. Second, the repair or replacement of damaged substructures can easily cost more than the original construction cost.

Substructure design demands manifold considerations because it is subject to various loadings not only from the superstructure, but also from surrounding materials such as water, backfill, and foundation soil. In addition to the various loadings, the substructure is subject to various failure modes, which include sliding failure, overturning failure, bearing capacity failure, overall stability failure, and structural failure.

Bridge engineers frequently spend more time in substructure design than in superstructure design due to its complex nature, which is created by numerous types of loadings, various limit states, unexpected geological conditions encountered during construction, and complicated geometrics involved in bridges with vertical or horizontal curve.

A properly designed substructure should prevent stability problems, structural failure, or excessive settlement, which may cause damage to the superstructure or poor rideability. In addition, the substructure should be designed considering water damages such as scour, traffic conditions during construction, and cooperation with existing structures or roads.

*This chapter written by Dr. John Sang Kim.

Detailed discussions on design considerations, bearings, and abutment design are presented in the following sections.

10.2   DESIGN CONSIDERATIONS

This section discusses various considerations involved in substructure design, which include site investigations, scour, and settlement.

10.2.1   Site Investigations

Prior to bridge design, a site investigation is conducted to determine surface and subsurface conditions. Surface condition information is important in deciding the construction method and the preliminary design of a bridge. Factors involved in the surface investigation include accessibility and workability of construction equipment, transportation of construction materials, availability of electric power or water, and various environmental regulations.

Accurate information on subsurface conditions is required for the reliable design of the substructure and the safe construction work below ground. Some information on the area can be obtained through surface topographic maps, or aerial photographs. The U. S. Geological Survey topographic maps, U. S. Department of Agriculture soil conservation maps, and state geologic maps are the sources available in the United States. The topographic maps display the elevation contours, water, forests, and constructed structures or facilities. The soil conservation maps show surface soils to a depth of 1.5–2.0 m. The geologic maps provide information on the extent of soil cover overlying the rock, and physical characteristics of the soil and the rock.

However, most design projects require performance of on-site subsurface explorations for the reliable design of the substructure. Test methods and the number of samples are decided based on site condition, the importance of the structure, or the financial situation. Subsurface information is usually obtained through borings, test pits, or by using geophysical investigative methods.

10.2.2   Scour

Bridge abutments and piers in streams, flood plains, or adjacent to water are subject to undermining by scouring action. This scouring action has caused many bridge failures.

Three types of scour in a river are shown in Figure 10.1 (Sowers, 1962). The first type of scour [Fig. 10.1(a)] is due to the lateral shifting of the channel. Continuous scouring occurs at the outside of each bend in a meandering river due to the higher velocity of the stream, while sedimentation occurs at the inside of each bend. An abutment close to the outside of a bend should be protected from undermining by placing concrete or asphaltic mats over the river bank or by founding the abutment below the greatest possible depth of scour.

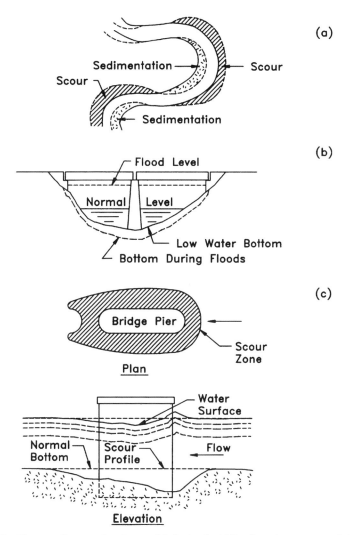

Fig. 10.1 Forms of scour in rivers. (a) Lateral shift of a stream caused by bank erosion and deposition; (b) normal bottom scour during floods; (c) accelerated scour caused by a bridge pier. [From Sowers, 1962.]

   The second type of scour [Fig. 10.1(b)] occurs due to the erosion of the river bed, which happens during periods of high flow. High velocity flow moves heavy bed materials so that the bed is lowered. The scouring actions are especially great where the width of the flow path is narrow. In general, this normal scour is proportional to the rise in the water surface and, therefore, the maximum depth of scour can be estimated by observation of the river bed during periods of high flow.
   The third type of scour [Fig. 10.1(c)] is the localized and accelerated scour resulting from such obstructions as bridge piers, which cause contraction of

the channel cross section and generate higher velocity of flow. The depth of scour in this case depends on factors such as the configuration of the pier, the angle between the flow and the pier, the contraction of the waterway, and the volume of debris that may be caught by the bridge.

Detailed information on scour prediction can be obtained from FHWA Technical Advisory T5140.20 titled "Scour at Bridges" (1988).

10.2.3 Settlement

The settlement of bridge foundations should be limited to a certain acceptable amount so that the bridge structure will not be damaged by excessive movements of the foundations or its functions impaired.

Design of foundations to prevent excessive settlement involves two considerations. One is determining the amount of movement that the supported structure can tolerate. The second is estimating the amount of movement induced by the applied loads. A number of investigators suggest various methods for estimating the amount of settlement.

> *Terzaghi and Peck's Method*   Terzaghi and Peck (1967) developed a procedure that uses Standard Penetration Test (SPT) results to estimate settlements of footings on sand. Their method, which has been widely used, is considered by some engineers as conservative in the sense that it often overestimates settlements. Studies by Tan and Duncan (1991) showed that about 85% of the time, Terzaghi and Peck's method estimated settlements larger than the actual settlements. Thus, while the method is not highly precise, it is considered reliable because it seldom underestimates settlements. Because settlements of footings on sand are highly erratic, methods that rarely underestimate settlements will often overestimate settlements by a fairly wide margin. The method proposed by Terzaghi and Peck provides a suitable compromise between accuracy and reliability. Although it is one of the simplest methods, it is as accurate as any other method with the same degree of reliablility.
>
> *The Method of D'Appolonia et al.*   D'Appolonia et al. (1970) developed a method for estimating settlement based on elastic theory. The method uses SPT blow count as the basis for estimating in situ soil compressibility. Tan and Duncan (1991) rated this method to be the most accurate method for estimating settlements of footings on sand. The average settlements estimated by this method are about equal to the average value of actual settlements for a large number of actual footings. However, it is important to note that their method underestimates settlements about one-half of the time. Values of settlement estimated using this method can be adjusted easily to provide the same level of reliability as Terzaghi and Peck's method.
>
> *Schmertmann's Method*   Schmertmann et al. (1977, 1978) developed a procedure that utilizes the Cone Penetration Test (CPT) results to esti-

mate settlements of footing on sand. This method has a rational basis and uses cone penetration resistance as a measure of in situ soil compressibility. Schmertmann's method proved to be a logical means for accounting for the variations in sand density and compressibility with depth. The computation process involves a rather lengthy procedure, and is warranted only when the site is characterized by pronounced and persistent layering.

Janbu's Method Janbu (1963, 1967) developed a unified approach for estimating settlements of footings on soil using tangent modulus values to characterize soil compressibility. In Janbu's method, the soil beneath the footing is divided into a number of sublayers, each characterized by a value of constrained tangent modulus.

Menard and Rousseau's Method Menard and Rousseau (1962) proposed a procedure for estimating settlements of footing using pressuremeter modulus to measure in situ soil compressibility.

10.3 BEARINGS OF BRIDGE

Bridge bearings are used to transfer forces from the superstructure to the substructure, allowing translational and rotational movements of the bridge. Until the middle of this century, the bearings used consisted of roller, rocker, or metal sliding bearings. With the advent of new materials and the need to satisfy more complicated design requirements, various advanced bearings were developed to accommodate movements, to resist loads, and to satisfy the requirements of the fatigue and fracture limit state. In the following sections, various types of bearing and a design illustration of a elastomeric bearing pad is discussed.

10.3.1 Types of Bearings

Because the failure of bridge bearings can cause deterioration or damage to the bridge, choosing the proper type of bearing is essential.

AASHTO (1994) LRFD Bridge Specifications provide a guide for selecting the suitable bearing type. Table 10.1 shows various bearing systems with the suitability ratings of each bearing for different types of movements and loads. Some of the bearings described in Table 10.1 are illustrated in Figure 10.2. A few decades ago, steel bearings consisting of steel plates, rockers, rollers, or pins were commonly used. Nowadays bearings made of new materials, such as elastomeric bearing pad, are more popular.

Elastomeric Bearings Two types of elastomeric bearings are available: plain elastomeric pads and reinforced elastomeric pads. Plain elastomeric pads are usually used in relatively small bridges where small shear and compressive strains are required. Reinforced elastomeric pads consist of multilayers of

TABLE 10.1 Bearing Suitability[a,b]

Type of Bearing	Movement		Rotation about bridge axis indicated			Resistance to Loads		
	Long	Trans	Long	Trans	Vert	Long	Trans	Vert
Plain elastomeric pad	L	L	S	S	L	L	L	L
Fiberglass reinforced pad	S	S	S	S	L	L	L	L
Cotton duck reinforced pad	U	U	U	U	U	L	L	S
Steel-reinforced elastomeric bearing	S	S	S	S	L	L	L	S
Plane sliding bearing	S	S	S	S	S	R	R	S
Curved sliding spherical bearing	R	R	S	S	S	R	R	S
Curved sliding cylindrical bearing	R	R	S	S	U	S	R	S
Disk bearing	R	R	S	S	L	R	R	S
Double cylindrical bearing	R	R	S	S	U	R	S	S
Pot bearing	R	R	S	S	L	U	R	S
Rocker bearing	S	U	U	S	U	S	R	S
Knuckle pinned bearing	U	U	U	S	U	S	R	S
Single roller bearing	S	U	S	U	U	U	R	S
Multiple roller bearing	S	U	U	U	U	U	U	S

[a] In AASHTO Table 14.6.2-1. [From *AASHTO LRFD Bridge Design Specifications*, Copyright © 1994 by the American Association of State Highway and Transportation Officials, Washington, DC. Used by Permission.]

[b] S = suitable, U = unsuitable, L = suitable for limited applications, R = may be suitable, but requires special considerations or additional elements such as slider or guideways.

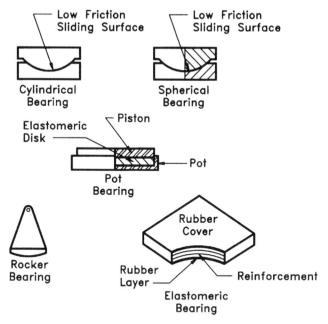

Fig. 10.2 Common bearing types. (AASHTO Fig. 14.6.2-1). [From *AASHTO LRFD Bridge Design Specifications,* Copyright © 1994 by the American Association of State Highway and Transportation Officials, Washington, DC. Used by permission.]

elastomer bonded in sandwich form to reinforcing materials such as fiberglass, cotton, or steel plate.

The elastomer should have enough shearing capacity to accommodate translational movement and sufficient rotational capacity to resist rotational movement. A design example of an elastomeric pad is presented in Example 10.1.

Sliding Bearings   Sliding bearings are made of two surfaces of similar or dissimilar material usually bronze sliding against steel with lubrication. Even though sliding bearings are simple and inexpensive, they have a shortcoming. For a century, it has been observed that the friction coefficient increased to unacceptable limits. The introduction of polytetrafluoroethylene (PTFE) revitalized the sliding bearings. The new material has excellent chemical resistance and provides a low coefficient of friction. Plain surfaces are used to accommodate horizontal sliding movement, and curved surfaces which can be cylindrical or spherical are used to permit uniaxial or multiaxial rotations, respectively.

Disk Bearings   Disk bearings consist of an unconfined elastomeric disk and a certain shear restriction mechanism to prevent rotational movement and to prevent translational movement. These bearings are more economical than steel spherical bearings and are used when small load capacity is required.

Pot Bearings   In pot bearings, a metal piston is supported by a plain elastomeric disk that is confined within a metal cylinder. These bearings are used when a multidirectional rotational capacity with medium range of load is demanded. Since the elastomeric disk is totally confined within a cylinder, no translational movement is allowed.

Rocker Bearings   In rocker bearings, a metal part with a curved surface is placed on top of another metal part with a flat or curved surface. The two metal parts are constrained by a dowel pin to prevent relative lateral movement. Rocker bearings are generally used at the fixed end of a bridge. The curved surface can be cylindrical or spherical to accommodate rotational movement about one or more axes.

Knuckle Pinned Bearings   This type of bearing consists of a pin encased by an upper and lower support that has a curved surface matching with the pin. A knuckle pinned bearing is used when large longitudinal forces are expected. A knuckle bearing has relatively high construction depths, and is easy to inspect and maintain.

Roller Bearings   Roller bearings are generally composed of one or more metal cylinders between pairs of parallel steel plates. Single roller bearings can accommodate rotational movement about the transverse axis and translational movement parallel to the longitudinal direction. Multiple roller bearings consist of a number of rollers between pairs of steel plates and accommodate multidirectional movements.

Other Bearing Types   In addition to the above bearings, other types may include leaf bearings and link bearings. Combinations of the different types of bearings can be used to obtain the required degree of freedom.

10.3.2   Example 10.1 Elastomeric Bearing Design

Design a steel reinforced elastomeric bearing pad according to AASHTO [A14.7.5]* that will support the concrete *T*-beam introduced in Example Problem 7.10.3.

---

*The article numbers in the AASHTO (1994) LRFD Specifications are enclosed in brackets and preceded by an A if specification and by C if commentary.

Design Loads

The elastomeric bearing layer should satisfy design criteria under service loads without impact [A14.7.5.3]. From Section 7.10.3, Parts G.5.d, H.1, and H.2, the vertical loads transmitted to the bearing can be calculated as

$$V_{LL} = mgr\ (V_{Tr} + V_{Ln})$$
$$V_{DL} = V_{DC} + V_{DW}$$

where

$$V_{LL} = \text{vertical load from live load}$$
$$V_{DL} = \text{vertical load from dead load}$$

mgr = live load distribution factor from Section 7.10.3, Parts G.4 and G.5

For shear distribution

$$mgr_V^I = 0.826 \times 1.115 = 0.921 \qquad \text{for interior beam}$$
$$mgr_V^E = 0.762 \times 1.115 = 0.850 \qquad \text{for exterior beam}$$

For moment distribution

$$mgr_M^I = 0.746 \times 0.948 = 0.707 \qquad \text{for interior beam}$$
$$mgr_M^E = 0.762 \times 0.948 = 0.722 \qquad \text{for exterior beam}$$

V_{Tr} = vertical load from truck load = 223.9 kN (Section 7.10.3, Part G.5.d)

V_{Ln} = vertical load from lane load = 45.2 kN (Section 7.10.3, Part G.5.d)

$$V_{DC} = \text{vertical load from concrete self weight}$$
$$= 73.7 \text{ kN for interior beam (Table E7.3-1)}$$
$$= 98.8 \text{ kN for exterior beam (Table E7.3-2)}$$
$$V_{DW} = \text{vertical load from wearing surface}$$
$$= 16.3 \text{ kN for interior beam (Table E7.3-1)}$$
$$= 11.1 \text{ kN for exterior beam (Table E7.3-2)}$$

The total vertical service load for an interior beam without impact is

$$V_{LL} = 0.921\ (223.9 + 45.2) = 247.8\ \text{kN}$$

$$\underline{V_{DL} = 73.7 + 16.3 \qquad\qquad = \quad 90.0\ \text{kN}}$$

$$337.8\ \text{kN}$$

and the total vertical service load for an exterior beam without impact is

$$V_{LL} = 0.850\ (223.9 + 45.2) = 228.7\ \text{kN}$$

$$\underline{V_{DL} = 98.8 + 11.1 \qquad\qquad = 109.9\ \text{kN}}$$

$$338.6\ \text{kN (governs)}$$

Therefore, design all bearing pads for a vertical reaction of 339 kN.

Maximum Longitudinal Movement at Abutment

The bridge is located in the low-temperature zone C, where the maximum number of consecutive freezing days is 14 [Table A14.7.5.2.2-2]. The concrete temperature ranges from -12 to 27°C in the zone with moderate climate [Table A3.12.2.1-1]. If the bridge is constructed during the summer, the temperature can be assumed to be 20°C and the temperature variation, ΔT, is

$$\Delta T = 20°\text{C} - (-12°\text{C}) = 32°\text{C}$$

The thermal coefficient, α, is 10.8×10^{-6} mm/mm/°C for normal density concrete [A5.4.2.2]. Then the temperature strain, ε_{temp}, is

$$\varepsilon_{temp} = \alpha\ \Delta T = 10.8 \times 10^{-6} \times 32°\text{C} = 0.000\ 346$$

The shrinkage strain, ε_{sh}, is [A5.4.2.3.1]

$$\varepsilon_{sh} = 0.0002 \text{ for 28 day and } 0.0005 \text{ for 1 year}$$

If the beams are lifted and reset after at least 28 days,

$$\varepsilon_{sh} = 0.0005 - 0.0002 = 0.0003$$

The maximum longitudinal movement, $_{max}\Delta_S$, can be obtained as (see Fig. 10.3)

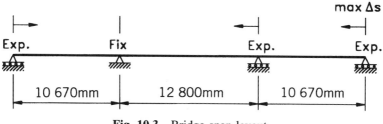

Fig. 10.3 Bridge span layout.

$$_{\max}\Delta_S = \gamma_{TU}L_e(\varepsilon_{temp} + \varepsilon_{sh})$$

where

$$\gamma_{TU} = \text{load factor for uniform temperature}$$

$$= 1.20 \text{ (Table 3.1) [Table A3.4.1-1]}$$

L_e = total expandable span length = 10 670 + 12 800 = 23 470 mm

Thus,

$$_{\max}\Delta_S = 1.20 \ (23 \ 470)(0.000 \ 346 + 0.000 \ 3) = 18.2 \text{ mm}$$

Preliminary Thickness of Bearing

The total elastomeric thickness, h_{rt}, should not be less than two times the maximum shear deformation, Δ_S, to prevent rollover at the edges and delamination due to fatigue [A14.7.5.3.4].

$$h_{rt} \geq 2 \ \Delta_S = 2 \ (18.2) = 36.4 \text{ mm}$$

Try $h_{rt} = 40$ mm and $h_{ri} = 10$ mm (see Fig. 10.4)

Preliminary Pad Area

Shape factor of an elastomeric layer, S_i, is the plan area divided by the area of perimeter free to bulge, expressed algebraically as [A14.7.5.1]

$$S_i = \frac{L \ W}{2h_{ri}(L + W)}$$

where

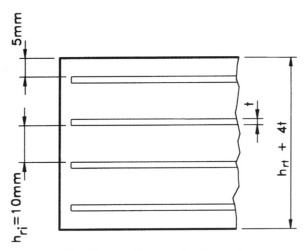

Fig. 10.4 Preliminary bearing pad.

h_{ri} = thickness of ith elastomeric layer (mm) = 10 mm

L = length of a rectangular pad (mm)

W = width of a rectangular pad in the transverse direction (mm)

Because the stem width of the T-beam is 350 mm, try W = 300 mm.

$$S_i = \frac{300L}{2(10)(L + 300)}$$

The compressive stresses of an elastomeric bearing layer subject to shear deformation should satisfy the following criteria [A14.7.5.3.2]:

$$\sigma_S \leq 1.66 \, GS \leq 11.0 \text{ MPa}$$

$$\sigma_L \leq 0.66 \, GS$$

where

σ_S = average compressive stress due to the total load (MPa)

σ_L = average compressive stress due to the live load (MPa)

G = shear modulus of elastomer (MPa) [Table A14.7.5.2-1]

G_L = low shear modulus

G_H = high shear modulus

S = shape factor of the thickest layer of the bearing

For the purpose of this example, the shear modulus is assumed to be between 0.95 and 1.20 MPa for the hardness 60 on the Shore A scale. Assuming σ_L is critical, solve for L.

$$\sigma_L = \frac{R_L}{L\,W} = \frac{0.66\,G_L\,L\,W}{2h_{ri}(L+W)}$$

$$\frac{247\,800}{300\,L} = \frac{0.66(0.95)(300L)}{2(10)(L+300)}$$

Solving quadratic, $L = 212$ mm.

Trial size is

$$W = 300 \text{ mm}, \; L = 220 \text{ mm}, \text{ and } h_{rt} = 40 \text{ mm}$$

For trial dimensions, check compressive stresses of the bearing layer.

$$S = \frac{L\,W}{2h_{ri}(L+W)} = \frac{(220)(300)}{2(10)(220+300)} = 6.35$$

$$\sigma_S = \frac{R}{L\,W} = \frac{339\,000}{(220)(300)} = 5.14 \text{ MPa} \le 11.0 \text{ MPa}$$

$$\le 1.66\,G_L\,S = 1.66(0.95)(6.35) = 10.01 \text{ MPa, OK}$$

$$\sigma_L = \frac{R_L}{L\,W} = \frac{247\,800}{(220)(300)} = 3.75 \text{ MPa}$$

$$\le 0.66\,G_L\,S = 0.66(0.95)(6.35) = 3.98 \text{ MPa, OK}$$

Instantaneous Compressive Deflection

Instantaneous deflection, δ, can be calculated as

$$\delta = \Sigma\,\varepsilon_i h_{ri}$$

where

ε_i = instantaneous compressive strain in ith elastomeric layer of a laminated bearing pad

h_{ri} = thickness of ith elastomeric layer in a laminated bearing pad (mm)

From Figure 10.5 with $\sigma_S = 5.14$ MPa, $S = 6.35$, read $\varepsilon_i = 0.033$ for 60 durometer reinforced elastomeric pad. Thus, the instantaneous deflection, δ, is

$$\delta = 4(0.033)(10) = 1.32 \text{ mm}$$

Rotation Capacity of Bearing

The rotation capacity of the bearing can be estimated as (see Fig. 10.6)

$$\theta_{max} = \frac{2\delta}{L} = \frac{2(1.32)}{220} = 0.0120 \text{ rad}$$

The design rotation, θ_S, at service limit state shall be obtained as [A14.4.2]

$$\theta_S = \theta_{DC} + \theta_L + \theta_{unk}$$

where

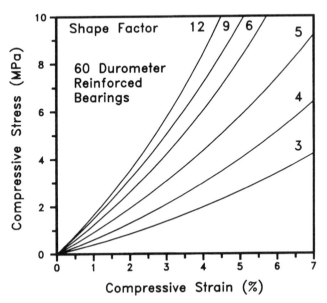

Fig. 10.5 Stress–strain curve for 60 durometer. (AASHTO Fig. C14.7.5.3.3-1). [From *AASHTO LRFD Bridge Design Specifications*, Copyright © 1994 by the American Association of State Highway and Transportation Officials, Washington, DC. Used by permission.]

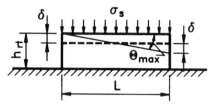

Fig. 10.6 Maximum rotational capacity of bearing.

θ_{DC} = rotation due to camber which reflects dead load deflection (negative value)

θ_L = rotation due to live load deflection

θ_{unk} = allowance for uncertainties = ± 0.005 rad

The instantaneous rotation due to dead load deflection, θ_{Di}, can be calculated by superposition as shown in Figure 10.7.

$$\theta_{Di} = \theta_{D1} - \theta_{D2}$$

where

$$\theta_{D1} = \frac{w_e L_s^3}{24EI} = \frac{(27.25)(10\ 670)^3}{24(610 \times 10^{12})} = 0.0023 \text{ rad}$$

$$\theta_{D2} = \frac{M_D L_s^3}{6EI} = \frac{(378 \times 10^6)(10\ 670)}{6(610 \times 10^{12})} = 0.0011 \text{ rad}$$

The values in these equations are taken from Section 7.10.3, Parts H.2 and I.7.d, w_e = uniformly distributed load = 24.49 + 2.76 = 27.25 N/mm, L_s = span length = 10 670 mm, EI = 610 × 10^{12} Nmm^2 and M_D = 13.866 × 10^6 (27.25) = 378 × 10^6 Nmm.

Thus,

$$\theta_{Di} = \theta_{D1} - \theta_{D2} = 0.0023 - 0.0011 = 0.0012 \text{ rad}$$

In addition to the instantaneous rotation due to dead load, θ_{Di}, a long-term rotation produced by creep effect should be considered in the calculation of camber. Thus, the rotation due to camber, θ_{DC}, is

$$\theta_{DC} = -(\theta_{Di} + \lambda\theta_{Di}) = -(1 + \lambda)\theta_{Di}$$

where the creep factor, λ = 2.47 from Section 7.10.3, Part I.7.d,

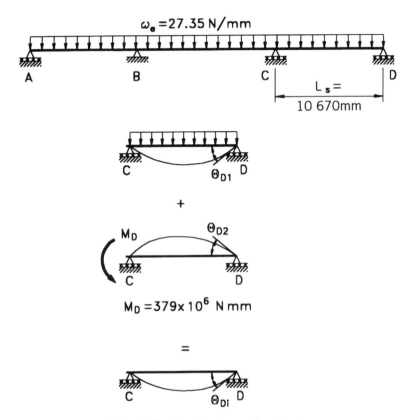

Fig. 10.7   Rotation due to dead load.

$$\theta_{DC} = -(1 + 2.47)(0.0012) = -0.0042 \text{ rad}$$

The rotation due to live load, θ_L, can be estimated by superposition as shown in Figure 10.8.

$$\theta_L = \theta_{L1} + \theta_{L2} + \theta_{L3}$$

where

$$\theta_{L1} = \frac{P_1 a_1 b_1}{6EI\,L_s}(a_1 + 2b_1)$$

$$= \frac{(145 \times 10^3)(8568)(2102)}{6(610 \times 10^{12})(10\,670)}[8568 + 2(2102)] = 0.85 \times 10^{-3}$$

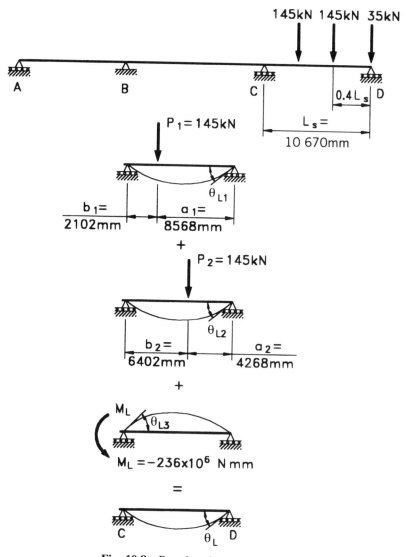

Fig. 10.8 Rotation due to live load.

$$\theta_{L2} = \frac{P_2 a_2 b_2}{6EI\,L_s}\,(a_2 + 2b_2)$$

$$= \frac{(145 \times 10^3)(4268)(6402)}{6(610 \times 10^{12})(10\ 670)}\,[4268 + 2(6402)] = 1.73 \times 10^{-3}$$

$$\theta_{L3} = \frac{M_L L_s}{6EI_s} = \frac{(-236 \times 10^6)(10\ 670)}{6(610 \times 10^{12})} = -0.69 \times 10^{-3}$$

Thus,

$$\theta_L = mgr(\Sigma \theta_{Li}) = 0.722(0.85 \times 10^{-3} + 1.73 \times 10^{-3} - 0.69 \times 10^{-3})$$

$$= 0.0014 \text{ rad}$$

The design rotation at service limit state is

$$\theta_S = \theta_{DC} + \theta_L + \theta_{unk}$$

$$= -0.0042 + 0.0014 \pm 0.005$$

$$= 0.0022 \quad \text{or} \quad -0.0078 \text{ rad (governs)}$$

Thus,

$$\theta_S = 0.0078 \text{ rad} < \theta_{\max} = 0.0120 \text{ rad, OK}$$

Combined Compression and Rotation

Bearings should be designed to avoid uplift at any point in the bearing and to prevent excessive compressive stress on an edge under any combined loads and corresponding rotations [A14.7.5.3.5].

Uplift Uplift requirement for rectangular bearing pads can be satisfied by the following condition:

$$\sigma_S > \sigma_{upmin} = 1.0\ G\ S\left(\frac{\theta_S}{n}\right)\left(\frac{B}{h_{ri}}\right)^2$$

where

θ_S = design rotation = 0.0078 rad

n = number of interior layers = 3

B = length in direction of rotation = 220 mm

G = shear modulus

    = G_H for uplift criterion = 1.20 MPa

    = G_L for shear criterion = 0.95 MPa

$$\sigma_{upmin} = 1.0(1.2)(6.35)\left(\frac{0.0078}{3}\right)\left(\frac{220}{10}\right)^2 = 9.59 \text{ MPa} > \sigma_S = 5.14 \text{ MPa, No Good}$$

Because the uplift criterion is not satisfied, the bearing pad must be redesigned. After several attempts, the pad is designed as follows:

$$L = 240 \text{ mm}$$

$$W = 350 \text{ mm}$$

Number of interior layers = 3

Thickness of each layer, $h_{ri} = 15$ mm

Total thickness of elastomeric pad, $h_{rt} = 60$ mm

Shape factor, $S = 4.75$

Total compressive stress, $\sigma_S = 4.04$ MPa $< \sigma_{\max} = 7.48$ MPa < 11.00 MPa, OK

Compressive stress due to live load, $\sigma_L = 2.95$ MPa $< \sigma_{L\max} = 2.98$ MPa, OK

Instantaneous compressive deflection, $\delta = 1.98$ mm

Rotational capacity, $\theta_{\max} = 0.0165$ rad $>$ design rotation,

$$\theta_S = 0.0078 \text{ rad, OK}$$

Uplift, $\sigma_{up\min} = 3.79$ MPa $< \sigma_S = 4.04$ MPa, OK

Compression   Compressive stress requirement for rectangular bearing pads subject to shear deformation can be satisfied by

$$\sigma_S < \sigma_{C\max} = 1.875 \, G \, S \left[1 - 0.20 \left(\frac{\theta_S}{n} \right) \left(\frac{B}{h_{ri}} \right)^2 \right]$$

$$\sigma_{C\max} = 1.875(0.95)(4.75) \left[1 - 0.20 \left(\frac{0.0078}{3} \right) \left(\frac{240}{15} \right)^2 \right]$$

$$= 7.33 \text{ MPa} > \sigma_S = 4.04 \text{ MPa, OK}$$

Stability of Elastomeric Bearings

The bearing pad should be designed to prevent instability at the service limit state load combinations by limiting the average compressive stress to one-half the estimated buckling stress [A14.7.5.3.6].

For the bridge deck free to translate horizontally,

$$\sigma_S \leq \sigma_{cr} = \frac{G}{2A - B}$$

where

$$A = \frac{1.92 \dfrac{h_{rt}}{L}}{S\sqrt{1 + \dfrac{2.0L}{W}}} = \frac{1.92(60/240)}{4.75\sqrt{1 + \dfrac{2.0(240)}{350}}} = 0.0656$$

$$B = \frac{2.67}{S(S + 2.0)\left(1 + \dfrac{L}{4.0W}\right)} = \frac{2.67}{4.75(4.75 + 2.0)\left(1 + \dfrac{240}{4.0(350)}\right)} = 0.0711$$

$$\sigma_{cr} = \frac{0.95}{2(0.0656) - 0.0711} = 15.81 \text{ MPa} > \sigma_S = 4.04 \text{ MPa, OK}$$

Reinforcement

The reinforcement should endure the tensile stresses produced by compression of the bearing pad. The thickness of the steel reinforcement, h_s, can be estimated as [A14.7.5.3.7]

At service limit state

$$h_s \geq \frac{3h_{max}\,\sigma_S}{F_y} = \frac{3(15)(4.04)}{345} = 0.53 \text{ mm}$$

At fatigue limit state

$$h_s \geq \frac{2h_{max}\,\sigma_L}{\Delta F_{TH}} = \frac{2(15)(2.95)}{165} = 0.54 \text{ mm} \qquad \text{(governs)}$$

where

h_{max} = maximum value of h_{ri} = 15 mm

F_y = yield strength of reinforcement = 345 MPa

ΔF_{TH} = fatigue threshold for Category A, given in Table 8.6 = 165 MPa

Thus, use h_s = 0.6 mm.

Summary of Example

The total thickness of the bearing pad is

$$3 \text{ interior layers} \times 15 \text{ mm} = 45 \text{ mm}$$
$$2 \text{ exterior layers} \times 7.5 \text{ mm} = 15 \text{ mm}$$
$$\underline{4 \text{ reinforcements} \times 0.6 \text{ mm} = 2.4 \text{ mm}}$$
$$\text{Total thickness} = 62.4 \text{ mm}$$

Use an elastomeric bearing pad with $L = 240$ mm, $W = 350$ mm, and total thickness $= 62.4$ mm.

10.4   ABUTMENTS

10.4.1   General

A retaining wall is used to hold back an earth embankment or water and to maintain a sudden change in elevation. An abutment is a particular type of retaining wall that supports the end of a bridge superstructure.

When an abutment is designed to resist the loads from the bridge superstructure, reaction pressures are developed across its foundation. The safety of the abutment is determined by the capacity of the foundation to resist these reaction forces. When an abutment is founded on a spread footing on soil or rock, the bearing capacity and sliding resistance of the foundation materials and overturning stability must be determined.

Types of abutments and the selection procedures are discussed in this section. A general overview of design considerations, estimation of earth pressures, stability criteria, design procedures, and a design example are also presented.

10.4.2   Types of Abutments

In the sixteenth edition of the AASHTO (1996) Standard Specifications, abutments are classified into four types: stub abutments, partial-depth abutments, full-depth abutments, and integral abutments. Peck et al. (1974) divide bridge abutments in a different way: gravity abutments, U-abutments, spill-through abutments, and pile-bent abutments. A gravity abutment with wing walls [Fig. 10.9(a)] is an abutment that consists of a bridge seat, wingwalls, backwall, and footing. A U-abutment [Fig. 10.9(b)] is an abutment whose wingwalls are perpendicular to the bridge seat. A spill-through abutment [Fig. 10.9(c)] consists of a beam that supports the bridge seat, two or more columns supporting the beam, and a footing supporting the columns. The columns are embedded up to the bottom of the beam in the fill, which extends on its natural slope in front of the abutment. A pile-bent abutment with stub wings [Fig. 10.9(d)] is another type of spill-through abutment, where a row of driven piles supports the beam.

Selection of Abutments   The procedure of selecting the most appropriate type of abutments can be based on the following considerations:

1. Construction and maintenance cost.
2. Cut or fill earthwork situation.
3. Traffic maintenance during construction.
4. Construction period.

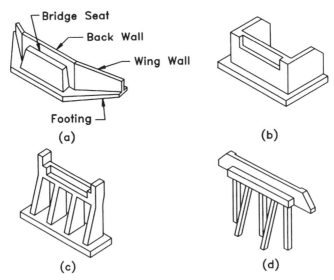

Fig. 10.9   Various types of abutments. (a) Gravity abutment with wing walls, (b) U abutment, (c) spill-through abutment, (d) pile-bent abutment with stub wings. [From *Foundation Engineering,* 2nd ed., by R. B. Peck, W. E. Hansen and T. H. Thornburn. Copyright © 1974 by John Wiley & Sons, Inc. Reprinted by permission.]

5. Safety of construction workers.
6. Availability and cost of backfill material.
7. Superstructure depth.
8. Size of abutment.
9. Horizontal and vertical alignment changes.
10. Area of excavation.
11. Aesthetics and similarity to adjacent structures.
12. Previous experience with the type of abutment.
13. Ease of access for inspection and maintenance.
14. Anticipated life, loading conditions, and acceptability of deformations.

The advantages and disadvantages of various types of abutments are discussed by Munfakh (1990) and Schnore (1990). The decision matrices that they provide considering the above factors can be very useful in selecting the proper abutment.

10.4.3   Limit States

When abutments fail to satisfy their intended design functions, they are considered to reach "limit states." Limit states can be categorized into two types: ultimate or strength limit states and serviceability limit states.

Ultimate Limit States   An abutment reaches an ultimate limit state when the
strength of at least one of its components is fully mobilized or when the
structure becomes unstable. In the ultimate limit state, an abutment may experience serious distress and structural damage, both local and global. In
addition, various failure modes in the soil that supports the abutment can also
be identified. These are also called ultimate limit states; they include bearing
capacity failure, sliding, overturning, and overall instability.

Serviceability Limit States   An abutment experiences a serviceability limit
state when it fails to perform its intended design function fully, due to excessive deformation or deterioration. Serviceability limit states include excessive total or differential settlement, lateral movement, fatigue, vibration, and
cracking.

10.4.4 Load and Performance Factors

The AASHTO (1994) bridge specifications require the use of the Load and
Resistance Factor Design (LRFD) method in the substructure design. A mathematical statement of LRFD can be expressed as

$$\phi R_n \geq \text{effect of } \Sigma\gamma_i Q_i \tag{10.1}$$

where

$$\phi = \text{performance or resistance factor}$$
$$R_n = \text{nominal resistance}$$
$$\gamma_i = \text{load factor for load component } i$$
$$Q_i = \text{load component } i$$

Load Factors   Load factors are applied to loads to account for uncertainties
in selecting loads and load effects. The load factors used in the first edition
of the AASHTO (1994) LRFD Bridge Specifications are shown in Tables 3.1
and 3.2.

Performance Factors   Performance or resistance factors are used to account
for uncertainties in structural properties, soil properties, variability in workmanship, and inaccuracies in the design equations used to estimate the capacity. These factors are used for design at the ultimate limit state. Suggested
values of performance factors for shallow foundations are listed in Table 10.2
[Table A10.5.4-1].

TABLE 10.2   Performance Factors for Shallow Foundations[a]

Type of Limit State	Performance Factor
1. Bearing capacity	
a. Sand	
Semiempirical procedure (SPT)	0.45
Semiempirical procedure (CPT)	0.55
Rational method	
Using ϕ_f estimated from SPT	0.35
Using ϕ_f estimated from CPT	0.45
b. Clay	
Semiempirical procedure (CPT)	0.50
Rational method	
Using shear strength in lab tests	0.60
Using shear strength from field vane tests	0.60
Using shear strength estimated from CPT data	0.50
c. Rock	
Semiempirical procedure	0.60
2. Sliding	
a. Precast concrete placed on sand	
Using ϕ_f estimated from SPT	0.90
Using ϕ_f estimated from CPT	0.90
b. Concrete cast in place on sand	
Using ϕ_f estimated from SPT	0.80
Using ϕ_f estimated from CPT	0.80
c. Clay (where shear strength is less than 0.5 times normal pressure)	
Using shear strength in lab	0.85
Using shear strength from field vane test	0.85
Using shear strength estimated from CPT data	0.80
d. Clay (where the strength is greater than 0.5 times normal pressure)	0.85

[a]Barker et al., 1991. Used by permission.

10.4.5   Forces on Abutments

Earth pressures exerted on an abutment can be classified into at-rest, active, and passive. Each of these earth pressures corresponds to different conditions with regard to the direction and the magnitude of the abutment movement. When a wall moves away from the backfill, the earth pressure decreases (active pressure), and when it moves toward the backfill, the earth pressure increases (passive pressure). Table 10.3, obtained through experimental data and finite element analyses (Clough and Duncan, 1991), gives approximate magnitudes of wall movements required to reach minimum active and maximum passive earth pressure conditions. Observations from the table can be summarized as follows (Clough and Duncan, 1991):

TABLE 10.3 Approximate Magnitudes of Movements Required to Reach Minimum Active and Maximum Passive Earth Pressure Conditions[a]

Type of Backfill	Values of Δ/H[b]	
	Active	Passive
Dense sand	0.001	0.01
Medium dense sand	0.002	0.02
Loose sand	0.004	0.04
Compacted silt	0.002	0.02
Compacted lean clay	0.01[c]	0.05[c]
Compacted fat clay	0.01[c]	0.05[c]

[a] After Clough and Duncan, 1991.
[b] Δ = movement of top of wall to reach minimum active or maximum passive pressure, by tilting or lateral translation.
H = height of wall.
[c] Under stress conditions close to the minimum active or maximum passive earth pressures, cohesive soils creep continually. The movements shown would produce active or passive pressures only temporarily. With time the movements would continue if pressures remain constant. If movement remains constant, active pressures will increase with time, approaching the at-rest pressure, and passive pressures will decrease with time, approaching values on the order of 40% of the maximum short-term passive pressure.

1. The required movements for the extreme conditions are approximately proportional to the wall height.
2. The movement required to reach the maximum passive pressure is about 10 times as great as that required to reach the minimum active pressure for walls of the same height.
3. The movement required to reach the extreme conditions for dense and incompressible soils is smaller than those for loose and compressible soils.

For any cohesionless backfill, conservative and simple guidelines for the maximum movements required to reach the extreme cases are provided by Clough and Duncan (1991). For minimum active pressure, the movement is no more than about 1 mm in 240 mm ($\Delta/H = 0.004$), and for maximum passive pressure about 1 mm in 24 mm ($\Delta/H = 0.04$).

As shown in Figure 10.10, the value of the earth pressure coefficient varies with wall displacement and eventually remains constant after sufficiently large displacements. When no wall displacements occur, the wall is said to be at rest and the at-rest earth pressure coefficient K_o applies. The change of pressures also varies with the type of soils, that is, the pressures in the dense sand change more quickly with wall movement.

The earth pressure coefficients shown in Figure 10.11 are for a backfill compacted to a medium dense condition. Compared with the values in Figure 10.10, the curve is shifted upward so that the required movement for the

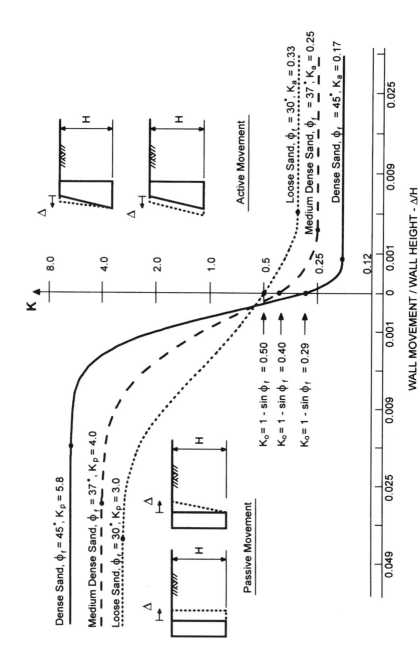

Fig. 10.10 Relationship between wall movement and earth pressure. [After Clough and Duncan, 1991.]

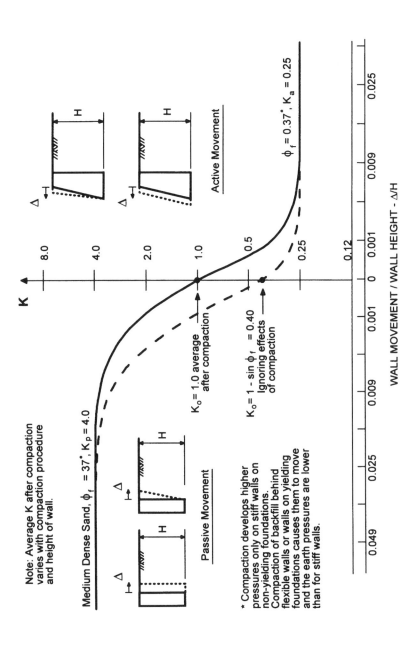

Fig. 10.11 Relationship between wall movement and earth pressure for a wall with compacted backfill. [After Clough and Duncan, 1991.]

minimum active earth pressure is increased, and the required movement for the maximum passive earth pressure is decreased. However, in spite of the effect of compaction, the conservative guidelines of 1 mm in 240 mm for active condition, and 1 mm in 24 mm for passive condition still can be used to estimate the required movements for the extreme earth pressure conditions.

Methods for Estimating K_a and K_p   Coulomb in 1776 and Rankine in 1856 developed simple methods for calculating the active and passive earth pressures exerted on retaining structures. Caquot and Kerisel (1948) developed the more generally applicable log spiral theory. Where the movements of walls are sufficiently large so that the shear strength of the backfill soil is fully mobilized, and where the strength properites of the backfill can be estimated with sufficient accuracy, these methods of calculation are useful for practical purposes.

Coulomb's trial wedge method can be used for irregular backfill configurations, and Rankine's theory and the log spiral analysis can be used for more regular configurations. Each of these methods will be discussed below.

Coulomb Theory   The Coulomb theory, the first rational solution to the earth pressure problem, is based on the concept that the lateral force exerted on a wall by the backfill can be evaluated by analysis of the equilibrium of a wedge-shaped mass of soil bounded by the back of the wall, the backfill surface, and a surface of sliding through the soil. The assumptions in this analysis are

1. The surface of sliding through the soil is a straight line.
2. The full strength of the soil is mobilized to resist sliding (shear failure) through the soil.

Active Pressure   A graphical illustration for the mechanism of active failure according to the Coulomb theory is shown in Figure 10.12(a). The active earth pressure force can be expressed as:

$$P_a = \frac{1}{2}\gamma H^2 \frac{\cos^2(\phi_f - \beta)}{\cos^2\beta \cos(\beta + \delta)\left[1 + \sqrt{\dfrac{\sin(\phi_f + \delta)\sin(\phi_f - i)}{\cos(\beta + \delta)\cos(\beta - i)}}\right]^2} \quad (10.2)$$

where

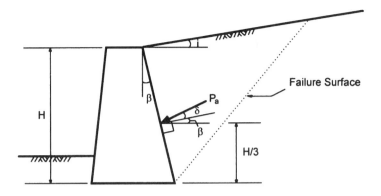

(a) Active Pressure Force

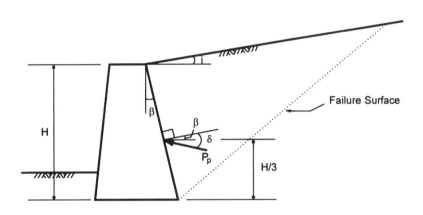

(b) Passive Pressure Force

Fig. 10.12   Coulomb theory for active and passive earth pressures.

P_a = active earth pressure force (force/length)

  = $\frac{1}{2}\,\gamma H^2 K_a$

K_a = coefficient of active earth pressure

  γ = unit weight of backfill soil (force/length^3)

H = wall height (length)

ϕ_f = the internal friction angle of soil (degrees)

β = the slope of stem face (degrees)

δ = the friction angle between wall and soil (degrees)

i = the slope of backfill surface (degrees)

Passive Pressure   The Coulomb theory can be used to evaluate passive resistance, using the same basic assumptions. Figure 10.12(b) shows the failure mechanism for the passive case. The passive earth pressure force, P_p, can be expressed as follows;

$$P_p = \frac{1}{2}\gamma H^2 \frac{\cos^2(\phi_f + \beta)}{\cos^2\beta \cos(\beta - \delta)\left[1 - \sqrt{\dfrac{\sin(\phi_f + \delta)\sin(\phi_f + i)}{\cos(\beta - \delta)\cos(\beta - i)}}\right]^2} \quad (10.3)$$

The basic assumption in the Coulomb theory is that the surface of sliding is a plane. This assumption does not affect appreciably the accuracy for the active case. However, for the passive case, values of P_p calculated by the Coulomb theory can be much larger than can actually be mobilized, especially when the value of δ exceeds about one-half of ϕ_f.

Wall Friction   Friction between the wall and backfill has an important effect on the magnitude of earth pressures and an even more important effect on the direction of the earth pressure force. Table 10.4 presents values of the maximum possible wall friction angle for various wall materials and soil types.

Rankine Theory   The Rankine theory is applicable to conditions where the wall friction angle (δ) is equal to the slope of the backfill surface (i). As in the case of the Coulomb theory, it is assumed that the strength of the soil is fully mobilized.

Active Pressure   The active earth pressure considered in the Rankine theory is illustrated in Figure 10.13(a) for a level backfill condition. The coefficient of active earth pressure, K_a, can be expressed as

$$K_a = \cos i \frac{\cos i - \sqrt{\cos^2 i - \cos^2 \phi_f}}{\cos i + \sqrt{\cos^2 i - \cos^2 \phi_f}} \quad (10.4)$$

When the ground surface is horizontal, that is, when $i = 0$, K_a can be expressed as

TABLE 10.4 Ultimate Friction Factors, Friction Angles and Adhesion for Dissimilar Materials[a]

Interface Materials	Friction Angle, δ (deg)
Mass concrete on the following foundation materials	
Clean sound rock	35
Clean gravel, gravel-sand mixtures, coarse sand	29–31
Clean fine to medium sand, silty medium to coarse sand, silty or clayey gravel	24–29
Clean fine sand, silty or clayey fine to medium sand	19–24
Fine sandy silt, nonplastic silt	17–19
Very stiff and hard residual or preconsolidated clay	22–26
Medium stiff and stiff clay and silty clay	17–19
Masonry on foundation materials has same friction factors	
Steel sheet piles against the following soils	
Clean gravel, gravel–sand mixtures, well-graded rock fill with spalls	22
Clean sand, silty sand–gravel mixture, single-size hard rock fill	17
Silty sand, gravel or sand mixed with silt or clay	14
Fine sandy silt, nonplastic silt	11
Formed or precast concrete or concrete sheet piling against the following soils	
Clean gravel, gravel–sand mixture, well-graded rock fill with spalls	22–26
Clean sand, silty sand–gravel mixture, single-size hard rock fill	17–22
Silty sand, gravel or sand mixed with silt or clay	17
Fine sandy silt, nonplastic silt	14
Various structural materials	
Masonry on masonry, igneous, and metamorphic rocks:	
Dressed soft rock on dressed soft rock	35
Dressed hard rock on dressed soft rock	33
Dressed hard rock on dressed hard rock	29
Masonry on wood in direction of cross grain	26
Steel on steel at sheet pile interlocks	17

[a]From U.S. Dept. of Navy, 1982b.

$$K_a = \frac{1 - \sin \phi_f}{1 + \sin \phi_f} \tag{10.5}$$

The active earth pressure, P_a, can be expressed as

$$P_a = K_a \gamma z - 2c\sqrt{K_a} \tag{10.6}$$

where

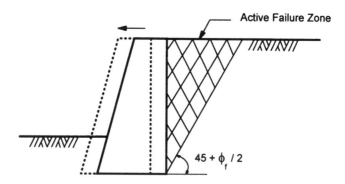

(a) Frictionless Wall Moves away from Backfill

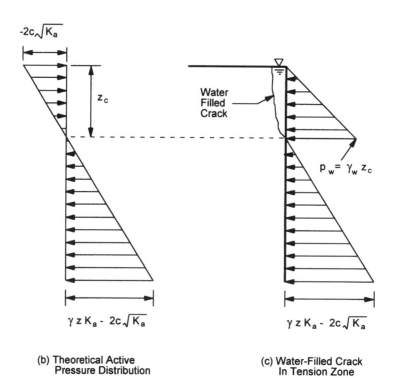

(b) Theoretical Active
Pressure Distribution

(c) Water-Filled Crack
In Tension Zone

Fig. 10.13 Rankine theory for active pressure, frictionless wall [After Clough and Duncan, 1991.]

P_a = the active pressure (force/length^2),

K_a = the active pressure coefficient

γ = the unit weight of soil (force/length^3)

c = the cohesion (force/length^2)

z = the depth below the ground surface (length)

The variation of active pressure with depth is linear, as shown in Figure 10.13(b). If the backfill is cohesive, the soil is theoretically in a tension zone down to a depth of $2c/\gamma(K_a)^{1/2}$. However, a tension crack is likely to develop in that zone and may be filled with water, so that hydrostatic pressure will be exerted on the wall, as shown in Figure 10.13(c).

Passive Pressure   The Rankine theory can also be applied to passive pressure conditions. The passive earth pressure coefficient (K_p) can be expressed as

$$K_p = \cos i \; \frac{\cos i + \sqrt{\cos^2 i - \cos^2 \phi_f}}{\cos i - \sqrt{\cos^2 i - \cos^2 \phi_f}} \tag{10.7}$$

When the ground surface is horizontal, K_p can be expressed as

$$K_p = \frac{1 + \sin \phi_f}{1 - \sin \phi_f} \tag{10.8}$$

The passive pressure at depth z can be expressed as

$$P_p = K_p \, \gamma \, z + 2c\sqrt{K_p} \tag{10.9}$$

where

P_p = the passive pressure (force/length^2)

K_p = the passive pressure coefficient

Log Spiral Analysis   The failure surface in most cases is more closely approximated by a log spiral than a straight line, as shown in Figure 10.14. Active and passive pressure coefficients, K_a and K_p obtained from analyses using log spiral surfaces are listed in Tables 10.5 and 10.6 (Caquot and Kerisel, 1948). Values of K_a and K_p for walls with level backfill and vertical stem are also shown in Figure 10.15. These values are also based on the log spiral analyses performed by Caquot and Kerisel.

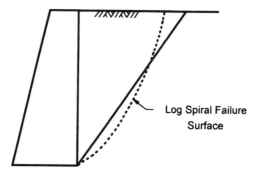

Fig. 10.14   Comparison of log spiral and straight line failure surfaces for active conditions [From Clough and Duncan, 1991.]

Selection of Earth Pressure Coefficients   Selecting a proper earth pressure coefficient is essential for successful wall design. A number of methods previously discussed can be used to decide the magnitude of the coefficients.

An alternative method is to use a hydrostatic fluid pressure, which is equivalent to the product of an earth pressure coefficient and the unit weight of the soil. A decision on what type of earth pressure coefficient should be used is based on the direction and the magnitude of the wall movement. Table 10.3, Figures 10.10 and 10.11 can be useful in this procedure.

The New Zealand Ministry of Works and Development (NZMWD, 1979) has recommended the following static earth pressure coefficients for use in design:

1. Counterfort or gravity walls founded on rock or piles: K_0.
2. Cantilever walls less than 4880-mm high founded on rock or piles: $0.5 (K_0 + K_a)$.
3. Cantilever walls higher than 4880-mm or any wall founded on a spread footing: K_a.

NZMWD (1979) also recommended that K_0 should be used for the types of abutments that are not mentioned above. Earth pressures on integral abutments or framed abutments may be higher than active due to thermal movements of the bridge superstructure.

Location of Horizontal Resultant   In conventional designs and analyses, the horizontal resultant is assumed to be located at one-third of total height from the bottom of the wall. However, several experimental tests performed by researchers, including Terzaghi (1934), Clausen and Johansen (1972), and Sherif et al. (1982), showed that the point of application is above the lower third point.

According to a study by Duncan et al. (1990), Terzaghi found that the resultant was applied at $0.40H$ to $0.45H$ from the bottom of the wall where

TABLE 10.5 Values of K_a for Log Spiral Failure Surface[a]

δ (deg)	i (deg)	β (deg)	ϕ_f (deg) 20	25	30	35	40	45
0	−15	−10	0.37	0.30	0.24	0.19	0.14	0.11
		0	0.42	0.35	0.29	0.24	0.19	0.16
		10	0.45	0.39	0.34	0.29	0.24	0.21
	0	−10	0.42	0.34	0.27	0.21	0.16	0.12
		0	0.49	0.41	0.33	0.27	0.22	0.17
		10	0.55	0.47	0.40	0.34	0.28	0.24
	15	−10	0.55	0.41	0.32	0.23	0.17	0.13
		0	0.65	0.51	0.41	0.32	0.25	0.20
		10	0.75	0.60	0.49	0.41	0.34	0.28
ϕ_f	−15	−10	0.31	0.26	0.21	0.17	0.14	0.11
		0	0.37	0.31	0.26	0.23	0.19	0.17
		10	0.41	0.36	0.31	0.27	0.25	0.23
	0	−10	0.37	0.30	0.24	0.19	0.15	0.12
		0	0.44	0.37	0.30	0.26	0.22	0.19
		10	0.50	0.43	0.38	0.33	0.30	0.26
	15	−10	0.50	0.37	0.29	0.22	0.17	0.14
		0	0.61	0.48	0.37	0.32	0.25	0.21
		10	0.72	0.58	0.46	0.42	0.35	0.31

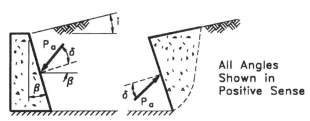

All Angles
Shown in
Positive Sense

[a]After Caquot and Kerisel, 1948.

H is the total height of the wall. Clausen and Johansen found the same range of locations as Terzaghi's, and Sherif et al. recommended $0.42H$ for a wall in static active condition. However, Duncan et al. (1990) suggested that the location of the resultant should be assumed to be $0.40H$ above the bottom of the wall.

Equivalent Fluid Pressure Equivalent fluid pressures provide a convenient means of estimating design earth pressures, especially when the backfill material is a clayey soil.

The lateral earth pressure at depth z can be expressed as

TABLE 10.6   Values of K_p for Log Spiral Failure Surface[a]

δ (deg)	i (deg)	β (deg)	ϕ_f (deg)					
			20	25	30	35	40	45
		−10	1.32	1.66	2.05	2.52	3.09	3.95
	−15	0	1.09	1.33	1.56	1.82	2.09	2.48
		10	0.87	1.03	1.17	1.30	1.33	1.54
		−10	2.33	2.96	3.82	5.00	6.68	9.20
0	0	0	2.04	2.46	3.00	3.69	4.59	5.83
		10	1.74	1.89	2.33	2.70	3.14	3.69
		−10	3.36	4.56	6.30	8.98	12.2	20.0
	15	0	2.99	3.86	5.04	6.72	10.4	12.8
		10	2.63	3.23	3.97	4.98	6.37	8.2
		−10	1.95	2.90	4.39	6.97	11.8	22.7
	−15	0	1.62	2.31	3.35	5.04	7.99	14.3
		10	1.29	1.79	2.50	3.58	5.09	8.86
		−10	3.45	5.17	8.17	13.8	22.5	52.9
ϕ_f	0	0	3.01	4.29	6.42	10.2	17.5	33.5
		10	2.57	3.50	4.98	7.47	12.0	21.2
		−10	4.95	7.95	13.5	24.8	50.4	11.5
	15	0	4.42	6.72	10.8	18.6	39.6	73.6
		10	3.88	5.62	8.51	13.8	24.3	46.9

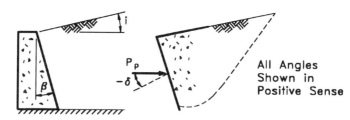

[a]After Caquot and Kerisel, 1948.

$$p_h = \gamma_{eq}\, z \qquad\qquad (10.10)$$

where

p_h = lateral earth pressure (force/length^2)

γ_{eq} = equivalent fluid unit weight (force/length^3);
unit weight of a fluid that would exert the same pressure
as the backfill soil

z = depth below the surface of backfill (length)

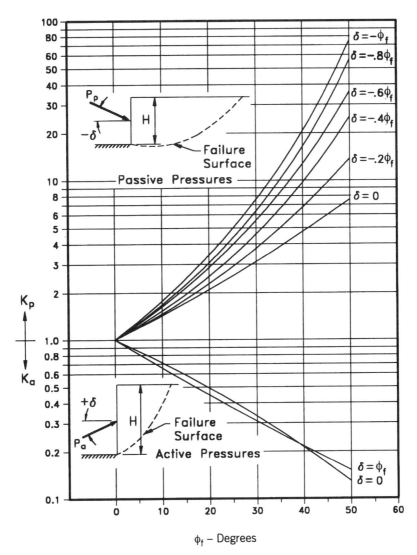

Fig. 10.15 Active and passive pressure coefficients for vertical wall and horizontal backfill—based on log spiral failure surfaces [After Caquot and Kerisel, 1948.]

Some typical equivalent fluid unit weights and corresponding pressure coefficients are presented in Table 10.7. These are appropriate for use in designing walls up to about 6100 mm in height. Values are presented for at-rest condition and for walls that can tolerate movements of 1 mm in 240 mm, and for level and sloped backfill.

When the equivalent fluid pressure is used in the estimation of horizontal earth pressure it is necessary to include vertical earth pressure acting on the wall to avoid an assumption that is too conservative. In the level backfill, the

TABLE 10.7 Coefficients and Unit Weights for Equivalent Fluid Pressures[a]

	Equivalent Fluid Unit Weights and Pressure Coefficients[b]							
	Level Backfill				Backfill 2(H) on 1(V)			
	At-Rest		$\Delta/H = 1/240$		At-Rest		$\Delta/H = 1/240$	
Type of Soil	γ_{eq} (kN/m³)	K	γ_{eq} (kN/m³)	K	γ_{eq} (kN/m³)	K	γ_{eq} (kN/m³)	K
Loose sand or gravel	8.64	0.45	6.28	0.35	10.21	0.55	7.86	0.45
Medium dense sand or gravel	7.86	0.40	5.50	0.25	9.43	0.50	7.07	0.35
Dense sand or gravel	7.07	0.35	4.71	0.20	8.64	0.45	6.28	0.30
Compacted silt (ML)	9.43	0.50	6.28	0.35	11.00	0.60	7.86	0.45
Compacted clay (CL)	11.00	0.60	7.07	0.40	12.57	0.70	8.64	0.50
Lean compacted fat clay (CH)	12.57	0.65	8.64	0.50	14.14	0.75	10.21	0.60

[a]After Clough and Duncan, 1991.
[b]$P_h = \gamma_{eq} z + K q_s$ where γ_{eq} = equivalent fluid unit weight, z = depth below ground surface, K = horizontal earth pressure coefficient, and q_s = uniform surcharge pressure.

amount of the vertical earth pressure acting on the wall can be taken as much as 10% of the soil weight.

Effect of Surcharges   When vertical loads act on a surface of the backfill near a retaining wall or an abutment, the lateral and vertical earth pressure used for the design of the wall should be increased.

Uniform Surcharge Load   A surcharge load uniformly distributed over a large ground surface area increases both the vertical and lateral pressures. The increase in the vertical pressure, Δp_v is the same as the applied surcharge pressure, q_s. That is,

$$\Delta p_v = q_s \tag{10.11}$$

and the amount of increase in the lateral pressure, Δp_h is

$$\Delta p_h = K \, q_s \tag{10.12}$$

where

K = an earth pressure coefficient (dimensionless)

$K = K_a$ for active pressure

$K = K_0$ for at-rest condition

$K = K_p$ for passive pressure

Because the applied area is infinitely large, the increases in both vertical and horizontal pressures are constant over the height of the wall. Therefore, the horizontal resultant force due to a surcharge load is located at midheight of the wall.

Point Load and Strip Loads   The theory of elasticity can be used to estimate the increased earth pressures induced by various types of surcharge loads. Equations for earth pressures due to point load and strip loads are presented in Figure 10.16.

Equivalent Height of Soil for Live Load Surcharge   In the AASHTO (1994) LRFD Bridge Specifications, the live load surcharge, *LS*, is specified in terms of an equivalent height of soil, h_{eq}, representing the vehicular loading. The values specified for h_{eq} vary with the height of the wall and are given in Table 10.8.

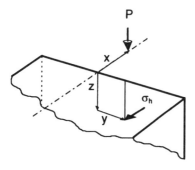

$$\sigma_h = \frac{P}{\pi R^2} \left[\frac{3r^2 z}{R^3} - \frac{(1-2v)\,R}{R+z} \right]$$

where   P = force

$$r = \sqrt{x^2 + y^2}$$

$$R = \sqrt{x^2 + y^2 + z^2}$$

$$v = \text{Poisson's ratio}$$

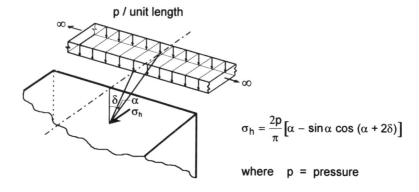

$$\sigma_h = \frac{2p}{\pi} \left[\alpha - \sin\alpha \, \cos(\alpha + 2\delta) \right]$$

where   p = pressure

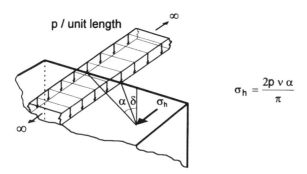

$$\sigma_h = \frac{2p\, v\, \alpha}{\pi}$$

Fig. 10.16   Earth pressure due to point load and strip loads.

10.4.6   Design Requirements for Abutments

Failure Modes for Abutments   Abutments are subject to various limit states or types of failure, as illustrated in Figure 10.17. Failures can occur within soils or the structural members. Sliding failure occurs when the lateral earth

TABLE 10.8 Equivalent Height of Soil for Live Load Surcharge[a]

Wall Height (mm)	h_{eq} (mm)
≤ 1500	1700
3000	1200
6000	760
≥ 9000	610

[a]In AASHTO Table 3.11.6.2-1. [From *AASHTO LRFD Bridge Design Specifications.* Copyright © 1994 by the American Association of State and Highway and Transportation Officials, Washington, DC. Used by permission.]

pressure exerted on the abutment exceeds the frictional sliding capacity of the foundation. If the bearing pressure is larger than the capacity of the foundation soil or rock, bearing failure results. Deep-seated sliding failure may develop in clayey soils. Structural failure also should be checked.

Basic Design Criteria for Abutments For design purposes, abutments on spread footings can be classified into three categories (Duncan et al., 1990):

1. Abutment with clayey soils in the backfill or foundations.
2. Abutment with granular backfills and foundations of sand or gravel.
3. Abutment with granular backfills and foundations on rock.

For each category, design procedures and stability criteria for the ASD method and the LRFD method are summarized in Figures 10.18–10.20. The uniform pressure distributions shown in the figures are based on the method presented by Meyerhof (1953). Gravity types of abutments are used in the figures, but the procedures and criteria are equally applicable to semigravity, cantilever, and other types of abutments that develop similar foundation pressures when subjected to lateral forces.

Procedure for Design of Abutments A series of steps must be followed to obtain a satisfactory design.

Step 1. Select preliminary proportions of the wall.
Step 2. Determine loads and earth pressures.
Step 3. Calculate magnitude of reaction forces on base.
Step 4. Check stability and safety criteria
    a. Location of normal component of reactions
    b. Adequacy of bearing pressure
    c. Safety against sliding.

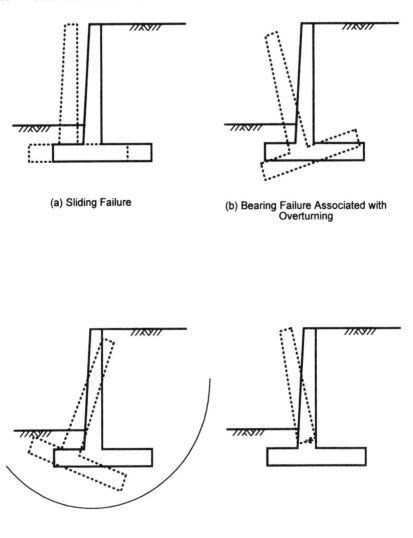

(a) Sliding Failure

(b) Bearing Failure Associated with Overturning

(c) Deep-seated Sliding Failure

(d) Structural Failure

Fig. 10.17   Failure modes of abutments.

Step 5. Revise proportions of wall and repeat Steps 2–4 until stability criteria are satisfied and then check
    a. Settlement within tolerable limits
    b. Safety against deep-seated foundation failure.
Step 6. If proportions become unreasonable, consider a foundation supported on driven piles or drilled shafts.
Step 7. Compare economics of completed design with other wall systems.

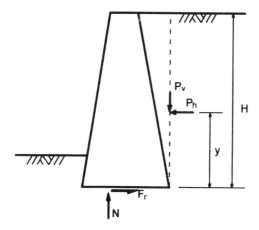

Earth Loads

P_v and P_h based on experience, with allowance for creep

$y = 0.4\,H$

Stability Criteria

| ASD Method | LRFD Method |

or

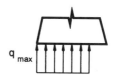

q_{max}

(1) N within middle third of base

(2) $R_I q_{ult} / FS \geq q_{max\,(unfactored)}$

(3) Safe against sliding

   $F_r / FS \geq P_{h(unfactored)}$

(4) Settlement within tolerable limits

(5) Safe against deep-seated foundation failure

(1) N within middle half of base

(2) $\phi R_I q_{ult} \geq q_{u\,max}$

(3) Safe against sliding

   $\phi_s F_r \geq \Sigma \gamma_i P_{hi}$

(4) Settlement within tolerable limits

(5) Safe against deep-seated foundation failure

Fig. 10.18   Earth loads and stability criteria for walls with clayey soils in the backfill of foundation. [After J. M. Duncan, G. W. Clough, and R. M. Ebeling (1990). "Behavior and Design of Gravity Earth Retaining Structures," *Proceedings of Conference on Design and Performance of Earth Retaining Structures,* ASCE, Cornell University, Ithaca, NY. Reproduced by permission of ASCE.]

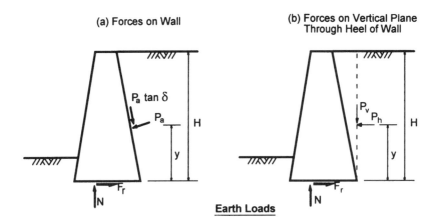

(a) Forces on Wall

(b) Forces on Vertical Plane
Through Heel of Wall

Earth Loads

P_a and P_h calculated using Coulomb active earth pressure theory
δ or P_a estimated using judgement, with allowance for movement
of backfill relative to wall.

$y = 0.4\,H$

Stability Criteria

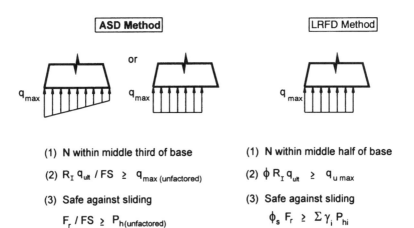

ASD Method	LRFD Method
or	

(1) N within middle third of base

(2) $R_I\,q_{ult}\,/\,FS \geq q_{max\,(unfactored)}$

(3) Safe against sliding

   $F_r\,/\,FS \geq P_{h(unfactored)}$

(4) Settlement within tolerable limits

(1) N within middle half of base

(2) $\phi\,R_I\,q_{ult} \geq q_{u\,max}$

(3) Safe against sliding

   $\phi_s\,F_r \geq \Sigma\,\gamma_i\,P_{hi}$

(4) Settlement within tolerable limits

Fig. 10.19 Earth loads and stability criteria for walls with granular backfills and foundations on sand or gravel. [After J. M. Duncan, G. W. Clough, and R. M. Ebeling (1990). "Behavior and Design of Gravity Earth Retaining Structures," *Proceedings of Conference on Design and Performance of Earth Retaining Structures,* ASCE, Cornell University, Ithaca, NY. Reproduced by permission of ASCE.]

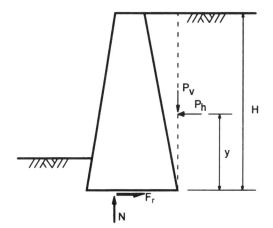

Earth Loads

P_h based on at-rest pressure

P_v estimated using judgement

$y = 0.4\,H$

Stability Criteria

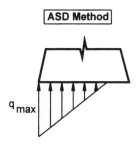

ASD Method

q_{max}

(1) N within middle half of base

(2) $R_I q_{ult} / FS \ge q_{max\,(unfactored)}$

(3) Safe against sliding:

$F_r / FS \ge P_{h(unfactored)}$

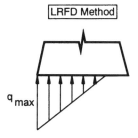

LRFD Method

q_{max}

(1) N within middle three quarters of base

(2) $\phi R_I q_{ult} \ge q_{u\,max}$

(3) Safe against sliding:

$\phi_s F_r \ge \Sigma \gamma_i P_{hi}$

Fig. 10.20 Earth loads and stability criteria for walls with granular backfills and foundations on rock. [After J. M. Duncan, G. W. Clough, and R. M. Ebeling (1990). "Behavior and Design of Gravity Earth Retaining Structures," *Proceedings of Conference on Design and Performance of Earth Retaining Structures,* ASCE, Cornell University, Ithaca, NY. Reproduced by permission of ASCE.]

Step 1. Preliminary Proportions   Figure 10.21 shows commonly used dimensions for a gravity-retaining wall and a cantilever wall. These proportions can be used when scour is not a concern to obtain dimensions for a first trial of the abutment.

Step 2. Loads and Earth Pressures   Design loads for an abutment are obtained by using group load combinations described in Tables 3.1 and 3.2. Methods for calculating earth pressures exerted on the wall are discussed in Section 10.4.5. The use of equivalent fluid pressures presented in Table 10.7 gives satisfactory earth pressures if conditions are not unusual.

Step 3. Reaction Forces on Base   Figure 10.22 illustrates a typical cantilever wall subjected to various loads causing reaction forces which are normal to the base (N) and tangent to the base (F_r). These reaction forces are determined by simple statics for each load combination being investigated.

Step 4. Stability Criteria

1. The location of the resultant on the base is determined by balancing moments about the toe of the wall. The criteria for foundations on soil for the location of the resultant is that it must lie within the middle half for LRFD (Figs. 10.18 and 10.19). This criterion replaces the check on the ratio of stabilizing moment to overturning moment. For foundations on rock, the acceptable location of the resultant has a greater range than for foundations on soil (Fig. 10.20).

   As shown in Figure 10.23, the location of the resultant, X_o, is obtained by

   $$X_o = \frac{\text{summation of moments about point o}}{N} \qquad (10.13)$$

   where N = the vertical resultant force (force/length).
   The eccentricity of the resultant, e, with respect to the centerline of the base is

   $$e = \frac{B}{2} - X_o \qquad (10.14)$$

   where B = base width (length).

2. Safety against bearing failure is obtained by applying a performance factor to the ultimate bearing capacity in the LRFD method. The ultimate bearing capacity can be calculated from in situ tests or semiempirical procedures as presented in an engineering manual for shallow foundations (Barker et al., 1991).

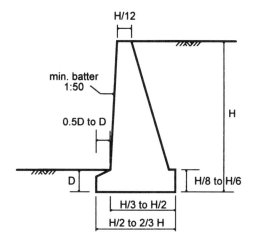

(a) Mass Concrete Wall

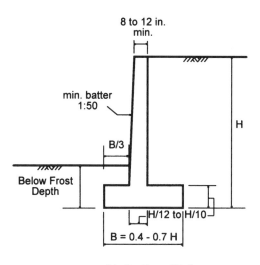

(b) Cantilever Wall

Fig. 10.21 Preliminary dimensions for gravity walls and cantilever walls. [After Clayton and Milititsky, 1986.]

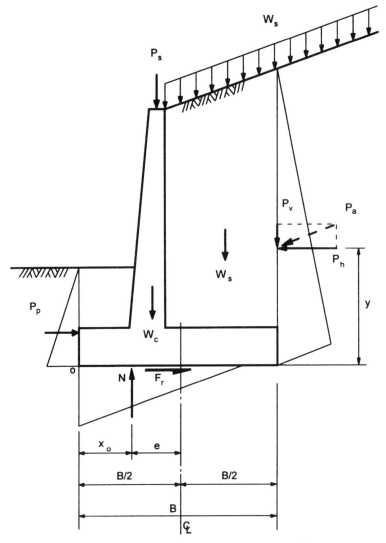

Fig. 10.22 Forces on a typical retaining wall or abutment.

Safety against bearing failure is checked by

$$\phi R_I \, q_{\text{ult}} \geq q_{u\text{max}} \tag{10.15}$$

where

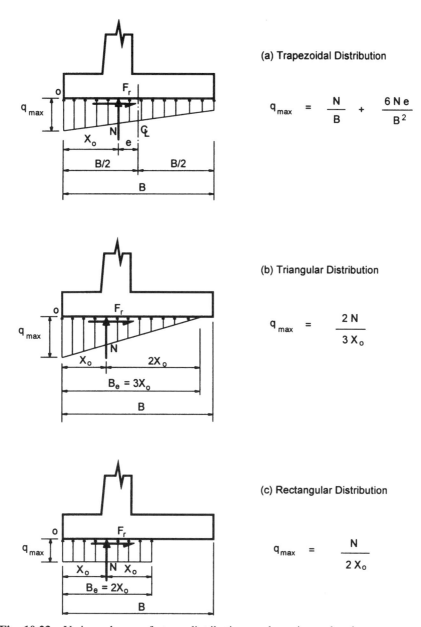

(a) Trapezoidal Distribution

$$q_{max} = \frac{N}{B} + \frac{6\,N\,e}{B^2}$$

(b) Triangular Distribution

$$q_{max} = \frac{2\,N}{3\,X_o}$$

(c) Rectangular Distribution

$$q_{max} = \frac{N}{2\,X_o}$$

Fig. 10.23 Various shapes of stress distributions and maximum bearing pressures.

q_{ult} = ultimate bearing capacity (force/length²)

R_I = reduction factor due to inclined loads = $(1 - H_n/V_n)^3$

H_n = unfactored horizontal force

V_n = unfactored vertical force

ϕ = performance or resistance factor

q_{umax} = maximum bearing pressure due to factored loads (force/length²)

Shape of Bearing Pressure Distribution The shape of the bearing pressure distribution for soil foundations can be treated as a triangle, a trapezoid or a rectangle, and for rock foundations treated as a triangle or a trapezoid (see Figs. 10.18–10.20). Therefore, the resultant, N, will pass through the centroid of a triangular or trapezoidal stress distribution, or the middle of a uniformly distributed stress block as illustrated in Figure 10.23.

Maximum Bearing Pressure The following equations are used to compute the maximum soil pressures, q_{umax} per unit length of a rigid footing.

For a triangular shape of bearing pressure:

a. when the resultant is within the middle third of base

$$q_{(u)max} = \frac{N_{(u)}}{B} - \frac{6N_{(u)}e}{B^2} \qquad (10.16)$$

b. when the resultant is outside of the middle third of base

$$q_{(u)max} = \frac{2N_{(u)}}{3X_o} \qquad (10.17)$$

For a uniform distribution of bearing pressure

$$q_{(u)max} = \frac{N_{(u)}}{2X_o} \qquad (10.18)$$

where

$N_{(u)}$ = unfactored (factored) vertical resultant (force/length),

X_o = location of the resultant measured from toe (length)

e = eccentricity of $N_{(u)}$ (length).

3. In the LRFD method, sliding stability is checked by

$$\phi_s F_{ru} \geq \Sigma \gamma_i P_{hi} \qquad (10.19)$$

where

ϕ_s = performance factor for sliding (values given in Table 10.2),

$F_{ru} = N_u \tan \delta_b + c_a B_e$

N_u = factored vertical resultant (force/length)

δ_b = friction angle between base and soil (degrees)

c_a = adhesion (force/length^2)

B_e = effective length of base in compression (length)

γ_i = load factor for force component i

P_{hi} = horizontal earth pressure force i causing sliding (force/length)

As shown in Figure 10.22, the lateral pressure, P_h, causes the wall to slide and is resisted by friction and adhesion between the base and foundation soil. According to the sixteenth edition of AASHTO (1996) Standard Specifications, the passive earth pressure, P_p, generated by the soil in front of the wall may be included to resist sliding if it is ensured that the soil in front of the wall will exist permanently (e.g., if the wall is embedded deeply below a covering such as a sidewalk or pavement). However, sliding failure occurs in many cases before the passive earth pressure is fully mobilized. Therefore, it is safer to ignore the effect of the passive earth pressure.

Step 5. Revise Proportions   When the preliminary wall dimensions are found inadequate, the wall dimensions should be adjusted by a trial and error method. A sensitivity study done by Kim (1995) shows that the stability can be improved by varying the location of the wall stem, the base width, and the wall height. Some suggestions for correcting each stability or safety problem are presented as follows:

1. Bearing Failure or Eccentricity Criterion Not Satisfied
   a. Increase the base width.
   b. Relocate the wall stem by moving it toward the heel.

    c. Minimize P_h by replacing a clayey backfill with granular material or by reducing porewater pressures behind the wall stem with a well-designed drainage system.

    d. Provide an adequately designed reinforced concrete approach slab supported at one end by the abutment so that no horizontal pressure due to live load surcharge need be considered.

2. Sliding Stability Criterion Not Satisfied
    a. Increase the base width.
    b. Minimize P_h as described above.
    c. Use an inclined base (heel side down) to increase horizontal resistance.
    d. Provide an adequately designed approach slab as mentioned above.
    e. Use a shear key.

The main function of a shear key is to generate additional passive soil pressure to increase sliding resistance. The considerations discussed earlier concerning the amount of displacement required to mobilize passive earth pressure are also important with regard to the resistance that can be achieved through the use of a key. Depending on local experience or job conditions, the design engineer must decide whether or not some passive earth pressure should be included.

3. Settlement and Overall Stability Check
Once the proportions of the wall have been selected to satisfy the bearing pressure, eccentricity and sliding criteria, then the requirements on settlement and overall slope stability must be checked.

    a. Settlement should be checked for walls founded on compressible soils to ensure that the predicted settlement is less than the settlement that the wall or structure it supports can tolerate. The magnitude of settlement can be estimated using the methods described in *Engineering Manual for Shallow Foundations* (Barker et al., 1991).

    b. The overall stability of slopes with regard to the most critical sliding surface should be evaluated if the wall is underlain by weak soil. This check is based on limiting equilibrium methods which employ the modified Bishop, simplified Janbu or Spenser analysis. This subject is discussed in a number of design manuals and papers including U.S. Dept. of Navy (1982a) and a manual by the Corps of Engineers (1989).

Step 6. Consider Deep Foundations   Driven piles and drilled shafts can be used when the configuration of the wall is unreasonable or uneconomical. Engineering manuals for driven piles and drilled shafts (Barker et al., 1991) may be consulted with regard to design of deep foundations to withstand vertical and lateral loads.

Step 7. Compare with Alternative Wall Systems   When a design is completed, it should be compared with other types of walls that may result in a more economical design. Detailed information can be found in Sections 2.2.1 and 10 of a manual by the Corps of Engineers (1989).

10.4.7   Example 10.2   Abutment Design

The design principles and procedures presented in the previous sections are demonstrated in this design example.

PROBLEM STATEMENT

Using the LRFD method, the stability and safety for the abutment shown in Figure 10.24 is to be checked. The abutment is founded on sandy gravel with an average Standard Penetration Test (SPT) blow count of 22. The ultimate bearing capacity of the foundation soil is estimated to be 1060 kPa.

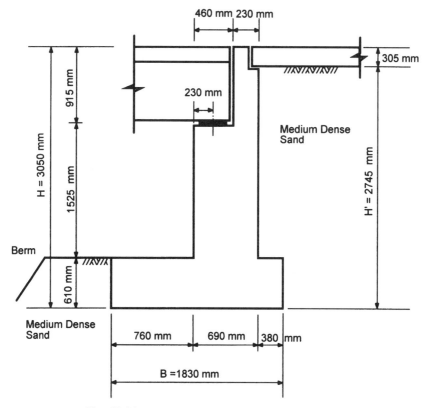

Fig. 10.24   A design example of bridge abutment.

The backfill material is medium dense sand with an angle of internal friction equal to 35°, and unit weight of 18.9 kN/m³. The density of the concrete is 23.6 kN/m³. The friction angle between the base and foundation soil, δ_b, is estimated to be 29°.

The passive pressure of the soil in front of the abutment will be ignored. From Table 10.8, the equivalent height of soil for live load surcharge is determined to be 1195 mm. The weight of the approach slab is considered dead load surcharge. (If the approach slab is designed to span across the backfill, both surcharges could be ignored.)

In this example, active pressure obtained from equivalent fluid pressure is used. From Table 10.7, for medium dense sand or gravel, the equivalent fluid unit weight, γ_{eq}, is 5.50 kN/m³ and the horizontal earth pressure coefficient, K, is 0.25. The equivalent vertical fluid pressure acting on the wall is taken as 10% of the soil weight or 1.89 kN/m³.

DETERMINATION OF LOADS AND EARTH PRESSURES

The loadings to be considered in this example are illustrated in Figure 10.25.

Loadings   The loadings from the superstructure are given as

$$DL = \text{dead load} = 109.4 \text{ kN/m}$$

$$LL = \text{live load} = 87.5 \text{ kN/m}$$

$$WS = \text{wind load on superstructure} = 2.9 \text{ kN/m}$$

$$WL = \text{wind load on live load} = 0.7 \text{ kN/m}$$

$$BR = \text{braking force} = 3.6 \text{ kN/m}$$

$$CR + SH + TU = \text{creep, shrinkage, and temperature}$$

$$= 10\% \text{ of } DL = 10.9 \text{ kN/m}$$

Pressures generated by the live load and dead load surcharges can be obtained as

$$\omega_L = h_{eq}\,\gamma = 1195 \text{ mm} \times 18.9 \text{ kN/m}^3 = 22.6 \text{ kN/m}^2$$

$$\omega_D = (\text{slab thickness})(\gamma_c) = 305 \text{ mm} \times 23.6 \text{ kN/m}^3 = 7.2 \text{ kN/m}^2$$

$$H_L = K\,\omega_L H' = 0.25 \times 22.6 \text{ kN/m}^2 \times 2743 \text{ mm} = 15.51 \text{ kN/m}$$

$$H_D = K\,\omega_D H' = 0.25 \times 7.2 \text{ kN/m}^2 \times 2743 \text{ mm} = 4.94 \text{ kN/m}$$

$$V_L = \omega_L \,(\text{heel width}) = 22.6 \text{ kN/m}^2 \times 380 \text{ mm} = 8.59 \text{ kN/m}$$

$$V_D = \omega_D \,(\text{heel width}) = 7.2 \text{ kN/m}^2 \times 380 \text{ mm} = 2.74 \text{ kN/m}$$

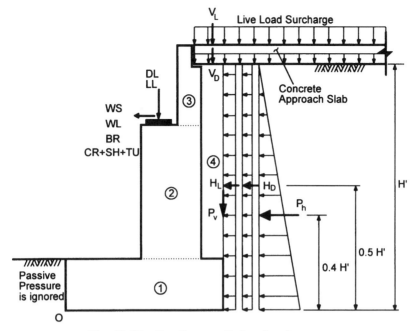

Fig. 10.25 Loadings applied to the abutment.

Pressures due to equivalent fluid pressure can be calculated as

$$P_h = \tfrac{1}{2}(\text{EFP}_h)H'^2 = \tfrac{1}{2}(5.50)(2.745)^2 = 20.72 \text{ kN/m}$$

$$P_v = \tfrac{1}{2}(\text{EFP}_v)H'^2 = \tfrac{1}{2}(1.89)(2.745)^2 = 7.12 \text{ kN/m}$$

Load Combinations From Table 3.1 [Table A3.4.1-1], the relevant load combinations are determined to be Strength *I* and Strength *III*. Considering the minimum and maximum load factors for permanent loads as shown in Table 3.2 [Table A3.4.1-2], load combinations can be expanded to four groups: Strength *I*, Strength *Ia*, Strength *III*, and Strength *IIIa*. The load factors and load combinations are summarized as follows:

	DC	EV	EH	LL	BR	LS	WS	WL	CR + SH + TU
Strength *I*	1.25	1.35	1.50	1.75	1.75	1.75	0	0	0.50
Strength *Ia*	0.90	1.00	1.50	1.75	1.75	1.75	0	0	0.50
Strength *III*	1.25	1.35	1.50	0	0	0	1.40	0	0.50
Strength *IIIa*	0.90	1.00	1.50	0	0	0	1.40	0	0.50

Unfactored Loads   Unfactored vertical and horizontal loads are summarized as follows:

Vertical Loads Items	V_n (kN)	Arm (mm)	Moment (kN m)
1. (1830)(610)(23.6)	26.34	915	24.10
2. (690)(1525)(23.6)	24.83	1105	27.44
3. (230)(915)(23.6)	4.97	1335	6.63
4. (380)(2440)(18.9)	17.52	1640	28.73
DL	109.40	990	108.31
LL	87.50	990	86.63
V_D	2.74	1640	4.49
V_L	8.59	1640	14.09
P_V	7.12	1830	13.03

Horizontal Loads Items	H_n (kN)	Arm (mm)	Moment (kN m)
P_h	20.72	1098	22.75
H_D	4.94	1373	6.78
H_L	15.51	1373	21.30
WS	2.90	2135	6.19
WL	0.70	2135	1.49
BR	3.60	2135	7.69
CR + SH + TU	10.90	2135	23.27

Design Loads   Factored design loads are summarized as follows:

Vertical Loads, V_u (kN/m)

Items Notations V_n	1 DC 26.34	2 DC 24.83	3 DC 4.97	4 EV 17.52	DL DC 109.40	LL LL 87.50	V_D DC 2.74	V_L LL 8.59	P_v EH 7.12	V_u Total
Strength I	32.93	31.04	6.21	23.65	136.75	153.13	3.43	15.03	10.68	412.85
Strength Ia	23.71	22.35	4.47	17.52	98.46	153.13	2.47	15.03	10.68	347.82
Strength III	32.93	31.04	6.21	23.65	136.75	0.00	3.43	0.00	10.68	244.69
Strength IIIa	23.71	22.35	4.47	17.52	98.46	0.00	2.47	0.00	10.68	179.66

Moment due to V_u (kN m/m)

Items Notations M_{Vn}	1 DC 24.10	2 DC 27.44	3 DC 6.63	4 EV 28.73	DL DC 108.31	LL LL 86.63	V_D DC 4.49	V_L LL 14.09	P_v EH 13.03	M_{Vu} Total
Strength I	30.13	34.30	8.29	38.79	135.39	151.60	5.61	24.66	19.55	448.32
Strength Ia	21.69	24.70	5.97	28.73	97.48	151.60	4.04	24.66	19.55	378.42
Strength III	30.13	34.30	8.29	38.79	135.39	0.00	5.61	0.00	19.55	272.06
Strength IIIa	21.69	24.70	5.97	28.73	97.48	0.00	4.04	0.00	19.55	202.16

Horizontal Loads, H_u(kN/m)

Items	P_h	H_D	H_L	WS	WL	BR	CR + SH + TU	
Notations	EH	EH	LS	WS	WL	BR	CR + SH + TU	H_u
H_n	20.72	4.94	15.51	2.90	0.70	3.60	10.90	Total
Strength I	31.08	7.41	27.14	0.00	0.00	6.30	5.45	77.38
Strength Ia	31.08	7.41	27.14	0.00	0.00	6.30	5.45	77.38
Strength III	31.08	7.41	0.00	4.06	0.00	0.00	5.45	48.00
Strength IIIa	31.08	7.41	0.00	4.06	0.00	0.00	5.45	48.00

Moment due to H_u(kN m/m)

Items	P_h	H_D	H_L	WS	WL	BR	CR + SH + TU	
Notations	EH	EH	LS	WS	WL	BR	CR + SH + TU	M_{Hu}
M_{Hn}	22.75	6.78	21.30	6.19	1.49	7.69	23.27	Total
Strength I	34.13	10.17	37.28	0.00	0.00	13.46	11.64	106.68
Strength Ia	34.13	10.17	37.28	0.00	0.00	13.46	11.64	106.68
Strength III	34.13	10.17	0.00	8.67	0.00	0.00	11.64	64.61
Strength IIIa	34.13	10.17	0.00	8.67	0.00	0.00	11.64	64.61

STABILITY AND SAFETY CRITERIA

Three design criteria should be satisfied: Eccentricity, Sliding, and Bearing Capacity. The last column of each table represents the design margin which is expressed as

$$\text{Design margin (\%)} = \frac{\text{provided} - \text{applied}}{\text{provided}} \times 100$$

Eccentricity   In the LRFD method, the eccentricity design criterion is ensured by keeping the resultant force within the middle half of the base width. In other words, the eccentricity should not exceed the maximum eccentricity, e_{max} (= $B/4$) in soil foundation. The results are summarized as follows:

	V_L	H_L	M_V	M_H	X_o	e	e_{max}	Design Margin (%)
Strength I	412.85	77.38	448.32	106.68	827.52	87.48	457.50	80.9
Strength Ia	347.82	77.38	378.42	106.68	781.27	133.73	457.50	70.8
Strength III	244.69	48.00	272.06	64.61	847.81	67.19	457.50	85.3
Strength IIIa	179.66	48.00	202.16	64.61	765.61	149.39	457.50	67.3

$$\text{where } \% = (e_{max} - e)/e_{max} \times 100$$

Sliding   The results of sliding design criterion are summarized as follows:

	V_L	Tan δ_b	F_r	ϕ_s	$\phi_s F_r$	H_L	Design Margin (%)
Strength *I*	412.85	0.55	227.07	0.8	181.66	77.38	57.40
Strength *Ia*	347.82	0.55	191.30	0.8	153.04	77.38	49.44
Strength *III*	244.69	0.55	134.58	0.8	107.66	48.00	55.42
Strength *IIIa*	179.66	0.55	98.81	0.8	79.05	48.00	39.28

$$\text{where } \% = (\phi_s F_r - H_L)/(\phi_s F_r) \times 100$$

Bearing Capacity   The results of bearing capacity criterion are summarized as follows:

	H_L	V_L	H_L/V_L	R_I	q_{ult}	$R_I q_{ult}$	$\phi R_I q_{ult}$	q_{max}	Design Margin (%)
Strength *I*	55.67	289.0	0.19	0.531	1060.0	562.86	253.29	249.45	1.52
Strength *Ia*	55.67	289.0	0.19	0.531	1060.0	562.86	253.29	222.60	12.12
Strength *III*	39.46	192.9	0.20	0.512	1060.0	542.72	244.22	144.31	40.91
Strength *IIIa*	39.46	192.9	0.20	0.512	1060.0	542.72	244.22	117.33	52.00

$$\text{where } \% = (\phi R_I q_{ult} - q_{max})/(\phi R_I q_{ult}) \times 100$$

CONCLUSIONS

Strength *I* governs the design with the bearing capacity criterion. If any design criterion is not satisfied, the abutment dimensions should be adjusted by increasing the base width, moving the stem location or changing wall thickness. Because an abutment is subject to numerous loadings and various limit states, it is not clear which dimensions should be adjusted to find the optimum design. Using a spreadsheet program may be one of the most effective ways to design the abutment.

REFERENCES

AASHTO (1996). *Standard Specifications for Highway Bridges,* 16th ed., American Association of State Highway and Transportation Officials, Washington, DC.

AASHTO (1994). *LRFD Bridge Design Specifications,* 1st ed., American Association of State Highway and Transportation Officials, Washington, DC.

Barker, R. M., J. M. Duncan, K. B. Rojiani, P. S. K. Ooi, C. K. Tan, and S. G. Kim (1991). *Manuals for the Design of Bridge Foundations,* NCHRP Report 343, Transportation Research Board, Washington, DC, 308 pp.

Caquot, A. and J. Kerisel (1948). *Tables for the Calculation of Passive Pressure, Active Pressure and Bearing Capacity of Foundations,* Gauthier-Villars, Imprimeur-Libraire, Libraire du Bureau des Longitudes, de L'Ecole Polytechnique, Paris, 120 pp.

Clausen, C. J. F. and S. Johansen (1972). "Earth Pressures Measured Against a Section of a Basement Wall," *Proceedings, 5th European Conference on SMFE,* Madrid, Spain, 1972, pp. 515–516.

Clayton, C. R. I. and J. Milititsky (1986). *Earth Pressure and Earth-Retaining Structures,* Surrey University Press, Glasgow, Scotland, 300 pp.

Clough, C. W. and J. M. Duncan (1991). *Foundation Engineering Handbook,* 2nd ed., edited by H. Y. Fang, Chapman & Hall, New York, pp. 223–235.

D'Appolonia, D. J., E. D'Appolonia, and R. F. Brisette (1970). "Settlement of Spread Footings on Sand," *Journal of Soil Mechanics and Foundation Division,* ASCE, Vol. 96, No. SM2, pp. 754–761.

Duncan, J. M., G. W. Clough, and R. M. Ebeling (1990). "Behavior and Design of Gravity Earth Retaining Structures," *Proceedings of Conference on Design and Performance of Earth Retaining Structures,* ASCE, Cornell University, Ithaca, New York, June, pp. 251–277.

Federal Highway Administration (1988). *Technical Advisory—Interim Procedures for Evaluating Scour at Bridges,* U.S. Department of Transportation, Office of Engineering, Bridge Division, 62 pp.

Janbu, N. (1963). "Soil Compressibility as Determined by Oedometer and Triaxial Tests," *Proceedings European Conference of Soil Mechanics and Foundation Engineering,* Vol. I, Wiesbaden, Germany.

Janbu, N. (1967). "Settlement Calculations Based on Tangent Modulus Concept," Bulletin No. 2, *Soil Mechanics and Foundation Engineering Series,* The Technical University of Norway, Trondheim, Norway, 57 pp.

Kim, J. S. (1995). Reliability-Based Design of a Retaining Wall, Ph.D Dissertation, Virginia Polytechnic Institute and State University, Blacksburg, VA, 236 pp.

Menard, L. and J. Rousseau (1962). "L'Evaluation des Tossements Tendances Nouvelles," *Sols-Soils,* Vol. 1, No. 1.

Meyerhof, G. G. (1953). "The Ultimate Bearing Capacity of Foundations under Eccentric and Inclined Loads," *Proceedings, 3rd International Conference on Soil Mechanics and Foundation Engineering,* Zurich, Switzerland, Vol. 1, pp. 440–445.

Munfakh, G. A. (1990). "Innovative Earth Retaining Structures: Selection, Design & Performance," *Proceedings of Conference on Design and Performance of Earth Retaining Structures,* ASCE, Cornell University, Ithaca, NY, pp. 85–118.

New Zealand Ministry of Works and Development (1979). *Retaining Wall Design Notes,* Wellington, New Zealand, 43 pp.

Peck, R. B., W. E. Hansen, and T. H. Thornburn (1974). *Foundation Engineering,* 2nd ed. Wiley, NY, 514 pp.

Schmertmann, J. H. (1977). *Guidelines for Cone Penetration Test—Performance and Design,* US Department of Transportation, Federal Highway Administration, pp. 54–55.

Schmertmann, J. H., J. P. Hartman, and P. R. Brown (1978). "Improved Strain Influence Factor Diagram," *Journal of Geotechnical Engineering Division,* ASCE, Vol. 104, No. GT8, pp. 1131–1135.

Schnore, A. R. (1990). "Selecting Retaining Wall Type and Specifying Proprietary Retaining Walls in NYSDOT Practice," *Proceedings of Conference on Design and*

Performance of Earth Retaining Structures, ASCE, Cornell University, Ithaca, New York, pp. 119–124.

Sherif, M. A., I. Ishibashi, and C. D. Lee (1982). "Earth Pressures Against Rigid Retaining Walls," *Journal of Geotechnical Engineering Division,* ASCE, Vol. 108, GT5, pp. 679–695.

Sowers, B. F. (1962)."Shallow Foundations," Chapter 6 in *Foundation Engineering,* edited by Leonards, G. A., McGraw-Hill, pp. 525–632.

Tan, C. K. and J. M. Duncan (1991). "Settlements of Footings on Sand—Accuracy and Reliablility," *Proceedings of Geotechnical Engineering Congress,* Boulder, CO.

Terzaghi, K. (1934). "Retaining Wall Design for Fifteen-Mile Falls Dam," *Engineering News Record,* May, pp. 632–636.

Terzaghi, K. and R. B. Peck (1967). *Soil Mechanics in Engineering Practice,* Wiley, New York, 729 pp.

U.S. Army Corps of Engineers (1949). "Report on Frost Penetration," Addendum No. 1, 1945–1947, Corps of Engineers, U.S. Army, New England Division, Boston, MA.

U.S. Army Corps of Engineers (1989). "Engineering and Design Retaining and Flood Walls," Manual EM No. 1110-2-2502, Washington, DC.

U.S. Dept. of Navy (1982a) "NAVFAC DM7.1, Soil Mechanics," Naval Facilities Engineering Command, VA, 348 pp.

U.S. Dept. of Navy (1982b). "NAVFAC DM7.2, Foundations and Earth Structures," Naval Facilities Engineering Command, VA, 244 pp.

APPENDIX A

INFLUENCE FUNCTIONS FOR DECK ANALYSIS

Throughout the book, several examples require the analysis of the deck for uniform and concentrated (line) loads. To facilitate analysis, influence functions were developed for a deck with five interior bays and two cantilevers. The widths (S) of the interior bays are assumed to be the same and the cantilevers are assumed to be of length (L). The required ordinates and areas are given in Table A1. The notes at the bottom of the table describe its use. Examples are given in Chapters 5, 7, and 9 to illustrate analysis using this table.

TABLE A1 Influence Functions for Deck Analysis[a]

Location		M_{200}	M_{204}	M_{205}	M_{300}	R_{200}[b]
C	100	−1.0000	−0.4920	−0.3650	0.2700	1 + 1.270L/S
A	101	−0.9000	−0.4428	−0.3285	0.2430	1 + 1.143L/S
N	102	−0.8000	−0.3936	−0.2920	0.2160	1 + 1.016L/S
T	103	−0.7000	−0.3444	−0.2555	0.1890	1 + 0.889L/S
I	104	−0.6000	−0.2952	−0.2190	0.1620	1 + 0.762L/S
L	105	−0.5000	−0.2460	−0.1825	0.1350	1 + 0.635L/S
E	106	−0.4000	−0.1968	−0.1460	0.1080	1 + 0.508L/S
V	107	−0.3000	−0.1476	−0.1095	0.0810	1 + 0.381L/S
E	108	−0.2000	−0.0984	−0.0730	0.0540	1 + 0.254L/S
R	109	−0.1000	−0.0492	−0.0365	0.0270	1 + 0.127L/S
110 or 200		0.0000	0.0000	0.0000	0.0000	1.0000
	201	0.0000	0.0494	0.0367	−0.0265	0.8735
	202	0.0000	0.0994	0.0743	−0.0514	0.7486
	203	0.0000	0.1508	0.1134	−0.0731	0.6269
	204	0.0000	0.2040	0.1150	−0.0900	0.5100
	205	0.0000	0.1598	0.1998	−0.1004	0.3996
	206	0.0000	0.1189	0.1486	−0.1029	0.2971
	207	0.0000	0.0818	0.1022	−0.0954	0.2044
	208	0.0000	0.0491	0.0614	−0.0771	0.1229
	209	0.0000	0.0217	0.0271	−0.0458	0.0542
210 or 300		0.0000	0.0000	0.0000	0.0000	0.0000
	301	0.0000	−0.0155	−0.0194	−0.0387	−0.0387
	302	0.0000	−0.0254	−0.0317	−0.0634	−0.0634
	303	0.0000	−0.0305	−0.0381	−0.0761	−0.0761
	304	0.0000	−0.0315	−0.0394	−0.0789	−0.0789
	305	0.0000	−0.0295	−0.0368	−0.0737	−0.0737
	306	0.0000	−0.0250	−0.0313	−0.0626	−0.0626
	307	0.0000	−0.0191	−0.0238	−0.0476	−0.0476
	308	0.0000	−0.0123	−0.0154	−0.0309	−0.0309
	309	0.0000	−0.0057	−0.0072	−0.0143	−0.0143
310 or 400		0.0000	0.0000	0.0000	0.0000	0.0000

401	0.0104	0.0104	0.0052	0.0042	0.0000
402	0.0171	0.0171	0.0086	0.0069	0.0000
403	0.0206	0.0206	0.0103	0.0083	0.0000
404	0.0214	0.0214	0.0107	0.0086	0.0000
405	0.0201	0.0201	0.0100	0.0080	0.0000
406	0.0171	0.0171	0.0086	0.0069	0.0000
407	0.0131	0.0131	0.0066	0.0053	0.0000
408	0.0086	0.0086	0.0043	0.0034	0.0000
409	0.0040	0.0040	0.0020	0.0016	0.0000
410 or 500	0.0000	0.0000	0.0000	0.0000	0.0000
501	-0.0031	-0.0031	-0.0015	-0.0012	0.0000
502	-0.0051	-0.0051	-0.0026	-0.0021	0.0000
503	-0.0064	-0.0064	-0.0032	-0.0026	0.0000
504	-0.0069	-0.0069	-0.0034	-0.0027	0.0000
505	-0.0067	-0.0067	-0.0033	-0.0027	0.0000
506	-0.0060	-0.0060	-0.0030	-0.0024	0.0000
507	-0.0049	-0.0049	-0.0024	-0.0020	0.0000
508	-0.0034	-0.0034	-0.0017	-0.0014	0.0000
509	-0.0018	-0.0018	-0.0009	-0.0007	0.0000
510	0.0000	0.0000	0.0000	0.0000	0.0000
Area + (w/o Cantilever)[c]	0.4464	0.0134	0.0982	0.0986	0.0000
Area − (w/o Cantilever)[c]	-0.0536	-0.1205	-0.0268	-0.0214	0.0000
Area Net (w/o Cantilever)[c]	0.3928	-0.1071	0.0714	0.0772	0.0000
Area + (Cantilever)[d]	1.0 + 0.635L/S	0.1350	0.0000	0.0000	0.0000
Area − (Cantilever)[d]	0.0000	0.0000	-0.1825	-0.2460	-0.5000
Area Net − (Cantilever)[d]	1.0 + 0.635L/S	0.1350	-0.1825	-0.2460	-0.5000

[a] Multiply coefficients by the span length where the load is applied, that is, L on cantilever and S in the other spans.
[b] DO NOT multiply by the cantilever span length, use formulae or values given.
[c] Multiply moment area coefficient by S^2, reaction area coefficient by S.
[d] Multiply moment area coefficient by L^2, reaction area coefficient by L.

APPENDIX B

METAL REINFORCEMENT INFORMATION

TABLE B.1 Standard Metric Reinforcing Bars

Designation Number	Nominal Diameter (mm)	Nominal Area (mm²)	Nominal Mass (kg/m)
10	11.3	100	0.785
15	16.0	200	1.870
20	19.5	300	2.356
25	25.2	500	3.925
30	29.9	700	5.495
35	35.7	1000	7.850
45	43.7	1500	11.775
55	56.4	2500	19.625

TABLE B.2   Standard Metric Prestressing Tendons

Type	Nominal Diameter (mm)	Nominal Area (mm^2)	Nominal Weight (kg/m)
Seven-wire strand	6.35	23.22	0.182
(Grade 250)	7.94	37.42	0.294
	9.53	51.61	0.405
	11.11	69.68	0.548
	12.70	92.90	0.730
	15.24	139.35	1.094
Seven-wire strand	9.53	54.84	0.432
(Grade 270)	11.11	74.19	0.582
	12.70	98.71	0.775
	15.24	140.00	1.102
Prestressing wire	4.88	18.7	0.146
	4.98	19.4	0.149
	6.35	32	0.253
	7.01	39	0.298
Prestressing bars	19	284	2.23
(plain)	22	387	3.04
	25	503	3.97
	29	639	5.03
	32	794	6.21
	35	955	7.52
Prestressing bars	15	181	1.46
(deformed)	20	271	2.22
	26	548	4.48
	32	806	6.54
	36	1019	8.28

TABLE B.3 Cross-Sectional Area (mm²) of Combinations of Metric Bars of the Same Size

Number of Bars	Bar Number							
	10	15	20	25	30	35	45	55
1	100	200	300	500	700	1000	1500	2500
2	200	400	600	1000	1400	2000	3000	5000
3	300	600	900	1500	2100	3000	4500	7500
4	400	800	1200	2000	2800	4000	6000	10 000
5	500	1000	1500	2500	3500	5000	7500	12 500
6	600	1200	1800	3000	4200	6000	9000	15 000
7	700	1400	2100	3500	4900	7000	10 500	17 500
8	800	1600	2400	4000	5600	8000	12 000	20 000
9	900	1800	2700	4500	6300	9000	13 500	22 500
10	1000	2000	3000	5000	7000	10 000	15 000	25 000

TABLE B.4 Cross-Sectional Area per Millimeter Width (mm^2/mm) of Metric Bars of the Same Size

Bar Spacing (mm)	Bar Number								
	10	15	20	25	30	35	45	55
100	1.000	2.000	3.000	5.000	7.000	10.000	15.000	20.000
125	0.800	1.600	2.400	4.000	5.600	8.000	12.000	16.667
150	0.667	1.333	2.000	3.333	4.667	6.667	10.000	14.286
175	0.571	1.143	1.714	2.857	4.000	5.714	8.571	12.500
200	0.500	1.000	1.500	2.500	3.500	5.000	7.500	11.111
225	0.444	0.889	1.333	2.222	3.111	4.444	6.667	10.000
250	0.400	0.800	1.200	2.000	2.800	4.000	6.000	8.333
300	0.333	0.667	1.000	1.667	2.333	3.333	5.000	7.143
350	0.286	0.571	0.857	1.429	2.000	2.857	4.286	6.250
400	0.250	0.500	0.750	1.250	1.750	2.500	3.750	5.555
450	0.222	0.444	0.667	1.111	1.555	2.222	3.333	

APPENDIX C

COMPUTER SOFTWARE FOR LRFD OF BRIDGES

At the time of this writing (1996), little software exits for the analysis and design of bridges according to the AASHTO LRFD Specification for Bridge Design. Those known to the authors are listed.

BT BEAM

Many of the numerical examples in this book were developed with the aid of BT Beam. This program is developed and maintained by Bridge Tech, Inc. and is available for a modest license fee. Capabilities include the analysis and load combination for the LRFD Specification. It does not include resistance computations.

Information can be obtained from Bridge Tech, Inc., 302 S. 2nd, Suite 200, Laramie, WY, 82070. Telephone (307)721-5070.

AASHTO BDS

AASHTO Bridge Design System (BDS) is software developed and supported by AASHTO in AASHTOware products. This program is a comprehensive program for the analysis, design, and rating of bridges. The most common types of bridges are addressed with the 15th ed. Specification. At the time of this writing, a live load module has been included for the analysis of AASHTO LRFD live loads. The incorporation of the LRFD Specification has been placed on hold pending a study on the program architecture and its ability and limitations to incorporate the new specification.

Information can be obtained from AASHTO, 444 N. Capitol St., NW, Washington, DC, 20001. Telephone (202)624-5800.

BRASS LRFD-SI

The program Bridge Analysis and Structural System (BRASS) is used by numerous departments of transportation for the analysis, design, and rating of bridges. At the time of this writing, the program is being updated for the incorporation of the LRFD loads and resistance articles. In addition, SI units are being incorporated. The program has the capability for the analysis, design review, and rating of steel, steel–concrete composite, conventional and pre-stressed concrete, and timber systems.

Information can be obtained from the Wyoming Department of Transportation, P.O. Box 1708, Cheyenne, WY, 82003-1708. Telephone (307)777-4427.

PENNDOT PROGRAMS

Recently, the Pennsylvania Department of Transportation contracted to develop a series of programs based upon the LRFD Specification. These programs include superstructure design review, pier design, culvert design, and abutment design. Other programs may also be forthcoming.

Information can be obtained from PennDOT, Bureau of Design, Bridge Division, Transportation and Safety Building, Room 1120, Harrisburg, PA, 17120. Telephone (717)787-2882.

APPENDIX D

NCHRP 12-33 PROJECT TEAM

John M. Kulicki, Principal Investigator
Dennis R. Mertz, Co-Principal Investigator
Scott A. Sabol, Program Officer (1992–1993)
Ian M. Friedland, Senior Program Officer (1988–1993)

Code Coordinating Committee
John M. Kulicki, Chair
John J. Ahlskog
Richard M. Barker
Robert C. Cassano
Paul F. Csagoly
James M. Duncan
Theodore V. Galambos
Andrzej S. Nowak
Frank D. Sears

*Editorial Committee**
John M. Kulicki, Chair
Paul F. Csagoly
Dennis R. Mertz
Frank D. Sears

**With appreciation to Modjeski &
Masters staff members:*
Diane M. Long
Scott R. Eshenaur
Chad M. Clancy
Robert P. Barrett
Donald T. Price
Nancy E. Kauhl
Malden B. Whipple
Raymond H. Rowand
Charles H. Johnson

NCHRP Panel
Veldo M. Goins, Chair
Roger Dorton
Steven J. Fenves
Richard S. Fountain
C. Stewart Gloyd
Stanley Gordon
Geerhard Haaijer
Clellon L. Loveall
Basile Rabbat
James E. Roberts
Arunprakash M. Shirole
James T. P. Yao
Luis Ybanez

TASK GROUPS

General Design Features
Frank D. Sears, Chair
Stanley R. Davis
Ivan M. Viest

Analysis and Evaluation
Paul F. Csagoly, Chair
Peter Buckland
Ian G. Buckle
Roy A. Imbsen
Jay A. Puckett
Wallace W. Sanders
Frieder Seible
William H. Walker

Loads and Load Factors
Paul F. Csagoly, Chair
Peter G. Buckland
Eugene Buth
James Cooper
C. Allin Cornell
James H. Gates
Michael A. Knott
Fred Moses
Andrzej S. Nowak
Robert Scanlan

Concrete Structures
Robert C. Cassano, Chair
John H. Clark
Michael P. Collins
Paul F. Csagoly
David P. Gustafson
Antonie E. Naaman
Paul Zia
Don W. Alden

Steel Structures
Frank D. Sears, Chair
John Barsom
Karl Frank
Wei Hsiong
William McGuire
Dennis R. Mertz
Roy L. Mion
Charles G. Schilling
Ivan M. Viest
Michael A. Grubb

Wood Structures
Andrzej S. Nowak, Chair
Baidar Bakht
R. Michael Caldwell
Donald J. Flemming
Hota V. S. Gangarao
Joseph F. Murphy
Michael A. Ritter
Raymond Taylor
Thomas G. Williamson

Bridge Railings
Ralph W. Bishop, Chair
Eugene Buth
James H. Hatton, Jr.
Teddy J. Hirsch
Robert A. Pege

Joints, Bearings and Accessories
Charles W. Purkiss, Chair
Ian G. Buckle
John J. Panak
David Pope
Charles W. Roeder
John F. Stanton

Earthquake Provisions Advisory Group
Ian Buckle, Chair
Robert Cassano
James Cooper
James Gates
Roy Imbsen
Geoffrey Martin

Aluminum Structures
Frank D. Sears, Chair
Teoman Pekoz

Foundations
J. Michael Duncan, Co-Chair
Richard M. Barker, Co-Chair

Deck Systems
Paul F. Csagoly, Chair
Barrington deVere Batchelor
Daniel H. Copeland
Gene R. Gilmore
Richard E. Klingner
Roman Wolchuk

Buried Structures
James Withiam, Chair
Edward P. Voytko

Walls, Piers and Abutments
J. Michael Duncan, Co-Chair
Richard M. Barker, Co-Chair
James Withiam

Calibration
Andrzej S. Nowak, Chair
C. Allin Cornell
Dan M. Frangopol
Theodore V. Galambos
Roger Green
Fred Moses
Kamal B. Rojiani

INDEX